AF716687

QUANTUM OSCILLATORS

QUANTUM OSCILLATORS

OLIVIER HENRI-ROUSSEAU and PAUL BLAISE

WILEY

A John Wiley & Sons, Inc., Publication

Copyright © 2011 by John Wiley & Sons, Inc. All rights reserved

Published by John Wiley & Sons, Inc., Hoboken, New Jersey
Published simultaneously in Canada

No part of this publication may be reproduced, stored in a retrieval system, or transmitted in any form or by any means, electronic, mechanical, photocopying, recording, scanning, or otherwise, except as permitted under Section 107 or 108 of the 1976 United States Copyright Act, without either the prior written permission of the Publisher, or authorization through payment of the appropriate per-copy fee to the Copyright Clearance Center, Inc., 222 Rosewood Drive, Danvers, MA 01923, (978) 750-8400, fax (978) 750-4470, or on the web at www.copyright.com. Requests to the Publisher for permission should be addressed to the Permissions Department, John Wiley & Sons, Inc., 111 River Street, Hoboken, NJ 07030, (201) 748-6011, fax (201) 748-6008, or online at http://www.wiley.com/go/permission.

Limit of Liability/Disclaimer of Warranty: While the publisher and author have used their best efforts in preparing this book, they make no representations or warranties with respect to the accuracy or completeness of the contents of this book and specifically disclaim any implied warranties of merchantability or fitness for a particular purpose. No warranty may be created or extended by sales representatives or written sales materials. The advice and strategies contained herein may not be suitable for your situation. You should consult with a professional where appropriate. Neither the publisher nor author shall be liable for any loss of profit or any other commercial damages, including but not limited to special, incidental, consequential, or other damages.

For general information on our other products and services or for technical support, please contact our Customer Care Department within the United States at (800) 762–2974, outside the United States at (317) 572–3993 or fax (317) 572–4002.

Wiley also publishes its books in a variety of electronic formats. Some content that appears in print may not be available in electronic formats. For more information about Wiley products, visit our web site at www.wiley.com.

Library of Congress Cataloging-in-Publication Data

Henri-Rousseau, Olivier.
Quantum oscillators / Olivier Henri-Rousseau and Paul Blaise.
p. cm.
Includes index.
ISBN 978-0-470-46609-4 (cloth)
1. Harmonic oscillators. 2. Spectrum analysis. 3. Wave mechanics. 4. Hydrogen bonding. I. Blaise, Paul. II. Title.
QC174.2.H45 2011
541′.224–dc22

2011008577

Printed in the United States of America

10 9 8 7 6 5 4 3 2 1

This book is dedicated to
Prof. Andrzej Witkowski of the Jagellonian University of Cracow,
on the occasion of his 80th birthday.

CONTENTS

PART II

SINGLE QUANTUM HARMONIC OSCILLATORS

PART III

ANHARMONICITY

CHAPTER 9 *ANHARMONIC OSCILLATORS*

CHAPTER 10 *OSCILLATORS INVOLVING ANHARMONIC COUPLINGS*

PART IV

OSCILLATOR POPULATIONS IN THERMAL EQUILIBRIUM

CHAPTER 11 *DYNAMICS OF A LARGE SET OF COUPLED OSCILLATORS*

CHAPTER 12 *DENSITY OPERATORS FOR EQUILIBRIUM POPULATIONS OF OSCILLATORS*

PART VII

VIBRATIONAL SPECTROSCOPY

CHAPTER 17 *APPLICATIONS TO OSCILLATOR SPECTROSCOPY*

CHAPTER 18 *APPENDIX*

LIST OF FIGURES

PREFACE

Quantum oscillators play a fundamental role in many area of physics and chemical physics, especially in infrared spectroscopy. They are encountered in molecular normal modes, or in solid-state physics with phonons, or in the quantum theory of light, with photons. Besides, quantum oscillators have the merit to be more easily exposed than the other physical systems interested by quantum mechanics because of their one-dimensional fundamental nature. However, despite the relative simplicity of quantum oscillators combined with their physical importance, there is a lack of monographs specifically devoted to them. Indeed it would be thereby of interest to dispose of a treatise widely covering the quantum properties of quantum harmonic oscillators at the following levels of increasing difficulty: (i) time-independent properties, (ii) reversible dynamics, (iii) thermal statistical equilibrium, and (iv) irreversible evolution toward equilibrium. And not only harmonic oscillators but also anharmonic ones, as well as single oscillators and anharmonically coupled oscillators.

As a matter of fact, such subjects are dispersed among different books of more or less difficulty and mixed with other physical systems. The aim of the present book is to remove that which would be considered as a lack. This book will start from an undergraduate level of knowledge and then will rise progressively to a graduate one. To allow that, it is divided into seven different parts of increasing conceptual difficulties.

Part I with Chapters 1–4 gives all the basic concepts required to study the different aspects of quantum oscillators. Part II, Chapters 5–8, is devoted to the properties of *single quantum harmonic oscillators*. Moreover, Part III deals with *anharmonicity*, either that of single anharmonic oscillator (Chapter 9) or that of anharmonically coupled harmonic oscillators (Chapter 10). Furthermore, Part IV, Chapters 11–13, treats the *thermal properties* of a large population of harmonic oscillators at statistical equilibrium. Part V concerns different kinds of *quantum normal modes* met either in light (Chapter 14) or in molecules and solids (Chapter 15). Finally, Part VI, Chapter 16, studies the irreversible behavior of *damped quantum oscillators*, whereas Part VII, Chapter 17, applies many of the results of the previous chapters to some *spectroscopic properties* of quantum oscillators. Its now time to be more precise with the contents of these parts.

Chapter 1 summarizes the minimal mathematical properties (specially those of Hilbert spaces and of noncommuting operator algebra) required to understand quantum principles. That is the aim of Chapter 2, which, after giving the postulates of quantum mechanics, treats quantum average values and dispersion, allowing one to get the Heisenberg uncertainty relations, and develops the basic consequences of the time-dependent Schrödinger equation. Then, Chapter 3 goes further by looking at the different representations of quantum mechanics, which makes tractable the quantum generalities exposed in the previous chapter, and which will be of great help in the further studies of quantum oscillators. These quantum descriptions are *matrix mechanics*, *wave mechanics*, and time-dependent representations, that is, *Schrödinger*, *Heisenberg*, and *interaction pictures*, and finally the *density operator* representation, which may be declined according to matrix mechanics or wave mechanics and also to different time-dependent pictures. Chapter 4 ends Part I, being devoted to three different but important physical models, which will enlighten the further studies of quantum oscillators. They are the *particle-in-a-box model*, which is a simple and didactic introduction to energy quantization that will be met for quantum oscillators, the *two-energy-level model,* which will be used when studying *Fermi resonances* appearing in vibrational spectroscopy, and the *Fermi golden rule*, involving concepts that will be used in the same area of vibrational spectroscopy.

Following Part 1, which deals with the basis required for quantum oscillators studies, Part II enters into the heart of the subject. Chapter 5 focuses attention on the quantum energetic representation of harmonic oscillators by solving their time-independent Schrödinger equation using *ladder operators (Boson operators)*, thus allowing one to determine the *quantized energy levels* and the corresponding *Hamiltonian eigenkets*, and also the action of the ladder operators on these eigenkets. It continues by obtaining the *oscillator excited wavefunctions,* from the corresponding ground state using the action of the ladder operators on the Hamiltonian eigenkets. After this Hamiltonian eigenket representation, Chapter 6 is concerned with *coherent states*, which minimize the Heisenberg uncertainty relations, and *translation operators,* the action of which on Hamiltonian ground states yields coherent states, by studying their properties, which are deeply interconnected, and then used to calculate *Franck–Condon factors* and to diagonalize the Hamiltonian of *driven harmonic oscillators*. Chapter 7 continues Part II by giving proofs of some *Boson operator theorems*, which are applied at its end to find the dynamics of a *driven harmonic oscillator* and which will be widely used in the following. Finally, Chapter 8 closes Part II by treating some more complicated topics such as *phase operators*, *squeezed states*, and *Bogoliubov–Valatin transformation*, which involve products of ladder operators.

The properties of single quantum harmonic oscillators found in Part II allow us to treat anharmonicity in Part III. That is first done in Chapter 9 by studying anharmonic oscillators such as those involving *Morse potentials*, which are more realistic than harmonic potentials for diatomic molecules or *double-well* potentials leading to *quantum tunneling*, and in Chapter 10 by studying several harmonic oscillators involving anharmonic coupling. In this last chapter of Part III, together with *Fermi resonances*, is studied the *strong anharmonic coupling theory* encountered in the quantum theory of weak H-bonded species and allowing the *adiabatic separation* between low- and high-frequency anharmonically coupled oscillators, which is studied in detail. Chapter 10 ends with a study of anharmonic coupling between four oscillators, which is used to model a centrosymmetric cyclic H-bonded dimer.

Parts II and III ignored the thermal properties of single or coupled quantum oscillators, considering them as isolated from the medium, what they may be, harmonic or anharmonic. The aim of Part IV is to address the thermal influence of the medium. Part IV begins this study with a somewhat unusual chapter (Chapter 11) dealing with the dynamics of very large populations of linearly coupled harmonic oscillators starting from an initial situation where the energy is found only on one of the oscillators. Moreover, having proven the *Boltzmann H-theorem* according to which the entropy increases until statistical equilibrium is attained, Chapter 12 applies the results of Chapter 11 to show how, after some characteristic time has elapsed, the statistical entropy reaches its maximum, in agreement with the Boltzmann theorem, whereas a *coarse-grained energy analysis* of the energy distribution of the oscillators sets reveals a *Boltzmann energy* distribution. Then, applying the principle of entropy maximization at statistical equilibrium, this chapter obtains the *microcanonical* and *canonical* density operators. Finally, Chapter 13 closes Part IV by studying the thermal properties of quantum harmonic oscillators (*thermal average energies, heat capacities, thermal energy fluctuations*) and ends with the demonstration of the expression of the thermal average of general functions of Boson operators, which contains as a special case the *Bloch theorem.*

Chapter 11 of Part IV studies the dynamics of a large population of coupled quantum harmonic oscillators that, as calculation intermediates, are considered to be *normal modes*, but without taking attention to them due to the dynamics preoccupations. Since normal modes of systems of many degrees of freedom are collective harmonic motions in which all the parts are moving at the same angular frequency and the same phase, it is possible, within classical physics, to extract for such systems the classical normal modes and then to quantize them to get quantum harmonic oscillators to which it is possible to apply all the results of Parts II–IV. This is the purpose of Part V, which starts (Chapter 14) with a study of the quantum normal modes of electromagnetic fields. That may be first performed with obtaining the classical normal modes of the fields by passing for the Maxwell equations in the vacuum, from the geometrical space to the reciprocal one, using Fourier transforms, and then introduce a commutation rule between the conjugate variables of the electromagnetic field, which are the potential vector and the electric field in the reciprocal space. Then, applying the thermal properties of quantum oscillators found in Chapter 13, it is possible to derive the *black-body radiation Planck law* and the *Stefan–Boltzmann law,* and also the ratio of the *Einstein coefficients*. Chapter 15 completes this part devoted to normal modes by determining the classical molecular normal modes and then quantizing them, and so obtaining the normal modes of a one-dimensional solid in the reciprocal space, allowing one, on application of the thermal properties of oscillators, to obtain the Einstein and the Debye results concerning the solids *heat capacity* of solids.

Continuing the work of Part IV devoted to thermal equilibrium, which was applied in Part V to find the thermal statistical properties of normal modes, Part VI, involving only Chapter 16, studies the irreversible behavior of harmonic oscillators, which are damped due to the influence of the medium. This irreversible influence is modeled by considering the medium, acting as a thermal bath, as a very large set of harmonic oscillators of variable angular frequencies, weakly coupled to the damped oscillator, and each constrained to remain in statistical thermal equilibrium. Then, solving within this approach the *Liouville equation*, and after performing the Markov approximation, the *master equations* governing the dynamics of the density operators of driven or undriven harmonic oscillators are obtained. This procedure allows one to derive in a subsequent section the *Fokker–Planck equation* for damped harmonic oscillators. Next, Chapter 16 continues, by aid of an approach similar to that used for the master equations by deriving the *Langevin equations* governing the time-dependent statistical averages of the Boson operators, and ends, using these Langevin equations, by obtaining the interaction picture time evolution operator of driven damped quantum harmonic oscillators, which allows one to get the corresponding time-dependent density operator, which may be envisaged as a consequence of the corresponding master equation governing the dynamics of damped oscillators.

The book ends with Part VII corresponding to the single Chapter 17, by applying many of the properties of quantum oscillators obtained in Parts II and III (Chapter 10), Part IV (Chapter 13), and Part VI (Chapter 16), to find some important results in vibrational spectroscopy, such as the IR selection rule for quantum harmonic oscillators, and to study using *linear response theory*, and after having proved it, the line shapes of some physical realistic situations involving anharmonically coupled damped quantum harmonic oscillators encountered in the area of H-bonded species.

Clearly, the topics studied in all these parts involve progressive levels of difficulty, varying from undergraduate to graduate.

It may be of interest to list the quantum theoretical tools necessary to treat the different subjects of the book. Essential tools are *kets*, *bras*, *scalar products*, *closure relation*, *linear Hermitian* and *unitary operators*, *commutators* and *eigenvalue equations*, as well as *quantum mechanical fundamentals*. There exist seven postulates, concerning the notions of quantum average values and of the corresponding fluctuations leading to the Heisenberg uncertainty relations. We list the time dependence of the quantum average values leading to the Ehrenfest theorem and to the virial theorem, the different representations of quantum mechanics involving wave mechanics, matrix representation, the different time-dependent representations, that is, the Schrödinger and Heisenberg ones and also the interaction picture, all using the time evolution operators and, finally, the various density operator representations. Furthermore, there are also mathematical tools that are not specific to the subject but necessary to the understanding of some developments and that will be treated in the Appendix (Chapter 18). Among them, some commutator algebra, particularly those dealing with the position and momentum operators, some theorems concerning exponential operators as the Baker–Campbell–Hausdorff relation or the Glauber–Weyl theorem, some information about Fourier transforms and distribution functions, the Lagrange multipliers method, complex results concerning vectorial analysis, and elements dealing with the point-group theory.

On the other hand, as it may be inferred from the presentation of the different parts of the book, the following quantum oscillator properties will be considered: Hamiltonian eigenkets of harmonic oscillators and their corresponding wavefunctions, ladder operators, action of these operators on the Hamiltonian eigenkets, coherent states, translation operators, squeezed states and corresponding squeezing operators, time dependence of the ladder operators, canonical transformations involving ladder operators, normal and antinormal ordering, Bogolyubov transformations, Boltzmann density operators of harmonic oscillators, and thermal quantum average values of operators, specially that of the translation operator leading to the Bloch theorem.

Despite the complexity of the project, our aim is to propose a progressive course where all the demonstrations, whatever their level may be, would present no particular difficulties, and thus would be readable at various levels ranging from undergraduate to postgraduate levels. In this end, we have applied our teaching experience, which used the Gestalt psychology, according to which the main operational principle of the mind is holistic, the whole being more important than the sum of its parts, that is particularly sensitive with respect to the visual recognition of figures and whole forms instead of just a collection of simple lines and curves: We have observed that this concept is very well verified to those unfamiliar with long equations involving many intricated symbols.

There are different ways to read this book. The first one concerns quantum mechanics, which, since considered from the viewpoint of oscillators, allows one to avoid all the mathematical difficulties related to the techniques for solving the second-order partial differential equations encountered in wave mechanics. The second one gives the elements required to understand the theories dealing with the line shapes met spectroscopy more specially in the area of H-bonded species. The third one may be viewed as a simple introduction to quantization of light. The fourth one may be considered as an introduction to quantum equilibrium statistical properties of oscillators, while the fifth focuses attention on the irreversible behavior of oscillators Finally, the sixth concerns chemists interested in molecular spectroscopy. The chapters may be considered as follows:

Domains	**Chapters**																
Quantum oscillators	1	2	3	4	5	6	7		9	10			13	14	15	16	
IR line shape spectra		2	3	4	5	6	7		9	10			13		15		17
Theory of light		2	3		5	6	7	8					13	14		16	
Statistical equilibrium		2	3		5	6	7					12	13				
Irreversibility		2	3		5	6	7				11					16	
Molecular spectroscopy	1	2			5				9	10							17

The cost to be paid will be the inclusion of many details in the demonstrations, which sometimes appear to the advanced readers to be superfluous. In addition, to make the equations more easily readable we have sometimes used unusual notations combined with the introduction of additive brackets, which would appear to be surprising and unnecessary for those indifferent to the didactic advantages of the Gestalt psychology, which is our option.

ACKNOWLEDGMENTS

Prof. W. Coffey (Dublin)
Prof. Ph. Durand (Toulouse)
Prof. J-L. Déjardin
Prof. Y. Kalmykov
Prof. H. Kachkachi
Dr. P. M. Déjardin
Dr. A. Velcescu-Ceasu
Dr. P. Villalongue
Dr. B. Boulil

PART I

BASIS REQUIRED FOR QUANTUM OSCILLATOR STUDIES

CHAPTER 1

BASIC CONCEPTS REQUIRED FOR QUANTUM MECHANICS

In order to summarize the quantum basis required for the study of oscillators, it is necessary to define some mathematical notions concerning the properties of state spaces, particularly the concepts of linear operators, kets, bras, Hermiticity, eigenvalues, and eigenvectors of linear operators involved in the formulation of the different postulates. The first two sections of this chapter are devoted to this. However, it is possible to pass directly to the third section leaving for later the lecture of the previous one.

1.1 BASIC CONCEPTS OF COMPLEX VECTORIAL SPACES

1.1.1 Kets, bras, and scalar products

Quantum mechanics deals with *state spaces*, that is, vectorial spaces involving complex scalar products that are generally of infinite dimension. Any element of these spaces is named a *ket* and symbolized $|\ldots\rangle|$ by inserting inside it a free notation allowing one to clearly identify this ket; for instance, $|\Psi_k\rangle$[1] or $|n\rangle$.

Since the space of states is vectorial, and if the kets $|\Psi_1\rangle$ and $|\Psi_2\rangle$ belong to the same state space, then the ket $|\Phi\rangle$ defined by the linear superposition

$$|\Phi\rangle = \lambda_1|\Psi_1\rangle + \lambda_2|\Psi_2\rangle$$

where λ_1 and λ_2 are two scalars, belongs also to the same state space.

Now, to some ket $|\Psi\rangle$ of the state space there exists a linear functional that associates with some another ket $|\Phi\rangle$ of this space a complex scalar $A_{\Phi\Psi}$, which is *the scalar product* of $|\Psi\rangle$ by $|\Phi\rangle$. This may be written

$$A_{\Phi\Psi} = \langle\Phi|\Psi\rangle \tag{1.1}$$

[1] All the notations inside the symbol $|\ldots\rangle|$ are designed to distinguish clearly the ket of interest. For instance, some Latin n or Greek Ψ letters lead to the writing $|n\rangle$ or $|\Psi\rangle$, but the notation may be as complex as required; for instance, $|n_l\rangle$ or $|\Psi_k\rangle$, the subscripts allowing to distinguish between two kets $|n_l\rangle$ and $|n_j\rangle$ of the same kind, and in a similar way to kets $|\Psi_k\rangle$ and $|\Psi_j\rangle$. In the following we shall use also as specification notations of the form: $|\{n\}\rangle$, $|(n)\rangle$, $|[n]\rangle$ in order to reserve the notations $|n_l\rangle$ or $|\Psi_k\rangle$ for kets belonging to the same basis.

Quantum Oscillators, First Edition. Olivier Henri-Rousseau and Paul Blaise.
© 2011 John Wiley & Sons, Inc. Published 2011 by John Wiley & Sons, Inc.

This linear functional, which is denoted $\langle\Phi|$, is named the *bra*, corresponding to the ket $|\Phi\rangle$. The bras may be viewed as belonging to a state space that is the *dual space* of the state space to which belong the kets, that is, the bras are the Hermitian conjugates of the corresponding kets, namely

$$\langle\Phi| = |\Phi\rangle^{\dagger} \tag{1.2}$$

where the superscript† denotes the Hermitian conjugation. The scalar products have the following properties:

$$(\langle\lambda_1\Phi_1 + \lambda_2\Phi_2|)|\Psi\rangle = \lambda_1^*\langle\Phi_1|\Psi\rangle + \lambda_2^*\langle\Phi_2|\Psi\rangle$$

$$\langle\Phi|(|\lambda_1\Psi_1 + \lambda_2\Psi_2\rangle) = \lambda_1\langle\Phi|\Psi_1\rangle + \lambda_2\langle\Phi|\Psi_2\rangle$$

$$\langle\Psi_k|\Phi_l\rangle = \langle\Phi_l|\Psi_k\rangle^* \tag{1.3}$$

$$\langle\Psi|\Psi\rangle > 0 \quad \text{if } |\Psi\rangle \neq 0$$

$$\boxed{\langle\Psi|\Psi\rangle = 0 \quad \text{if and only if } |\Psi\rangle = 0} \tag{1.4}$$

In addition, if this scalar product is *normalized*, we have

$$\langle\Psi|\Psi\rangle = 1$$

If the scalar product of two kets $|\Psi\rangle$ and $|\Phi\rangle$ is zero, the two kets $|\Psi\rangle$ and $|\Phi\rangle$ are said to be *orthogonal*:

$$\langle\Psi|\Phi\rangle = 0$$

1.1.2 Linear transformations

Let us consider the action of a linear operator $\mathbf{A}$ on a *ket* $|\xi\rangle$ belonging to the state space. This action leads to another ket $|\Phi\rangle$ according to

$$\mathbf{A}|\xi\rangle = |\Phi\rangle \tag{1.5}$$

Consider now the action of an another linear operator $\mathbf{B}$ acting on the same ket $|\xi\rangle$. Generally, it will yield another ket $|\Psi\rangle$:

$$\mathbf{B}|\xi\rangle = |\Psi\rangle$$

In most situations, the product of two operators $\mathbf{A}$ and $\mathbf{B}$ does not commute, that is,

$$\mathbf{AB} \neq \mathbf{BA}$$

The *commutator* of two operators $\mathbf{A}$ and $\mathbf{B}$ is symbolized[2] by

$$[\mathbf{A}, \mathbf{B}] \equiv \mathbf{AB} - \mathbf{BA}$$

[2]The standard notation for a commutator is $[\ldots, \ldots]$ where the comma separates the two operators involved. Since the comma risks being unnoticed, in order to avoid this risk we have chosen to reserve as far as, the notation involving [..,..] to commutators, and to use for other situations notations of the kinds $(\ldots)$ or $\{\ldots\}$.

In some situations, a linear operator $\mathbf{A}$ may act on different kets $|\Psi_1\rangle$, $|\Psi_2\rangle$, ..., in such a way as it multiplies them by scalars A_1, A_2, ..., more generally

$$\mathbf{A}|\Psi_l\rangle = A_l|\Psi_l\rangle \tag{1.6}$$

The kets $|\Psi_l\rangle$ corresponding to these special situations are the *eigenvectors* of the operator $\mathbf{A}$ while the scalars A_l are the corresponding *eigenvalues*. Equation (1.6) is called an *eigenvalue equation*. The scalar A_l is generally complex.

When different eigenvectors exist corresponding to a same eigenvalue, then a *degeneracy* exists, its degree being the number of eigenvectors associated with this same eigenvalue. In the following, we shall not encounter degeneracy except for very special situations so that we shall ignore the particular treatment of this case.

1.1.2.1 Hermitian conjugate of a linear transformation

The Hermitian conjugate of the linear operator $\mathbf{A}$ is $\mathbf{A}^\dagger$. Consider a linear transformation of the form (1.5)

$$\mathbf{A}|\Psi\rangle = |\Phi\rangle \tag{1.7}$$

Its Hermitian conjugate is the bra $\langle\Phi|$:

$$\{\mathbf{A}|\Psi\rangle\}^\dagger = \langle\Phi|$$

Now, the Hermitian conjugate of the linear transformation (1.7) is

$$\boxed{\{\mathbf{A}|\Psi\rangle\}^\dagger = \langle\Psi|\mathbf{A}^\dagger} \tag{1.8}$$

which is equivalent to

$$\langle\Phi| = \langle\Psi|\mathbf{A}^\dagger \tag{1.9}$$

Consider now an eigenvalue equation of the form (1.6)

$$\mathbf{A}|\Psi\rangle = A|\Psi\rangle \tag{1.10}$$

Then, owing to Eq. (1.8), and because the Hermitian conjugate of a scalar is its complex conjugate, the Hermitian conjugate of Eq. (1.10) is

$$\langle\Psi|\mathbf{A}^\dagger = \langle\Psi|A^* \tag{1.11}$$

1.1.3 Basis and closure relation

A set $\{|\Phi_n\rangle\}$ of kets $|\Phi_n\rangle$ of the state space is said to be orthonormal if these states satisfy

$$\boxed{\langle\Phi_n|\Phi_m\rangle = \delta_{nm}} \tag{1.12}$$

where δ_{nm} is the Kronecker symbol given by

$$\delta_{mm} = 1 \qquad \text{and} \qquad \delta_{mn} = 0 \quad \text{if} \quad m \neq n \tag{1.13}$$

Again, such a set $\{|\Phi_n\rangle\}$ forms a basis in the state space, provided all kets $|\Psi_k\rangle$ belonging to this space may be expanded according to

$$|\Psi_k\rangle = \sum_{n=1}^{\infty} C_{nk}|\Phi_n\rangle \tag{1.14}$$

where the C_{nk} are the expansion coefficients, which may be complex.

Now, premultiply both members of Eq. (1.14) by a bra $\langle\Phi_m|$ corresponding to some ket belonging to the basis $\{|\Phi_n\rangle\}$. It reads

$$\langle\Phi_m|\Psi_k\rangle = \langle\Phi_m|\sum_{n=1}^{\infty} C_{nk}|\Phi_n\rangle$$

or

$$\langle\Phi_m|\Psi_k\rangle = \sum_{n=1}^{\infty} C_{nk}\langle\Phi_m|\Phi_n\rangle$$

Therefore, in view of Eq. (1.12), it transforms to

$$\langle\Phi_m|\Psi_k\rangle = \sum_{n=1}^{\infty} C_{nk}\delta_{nm}$$

or, in view of Eq. (1.13),

$$\langle\Phi_m|\Psi_k\rangle = C_{mk} \tag{1.15}$$

Then, introducing Eq. (1.15) into Eq. (1.14) we have

$$|\Psi_k\rangle = \sum_{n=1}^{\infty}\langle\Phi_n|\Psi_k\rangle|\Phi_n\rangle$$

Furthermore, after commuting the scalar product with the ket in the second member of this equation, we have

$$|\Psi_k\rangle = \sum_{n=1}^{\infty}\{|\Phi_n\rangle\langle\Phi_n|\}\,|\Psi_k\rangle \tag{1.16}$$

Now, in order for Eq. (1.16) to be satisfied, whatever may be the ket $|\Psi_k\rangle$ appearing on both sides of this equation, it is necessary that

$$\boxed{\sum_{n=1}^{\infty}|\Phi_n\rangle\langle\Phi_n| = \mathbf{1}} \tag{1.17}$$

Equation (1.17) is known as *the closure relation*. The closure relation (1.17) together with the orthonormality condition (1.12) are the two important properties of a basis, the first being the consequence of the second one.

Now, consider the following operation:

$$\{|\Phi_n\rangle\langle\Phi_n|\}\,|\Psi_k\rangle$$

Then, using the expansion (1.16), this expression reads

$$\{|\Phi_n\rangle\langle\Phi_n|\}|\Psi_k\rangle = \{|\Phi_n\rangle\langle\Phi_n|\}\sum_{m=1}^{\infty} C_{mk}|\Phi_m\rangle$$

or

$$\{|\Phi_n\rangle\langle\Phi_n|\}|\Psi_k\rangle = |\Phi_n\rangle\sum_{m=1}^{\infty} C_{mk}\langle\Phi_n|\Phi_m\rangle$$

and thus, using the orthonormality properties (1.12)

$$\{|\Phi_n\rangle\langle\Phi_n|\}|\Psi_k\rangle = |\Phi_n\rangle\sum_{m=1}^{\infty} C_{mk}\delta_{nm}$$

so that

$$\{|\Phi_n\rangle\langle\Phi_n\}|\Psi_k\rangle = |\Phi_n\rangle C_{nk}$$

Thus, $|\Phi_n\rangle\langle\Phi_n|$ acts on the ket $|\Psi_k\rangle$ as an operator, projecting it on to the state $|\Phi_n\rangle$. Thus it is called a *projector*

1.1.4 Schwarz inequality

Consider a ket $|\Psi\rangle$ that is the superposition of two different kets $|\Phi\rangle$ and $|\xi\rangle$:

$$|\Psi\rangle = |\Phi\rangle + \lambda|\xi\rangle \tag{1.18}$$

where λ is a complex scalar number. The Hermitian conjugate of this equation is

$$\langle\Psi| = \langle\Phi| + \lambda^*\langle\xi| \tag{1.19}$$

Consider now the norm of this ket, which cannot be negative, so that it must be written

$$\langle\Psi|\Psi\rangle \geqslant 0 \tag{1.20}$$

Then, using Eqs. (1.18) and (1.19) the norm becomes

$$\langle\Psi|\Psi\rangle = \langle\Phi|\Phi\rangle + \lambda\langle\Phi|\xi\rangle + \lambda^*\langle\xi|\Phi\rangle + \lambda\lambda^*\langle\xi|\xi\rangle \tag{1.21}$$

Now, suppose that the scalar λ is given by

$$\lambda = -\frac{\langle\xi|\Phi\rangle}{\langle\xi|\xi\rangle} \tag{1.22}$$

Then, according to Eq. (1.3), the complex conjugate of λ is

$$\lambda^* = -\frac{\langle\Phi|\xi\rangle}{\langle\xi|\xi\rangle} \tag{1.23}$$

Again, introducing Eqs. (1.22) and (1.23) in (1.21), one obtains

$$\langle\Psi|\Psi\rangle = \langle\Phi|\Phi\rangle - \frac{\langle\xi|\Phi\rangle}{\langle\xi|\xi\rangle}\langle\Phi|\xi\rangle - \frac{\langle\Phi|\xi\rangle}{\langle\xi|\xi\rangle}\langle\xi|\Phi\rangle + \frac{\langle\xi|\Phi\rangle}{\langle\xi|\xi\rangle}\frac{\langle\Phi|\xi\rangle}{\langle\xi|\xi\rangle}\langle\xi|\xi\rangle \tag{1.24}$$

yielding, after an initial simplification

$$\langle\Psi|\Psi\rangle = \langle\Phi|\Phi\rangle - \frac{\langle\xi|\Phi\rangle}{\langle\xi|\xi\rangle}\langle\Phi|\xi\rangle - \frac{\langle\Phi|\xi\rangle}{\langle\xi|\xi\rangle}\langle\xi|\Phi\rangle + \frac{\langle\xi|\Phi\rangle}{\langle\xi|\xi\rangle}\langle\Phi|\xi\rangle$$

Then, after cancellation of the two last right-hand terms, this last equation becomes

$$\langle\Psi|\Psi\rangle = \langle\Phi|\Phi\rangle - \frac{\langle\xi|\Phi\rangle\langle\Phi|\xi\rangle}{\langle\xi|\xi\rangle}$$

or, in view of the inequality (1.20)

$$\langle\Phi|\Phi\rangle\langle\xi|\xi\rangle - \langle\xi|\Phi\rangle\langle\Phi|\xi\rangle \geqslant 0$$

leading to a result that is known as the *Schwarz inequality*:

$$\langle\Phi|\Phi\rangle\langle\xi|\xi\rangle \geqslant \langle\xi|\Phi\rangle\langle\Phi|\xi\rangle \tag{1.25}$$

1.2 HERMITIAN CONJUGATION

1.2.1 Theorem dealing with Hermitian conjugates

Consider the linear transformation

$$\mathbf{B}|\Psi\rangle = |\xi\rangle \tag{1.26}$$

Again, owing to Eq. (1.8), its Hermitian conjugate is

$$\langle\Psi|\mathbf{B}^\dagger = \langle\xi| \tag{1.27}$$

Then, premultiplying Eq. (1.26) by $\langle\Phi|$ and postmultiplying Eq. (1.27) by $|\Phi\rangle$, one obtains, respectively,

$$\langle\Phi|\mathbf{B}|\Psi\rangle = \langle\Phi|\xi\rangle \tag{1.28}$$

$$\langle\Psi|\mathbf{B}^\dagger|\Phi\rangle = \langle\xi|\Phi\rangle \tag{1.29}$$

Thus, owing to Eq. (1.3), it appears that, in the present situation

$$\langle\Phi|\xi\rangle = \langle\xi|\Phi\rangle^*$$

Thus, Eqs. (1.28) and (1.29) yield

$$\boxed{\langle\Psi|\mathbf{B}^\dagger|\Phi\rangle = \langle\Phi|\mathbf{B}|\Psi\rangle^*} \tag{1.30}$$

1.2.2 Hermitian conjugate of $\mathbf{A}^\dagger$

Consider the Hermitian conjugate $(\mathbf{A}^\dagger)^\dagger$ of the Hermitian conjugate $\mathbf{A}^\dagger$ of the linear operator $\mathbf{A}$. First, we may write that the Hermitian conjugate of the operator $\mathbf{A}$ is a new operator $\mathbf{B}$:

$$\mathbf{B} = \mathbf{A}^\dagger \tag{1.31}$$

Then, the Hermitian conjugate of $\mathbf{A}^\dagger$ is $\mathbf{B}^\dagger$:

$$(\mathbf{A}^\dagger)^\dagger = \mathbf{B}^\dagger \tag{1.32}$$

Now, premultiply the two members of Eq. (1.32) by some bra $\langle\Psi|$ and postmultiply them by some ket $|\Phi\rangle$. Then, one obtains

$$\langle\Psi|(\mathbf{A}^\dagger)^\dagger|\Phi\rangle = \langle\Psi|\mathbf{B}^\dagger|\Phi\rangle$$

Owing to Eq. (1.30), this last expression becomes

$$\langle\Psi|(\mathbf{A}^\dagger)^\dagger|\Phi\rangle = \langle\Phi|\mathbf{B}|\Psi\rangle^*$$

Again, introduce Eq. (1.31) on the right-hand side of this last result. Then, one finds

$$\langle\Psi|(\mathbf{A}^\dagger)^\dagger|\Phi\rangle = \langle\Phi|\mathbf{A}^\dagger|\Psi\rangle^*$$

Moreover, using again theorem (1.30), one gets

$$\langle\Psi|(\mathbf{A}^\dagger)^\dagger|\Phi\rangle = \langle\Psi|\mathbf{A}|\Phi\rangle$$

Finally, since the latter must be true whatever $\langle\Psi|$ and $|\Phi\rangle$ are, it follows that

$$\boxed{(\mathbf{A}^\dagger)^\dagger = \mathbf{A}} \tag{1.33}$$

1.2.3 Successive Hermitian conjugations of a linear transformation

Consider the Hermitian conjugate of a linear transformation (1.8). It is

$$\{\{\mathbf{A}|\Psi\rangle\}^\dagger\}^\dagger = \{\langle\Psi|\mathbf{A}^\dagger\}^\dagger \tag{1.34}$$

Now, let

$$\mathbf{A}|\Psi\rangle = |\Phi\rangle \qquad \text{and} \qquad \langle\Psi|\mathbf{A}^\dagger = \langle\Phi| \tag{1.35}$$

Then, due to this last equation, Eq. (1.34) reads

$$\{\{\mathbf{A}|\Psi\rangle\}^\dagger\}^\dagger = \langle\Phi|^\dagger$$

or, in view of Eq. (1.2)

$$\{\{\mathbf{A}|\Psi\rangle\}^\dagger\}^\dagger = |\Phi\rangle$$

Moreover, due to the first equation of (1.35), we also have

$$\{\{\mathbf{A}|\Psi\rangle\}^\dagger\}^\dagger = \mathbf{A}|\Psi\rangle$$

1.2.4 Hermitian conjugate of $|\xi\rangle\langle\zeta|$

Consider the following operator and its Hermitian conjugate:

$$\mathbf{A} = |\xi\rangle\langle\zeta| \qquad \text{and} \qquad \mathbf{A}^\dagger = \{|\xi\rangle\langle\zeta|\}^\dagger \tag{1.36}$$

What is the relation between $\mathbf{A}$ and $\mathbf{A}^\dagger$? To answer this question, premultiply both the operator and its Hermitian conjugate by the bra $\langle\Phi|$ and postmultiply both of them by the ket $|\Psi\rangle$ leading, respectively, to

$$\langle\Phi|\mathbf{A}|\Psi\rangle = \langle\Phi|\{|\xi\rangle\langle\zeta|\}|\Psi\rangle$$

and

$$\langle\Phi|\mathbf{A}^\dagger|\Psi\rangle = \langle\Phi|\,\{|\xi\rangle\langle\zeta|\}^\dagger\,|\Psi\rangle$$

Now, according to Eq. (1.30), the operator defined by Eq. (1.36) must obey

$$\langle\Phi|\mathbf{A}^\dagger|\Psi\rangle = \langle\Psi|\mathbf{A}|\Phi\rangle^*$$

Thus, in the present situation, due to the expressions (1.36), the latter takes on the form

$$\langle\Phi|\left\{|\xi\rangle\langle\zeta|\right\}^{\dagger}|\Psi\rangle = \left\{\langle\Psi|\left\{|\xi\rangle\langle\zeta|\right\}|\Phi\rangle\right\}^{*}$$

After simplifying the notation in the more usual form, we have

$$\langle\Phi|\left\{|\xi\rangle\langle\zeta|\right\}^{\dagger}|\Psi\rangle = \left\{\langle\Psi|\xi\rangle\langle\zeta|\Phi\rangle\right\}^{*} \qquad (1.37)$$

Again, the two terms of the right-hand side of this last equation are scalars obeying

$$\langle\Psi|\xi\rangle^{*} = \langle\xi|\Psi\rangle \qquad \text{and} \qquad \langle\zeta|\Phi\rangle^{*} = \langle\Phi|\zeta\rangle$$

Thus, Eq. (1.37) transforms to

$$\langle\Phi|\left\{|\xi\rangle\langle\zeta|\right\}^{\dagger}|\Psi\rangle = \langle\xi|\Psi\rangle\langle\Phi|\zeta\rangle$$

Now, the two right scalars appearing on the right-hand side of this last expression do commute, so that

$$\langle\Phi|\left\{|\xi\rangle\langle\zeta|\right\}^{\dagger}|\Psi\rangle = \langle\Phi|\zeta\rangle\langle\xi|\Psi\rangle$$

Finally, since this last equation must be satisfied, whatever $\langle\Psi|$ and $|\Phi\rangle$ are, one obtains the final result

$$\boxed{\left\{|\xi\rangle\langle\zeta|\right\}^{\dagger} = |\zeta\rangle\langle\xi|} \qquad (1.38)$$

1.2.5 Hermitian conjugate of a product of operators that do not commute

Now, consider two noncommuting linear operators **A** and **B** the product of which is **C**, that is,

$$\mathbf{AB} = \mathbf{C} \qquad \text{and} \qquad [\mathbf{A}, \mathbf{B}] \neq 0$$

Then, seek the Hermitian conjugate $(\mathbf{AB})^{\dagger}$ of their product **AB**. Hence, premultiply the product **AB** by the bra $\langle\Psi|$ and postmultiply it by the ket $|\Phi\rangle$. Then, considering the product **AB** as a new operator **C**, and applying the theorem (1.30), that is,

$$\langle\Psi|\mathbf{C}^{\dagger}|\Phi\rangle = \langle\Phi|\mathbf{C}|\Psi\rangle^{*}$$

we have

$$\langle\Psi|(\mathbf{AB})^{\dagger}|\Phi\rangle = \langle\Phi|\mathbf{AB}|\Psi\rangle^{*} \qquad (1.39)$$

Now, observe that the action of the operator **B** on the ket $|\Psi\rangle$ and that of the operator **A** on the bra $\langle\Phi|$ are linear transformations of the type

$$\mathbf{B}|\Psi\rangle = |\chi\rangle \qquad \text{and} \qquad \langle\Phi|\mathbf{A} = \langle\mu| \qquad (1.40)$$

Then, owing to these linear transformations defining the ket $|\chi\rangle$ and the bra $\langle\mu|$, Eq. (1.39) reads

$$\langle\Psi|(\mathbf{AB})^{\dagger}|\Phi\rangle = \langle\mu|\chi\rangle^{*} \qquad (1.41)$$

Again, due to the relation (1.3) defining the scalar product and its complex conjugate, there is

$$\langle \mu | \chi \rangle^* = \langle \chi | \mu \rangle$$

Hence, Eq. (1.41) takes the form

$$\langle \Psi | (\mathbf{AB})^\dagger | \Phi \rangle = \langle \chi | \mu \rangle \tag{1.42}$$

Moreover, the Hermitian conjugate of the linear transformations (1.40) is

$$\langle \Psi | \mathbf{B}^\dagger = \langle \chi | \qquad \text{and} \qquad \mathbf{A}^\dagger | \Phi \rangle = | \mu \rangle$$

Thus, the corresponding scalar product yields

$$\langle \chi | \mu \rangle = \langle \Psi | \mathbf{B}^\dagger \mathbf{A}^\dagger | \Phi \rangle$$

As a consequence, Eq. (1.42) becomes

$$\langle \Psi | (\mathbf{AB})^\dagger | \Phi \rangle = \langle \Psi | \mathbf{B}^\dagger \mathbf{A}^\dagger | \Phi \rangle$$

Of course, this last equation must be true for all $\langle \Psi |$ and $| \Phi \rangle$ so that

$$\boxed{(\mathbf{AB})^\dagger = \mathbf{B}^\dagger \mathbf{A}^\dagger} \tag{1.43}$$

1.2.6 Hermitian conjugate of a general expression involving kets, bra operators, and scalars

We may summarize here the present results obtained previously and that dealt with the Hermitian conjugate in special situations given, respectively, by Eqs. (1.33), (1.38), and (1.43). For operators we have

$$(\mathbf{A}^\dagger)^\dagger = \mathbf{A}; \qquad \{|\xi\rangle\langle\zeta|\}^\dagger = |\zeta\rangle\langle\xi| \qquad \text{and} \qquad (\mathbf{AB})^\dagger = \mathbf{B}^\dagger \mathbf{A}^\dagger$$

For linear transformations, we have

$$\text{If} \qquad \mathbf{A}|\Psi\rangle = A|\Psi\rangle \quad \text{then} \quad \{\mathbf{A}|\Psi\rangle\}^\dagger = \langle\Psi|\mathbf{A}^\dagger \quad \text{with} \quad \langle\Psi|\mathbf{A}^\dagger = \langle\Psi|A^*$$

Finally, for scalars, we have

$$\langle \Phi | \Psi \rangle = \langle \Psi | \Phi \rangle^* \qquad \text{and} \qquad \langle \Psi | \mathbf{B}^\dagger | \Phi \rangle = \langle \Phi | \mathbf{B} | \Psi \rangle^*$$

Thus, it is possible to deduce general rules allowing one to find the Hermitian conjugate of a general expression involving linear operators kets, bra, and scalars, that is,

1. Replace
   (a) scalars by their complex conjugates
   (b) kets by the corresponding bras and vice versa
   (c) linear operators by their Hermitian conjugates
2. Invert the order of the different terms, recalling that the position of the scalar is irrelevant.

As a first example, consider the following expression, which is a scalar:

$$\langle\Psi|\mathbf{A}|\Psi\rangle = A$$

Since the Hermitian conjugate of $|\Psi\rangle$ is $\langle\Psi|$ and vice versa and since the Hermitian conjugate of the scalar A is its complex conjugate A^*, the Hermitian conjugate of this expression is

$$\langle\Psi|\mathbf{A}^\dagger|\Psi\rangle = A^*$$

Now, consider the following operator:

$$\mathbf{B} = \lambda\langle\Phi|\mathbf{A}|\chi\rangle|\Phi\rangle\langle\mu|$$

Applying the above rules, its Hermitian conjugate is given by

$$\mathbf{B}^\dagger = \lambda^*|\mu\rangle\langle\Phi|\langle\chi|\mathbf{A}^\dagger|\Phi\rangle$$

Finally, consider the operator **C**, which consists of an exponential of another operator **A**:

$$\mathbf{C} = e^{i\mathbf{A}} \quad \text{with} \quad i^2 = -1$$

Since the complex conjugate of the scalar i is $-i$, the Hermitian conjugate of the operator **C** is

$$\mathbf{C}^\dagger = (e^{i\mathbf{A}})^\dagger = e^{-i\mathbf{A}^\dagger} \tag{1.44}$$

1.3 HERMITICITY AND UNITARITY

1.3.1 Hermitian operators

If certain linear operators **A** are equal to their Hermitian conjugate $\mathbf{A}^\dagger$, then they are said to be *Hermitian*:

$$\mathbf{A} = \mathbf{A}^\dagger \tag{1.45}$$

1.3.1.1 Reality of eigenvalues and orthonormality of the eigenvectors In order to show that the eigenvalues of Hermitian operators are real, let us write the eigenvalue equation of a linear operator

$$\mathbf{A}|\Psi_i\rangle = A_i|\Psi_i\rangle \tag{1.46}$$

where A_i is one of the eigenvalues of this operator and $|\Psi_i\rangle$ the corresponding eigenvector. Premultiply the two members of this equation by the bra $\langle\Psi_i|$, conjugate to the ket $|\Psi_i\rangle$:

$$\langle\Psi_i|\mathbf{A}|\Psi_i\rangle = \langle\Psi_i|A_i|\Psi_i\rangle$$

The eigenvalue A_i being a scalar, must commute with the bra so that one may write

$$\langle\Psi_i|\mathbf{A}|\Psi_i\rangle = A_i\langle\Psi_i|\Psi_i\rangle \tag{1.47}$$

Next, assume that the eigenvector $|\Psi_i\rangle$ is normalized, that is,

$$\langle\Psi_i|\Psi_i\rangle = 1$$

Then, Eq. (1.47) simplifies to

$$\langle\Psi_i|\mathbf{A}|\Psi_i\rangle = A_i \tag{1.48}$$

On the other hand, the Hermitian conjugate of this equation is

$$\langle\Psi_i|\mathbf{A}^\dagger|\Psi_i\rangle = A_i^* \tag{1.49}$$

Again, since we have assumed that the linear operator is Hermitian, it obeys Eq. (1.45), so that

$$\langle\Psi_i|\mathbf{A}^\dagger|\Psi_i\rangle = \langle\Psi_i|\mathbf{A}|\Psi_i\rangle$$

Thus, it appears from Eqs. (1.48) and (1.49) that the eigenvalue A_i of the Hermitian operator is equal to its complex conjugate A_i^*, that is, it is real since it obeys

$$A_i = A_i^* \tag{1.50}$$

Thus, we have the following property:

$$\boxed{\text{If} \qquad \mathbf{A} = \mathbf{A}^\dagger \quad \text{then} \quad A_i = A_i^*} \tag{1.51}$$

Now, in order to show that the eigenvectors of an Hermitian operator are orthogonal, let us write the eigenvalue equation of a linear operator for two distinct eigenvalues and eigenvectors:

$$\mathbf{A}|\Psi_i\rangle = A_i|\Psi_i\rangle \qquad \text{and} \qquad \mathbf{A}|\Psi_k\rangle = A_k|\Psi_k\rangle$$

The Hermitian conjugate of the first expression in this eigenvalue equation, is

$$\langle\Psi_i|\mathbf{A}^\dagger = A_i^*\langle\Psi_i|$$

Besides, if we assume that the operator $\mathbf{A}$ is Hermitian, then the eigenvalue A_i is real; then, according to Eq. (1.50), the following results hold:

$$\mathbf{A}|\Psi_k\rangle = A_k|\Psi_k\rangle \qquad \text{and} \qquad \langle\Psi_i|\mathbf{A} = A_i\langle\Psi_i| \quad \text{if } \mathbf{A} = \mathbf{A}^\dagger$$

Now, premultiply the two members of the first eigenvalue equation by the bra $\langle\Psi_i|$, and postmultiply the two members of its Hermitian conjugate by the ket $|\Psi_k\rangle$. Then, after commuting the eigenvalues that are scalars, one obtains the two expressions

$$\langle\Psi_i|\mathbf{A}|\Psi_k\rangle = A_k\langle\Psi_i|\Psi_k\rangle \qquad \text{and} \qquad \langle\Psi_i|\mathbf{A}|\Psi_k\rangle = A_i\langle\Psi_i|\Psi_k\rangle$$

Now, substract the second expression from the first one, that is,

$$\langle\Psi_i|\mathbf{A}|\Psi_k\rangle - \langle\Psi_i|\mathbf{A}|\Psi_k\rangle = (A_k - A_i)\langle\Psi_i|\Psi_k\rangle$$

yielding

$$(A_k - A_i)\langle\Psi_i|\Psi_k\rangle = 0$$

Thus, since we have assumed that the two eigenvalues are different, that is,

$$(A_k - A_i) \neq 0$$

hence it appears that the corresponding eigenvectors of Hermitian operators are orthogonal, leading us to write

$$\boxed{\langle \Psi_i | \Psi_k \rangle = 0 \quad \text{if} \quad \mathbf{A} = \mathbf{A}^\dagger \quad \text{with} \quad \mathbf{A}|\Psi_k\rangle = A_k|\Psi_k\rangle} \tag{1.52}$$

1.3.1.2 Trace and invariance of the trace

By definition the trace operation, denoted **tr**, over any operator **C** is[3]

$$\mathbf{tr}\{\mathbf{C}\} = \sum_n \langle \Phi_n | \mathbf{C} | \Phi_n \rangle \tag{1.53}$$

where the $|\Phi_n\rangle$ involved in the infinite sum belong to the basis $\{|\Phi_n\rangle\}$. Next, suppose that the operator **C** is the product of two operators **A** and **B**, which do not commute, that is,

$$\mathbf{C} = \mathbf{AB} \quad \text{with} \quad [\mathbf{A}, \mathbf{B}] \neq 0$$

Then, the trace takes the form

$$\mathbf{tr}\{\mathbf{AB}\} = \sum_n \langle \Phi_n | \mathbf{AB} | \Phi_n \rangle \tag{1.54}$$

Introduce between **A** and **B** the closure relation built up from the basis $\{|\Phi_m\rangle\}$. This procedure leads to a double summation not only over n but also over m:

$$\mathbf{tr}\{\mathbf{AB}\} = \sum_n \sum_m \langle \Phi_n | \mathbf{A} | \Phi_m \rangle \langle \Phi_m | \mathbf{B} | \Phi_n \rangle$$

Since the terms involved in the double summation are scalar, they commute, so that

$$\mathbf{tr}\{\mathbf{AB}\} = \sum_n \sum_m \langle \Phi_m | \mathbf{B} | \Phi_n \rangle \langle \Phi_n | \mathbf{A} | \Phi_m \rangle$$

Then, one may omit between **B** and **A** the closure relation involving the summation over n to give

$$\mathbf{tr}\{\mathbf{AB}\} = \sum_m \langle \Phi_m | \mathbf{BA} | \Phi_m \rangle \tag{1.55}$$

However, owing to the definition (1.53) of the trace, the right-hand side of Eq. (1.56) yields

$$\sum_m \langle \Phi_m | \mathbf{BA} | \Phi_m \rangle = \mathbf{tr}\{\mathbf{BA}\} \tag{1.56}$$

Hence, comparison of Eq. (1.54) and (1.56) shows that

$$\mathbf{tr}\{\mathbf{AB}\} = \mathbf{tr}\{\mathbf{BA}\}$$

so that the trace operation is invariant with respect to a permutation of **A** and **B**.

[3] In order to make clear what is meant by the trace operation, in the following we shall denote it, by $\mathbf{tr}\{\ \}$ where all the operators involved **A**, **B**... will be inside the notation $\{\ldots\}$. For instance, $\mathbf{tr}\{\mathbf{AB}\}$.

1.3.1.3 Hermitization of the product **AB** ***of Hermitian operators when*** $[\mathbf{A}, \mathbf{B}] \neq 0$ Consider the product **C** of two linear operators **A** and **B**:

$$\mathbf{C} = \mathbf{AB} \tag{1.57}$$

Again, assume that **A** and **B** are Hermitian operators that do not commute, that is,

$$\mathbf{A} = \mathbf{A}^\dagger \qquad \mathbf{B} = \mathbf{B}^\dagger \qquad [\mathbf{A}, \mathbf{B}] \neq 0$$

As we shall see, their product **C** is not Hermitian so that it is necessary to convert it to Hermitian form. To show that the product **C** is not Hermitian let us write with the aid of Eq. (1.43) the Hermitian of **C**:

$$\mathbf{C}^\dagger = (\mathbf{AB})^\dagger = \mathbf{B}^\dagger \mathbf{A}^\dagger$$

Since both operators are Hermitian, it is possible to write

$$\mathbf{C}^\dagger = \mathbf{BA} \quad \text{if} \quad \mathbf{B} = \mathbf{B}^\dagger \quad \text{and} \quad \mathbf{A} = \mathbf{A}^\dagger \tag{1.58}$$

Thus, since by hypothesis the two operators do not commute, the comparison of Eqs. (1.57) and (1.58) shows that the product **C** is not Hermitian. Hence, it is necessary to recall that

$$\mathbf{C}^\dagger \neq \mathbf{C} \quad \text{if} \quad [\mathbf{A}, \mathbf{B}] \neq 0 \quad \text{when} \quad \mathbf{A} = \mathbf{A}^\dagger \quad \text{and} \quad \mathbf{B} = \mathbf{B}^\dagger \tag{1.59}$$

In order to write the product in Hermitian form, we consider the linear combination of **C** and its Hermitian conjugate, namely

$$\mathbf{D} = \tfrac{1}{2}(\mathbf{C} + \mathbf{C}^\dagger)$$

Then, the Hermitian conjugate $\mathbf{D}^\dagger$ of **D** is Hermitian since

$$\mathbf{D}^\dagger = \tfrac{1}{2}(\mathbf{C}^\dagger + \mathbf{C}) = \mathbf{D}$$

As a consequence, it appears that the linear combination of the products **AB** and **BA** is Hermitian. Hence, important property of Hermitization of the product of two Hermitian operators follows, namely

$$\boxed{\mathbf{D} = \tfrac{1}{2}\{\mathbf{AB} + \mathbf{BA}\} = \mathbf{D}^\dagger \quad \text{if} \quad \mathbf{A} = \mathbf{A}^\dagger \quad \mathbf{B} = \mathbf{B}^\dagger \quad [\mathbf{A}, \mathbf{B}] \neq 0} \tag{1.60}$$

1.3.2 Eigenkets of two commuting Hermitian operators

1.3.2.1 First theorem Consider two operators **A** and **B** that commute, that is,

$$[\mathbf{A}, \mathbf{B}] = 0 \tag{1.61}$$

The eigenvalue equation of **A** is

$$\mathbf{A}|\Psi_i\rangle = A_i|\Psi_i\rangle \tag{1.62}$$

where A_i is the scalar eigenvalue of the operator **A**. Now, consider the action of the product **BA** of the two operators on any eigenket of **A**

$$\mathbf{BA}|\Psi_i\rangle = \mathbf{B}A_i|\Psi_i\rangle = A_i\mathbf{B}|\Psi_i\rangle = A_i|\mathbf{B}\Psi_i\rangle$$

In addition, owing to the nullity of the commutator (1.61), we have

$$\mathbf{BA}|\Psi_i\rangle = \mathbf{AB}|\Psi_i\rangle = \mathbf{A}|\mathbf{B}\Psi_i\rangle \tag{1.63}$$

where $|\mathbf{B}\Psi_i\rangle$ is the ket obtained by the linear transformation of $\mathbf{B}$ over $|\Psi_i\rangle$. Thus, by identification of the two last equations, it appears that

$$\mathbf{A}|\mathbf{B}\Psi_i\rangle = A_i|\mathbf{B}\Psi_i\rangle \tag{1.64}$$

This result shows that when $\mathbf{A}$ and $\mathbf{B}$ commute, and that, according to Eq. (1.62) if $|\Psi_i\rangle$ is an eigenket of $\mathbf{A}$, then, due to Eq. (1.64), $|\mathbf{B}\Psi_i\rangle$ is also an eigenket of $\mathbf{A}$.

In a like manner, if the eigenvalue equation of $\mathbf{B}$ is

$$\mathbf{B}|\Phi_k\rangle = B_k|\Phi_k\rangle$$

then one obtains

$$\mathbf{B}|\mathbf{A}\Phi_k\rangle = B_k|\mathbf{A}\Phi_k\rangle \tag{1.65}$$

showing that when $\mathbf{A}$ and $\mathbf{B}$ commute, if $|\Phi_k\rangle$ is an eigenket of $\mathbf{B}$, $|\mathbf{A}\Phi_k\rangle$ is also an eigenket of $\mathbf{B}$.

1.3.2.2 Second theorem Consider the two following eigenvalue equations of the same linear Hermitian operator $\mathbf{A}$:

$$\mathbf{A}|\Psi_1\rangle = A_1|\Psi_1\rangle \qquad \text{and} \qquad \mathbf{A}|\Psi_2\rangle = A_2|\Psi_2\rangle \tag{1.66}$$

where A_1 and A_2 are two different eigenvalues of $\mathbf{A}$, that is,

$$A_1 - A_2 \neq 0 \tag{1.67}$$

Now, consider another linear operator $\mathbf{B}$, which commutes with $\mathbf{A}$, but which is not necessarily Hermitian, that is,

$$[\mathbf{A}, \mathbf{B}] = 0$$

Then, owing to the nullity of this commutator, we have

$$\langle\Psi_1|[\mathbf{A}, \mathbf{B}]|\Psi_2\rangle = 0 \tag{1.68}$$

Expanding the commutator gives

$$\langle\Psi_1|[\mathbf{A}, \mathbf{B}]|\Psi_2\rangle = \langle\Psi_1|\mathbf{AB}|\Psi_2\rangle - \langle\Psi_1|\mathbf{BA}|\Psi_2\rangle$$

Then, using the first equation of (1.66) or the Hermitian conjugate of the second, one reads

$$\langle\Psi_1|\,[\mathbf{A}, \mathbf{B}]\,|\Psi_2\rangle = (A_1 - A_2)\,\langle\Psi_1|\mathbf{B}|\Psi_2\rangle \tag{1.69}$$

Hence, owing to Eqs. (1.67)–(1.69), it appears that

$$\langle\Psi_1|\mathbf{B}|\Psi_2\rangle = 0 \tag{1.70}$$

Thus, if $|\Psi_1\rangle$ and $|\Psi_2\rangle$ are eigenkets of any Hermitian operator $\mathbf{A}$, then Eq. (1.70) holds for any operator $\mathbf{B}$ that commutes with $\mathbf{A}$.

1.3.3 Eigenvalue equation of an exponential operator

Consider an exponential operator $e^{\xi\mathbf{A}}$, which is a function of the scalar ξ, and another operator $\mathbf{A}$ obeying the eigenvalue equation

$$\mathbf{A}|\Phi_n\rangle = A_n|\Phi_n\rangle \tag{1.71}$$

We search what is the effect of this operator on an eigenstate $|\Phi_n\rangle$. For this purpose, we may expand on the right-hand side of this last equation the exponential operator in Taylor series, to give

$$e^{\xi\mathbf{A}}|\Phi_n\rangle = \sum_k \frac{\xi^k}{k!}\mathbf{A}^k|\Phi_n\rangle \tag{1.72}$$

Now, observe that

$$\mathbf{A}^k|\Phi_n\rangle = \mathbf{A}^{k-1}\mathbf{A}|\Phi_n\rangle$$

or, in view of Eq. (1.71)

$$\mathbf{A}^k|\Phi_n\rangle = \mathbf{A}^{k-1}A_n|\Phi_n\rangle$$

Again, after commuting the scalar A_n with the operator $\mathbf{A}^{k-1}$ written $\mathbf{A}^{k-2}\mathbf{A}$, one obtains

$$\mathbf{A}^k|\Phi_n\rangle = A_n\mathbf{A}^{k-2}\mathbf{A}|\Phi_n\rangle$$

or

$$\mathbf{A}^k|\Phi_n\rangle = A_nA_n\mathbf{A}^{k-2}|\Phi_n\rangle$$

Proceeding in the same way for each power of $\mathbf{A}$, one gets finally

$$\mathbf{A}^k|\Phi_n\rangle = A_n^k|\Phi_n\rangle$$

Then, using this result, Eq. (1.72) becomes

$$e^{\xi\mathbf{A}}|\Phi_n\rangle = \sum_k \frac{\xi^k}{k!}A_n^k|\Phi_n\rangle$$

Again, return to the expansion appearing on the right-hand side of this last equation to the exponential, and one obtains

$$\boxed{e^{\xi\mathbf{A}}|\Phi_n\rangle = e^{\xi A_n}|\Phi_n\rangle} \tag{1.73}$$

1.3.4 Unitary operators

Consider the inverse $\mathbf{U}^{-1}$ of a linear operator $\mathbf{U}$. This inverse is defined by

$$\mathbf{U}\mathbf{U}^{-1} = \mathbf{U}^{-1}\mathbf{U} = \mathbf{1}$$

Next, assume that the inverse $\mathbf{U}^{-1}$ of the linear operator $\mathbf{U}$ is the Hermitian conjugate of $\mathbf{U}$:

$$\mathbf{U}^{-1} = \mathbf{U}^{\dagger}$$

Then the operator **U**, which is said to be *unitary*, obeys the following relation:

$$\boxed{\mathbf{UU}^\dagger = \mathbf{U}^\dagger\mathbf{U} = 1 \quad \text{if} \quad \mathbf{U}^{-1} = \mathbf{U}^\dagger} \tag{1.74}$$

As an example of unitary operator, consider the following expression for the linear operator **U**, which is an exponential of the Hermitian operator **B** times a real scalar λ times the imaginary number i:

$$\mathbf{U} = e^{i\lambda\mathbf{B}} \quad \text{with} \quad \mathbf{B} = \mathbf{B}^\dagger \qquad \lambda = \lambda^* \qquad \text{and} \qquad i^2 = -1$$

Then, using Eq. (1.44), the Hermitian conjugate of **U** appears to be given by

$$\mathbf{U}^\dagger = e^{-i\lambda\mathbf{B}}$$

On the other hand, it is obvious that the inverse of **U** is

$$\mathbf{U}^{-1} = e^{-i\lambda\mathbf{B}}$$

As a consequence, comparison of the two above equations shows that the Hermitian conjugate of **U** is its inverse, showing that **U** is unitary:

$$\mathbf{U}^\dagger = \mathbf{U}^{-1}$$

1.4 ALGEBRA OPERATORS

Here, we give some important results dealing with the algebra of operators, which are proved in Appendices 1–5. They are

- The commutator involving three noncommuting operators. **A**, **B**, and **C**:

$$\boxed{[\mathbf{A}, \mathbf{BC}] = [\mathbf{A}, \mathbf{B}]\mathbf{C} + \mathbf{B}[\mathbf{A}, \mathbf{C}]} \tag{1.75}$$

- The transformations

$$\boxed{e^{\xi\mathbf{A}}\mathbf{B}e^{-\xi\mathbf{A}} = \mathbf{B} + [\mathbf{A}, \mathbf{B}]\xi + \frac{1}{2}[\mathbf{A}, [\mathbf{A}, \mathbf{B}]]\xi^2 + \frac{1}{3!}[\mathbf{A}, [\mathbf{A}, [\mathbf{A}, \mathbf{B}]]]\xi^3 + \ldots} \tag{1.76}$$

$$\boxed{e^{\xi\mathbf{A}}\mathbf{F}(\mathbf{B})e^{-\xi\mathbf{A}} = \mathbf{F}(e^{\xi\mathbf{A}}\mathbf{B}e^{-\xi\mathbf{A}})} \tag{1.77}$$

  where ξ is a scalar and **A** and **B** are two independent linear operators that do not depend on ξ and that do not commute.
- the *Glauber* or *Glauber–Weyl relation*

$$\boxed{e^{\xi\mathbf{A}}e^{\xi\mathbf{B}} = e^{(\mathbf{A}+\mathbf{B})\xi}e^{+[\mathbf{A},\mathbf{B}]\xi^2/2} \text{ with } [\mathbf{A}, [\mathbf{A}, \mathbf{B}]] = 0 \quad \text{and} \quad [\mathbf{B}, [\mathbf{A}, \mathbf{B}]] = 0} \tag{1.78}$$

  where ξ is a scalar and may be also written

$$e^{(\mathbf{A}+\mathbf{B})\xi} = e^{\xi\mathbf{A}}e^{\xi\mathbf{B}}e^{-[\mathbf{A},\mathbf{B}]\xi^2/2} = e^{(\mathbf{B}+\mathbf{A})\xi} \tag{1.79}$$

$$e^{(\mathbf{A}+\mathbf{B})\xi} = e^{\xi\mathbf{B}\xi\mathbf{A}}e^{-[\mathbf{B},\mathbf{A}]\xi^2/2} = e^{(\mathbf{B}+\mathbf{A})\xi} \tag{1.80}$$

In the latter equations, the last terms on the right-hand side have been introduced in order to focus attention on the fact that

$$e^{(\mathbf{B}+\mathbf{A})\xi} = e^{(\mathbf{A}+\mathbf{B})\xi}$$

Now, we may summarize the most important results as follows:

Basic equations for quantum mechanics
Linear transformations and their Hermitian conjugates:
$\mathbf{A}\vert\Psi\rangle = \vert\Phi\rangle \qquad \langle\Psi\vert\mathbf{A}^\dagger = \langle\Phi\vert$ **Hermitian operators A, unitary operators U, commutators:** $\mathbf{A} = \mathbf{A}^\dagger \qquad \mathbf{U}^{-1} = \mathbf{U}^\dagger \quad \text{with} \quad \mathbf{U}\mathbf{U}^{-1} = \mathbf{U}^{-1}\mathbf{U} = 1 \qquad [\mathbf{A}, \mathbf{B}] = \mathbf{AB} - \mathbf{BA}$
Eigenvalue equations and their Hermitean conjugates:
$\mathbf{A}\vert\Psi_i\rangle = A_i\vert\Psi_i\rangle \qquad \langle\Psi_i\vert\mathbf{A}^\dagger = \langle\Psi_i\vert A_i^*$
Eigenvalue equations of Hermitian operators and their Hermitean conjugates:
$\mathbf{A}\vert\Psi_i\rangle = A_i\vert\Psi_i\rangle \qquad \langle\Psi_i\vert\mathbf{A} = \langle\Psi_i\vert A_i$ with $\langle\Psi_i\vert\Psi_k\rangle = \delta_{ik}$ and $\sum \vert\Psi_k\rangle\langle\Psi_k\vert = 1$
An important relation:
$\langle\Psi\vert\mathbf{B}^\dagger\vert\Phi\rangle = \langle\Phi\vert\mathbf{B}\vert\Psi\rangle^*$
Invariance of the trace:
$\sum\langle\Psi_k\vert\mathbf{AB}\vert\Psi_k\rangle = \sum\langle\Psi_k\vert\mathbf{BA}\vert\Psi_k\rangle \quad \text{even if} \quad [\mathbf{A}, \mathbf{B}] \neq 0$

CHAPTER 2

BASIS FOR QUANTUM APPROACHES OF OSCILLATORS

INTRODUCTION

Using the mathematical basis treated in this chapter, it will be possible to discuss the quantum mechanics tools necessary for the study of the behavior of oscillators. We begin with an exposition of the postulates of quantum mechanics, which will be the purpose of Section 2.1. An important place will be given to the notions of *quantum average values* and to *quantum fluctuations*, allowing one to deduce from quantum principles the *Heisenberg uncertainty relations* according to which it is not possible to simultaneously know with arbitrary accuracy both the position and the momentum of any particle. In a subsequent section, some dynamic aspects will be developed allowing one both to determine the time dependence of the quantum average values and show that the Heisenberg uncertainty relations introduce a limit to the perfect knowledge assumed by classical mechanics. However, the quantum principles lead to the *Ehrenfest equations*, which nearly behave as the Newton equations, save that they are dealing with average values and not with exact ones, as for the classical equations. Related to these dynamic aspects, we shall prove the *energy conservation*, in a quantum averaged form, and the *virial theorem* relating the quantum average value of the kinetic and potential energies to the total energy, which holds also in classical mechanics. The last section will be devoted to some developments dealing with quantum concepts related to the connection between the position and the momentum, which will be used in Chapter 3 to relate quantum mechanics to wave mechanics.

2.1 OSCILLATOR QUANTIZATION AT THE HISTORICAL ORIGIN OF QUANTUM MECHANICS

2.1.1 Ultraviolet catastrophe

Measurements of thermal capacity of solids were discovered at the beginning of the twentieth century to be in contradiction with the principles of statistical physics

Quantum Oscillators, First Edition. Olivier Henri-Rousseau and Paul Blaise.
© 2011 John Wiley & Sons, Inc. Published 2011 by John Wiley & Sons, Inc.

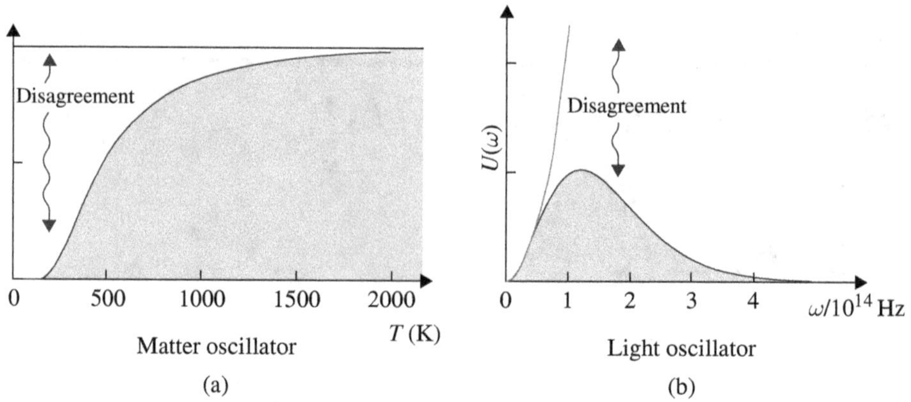

Figure 2.1 Contradiction between experiment (shaded areas) and classical prediction (lines).

based on classical mechanics: The experiments show that these thermal capacities are temperature dependent, whereas the theory assuming that they result from the partial derivative with respect to the temperature of the average oscillation energy of the atoms within the solids predicted that they ought to be constant, due to the equipartition theorem of statistical mechanics applied to classical mechanics, according to which each degree of freedom of vibration of the solid contributes the same energy amount k_BT (where k_B is the Boltzmann constant and T the absolute temperature). See, for instance, Fig. 2.1a. In addition, the study of the frequency distribution of the intensity of the electromagnetic radiations enclosed in a heated cavity at thermal equilibrium (black-body radiations) lead to the results that this intensity narrows to zero as the frequency increases, in utter contradiction with the classical statistical mechanics predictions (applied to Maxwell electromagnetic modes of vibration) by Rayleigh and Jeans, according to which the intensity ought to tend to infinity (*the ultraviolet catastrophe*). See Fig. 2.1b.

2.1.2 Planck, Einstein, and Bohr's old quantum mechanics

To reconcile the ultraviolet catastrophe with physics, Planck (1858–1947) assumed that the walls of the black body responsible for the absorption and emission of ultraviolet light are made of small oscillators of various frequencies, the energy of which cannot vary continuously as in Newtonian mechanics, but is quantized, the energy levels E_n obeying

$$E_n(\text{matter oscillator}) = nh\nu$$

where n is an integer, ν the frequency of the microscopic oscillator, and h *Planck's constant*. With this assumption of the oscillator energy quantization, Planck was able in 1901 to reproduce with great accuracy the experimental results.

Moreover, some time later, Einstein proposed (1905) a theoretical interpretation of the *photoelectric effect*, a recent and unexplained laboratory result: It had been discovered that an electron can be expelled from a material by a light radiation, when its frequency is greater than a threshold characteristic of the material, the kinetic energy of the emitted electron increasing linearly with the light frequency beyond the threshold. To interpret that Einstein assumed that light, considered at this time by the physicists as of wave nature, has also to be considered as consisting of a grain of light, *the photon*, the energy of which is proportional to the angular frequency ω of the light, the proportionality constant being that introduced by Planck in his theory.

$$E_n(\text{light oscillator}) = n\hbar\omega \quad \text{with} \quad \hbar = \frac{h}{2\pi}$$

A few years later, in 1913, Bohr (1885–1962), a Danish physicist, attacked the problem raised by the absorption and emission of light rays by hydrogen atoms. The frequencies of these lines, which are the same for both processes, were found by Balmer (1825–1898) to obey with a perfect precision an empirical formula, the *Balmer formula*, involving integer numbers. Bohr was able to theoretically reproduce the empirical Balmer formula by assuming that the angular momentum of the electron generated by its circular orbit motion around the proton is quantized, being an integer multiple of Planck's constant already introduced in the Planck and in the Einstein theoretical approaches.

Moreover, Bohr assumed that when the electron moves from one orbit to another, it performs that in a sudden and unrepresentative manner, by emitting or absorbing a quantum of light (photon) of frequency given by the absolute difference between the orbit's energies divided by Planck's constant. In addition, to link his theoretical approach with classical mechanics, Bohr introduced a *correspondence principle*, claiming that when the quantized energy levels of the electronic orbits are higher and higher, the transitions between successive energy levels involve a dynamics that approaches more and more closely the classical circular motion.

2.1.3 Heisenberg and matrix mechanics

All these works of Planck, Einstein, and Bohr called into question the continuous variation of the energy level of atoms since they assumed that energy may change only by small packets, the energy quanta involving the Planck constant. These works aroused passionate debates, some scientists thinking with Bohr that the classical mechanics of Newton would have to be rethought from top to bottom in order to deeply reflect the new realities at the scale of molecules, atoms, and their elementary constituents.

Among these young physicists, Heisenberg (1901–1976) played a pioneering role, building during his thesis in 1924 a new theory. He focused on quantizing the energy of microscopic oscillators proposed by Planck. His ideas were primarily based on two kinds of square noncommutative matrices, one of which was intended to represent the position coordinate and the other one the conjugated momentum. Heisenberg completed this assumption by introducing Planck's constant in these matrices.

Heisenberg justified his assumption of noncommutative matrices representing position coordinates and momentum by the positivist postulate that it is impossible on the atomic scale to measure the position of a particle without changing instantly ipso facto speed and therefore its momentum. Heisenberg was able, from his noncommutative matrices (recognized as such by Jordan), to find the formulas postulated by Planck for the quantization of the energy of small oscillators belonging to the atomic scale. This work may be regarded as the foundation stone of the new quantum mechanics.

2.1.4 De Broglie and wave mechanics

As seen above, Einstein introduced in his interpretation of the photoelectric effect the necessity to add corpuscular properties to the wave ones assumed for light, following the interference experiments of Young and others. This dual nature of light, Louis de Broglie (1892–1987) has extended it to *matter*, that is, all entities involving mass, which comprise the physical realities around us: At the same time Heisenberg was working on his thesis on the matrix mechanics, de Broglie, starting from intuitions of the Irish physicist Hamilton (1805–1865). proposed a new mechanism applying to the microscopic scale in which a wave is associated with the particle dynamics. In this new mechanics, the wavelength λ (*de Broglie wavelength*) of free particles (particles moving in a straight line in the absence of potential) is equal to Planck's constant divided by the momentum p of the particles (*de Broglie relation*):

$$\lambda = \frac{h}{p}$$

Hence, since the momentum is proportional to the product of mass times velocity, the de Broglie wavelength becomes smaller the greater the mass, so that it becomes negligible when going from the atomic and molecular scales to the human one and, a fortiori, to those of the planets and stars.

In this wave mechanics, the corpuscular properties of matter are linked with the position coordinates, while the wave properties are linked to the momentum through the de Broglie wavelength. That is the origin of the term *wave mechanics* given to this new discipline of physics. One of the famous theoreticians of the time, the Austrian physicist Schrödinger (1887–1961), who initially despised the ideas of the young French physicist de Broglie, thereafter applied them to the hydrogen atom. By solving the partial differential equation governing in wave mechanics the electron behavior of the hydrogen atom, Schrödinger retrieved the results of Bohr concerning the empirical Balmer formula.

Wave mechanics was soon experimentally confirmed by Davisson (1881–1958) and Germer (1896–1971) in connection with diffraction observations on crystals, allowing to verify the validity of the de Broglie relation. In addition the wave particle duality nature became evident via new experiments where particles having crossed separately a dispersion pattern, strike a screen by exhibit an interference pattern, thus suggesting that each isolated particle *interferes with itself*. This phenomenon was observed for light (photons) and also for material particles such as atoms.

2.2 QUANTUM MECHANICS POSTULATES AND NONCOMMUTATIVITY

2.2.1 The principles

2.2.1.1 First postulate At a given time, the physical state of a system is described by a *ket* $|\Psi_j(t)\rangle$ belonging to the *state space*, that is, to a vector space of infinite dimension involving complex scalar products.

2.2.1.2 Second postulate With each classical physical variable A is associated a linear operator $\mathbf{A}$ acting in the state space, which must be *Hermitian* (*observable*), and obeying, therefore,

$$\mathbf{A} = \mathbf{A}^\dagger$$

2.2.1.3 Third postulate The possible measurements of an observable $\mathbf{A}$ are given by the *eigenvalues* A_n of this operator, that is

$$\mathbf{A}|\Phi_n\rangle = A_n|\Phi_n\rangle$$

where $|\Phi_n\rangle$ are the corresponding eigenkets of the eigenvectors of $\mathbf{A}$. Owing to the Hermiticity of the observables, their eigenvalues are real:

$$A_n = A_n^*$$

This constraint of Hermiticity on linear operators, which describe the physical variables, avoids the possibility of complex expressions involving an imaginary part in measurements of many physical variables.

2.2.1.4 Fourth postulate The transition of a system from any ket $|\Phi_n\rangle$ to another $|\Psi_j\rangle$ cannot be predicted in a deterministic way but only in a probabilistic one defined by a probability $\mathbb{P}_{nj}$, which may be calculated from the squared modulus of the scalar product of the initial and final kets, that is, by

$$\boxed{\mathbb{P}_{nj} = |\langle\Psi_j|\Phi_n\rangle|^2} \tag{2.1}$$

2.2.1.5 Fifth postulate This postulate concerns situations where the eigenvalues are *degenerate*, which we shall not encounter here. Thus, in order to simplify, we do not give it here.

2.2.1.6 Sixth postulate There are two different equivalent ways to obtain the dynamics of a quantum system. In the first one, the kets and bras are time dependent and the operators are constants. This is the *Schrödinger picture*. In the second one, the kets and bras are constants, and it is the operators that are time dependent. The latter is the *Heisenberg picture*.

The sixth postulate, in the Schrödinger picture, states that the kets describing a physical system evolve with time between two quantum jumps in a deterministic way, which is given by the following equation named the *time-dependent Schrödinger equation* or more shortly the *Schrödinger equation*:

$$\boxed{i\hbar \frac{\partial}{\partial t}|\Psi_j(t)\rangle = \mathbf{H}|\Psi_j(t)\rangle \quad \text{with} \quad i^2 = -1} \tag{2.2}$$

where $\mathbf{H}$ is the total quantum Hamiltonian describing the system, whereas $\hbar$ is the Planck constant divided by 2π. Note that in Eq. (2.2) the partial derivative with respect to time is sometimes replaced by a time derivative. However, as the ket may be affected by transformations other than that of time, for instance, translations of the origin (vide infra the translation operators), we prefer the partial derivative notation.

2.2.1.7 Seventh postulate The quantum operator $\mathbf{A}$ describing a classical physical variable A may be obtained as follows:

1. Express the classical variable A in terms of the space variables Q_k related to the different freedom degrees k of the system, and of their corresponding conjugate momentum P_k, that is, write $A(P_k, Q_k)$.
2. Associate the Hermitian operators $\mathbf{Q}_i$ and $\mathbf{P}_i$, respectively, to each space variable Q_k and to its corresponding conjugate momentum P_k, in order to pass from the classical expression $A(P_k, Q_k)$ to the corresponding quantum Hermitian operator $\mathbf{A}(\mathbf{P}_k, \mathbf{Q}_k)$, that is,

$$A(P_k, Q_k) \rightarrow \mathbf{A}(\mathbf{P}_k, \mathbf{Q}_k)$$

3. Require that the $\mathbf{Q}_k$ and $\mathbf{P}_k$ operators obey the commutation rule

$$\boxed{[\mathbf{Q}_k, \mathbf{P}_l] = i\hbar\delta_{kl} \quad \text{with} \quad i^2 = -1} \tag{2.3}$$

where $\hbar$ is the Planck constant given by

$$\hbar = \frac{h}{2\pi} = \frac{6.62}{2\pi} \times 10^{-34}\ \mathrm{J \cdot S}$$

Some further information concerning the commutation rules are given in Section 18.5.

The third and fourth postulates lead to the following important remarks:

The third postulate leads one to distinguish, in the measurement of an observable, two different possibilities according to whether or not before any measurement of one of its observables, the system was in an eigenket of the measured operator. This postulate gives directly a response only in the specific situation where the system was in an eigenket of this last one. Owing to Eq. (2.3), noting that the basic physical variables $\mathbf{P}$ and $\mathbf{Q}$ do not commute, different Hermitian operators $\mathbf{A}(\mathbf{P}_k, \mathbf{Q}_k)$ and $\mathbf{B}(\mathbf{P}_l, \mathbf{Q}_l)$, both functions of $\mathbf{P}$ and $\mathbf{Q}$, have no reasons to commute:

$$[\mathbf{A}(\mathbf{P}_k, \mathbf{Q}_k), \mathbf{B}(\mathbf{P}_l, \mathbf{Q}_l)] \neq 0 \tag{2.4}$$

To make clear the discussion, write the eigenvalue equations of these two Hermitian operators:

$$\mathbf{A}(\mathbf{P}_k, \mathbf{Q}_k)|\Phi_\nu\rangle = A_\nu|\Phi_\nu\rangle \qquad \text{and} \qquad \mathbf{B}(\mathbf{P}_k, \mathbf{Q}_k)|\Psi_\mu\rangle = B_\mu|\Psi_\mu\rangle$$

where $|\Phi_\nu\rangle$ and $|\Psi_\mu\rangle$ are, respectively, the eigenkets of $\mathbf{A}(\mathbf{P}_k, \mathbf{Q}_k)$ and $\mathbf{B}(\mathbf{P}_k, \mathbf{Q}_k)$, whereas A_ν and B_μ are the corresponding eigenvalues. Of course, since these operators do not commute, they do not admit the same set of eigenvectors. Moreover, since they are Hermitian, each eigenket of one of these operators may be linearly expanded in the set of eigenkets of the other operator. For instance,

$$|\Psi_\mu\rangle = \sum_\nu a_{\nu\mu}|\Phi_\nu\rangle \quad \text{with} \quad a_{\nu\mu} = \langle\Phi_\nu|\Psi_\mu\rangle \tag{2.5}$$

Now, suppose that at an initial time the system is in one of the eigenstates $|\Psi_\mu\rangle$ of the $\mathbf{B}(\mathbf{P}_k, \mathbf{Q}_k)$ operator. Next, if a measurement of the Hermitian operator $\mathbf{A}(\mathbf{P}_k, \mathbf{Q}_k)$ is performed on this system, then, according to the third postulate, this measurement will yield, for instance, A_η of the different eigenvalues and A_ν of the Hermitian operator $\mathbf{A}(\mathbf{P}_k, \mathbf{Q}_k)$. That implies that, after such a measurement, the system is now in the ket $|\Phi_\eta\rangle$ corresponding to the eigenvalue A_η. It appears, therefore, that measurement of the operator $\mathbf{A}(\mathbf{P}_k, \mathbf{Q}_k)$ of the system, which was initially in the ket $|\Psi_\mu\rangle$, has induced a jump in the ket $|\Phi_\eta\rangle$. Hence, according to the fourth postulate, this jump is not deterministic but occurs with probability

$$\mathbb{P}_{\mu\eta} = |\langle\Psi_\mu|\Phi_\eta\rangle|^2$$

or, compare, Eq. (2.5),

$$\mathbb{P}_{\mu\eta} = \left|\sum_\nu a_{\nu\mu}\langle\Phi_\nu|\Phi_\eta\rangle\right|^2$$

so that due to the orthonormality of the eigenkets of a Hermitian operator $\mathbf{A}(\mathbf{P}_k, \mathbf{Q}_k)$

$$\mathbb{P}_{\mu\eta} = \left|\sum_\nu a_{\nu\mu}\delta_{\eta\nu}\right|^2 = |a_{\mu\eta}|^2$$

Thus, the measurement of $\mathbf{A}(\mathbf{P}_k, \mathbf{Q}_k)$ has induced the abrupt change

$$\sum_\nu a_{\nu\mu}|\Phi_\nu\rangle \rightarrow |\Phi_\eta\rangle$$

with the probability equal to the squared absolute value of the coefficients $a_{\mu\eta}$ of the expansion appearing on the left-hand side of this last equation. As a matter of fact, the measurement of $\mathbf{A}(\mathbf{P}_k, \mathbf{Q}_k)$ has induced a reduction of the left-hand-side expansion, which is called the *wave packet reduction*, for historical reasons related to the fact that in wave mechanics the $|\Phi_\nu\rangle$ may be related to different orthogonal wavefunctions (see the discussion in Chapter 3 dealing with wave mechanics).

2.2.2 Classical mechanics as special limit of quantum mechanics

Despite its very formal character, which is far from classical mechanics, quantum mechanics is not without a link with it. As we shall see, it is possible from the postulates of quantum mechanics, to demonstrate the following equations, named the *Ehrenfest equations*, which govern the dynamics of a system:

$$\boxed{\frac{d\langle\mathbf{Q}(t)\rangle_\Psi}{dt} = \frac{\langle\mathbf{P}(t)\rangle_\Psi}{m} \quad \text{and} \quad \frac{d\langle\mathbf{P}(t)\rangle_\Psi}{dt} = -\left\langle\frac{\partial\mathbf{V}}{\partial\mathbf{Q}}\right\rangle_\Psi} \tag{2.6}$$

Here, $\langle \mathbf{Q}(t) \rangle_\Psi$, $\langle \mathbf{P}(t) \rangle_\Psi$, and $\langle (\partial \mathbf{V}/\partial \mathbf{Q}) \rangle_\Psi$ are, respectively, the average values of the position, momentum, and potential when the system is in the quantum state characterized by the ket $|\Psi\rangle$. Now, these equations are very similar to Newton's equations:

$$\boxed{\frac{d\vec{Q}(t)}{dt} = \frac{\vec{P}(t)}{m} \quad \text{and} \quad \frac{d\vec{P}(t)}{dt} = -\frac{\partial V}{\partial \vec{Q}}} \tag{2.7}$$

Letters with arrow mean vectorial entities in classical mechanics, at the opposite of bold letters appearing above and meaning quantum mechanical operators. However, an important difference exists because the quantum Eqs. (2.6) govern average values, whereas the classical Eqs. (2.7) govern exact ones. Hence, if they are average values $\langle \mathbf{A} \rangle_\Psi$, they indicate that dispersion of the possible values around the average values exists, which may be analyzed using the *variance* $\Delta\mathbf{A}$, namely

$$\Delta\mathbf{A} = \sqrt{\langle \mathbf{A}^2 \rangle_\Psi - \langle \mathbf{A} \rangle_\Psi^2} \tag{2.8}$$

where $\langle \mathbf{A}^2 \rangle_\Psi$ is the average of the square of $\mathbf{A}$. For the position and the momentum, the time-dependent average values are governed by Eqs. (2.6), whereas the corresponding variance are governed by the *Heisenberg uncertainty relation* (which will be demonstrated later). Figure 2.2 shows two situations occurring for the relative variance $\Delta\mathbf{A}/\langle \mathbf{A} \rangle$, the left-hand-side showing a quantum behavior, whereas the right-hand-side exhibits classical behavior.

$$\boxed{(\Delta\mathbf{P}(t))_\Psi (\Delta\mathbf{Q}(t))_\Psi \geqslant \frac{\hbar}{2}} \tag{2.9}$$

The passage from the quantum mechanics to the classical mechanics occurs when

$$\boxed{\frac{(\Delta\mathbf{P}(t))_\Psi}{\langle \mathbf{P}(t) \rangle_\Psi} \to 0} \quad \text{and} \quad \boxed{\frac{(\Delta\mathbf{Q}(t))_\Psi}{\langle \mathbf{Q}(t) \rangle_\Psi} \to 0} \tag{2.10}$$

When the size of the system is very small, of the order of the size of molecules or atoms or smaller, the quantum mechanics of Eqs. (2.6) holds. However, when this

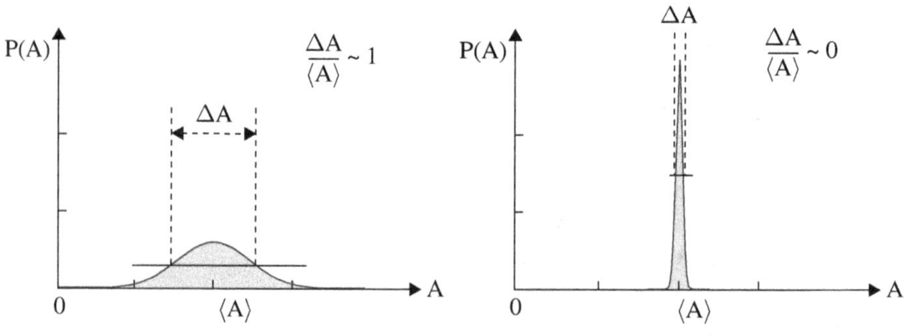

Figure 2.2 Quantum and classical relative variance $\Delta\mathbf{A}/\langle \mathbf{A} \rangle$.

size is progressively increasing, the conditions (2.10) are more and more verified so that the quantum mechanics of Eqs. (2.6) transforms to classical mechanics (2.7).

The specific physical behavior as the size of the system decreases is linked to the basic uncertainty characterizing the fundamental physical variables of small particles manifested via the following probability passage from any state of position to one of momentum, which may be the initial position and the final momentum, or vice versa, through the following relation, which will be demonstrated later

$$\boxed{|\langle\{P\}|\{Q\}\rangle|^2 = \frac{1}{2\pi\hbar}} \tag{2.11}$$

where $|\{Q\}\rangle$ and $|\{P\}\rangle$ are, respectively, the eigenkets of the position $\mathbf{Q}$ and momentum $\mathbf{P}$ operators defined, according to the third postulate by the continuous eigenvalue equation

$$\mathbf{Q}|\{Q\}\rangle = Q|\{Q\}\rangle \qquad \text{and} \qquad \mathbf{P}|\{P\}\rangle = P|\{P\}\rangle$$

where Q and P are, respectively, the eigenvalues of $\mathbf{Q}$ and $\mathbf{P}$, and thus the respective measured values of these operators, when the system is either in $|\{Q\}\rangle$ or in $|\{P\}\rangle$, Eq. (2.11) implies therefore that, after a measurement of the position yielding Q, another measurement of the momentum may lead to all the possible values P, with the same probability, is given by

$$\mathbb{P}_{Q\rightarrow P} = \mathbb{P}_{P\rightarrow Q} = \frac{1}{2\pi\hbar}$$

The Heisenberg uncertainty relation (2.9) and the jump probability (2.11) are consequence of the fundamental commutator

$$[\mathbf{Q}, \mathbf{P}] = i\hbar \tag{2.12}$$

Thus, the noncommutativity properties of observables, which are very general, play a fundamental role in the knowledge of the possible measurement of a physical variable. In order to appreciate the role played by the Hermitian operators in quantum mechanics, it is necessary to find the expression of the commutators $[\mathbf{Q}, \mathbf{F}(\mathbf{P})]$ and $[\mathbf{P}, \mathbf{F}(\mathbf{Q})]$, which are deeply linked to their behavior. In Appendix 5 are demonstrated some expressions dealing with commutators that are functions of $\mathbf{P}$ and $\mathbf{Q}$, and that result from the basic commutator (2.12). They are the following:

$$\boxed{[\mathbf{Q}, \mathbf{P}^n] = n(i\hbar)\mathbf{P}^{n-1}} \tag{2.13}$$

$$\boxed{[\mathbf{Q}, \mathbf{F}(\mathbf{P})] = (i\hbar)\left(\frac{\partial \mathbf{F}(\mathbf{P})}{\partial \mathbf{P}}\right)} \tag{2.14}$$

$$\boxed{[\mathbf{P}, \mathbf{Q}^n] = -n(i\hbar)\mathbf{Q}^{n-1}}$$

$$\boxed{[\mathbf{P}, \mathbf{F}(\mathbf{Q})] = -(i\hbar)\left(\frac{\partial \mathbf{F}(\mathbf{Q})}{\partial \mathbf{Q}}\right)} \tag{2.15}$$

2.3 HEISENBERG UNCERTAINTY RELATIONS

2.3.1 Mean values

Clearly, according to the third postulate, if a system is in a state $|\Phi_n\rangle$, which is an eigenvector of some Hermitian operator **A**, the measurement of the physical variable associated to this operator is given by the corresponding real eigenvalue A_n of this operator, that is,

$$\mathbf{A}|\Phi_n\rangle = A_n|\Phi_n\rangle \quad \text{with} \quad \mathbf{A} = \mathbf{A}^\dagger \tag{2.16}$$

However, if the system is described by a state $|\Psi_k\rangle$ that is not an eigenvector of the operator, we have seen that, according to the fourth postulate, there are as many possibilities to get measurements of the physical variable associated to **A** as there are eigenvalues of **A**. Then the only possibility for a measurement of **A** is an average value $\langle\mathbf{A}\rangle_{\Psi_k}$ given by

$$\langle\mathbf{A}\rangle_{\Psi_k} = \langle\Psi_k|\mathbf{A}|\Psi_k\rangle \tag{2.17}$$

To show that $\langle\mathbf{A}\rangle_{\Psi_k}$ is an average value, use the closure relation of the eigenkets of the Hermitian operator **A**:

$$1 = \sum_n |\Phi_n\rangle\langle\Phi_n| \tag{2.18}$$

Then insert it on the right-hand side of Eq. (2.17) just after **A**:

$$\langle\mathbf{A}\rangle_{\Psi_k} = \langle\Psi_k|\mathbf{A}\left\{\sum_n |\Phi_n\rangle\langle\Phi_n|\right\}|\Psi_k\rangle$$

This last expression reads in the more usual form on commuting the sum

$$\langle\mathbf{A}\rangle_{\Psi_k} = \sum_n \langle\Psi_k|\mathbf{A}|\Phi_n\rangle\langle\Phi_n|\Psi_k\rangle$$

Again, according to the eigenvalue Eq. (2.16), this equation transforms to

$$\langle\mathbf{A}\rangle_{\Psi_k} = \sum_n \langle\Psi_k|A_n|\Phi_n\rangle\langle\Phi_n|\Psi_k\rangle$$

or, on commuting the eigenvalues A_n that are scalars

$$\langle\mathbf{A}\rangle_{\Psi_k} = \sum_n A_n\langle\Psi_k|\Phi_n\rangle\langle\Phi_n|\Psi_k\rangle$$

Moreover, using the fact that the two right-hand-side scalar products are complex conjugates, we have

$$\langle\mathbf{A}\rangle_{\Psi_k} = \sum_n A_n|\langle\Psi_k|\Phi_n\rangle|^2 \tag{2.19}$$

Finally, due to the fourth postulate, the right-hand-side squared modulus is the transition probability to pass from the ket $|\Psi_k\rangle$ in which the system was initially before the measurement of **A** to the eigenket $|\Phi_n\rangle$ of this operator **A** associated with the eigenvalue A_n to which the measurement of **A** has lead.

$$|\langle\Psi_k|\Phi_n\rangle|^2 = \mathbb{P}_{kn} \tag{2.20}$$

Thus, the left-hand side of Eq. (2.19), which is defined by Eq. (2.17), appears to be given by

$$\boxed{\langle \mathbf{A}\rangle_{\Psi_k} = \sum_n \mathbb{P}_{kn} A_n = \langle\Psi_k|\mathbf{A}|\Psi_k\rangle} \tag{2.21}$$

Examination of this last result shows that $\langle \mathbf{A}\rangle_{\Psi_k}$ has the properties of a statistical average value since it is the sum of the possible values of the observable **A** weighted by their corresponding probabilities.

2.3.2 Variation theorem

It is now possible to prove the variation theorem. The sixth postulate attributes to the Hamiltonian a privileged role. Dealing with the Hamiltonian, there is, in quantum mechanics, a theorem that is of great interest concerning the energy of physical systems. Let us write the eigenvalue equation of the Hamiltonian **H**:

$$\mathbf{H}|\Psi_i\rangle = E_i\,|\Psi_i\rangle \quad \text{with} \quad \langle\Psi_i|\Psi_j\rangle = \delta_{ij} \tag{2.22}$$

where E_i are the eigenvalues and $|\Psi_i\rangle$ the corresponding eigenvectors. Now, consider the average value over any ket $|\Phi_l\rangle$ of the difference between the Hamiltonian and the lowest eigenvalue E_0:

$$\langle\Phi_l|\,(\mathbf{H}-E_0)\,|\Phi_l\rangle = \langle\Phi_l|\,\mathbf{H}|\,\Phi_l\rangle - E_0\langle\Phi_l|\Phi_l\rangle \tag{2.23}$$

Next, assume that the ket $|\Phi_l\rangle$ is given by the following expansion over the eigenkets $|\Psi_i\rangle$ of the Hamiltonian, that is,

$$|\Phi_l\rangle = \sum_j a_{jl}|\Psi_j\rangle \qquad \text{and} \qquad \langle\Phi_l| = \sum_i a_{li}\langle\Psi_i|$$

Then, Eq. (2.23) becomes

$$\langle\Phi_l|\,(\mathbf{H}-E_0)\,|\Phi_l\rangle = \sum_i\sum_j a_{jl}a_{li}\,\{\langle\Psi_i|\,\mathbf{H}\,|\,\Psi_j\rangle - E_0\langle\Psi_i|\Psi_j\rangle\}$$

Furthermore, due to the eigenvalue equation (2.22) by orthogonality properties we have

$$\langle\Phi_l|\,(\mathbf{H}-E_0)|\Phi_l\rangle = \sum_i\sum_j a_{jl}a_{li}(E_j - E_0)\delta_{ij}$$

and thus

$$\langle\Phi_l|\,(\mathbf{H}-E_0)|\Phi_l\rangle = \sum_j |a_{jl}|^2(E_j - E_0) \tag{2.24}$$

Now, observe that, since E_0 is the lowest eigenvalue, the right-hand-side differences are positive in the same way as the squared modulus of the expansion coefficients, that is,

$$(E_j - E_0) \geqslant 0 \qquad \text{and} \qquad |a_{jl}|^2 \geqslant 0$$

Thus, we have from the left-hand side of Eq. (2.24), the following inequality:

$$\langle \Phi_l | (\mathbf{H} - E_0) | \Phi_l \rangle \geqslant 0$$

Since E_0 is a scalar, one then obtains the following fundamental result:

$$\boxed{\langle \Phi_l | \mathbf{H} | \Phi_l \rangle \geqslant E_0} \tag{2.25}$$

Hence, the average value of the Hamiltonian performed over any one ket cannot be smaller than the lowest energy E_0. That gives the possibility to approach E_0 by variational methods if it is not possible to solve exactly the eigenvalue equation (2.22) of the Hamiltonian.

2.3.3 Variance

Now, observe that in probability theory and statistics, the variance ΔA of a random variable is a measurement of the statistical dispersion averaging the squared distance of its possible values from the mean value $\langle A \rangle$. Start from the variance (2.8):

$$\Delta \mathbf{A}_{\Psi_k} = \sqrt{\langle \mathbf{A}^2 \rangle_{\Psi_k} - \langle \mathbf{A} \rangle^2_{\Psi_k}} \tag{2.26}$$

In this last equation, the second term under the square root is given by Eq. (2.21). In order to get the first one, we may begin by using the definition (2.17) of the average of some operator, by taking $\mathbf{A}^2$ in place of $\mathbf{A}$:

$$\langle \mathbf{A}^2 \rangle_{\Psi_k} = \langle \Psi_k | \mathbf{A}^2 | \Psi_k \rangle$$

which may also be written

$$\langle \mathbf{A}^2 \rangle_{\Psi_k} = \langle \Psi_k | \mathbf{A}\mathbf{A} | \Psi_k \rangle \tag{2.27}$$

Again, write the eigenvalue equation of Hermitian operators and the corresponding closure relation:

$$\mathbf{A} | \Phi_n \rangle = A_n | \Phi_n \rangle \qquad \text{and} \qquad \sum_n | \Phi_n \rangle \langle \Phi_n | = 1 \tag{2.28}$$

Then, introduce in Eq. (2.27) between $\mathbf{A}$ and $|\Psi_k\rangle$ this last closure relation

$$\langle \mathbf{A}^2 \rangle_{\Psi_k} = \langle \Psi_k | \mathbf{A}\mathbf{A} \left(\sum_n | \Phi_n \rangle \langle \Phi_n | \right) | \Psi_k \rangle$$

Using the eigenvalue equation appearing in (2.28), one obtains

$$\langle \mathbf{A}^2 \rangle_{\Psi_k} = \sum_n \langle \Psi_k | \mathbf{A} A_n | \Phi_n \rangle \langle \Phi_n | \Psi_k \rangle$$

Next, commuting the scalar A_n, this equation becomes

$$\langle \mathbf{A}^2 \rangle_{\Psi_k} = \sum_n A_n \langle \Psi_k | \mathbf{A} | \Phi_n \rangle \langle \Phi_n | \Psi_k \rangle$$

Again using in turn the eigenvalue equation (2.28), one finds

$$\langle \mathbf{A}^2 \rangle_{\Psi_k} = \sum_n A_n \langle \Psi_k | A_n | \Phi_n \rangle \langle \Phi_n | \Psi_k \rangle$$

or, A_n being a scalar

$$\langle \mathbf{A}^2 \rangle_{\Psi_k} = \sum_n A_n^2 |\langle \Psi_k | \Phi_n \rangle|^2$$

Finally, using the fourth postulate given in the present context by Eq. (2.20) leads to

$$\langle \mathbf{A}^2 \rangle_{\Psi_k} = \sum_n \mathbb{P}_{kn} A_n^2 \quad \text{with} \quad \mathbb{P}_{kn} = |\langle \Psi_k | \Phi_n \rangle|^2$$

Thus, the variance (2.26) takes the form

$$\Delta \mathbf{A}_{\Psi_k} = \sqrt{\left(\sum_n \mathbb{P}_{kn} A_n^2 \right) - \left(\sum_n \mathbb{P}_{kn} A_n \right)^2}$$

2.3.4 Product of two variances

2.3.4.1 Variances of two different operators before and after some shift

Consider two Hermitian operators **A** and **B**, the commutator of which is obeying

$$[\mathbf{A}, \mathbf{B}] = i\mathbf{C} \quad \text{with} \quad i^2 = -1 \tag{2.29}$$

and

$$\mathbf{A} = \mathbf{A}^\dagger \qquad \mathbf{B} = \mathbf{B}^\dagger \qquad \mathbf{C} = \mathbf{C}^\dagger$$

Again, consider the average values of these operators **A** and **B**, respectively, calculated on some ket $|\Psi\rangle$ that we shall suppose normalized:

$$\langle \mathbf{A} \rangle_\Psi = \langle \Psi | \mathbf{A} | \Psi \rangle \quad \text{and} \quad \langle \mathbf{B} \rangle_\Psi = \langle \Psi | \mathbf{B} | \Psi \rangle \quad \text{with} \quad \langle \Psi | \Psi \rangle = 1 \tag{2.30}$$

Next, consider the following transformed operators:

$$\tilde{\mathbf{A}} = \{\mathbf{A} - \langle \mathbf{A} \rangle_\Psi\} \qquad \tilde{\mathbf{B}} = \{\mathbf{B} - \langle \mathbf{B} \rangle_\Psi\} \tag{2.31}$$

with average values on $|\Psi\rangle$

$$\langle \tilde{\mathbf{A}} \rangle_\Psi = \langle \Psi | \tilde{\mathbf{A}} | \Psi \rangle \quad \text{and} \quad \langle \tilde{\mathbf{B}} \rangle_\Psi = \langle \Psi | \tilde{\mathbf{B}} | \Psi \rangle \tag{2.32}$$

Now, write explicitly the first of the average values (2.32), using the first equation of (2.31):

$$\langle \tilde{\mathbf{A}} \rangle_\Psi = \langle \Psi | \{\mathbf{A} - \langle \mathbf{A} \rangle_\Psi\} | \Psi \rangle$$

That gives

$$\langle \tilde{\mathbf{A}} \rangle_\Psi = \langle \mathbf{A} \rangle_\Psi - \langle \mathbf{A} \rangle_\Psi \langle \Psi | \Psi \rangle = \{\langle \mathbf{A} \rangle_\Psi - \langle \mathbf{A} \rangle_\Psi\} = 0 \tag{2.33}$$

where the normalization of $|\Psi\rangle$ has been used. In like manner one may find that $\langle \tilde{\mathbf{B}} \rangle_\Psi$ is zero. Hence, one may write

$$\langle \tilde{\mathbf{A}} \rangle_\Psi = 0 \quad \text{and} \quad \langle \tilde{\mathbf{B}} \rangle_\Psi = 0 \tag{2.34}$$

Now, consider the corresponding squares of the variances concerning the two operators (2.31), that is,

$$\Delta\tilde{\mathbf{A}}_{\Psi}^2 = \{\langle\tilde{\mathbf{A}}^2\rangle_\Psi - \langle\tilde{\mathbf{A}}\rangle_\Psi^2\} \qquad \text{and} \qquad \Delta\tilde{\mathbf{B}}_{\Psi}^2 = \{\langle\tilde{\mathbf{B}}^2\rangle_\Psi - \langle\tilde{\mathbf{B}}\rangle_\Psi^2\}$$

or, owing to Eq. (2.34)

$$\Delta\tilde{\mathbf{A}}_{\Psi}^2 = \langle\tilde{\mathbf{A}}^2\rangle_\Psi \qquad \text{and} \qquad \Delta\tilde{\mathbf{B}}_{\Psi}^2 = \langle\tilde{\mathbf{B}}^2\rangle_\Psi \tag{2.35}$$

Furthermore, the quantum averages over $|\Psi\rangle$ of the operators (2.31) are

$$\langle\tilde{\mathbf{A}}^2\rangle_\Psi = \langle\Psi|\tilde{\mathbf{A}}^2|\Psi\rangle \qquad \text{and} \qquad \langle\tilde{\mathbf{B}}^2\rangle_\Psi = \langle\Psi|\tilde{\mathbf{B}}^2|\Psi\rangle \tag{2.36}$$

Thus, Eq. (2.35) reads

$$\Delta\tilde{\mathbf{A}}_{\Psi}^2 = \langle\Psi|\tilde{\mathbf{A}}^2|\Psi\rangle \qquad \text{and} \qquad \Delta\tilde{\mathbf{B}}_{\Psi}^2 = \langle\Psi|\tilde{\mathbf{B}}^2|\Psi\rangle$$

Next, owing to the first equation of (2.31), the first equation of (2.36) transforms, after expanding the squared expression, into

$$\langle\tilde{\mathbf{A}}^2\rangle_\Psi = \langle\Psi|\{\mathbf{A}^2 + \langle\mathbf{A}\rangle_\Psi^2 - 2\mathbf{A}\langle\mathbf{A}\rangle_\Psi\}|\Psi\rangle$$

or

$$\langle\tilde{\mathbf{A}}^2\rangle_\Psi = \langle\Psi|\mathbf{A}^2|\Psi\rangle + \langle\Psi|\langle\mathbf{A}\rangle_\Psi^2|\Psi\rangle - 2\langle\Psi|\mathbf{A}|\Psi\rangle\langle\mathbf{A}\rangle_\Psi$$

Since $|\Psi\rangle$ is normalized and due to the first equation of (2.30), we then have

$$\langle\tilde{\mathbf{A}}^2\rangle_\Psi = \langle\mathbf{A}^2\rangle_\Psi + \langle\mathbf{A}\rangle_\Psi^2 - 2\langle\mathbf{A}\rangle_\Psi\langle\mathbf{A}\rangle_\Psi$$

or

$$\langle\tilde{\mathbf{A}}^2\rangle_\Psi = \langle\mathbf{A}^2\rangle_\Psi - \langle\mathbf{A}\rangle_\Psi^2 \tag{2.37}$$

Now, observe that the right-hand side of Eq. (2.37) is the dispersion of the operator $\mathbf{A}$ averaged on the ket $|\Psi\rangle$:

$$\langle\mathbf{A}^2\rangle_\Psi - \langle\mathbf{A}\rangle_\Psi^2 = \Delta\mathbf{A}_\Psi^2$$

Thus, Eq. (2.37) yields

$$\langle\tilde{\mathbf{A}}^2\rangle_\Psi = \Delta\mathbf{A}_\Psi^2$$

Thus, compare Eq. (2.35), and working in the same way for $\tilde{\mathbf{B}}$ as for $\tilde{\mathbf{A}}$, one obtains, respectively,

$$\boxed{\Delta\tilde{\mathbf{A}}_{\Psi}^2 = \Delta\mathbf{A}_\Psi^2 \qquad \text{and} \qquad \Delta\tilde{\mathbf{B}}_{\Psi}^2 = \Delta\mathbf{B}_\Psi^2} \tag{2.38}$$

which we shall use later on.

2.3.4.2 Product of variances of A and B and Heisenberg uncertainty relations Now, consider the product of the variances of $\tilde{\mathbf{A}}$ and $\tilde{\mathbf{B}}$, over $|\Psi\rangle$. In view of Eqs. (2.30) and (2.35), it is

$$\Delta\tilde{\mathbf{A}}_{\Psi}^2\Delta\tilde{\mathbf{B}}_{\Psi}^2 = \langle\Psi|\tilde{\mathbf{A}}^2|\Psi\rangle\langle\Psi|\tilde{\mathbf{B}}^2|\Psi\rangle$$

or

$$\Delta\tilde{\mathbf{A}}_{\Psi}^{2}\Delta\tilde{\mathbf{B}}_{\Psi}^{2} = (\langle\Psi|\tilde{\mathbf{A}})(\tilde{\mathbf{A}}|\Psi\rangle)(\langle\Psi|\tilde{\mathbf{B}})(\tilde{\mathbf{B}}|\Psi\rangle) \tag{2.39}$$

Next, observe that the linear action of the operators $\tilde{\mathbf{A}}$ and $\tilde{\mathbf{B}}$ transforms, respectively, the ket $|\Psi\rangle$ into the new kets $|\Phi\rangle$ and $|\xi\rangle$ according to

$$\tilde{\mathbf{A}}|\Psi\rangle = |\Phi\rangle \qquad \text{and} \qquad \tilde{\mathbf{B}}|\Psi\rangle = |\xi\rangle \tag{2.40}$$

The Hermitian conjugates of these two linear transformations are

$$\langle\Psi|\tilde{\mathbf{A}} = \langle\Phi| \qquad \text{and} \qquad \langle\Psi|\tilde{\mathbf{B}} = \langle\xi| \tag{2.41}$$

Thus, Eq. (2.39) becomes

$$\Delta\tilde{\mathbf{A}}_{\Psi}^{2}\Delta\tilde{\mathbf{B}}_{\Psi}^{2} = \langle\Phi|\Phi\rangle\langle\xi|\xi\rangle \tag{2.42}$$

Now, observe that the Schwarz inequality (1.25) stipulates that

$$\langle\Phi|\Phi\rangle\langle\xi|\xi\rangle \geqslant \langle\xi|\Phi\rangle\langle\Phi|\xi\rangle$$

Hence, the product of uncertainties (2.42) transforms to the following inequality:

$$\Delta\tilde{\mathbf{A}}_{\Psi}^{2}\Delta\tilde{\mathbf{B}}_{\Psi}^{2} \geqslant \langle\xi|\Phi\rangle\langle\Phi|\xi\rangle \tag{2.43}$$

Again, in view of the linear transformations (2.40) and (2.41), the scalar products involved on the right-hand side of this last inequality are given by

$$\langle\Phi|\xi\rangle = \langle\Psi|\tilde{\mathbf{A}}\tilde{\mathbf{B}}|\Psi\rangle \qquad \text{and} \qquad \langle\xi|\Phi\rangle = \langle\Psi|\tilde{\mathbf{B}}\tilde{\mathbf{A}}|\Psi\rangle$$

Thus, the product of uncertainties (2.43) transforms to

$$\Delta\tilde{\mathbf{A}}_{\Psi}^{2}\Delta\tilde{\mathbf{B}}_{\Psi}^{2} \geqslant \langle\Psi|\tilde{\mathbf{A}}\tilde{\mathbf{B}}|\Psi\rangle\langle\Psi|\tilde{\mathbf{B}}\tilde{\mathbf{A}}|\Psi\rangle \tag{2.44}$$

Moreover, keeping in mind Eq. (1.60), and since neither $\tilde{\mathbf{A}}\tilde{\mathbf{B}}$ nor $\tilde{\mathbf{B}}\tilde{\mathbf{A}}$ are Hermitian, it is suitable to express these products in terms of symmetric and antisymmetric combinations according to

$$\tilde{\mathbf{A}}\tilde{\mathbf{B}} = \tfrac{1}{2}(\tilde{\mathbf{A}}\tilde{\mathbf{B}} + \tilde{\mathbf{B}}\tilde{\mathbf{A}}) + \tfrac{1}{2}(\tilde{\mathbf{A}}\tilde{\mathbf{B}} - \tilde{\mathbf{B}}\tilde{\mathbf{A}}) \tag{2.45}$$

Remark that the antisymmetric part is just the commutator of $\tilde{\mathbf{A}}$ and $\tilde{\mathbf{B}}$:

$$(\tilde{\mathbf{A}}\tilde{\mathbf{B}} - \tilde{\mathbf{B}}\tilde{\mathbf{A}}) = [\tilde{\mathbf{A}}, \tilde{\mathbf{B}}]$$

Now, this commutator involving the transformed operators may be expressed in terms of the initial ones using Eq. (2.31), so that

$$[\tilde{\mathbf{A}}, \tilde{\mathbf{B}}] = [(\mathbf{A} - \langle\mathbf{A}\rangle_{\Psi}), (\mathbf{B} - \langle\mathbf{B}\rangle_{\Psi})]$$

Then, since the average values involved in this last equation are scalars, the commutator of the transformed operators appears to be that of the nontransformed ones:

$$[\tilde{\mathbf{A}}, \tilde{\mathbf{B}}] = [\mathbf{A}, \mathbf{B}]$$

Thus, Eq. (2.45) transforms to

$$\tilde{\mathbf{A}}\tilde{\mathbf{B}} = \tfrac{1}{2}(\tilde{\mathbf{A}}\tilde{\mathbf{B}} + \tilde{\mathbf{B}}\tilde{\mathbf{A}}) + \tfrac{1}{2}[\mathbf{A}, \mathbf{B}]$$

Again, owing to the assumption (2.29) we have performed for the commutators of the initial operators **A** and **B**, this last equation leads to

$$\tilde{\mathbf{A}}\tilde{\mathbf{B}} = \tfrac{1}{2}\{\tilde{\mathbf{A}}\tilde{\mathbf{B}} + \tilde{\mathbf{B}}\tilde{\mathbf{A}} + i\mathbf{C}\} \tag{2.46}$$

Now, consider $\tilde{\mathbf{B}}\tilde{\mathbf{A}}$, which reads

$$(\tilde{\mathbf{B}}\tilde{\mathbf{A}} - \tilde{\mathbf{A}}\tilde{\mathbf{B}}) = [\tilde{\mathbf{B}}, \tilde{\mathbf{A}}] = -[\mathbf{A}, \mathbf{B}]$$

Then, due to Eq. (2.29), we have

$$\tilde{\mathbf{B}}\tilde{\mathbf{A}} = \tfrac{1}{2}\{\tilde{\mathbf{A}}\tilde{\mathbf{B}} + \tilde{\mathbf{B}}\tilde{\mathbf{A}} - i\mathbf{C}\} \tag{2.47}$$

As a consequence of Eqs. (2.46) and (2.47), Eq. (2.44) becomes

$$\Delta\tilde{\mathbf{A}}^2_\Psi \Delta\tilde{\mathbf{B}}^2_\Psi \geqslant \tfrac{1}{4}\langle\Psi|\{\tilde{\mathbf{A}}\tilde{\mathbf{B}} + \tilde{\mathbf{B}}\tilde{\mathbf{A}} + i\mathbf{C}\}|\Psi\rangle\langle\Psi|\{\tilde{\mathbf{A}}\tilde{\mathbf{B}} + \tilde{\mathbf{B}}\tilde{\mathbf{A}} - i\mathbf{C}\}|\Psi\rangle$$

Thus

$$\Delta\tilde{\mathbf{A}}^2_\Psi \Delta\tilde{\mathbf{B}}^2_\Psi \geqslant \tfrac{1}{4}(\langle\Psi|(\tilde{\mathbf{A}}\tilde{\mathbf{B}} + \tilde{\mathbf{B}}\tilde{\mathbf{A}})|\Psi\rangle + i\langle\Psi|\mathbf{C}|\Psi\rangle)(\langle\Psi|(\tilde{\mathbf{A}}\tilde{\mathbf{B}} + \tilde{\mathbf{B}}\tilde{\mathbf{A}})|\Psi\rangle - i\langle\Psi|\mathbf{C}|\Psi\rangle)$$

or

$$\Delta\tilde{\mathbf{A}}^2_\Psi \Delta\tilde{\mathbf{B}}^2_\Psi \geqslant \tfrac{1}{4}\{\langle\Psi|(\tilde{\mathbf{A}}\tilde{\mathbf{B}} + \tilde{\mathbf{B}}\tilde{\mathbf{A}})|\Psi\rangle^2 + \langle\Psi|\mathbf{C}|\Psi\rangle^2\} \tag{2.48}$$

Now, as it appears by inspection of this last inequality, each member of the right-hand-side, however small it may be, cannot be negative since it is a squared average value. Moreover, the inequality is also satisfied when one substracts from the smallest right-hand-side term, its first squared term involving the products $\tilde{\mathbf{A}}\tilde{\mathbf{B}}$ and $\tilde{\mathbf{B}}\tilde{\mathbf{A}}$. Hence, if the inequality (2.48) is satisfied, the following one will be a fortiori satisfied:

$$\Delta\tilde{\mathbf{A}}^2_\Psi \Delta\tilde{\mathbf{B}}^2_\Psi \geqslant \tfrac{1}{4}\langle\Psi|\mathbf{C}|\Psi\rangle^2$$

Hence, owing to Eq. (2.38), it appears that the same inequality for the dispersions dealing with the initial operators **A** and **B** exists, so that it reads

$$\Delta\mathbf{A}^2_\Psi \Delta\mathbf{B}^2_\Psi \geqslant \tfrac{1}{4}\langle\Psi|\mathbf{C}|\Psi\rangle^2$$

or, in view of Eq. (2.29)

$$\boxed{\Delta\mathbf{A}^2_\Psi \Delta\mathbf{B}^2_\Psi \geqslant -i\tfrac{1}{4}\langle\Psi|[\mathbf{A}, \mathbf{B}]|\Psi\rangle^2} \tag{2.49}$$

Now, apply the inequality (2.49) to the coordinate operator **Q** and its conjugate momentum **P**. Then take

$$\mathbf{A} = \mathbf{Q} \qquad \mathbf{B} = \mathbf{P} \quad \text{and} \quad [\mathbf{A}, \mathbf{B}] = [\mathbf{Q}, \mathbf{P}] = i\hbar$$

Besides, due to Eq. (2.29), that is,

$$i\mathbf{C} = [\mathbf{Q}, \mathbf{P}] = i\hbar$$

Then, the inequality (2.49) takes the form

$$\Delta\mathbf{P}^2_\Psi \Delta\mathbf{Q}^2_\Psi \geqslant \frac{1}{4}\langle\Psi|\hbar|\Psi\rangle^2 = \left(\frac{\hbar}{2}\right)^2$$

so that the product of variances of the coordinate **Q** and its conjugate momentum **P** appears, whatever the ket $|\Psi\rangle$ describing the system

$$\boxed{\Delta \mathbf{P}_\Psi \Delta \mathbf{Q}_\Psi \geqslant \frac{\hbar}{2}} \tag{2.50}$$

That is the Heisenberg uncertainty relation. Of course, this relation holds for the three Cartesian coordinates of some particle, so that

$$\boxed{(\Delta P_x)_\Psi (\Delta Q_x)_\Psi \geqslant \frac{\hbar}{2}}$$

$$\boxed{(\Delta P_y)_\Psi (\Delta Q_y)_\Psi \geqslant \frac{\hbar}{2}}$$

$$\boxed{(\Delta P_z)_\Psi (\Delta Q_z)_\Psi \geqslant \frac{\hbar}{2}}$$

An important consequence of these uncertainty relations is that the *trajectory*, which is fundamental in classical mechanics, has no meaning in quantum mechanics. The reason is that, to define the trajectory of some particle, it is necessary to know exactly both its position and momentum at all times. This impossibility of a precise trajectory indicates that two particles of the same kind, such as, for instance, two electrons, or two protons, or two hydrogen atoms, are *indistinguishable* because the only possibility to distinguish them would be their individual trajectories, which is impossible because of the uncertainty relations.

2.4 SCHRÖDINGER PICTURE DYNAMICS

Now, we shall consider some dynamic behaviors appearing in quantum mechanics as a consequence of the Schrödinger equation appearing in the sixth postulate. Recall that according to this equation, the kets and the corresponding bras are time dependent, whereas the operators are constant. Such a time description in which kets and bras are time dependent whereas the operators are constant is called the *Schrödinger picture* (SP) in order to differentiate it from another description named *Heisenberg picture* (HP) in which the operators are changing with time whereas the kets and bras remain constant.

We shall first show that the Schrödinger equation preserves the conservation of the norm that is required from the physical viewpoint. Then, we shall demonstrate some fundamental dynamic equations, and, finally, two theorems, one of which is the Ehrenfest theorem, which resembles the basic Newtonian equations of classical mechanics, the only difference being that the Ehrenfest theorem governs average values instead of exact ones.

2.4.1 Norm conservation and average values time dependence

2.4.1.1 Norm conservation Consider the Schrödinger equation

$$i\hbar\left(\frac{\partial|\Psi(t)\rangle}{\partial t}\right) = \mathbf{H}|\Psi(t)\rangle \tag{2.51}$$

where **H** is the Hamiltonian operator, which is, of course, Hermitian. If it is normalized, the norm of the ket$|\Psi(t^\circ)\rangle$ at time t° is

$$\langle\Psi(t^\circ)|\Psi(t^\circ)\rangle = 1$$

Of course, if the norm has to be conserved, it must be given at any time $t \neq t^\circ$ by

$$\langle\Psi(t)|\Psi(t)\rangle = 1$$

We shall show that this last equation is in agreement with the Schrödinger equation. For this purpose, we write explicitly the time derivative of the norm

$$\left(\frac{\partial\langle\Psi(t)|\Psi(t)\rangle}{\partial t}\right) = \left(\frac{\partial\langle\Psi(t)|}{\partial t}\right)|\Psi(t)\rangle + \langle\Psi(t)|\left(\frac{\partial|\Psi(t)\rangle}{\partial t}\right) \tag{2.52}$$

To calculate the time derivative of the bra involved on the first right-hand-side term of this last equation, we consider the Hermitian conjugate of Eq. (2.51)

$$-i\hbar\left(\frac{\partial\langle\Psi(t)|}{\partial t}\right) = \langle\Psi(t)|\mathbf{H}^\dagger$$

Then, since the Hamiltonian is Hermitian, that is, $\mathbf{H}^\dagger = \mathbf{H}$, this last equation becomes

$$-i\hbar\left(\frac{\partial\langle\Psi(t)|}{\partial t}\right) = \langle\Psi(t)|\mathbf{H} \tag{2.53}$$

As a consequence of Eqs. (2.51) and (2.53), the time derivative of the norm (2.52) becomes

$$\left(\frac{\partial\langle\Psi(t)|\Psi(t)\rangle}{\partial t}\right) = -\frac{1}{i\hbar}\langle\Psi(t)|\mathbf{H}|\Psi(t)\rangle + \frac{1}{i\hbar}\langle\Psi(t)|\mathbf{H}|\Psi(t)\rangle$$

This last result simplifies to

$$\boxed{\left(\frac{\partial\langle\Psi(t)|\Psi(t)\rangle}{\partial t}\right) = 0}$$

showing, as required, that the norm is conserved along the time.

2.4.2 Time evolution of operator average value

2.4.2.1 General expression We shall now consider how the average value of some operator **A** calculated over any ket $|\Psi(t)\rangle$ evolves, which is time dependent because of the Schrödinger equation.

In the Schrödinger time-dependent picture, any operator **A** does not depend on time so that the time derivative of the average value of **A** over $|\Psi(t)\rangle$ is

$$\left(\frac{\partial\langle\Psi(t)|\mathbf{A}|\Psi(t)\rangle}{\partial t}\right) = \left(\frac{\partial\langle\Psi(t)|}{\partial t}\right)\mathbf{A}|\Psi(t)\rangle + \langle\Psi(t)|\mathbf{A}\left(\frac{\partial|\Psi(t)\rangle}{\partial t}\right) \tag{2.54}$$

Using Eqs. (2.51) and (2.53), this equation transforms to

$$\left(\frac{\partial\langle\Psi(t)|\mathbf{A}|\Psi(t)\rangle}{\partial t}\right) = -\frac{1}{i\hbar}\langle\Psi(t)|\mathbf{HA}|\Psi(t)\rangle + \frac{1}{i\hbar}\langle\Psi(t)|\mathbf{AH}|\Psi(t)\rangle$$

or, in term of the commutator of **H** and **A**

$$\left(\frac{\partial\langle\Psi(t)|\mathbf{A}|\Psi(t)\rangle}{\partial t}\right) = \frac{i}{\hbar}\langle\Psi(t)|[\mathbf{H},\mathbf{A}]|\Psi(t)\rangle \tag{2.55}$$

which may be written in the compact form

$$\boxed{\left(\frac{\partial\langle\mathbf{A}(t)\rangle_\Psi}{\partial t}\right) = \frac{i}{\hbar}\langle[\mathbf{H},\mathbf{A}]\rangle_\Psi} \tag{2.56}$$

with

$$\langle\mathbf{A}(t)\rangle_\Psi \equiv \langle\Psi(t)|\mathbf{A}|\Psi(t)\rangle \qquad \langle[\mathbf{H},\mathbf{A}]\rangle_\Psi = \langle\Psi(t)|[\mathbf{H},\mathbf{A}]|\Psi(t)\rangle \tag{2.57}$$

We remark that the notation $\langle\mathbf{A}(t)\rangle_\Psi$ does not imply that **A** depends on time but only means that the average value $\langle\mathbf{A}(t)\rangle_\Psi$ of **A** depends on time. Besides, observe that, in the Schrödinger time-dependent picture, some physical systems that have to be studied may appear quantum mechanically for one part and classically for another one. In such systems, which are said to be *hemiquantal*, there is then the possibility for any operator **A** to present a time dependence through its classical part. Then, Eq. (2.55) has to be generalized into

$$\left(\frac{\partial\langle\Psi(t)|\mathbf{A}(t)|\Psi(t)\rangle}{\partial t}\right) = \frac{i}{\hbar}\langle\Psi(t)|[\mathbf{H},\mathbf{A}(t)]|\Psi(t)\rangle + \langle\Psi(t)|\left(\frac{\partial\mathbf{A}(t)}{\partial t}\right)|\Psi(t)\rangle \tag{2.58}$$

2.4.2.2 Conservation of the total energy and exchange of energies In the special situation where the operator is the Hamiltonian, and what may be the ket $|\Psi(t)\rangle$ describing the system, Eq. (2.56) reads

$$\left(\frac{\partial\langle\mathbf{H}\rangle_\Psi}{\partial t}\right) = \frac{i}{\hbar}\langle[\mathbf{H},\mathbf{H}]\rangle_\Psi = 0 \tag{2.59}$$

Hence, the average value of the total Hamiltonian, that is, the total energy, remains constant whichever ket $|\Psi(t)\rangle$ describes the system.

However, if the total energy is conserved, it is not true for the energies of subsystems from which any physical system is built up. Suppose, for instance, that the total Hamiltonian is the sum of two Hamiltonians, which do not mutually commute:

$$\mathbf{H} = \mathbf{H}_1 + \mathbf{H}_2 \quad \text{with} \quad [\mathbf{H}_1,\mathbf{H}_2] \neq 0$$

Then, the commutators of $\mathbf{H}_1$ and $\mathbf{H}_2$ with **H** are

$$[\mathbf{H}_1,\mathbf{H}] = [\mathbf{H}_1,\mathbf{H}_2] \qquad \text{and} \qquad [\mathbf{H}_2,\mathbf{H}] = [\mathbf{H}_2,\mathbf{H}_1] = -[\mathbf{H}_1,\mathbf{H}_2] \tag{2.60}$$

Thus, due to Eqs. (2.56) and (2.60), the time dependences of the averages of the Hamiltonians of the two subsystems obey

$$\left(\frac{\partial\langle\mathbf{H}_1(t)\rangle_\Psi}{\partial t}\right) = \frac{i}{\hbar}\langle[\mathbf{H}_2,\mathbf{H}_1]\rangle_\Psi \tag{2.61}$$

$$\left(\frac{\partial \langle \mathbf{H}_2(t)\rangle_\Psi}{\partial t}\right) = -\frac{i}{\hbar}\langle [\mathbf{H}_2, \mathbf{H}_1]\rangle_\Psi \tag{2.62}$$

We emphasize that in these last equations, they are the average values of $\mathbf{H}_1$ and $\mathbf{H}_2$, which depend on time through $\Psi(t)$, although the operators $\mathbf{H}_1$ and $\mathbf{H}_2$ do not depend on time. Equations (2.61) and (2.62) show that the energy moves back and forth between the two subsystems according to

$$\left(\frac{\partial \langle \mathbf{H}_1(t)\rangle_\Psi}{\partial t}\right) = -\left(\frac{\partial \langle \mathbf{H}_2(t)\rangle_\Psi}{\partial t}\right)$$

in such a way as their sum remains constant according to Eq. (2.59). Of course, if one considers, respectively, in place of $\mathbf{H}_1$ and $\mathbf{H}_2$ the kinetic and potential energies, $\mathbf{T}$ and $\mathbf{V}$ of the system, the Eqs. (2.61) and (2.62) become

$$\left(\frac{\partial \langle \mathbf{T}(t)\rangle_\Psi}{\partial t}\right) = \frac{i}{\hbar}\langle [\mathbf{V}, \mathbf{T}]\rangle_\Psi$$

$$\left(\frac{\partial \langle \mathbf{V}(t)\rangle_\Psi}{\partial t}\right) = -\frac{i}{\hbar}\langle [\mathbf{V}, \mathbf{T}]\rangle_\Psi$$

so that the kinetic and potential energies exchange themselves with time according to

$$\left(\frac{\partial \langle \mathbf{T}(t)\rangle_\Psi}{\partial t}\right) = -\left(\frac{\partial \langle \mathbf{V}(t)\rangle_\Psi}{\partial t}\right)$$

2.4.2.3 Stationary states By definition, a stationary state is an eigenstate of the Hamiltonian, that is, it obeys the eigenvalue equation

$$\mathbf{H}|\Psi_k(t)\rangle = E_k|\Psi_k(t)\rangle$$

The time-dependent Schrödinger equation is

$$i\hbar\left(\frac{\partial |\Psi_k(t)\rangle}{\partial t}\right) = \mathbf{H}|\Psi_k(t)\rangle$$

For a stationary state, it reads

$$i\hbar\left(\frac{\partial |\Psi_k(t)\rangle}{\partial t}\right) = E_k|\Psi_k(t)\rangle$$

so that, by integration

$$|\Psi_k(t)\rangle = |\Psi_k(0)\rangle e^{-iE_kt/\hbar} \tag{2.63}$$

Now, consider the average value of any operator over a stationary state. At an initial time it is

$$\langle \mathbf{A}(0)\rangle_{\Psi_k} = \langle \Psi_k(0)|\mathbf{A}|\Psi_k(0)\rangle \tag{2.64}$$

At time t, it is given by

$$\langle \mathbf{A}(t)\rangle_{\Psi_k} = \langle \Psi_k(t)|\mathbf{A}|\Psi_k(t)\rangle \tag{2.65}$$

Then, owing to Eq. (2.63) and to its Hermitian conjugate, one has

$$\langle \mathbf{A}(t)\rangle_{\Psi_k} = e^{iE_kt/\hbar}\langle \Psi_k(0)|\mathbf{A}|\Psi_k(0)\rangle e^{-iE_kt/\hbar}$$

and, thus, after simplification

$$\langle \mathbf{A}(t)\rangle_{\Psi_k} = \langle \Psi_k(0)|\mathbf{A}|\Psi_k(0)\rangle \tag{2.66}$$

Comparison of Eqs. (2.64) and (2.66) shows that

$$\left(\frac{\partial \langle \mathbf{A}(t)\rangle_{\Psi_k}}{\partial t}\right) = 0 \quad \text{for any stationary state} \tag{2.67}$$

2.4.3 Ehrenfest equations

Now, we are able to demonstrate the Ehrenfest equations governing the dynamics of the operators **Q** and **P**. Applying Eq. (2.56), one obtains, respectively,

$$\left(\frac{\partial \langle \mathbf{Q}(t)\rangle_{\Psi_k}}{\partial t}\right) = \frac{i}{\hbar}\langle [\mathbf{H},\mathbf{Q}]\rangle_{\Psi_k} \tag{2.68}$$

$$\left(\frac{\partial \langle \mathbf{P}(t)\rangle_{\Psi_k}}{\partial t}\right) = \frac{i}{\hbar}\langle [\mathbf{H},\mathbf{P}]\rangle_{\Psi_k} \tag{2.69}$$

When the system involves only forces that are the derivative of a potential, the Hamiltonian $\mathbf{H}(\mathbf{P},\mathbf{Q})$ is as above the sum of the kinetic $\mathbf{T}(\mathbf{P})$ and potential $\mathbf{V}(\mathbf{Q})$ operators, the first one depending on **P** and the last one on **Q**:

$$\mathbf{H}(\mathbf{P},\mathbf{Q}) = \mathbf{T}(\mathbf{P}) + \mathbf{V}(\mathbf{Q})$$

For a single particle, the kinetic operator is simply

$$\mathbf{T}(\mathbf{P}) = \frac{\mathbf{P}^2}{2m}$$

Of course, the commutators of the kinetic momentum operators and that of the potential and coordinate operators, are, respectively, zero, that is

$$[\mathbf{T}(\mathbf{P}),\mathbf{P}] = [\mathbf{V}(\mathbf{Q}),\mathbf{Q}] = 0$$

Thus, the commutators of the Hamiltonian with the coordinate and momentum operators are, respectively,

$$[\mathbf{H},\mathbf{Q}] = \frac{1}{2m}[\mathbf{P}^2,\mathbf{Q}] \tag{2.70}$$

$$[\mathbf{H},\mathbf{P}] = [\mathbf{V}(\mathbf{Q}),\mathbf{P}] \tag{2.71}$$

Besides, owing to Eqs. (2.14) and (2.15), the commutators appearing on the right-hand sides of these two last equations are, respectively,

$$[\mathbf{P}^2,\mathbf{Q}] = -2i\hbar\mathbf{P}$$

$$[\mathbf{V}(\mathbf{Q}),\mathbf{P}] = i\hbar\left(\frac{\partial \mathbf{V}}{\partial \mathbf{Q}}\right)$$

Then, using for the two commutators (2.70) and (2.71), these two last equations, and introducing them into Eqs. (2.68) and (2.69), one obtains the final results, which are known as the *Ehrenfest equations*, and which hold whatever the ket $|\Psi_k(t)\rangle$ considered for the calculation:

$$\boxed{\langle \mathbf{P}(t)\rangle_{\Psi_k} = m\left(\frac{d\langle \mathbf{Q}(t)\rangle_{\Psi_k}}{dt}\right)} \tag{2.72}$$

$$\boxed{\frac{\partial}{\partial t}\langle \mathbf{P}(t)\rangle_{\Psi_k} = -\left\langle\left(\frac{\partial \mathbf{V}}{\partial \mathbf{Q}}\right)\right\rangle_{\Psi_k}} \tag{2.73}$$

Thus, the first Ehrenfest equation looks like the Newton equation defining the momentum in terms of the velocity, whereas the second one looks like that relating the time derivative of the momentum (i.e., the acceleration) to the gradient of the potential (i.e., the force). However, the ket $|\Psi_k(t)\rangle$ considered for the calculation can never be simultaneously an eigenket of $\mathbf{P}$ and $\mathbf{Q}$ because

$$[\mathbf{Q}, \mathbf{P}] = i\hbar$$

thus, the uncertainty relations must be retained so that the Ehrenfest equations [(2.72) and (2.73)] have always to be considered mindful of the Heisenberg uncertainty relation:

$$\boxed{(\Delta \mathbf{P})_{\Psi_k}(\Delta \mathbf{Q})_{\Psi_k} \geqslant \frac{\hbar}{2}}$$

2.4.4 Virial theorem

2.4.4.1 Demonstration of the virial theorem Now, we shall prove the virial theorem, which relates the average values of the kinetic and potential operators, when the averages are performed over stationary states $|\Psi_k\rangle$, that is, eigenstates of the Hamiltonian $\mathbf{H}$ obeying the eigenvalue equation

$$\mathbf{H}|\Psi_k\rangle = E_k|\Psi_k\rangle \tag{2.74}$$

Apply Eq. (2.56) to the product $\mathbf{QP}$ of the coordinate and momentum operators $\mathbf{Q}$ and $\mathbf{P}$. Hence

$$i\hbar\left(\frac{\partial\langle \mathbf{QP}\rangle_{\Psi_k}}{\partial t}\right) = \langle[\mathbf{QP}, \mathbf{H}]\rangle_{\Psi_k} \tag{2.75}$$

The Hamiltonian $\mathbf{H}$ may be written, as above, as the sum of the kinetic $\mathbf{T}$ and potential $\mathbf{V}$ operators, the first only a function of $\mathbf{P}$ and the latter of $\mathbf{Q}$.

$$\mathbf{H} = \mathbf{T}(\mathbf{P}) + \mathbf{V}(\mathbf{Q})$$

Of course, as above, the following commutators are zero:

$$[\mathbf{T}(\mathbf{P}), \mathbf{P}] = [\mathbf{V}(\mathbf{Q}), \mathbf{Q}] = 0 \tag{2.76}$$

Thus, the commutator appearing on the right-hand side of Eq. (2.75) reads

$$[\mathbf{QP}, \mathbf{H}] = [\mathbf{QP}, \mathbf{T}(\mathbf{P})] + [\mathbf{QP}, \mathbf{V}(\mathbf{Q})] \tag{2.77}$$

For a single particle, the kinetic operator is

$$\mathbf{T}(\mathbf{P}) = \frac{1}{2m}\mathbf{P}^2 \tag{2.78}$$

Hence, the first commutator appearing on the right-hand side of Eq. (2.77) is

$$[\mathbf{QP}, \mathbf{T}(\mathbf{P})] = \frac{1}{2m}[\mathbf{QP}, \mathbf{P}^2] \tag{2.79}$$

The commutator appearing on the right member of this last equation may be written

$$[\mathbf{QP}, \mathbf{P}^2] = (\mathbf{QP}^2 - \mathbf{P}^2\mathbf{Q})\mathbf{P}$$

or

$$[\mathbf{QP}, \mathbf{P}^2] = [\mathbf{Q}, \mathbf{P}^2]\mathbf{P}$$

Thus, in view of Eq. (2.13), it transforms to

$$[\mathbf{QP}, \mathbf{P}^2] = (i\hbar)2\mathbf{P}^2$$

Hence, the commutator (2.79) becomes

$$[\mathbf{QP}, \mathbf{T}(\mathbf{P})] = i\hbar\frac{\mathbf{P}^2}{m} \tag{2.80}$$

Now, consider the second commutator appearing on the right-hand side of Eq. (2.77):

$$[\mathbf{QP}, \mathbf{V}(\mathbf{Q})] = \mathbf{QPV}(\mathbf{Q}) - \mathbf{V}(\mathbf{Q})\mathbf{QP}$$

which, since $\mathbf{Q}$ commutes with $\mathbf{V}(\mathbf{Q})$, transforms to

$$[\mathbf{QP}, \mathbf{V}(\mathbf{Q})] = \mathbf{QPV}(\mathbf{Q}) - \mathbf{QV}(\mathbf{Q})\mathbf{P}$$

so that Eq. (2.76) reads

$$[\mathbf{QP}, \mathbf{V}(\mathbf{Q})] = \mathbf{Q}[\mathbf{P}, \mathbf{V}(\mathbf{Q})]$$

Then, using Eq. (2.15), we have

$$[\mathbf{QP}, \mathbf{V}(\mathbf{Q})] = -(i\hbar)\mathbf{Q}\left(\frac{\partial \mathbf{V}}{\partial \mathbf{Q}}\right) \tag{2.81}$$

Now, using Eqs. (2.80) and (2.81) the commutator (2.77) appears to be given by

$$[\mathbf{QP}, \mathbf{H}] = (i\hbar)\left\{\frac{1}{m}\mathbf{P}^2 - \mathbf{Q}\left(\frac{\partial \mathbf{V}}{\partial \mathbf{Q}}\right)\right\}$$

Moreover, after averaging over $|\Psi_k\rangle$ one obtains

$$\langle[\mathbf{QP}, \mathbf{H}]\rangle_{\Psi_k} = (i\hbar)\left\{\frac{1}{m}\langle\mathbf{P}^2\rangle_{\Psi_k} - \left\langle\mathbf{Q}\left(\frac{\partial \mathbf{V}}{\partial \mathbf{Q}}\right)\right\rangle_{\Psi_k}\right\}$$

Finally, using this result into Eq. (2.75), we have

$$\left(\frac{\partial\langle\mathbf{Q}\,\mathbf{P}\rangle_{\Psi_k}}{\partial t}\right) = 2\left\langle\frac{\mathbf{P}^2}{2m}\right\rangle_{\Psi_k} - \left\langle\mathbf{Q}\left(\frac{\partial \mathbf{V}}{\partial \mathbf{Q}}\right)\right\rangle_{\Psi_k} \tag{2.82}$$

When the ket, over which the average value is performed, is stationary, Eq. (2.67) holds, so that the time dependence of the average value of the correlation between $\mathbf{Q}$ and $\mathbf{P}$ is zero:

$$\left(\frac{\partial\langle \mathbf{Q}\,\mathbf{P}\rangle_{\Psi_k}}{\partial t}\right) = 0$$

Hence, for stationary situations, Eq. (2.82) simplifies to

$$2\left\langle \frac{\mathbf{P}^2}{2m}\right\rangle_{\Psi_k} = \left\langle \mathbf{Q}\left(\frac{\partial \mathbf{V}}{\partial \mathbf{Q}}\right)\right\rangle_{\Psi_k} \tag{2.83}$$

Observe that the gradient of the potential may be written as a force $\mathbf{F}$ according to

$$\left(\frac{\partial \mathbf{V}}{\partial \mathbf{Q}}\right) = -\mathbf{F}$$

so that, Eq. (2.83) yields

$$2\left\langle \frac{\mathbf{P}^2}{2m}\right\rangle_{\Psi_k} = -\langle \mathbf{Q}\,\mathbf{F}\rangle_{\Psi_k} \tag{2.84}$$

This equation may be generalized for many degrees of freedom j. We have

$$2\sum_{j=1}^{N}\left\langle \frac{\mathbf{P}_j^2}{2m}\right\rangle_{\Psi_k} = -\sum_{j=1}^{N}\langle \mathbf{Q}_j\,\mathbf{F}_j\rangle_{\Psi_k}$$

2.4.4.2 *Applications of the virial theorem*

2.4.4.2.1 Systems involving harmonic potential Now, apply Eq. (2.83) to a quantum harmonic oscillator where the potential obeys

$$\mathbf{V}(\mathbf{Q}) = \tfrac{1}{2}k\mathbf{Q}^2 \tag{2.85}$$

where k is the force constant of the potential, which is a scalar. Then, deriving Eq. (2.85) leads to

$$\left(\frac{\partial \mathbf{V}}{\partial \mathbf{Q}}\right) = k\mathbf{Q}$$

Besides, multiplying both terms by $\mathbf{Q}$ gives

$$\mathbf{Q}\left(\frac{\partial \mathbf{V}}{\partial \mathbf{Q}}\right) = k\mathbf{Q}^2$$

Again, averaging over the ket $|\Psi_k\rangle$ and in view of Eq. (2.85),

$$\left\langle \mathbf{Q}\left(\frac{\partial \mathbf{V}}{\partial \mathbf{Q}}\right)\right\rangle_{\Psi_k} = 2\langle \mathbf{V}(\mathbf{Q})\rangle_{\Psi_k} \tag{2.86}$$

At last, owing to Eqs. (2.78) and (2.83), Eq. (2.86) yields

$$\langle \mathbf{T}\rangle_{\Psi_k} = \langle \mathbf{V}(\mathbf{Q})\rangle_{\Psi_k}$$

On the other hand, the average value of the Hamiltonian is the sum of the kinetic and potential operators, that is,

$$\langle \mathbf{H} \rangle_{\Psi_k} = \langle \mathbf{T} \rangle_{\Psi_k} + \langle \mathbf{V}(\mathbf{Q}) \rangle_{\Psi_k} \tag{2.87}$$

However, since the average value of the Hamiltonian is performed over one of its eigenstates obeying Eq. (2.74), this is just the corresponding eigenvalue E_k so that Eq. (2.87) gives

$$E_k = \langle \mathbf{T} \rangle_{\Psi_k} + \langle \mathbf{V}(\mathbf{Q}) \rangle_{\Psi_k} \tag{2.88}$$

Hence, one may determine the average value of the kinetic and potential operators from the value of the corresponding energy levels via

$$\langle \mathbf{T} \rangle_{\Psi_k} = \frac{E_k}{2} = \langle \mathbf{V}(\mathbf{Q}) \rangle_{\Psi_k} \tag{2.89}$$

2.4.4.2.2 Systems involving Coulomb potential Now consider, as a second example, a Coulomb potential involving two electrical charges q and q' obeying

$$\mathbf{V}(\mathbf{Q}) = -K\frac{1}{\mathbf{Q}} \quad \text{with} \quad K = \frac{qq'}{4\pi\varepsilon^\circ} \tag{2.90}$$

where ε° is the vacuum permittivity, which is a scalar. Then, after deriving $\mathbf{V}$ with respect to $\mathbf{Q}$ and rearranging, it reads

$$\mathbf{Q}\left(\frac{\partial \mathbf{V}}{\partial \mathbf{Q}}\right) = K\frac{1}{\mathbf{Q}}$$

Furthermore, the quantum average over $|\Psi_k\rangle$ leads, by aid of Eq. (2.90), to

$$\left\langle \mathbf{Q}\left(\frac{\partial \mathbf{V}}{\partial \mathbf{Q}}\right)\right\rangle_{\Psi_k} = -\langle \mathbf{V}(\mathbf{Q}) \rangle_{\Psi_k}$$

Now, with Eq. (2.78), the virial theorem (2.83), takes the form

$$2\langle \mathbf{T} \rangle_{\Psi_k} = -\langle \mathbf{V}(\mathbf{Q}) \rangle_{\Psi_k}$$

Of course, since Eqs. (2.87) and (2.88) continue to apply, one may obtain from the expression for the energy levels the corresponding average values of the potential and kinetic operators by aid of

$$\langle \mathbf{V}(\mathbf{Q}) \rangle_{\Psi_k} = 2E_k \qquad \text{and} \qquad \langle \mathbf{T} \rangle_{\Psi_k} = -E_k$$

2.5 POSITION OR MOMENTUM TRANSLATION OPERATORS

2.5.1 Eigenvalue equations of the position and momentum operators

Consider the eigenvalue equation of the coordinate operator $\mathbf{Q}$:

$$\mathbf{Q}|\{Q\}\rangle = Q|\{Q\}\rangle \tag{2.91}$$

The meaning of this eigenvalue equation is that when a system is in an eigenket $|\{Q\}\rangle$[1] of the coordinate operator $\mathbf{Q}$, the measurement of its position is given by the corresponding eigenvalue Q. Of course, since the $\mathbf{Q}$ operator is Hermitian, the Hermitian conjugate of Eq. (2.91) is

$$\langle\{Q\}|\mathbf{Q} = \langle\{Q\}|Q$$

Now, observe that the possible measurements of the Q coordinate are continuous. Thus, the orthormality properties of the $\mathbf{Q}$ operator involving two different kets $|\{Q\}\rangle$ and $|\{Q'\}\rangle$ must be written according to this continuous property. Hence

$$\langle\{Q\}|\{Q'\}\rangle = \delta(Q - Q') \tag{2.92}$$

Furthermore, the eigenvectors of the Hermitian operator $\mathbf{Q}$ form a basis that must be continuous, owing to this continuity of Q. Thus, the usual closure relation (1.17) built up on the eigenvectors, must be replaced by a new one where an infinite integral takes the place of the sum over the eigenkets. That leads to

$$\int_{-\infty}^{+\infty} |\{Q\}\rangle\langle\{Q\}|\, dQ = \mathbf{1}$$

In a similar way, we may write the eigenvalue equation of the momentum operator $\mathbf{P}$ and its Hermitian conjugate

$$\mathbf{P}|\{P\}\rangle = P|\{P\}\rangle \qquad \text{and} \qquad \langle\{P\}|\mathbf{P} = \langle\{P\}|P \tag{2.93}$$

Here, $|\{P\}\rangle$[2] is an eigenket of the momentum operator $\mathbf{P}$ with the eigenvalue P. Besides, owing to the continuity of the eigenvalues of the operator $\mathbf{P}$, that is, of its possible measured values, the orthonormality of the eigenkets of $\mathbf{P}$ and the closure relation are similar to those dealing with $\mathbf{Q}$, that is,

$$\langle\{P\}|\{P'\}\rangle = \delta(P - P') \qquad \text{and} \qquad \int_{-\infty}^{+\infty} |\{P\}\rangle\langle\{P\}|\, dP = \mathbf{1} \tag{2.94}$$

In the following, we shall show that the scalar product of any eigenket of the position operator $\mathbf{Q}$ by some eigenket of its momentum conjugate $\mathbf{P}$ is the same irrespective of the corresponding eigenvalues Q and P:

$$\boxed{\langle\{P\}|\{Q\}\rangle = \frac{1}{\sqrt{2\pi\hbar}} \exp\left\{-\frac{iPQ}{\hbar}\right\}}$$

which is consistent with the Heisenberg uncertainty relation

$$(\Delta P)_\Psi (\Delta Q)_\Psi \geqslant \frac{\hbar}{2}$$

[1]We shall use for the eigenkets of $\mathbf{Q}$, the notation $|\{Q\}\rangle$ in place of the usual one $|Q\rangle$ in order to make clearer some equations (see later).

[2]In a similar way we shall use for the eigenkets of $\mathbf{P}$, the notation $|\{P\}\rangle$ in place of the usual one $|P\rangle$ in order to make clearer some equations (see later).

and with the basic postulate commutator

$$[\mathbf{Q}, \mathbf{P}] = i\hbar$$

Now, one has to get the expression of the unitary operators, allowing one to translate the origin of the position and momentum operators.

2.5.2 Position operator translation

First, consider the following linear operator:

$$\boxed{\mathbf{A}(\mathbf{P}, Q^\circ) \equiv \mathbf{A}(Q^\circ) = \exp\left\{-\frac{iQ^\circ\mathbf{P}}{\hbar}\right\}} \tag{2.95}$$

where Q° is a real scalar having the dimension of a length, and $\mathbf{P}$ the momentum operator, conjugate to the position operator $\mathbf{Q}$. Its Hermitian conjugate is

$$\mathbf{A}(\mathbf{P}, Q^\circ)^\dagger = \exp\left\{\frac{iQ^\circ\mathbf{P}^\dagger}{\hbar}\right\}$$

Since $\mathbf{P}$ is Hermitian, that is, $\mathbf{P} = \mathbf{P}^\dagger$, this equation transforms to

$$\mathbf{A}(\mathbf{P}, Q^\circ)^\dagger = \exp\left\{\frac{iQ^\circ\mathbf{P}}{\hbar}\right\} \tag{2.96}$$

Thus

$$\mathbf{A}(\mathbf{P}, Q^\circ)^\dagger = \mathbf{A}(\mathbf{P}, Q^\circ)^{-1} \tag{2.97}$$

Now, the operator $\mathbf{A}(\mathbf{P}, Q^\circ)$ is unitary, so that

$$\mathbf{A}(\mathbf{P}, Q^\circ)^\dagger\mathbf{A}(\mathbf{P}, Q^\circ) = 1 \tag{2.98}$$

Now, calculate the commutator of the operator (2.95) with $\mathbf{Q}$. Then, since $\mathbf{A}(Q^\circ, \mathbf{P})$ is a function of $\mathbf{P}$, in view of Eq. (2.14) it takes the form

$$[\mathbf{Q}, \mathbf{A}(\mathbf{P}, Q^\circ)] = i\hbar\left(\frac{\partial\mathbf{A}(\mathbf{P}, Q^\circ)}{\partial\mathbf{P}}\right) \tag{2.99}$$

The right-hand side of this last equation may be obtained differentiating Eq. (2.95) to give

$$\left(\frac{\partial\mathbf{A}(\mathbf{P}, Q^\circ)}{\partial\mathbf{P}}\right) = -\left(\frac{i}{\hbar}\right)Q^\circ\mathbf{A}(\mathbf{P}, Q^\circ)$$

As a consequence, Eq. (2.99) becomes

$$[\mathbf{Q}, \mathbf{A}(\mathbf{P}, Q^\circ)] = Q^\circ\mathbf{A}(\mathbf{P}, Q^\circ) \tag{2.100}$$

Next, writing explicitly the left-hand side of Eq. (2.100) yields

$$\mathbf{Q}\mathbf{A}(Q^\circ) = \mathbf{A}(\mathbf{P}, Q^\circ)\mathbf{Q} + Q^\circ\mathbf{A}(\mathbf{P}, Q^\circ) \tag{2.101}$$

Again, premultiply each member of Eq. (2.101) by the inverse of **A**, that is,

$$\mathbf{A}(\mathbf{P}, Q^\circ)^{-1}\mathbf{Q}\mathbf{A}(\mathbf{P}, Q^\circ) = \mathbf{A}(\mathbf{P}, Q^\circ)^{-1}\mathbf{A}(\mathbf{P}, Q^\circ)\mathbf{Q} + Q^\circ\mathbf{A}(\mathbf{P}, Q^\circ)^{-1}\mathbf{A}(\mathbf{P}, Q^\circ)$$

Then, after simplifying, by aid of Eqs. (2.97) and (2.98), this last expression reduces to

$$\boxed{\mathbf{A}(\mathbf{P}, Q^\circ)^{-1}\mathbf{Q}\mathbf{A}(\mathbf{P}, Q^\circ) = \mathbf{Q} + Q^\circ} \tag{2.102}$$

Thus, Eq. (2.102), which is called *a canonical transformation* on the coordinate operator **Q**, translates the origin of **Q** by the scalar amount Q°. Furthermore, for an infinitesimal scalar displacement dQ°, Eq. (2.102) transforms to

$$\mathbf{A}(\mathbf{P}, dQ^\circ)^{-1}\mathbf{Q}\mathbf{A}(\mathbf{P}, dQ^\circ) = \mathbf{Q} + dQ^\circ \tag{2.103}$$

2.5.3 Momentum operator translation

Now, consider the linear operator $\mathbf{B}(\mathbf{Q}, P^\circ)$:

$$\boxed{\mathbf{B}(\mathbf{Q}, P^\circ) = \exp\left\{\frac{iP^\circ\mathbf{Q}}{\hbar}\right\}} \tag{2.104}$$

where P° is a scalar having the dimension of a momentum and **Q** being the Hermitian coordinate operator. The inspection of its expression shows that the operator $\mathbf{B}(P^\circ)$ is unitary since its inverse $\mathbf{B}(\mathbf{Q}, P^\circ)^{-1}$ is equal to its Hermitian conjugate $\mathbf{B}(\mathbf{Q}, P^\circ)^\dagger$:

$$\mathbf{B}(\mathbf{Q}, P^\circ)^\dagger = \mathbf{B}(\mathbf{Q}, P^\circ)^{-1}$$

Calculate the commutator of this operator with the momentum operator **P**. Since $\mathbf{B}(P^\circ, \mathbf{Q})$ is a function of **Q**, one may use Eq. (2.15), which leads to

$$[\mathbf{P}, \mathbf{B}(\mathbf{Q}, P^\circ)] = -i\hbar\left(\frac{\partial\mathbf{B}(\mathbf{Q}, P^\circ)}{\partial\mathbf{Q}}\right)$$

Differentiating Eq. (2.104), and after identification, one obtains

$$[\mathbf{P}, \mathbf{B}(\mathbf{Q}, P^\circ)] = P^\circ\mathbf{B}(\mathbf{Q}, P^\circ) \tag{2.105}$$

Then, writing explicitly the commutator, Eq. (2.105) reads

$$\mathbf{P}\mathbf{B}(\mathbf{Q}, P^\circ) = \mathbf{B}(\mathbf{Q}, P^\circ)\mathbf{P} + P^\circ\mathbf{B}(\mathbf{Q}, P^\circ) \tag{2.106}$$

Moreover, premultiply this equation by the inverse of **B** to get

$$\mathbf{B}(\mathbf{Q}, P^\circ)^{-1}\mathbf{P}\mathbf{B}(\mathbf{Q}, \mathbf{P}^\circ) = \mathbf{B}(\mathbf{Q}, P^\circ)^{-1}\mathbf{B}(\mathbf{Q}, P^\circ)\mathbf{P} + \mathbf{B}(\mathbf{Q}, P^\circ)^{-1}P^\circ\,\mathbf{B}(\mathbf{Q}, P^\circ) \tag{2.107}$$

On simplification, this expression reduces to

$$\boxed{\mathbf{B}(\mathbf{Q}, P^\circ)^{-1}\mathbf{P}\mathbf{B}(\mathbf{Q}, P^\circ) = \mathbf{P} + P^\circ} \tag{2.108}$$

Clearly, this canonical transformation allows to translate **P** by the scalar amount P°.

2.5.4 Quantum Galilean transformation

One may define the Galilean transformation operator according to

$$\boxed{\mathbf{S}(v) = \exp\left\{\frac{i}{\hbar}(mv\mathbf{Q} - \mathbf{P}vt)\right\}} \tag{2.109}$$

where v is the scalar velocity. Observe that this operator is Hermitian since

$$\mathbf{S}^{\dagger}(v) = \exp\left\{-\frac{i}{\hbar}(mv\mathbf{Q} - \mathbf{P}vt)\right\} = \mathbf{S}^{-1}(v)$$

Using the Glauber theorem (1.78), the operator (2.109) and its inverse take, respectively, the forms

$$\mathbf{S}(v) = \exp\left\{\frac{i}{\hbar}mv\mathbf{Q}\right\}\exp\left\{-\frac{i}{\hbar}\mathbf{P}vt\right\}e^{\zeta}$$

$$\mathbf{S}^{-1}(v) = \exp\left\{\frac{i}{\hbar}\mathbf{P}vt\right\}\exp\left\{-\frac{i}{\hbar}mv\mathbf{Q}\right\}e^{-\zeta}$$

with

$$\zeta = -\frac{1}{2}\left[\frac{i}{\hbar}mv\mathbf{Q}, -\frac{i}{\hbar}\mathbf{P}v\right]$$

Next, perform the following transformation on the position coordinate according to

$$\mathbf{S}(v)^{-1}\mathbf{Q}\mathbf{S}(v) = \exp\left\{\frac{i}{\hbar}\mathbf{P}vt\right\}\exp\left\{-\frac{i}{\hbar}mv\mathbf{Q}\right\}\mathbf{Q}\,\exp\left\{\frac{i}{\hbar}mv\mathbf{Q}\right\}\exp\left\{-\frac{i}{\hbar}\mathbf{P}vt\right\}$$

Hence

$$\mathbf{S}(v)^{-1}\mathbf{Q}\mathbf{S}(v) = \exp\left\{\frac{i}{\hbar}\mathbf{P}vt\right\}\mathbf{Q}\,\exp\left\{-\frac{i}{\hbar}\mathbf{P}vt\right\} \tag{2.110}$$

Next, taking vt in place of $P°$, and using Eq. (2.108), with the aid of Eq. (2.104), Eq. (2.110) reads

$$\boxed{\mathbf{S}(v)^{-1}\mathbf{Q}\mathbf{S}(v) = \mathbf{Q} - vt} \tag{2.111}$$

On the other hand, the transformation on the **P** coordinate involving the unitary operator (2.109) takes the form

$$\mathbf{S}(v)^{-1}\mathbf{P}\mathbf{S}(v) = \exp\left\{\frac{i}{\hbar}\mathbf{P}vt\right\}\exp\left\{-\frac{i}{\hbar}mv\mathbf{Q}\right\}\mathbf{P}\,\exp\left\{\frac{i}{\hbar}mv\mathbf{Q}\right\}\exp\left\{-\frac{i}{\hbar}\mathbf{P}vt\right\}$$

Again, taking mv in place of $Q°$, and then using Eqs. (2.102) and (2.95), yields

$$\mathbf{S}(v)^{-1}\mathbf{P}\mathbf{S}(v) = \exp\left\{\frac{i}{\hbar}\mathbf{P}vt\right\}(\mathbf{P} + mv)\exp\left\{-\frac{i}{\hbar}\mathbf{P}vt\right\}$$

or, after simplification

$$\boxed{\mathbf{S}(v)^{-1}\mathbf{P}\mathbf{S}(v) = \mathbf{P} + mv} \tag{2.112}$$

Equations (2.111) and (2.112) are the quantum Galilean transformations dealing with the **Q** and **P** operators

2.5.5 Action of translation operators on the Q eigenkets

Start from Eq. (2.101), that is, omitting the dependence of the translation operator on **P**

$$\mathbf{QA}(Q^\circ) = \mathbf{A}(Q^\circ)\mathbf{Q} + Q^\circ\mathbf{A}(Q^\circ) \quad \text{with} \quad \mathbf{A}(Q^\circ) = \exp\left\{-\frac{iQ^\circ\mathbf{P}}{\hbar}\right\} \tag{2.113}$$

where $\mathbf{A}(Q^\circ)$ is the translation operator, Q° a scalar having the dimension of a length, and $\mathbf{Q}$ and $\mathbf{P}$ having their usual meaning. Now, postmultiply both members of the first equation appearing in (2.113) by an eigenket $|\{Q\}\rangle$ of the position operator $\mathbf{Q}$:

$$\mathbf{QA}(Q^\circ)|\{Q\}\rangle = \mathbf{A}(Q^\circ)\mathbf{Q}|\{Q\}\rangle + Q^\circ\mathbf{A}(Q^\circ)|\{Q\}\rangle \tag{2.114}$$

Owing to the eigenvalue equation (2.91), this equation transforms to

$$\mathbf{QA}(Q^\circ)|\{Q\}\rangle = \mathbf{A}(Q^\circ)Q|\{Q\}\rangle + Q^\circ\mathbf{A}(Q^\circ)|\{Q\}\rangle$$

or, after commuting the scalar Q, with the translation operator

$$\mathbf{QA}(Q^\circ)|\{Q\}\rangle = (Q + Q^\circ)\mathbf{A}(Q^\circ)|\{Q\}\rangle \tag{2.115}$$

Now, using the notation

$$\mathbf{A}(Q^\circ)|\{Q\}\rangle \equiv |\{\mathbf{A}(Q^\circ)Q\}\rangle$$

Eq. (2.115) yields

$$\mathbf{Q}|\{\mathbf{A}(Q^\circ)Q\}\rangle = (Q + Q^\circ)|\{\mathbf{A}(Q^\circ)Q\}\rangle \tag{2.116}$$

On the other hand, the eigenvalue equation Eq. (2.91) reads

$$\mathbf{Q}\;|\{Q + Q^\circ\}\rangle = (Q + Q^\circ)|\{Q + \;Q^\circ\}\rangle \tag{2.117}$$

where $|\{Q+Q^\circ\}\rangle$ is the corresponding eigenvector of $\mathbf{Q}$. Then, by comparison of Eqs. (2.115) and (2.117) and ignoring a phase factor without interest, it appears that

$$\boxed{\mathbf{A}(Q^\circ)|\{Q\}\rangle = |\{Q + Q^\circ\}\rangle} \tag{2.118}$$

or, in view of Eq. (2.95)

$$\exp\left\{-\frac{iQ^\circ\mathbf{P}}{\hbar}\right\}|\{Q\}\rangle = |\{Q + Q^\circ\}\rangle \tag{2.119}$$

Of course, since $\mathbf{P}$ is Hermitian and Q° a real scalar, the Hermitian conjugate of this last expression is

$$\langle\{Q\}|\exp\left\{\frac{iQ^\circ\mathbf{P}}{\hbar}\right\} = \langle\{Q + Q^\circ\}| \tag{2.120}$$

Now, remark that for the infinitesimal transformation (2.103), the translation operator (2.95) may be expanded up to first order in dQ° to give

$$\mathbf{A}(dQ^\circ) = 1 - \frac{i}{\hbar}dQ^\circ\mathbf{P}$$

Next, the action of this infinitesimal translation operator on an eigenket of $\mathbf{Q}$ takes the form

$$\mathbf{A}(dQ^\circ)|\{Q\}\rangle = \left(1 - \frac{i}{\hbar}dQ^\circ\mathbf{P}\right)|\{Q\}\rangle$$

or, due to Eq. (2.118), with dQ° in place of Q°

$$\mathbf{A}(dQ^\circ)|\{Q\}\rangle = |\{Q + dQ^\circ\}\rangle$$

Thus, by identification of these two last equations, one gets

$$\boxed{|\{Q + dQ^\circ\}\rangle = \left(1 - \frac{i}{\hbar}dQ^\circ\mathbf{P}\right)|\{Q\}\rangle} \tag{2.121}$$

Next, let $|\{0\}_Q\rangle$ be the eigenket of the coordinate operator $\mathbf{Q}$, corresponding to the zero eigenvalue

$$\mathbf{Q}|\{0\}_Q\rangle = 0\ |\{0\}_Q\rangle \tag{2.122}$$

Then, Eq. (2.118) reads

$$\mathbf{A}(Q^\circ)|\{0\}_Q\rangle = |\{0 + Q^\circ\}\rangle$$

or

$$\mathbf{A}(Q^\circ)|\{0\}_Q\rangle = |\{Q^\circ\}\rangle \tag{2.123}$$

Again, writing explicitly the translation operator by the aid of Eq. (2.95), and substituting the notation Q° by the more general one Q, without modifying anything, one obtains

$$\boxed{\exp\left\{-\frac{i}{\hbar}Q\mathbf{P}\right\}|\{0\}_Q\rangle = |\{Q\}\rangle} \tag{2.124}$$

On the other hand, recall Eq. (2.106), that is,

$$\mathbf{PB}(P^\circ) = \mathbf{B}(P^\circ)\mathbf{P} + P^\circ\mathbf{B}(P^\circ) \tag{2.125}$$

with

$$\mathbf{B}(P^\circ) = \exp\left\{\frac{iP^\circ\mathbf{Q}}{\hbar}\right\}$$

where $\mathbf{B}(P^\circ)$ is the translation operator and where P° is a scalar having the dimension of an impulsion. Now, postmultiply Eq. (2.125) by an eigenket $|\{P\}\rangle$ of $\mathbf{P}$

$$\mathbf{PB}(P^\circ)|\{P\}\rangle = \mathbf{B}(P^\circ)\mathbf{P}|\{P\}\rangle + P^\circ\mathbf{B}(P^\circ)|\{P\}\rangle$$

Then, by an inference very similar to that allowing one to pass from Eq. (2.114) to (2.118), one finds

$$\mathbf{B}(P^\circ)|\{P\}\rangle = |\{P + P^\circ\}\rangle \tag{2.126}$$

Now, consider the eigenvalue equation of the momentum operator corresponding to the zero eigenvalue, that is,

$$\mathbf{P}|\{0\}_P\rangle = 0|\{0\}_P\rangle$$

Then, Eq. (2.126) yields

$$\mathbf{B}(P^\circ)|\{0\}_P\rangle = |\{P^\circ\}\rangle$$

Finally, explicitly writing the translation operator $\mathbf{B}(P^\circ)$, using Eq. (2.104), and taking P in place of P°, and $|\{P\}\rangle$ in place of $|\{P^\circ\}\rangle$, this last equation becomes

$$\boxed{\exp\left\{\frac{iP\mathbf{Q}}{\hbar}\right\}|\{0\}_P\rangle = |\{P\}\rangle} \tag{2.127}$$

2.5.6 Scalar products $\langle\{P\}|\{Q\}\rangle$

We have now to find the expression of the scalar product between an eigenket of $\mathbf{Q}$ and one of $\mathbf{P}$.

2.5.6.1 A first expression To this aim, premultiply Eq. (2.124) by any bra $\langle\{P\}|$:

$$\langle\{P\}|\{Q\}\rangle = \langle\{P\}|\exp\left\{-\frac{iQ\mathbf{P}}{\hbar}\right\}|\{0\}_Q\rangle$$

Using Eq. (2.93) and the action of the exponential operator on the left bra, which is an eigenbra of the $\mathbf{P}$ operator with the eigenvalue P, one obtains from (2.93)

$$\langle\{P\}|\{Q\}\rangle = \exp\left\{-\frac{iQP}{\hbar}\right\}\langle\{P\}|\{0\}_Q\rangle \tag{2.128}$$

Now, observe that in this last equation, the bra $\langle\{P\}|$ may be obtained via the Hermitian conjugate of Eq. (2.127), that is,

$$\langle\{P\}| = \langle\{0\}_P|\exp\left\{-\frac{iP\mathbf{Q}}{\hbar}\right\}$$

Then, using this expression for the bra $\langle\{P\}|$, Eq. (2.128) transforms to

$$\langle\{P\}|\{Q\}\rangle = \exp\left\{-\frac{iQP}{\hbar}\right\}\left\langle\{0\}_P\left|\exp\left\{-\frac{iP\mathbf{Q}}{\hbar}\right\}\right|\{0\}_Q\right\rangle \tag{2.129}$$

Next, owing to Eq. (2.122), we may remark that the eigenvalue of $\mathbf{Q}$ corresponding to the right-hand-side ket of this last equation, is zero, so that the corresponding eigenvalue equation involving the exponential of $\mathbf{Q}$ reduces to

$$\exp\left\{-\frac{iP\mathbf{Q}}{\hbar}\right\}|\{0\}_Q\rangle = |\{0\}_Q\rangle$$

Thus, the scalar product (2.129) yields

$$\boxed{\langle\{P\}|\{Q\}\rangle = \exp\left\{-\frac{iQP}{\hbar}\right\}\langle\{0\}_P|\{0\}_Q\rangle} \tag{2.130}$$

2.5.6.2 Scalar products involved on the right-hand side of Eq. (2.130) To further utilize Eq. (2.130), we require the following scalar product:

$$\langle\{0\}_P|\{0\}_Q\rangle$$

For this purpose, let us first consider the scalar product $\langle\{P'\}|\{P''\}\rangle$ between two different eigenkets of the momentum operator, which obeys Eq. (2.94), that is,

$$\langle\{P'\}|\{P''\}\rangle = \delta(P' - P'')$$

Introduce between the ket and the bra the closure relation on the eigenkets of the coordinate operator:

$$\langle\{P'\}|\left\{\int_{-\infty}^{+\infty} |\{Q\}\rangle\langle\{Q\}|\, dQ\right\}|\{P''\}\rangle = \delta(P' - P'')$$

or

$$\int_{-\infty}^{+\infty} \langle\{P'\}|\{Q\}\rangle\langle\{Q\}|\{P''\}\rangle dQ' = \delta(P' - P'')$$

On the other hand, using Eq. (2.130) and its complex conjugate, this last expression yields

$$|\langle\{0\}_P|\{0\}_Q\rangle|^2 \int_{-\infty}^{+\infty} \exp\left\{-\frac{iQP'}{\hbar}\right\} \exp\left\{\frac{iQP''}{\hbar}\right\} dQ = \delta(P' - P'')$$

or

$$|\langle\{0\}_P|\{0\}_Q\rangle|^2 \int_{-\infty}^{+\infty} \exp\left\{-\frac{iQ(P' - P'')}{\hbar}\right\} dQ = \delta(P' - P'') \tag{2.131}$$

Now, observe that according to Eq. (18.60) and keeping in mind the fact that the dimension of P is that of $Q/\hbar$, the integral appearing in Eq. (2.131) reads

$$\int_{-\infty}^{+\infty} \exp\left\{-\frac{iQ(P' - P'')}{\hbar}\right\} dQ = 2\pi\hbar\delta(P' - P'')$$

Thus, Eq. (2.131) simplifies to

$$|\langle\{0\}_P|\{0\}_Q\rangle|^2 = \frac{1}{2\pi\hbar}$$

Therefore, ignoring the unknown phase factor, which is without interest,

$$\langle\{0\}_P|\{0\}_Q\rangle = \sqrt{\frac{1}{2\pi\hbar}}$$

Eq. (2.130) becomes

$$\boxed{\langle\{P\}|\{Q\}\rangle = \sqrt{\frac{1}{2\pi\hbar}} \exp\left\{-\frac{iPQ}{\hbar}\right\}} \tag{2.132}$$

At last, the probability of passage from some ket $|\{Q\}\rangle$ to any ket $|\{P\}\rangle$ or vice versa is, according to the fourth postulate

$$\boxed{|\langle\{P\}|\{Q\}\rangle|^2 = \frac{1}{2\pi\hbar}} \tag{2.133}$$

That shows that whatever the value observed for the position coordinate before a measurement of the momentum, the probability to find after such a measurement some value of the momentum is the same whatever this last value and vice versa. Equation (2.133) may be viewed as the expression of the basic contingency affecting the most simple and fundamental variables appearing in physics through Lagrange's equations.

2.6 CONCLUSION

This chapter has considered the presentation of the principles of quantum mechanics. We have introduced the important concepts of bras and kets describing the quantum states, that of Hermitean operators describing the physical variables, that of the measurement of physical variables through the eigenvalues of the corresponding Hermitian operators, and the notions of quantum average values generally relating kets and Hermitian operators. In discussing the quantum principles, large parts have been devoted to the time-dependent Schrödinger equation and to quantum averages and to their corresponding fluctuations. The quantum principles were shown to lead to a limitation of the knowledge of some physical conjugated variables, which is illustrated by the Heisenberg uncertainty relations, forbidding one to know simultaneously and exactly the position and momentum, however, preserving the main features of Newton's laws of classical mechanics, the cost to be paid to the Heisenberg uncertainty relations being the fact that these laws govern average values of the position and momentum in place of exact ones. Now, to be useful applied to particular situations such as, for instance, oscillators, quantum mechanics requires different equivalent representations such as matrix mechanics, wave mechanics, density operator approach, and also equivalent different time-dependent representations such as the Schrödinger, the Heisenberg, and the interaction pictures. The most important results of this chapter are summarized below:

Basic equations for quantum mechanics
Deterministic and probalistic changes:
$i\hbar\left(\dfrac{\partial|\Psi(t)\rangle}{\partial t}\right) = \mathbf{H}|\Psi(t)\rangle \qquad \mathbb{P}_{kl} = |\langle\Phi_k|\Psi_l\rangle|^2$
Average values, dispersions, and their dynamics:
$\langle\mathbf{A}\rangle_\Psi = \langle\Psi|\mathbf{A}|\Psi\rangle \qquad \Delta\mathbf{A}_\Psi = \sqrt{\langle\mathbf{A}^2\rangle_\Psi - \langle\mathbf{A}\rangle_\Psi^2}$ $\left(\dfrac{\partial\langle\Psi(t)|\mathbf{A}|\Psi(t)\rangle}{\partial t}\right) = \dfrac{i}{\hbar}\langle\Psi(t)|[\mathbf{H},\mathbf{A}]|\Psi(t)\rangle$

Eigenvalue equations of the Hermitian operators and their eigenvalues and eigenvectors:
$\mathbf{A}\|\Phi_n\rangle = A_n\|\Phi_n\rangle$ since $\mathbf{A} = \mathbf{A}^\dagger$, $A_n = A_n^*$, $\langle\Phi_n\|\Phi_m\rangle = \delta_{nm}$, $\sum \|\Phi_n\rangle\langle\Phi_n\| = 1$
Important relations resulting from the commutation rule $[\mathbf{Q}, \mathbf{P}] = i\hbar$:
$[\mathbf{Q}, \mathbf{F}(\mathbf{P})] = (i\hbar)\left(\frac{\partial \mathbf{F}(\mathbf{P})}{\partial \mathbf{P}}\right)$ $[\mathbf{P}, \mathbf{F}(\mathbf{Q})] = -(i\hbar)\left(\frac{\partial \mathbf{F}(\mathbf{Q})}{\partial \mathbf{Q}}\right)$ $(e^{iQ^\circ \mathbf{P}/\hbar})\mathbf{Q}\,(e^{-iQ^\circ \mathbf{P}/\hbar}) = \mathbf{Q} + Q^\circ$ $(e^{-iP^\circ \mathbf{Q}/\hbar})\mathbf{P}(e^{iP^\circ \mathbf{Q}/\hbar}) = \mathbf{P} + P^\circ$ if $\mathbf{Q}\|\{Q\}\rangle = Q\|\{Q\}\rangle$ and $\mathbf{P}\|\{P\}\rangle = P\|\{P\}\rangle$ $\langle\{P\}\|\{Q\}\rangle = \sqrt{\frac{1}{2\pi\hbar}}e^{-iQP/\hbar}$ $\Delta P_\psi\, \Delta Q_\psi \geqslant \hbar$

BIBLIOGRAPHY

C. Cohen-Tannoudji, B. Diu, and F. Laloe. *Quantum Mechanics.* Wiley-Interscience: New York, 2006.

P. A. M. Dirac. *The Principles of Quantum Mechanics*, 4th ed. Oxford University Press: 1982.

A. Messiah. *Quantum Mechanics.* Dover Publications, New York, 1999.

L. I. Schiff. *Quantum Mechanics*, 3rd ed. McGraw-Hill: New York, 1968.

CHAPTER 3

QUANTUM MECHANICS REPRESENTATIONS

INTRODUCTION

In the previous chapter we obtained different simple but important results following from the postulates of quantum mechanics such as the Ehrenfest and the virial theorems, the Heisenberg uncertainty relations, and the scalar products between any eigenket of **Q** and another one of **P**, the modulus of them being the same whatever the corresponding eigenvalues. But, in order to become tractable for the study of concrete situations, it is necessary to adapt the postulates. That is the aim of what is termed different representations of quantum mechanics. Among them there are the *matrix mechanics*, due initially to Heisenberg, Born, Jordan, and Pauli, and the *wave mechanics* of Louis de Broglie and Schrödinger. There are also different time-dependent representations besides those of Schrödinger, that is, the *Heisenberg picture* and the different *interaction pictures*, which deal with time evolution operators. Finally, there are the density operator representations in which the informations dealing with the kets or the wavefunctions are introduced into an operator and which are very useful when working on many-particle systems. All these representations will be studied in the present chapter.

3.1 MATRIX REPRESENTATION

Because the postulates of quantum mechanics concern the state space, which is a vector space, the matrices play a fundamental role in quantum mechanics leading to the fact that there are matrix representations for all theoretical entities involved in the postulates, that is, for kets, bras, linear transformations, eigenvalue equations, and so on. The purpose of the present section is to consider that subject more deeply.

3.1.1 Kets and bras

First, consider the eigenvalue equation of a Hermitian operator **A**:

$$\mathbf{A}|\Phi_l\rangle = A_l|\Phi_l\rangle \quad \text{with} \quad \mathbf{A} = \mathbf{A}^\dagger \qquad \text{and thus} \qquad \langle\Phi_l|\Phi_k\rangle = \delta_{lk}$$

Quantum Oscillators, First Edition. Olivier Henri-Rousseau and Paul Blaise.
© 2011 John Wiley & Sons, Inc. Published 2011 by John Wiley & Sons, Inc.

We remember that the eigenvectors $|\Phi_l\rangle$ of $\mathbf{A}$ obey the closure relation

$$\sum_i |\Phi_i\rangle\langle\Phi_i| = 1$$

Then consider a ket $|\Psi_k\rangle$ of the state space that does not belong to the set $\{\Phi_l\}$ of the eigenvectors $|\Phi_l\rangle$, and multiply it by the above closure relation:

$$|\Psi_k\rangle = \mathbf{1}|\Psi_k\rangle = \left\{\sum_i |\Phi_i\rangle\langle\Phi_i|\right\}|\Psi_k\rangle$$

Hence

$$|\Psi_k\rangle = \sum_i \langle\Phi_i|\Psi_k\rangle|\Phi_i\rangle$$

or

$$|\Psi_k\rangle = \sum_i a_{ik}|\Phi_i\rangle \quad \text{with} \quad a_{ik} = \langle\Phi_i|\Psi_k\rangle \tag{3.1}$$

Owing to the convention for matrix notation in which the first index corresponds to the row and the second one to the column, and in view of Eq. (3.1), a ket $|\Psi_k\rangle$ may be represented, in a basis $\{|\Phi_i\rangle\}$, by a column vector, the components of which are the coefficients a_{ik}:

$$|\Psi_k\rangle \Leftrightarrow \begin{pmatrix} a_{1k} \\ a_{2k} \\ \ldots \\ a_{ik} \\ \ldots \end{pmatrix} \tag{3.2}$$

Next, one may proceed in a similar way for the bra $\langle\Psi_j|$ corresponding to the above ket. Then, the Hermitian conjugate of Eq. (3.1) is

$$\langle\Psi_j| = \sum_i a_{ji}\langle\Phi_i| \quad \text{with} \quad a_{ji} = \langle\Psi_j|\Phi_i\rangle$$

This last result shows that the matrix representation of the bra $\langle\Psi_j|$ is a row vector, the components of which are the expansion coefficients of the above equation:

$$\langle\Psi_j| \quad \Leftrightarrow \quad (a_{j1} \quad a_{j2} \quad \ldots \quad a_{ji} \quad \ldots) \tag{3.3}$$

Note that the expansion coefficients a_{ik} and a_{ki} are complex conjugates since they are the expressions of the complex conjugate scalar products, tha is,

$$a_{ij} = a_{ji}^* \quad \text{because} \quad \langle\Phi_i|\Psi_j\rangle = \langle\Psi_j|\Phi_i\rangle^*$$

3.1.2 Scalar products

Consider the following expansions of the ket $|\Psi_k\rangle$ and of the bra $\langle\xi_j|$ in the basis $\{|\Phi_i\rangle\}$ obeying

$$\langle\Phi_i|\Phi_l\rangle = \delta_{il} \tag{3.4}$$

$$|\Psi_k\rangle = \sum_i a_{ik}|\Phi_i\rangle \quad \text{with} \quad a_{ik} = \langle\Phi_i|\Psi_k\rangle$$

$$\langle\xi_j| = \sum_l b_{jl}\langle\Phi_l| \quad \text{with} \quad b_{jl} = \langle\xi_j|\Phi_l\rangle$$

Now, the scalar product $\langle\xi_j|\Psi_k\rangle$ reads

$$\langle\xi_j|\Psi_k\rangle = \left\{\sum_l b_{jl}\langle\Phi_l|\right\}\left\{\sum_i a_{ik}|\Phi_i\rangle\right\}$$

or

$$\langle\,\xi_j|\Psi_k\rangle = \sum_l \sum_i b_{jl}a_{ik}\langle\Phi_l|\Phi_i\rangle$$

so that, due to the orthonormality properties (3.4)

$$\langle\xi_j|\Psi_k\rangle = \sum_l b_{jl}a_{lk} \tag{3.5}$$

Owing to the matrix convention according to which the first index refers to the row and the second to the column, expression (3.5) appears to be the matrix product between the jth line and the kth column vectors constructed, respectively, from the set of b_{jl} and a_{lk} coefficients.

$$\langle\xi_j|\Psi_k\rangle \Leftrightarrow (b_{j1} \quad b_{j2} \quad \cdots \quad b_{jl} \quad \cdots)\begin{pmatrix} a_{1k} \\ a_{2k} \\ \cdots \\ a_{lk} \\ \cdots \end{pmatrix}$$

3.1.3 Operators

Consider a linear operator $\mathbf{A}$. Premultiply it by the bra $\langle\Phi_i|$ and postmultiply it by the ket $|\Phi_k\rangle$ belonging to the same basis as the ket $|\Phi_l\rangle$, the Hermitian conjugate of which is $\langle\Phi_i|$. The linear operation of $\mathbf{A}$ on $|\Phi_k\rangle$ gives a new ket $|\Psi_k\rangle$ on which the action of $\langle\Phi_i|$ corresponds to a scalar product, the result of which is the double index scalar A_{ik}:

$$\langle\Phi_i|\mathbf{A}|\Phi_k\rangle = \langle\Phi_i|\Psi_k\rangle = A_{ik} \tag{3.6}$$

The different scalars A_{ik} (which may be obtained by allowing the indexes of the ket and of the bra to run over the different terms of the basis) appear to be the matrix elements of a square matrix the dimension of which is generally infinite. Observe that, owing to Eq. (1.30),

$$\langle\Phi_i|\mathbf{A}|\Phi_k\rangle = \langle\Phi_k|\mathbf{A}^\dagger|\Phi_i\rangle^* \tag{3.7}$$

3.1.3.1 Hermitian operators If the linear operator is Hermitian, that is,

$$\mathbf{A} = \mathbf{A}^\dagger$$

then the matrix elements (3.7) simplify to

$$\langle \Phi_i|\mathbf{A}|\Phi_k\rangle = \langle \Phi_k|\mathbf{A}|\Phi_i\rangle^*$$

Hence, from Eq. (3.6), it appears that the matrix elements are complex conjugate with respect to the diagonal part, which is real, so that

$$A_{ik} = A_{ki}^* \quad \text{and thus} \quad A_{kk} = A_{kk}^* \quad \text{so that} \quad A_{kk} \text{ is real} \tag{3.8}$$

with

$$A_{ki}^* = \langle \Phi_k|\mathbf{A}|\Phi_i\rangle^*$$

A matrix the elements defined by condition (3.8) is a Hermitian matrix.

3.1.3.2 Unitary operators Consider the linear unitary operator **U** satisfying

$$\mathbf{U}^{-1} = \mathbf{U}^\dagger \quad \text{with} \quad \mathbf{U}^{-1}\mathbf{U} = 1 \tag{3.9}$$

Now, consider a matrix element of this operator

$$U_{ik} = \langle \Phi_i|\mathbf{U}|\Phi_k\rangle$$

and the corresponding matrix elements of its inverse and of its Hermitian conjugate. They must be equal owing to the fact that the inverse of the unitary operator is equal to its Hermitian conjugate. Hence

$$\langle \Phi_i|\mathbf{U}^{-1}|\Phi_k\rangle = \langle \Phi_i|\mathbf{U}^\dagger|\Phi_k\rangle \tag{3.10}$$

Of course, owing to the general property (1.30), the following relation for the right-hand-side matrix element of the latter equation exists:

$$\langle \Phi_i|\mathbf{U}^\dagger|\Phi_k\rangle = \langle \Phi_k|\mathbf{U}|\Phi_i\rangle^* \tag{3.11}$$

Thus, because of this last equation, Eq. (3.10) becomes

$$\langle \Phi_i|\mathbf{U}^{-1}|\Phi_k\rangle = \langle \Phi_k|\mathbf{U}|\Phi_i\rangle^*$$

Thus, the following general relation between the matrix elements of the unitary operator and those of its inverse exists:

$$\boxed{U_{ik}^{-1} = U_{ki}^*} \tag{3.12}$$

with

$$U_{ki} = \langle \Phi_k|\mathbf{U}|\Phi_i\rangle \qquad U_{ik}^{-1} = \langle \Phi_i|\mathbf{U}^{-1}|\Phi_k\rangle \qquad \text{and} \qquad U_{ki}^* = \langle \Phi_k|\mathbf{U}|\Phi_i\rangle^* \tag{3.13}$$

Next, consider the matrix element built up from the definition of an inverse operator:

$$\langle \Phi_i|\mathbf{U}^{-1}\mathbf{U}|\Phi_k\rangle = \langle \Phi_i|\mathbf{1}|\Phi_k\rangle \tag{3.14}$$

Since the ket $|\Phi_k\rangle$ and the ket $|\Phi_i\rangle$ (which is the Hermitian conjugate of the bra $\langle \Phi_i|$) belong to the same basis, they are orthogonal, that is,

$$\langle \Phi_i|\mathbf{1}|\Phi_k\rangle = \langle \Phi_i|\Phi_k\rangle = \delta_{ik}$$

so that the matrix element (3.14) obeys

$$\langle \Phi_i | \mathbf{U}^{-1} \mathbf{U} | \Phi_k \rangle = \delta_{ik} \tag{3.15}$$

Next, use the closure relation on the basis $\{|\Phi_l\rangle\}$

$$\sum_l |\Phi_l\rangle\langle\Phi_l| = \mathbf{1}$$

By inserting it in Eq. (3.15) between the unitary operator and its inverse, it yields

$$\langle \Phi_i | \mathbf{U}^{-1} \left\{ \sum_l |\Phi_l\rangle\langle\Phi_l| \right\} \mathbf{U} | \Phi_k \rangle = \delta_{ik}$$

Then, using the properties (3.9) of the unitary operator, we have

$$\sum_l \langle \Phi_i | \mathbf{U}^\dagger | \Phi_l \rangle \langle \Phi_l | \mathbf{U} | \Phi_k \rangle = \delta_{ik} \tag{3.16}$$

or, due to Eq. (3.11)

$$\sum_l \langle \Phi_l | \mathbf{U} | \Phi_i \rangle^* \langle \Phi_l | \mathbf{U} | \Phi_k \rangle = \delta_{ik}$$

so that owing to Eq. (3.13)

$$\boxed{\sum_l U_{li}^* U_{lk} = \delta_{ik}}$$

This last expression may be split into two equations, the first of which shows that any column labeled i of some unitary matrix is normalized and that two different columns labeled i and k of such a matrix are orthogonal:

$$\sum_l |U_{li}|^2 = 1 \qquad \text{and} \qquad \sum_l U_{li}^* U_{lk} = 0 \quad \text{if} \quad i \neq k \tag{3.17}$$

All the matrix elements U_{li}, with l running over the elements of the basis, form therefore a column vector so that Eq. (3.17) may be visualized as the orthonormality properties of the column vectors from which the unitary matrix is built up. Now, taking the Hermitian conjugate of Eq. (3.16) and proceeding in a similar way, one would obtain the two following equations, expressing, respectively, that any row i of a unitary matrix is normalized and that two different rows i and k of such a matrix are orthogonal:

$$\sum_l |U_{il}|^2 = 1 \qquad \text{and} \qquad \sum_l U_{il}^* U_{kl} = 0 \quad \text{if} \quad i \neq k$$

Observe that some unitary matrices are real so that, owing to Eq. (3.12), their matrix elements O_{ik} obey

$$O_{ik}^{-1} = O_{ki}$$

For such matrices, which are said to be *orthogonal*, and, owing to Eq. (3.17), their columns obey the simplified orthonormality properties (which are at the origin of their name):

$$\sum_l O_{lk}^2 = 1 \qquad \text{and} \qquad \sum_l O_{li}O_{lk} = 0 \quad \text{if} \quad l \neq k$$

3.1.4 Linear transformations

3.1.4.1 Simple linear transformations Consider the eigenvalue equation of the Hermitian operator $\mathbf{B}$:

$$\mathbf{B}|\Phi_l\rangle = B_l|\Phi_l\rangle$$

Since it is Hermitian, its eigenkets $|\Phi_l\rangle$ are orthonormal so that the following basis $\{|\Phi_l\rangle\}$ can be constructed:

$$\sum_i |\Phi_i\rangle\langle\Phi_i| = \mathbf{1} \qquad \text{and} \qquad \langle\Phi_i|\Phi_j\rangle = \delta_{ij} \tag{3.18}$$

Next, consider the following linear transformation involving the linear operator $\mathbf{A}$, which does not commute with $\mathbf{B}$ and which transforms a ket $|\Psi_k\rangle$ into any another one $|\xi_q\rangle$:

$$\mathbf{A}|\Psi_k\rangle = |\xi_q\rangle \quad \text{with} \quad [\mathbf{A}, \mathbf{B}] \neq 0 \tag{3.19}$$

Now, introduce the closure relation appearing in (3.18) in this linear transformation according to

$$\mathbf{A}\sum_i |\Phi_i\rangle\langle\Phi_i|\Psi_k\rangle = |\xi_q\rangle$$

Again, premultiply both sides of this equation by the bra $\langle\Phi_r|$:

$$\boxed{\sum_i \langle\Phi_r|\mathbf{A}|\Phi_i\rangle\langle\Phi_i|\Psi_k\rangle = \langle\Phi_r|\xi_q\rangle}$$

which reads

$$\sum_i A_{ri}\, b_{ik} = a_{rk} \tag{3.20}$$

with, respectively,

$$A_{ri} = \langle\Phi_r|\mathbf{A}|\Phi_i\rangle \qquad b_{ik} = \langle\Phi_i|\Psi_k\rangle \qquad \text{and} \qquad a_{rq} = \langle\Phi_r|\xi_q\rangle$$

Owing to the matrix convention, and within the representation defined by the basis (3.18), Eq. (3.19) appears to be the matrix linear transformation (3.20) through

$$\begin{pmatrix} a_{1q} \\ a_{2q} \\ \cdots \\ a_{lq} \\ \cdots \end{pmatrix} = \begin{pmatrix} A_{11} & A_{12} & \cdots & A_{1i} & \cdots \\ A_{21} & A_{22} & & & \\ \cdots & & & & \\ A_{r1} & & & & \\ \cdots & & & & \end{pmatrix} \begin{pmatrix} b_{1k} \\ b_{2k} \\ \cdots \\ b_{lk} \\ \cdots \end{pmatrix}$$

3.1.4.2 Inverse transformations Now, we shall consider the inverse of the transformation (3.19). Hence, premultiply both members of this equation by the inverse $\mathbf{A}^{-1}$ of the operator $\mathbf{A}$:

$$\mathbf{A}^{-1}\mathbf{A}|\Psi_k\rangle = \mathbf{A}^{-1}|\xi_q\rangle$$

Then, after simplification, we have

$$|\Psi_k\rangle = \mathbf{A}^{-1}|\xi_q\rangle$$

Now, insert the closure relation (3.18) in the following way:

$$|\Psi_k\rangle = \mathbf{A}^{-1}\sum_i |\Phi_i\rangle\langle\Phi_i|\xi_q\rangle$$

Premultiplying by the bra $\langle\Phi_r|$ reads

$$\langle\Phi_r|\Psi_k\rangle = \sum_i \langle\Phi_r|\mathbf{A}^{-1}|\Phi_i\rangle\langle\Phi_i|\xi_q\rangle$$

leading to the following matrix representation of the inverse linear transformation:

$$\begin{pmatrix} b_{1k} \\ b_{2k} \\ \cdots \\ b_{lk} \\ \cdots \end{pmatrix} = \begin{pmatrix} A_{11}^{-1} & A_{12}^{-1} & \cdots & A_{1i}^{-1} & \cdots \\ A_{21}^{-1} & A_{22}^{-1} & & & \\ \cdots & & & & \\ A_{r1}^{-1} & & & & \\ \cdots & & & & \end{pmatrix} \begin{pmatrix} a_{1q} \\ a_{2q} \\ \cdots \\ a_{iq} \\ \cdots \end{pmatrix}$$

that may be also written

$$b_{rk} = \sum_i A_{ri}^{-1} a_{iq}$$

with

$$a_{iq} = \langle\Phi_i|\xi_q\rangle \qquad \text{and} \qquad A_{ri}^{-1} = \langle\Phi_r|\mathbf{A}^{-1}|\Phi_i\rangle$$

respectively.

3.1.4.3 Unitary transformations Consider a matrix element of a matrix representation of a linear operator $\mathbf{A}$ in some basis $\{|\Psi_k\rangle\}$ defined by the eigenvalue equation of a Hermitian operator $\mathbf{C}$ that does not commute with $\mathbf{A}$:

$$\mathbf{C}|\Psi_k\rangle = C_k|\Psi_k\rangle \quad \text{with} \quad \mathbf{C} = \mathbf{C}^\dagger \qquad \text{and} \qquad [\mathbf{C}, \mathbf{A}] \neq 0$$

$$\sum_k |\Psi_k\rangle\langle\Psi_k| = \mathbf{1} \qquad \text{and} \qquad \langle\Psi_k|\Psi_l\rangle = \delta_{kl} \tag{3.21}$$

This element is

$$\langle\Psi_l|\mathbf{A}|\Psi_k\rangle = A_{lk} \tag{3.22}$$

Now, seek the representation of this operator within a new basis $\{|\Phi_q\rangle\}$ defined by the eigenvalue equation of another Hermitian operator **B**, which commutes neither with **A** nor with **C**:

$$\mathbf{B}|\Phi_q\rangle = B_q|\Phi_q\rangle \quad \text{with} \quad \mathbf{B} = \mathbf{B}^\dagger \qquad \text{and} \qquad [\mathbf{B}, \mathbf{A}] \neq 0$$

$$\sum_q |\Phi_q\rangle\langle\Phi_q| = \mathbf{1} \qquad \text{and} \qquad \langle\Phi_p|\Phi_q\rangle = \delta_{pq} \tag{3.23}$$

To this end, introduce twice the unity operator inside the matrix element (3.22):

$$\langle\Psi_l|\mathbf{A}|\Psi_k\rangle = \langle\Psi_l|\mathbf{1}\ \mathbf{A}\ \mathbf{1}|\Psi_k\rangle$$

Then, using for the unity operator of the closure relation appearing in Eq. (3.23), it reads

$$A_{lk} = \langle\Psi_l|\mathbf{A}|\Psi_k\rangle = \langle\Psi_l| \left\{ \sum_q |\Phi_q\rangle\langle\Phi_q| \right\} \mathbf{A} \left\{ \sum_p |\Phi_p\rangle\langle\Phi_p| \right\} |\Psi_k\rangle$$

or

$$A_{lk} = \sum_q \sum_p \langle\Psi_l|\Phi_q\rangle\langle\Phi_q|\mathbf{A}|\Phi_p\rangle\langle\Phi_p|\Psi_k\rangle$$

and thus

$$\boxed{A_{lk} = \sum_q \sum_p a_{lq}\tilde{A}_{qp}a_{pk}} \tag{3.24}$$

with

$$a_{lq} = \langle\Psi_l|\Phi_q\rangle \qquad \tilde{A}_{qp} = \langle\Phi_q|\mathbf{A}|\Phi_p\rangle \qquad \text{and} \qquad a_{pk} = \langle\Phi_p|\Psi_k\rangle \tag{3.25}$$

Owing to the matrix notation conventions, Eq. (3.24) appears as the following product of matrices:

$$\begin{pmatrix} A_{11} & A_{12} & & A_{1k} & \cdots \\ A_{21} & A_{22} & & & \\ \vdots & & \cdots & & \\ A_{k1} & A_{k2} & & A_{kk} & \\ & & & & \cdots \end{pmatrix}$$

$$= \begin{pmatrix} a_{11} & a_{12} & \cdots & \cdots & \\ a_{21} & a_{22} & \cdots & \cdots & \\ \cdots & \cdots & & & \\ a_{l1} & a_{l2} & & a_{ll} & \\ & & & & \cdots \end{pmatrix} \begin{pmatrix} \tilde{A}_{11} & \tilde{A}_{12} & & \cdots & \cdots \\ \tilde{A}_{21} & \tilde{A}_{22} & & & \\ \cdots & \cdots & \cdots & & \\ \tilde{A}_{q1} & \tilde{A}_{q2} & & \cdots & \\ & & & & \cdots \end{pmatrix} \begin{pmatrix} a_{11} & a_{12} & \cdots & \cdots & \\ a_{21} & a_{22} & \cdots & \cdots & \\ \cdots & \cdots & & & \\ a_{p1} & a_{p2} & & \cdots & \\ & & & & \cdots \end{pmatrix} \tag{3.26}$$

Now, observe that the first and third right-hand-side matrices are unitary, which may be proved by first observing that due to Eq. (3.25), it is always possible to write

$$\sum_q a_{lq} a_{qk} = \sum_q \langle \Psi_l | \Phi_q \rangle \langle \Phi_q | \Psi_k \rangle$$

Again, using the closure relation appearing in Eq. (3.23), and the orthonormality properties (3.21), we have

$$\sum_q a_{lq} a_{qk} = \langle \Psi_l | \Psi_k \rangle = \delta_{lk}$$

On the other hand, the unitary transformation (3.26) may be denoted

$$\boxed{\mathbf{A}} = \boxed{\mathbf{U}}^{-1} \boxed{\tilde{\mathbf{A}}} \boxed{\mathbf{U}}$$

where $\boxed{\mathbf{A}}$ is the matrix representation of the operator $\mathbf{A}$ in the basis (3.21), $\boxed{\tilde{\mathbf{A}}}$ is that of the same operator in the basis (3.23), and $\boxed{\mathbf{U}}$ is the unitary matrix whose elements are given by (3.25).

3.1.4.4 Eigenvalue equations Now, write the eigenvalue equations of a linear operator $\mathbf{A}$:

$$\mathbf{A} | \Psi_k \rangle = A_k | \Psi_k \rangle \tag{3.27}$$

Now, seek the matrix representation of this equation in the basis $\{|\Phi_i\rangle\}$ of the eigenkets of a Hermitian operator $\mathbf{B}$, which does not commute with $\mathbf{A}$:

$$\mathbf{B} | \Phi_q \rangle = B_q | \Phi_q \rangle \quad \text{with} \quad \mathbf{B} = \mathbf{B}^\dagger \qquad \text{and} \qquad [\mathbf{B}, \mathbf{A}] \neq 0$$

$$\sum_q | \Phi_q \rangle \langle \Phi_q | = \mathbf{1} \qquad \text{and} \qquad \langle \Phi_q | \Phi_p \rangle = \delta_{qp} \tag{3.28}$$

Now, introduce this closure relation on both sides of the eigenvalue equation (3.27) according to

$$\mathbf{A} \left\{ \sum_q | \Phi_q \rangle \langle \Phi_q | \right\} | \Psi_k \rangle = A_k \left\{ \sum_q | \Phi_q \rangle \langle \Phi_q | \right\} | \Psi_k \rangle$$

so that

$$\sum_q \mathbf{A} | \Phi_q \rangle \langle \Phi_q | \Psi_k \rangle = \sum_q A_k | \Phi_q \rangle \langle \Phi_q | \Psi_k \rangle$$

Again, premultiply both sides of this last equation by a bra $\langle \Phi_p |$:

$$\langle \Phi_p | \sum_q \mathbf{A} | \Phi_q \rangle \langle \Phi_q | \Psi_k \rangle = \langle \Phi_p | \sum_q A_k | \Phi_q \rangle \langle \Phi_q | \Psi_k \rangle$$

which may be written

$$\sum_q \langle \Phi_p | \mathbf{A} | \Phi_q \rangle \langle \Phi_q | \Psi_k \rangle = \sum_q A_k \langle \Phi_p | \Phi_q \rangle \langle \Phi_q | \Psi_k \rangle$$

which, owing to the orthonormality properties (3.28) of the basis $\{|\Phi_i\rangle\}$, transforms to

$$\sum_q \langle\Phi_p|\mathbf{A}|\Phi_q\rangle\langle\Phi_q|\Psi_k\rangle = \sum_q A_k\delta_{pq}\langle\Phi_q|\Psi_k\rangle$$

or

$$\sum_q \langle\Phi_p|\mathbf{A}|\Phi_q\rangle\langle\Phi_q|\Psi_k\rangle = A_k\langle\Phi_p|\Psi_k\rangle$$

and thus

$$\boxed{\sum_q A_{pq}a_{qk} = A_k a_{pk}} \tag{3.29}$$

with

$$A_{pq} = \langle\Phi_p|\mathbf{A}|\Phi_q\rangle \qquad \text{and} \qquad a_{pk} = \langle\Phi_p|\Psi_k\rangle$$

Equation (3.29) leads to the following matrix representation in the basis (3.28) of the eigenvalue equation (3.27):

$$\begin{pmatrix} A_{11} & A_{12} & & A_{1q} & \cdots \\ A_{21} & A_{22} & & & \\ \cdots & \cdots & \cdots & & \\ A_{q1} & A_{q2} & & A_{qq} & \\ & & & & \cdots \end{pmatrix}\begin{pmatrix} a_{1k} \\ a_{2k} \\ \cdots \\ a_{qk} \\ \cdots \end{pmatrix} = A_k\begin{pmatrix} 1 & & & & \\ & 1 & & & \\ & & 1 & & \\ & & & 1 & \\ & & & & 1 \end{pmatrix}\begin{pmatrix} a_{1k} \\ a_{2k} \\ \cdots \\ a_{pk} \\ \cdots \end{pmatrix} \tag{3.30}$$

3.1.5 Block matrix representation and symmetry

When some symmetry in a system exists, the matrix representation of the Hamiltonian operator takes the form of a block matrix, the study of which is the aim of the present section. As shown in section 18.9, the symmetry operations all have an inverse, so that the operators $\mathbf{S}$ describing them must also have an inverse $\mathbf{S}^{-1}$ obeying

$$\mathbf{S}^{-1}\mathbf{S} = \mathbf{S}\mathbf{S}^{-1} = \mathbf{1} \tag{3.31}$$

Furthermore, since the Hamiltonian operator $\mathbf{H}$ of a system cannot be modified by symmetry operations in the same way as its corresponding classical scalar form, the action of any symmetry operator on it cannot modify it so that one may write

$$\mathbf{SH} = \mathbf{H} \qquad \text{and} \qquad \mathbf{S}^{-1}\mathbf{H} = \mathbf{H}$$

Hence, the following canonical transformation yields

$$\mathbf{S}^{-1}\mathbf{HS} = \mathbf{H} \tag{3.32}$$

demonstrating that the symmetry operators $\mathbf{S}$ commute with the Hamiltonians $\mathbf{H}$, that is,

$$[\mathbf{H}, \mathbf{S}] = 0 \tag{3.33}$$

Now, consider a basis $\{|\Phi_l\rangle\}$ yielding a matrix representation of the Hamiltonian. Then, one may form linear combinations $|\Psi_k^{\{g\}}\rangle$ or $|\Psi_j^{\{u\}}\rangle$ of the kets $|\Phi_l\rangle$ belonging to this basis, which are such that they will be symmetric or antisymmetric with respect to the symmetry operation corresponding to the **S** operator, that is, constructed from the following linear combinations:

$$|\Psi_k^{\{g\}}\rangle = \sum_l \{C_{lk}^{\{g\}}\}|\Phi_l\rangle$$

$$|\Psi_j^{\{u\}}\rangle = \sum_l \{C_{lj}^{\{u\}}\}|\Phi_l\rangle$$

obeying

$$\mathbf{S}^{-1}|\Psi_k^{\{g\}}\rangle = |\Psi_k^{\{g\}}\rangle \qquad \text{and} \qquad \langle\Psi_k^{\{g\}}|\mathbf{S} = \langle\Psi_k^{\{g\}}| \tag{3.34}$$

$$\mathbf{S}^{-1}|\Psi_j^{\{u\}}\rangle = -|\Psi_j^{\{u\}}\rangle \qquad \text{and} \qquad \langle\Psi_j^{\{u\}}|\mathbf{S} = -\langle\Psi_j^{\{u\}}| \tag{3.35}$$

Here the symbols $\{g\}$ (*gerade*) and $\{u\}$ (*ungerade*) have been used to distinguish between the symmetric and antisymmetric linear combinations.

Moreover, consider a matrix element of the Hamiltonian built up from a gerade ket and an ungerade bra. Then, insert the unity operator defined by Eq. (3.31) before and after **H** in such a way that

$$\langle\Psi_j^{\{u\}}|\mathbf{H}|\Psi_k^{\{g\}}\rangle = \langle\Psi_j^{\{u\}}|\mathbf{SS^{-1}HSS^{-1}}|\Psi_k^{\{g\}}\rangle$$

which, because of Eq. (3.32), simplifies to

$$\langle\Psi_j^{\{u\}}|\mathbf{H}|\Psi_k^{\{g\}}\rangle = \langle\Psi_j^{\{u\}}|\mathbf{SHS^{-1}}|\Psi_k^{\{g\}}\rangle$$

a result that, owing to Eqs. (3.34) and (3.35), reads

$$\langle\Psi_j^{\{u\}}|\mathbf{H}|\Psi_k^{\{g\}}\rangle = -\langle\Psi_j^{\{u\}}|\mathbf{H}|\Psi_k^{\{g\}}\rangle$$

so that

$$\boxed{\langle\Psi_j^{\{u\}}|\mathbf{H}|\Psi_k^{\{g\}}\rangle = \langle\Psi_k^{\{g\}}|\mathbf{H}|\Psi_j^{\{u\}}\rangle = 0} \tag{3.36}$$

where, in the last step, has used Eq. (1.30) and the Hermiticity of **H**. Equation (3.36) expresses the fact that the matrix element of a Hamiltonian between two kets of different symmetry is zero.

As an illustration, if, for instance, a subspace spanned by two gerade and two ungerade kets exists, then, according to Eq. (3.36), the matrix representation of the Hamiltonian takes on the following block form:

	$\lvert\Psi_1^{\{g\}}\rangle$	$\lvert\Psi_2^{\{g\}}\rangle$	$\lvert\Psi_1^{\{u\}}\rangle$	$\lvert\Psi_2^{\{u\}}\rangle$
$\langle\Psi_1^{\{g\}}\rvert$	$\{H_{11}^{\{g\}}\}$	$\{H_{12}^{\{g\}}\}$	0	0
$\langle\Psi_2^{\{g\}}\rvert$	$\{H_{21}^{\{g\}}\}$	$\{H_{22}^{\{g\}}\}$	0	0
$\langle\Psi_1^{\{u\}}\rvert$	0	0	$\{H_{11}^{\{u\}}\}$	$\{H_{12}^{\{u\}}\}$
$\langle\Psi_2^{\{u\}}\rvert$	0	0	$\{H_{21}^{\{u\}}\}$	$\{H_{22}^{\{u\}}\}$

(3.37)

The interest of the symmetry is to allow size reducing of Hamiltonian matrix representations to be diagonalized.

Now, consider any ket $|\{\xi\}\rangle$ describing the state of a system of the given symmetry characterized by the symmetry operation **S**, which may be expressed by a linear combination of g and u state, according to

$$|\{\xi_l\}\rangle = \sum_k \{b_{kl}^{\{g\}}\}|\Psi_k^{\{g\}}\rangle + \sum_j \{b_{jl}^{\{u\}}\}|\Psi_j^{\{u\}}\rangle \tag{3.38}$$

Then, due to the first expressions of (3.34) and (3.35), it reads, respectively

$$\tfrac{1}{2}\{1+\mathbf{S}\}|\Psi_k^{\{g\}}\rangle = \tfrac{1}{2}\{|\Psi_k^{\{g\}}\rangle + |\Psi_k^{\{g\}}\rangle\} = |\Psi_k^{\{g\}}\rangle$$

$$\tfrac{1}{2}\{1-\mathbf{S}\}|\Psi_k^{\{u\}}\rangle = \tfrac{1}{2}\{|\Psi_k^{\{u\}}\rangle + |\Psi_k^{\{u\}}\rangle\} = |\Psi_k^{\{u\}}\rangle$$

$$\tfrac{1}{2}\{1-\mathbf{S}\}|\Psi_k^{\{g\}}\rangle = \tfrac{1}{2}\{|\Psi_k^{\{g\}}\rangle - |\Psi_k^{\{g\}}\rangle\} = 0$$

$$\tfrac{1}{2}\{1+\mathbf{S}\}|\Psi_k^{\{u\}}\rangle = \tfrac{1}{2}\{|\Psi_k^{\{u\}}\rangle - |\Psi_k^{\{u\}}\rangle\} = 0$$

As a consequence of these results and of Eq. (3.38), one obtains, respectively

$$\boxed{\frac{1}{2}\{1+\mathbf{S}\}|\{\xi_l\}\rangle = \sum_k \{b_{kl}^{\{g\}}\}|\Psi_k^{\{g\}}\rangle} \tag{3.39}$$

$$\boxed{\frac{1}{2}\{1-\mathbf{S}\}|\{\xi_l\}\rangle = \sum_j \{b_{jl}^{\{u\}}\}|\Psi_j^{\{u\}}\rangle} \tag{3.40}$$

3.2 WAVE MECHANICS

Following the above exposition of the matrix representation of quantum mechanics, we now pass to wave mechanics, that is, to the representation of quantum mechanics in the basis of the eigenkets of the **Q** operator, which is sometimes called the *Q representation* of quantum mechanics. The precise foundation of wave mechanics by Louis de Broglie in 1924 was completely independent from that of quantum matrix mechanics by Heisenberg, the deep link between the two approaches being later discovered.

3.2.1 Quantum mechanics in representation $\{|\{Q\}\rangle\}$

In order to introduce wave mechanics, we start from the eigenvalue equation of the coordinate operator **Q** and its Hermitian conjugate:

$$\mathbf{Q}|\{Q\}\rangle = Q|\{Q\}\rangle \qquad \text{and} \qquad \langle\{Q\}|\mathbf{Q} = Q\langle\{Q\}| \tag{3.41}$$

together with the closure relation over the eigenstates of **Q** and the corresponding orthonormality relations

$$\int_{-\infty}^{+\infty} |\{Q\}\rangle\langle\{Q\}|dQ = \mathbf{1} \qquad \text{and} \qquad \langle\{Q\}|\{Q'\}\rangle = \delta(Q-Q') \tag{3.42}$$

Now, consider the following scalar product of a ket $|\Psi\rangle$ by any eigenket $|\{Q\}\rangle$ of $\mathbf{Q}$ and its complex conjugate, that is,

$$\langle\{Q\}|\Psi\rangle = \Psi(Q) \qquad \text{and} \qquad \langle\Psi|\{Q\}\rangle = \Psi^*(Q) \tag{3.43}$$

Here, the scalar $\Psi(Q)$, which is by definition the *representation* $\{|\{Q\}\rangle\}$ of the ket $|\Psi\rangle$, is named the *wavefunction* associated with this ket at the measured position Q. It is generally complex. The squared modulus of this scalar product is

$$|\langle\{Q\}|\Psi\rangle|^2 = |\Psi(Q)|^2 \tag{3.44}$$

Owing to the fourth postulate, the left-hand side of Eq. (3.44) corresponds to the probability for the system to jump from the ket $|\Psi\rangle$ into the ket $|\{Q\}\rangle$, which is an eigenket of the position operator $\mathbf{Q}$ with the corresponding eigenvalue Q. Thus, on the right-hand side of Eq. (3.44), $|\Psi(Q)|^2$ *is the probability for the system described by the scalar function* $\Psi(Q)$ *to be found at the position* Q.

Now, consider the scalar products of two different kets $|\Psi\rangle$ and $|\Phi\rangle$, and introduce inside the scalar product the closure relation (3.42)

$$\langle\Psi|\Phi\rangle = \langle\Psi| \left\{ \int_{-\infty}^{+\infty} |\{Q\}\rangle\langle\{Q\}| dQ \right\} |\Phi\rangle$$

Since the integration operation commutes with the kets or the bras, the scalar product simply reads

$$\langle\Psi|\Phi\rangle = \int_{-\infty}^{+\infty} \langle\Psi|\{Q\}\rangle\langle\{Q\}|\Phi\rangle dQ$$

Thus, in view of Eq. (3.43), it takes the form

$$\langle\Psi|\Phi\rangle = \int_{-\infty}^{+\infty} \Psi^*(Q)\Phi(Q)\, dQ$$

Next, if the two kets involved in the scalar product belong to a given orthonormal basis, we have

$$\langle\Psi_k|\Psi_l\rangle = \int_{-\infty}^{+\infty} \Psi_k^*(Q)\Psi_l(Q)\, dQ = \delta_{kl} \tag{3.45}$$

When applied to the norm of any ket, Eq. (3.45) reduces to the *normalization* condition

$$\int_{-\infty}^{+\infty} \Psi_k^*(Q)\Psi_k(Q)\, dQ = 1$$

3.2.2 Many-particle systems

The fourth postulate allows one to find the ket of a system formed by many particles, each of them being characterized by their own ket. We illustrate as follows:

Consider the value of the total wavefunction $\Psi_{\text{Tot}}(Q)$ of two particles at any value Q of the position, the individual wavefunction of each particle being, respectively, $\Phi^{\{1\}}(Q)$ and $\Phi^{\{2\}}(Q)$. The probability to find the two particles at position Q may be obtained by

$$\mathbb{P}_{\text{Tot}}(Q) = |\Psi_{\text{Tot}}(Q)|^2$$

Again, since the probabilities multiply, one has

$$|\Psi_{\text{Tot}}(Q)|^2 = |\Phi^{\{1\}}(Q)|^2 \, |\Phi^{\{2\}}(Q)|^2$$

As a consequence of each wavefunction working within its own state space, the total wavefunction may be written

$$\Psi_{\text{Tot}}(Q)\Psi_{\text{Tot}}(Q)^* = (\Phi^{\{1\}}(Q)\Phi^{\{1\}}(Q)^*)(\Phi^{\{2\}}(Q)\Phi^{\{2\}}(Q)^*)$$

so that

$$\Psi_{\text{Tot}}(Q) = \Phi^{\{1\}}(Q)\Phi^{\{2\}}(Q)$$

Hence, since the probabilities multiply, the meaning of the wavefunction implies that the total wavefunction of a system composed of two particles may be written as the product of the wavefunctions of each particle. By generalization to N particles, we have

$$\boxed{\Psi_{\text{Tot}}(Q) = \prod_{k=1}^{N} \Phi^{\{k\}}(Q)} \tag{3.46}$$

Furthermore, the wavefunctions $\Psi_{\text{Tot}}(Q)$ and $\Phi^{\{k\}}(Q)$ are given, respectively, by the following scalar products:

$$\Psi_{\text{Tot}}(Q) = \langle\{Q\}|\Psi_{\text{Tot}}\rangle$$

$$\Phi^{\{k\}}(Q) = \langle\{Q\}|\Phi^{\{k\}}\rangle$$

Then, Eq. (3.46) leads to

$$\langle\{Q\}|\ \Psi_{\text{Tot}}\rangle = \prod_{k=1}^{N} \langle\{Q\}|\ \Phi^{\{k\}}\rangle$$

This equation may be also written

$$\langle\{Q\}|\ \Psi_{\text{Tot}}\rangle = \langle\{Q\}| \prod_{k=1}^{N} |\Phi^{\{k\}}\rangle$$

Of course, this expression holds what may be the bra involved in the scalar products. Thus, it is possible to write

$$\boxed{|\Psi_{\text{Tot}}\rangle = \prod_{k=1}^{N} |\Phi^{\{k\}}\rangle} \tag{3.47}$$

That shows that the total ket of a system formed by several particles is the product of the kets of the different particles.

3.2.3 Momentum operator in representation $\{|\{Q\}\rangle\}$

In order to get the action of the momentum operator **P** on any ket $|\Psi\rangle$ within the $\{Q\}$ representation, introduce between **P** and $|\Psi\rangle$ the closure relation on the eigenkets of **P**:

$$\langle\{Q\}|\mathbf{P}|\Psi\rangle = \langle\{Q\}|\mathbf{P}\left\{\int_{-\infty}^{+\infty} |\{P\}\rangle\langle\{P\}|dP\right\}|\Psi\rangle$$

which due to the eigenvalue equation of **P** becomes

$$\langle\{Q\}|\mathbf{P}|\Psi\rangle = \langle\{Q\}|\int_{-\infty}^{+\infty} P|\{P\}\rangle\langle\{P\}|\{\Psi\}\rangle dP$$

or, using

$$\langle\{P\}|\{\Psi\}\rangle = \Psi(P)$$

$$\langle\{Q\}|\mathbf{P}|\Psi\rangle = \int_{-\infty}^{+\infty} P\langle\{Q\}|\{P\}\rangle\Psi(P)dP \tag{3.48}$$

Moreover, observe that we have shown that the scalar product of an eigenket of **Q** by another one of **P** is given by Eq. (2.132), that is,

$$\langle\{P\}|\{Q\}\rangle = \frac{1}{\sqrt{2\pi\hbar}}\exp\left\{-\frac{iPQ}{\hbar}\right\}$$

so that Eq. (3.48) transforms to

$$\langle\{Q\}|\mathbf{P}|\Psi\rangle = \frac{1}{\sqrt{2\pi\hbar}}\int_{-\infty}^{+\infty} P\exp\left\{-\frac{iPQ}{\hbar}\right\}\Psi(P)dP \tag{3.49}$$

Now, using Eq. (18.49), that is,

$$\frac{1}{\sqrt{2\pi\hbar}}\int_{-\infty}^{+\infty} Pf(P)e^{iQP/\hbar}dQ = i\hbar\left(\frac{\partial f(Q)}{\partial Q}\right) \quad \text{with} \quad \Psi(Q) = f(Q)$$

Eq. (3.49) takes the form

$$\boxed{\langle\{Q\}|\mathbf{P}|\Psi\rangle = i\hbar\left(\frac{\partial\Psi(Q)}{\partial Q}\right)} \tag{3.50}$$

This last result shows that in the quantum representation $\{|\{Q\}\rangle\}$, the momentum operator is acting on a wavefunction as a partial derivative with respect to the scalar Q times $\hbar/i$, which may be written formally as

$$\boxed{P = \frac{\hbar}{i}\frac{\partial}{\partial Q} = -i\hbar\frac{\partial}{\partial Q}} \tag{3.51}$$

Of course, in the quantum representation $\{|\{Q\}\rangle\}$, the action of the operator $\mathbf{Q}$ over some ket $|\Psi\rangle$ reads

$$\langle\{Q\}|\mathbf{Q}|\Psi\rangle = Q\langle\{Q\}|\Psi\rangle = Q\Psi(Q) \tag{3.52}$$

Observe that the following commutator reads

$$[Q, P] = \frac{\hbar}{i}\left(Q\frac{\partial}{\partial Q} - \frac{\partial}{\partial Q}Q\right)$$

thereby taking into account the fact that the right-hand side of this last equation is acting on any function. Thus one has to write

$$\frac{\partial}{\partial Q}Q = Q\frac{\partial}{\partial Q} + 1$$

and, after simplification, the above commutator becomes

$$\boxed{[Q, P] = i\hbar}$$

That is the equivalent in the quantum representation $\{|\{Q\}\rangle\}$ of the fundamental commutator given by the last postulate of quantum mechanics:

$$[\mathbf{Q}, \mathbf{P}] = i\hbar$$

3.2.4 Time-independent Schrödinger equation

Consider some operator function $\mathbf{F}(\mathbf{Q}, \mathbf{P})$ of $\mathbf{P}$ and $\mathbf{Q}$ that may be separately expanded in powers of $\mathbf{P}$ and $\mathbf{Q}$ according to

$$\mathbf{F}(\mathbf{Q}, \mathbf{P}) = \sum_n \{C_n\mathbf{P}^n + B_n\mathbf{Q}^n\} \tag{3.53}$$

where C_n and B_n are, respectively, the expansion coefficients that are scalars.

Now, consider matrix elements of this operator:

$$\langle\{Q\}|\mathbf{F}(\mathbf{Q}, \mathbf{P})|\Psi\rangle = \sum_n \langle\{Q\}|\{C_n\mathbf{P}^n + B_n\mathbf{Q}^n\}|\Psi\rangle \tag{3.54}$$

Again, owing to Eqs. (3.52), and (3.50), it appears that

$$\langle\{Q\}|\mathbf{Q}^n|\Psi\rangle = Q^n\Psi(Q) \tag{3.55}$$

$$\langle\{Q\}|\mathbf{P}^n|\Psi\rangle = \left(\frac{\hbar}{i}\frac{\partial}{\partial Q}\right)^n \Psi(Q) \tag{3.56}$$

Hence, with Eqs. (3.55) and (3.56), Eq. (3.54) transforms to

$$\langle\{Q\}|\mathbf{F}(\mathbf{Q}, \mathbf{P})|\Psi\rangle = \sum_n \left\{C_n\left(\frac{\hbar}{i}\frac{\partial}{\partial Q}\right)^n + B_nQ^n\right\}\Psi(Q) \tag{3.57}$$

When the operator $\mathbf{F}(\mathbf{Q}, \mathbf{P})$ is the Hamiltonian $\mathbf{H}(\mathbf{Q}, \mathbf{P})$

$$\mathbf{H}(\mathbf{Q}, \mathbf{P}) = \mathbf{T}(\mathbf{P}) + \mathbf{V}(\mathbf{Q})$$

with

$$\mathbf{T(P)} = \frac{\mathbf{P}^2}{2m} \qquad \text{and} \qquad \mathbf{V(Q)} = \sum_n B_n \mathbf{Q}^n$$

Eq. (3.57) takes the form

$$\langle \{Q\} | \mathbf{H(Q, P)} | \Psi \rangle = \hat{H} \Psi(Q) \tag{3.58}$$

with

$$\hat{H} = \hat{T} + \hat{V}(Q) \quad \text{with} \quad \hat{T} = -\frac{\hbar^2}{2m} \frac{\partial^2}{\partial Q^2} \tag{3.59}$$

and

$$\hat{V}(Q) = B_n Q^n$$

so that

$$\hat{H} = -\frac{\hbar^2}{2m} \frac{\partial^2}{\partial Q^2} + \hat{V}(Q) \tag{3.60}$$

This equation is the wave mechanics representation of the Hamiltonian, the eigenvalue equation of which is

$$\hat{H} \Phi_k(Q) = E_k \Phi_k(Q)$$

or

$$-\frac{\hbar^2}{2m} \frac{\partial^2}{\partial Q^2} \Phi_k(Q) + \hat{V}(Q) \Phi_k(Q) = E_k \Phi_k(Q) \tag{3.61}$$

This is the time-independent Schrödinger equation, that is, the wave mechanics representation of the Hamiltonian eigenvalue equation

$$\mathbf{H(Q, P)} | \Phi_k \rangle = E_k | \Phi_k \rangle \tag{3.62}$$

3.2.5 Wavefunction boundary conditions

The eigenvalues E_k are the same in both Eqs. (3.61) and (3.62), whereas the connection between the eigenfunction $\Phi_k(Q)$ of H and the eigenket $|\Phi_k\rangle$ of $\mathbf{H(Q, P)}$ is through the following scalar product:

$$\Phi_k(Q) = \langle \{Q\} | \Phi_k \rangle \tag{3.63}$$

Recall that, the fourth postulate allows one to write

$$|\langle \{Q\} | \Phi_k \rangle|^2 = |\Phi_k(Q)|^2 \equiv \mathbb{P}(Q) \tag{3.64}$$

Observe that $\mathbb{P}(Q)$ may be regarded as the probability density, which is also denoted $\rho(Q)$. Again, since the probabilities $\mathbb{P}(Q)$ must obey

$$\int_{-\infty}^{+\infty} \mathbb{P}(Q) dQ = 1$$

thus Eq. (3.64) implies that the wavefunction $\Phi_k(Q)$ has the normalization property

$$\boxed{\int_{-\infty}^{+\infty} |\Phi_k(Q)|^2 dQ = 1} \tag{3.65}$$

Furthermore, if the wavefunction is not normalized, it may be written as $\tilde{\Phi}_k(Q)$. In order to be square summable, its integral must be finite according to

$$\int_{-\infty}^{+\infty} |\tilde{\Phi}_k(Q)|^2 dQ = k^2 \quad \text{with } k^2 \text{ finite}$$

Then, in order to satisfy Eq. (3.65), one has

$$\Phi_k(Q) = \frac{1}{k}\tilde{\Phi}_k(Q)$$

where $1/k$ is the normalization constant. The normalization condition implies that at infinity the wavefunction must vanish, that is,

$$\boxed{\Phi_k(Q \to \pm\infty) \to 0} \tag{3.66}$$

This is an essential boundary condition for the time-independent Schrödinger equation (3.61). Such a condition leads to quantized eigenvalues and thus to quantized energy levels, not only for the eigenvalue equation (3.61) but also for that (3.62), which is equivalent.

Since the eigenvalue equation (3.61) has the structure of a wave equation, the $\{Q\}$ representation (3.61) of the eigenvalue equation (3.62) of the Hamiltonian may be viewed as a *wave mechanics equation.*

3.2.6 Time-dependent Schrödinger equation

From the Schrödinger equation it is possible, with help from the sixth postulate, to find the linear time-dependent operator that transforms some ket at initial time $|\Psi(0)\rangle$ into the corresponding one $|\Psi(t)\rangle$ at time t.

To get this operator, we start from the Schrödinger equation

$$i\hbar\left(\frac{\partial|\Psi(t)\rangle}{\partial t}\right) = \mathbf{H}|\Psi(t)\rangle$$

In order to solve this equation, premultiply it by some eigenbra of $\mathbf{Q}$ leading to

$$i\hbar\left(\frac{\partial\langle\{Q\}|\Psi(t)\rangle}{\partial t}\right) = \langle\{Q\}|\mathbf{H}|\Psi(t)\rangle$$

or, due to Eq. (3.58),

$$\boxed{i\hbar\left(\frac{\partial\Psi(Q,t)}{\partial t}\right) = \hat{H}\Psi(Q,t)} \tag{3.67}$$

By omitting the Q dependence, after integration between $t = 0$ and t one obtains

$$\ln\left(\frac{\Psi(t)}{\Psi(0)}\right) = -\frac{i}{\hbar}\hat{H}t$$

Passing to the exponential, that reads

$$\frac{\Psi(t)}{\Psi(0)} = e^{-i\hat{H}t/\hbar}$$

or

$$\boxed{\Psi(t) = e^{-i\hat{H}t/\hbar}\Psi(0)} \tag{3.68}$$

3.2.7 Current density and continuity equation

Now, let us define the *current density* operator according to the Hermitian product of the two Hermitian operators $\mathbf{P}|\{Q\}\rangle\langle\{Q\}|$ and $|\{Q\}\rangle\langle\{Q\}|\mathbf{P}$:

$$\mathbf{J} \equiv \frac{1}{2m}\{\mathbf{P}|\{Q\}\rangle\langle\{Q\}| + |\{Q\}\rangle\langle\{Q\}|\mathbf{P}\}$$

Now, consider the diagonal matrix elements of this operator built up from some eigenkets of any Hermitian operator, that is,

$$J = \langle\Psi|\mathbf{J}|\Psi\rangle$$

This representation of the operator is therefore

$$J = \left\{\frac{1}{2m}\langle\Psi|\mathbf{P}|\{Q\}\rangle\langle\{Q\}|\Psi\rangle\right\} + \text{hc}$$

where hc denotes the Hermitian conjugate. Then, using Eq. (3.50) and its Hermitian conjugate, one obtains

$$\boxed{J = -\left(i\frac{\hbar}{2m}\right)\left(\Psi(Q)^*\frac{\partial}{\partial Q}\Psi(Q) - \Psi(Q)\frac{\partial}{\partial Q}\Psi(Q)^*\right)} \tag{3.69}$$

Now, in order to find the continuity equation governing the wavefunction, differentiate the current density (3.69) with respect to Q

$$\left(\frac{\partial J}{\partial Q}\right) = -\left(i\frac{\hbar}{2m}\right)\frac{\partial}{\partial Q}\left(\Psi^*\frac{\partial}{\partial Q}\Psi - \Psi\frac{\partial}{\partial Q}\Psi^*\right) \tag{3.70}$$

One obtains, respectively,

$$\frac{\partial}{\partial Q}\left(\Psi^*\frac{\partial}{\partial Q}\Psi\right) = \left(\frac{\partial}{\partial Q}\Psi^*\frac{\partial}{\partial Q}\Psi + \Psi^*\frac{\partial^2}{\partial Q^2}\Psi\right)$$

$$\frac{\partial}{\partial Q}\left(\Psi\frac{\partial}{\partial Q}\Psi^*\right) = \left(\frac{\partial}{\partial Q}\Psi\frac{\partial}{\partial Q}\Psi^* + \Psi\frac{\partial^2}{\partial Q^2}\Psi^*\right)$$

Thus, Eq. (3.70) transforms to

$$\left(\frac{\partial J}{\partial Q}\right) = -\left(i\frac{\hbar}{2m}\right)\left(\Psi^*\frac{\partial^2}{\partial Q^2}\Psi - \Psi\frac{\partial^2}{\partial Q^2}\Psi^*\right) \tag{3.71}$$

On the other hand, consider the probability density related to the wavefunction

$$\boxed{\rho = \Psi^*\Psi} \tag{3.72}$$

By differentiation with respect to time, we get

$$\left(\frac{\partial \rho}{\partial t}\right) = \left(\frac{\partial \Psi^*}{\partial t}\right)\Psi + \Psi^*\left(\frac{\partial \Psi}{\partial t}\right) \tag{3.73}$$

Besides, the two derivatives of the wavefunction are governed by the time-dependent Schrödinger equation (3.67) and its Hermitian conjugate, that is,

$$i\hbar\left(\frac{\partial \Psi}{\partial t}\right) = H\Psi \qquad \text{and} \qquad -i\hbar\left(\frac{\partial \Psi^*}{\partial t}\right) = \hat{H}\Psi^* \tag{3.74}$$

Moreover, according to Eq. (3.60), the Hamiltonian is

$$\hat{H} = -\frac{\hbar^2}{2m}\left(\frac{\partial^2}{\partial Q^2}\right) + \hat{V}$$

Hence, Eqs. (3.74) lead to

$$i\hbar\left(\frac{\partial \Psi}{\partial t}\right) = -\frac{\hbar^2}{2m}\left(\frac{\partial^2}{\partial Q^2}\right)\Psi + \hat{V}\Psi \quad \text{and} \quad -i\hbar\left(\frac{\partial \Psi^*}{\partial t}\right) = -\frac{\hbar^2}{2m}\left(\frac{\partial^2}{\partial Q^2}\right)\Psi^* + \hat{V}\Psi^*$$

These two last equations allow one to transform Eq. (3.73) into

$$\left(\frac{\partial \rho}{\partial t}\right) = -i\left(\frac{\hbar}{2m}\right)\left\{\Psi\frac{\partial^2}{\partial Q^2}\Psi^* - \Psi^*\frac{\partial^2}{\partial Q^2}\Psi\right\}$$

Finally, it appears from comparison with Eq. (3.71) that the following one-dimensional equation is verified:

$$\left(\frac{\partial \rho}{\partial t}\right) = -\left(\frac{\partial J}{\partial Q}\right)$$

By generalization to the three-dimensional equation, one obtains the *continuity equation*

$$\boxed{\left(\frac{\partial \rho}{\partial t}\right) + \mathbf{Div}\ \vec{J} = 0} \tag{3.75}$$

where the arrow indicates a vectorial entity.

3.3 EVOLUTION OPERATORS

As we have seen, when considering the sixth postulate of quantum mechanics dealing with the dynamics involved in quantum mechanics, there are several time-dependent descriptions of quantum mechanics. In the *Schrödinger picture (SP)*, the kets depend on time, whereas the operators do not change with it. However, another time-dependent description, the *Heisenberg picture (HP)* exists, where the operators depend on time whereas the kets remain constant. Finally, many other time-dependent representations of quantum mechanics exist, which are intermediate between the Schrödinger and the Heisenberg pictures, in which both the kets and the operators depend on time in subtle ways. They are named the *interaction pictures*. We shall first consider the *time evolution operator* within the Schrödinger picture.

3.3.1 Schrödinger picture

Starting from the Schrödinger equation defined by the sixth postulate and governing the dynamics of some time-dependent ket $|\Psi(t)\rangle$

$$i\hbar\frac{\partial}{\partial t}|\Psi(t)\rangle = \mathbf{H}|\Psi(t)\rangle \tag{3.76}$$

where $\mathbf{H}$ is the total Hamiltonian of the system. Then, we introduce a linear operator $\mathbf{U}(t)$, the time evolution operator, allowing one to transform the ket $|\Psi(0)\rangle$ at an initial time $t = 0$ into one at time t according to

$$\boxed{|\Psi(t)\rangle = \mathbf{U}(t)|\Psi(0)\rangle} \tag{3.77}$$

with the obvious condition

$$\mathbf{U}(0) = 1$$

Then, time differentiation of Eq. (3.77) yields

$$\frac{\partial}{\partial t}|\Psi(t)\rangle = \left(\frac{\partial \mathbf{U}(t)}{\partial t}\right)|\Psi(0)\rangle \tag{3.78}$$

Again, introduce on the right-hand side of Eq. (3.78)

$$\mathbf{U}(t)^{-1}\mathbf{U}(t) = 1 \tag{3.79}$$

in such a way as to write

$$\frac{\partial}{\partial t}|\Psi(t)\rangle = \left(\frac{\partial \mathbf{U}(t)}{\partial t}\right)\mathbf{U}(t)^{-1}\mathbf{U}(t)|\Psi(0)\rangle$$

leading with the help of Eq. (3.77) and after multiplying by $i\hbar$, to

$$i\hbar\frac{\partial}{\partial t}|\Psi(t)\rangle = i\hbar\left(\frac{\partial \mathbf{U}(t)}{\partial t}\right)\mathbf{U}(t)^{-1}|\Psi(t)\rangle \tag{3.80}$$

Thus, identification of Eqs. (3.76) and (3.80) yields

$$\mathbf{H}|\Psi(t)\rangle = i\hbar\left(\frac{\partial \mathbf{U}(t)}{\partial t}\right)\mathbf{U}(t)^{-1}|\Psi(t)\rangle$$

Hence, since this latter result holds irrespective of $|\Psi(t)\rangle$, it appears that the following relation between the operators $\mathbf{U}(t)$ and $\mathbf{H}$ exists:

$$\boxed{i\hbar\left(\frac{\partial \mathbf{U}(t)}{\partial t}\right) = \mathbf{H}\mathbf{U}(t)} \tag{3.81}$$

The foregoing partial differential equation reads

$$\frac{d\mathbf{U}(t)}{\mathbf{U}(t)} = -\frac{i}{\hbar}\mathbf{H}\,dt$$

which, by integration yields

$$\ln\frac{\mathbf{U}(t)}{\mathbf{U}(0)} = -\frac{i}{\hbar}\mathbf{H}t$$

or due to the boundary condition (3.79)

$$\boxed{\mathbf{U}(t) = e^{-i\mathbf{H}t/\hbar}} \tag{3.82}$$

Observe that the Hamiltonian **H** being Hermitian, the evolution operator $\mathbf{U}(t)$ is unitary since its inverse is equal to its Hermitian conjugate, that is,

$$\mathbf{U}(t)^{-1} = \mathbf{U}(t)^{\dagger} = e^{i\mathbf{H}t/\hbar} \tag{3.83}$$

so that

$$\mathbf{U}(t)^{\dagger}\mathbf{U}(t) = \mathbf{U}(t)^{-1}\mathbf{U}(t) = 1 \tag{3.84}$$

Moreover, due to Eq. (3.82), Eq. (3.77) becomes

$$|\Psi(t)\rangle = (e^{-i\mathbf{H}t/\hbar})|\Psi(0)\rangle \tag{3.85}$$

We remark that Eq. (3.68) is the wave mechanics representation of the quantum relation (3.85).

Sometimes the Hamiltonian may depend on time, so that one has to solve a dynamic equation that is more complicated than (3.81) and of the form

$$i\hbar\left(\frac{\partial \mathbf{U}(t)}{\partial t}\right) = \mathbf{H}(t)\mathbf{U}(t) \tag{3.86}$$

Here, the Hamiltonians at different times do not commute:

$$[\mathbf{H}(t), \mathbf{H}(t')] \neq 0$$

Moreover, it is possible to write formally a solution of Eq. (3.86) in the same way as (3.81) according to

$$\boxed{\mathbf{U}(t) = \hat{\mathbf{P}}\exp\left\{-\frac{i}{\hbar}\int_0^t \mathbf{H}(t')\,dt'\right\}} \tag{3.87}$$

where $\hat{\mathbf{P}}$ is the *Dyson time-ordering operator.*

3.3.2 Heisenberg picture

Now, it is suitable to introduce a new time-independent picture in which (in contrast to the Schrödinger picture where the kets are time dependent and the operators constant) the kets are constant and the operators time dependent. This is the *Heisenberg picture*. For this purpose, we start from the Schrödinger picture equation (2.65) yielding the mean value of some operator **A** averaged over the time-dependent states, that is,

$$\langle \mathbf{A}(t)\rangle_{\Psi_k} = \langle \Psi_k(t)^{\mathrm{SP}}|\mathbf{A}^{\mathrm{SP}}|\Psi_k(t)^{\mathrm{SP}}\rangle$$

where the superior index SP indicates that the Schrödinger picture has been used. Next, due to Eq. (3.85) and to its Hermitian conjugate, this average value reads

$$\langle \mathbf{A}(t)\rangle_{\Psi_k} = \langle \Psi_k(0)|(e^{i\mathbf{H}t/\hbar})\mathbf{A}^{\mathrm{SP}}(e^{-i\mathbf{H}t/\hbar})|\Psi_k(0)\rangle$$

which may be written

$$\langle \mathbf{A}(t)\rangle_{\Psi_k} = \langle \Psi_k^{\mathrm{HP}}|\mathbf{A}(t)^{\mathrm{HP}}|\Psi_k^{\mathrm{HP}}\rangle$$

where $|\Psi_k^{\mathrm{HP}}\rangle$ is the time-independent ket in the Heisenberg picture, whereas $\mathbf{A}(t)^{\mathrm{HP}}$ is the time-dependent operator in this same Heisenberg picture given by

$$\boxed{\mathbf{A}(t)^{\mathrm{HP}} = (e^{i\mathbf{H}t/\hbar})\mathbf{A}^{\mathrm{SP}}(e^{-i\mathbf{H}t/\hbar})} \tag{3.88}$$

or, due to Eqs. (3.82) and (3.83), by

$$\boxed{\mathbf{A}(t)^{\mathrm{HP}} = \mathbf{U}(t)^{\dagger}\mathbf{A}^{\mathrm{SP}}\mathbf{U}(t)} \tag{3.89}$$

Since the operator **A** in the Schrödinger picture is time independent, the time derivation of each members of Eq. (3.89) gives after writing $\mathbf{A}^{\mathrm{SP}} = \mathbf{A}$

$$\left(\frac{\partial \mathbf{A}(t)^{\mathrm{HP}}}{\partial t}\right) = \left(\frac{\partial \mathbf{U}(t)^{\dagger}}{\partial t}\right)\mathbf{A}\mathbf{U}(t) + \mathbf{U}(t)^{\dagger}\mathbf{A}\left(\frac{\partial \mathbf{U}(t)}{\partial t}\right) \tag{3.90}$$

Besides, since the Hamiltonian **H** is Hermitian, note that the derivative with respect to time of the evolution operator (3.82) and that of its Hermitian conjugate (3.83) are, respectively

$$\left(\frac{\partial \mathbf{U}(t)}{\partial t}\right) = -\left(\frac{i}{\hbar}\right)\mathbf{H}\mathbf{U}(t)$$

$$\left(\frac{\partial \mathbf{U}(t)^{\dagger}}{\partial t}\right) = \left(\frac{i}{\hbar}\right)\mathbf{U}(t)^{\dagger}\mathbf{H}$$

Using these two equations, Eq. (3.90) becomes

$$\left(\frac{\partial \mathbf{A}(t)^{\mathrm{HP}}}{\partial t}\right) = \left(\frac{i}{\hbar}\right)\mathbf{U}(t)^{\dagger}\mathbf{H}\mathbf{A}\mathbf{U}(t) - \left(\frac{i}{\hbar}\right)\mathbf{U}(t)^{\dagger}\mathbf{A}\mathbf{H}\mathbf{U}(t) \tag{3.91}$$

Next, using Eq. (3.84), that is,

$$\mathbf{1} = \mathbf{U}(t)\mathbf{U}(t)^{\dagger} \tag{3.92}$$

and, inserting on the right-hand-side term of Eq. (3.91) this unity operator, first between the Hamiltonian **H** and the operator **A**, and then between the operator **A** and the Hamiltonian **H**, one obtains

$$\left(\frac{\partial \mathbf{A}(t)^{\mathrm{HP}}}{\partial t}\right) = \left(\frac{i}{\hbar}\right)\mathbf{U}(t)^{\dagger}\mathbf{H}\{\mathbf{U}(t)\mathbf{U}(t)^{\dagger}\}\mathbf{A}\mathbf{U}(t) - \left(\frac{i}{\hbar}\right)\mathbf{U}(t)^{\dagger}\mathbf{A}\{\mathbf{U}(t)\mathbf{U}(t)^{\dagger}\}\mathbf{H}\mathbf{U}(t)$$

Thus, by changing the position of the brackets, we have

$$\left(\frac{\partial \mathbf{A}(t)^{\mathrm{HP}}}{\partial t}\right) = \left(\frac{i}{\hbar}\right)(\{\mathbf{U}(t)^{\dagger}\mathbf{H}\mathbf{U}(t)\}\{\mathbf{U}(t)^{\dagger}\mathbf{A}\mathbf{U}(t)\} - \{\mathbf{U}(t)^{\dagger}\mathbf{A}\mathbf{U}(t)\}\{\mathbf{U}(t)^{\dagger}\mathbf{H}\mathbf{U}(t)\}) \tag{3.93}$$

Now, observe that, according to Eqs. (3.82) and (3.83),

$$\{\mathbf{U}(t)^{\dagger}\mathbf{H}\mathbf{U}(t)\} = (e^{i\mathbf{H}t/\hbar})\mathbf{H}(e^{-i\mathbf{H}t/\hbar})$$

Again, since the exponential depends on the Hamiltonian, it must commute with it, so that after simplification this unitary transformation reduces to

$$\{\mathbf{U}(t)^{\dagger}\mathbf{H}\mathbf{U}(t)\} = \mathbf{H}$$

Hence, Eq. (3.93) simplifies to

$$i\hbar\left(\frac{\partial \mathbf{A}(t)^{\mathrm{HP}}}{\partial t}\right) = -(\mathbf{H}\{\mathbf{U}(t)^{\dagger}\mathbf{A}\mathbf{U}(t)\} - \{\mathbf{U}(t)^{\dagger}\mathbf{A}\mathbf{U}(t)\}\mathbf{H})$$

Thus, due to the definition (3.89), this equation transforms to the final result, which is the Heisenberg equation governing the dynamics of any operator in the Heisenberg picture:

$$\boxed{i\hbar\left(\frac{\partial \mathbf{A}(t)^{\mathrm{HP}}}{\partial t}\right) = [\mathbf{A}^{\mathrm{HP}}(t), \mathbf{H}]} \tag{3.94}$$

This equation contains the same information as the Schrödinger time-dependent equation in the Schrödinger picture. It may also be of interest to take the average of this equation in a state $|\Psi\rangle$, in postmultiplying both terms of this equation by $|\Psi\rangle$, and premultiplying them by $\langle\Psi|$

$$i\hbar\langle\Psi|\left(\frac{\partial \mathbf{A}(t)^{\mathrm{HP}}}{\partial t}\right)|\Psi\rangle = \langle\Psi|[\mathbf{A}^{\mathrm{HP}}(t), \mathbf{H}]|\Psi\rangle \tag{3.95}$$

This time dependence of the average value in the Heisenberg picture may be compared to that (2.55) we have obtained above in the Schrödinger picture, that is,

$$i\hbar\langle\Psi(t)|\left(\frac{\partial \mathbf{A}}{\partial t}\right)|\Psi(t)\rangle = \langle\Psi(t)|[\mathbf{A}, \mathbf{H}]|\Psi(t)\rangle \tag{3.96}$$

Comparison of the Heisenberg picture (3.95) and Schödinger picture (3.96) shows clearly the exchange of the time dependence between the operator and the kets.

3.3.3 Hamilton equations

Consider the position and momentum operators in the Heisenberg picture. To simplify the notation, we shall write

$$\mathbf{Q}(t)^{\mathrm{HP}} \equiv \mathbf{Q}(t) \qquad \text{and} \qquad \mathbf{P}(t)^{\mathrm{HP}} \equiv \mathbf{P}(t)$$

These operators are given in the Heisenberg picture by

$$\mathbf{Q}(t) = \mathbf{U}(t)^{\dagger}\mathbf{Q}\mathbf{U}(t) \qquad \text{and} \qquad \mathbf{P}(t) = \mathbf{U}(t)^{\dagger}\mathbf{P}\mathbf{U}(t) \tag{3.97}$$

First, verify that the commutators of the two operators remain the same in the Heisenberg picture, where they are time dependent, as in the Schrödinger picture where they are not so. To verify that, use Eq. (3.97) to write explicitly the commutator appearing on the left-hand side, yielding

$$[\mathbf{Q}(t), \mathbf{P}(t)] = \mathbf{U}(t)^{\dagger}\mathbf{Q}\mathbf{U}(t)\mathbf{U}(t)^{\dagger}\mathbf{P}\mathbf{U}(t) - \mathbf{U}(t)^{\dagger}\mathbf{P}\mathbf{U}(t)\mathbf{U}(t)^{\dagger}\mathbf{Q}\mathbf{U}(t)$$

After simplification using Eq. (3.92), we have

$$[\mathbf{Q}(t), \mathbf{P}(t)] = \mathbf{U}(t)^{\dagger}[\mathbf{Q}, \mathbf{P}]\mathbf{U}(t)$$

so that, after using Eq. (3.92), it appears that

$$[\mathbf{Q}(t), \mathbf{P}(t)] = [\mathbf{Q}, \mathbf{P}]$$

or, due to the basic commutator (2.3), of **Q** and **P**

$$[\mathbf{Q}(t), \mathbf{P}(t)] = i\hbar \tag{3.98}$$

Now, from the Heisenberg dynamic equation (3.94), it is possible to obtain the dynamics governing the time dependence of the position operator and its conjugate momentum.

First consider that of the $\mathbf{Q}(t)$ coordinate. Keeping in mind that $\mathbf{Q}(t)$ depends only on time, Eq. (3.94) allows one to write the differential equation

$$i\hbar\left(\frac{d\mathbf{Q}(t)}{dt}\right) = [\mathbf{Q}, \mathbf{H}(\mathbf{Q}, \mathbf{P})] \tag{3.99}$$

Next, use the theorem (2.14), which in the present situation reads as follows:

$$[\mathbf{Q}, \mathbf{H}(\mathbf{Q}, \mathbf{P})] = i\hbar\left(\frac{\partial \mathbf{H}(\mathbf{Q}, \mathbf{P})}{\partial \mathbf{P}}\right)$$

Then, in view of this result, Eq. (3.99) takes the form

$$\left(\frac{d\mathbf{Q}(t)}{dt}\right) = \left(\frac{\partial \mathbf{H}(\mathbf{Q}, \mathbf{P})}{\partial \mathbf{P}}\right) \tag{3.100}$$

Now, consider the Heisenberg equation governing the dynamics of the momentum

$$i\hbar\left(\frac{d\mathbf{P}(t)}{dt}\right) = [\mathbf{P}, \mathbf{H}(\mathbf{Q}, \mathbf{P})] \tag{3.101}$$

Next, in view of Eq. (2.15), the commutator involved in this equation reads

$$[\mathbf{P}, \mathbf{H}(\mathbf{Q}, \mathbf{P})] = -i\hbar\left(\frac{\partial \mathbf{H}(\mathbf{Q}, \mathbf{P})}{\partial \mathbf{Q}}\right)$$

Thus, Eq. (3.101) becomes

$$\left(\frac{d\mathbf{P}(t)}{dt}\right) = -\left(\frac{\partial \mathbf{H}(\mathbf{Q}, \mathbf{P})}{\partial \mathbf{Q}}\right) \tag{3.102}$$

Both Eqs. (3.100) and (3.102), which satisfy the quantum commutator (3.98), are the quantum Hamilton equations of motion, the classical limits of which are the classical Hamilton equations

$$\left(\frac{d\vec{P}}{dt}\right) = -\left(\frac{\partial H}{\partial \vec{Q}}\right) \quad \text{and} \quad \left(\frac{d\vec{Q}}{dt}\right) = \left(\frac{\partial H}{\partial \vec{P}}\right) \quad \text{with} \quad [\vec{Q}, \vec{P}] = 0$$

3.3.4 Interaction picture

Now, consider a new time-dependent picture of quantum mechanics, the *interaction picture (IP)*, which is intermediate between the Schrödinger and Heisenberg pictures. This picture is sometimes more practical than the pure Schrödinger and Heisenberg representations.

3.3.4.1 Operators and kets in the interaction picture Suppose that the Hamiltonian $\mathbf{H}$ of a system may be split into two parts $\mathbf{H}^{\circ}$ and $\mathbf{V}$

$$\mathbf{H} = \mathbf{H}^{\circ} + \mathbf{V} \tag{3.103}$$

Now, we may introduce an IP time-dependent ket through an action on a time-dependent ket in the Schrödinger picture by aid of the Hermitian of the time evolution operator obtained from $\mathbf{H}^{\circ}$

$$\boxed{|\Psi(t)^{\mathrm{IP}}\rangle \equiv (e^{i\mathbf{H}^{\circ}t/\hbar})|\Psi(t)^{\mathrm{SP}}\rangle} \tag{3.104}$$

$|\Psi(t)^{\mathrm{SP}}\rangle$ is the ket at time t in the Schrödinger representation, whereas $|\Psi(t)^{\mathrm{IP}}\rangle$ is the corresponding ket at the same time t in the interaction picture.

Now, premultiply each member of this last equation by the inverse of the time evolution operator involved in the previous equation.

$$(e^{-i\mathbf{H}^{\circ}t/\hbar})|\Psi(t)^{\mathrm{IP}}\rangle = (e^{-i\mathbf{H}^{\circ}t/\hbar})(e^{i\mathbf{H}^{\circ}t/\hbar})|\Psi(t)^{\mathrm{SP}}\rangle$$

After simplification, which leads to the equation inverse of (3.104) which allows us to pass from the IP to the SP for all kets

$$|\Psi(t)^{\mathrm{SP}}\rangle = (e^{-i\mathbf{H}^{\circ}t/\hbar})|\Psi(t)^{\mathrm{IP}}\rangle \tag{3.105}$$

Next, take the partial time derivative of Eq. (3.104), that is,

$$\left(\frac{\partial|\Psi(t)^{\mathrm{IP}}\rangle}{\partial t}\right) = (e^{i\mathbf{H}^{\circ}t/\hbar})\left(\frac{\partial|\Psi(t)^{\mathrm{SP}}\rangle}{\partial t}\right) + \left(\frac{\partial e^{i\mathbf{H}^{\circ}t/\hbar}}{\partial t}\right)|\Psi(t)^{\mathrm{SP}}\rangle$$

The last partial time derivative appearing on the right-hand side of this equation is

$$\left(\frac{\partial e^{i\mathbf{H}^{\circ}t/\hbar}}{\partial t}\right) = \left(\frac{i}{\hbar}\right)\mathbf{H}^{\circ}(e^{i\mathbf{H}^{\circ}t/\hbar})$$

whereas the first one is given by the time-dependent Schrödinger equation defined by the sixth postulate, that is,

$$\left(\frac{\partial|\Psi(t)^{\mathrm{SP}}\rangle}{\partial t}\right) = \left(\frac{1}{i\hbar}\right)\mathbf{H}|\Psi(t)^{\mathrm{SP}}\rangle$$

Thus, the time derivative of the IP ket becomes

$$i\hbar\left(\frac{\partial|\Psi(t)^{\mathrm{IP}}\rangle}{\partial t}\right) = (e^{i\mathbf{H}^{\circ}t/\hbar})\mathbf{H}|\Psi(t)^{\mathrm{SP}}\rangle - \mathbf{H}^{\circ}(e^{i\mathbf{H}^{\circ}t/\hbar})|\Psi(t)^{\mathrm{SP}}\rangle$$

Next, use for the right-hand-side SP kets, Eq. (3.105), in order to obtain an equation involving only IP kets. Hence

$$i\hbar\left(\frac{\partial|\Psi(t)^{\mathrm{IP}}\rangle}{\partial t}\right) = (e^{i\mathbf{H}^{\circ}t/\hbar})\mathbf{H}(e^{-i\mathbf{H}^{\circ}t/\hbar})|\Psi(t)^{\mathrm{IP}}\rangle - \mathbf{H}^{\circ}|\Psi(t)^{\mathrm{IP}}\rangle \tag{3.106}$$

where we have performed a simplification on the right-hand-side because the Hamiltonian $\mathbf{H}^{\circ}$ commutes with all function of it, that is,

$$\mathbf{H}^{\circ} = (e^{i\mathbf{H}^{\circ}t/\hbar})\mathbf{H}^{\circ}(e^{-i\mathbf{H}^{\circ}t/\hbar})$$

Thus, one may use in Eq. (3.106) the right-hand side of this last equation in place of $\mathbf{H}^\circ$, which reads

$$i\hbar\left(\frac{\partial|\Psi(t)^{\mathrm{IP}}\rangle}{\partial t}\right) = (e^{i\mathbf{H}^\circ t/\hbar})\mathbf{H}(e^{-i\mathbf{H}^\circ t/\hbar})|\Psi(t)^{\mathrm{IP}}\rangle - (e^{i\mathbf{H}^\circ t/\hbar})\mathbf{H}^\circ(e^{-i\mathbf{H}^\circ t/\hbar})|\Psi(t)^{\mathrm{IP}}\rangle$$

Then, rearranging, one obtains

$$i\hbar\left(\frac{\partial|\Psi(t)^{\mathrm{IP}}\rangle}{\partial t}\right) = (e^{i\mathbf{H}^\circ t/\hbar})(\mathbf{H} - \mathbf{H}^\circ)(e^{-i\mathbf{H}^\circ t/\hbar})|\Psi(t)^{\mathrm{IP}}\rangle$$

Again, in view of the partition (3.103), the previous expression reduces to

$$i\hbar\left(\frac{\partial|\Psi(t)^{\mathrm{IP}}\rangle}{\partial t}\right) = (e^{i\mathbf{H}^\circ t/\hbar})\mathbf{V}(e^{-i\mathbf{H}^\circ t/\hbar})|\Psi(t)^{\mathrm{IP}}\rangle$$

which may be written

$$\boxed{i\hbar\left(\frac{\partial|\Psi(t)^{\mathrm{IP}}\rangle}{\partial t}\right) = \mathbf{V}(t)^{\mathrm{IP}}|\Psi(t)^{\mathrm{IP}}\rangle} \tag{3.107}$$

where $\mathbf{V}(t)^{\mathrm{IP}}$ is the perturbation $\mathbf{V}$ in the interaction picture, which is given by

$$\boxed{\mathbf{V}(t)^{\mathrm{IP}} = (e^{i\mathbf{H}^\circ t/\hbar})\mathbf{V}(e^{-i\mathbf{H}^\circ t/\hbar})} \tag{3.108}$$

Observe that in the IP both the perturbation operator and the ket are time dependent at the difference of the SP and HP where it is either the ket or the operator, which evolves with time. More generally, under partition (3.103), the IP time dependence of an operator is given by

$$\mathbf{A}(t)^{\mathrm{IP}} = (e^{i\mathbf{H}^\circ t/\hbar})\mathbf{A}(e^{-i\mathbf{H}^\circ t/\hbar})$$

3.3.4.2 Dynamics of IP time evolution operators Now, we may introduce an interaction picture operator $\mathbf{U}(t)^{\mathrm{IP}}$, which transforms any SP ket at initial time t_0 into the corresponding IP ket at time t, according to

$$\boxed{|\Psi(t)^{\mathrm{IP}}\rangle \equiv \mathbf{U}(t - t_0)^{\mathrm{IP}}|\Psi(t_0)^{\mathrm{SP}}\rangle} \tag{3.109}$$

with the stipulation that

$$\mathbf{U}(t_0)^{\mathrm{IP}} = 1 \tag{3.110}$$

Now, observe that Eq. (3.104) may be written

$$|\Psi^{\mathrm{IP}}(t)\rangle \equiv \mathbf{U}^\circ(t - t_0)^{-1}|\Psi^{\mathrm{SP}}(t)\rangle \tag{3.111}$$

where $\mathbf{U}^\circ(t - t_0)$ is the time evolution operator given by

$$\boxed{\mathbf{U}^\circ(t - t_0) = e^{-i\mathbf{H}^\circ(t-t_0)/\hbar}} \tag{3.112}$$

with, of course,

$$\mathbf{U}^\circ(t_0) = 1$$

Moreover, the inverse transformation of Eq. (3.109) may be obtained by premultiplying in it both members of the inverse of the IP time evolution operator. On simplification, we have

$$|\Psi(t_0)^{SP}\rangle = \mathbf{U}(t-t_0)^{IP-1}|\Psi(t)^{IP}\rangle \tag{3.113}$$

On the other hand, taking the partial derivative of both terms of Eq. (3.109) reads

$$\left(\frac{\partial|\Psi(t)^{IP}\rangle}{\partial t}\right) = \left(\frac{\partial \mathbf{U}(t)^{IP}}{\partial t}\right)|\Psi(t_0)^{SP}\rangle$$

Thus, in view of Eq. (3.113) allowing to pass from any SP ket at initial time t_0 to the corresponding IP ket at time t, the equation transforms to

$$\left(\frac{\partial|\Psi(t)^{IP}\rangle}{\partial t}\right) = \left(\frac{\partial \mathbf{U}(t)^{IP}}{\partial t}\right)\mathbf{U}(t-t_0)^{IP-1}|\Psi(t)^{IP}\rangle$$

Then, one may replace the left-hand side of this last equation by its expression given by Eq. (3.107). After rearranging, we have

$$\mathbf{V}(t-t_0)^{IP}|\Psi(t)^{IP}\rangle = i\hbar\left(\frac{\partial \mathbf{U}(t)^{IP}}{\partial t}\right)\mathbf{U}(t-t_0)^{IP-1}|\Psi(t)^{IP}\rangle$$

Then, since this last linear transformation is satisfied irrespective of the IP ket at any time, we have

$$i\hbar\left(\frac{\partial \mathbf{U}(t)^{IP}}{\partial t}\right)\mathbf{U}(t-t_0)^{IP-1} = \mathbf{V}(t-t_0)^{IP}$$

Finally, postmultiply both member of this last equation by $\mathbf{U}(t-t_0)^{IP}$. Hence, after simplification using the operator property

$$\boxed{i\hbar\left(\frac{\partial \mathbf{U}(t)^{IP}}{\partial t}\right) = \mathbf{V}(t-t_0)^{IP}\mathbf{U}(t-t_0)^{IP}} \tag{3.114}$$

which, when $t_0 = 0$ simplifies to

$$i\hbar\left(\frac{\partial \mathbf{U}(t)^{IP}}{\partial t}\right) = \mathbf{V}(t)^{IP}\mathbf{U}(t)^{IP} \tag{3.115}$$

and, due to Eq. (3.110)

$$\mathbf{U}(0)^{IP} = 1 \tag{3.116}$$

3.3.4.3 Relation between IP and SP time evolution operators Now, observe that the linear transformation, which is inverse of that given by Eq. (3.111), may be obtained by premultiplying both terms by $\mathbf{U}^\circ(t)$ and then simplifying the result using $\mathbf{U}^\circ(t-t_0)\mathbf{U}^\circ(t-t_0)^{-1} = 1$, leading to

$$|\Psi(t)^{SP}\rangle = \mathbf{U}^\circ(t-t_0)|\Psi(t)^{IP}\rangle \tag{3.117}$$

Then, premultiplying both members of this last equation by $\mathbf{U}^\circ(t-t_0)^{-1}$, we have on simplification

$$\mathbf{U}^\circ(t-t_0)^{-1}\mathbf{U}^\circ(t-t_0) = 1$$

and

$$\mathbf{U}^{\circ}(t-t_0)^{-1}|\Psi(t)^{\mathrm{SP}}\rangle = |\Psi(t)^{\mathrm{IP}}\rangle \tag{3.118}$$

Next, owing to Eq. (3.109) relating the IP ket at time t with the SP one at initial time t_0, the last equation becomes

$$|\Psi(t)^{\mathrm{SP}}\rangle \equiv \mathbf{U}^{\circ}(t-t_0)\mathbf{U}(t-t_0)^{\mathrm{IP}}|\Psi(t_0)^{\mathrm{SP}}\rangle \tag{3.119}$$

Remark that, according to Eq. (3.85), the SP ket at time t is related to the corresponding one on initial time t_0 via

$$|\Psi(t)^{\mathrm{SP}}\rangle \equiv \mathbf{U}(t-t_0)|\Psi(t_0)^{\mathrm{SP}}\rangle \tag{3.120}$$

with

$$\mathbf{U}(t-t_0) = e^{-i\mathbf{H}(t-t_0)/\hbar} \tag{3.121}$$

Thus, comparison of Eqs. (3.119) and (3.120) shows that

$$\boxed{\mathbf{U}(t-t_0) = \mathbf{U}^{\circ}(t-t_0)\mathbf{U}(t-t_0)^{\mathrm{IP}}} \tag{3.122}$$

Equation (3.122) shows that the full time evolution operator $\mathbf{U}(t-t_0)$, which is given by Eq. (3.121), is equal to the unperturbed time evolution operator $\mathbf{U}^{\circ}(t-t_0)$ given by Eq. (3.112) times the IP time evolution operator governed by the partial differential equation (3.114).

3.3.4.4 Perturbation expansion of the time evolution operator We shall now obtain the full time evolution operator $\mathbf{U}(t)$ when it is only easy to find its corresponding unperturbed time evolution operator $\mathbf{U}^{\circ}(t)$. The solution of the problem requires one to get the IP time evolution operator by solution of Eq. (3.115) with the boundary condition (3.116), that is,

$$i\hbar\left(\frac{\partial \mathbf{U}(t)^{\mathrm{IP}}}{\partial t}\right) = \mathbf{V}(t)^{\mathrm{IP}}\mathbf{U}(t)^{\mathrm{IP}} \quad \text{with} \quad \mathbf{U}(0)^{\mathrm{IP}} = 1$$

On integration between $t' = 0$ and $t' = t$, and using the boundary condition, we have

$$\mathbf{U}(t)^{\mathrm{IP}} = 1 + \left(\frac{1}{i\hbar}\right)\int_0^t \mathbf{V}(t')^{\mathrm{IP}}\,\mathbf{U}(t')^{\mathrm{IP}}dt' \tag{3.123}$$

Now, in order to solve the integral equation (3.123), one may write for $\mathbf{U}(t')^{\mathrm{IP}}$ on its right-hand side, an expression that may be obtained from Eq. (3.123), by the replacements $t' \to t$, and $t'' \to t'$, namely

$$\mathbf{U}(t')^{\mathrm{IP}} = 1 + \left(\frac{1}{i\hbar}\right)\int_0^{t'} \mathbf{V}(t'')^{\mathrm{IP}}\,\mathbf{U}(t'')^{\mathrm{IP}}dt''$$

Hence, Eq. (3.123) yields

$$\mathbf{U}(t)^{\mathrm{IP}} = 1+\left(\frac{1}{i\hbar}\right)\int_0^t \mathbf{V}(t')^{\mathrm{IP}}dt'+\left(\frac{1}{i\hbar}\right)^2\int_0^t\int_0^{t'} \mathbf{V}^{\mathrm{IP}}(t')\mathbf{V}^{\mathrm{IP}}(t'')\mathbf{U}^{\mathrm{IP}}(t'')dt'dt'' \tag{3.124}$$

The foregoing equation may be iterated as many times as required. If the perturbation $\mathbf{V}$ is very small with respect to $\mathbf{H}^\circ$, the third term on the right-hand side of this last equation, which is quadratic in $\mathbf{V}$, may be neglected with respect to the second term, which is linear in $\mathbf{V}$, leading to the first-order expansion of the IP time evolution operator given by

$$\mathbf{U}(t)^{\mathrm{IP}} = 1 + \left(\frac{1}{i\hbar}\right)\int_0^t \mathbf{V}(t')^{\mathrm{IP}} dt' \tag{3.125}$$

To simplify, limit the iteration by truncating the IP time evolution operator $\mathbf{U}(t'')^{\mathrm{IP}}$ at time t'' appearing in Eq. (3.124), to the first term unity

$$\mathbf{U}(t'')^{\mathrm{IP}} = 1 + \left(\frac{1}{i\hbar}\right)\int_0^{t''} \mathbf{V}(t''')^{\mathrm{IP}}\mathbf{U}(t''')^{\mathrm{IP}} dt''' \simeq 1$$

That leads to the following second-order perturbative expansion for the IP time evolution operator:

$$\boxed{\mathbf{U}(t)^{\mathrm{IP}} \simeq 1 + \left(\frac{1}{i\hbar}\right)\int_0^t \mathbf{V}(t')^{\mathrm{IP}} dt' + \left(\frac{1}{i\hbar}\right)^2 \int_0^t\int_0^{t'} \mathbf{V}^{\mathrm{IP}}(t')\mathbf{V}^{\mathrm{IP}}(t'')dt'dt''}$$

Next, owing to Eqs. (3.108) and (3.112), the IP time evolution operator reads

$$\mathbf{V}(t)^{\mathrm{IP}} = \mathbf{U}^\circ(t)^{-1}\mathbf{V}\mathbf{U}^\circ(t) \tag{3.126}$$

Then, using (3.126) and also Eq. (3.122) allowing to pass from $\mathbf{U}(t)^{\mathrm{IP}}$ to $\mathbf{U}(t)$, the full time evolution operator appears to be given by

$$\begin{aligned}\mathbf{U}(t) &\simeq \mathbf{U}^\circ(t)\\ &+ \left(\frac{1}{i\hbar}\right)\mathbf{U}^\circ(t)\int_0^t \mathbf{U}^\circ(t')^{-1}\mathbf{V}\mathbf{U}^\circ(t')dt'\\ &+ \left(\frac{1}{i\hbar}\right)^2\mathbf{U}^\circ(t)\int_0^t\int_0^{t'} \mathbf{U}^\circ(t')^{-1}\mathbf{V}\mathbf{U}^\circ(t')\mathbf{U}^\circ(t'')^{-1}\mathbf{V}\mathbf{U}^\circ(t'')dt'dt''\end{aligned}$$

This result must be considered, keeping in mind Eqs. (3.82) and (3.112), that is,

$$\mathbf{U}(t) = (e^{-i\mathbf{H}t/\hbar}) \qquad \text{and} \qquad \mathbf{U}^\circ(t) = (e^{-i\mathbf{H}^\circ t/\hbar})$$

3.3.5 Formal expression to make Eq. (3.123) tractable

Observe that the time evolution operators allow one to pass from a ket at initial time $t = 0$ to another at time t.

$$|\Psi(t)\rangle = \mathbf{U}(t)|\Psi(0)\rangle$$

Now, we may replace the initial time $t = 0$ by any time t°. Then, this equation may be written

$$|\Psi(t)\rangle = \mathbf{U}(t, t^\circ)|\Psi(t^\circ)\rangle \tag{3.127}$$

In this equation, the time evolution operator appears to be a conditional operator, which, if the ket is $|\Psi(t^\circ)\rangle$ at time $t = t^\circ$, transforms this ket into $|\Psi(t)\rangle$ at time t. Note that, $\mathbf{U}(t, t^\circ)$ is by definition given by

$$\mathbf{U}(t, t^\circ) = e^{-i(t-t^\circ)\mathbf{H}/\hbar}$$

Next, consider the following time evolution operators:

$$\mathbf{U}(t_2, t_1) = e^{-i(t_2-t_1)\mathbf{H}/\hbar} \quad \text{and} \quad \mathbf{U}(t_1, t^\circ) = e^{-i(t_1-t^\circ)\mathbf{H}/\hbar} \tag{3.128}$$

Again, consider the third evolution operator

$$\mathbf{U}(t_2, t^\circ) = e^{-i(t_2-t^\circ)\mathbf{H}/\hbar}$$

Of course, this operator may be written

$$\mathbf{U}(t_2, t^\circ) = e^{-i\{(t_2-t_1)+(t_1-t^\circ)\}\mathbf{H}/\hbar} \tag{3.129}$$

Then, owing to the equation appearing in Eq. (3.128), the time evolution operator (3.129) appears to be

$$\mathbf{U}(t_2, t^\circ) = \mathbf{U}(t_2, t_1)\,\mathbf{U}(t_1, t^\circ) \tag{3.130}$$

It may be observed that Eqs. (3.127) and (3.130) are true for all kinds of time evolution operators, that is, for full, unperturbed, and IP time evolution operators.

Keeping that in mind, we may return to Eq. (3.123).

$$\mathbf{U}^{\mathrm{IP}}(t, t^\circ) = 1 + \left(\frac{1}{i\hbar}\right)\int_{t^\circ}^{t} \mathbf{V}^{\mathrm{IP}}(\tau, t^\circ)\mathbf{U}^{\mathrm{IP}}(\tau, t^\circ)\, d\tau \tag{3.131}$$

Next, by inversion of Eq. (3.122), one obtains

$$\mathbf{U}^{\mathrm{IP}}(t, t^\circ) = \mathbf{U}^\circ(t, t^\circ)^{-1}\mathbf{U}(t, t^\circ)$$

This equation allows to transform Eq. (3.131) into

$$\mathbf{U}^\circ(t, t^\circ)^{-1}\mathbf{U}(t, t^\circ) = 1 + \left(\frac{1}{i\hbar}\right)\int_{t^\circ}^{t} \mathbf{V}^{\mathrm{IP}}(\tau, t^\circ)\mathbf{U}^{\mathrm{IP}}(\tau, t^\circ)\, d\tau \tag{3.132}$$

Next, we may use

$$\mathbf{U}^\circ(t, t^\circ)^{-1}\mathbf{U}^\circ(t, t^\circ) = 1 \tag{3.133}$$

Then, premultiplying the right-hand side of Eq. (3.132) by this last equation leads to

$$\mathbf{U}^\circ(t, t^\circ)^{-1}\mathbf{U}(t, t^\circ) = \mathbf{U}^\circ(t, t^\circ)^{-1}\mathbf{U}^\circ(t, t^\circ)\left(1 + \left(\frac{1}{i\hbar}\right)\int_{t^\circ}^{t} \mathbf{V}^{\mathrm{IP}}(\tau, t^\circ)\mathbf{U}^{\mathrm{IP}}(\tau, t^\circ)\, d\tau\right) \tag{3.134}$$

Next, premultiply both members of this last equation by $\mathbf{U}^\circ(t, t^\circ)$. Then, owing to Eq. (3.133), Eq. (3.134) reduces to

$$\mathbf{U}(t,t^\circ) = \mathbf{U}^\circ(t,t^\circ)\left(1 + \left(\frac{1}{i\hbar}\right)\int_{t^\circ}^{t} \mathbf{V}^{\mathrm{IP}}(\tau,t^\circ)\mathbf{U}^{\mathrm{IP}}(\tau,t^\circ)\,d\tau\right)$$

which may be written as

$$\mathbf{U}(t,t^\circ) = \mathbf{U}^\circ(t,t^\circ) + \left(\frac{1}{i\hbar}\right)\int_{t^\circ}^{t} \mathbf{U}^\circ(t,t^\circ)\mathbf{V}^{\mathrm{IP}}(\tau,t^\circ)\mathbf{U}^{\mathrm{IP}}(\tau,t^\circ)\,d\tau \tag{3.135}$$

Now, observe that the unperturbed time evolution operator, just after the integral allowing one to pass from $t = 0$ to t may be viewed as the product:

$$\mathbf{U}^\circ(t,t^\circ) = [\mathbf{U}^\circ(t,\tau)\mathbf{U}^\circ(\tau,t^\circ)]$$

Then, Eq. (3.135) may be written

$$\mathbf{U}(t,t^\circ) = \mathbf{U}^\circ(t,t^\circ) + \left(\frac{1}{i\hbar}\right)\int_{t^\circ}^{t} [\mathbf{U}^\circ(t,\tau)\mathbf{U}^\circ(\tau,t^\circ)]\{\mathbf{V}^{\mathrm{IP}}(\tau,t^\circ)\}\mathbf{U}^{\mathrm{IP}}(\tau,t^\circ)\,d\tau$$

Again, for the perturbation Hamiltonian in the interaction picture use Eq. (3.126):

$$\begin{aligned}\mathbf{U}(t,t^\circ) &= \mathbf{U}^\circ(t,t^\circ)\\ &\quad + \left(\frac{1}{i\hbar}\right)\int_{t^\circ}^{t} \mathbf{U}^\circ(t,\tau)\mathbf{U}^\circ(\tau,t^\circ)\mathbf{U}^\circ(\tau,t^\circ)^{-1}\mathbf{V}\mathbf{U}^\circ(\tau,t^\circ)\mathbf{U}^\circ(\tau,t^\circ)^{-1}\mathbf{U}^\circ(\tau,t^\circ)\,d\tau\end{aligned}$$

Finally, in order to simplify this last result, we may use the property of a time evolution operator and of its inverse in the following way:

$$\mathbf{U}^\circ(\tau,t^\circ)\mathbf{U}^\circ(\tau,t^\circ)^{-1} = 1 \qquad \text{and} \qquad \mathbf{U}^\circ(\tau,t^\circ)^{-1}\mathbf{U}^\circ(\tau,t^\circ) = 1$$

That leads to the final result of importance:

$$\boxed{\mathbf{U}(t,t^\circ) = \mathbf{U}^\circ(t,t^\circ) + \left(\frac{1}{i\hbar}\right)\int_{t^\circ}^{t} \mathbf{U}^\circ(t,\tau)\mathbf{V}\mathbf{U}^\circ(\tau,t^\circ)\,d\tau} \tag{3.136}$$

Note in this last equation the respective places of the times t°, τ, and t,

3.4 DENSITY OPERATORS

After studying the time dependence of quantum mechanics, through the Schrödinger, Heisenberg, and interaction pictures using the time evolution operator, it is now appropriate to introduce the fundamental concept of the *density operator*, which is a very powerful tool in quantum mechanics.

3.4.1 Basic properties

3.4.1.1 Definition By definition, the density operator $\boldsymbol{\rho}$ of a statistical mixture is

$$\boxed{\rho = \sum_i W_i |\Phi_i\rangle\langle\Phi_i|} \tag{3.137}$$

Here the W_i are the probabilities for the states to be occupied, which are therefore real and must obey

$$\sum_i W_i = 1 \qquad \text{and} \qquad 0 \leq W_i \leq 1$$

whereas the kets $|\Phi_i\rangle$ belong to an arbitrary basis in the state space, and thus obey

$$\sum_i |\Phi_i\rangle\langle\Phi_i| = 1 \qquad \text{and} \qquad \langle\Phi_i|\Phi_k\rangle = \delta_{ik} \tag{3.138}$$

Note that for a pure state, all the operations are zero except one, which is equal to unity. Then the density operator expression (3.137) reduces to

$$\rho = |\Phi_i\rangle\langle\Phi_i| \tag{3.139}$$

3.4.1.2 Trace of the density operator Consider now the trace of the density operator. It is in the basis used for its description:

$$\mathbf{tr}\{\rho\} = \sum_k \langle\Phi_k| \left\{\sum_i W_i|\Phi_i\rangle\langle\Phi_i|\right\} |\Phi_k\rangle$$

Then, since the W_i are scalars, owing to the orthonormality properties of the basis, this last equation transforms to

$$\mathbf{tr}\{\rho\} = \sum_k \sum_i W_i \langle\Phi_k|\Phi_i\rangle\langle\Phi_i|\Phi_k\rangle$$

Again, owing to the orthonormality properties (3.138) of the basis, that reduces to

$$\mathbf{tr}\{\rho\} = \sum_i W_i$$

At last, since the sum of the probabilities W_i is equal to unity, the trace appears to be simply given as

$$\mathbf{tr}\{\rho\} = 1 \tag{3.140}$$

3.4.1.3 Hermiticity of the density operator The Hermitian conjugate of the density operator (3.137) is

$$\rho^\dagger = \left\{\sum_i W_i|\Phi_i\rangle\langle\Phi_i|\right\}^\dagger \tag{3.141}$$

Again, using the rules of this section governing Hermitian conjugation, the right-hand side of this last equation is

$$\left\{\sum_i W_i|\Phi_i\rangle\langle\Phi_i|\right\}^\dagger = \sum_i W_i^*|\Phi_i\rangle\langle\Phi_i|$$

or, since the probabilities are real,

$$\left\{\sum_i W_i|\Phi_i\rangle\langle\Phi_i|\right\}^{\dagger} = \sum_i W_i|\Phi_i\rangle\langle\Phi_i|$$

Hence, Eq. (3.141) becomes

$$\rho^{\dagger} = \left\{\sum_i W_i|\Phi_i\rangle\langle\Phi_i|\right\} \tag{3.142}$$

Thus, by comparison of Eq. (3.142) with Eq. (3.137), it appears that

$$\rho^{\dagger} = \rho$$

showing that the density operator is Hermitian.

3.4.1.4 Inequality governing the density operator in the general case of mixed states Consider the square of the density operator, which, owing to Eq. (3.137), reads

$$\rho^2 = \sum_i W_i|\Phi_i\rangle\langle\Phi_i| \sum_k W_k|\Phi_k\rangle\langle\Phi_k|$$

or, since the probabilities W_i are scalars,

$$\rho^2 = \sum_i \sum_k W_i W_k|\Phi_i\rangle\langle\Phi_i|\Phi_k\rangle\langle\Phi_k|$$

so that due to the orthonormality properties (3.138)

$$\rho^2 = \sum_i \sum_k W_i W_k|\Phi_i\rangle\delta_{ik}\langle\Phi_k|$$

and, thus, after simplification using the properties of the Kronecker symbol, it is found that

$$\rho^2 = \sum_i W_i^2|\Phi_i\rangle\langle\Phi_i| \tag{3.143}$$

Moreover, since the probabilities are smaller than unity, their squares obey the inequality

$$W_i^2 < W_i^2$$

and it appears by comparison of Eq. (3.143) with (3.137) that

$$\rho^2 < \rho \tag{3.144}$$

For a pure state verifying Eq. (3.139), the square of the density operator reduces to

$$\rho^2 = |\Phi_i\rangle\langle\Phi_i|\Phi_i\rangle\langle\Phi_i|$$

or, because of the orthonormality properties (3.138), to

$$\rho^2 = |\Phi_i\rangle\langle\Phi_i|$$

so that

$$\rho^2 = \rho \quad \text{(pure state)}$$

3.4.2 Density operator for many particles

Now, consider the density operator of a set of N particles. Then, according to Eq. (3.47), a ket characterizing a whole system is given by the product

$$|\Psi_{k(1),l(2)..f(N)}\rangle = \prod_{(r)}^{(N)} |\Phi_{j(r)}\rangle \tag{3.145}$$

where $|\Phi_{l(r)}\rangle$ is the lth ket $|\Phi_l\rangle$ of the rth particle.

Next, for a pure case, the full density operator $\boldsymbol{\rho}_{\text{Tot}}$ of the set of N particles is given by an expression of the same form as that in Eq. (3.139) in which the ket given by Eq. (3.145) plays the role of $|\Phi_i\rangle$ in Eq. (3.139), that is,

$$\boldsymbol{\rho}_{\text{Tot}} = |\Psi_{k(1)l(2)..f(N)}\rangle\langle\Psi_{k(1)l(2)...f(N)}|$$

or

$$\boldsymbol{\rho}_{\text{Tot}} = \prod_{(r)}^{(N)} |\Phi_{j(r)}\rangle\langle\Phi_{j(r)}| \tag{3.146}$$

Again, for a mixed situation, the generalization to N particles of Eq. (3.137), leads to

$$\boldsymbol{\rho}_{\text{Tot}} = \sum_{k(1)}\sum_{l(2)}\cdots\sum_{f(N)} \mathbb{W}_{k(1)l(2)..f(N)}|\Psi_{k(1)l(2)..f(N)}\rangle\ \langle\Psi_{k(1)l(2)...f(N)}|$$

where the $\mathbb{W}_{k(1)l(2)..f(N)}$ are the joint probabilities to find the first particle (1) in the kth state $|\Phi_k\rangle$, with the probability $W_{k(1)}$, the second particle (2) in the lth state $|\Phi_l\rangle$ with the probability $W_{l(2)}$, and so on given by

$$\mathbb{W}_{k(1)l(2)..f(N)} = W_{k(1)}\ W_{l(2)}\ \cdots$$

On the other hand, consider a physical system that may be divided into two different subsystems. Then, the *full density operator* of this system may be written as the product of the density operators of the two subsystems:

$$\boldsymbol{\rho}_{\text{Tot}} = \boldsymbol{\rho}_{(1)}\boldsymbol{\rho}_{(2)}$$

By definition, the *reduced density operator* of one of the two subsystems is the partial trace over the subspace spanned by the other subsystem over the full density operator:

$$\boldsymbol{\rho}_{\text{Red}(2)} = \mathbf{tr}_{(1)}\{\boldsymbol{\rho}_{\text{Tot}}\} \qquad \boldsymbol{\rho}_{\text{Red}(1)} = \mathbf{tr}_{(2)}\{\boldsymbol{\rho}_{\text{Tot}}\}$$

3.4.3 Average values

We now show that the average value of an operator $\mathbf{A}$ performed over the density operator of a statistical mixed state is

$$\boxed{\langle\mathbf{A}\rangle_\rho = \mathbf{tr}\{\boldsymbol{\rho}\mathbf{A}\}} \tag{3.147}$$

In order to prove this equation, recall that, according to Eqs. (3.137) and (3.140), the density operator obeys

$$\boldsymbol{\rho} = \sum_i W_i|\Phi_i\rangle\langle\Phi_i| \qquad \text{and} \qquad \mathbf{tr}\{\boldsymbol{\rho}\} = 1$$

Then, Eq. (3.147) becomes

$$\langle \mathbf{A} \rangle_\rho = \mathbf{tr}\left\{\sum_i W_i |\Phi_i\rangle\langle\Phi_i|\mathbf{A}\right\}$$

Perform the trace over the basis $\{|\Phi_i\rangle\}$, which reads

$$\langle \mathbf{A} \rangle_\rho = \sum_k \sum_i \langle\Phi_k|W_i|\Phi_i\rangle\langle\Phi_i|\mathbf{A}|\Phi_k\rangle$$

Or, since the probabilities W_i are scalars,

$$\langle \mathbf{A} \rangle_\rho = \sum_k \sum_i W_i \langle\Phi_k|\Phi_i\rangle\langle\Phi_i|\mathbf{A}|\Phi_k\rangle$$

Finally, owing to the orthonormality properties (3.138) of the basis $\{|\Phi_i\rangle\}$, that simplifies to

$$\langle \mathbf{A} \rangle_\rho = \sum_k \sum_i W_i \delta_{ki} \langle\Phi_i|\mathbf{A}|\Phi_k\rangle$$

and thus

$$\langle \mathbf{A} \rangle_\rho = \sum_i W_i \langle\Phi_i|\mathbf{A}|\Phi_i\rangle \tag{3.148}$$

Hence, the average of operator $\mathbf{A}$ over density operator $\boldsymbol{\rho}$ is the sum of all the quantum average values of operator $\mathbf{A}$ over the kets $|\Phi_i\rangle$ belonging to the basis $\{|\Phi_i\rangle\}$, times the corresponding probabilities W_i.

Of course, for a pure density state where all the probabilities are zero, except one which is unity, the average value over the density operator (3.148) reduces to the simple quantum average value (2.21), that is,

$$\langle \mathbf{A} \rangle_\rho = \langle\Phi_i|\mathbf{A}|\Phi_i\rangle$$

3.4.4 Entropy and density operators

Introduce the statistical entropy function through

$$S = -k_B \langle \ln \boldsymbol{\rho} \rangle_\rho$$

where k_B is the Boltzmann constant. Now, keeping in mind that the average of an operator over the density operator is given by Eq. (3.147), the statistical entropy becomes

$$S = -k_B \mathbf{tr}\{\boldsymbol{\rho} \ln \boldsymbol{\rho}\} \tag{3.149}$$

Again, writing explicitly the trace involved in Eq. (3.149) by performing the trace over the basis $\{|\Phi_i\rangle\}$ obeying Eq. (3.138), that is,

$$\langle\Phi_k|\Phi_l\rangle = \delta_{kl} \tag{3.150}$$

we have

$$S = -k_B \sum_i \langle\Phi_i|\boldsymbol{\rho} \ln \boldsymbol{\rho}|\Phi_i\rangle \tag{3.151}$$

Moreover, since, owing to Eq. (3.137), the density operator of a mixed state expressed in the basis $\{|\Phi_i\rangle\}$ is given by

$$\rho = \sum_k W_k |\Phi_k\rangle\langle\Phi_k|$$

the statistical entropy (3.151) yields

$$S = -\sum_i \sum_k W_k \langle\Phi_i|\Phi_k\rangle\langle\Phi_k| \ln\left\{\sum_l W_l|\Phi_l\rangle\langle\Phi_l|\right\}|\Phi_i\rangle$$

or, due to the orthonormality properties (3.138)

$$S = -k_B \sum_i W_i \langle\Phi_i| \ln\left\{\sum_l W_l|\Phi_l\rangle\langle\Phi_l|\right\}|\Phi_i\rangle \tag{3.152}$$

Next, to calculate the operator ln **A** averaged over $|\Phi_i\rangle$, where

$$\mathbf{A} = \sum_l W_l|\Phi_l\rangle\langle\Phi_l|$$

we use the following formal expansion of the logarithm of some function A given by

$$\ln A = \sum_k C_k A^k \tag{3.153}$$

where C_k are the coefficients involved in the expansion of the logarithm. Then, the logarithm involved on the right-hand side of Eq. (3.152) expands as

$$\ln\left\{\sum_l W_l|\Phi_l\rangle\langle\Phi_l|\right\} = \sum_k C_k \left(\sum_l W_l|\Phi_l\rangle\langle\Phi_l|\right)^k \tag{3.154}$$

Next, observe that, when $k=2$, it reads

$$\left(\sum_l W_l|\Phi_l\rangle\langle\Phi_l|\right)^2 = \sum_l \sum_s W_l|\Phi_l\rangle\langle\Phi_l|W_s|\Phi_s\rangle\langle\Phi_s|$$

or, since W_s is a scalar,

$$\left(\sum_l W_l|\Phi_l\rangle\langle\Phi_l|\right)^2 = \sum_l \sum_s W_l W_s|\Phi_l\rangle\langle\Phi_l|\Phi_s\rangle\langle\Phi_s|$$

so that, due to the orthonormality property (3.150),

$$\left(\sum_l W_l|\Phi_l\rangle\langle\Phi_l|\right)^2 = \sum_l \sum_s W_l|\Phi_l\rangle W_s \delta_{ls}\langle\Phi_s|$$

After simplification using the orthonormality properties (3.138) of the basis, that simplifies to

$$\left(\sum_l W_l|\Phi_l\rangle\langle\Phi_l|\right)^2 = \sum_l (W_l)^2(|\Phi_l\rangle\langle\Phi_l|)^2$$

Again, by recurrence one obtains for any value of k

$$\left(\sum_l W_l|\Phi_l\rangle\langle\Phi_l|\right)^k = \sum_l (W_l)^k|\Phi_l\rangle\langle\Phi_l|$$

so that Eq. (3.154) takes the form

$$\ln\left\{\sum_l W_l|\Phi_l\rangle\langle\Phi_l|\right\} = \sum_k C_k \sum_l (W_l)^k|\Phi_l\rangle\langle\Phi_l|$$

Moreover, due to the latter result, the diagonal matrix elements of Eq. (3.154) read

$$\langle\Phi_i|\ln\left\{\sum_l W_l|\Phi_l\rangle\langle\Phi_l|\right\}|\Phi_i\rangle = \sum_k C_k\langle\Phi_i|\sum_l (W_l)^k(|\Phi_l\rangle\langle\Phi_l|)|\Phi_i\rangle \tag{3.155}$$

or, in view of the orthonormality properties (3.138), one has

$$\langle\Phi_i|(W_l)^k(|\Phi_l\rangle\langle\Phi_l|)|\Phi_i\rangle = (W_l)^k\delta_{il}$$

Hence, after simplification using the property of δ_{il}, Eq. (3.155) becomes

$$\langle\Phi_i|\ln\left\{\sum_l W_l|\Phi_l\rangle\langle\Phi_l|\right\}|\Phi_i\rangle = \sum_k C_k(W_i)^k$$

Hence, according to the formal expression of the expansion (3.153), in which W_l plays now the role of the function A, we have

$$\langle\Phi_i|\ln\left\{\sum_l W_l|\Phi_l\rangle\langle\Phi_l|\right\}|\Phi_i\rangle = \ln W_i$$

Thus, the entropy given by Eq. (3.152) transforms to the simple form

$$\boxed{S = -k_B\sum_i W_i \ln W_i} \tag{3.156}$$

which is the usual statistical expression of entropy in information theory. Of course, the probabilities may depend on time, so that the statistical entropy depends also on time. Thus, Eq. (3.156) may be written for any time

$$S = -k_B\sum_i W_i(t)\ln W_i(t) \tag{3.157}$$

3.4.5 Density operator representations

Start from the general expression (3.137) of the density operator $\boldsymbol{\rho}$ of a mixed state, that is,

$$\boldsymbol{\rho} = \sum_i W_i|\Phi_i\rangle\langle\Phi_i| \tag{3.158}$$

where W_i is the probability for the ket $|\Phi_i\rangle$ to be occupied. This operator may be expressed in the basis $\{|\{Q\}\rangle\}$ of the eigenstates of the position operator as it will be now seen.

3.4.5.1 *Position representation* For this purpose, write the eigenvalue equation of the position operator $\mathbf{Q}$, that is,

$$\mathbf{Q}|\{Q\}\rangle = Q|\{Q\}\rangle$$

In the basis $\{|\{Q\}\rangle\}$, the matrix elements of the density operator (3.158) read

$$\langle\{Q\}|\boldsymbol{\rho}|\{Q'\}\rangle = \sum_i W_i \langle\{Q\}|\Phi_i\rangle\langle\Phi_i|\{Q'\}\rangle$$

The scalar products involved in this last equation are the wavefunctions given by

$$\langle\{Q\}|\Phi_i\rangle = \Phi_i(Q) \qquad \text{and} \qquad \langle\Phi_i|\{Q'\}\rangle = \Phi_i^*(Q')$$

Hence, the matrix elements, which may be denoted as $\rho(Q, Q')$, become

$$\langle\{Q\}|\boldsymbol{\rho}|\{Q'\}\rangle = \sum_i W_i \Phi_i(Q)\Phi_i^*(Q') \equiv \rho(Q, Q') \tag{3.159}$$

the corresponding diagonal matrix elements denoted $\rho(Q, Q)$ reduce to

$$\langle\{Q\}|\boldsymbol{\rho}|\{Q\}\rangle = \sum_i W_i|\Phi_i(Q)|^2 \equiv \rho(Q, Q) \tag{3.160}$$

3.4.5.2 *Momentum representation* Now, write the eigenvalue equation of the momentum P as

$$\mathbf{P}|\{P\}\rangle = P|\{P\}\rangle$$

In the basis of the eigenstates of the position operator, the matrix elements of the density operator are, comparing Eq. (3.158),

$$\langle\{P\}|\boldsymbol{\rho}|\{P'\}\rangle = \sum_i W_i \langle\{P\}|\Phi_i\rangle\langle\Phi_i|\{P'\}\rangle$$

The scalar products involved here are the wavefunctions in the momentum representation, that is,

$$\langle\{P\}|\Phi_i\rangle = \Phi_i(P) \qquad \text{and} \qquad \langle\Phi_i|\{P'\}\rangle = \Phi_i^*(P')$$

Thus, the matrix elements $\rho(P, P')$ become

$$\rho(P, P') = \langle\{P\}|\boldsymbol{\rho}|\{P'\}\rangle = \sum_i W_i \Phi_i(P)\Phi_i^*(P') \tag{3.161}$$

the corresponding diagonal matrix elements being

$$\rho(P, P) = \langle\{P\}|\boldsymbol{\rho}|\{P\}\rangle = \sum_i W_i|\Phi_i(P)|^2 \tag{3.162}$$

3.4.5.3 *Wigner distribution function* Now, consider for one dimension in the position representation the following off-diagonal matrix elements of the density operator:

$$\rho\left(Q + \frac{\eta}{2}, Q - \frac{\eta}{2}\right) = \left\langle\left\{Q + \frac{\eta}{2}\right\}|\boldsymbol{\rho}|\left\{Q - \frac{\eta}{2}\right\}\right\rangle$$

Then it is possible to write from this matrix element the following function:

$$f_w(P,Q) = \int_{-\infty}^{+\infty} \rho\left(Q+\frac{\eta}{2}, Q-\frac{\eta}{2}\right) \exp\left\{\frac{-iP\eta}{\hbar}\right\} d\eta \tag{3.163}$$

Next, multiply both members of this last equation by $\frac{1}{2}\pi\hbar$ and integrate over all the momentum:

$$\frac{1}{2\pi\hbar}\int_{-\infty}^{+\infty} f_w(P,Q)\,dP = \frac{1}{2\pi\hbar}\int_{-\infty}^{+\infty}\int_{-\infty}^{+\infty} \rho\left(Q+\frac{\eta}{2}, Q-\frac{\eta}{2}\right) \exp\left\{\frac{-iP\eta}{\hbar}\right\} d\eta dP \tag{3.164}$$

or

$$\frac{1}{2\pi\hbar}\int_{-\infty}^{+\infty} f_w(P,Q)\,dP = \int_{-\infty}^{+\infty} \rho\left(Q+\frac{\eta}{2}, Q-\frac{\eta}{2}\right) \frac{1}{2\pi\hbar}\left\{\int_{-\infty}^{+\infty} \exp\left\{\frac{-iP\eta}{\hbar}\right\} dP\right\} d\eta \tag{3.165}$$

Next, owing to the distribution theory leading to Eq. (18.60), the last integral of the right-hand part of Eq. (3.165) reads

$$\frac{1}{2\pi\hbar}\left\{\int_{-\infty}^{+\infty} \exp\left\{\frac{-iP\eta}{\hbar}\right\} dP\right\} = \delta(\eta)$$

so that Eq. (3.164) becomes

$$\boxed{\frac{1}{2\pi\hbar}\int_{-\infty}^{+\infty} f_w(P,Q)\,dP = \int_{-\infty}^{+\infty} \rho\left(Q+\frac{\eta}{2}, Q-\frac{\eta}{2}\right)\delta(\eta)\,d\eta}$$

Therefore, according to the fact that $\delta(\eta)$ is zero, except if $\eta = 0$, for which $\delta(\eta) = 1$, and keeping in mind Eq. (3.160), this last expression reduces to

$$\boxed{\frac{1}{2\pi\hbar}\int_{-\infty}^{+\infty} f_w(P,Q)\,dP = \rho(Q,Q) = f(Q)} \tag{3.166}$$

The function $f_w(P,Q)$ (3.163), known as the *Wigner distribution function*, may be viewed as corresponding from quantum mechanics to the classical distribution function in the phase space $f(P,Q)$. However, it must be observed that the Wigner distribution function may be negative, that is, impossible for the classical distribution function $f(P,Q)$. This aspect is the cost to be paid by the requirement to save the Heisenberg uncertainty relations, which forbid the simultaneous knowledge of the position and of the momentum.

3.4.6 Dynamics

3.4.6.1 Schrödinger picture At the difference of the other operators, which do not depend on time in the Schrödinger picture, the density operator is time dependent in this representation because it is built up from the kets and the corresponding bras, which evolve with time according to the time-dependent Schrödinger equation.

3.4.6.1.1 Populations and coherences Start from the general expression (3.137) of the density operator of a mixed state. In the Schrödinger picture, the kets and bras are time dependent, so that at the difference of the other operators of quantum mechanics, the density operators must be time dependent, that is, when time is taken into account, Eq. (3.137) must read

$$\rho(t)^{\text{SP}} = \sum_i W_i |\Phi_i(t)\rangle\langle\Phi_i(t)| \tag{3.167}$$

where the W_i are time-dependent probabilities. Now, consider the eigenvalue equation of a Hermitian operator $\mathbf{A}$

$$\mathbf{A}|\Psi_n\rangle = A_n|\Psi_n\rangle$$

Next, consider a matrix element of the density operator in the basis $\{|\Psi_n\rangle\}$:

$$\rho_{nm}(t)^{\text{SP}} = \langle\Psi_n|\rho(t)^{\text{SP}}|\Psi_m\rangle \tag{3.168}$$

The time-dependent off-diagonal matrix elements of the density operator are known as *coherences*, whereas the diagonal corresponding ones are known as *populations*. Using Eq. (3.167) gives

$$\rho_{nm}(t)^{\text{SP}} = \sum_i W_i \langle\Psi_n|\ \Phi_i(t)\rangle\langle\Phi_i(t)|\Psi_m\rangle$$

This latter result may be also written for the coherences and for the populations, respectively

$$\rho_{nm}(t)^{\text{SP}} = \sum_i W_i C_{ni}(t) C_{im}(t)$$

$$\rho_{nn}(t)^{\text{SP}} = \sum_i W_i |C_{ni}(t)|^2$$

with

$$C_{ni}(t) = \langle\Psi_n|\ \Phi_i(t)\rangle$$

3.4.6.1.2 Liouville equation In order to get the time dependence of the density operator, first start from its expression (3.167) for a mixed state. Since the W_i are time independent, the partial derivative of Eq. (3.167) is

$$\left(\frac{\partial\rho(t)^{\text{SP}}}{\partial t}\right) = \sum_i W_i \left(\frac{\partial|\Phi_i(t)\rangle\langle\Phi_i(t)|}{\partial t}\right) \tag{3.169}$$

The time derivative of the right-hand side of this last equation is, of course,

$$\left(\frac{\partial|\Phi_i(t)\rangle\langle\Phi_i(t)|}{\partial t}\right) = \left(\frac{\partial|\Phi_i(t)\rangle}{\partial t}\right)\langle\Phi_i(t)| + |\Phi_i(t)\rangle\left(\frac{\partial\langle\Phi_i(t)|}{\partial t}\right)$$

Again, recall that thetime-dependent Schrödinger equation and its Hermitian conjugate are

$$i\hbar\left(\frac{\partial|\Phi_i(t)\rangle}{\partial t}\right) = \mathbf{H}|\Phi_i(t)\rangle \qquad \text{and} \qquad -i\hbar\left(\frac{\partial\langle\Phi_i(t)|}{\partial t}\right) = \langle\Phi_i(t)|\mathbf{H}$$

where **H** is the Hamiltonian. Thus, Eq. (3.169) becomes

$$i\hbar\left(\frac{\partial\rho(t)^{\text{SP}}}{\partial t}\right) = \sum_i W_i\mathbf{H}|\Phi_i(t)\rangle\langle\Phi_i(t)| - \sum_i W_i|\Phi_i(t)\rangle\langle\Phi_i(t)|\mathbf{H}$$

or, owing to Eq. (3.167) and since W_i commutes with **H**,

$$i\hbar\left(\frac{\partial\rho(t)^{\text{SP}}}{\partial t}\right) = \mathbf{H}\rho(t)^{\text{SP}} - \rho(t)^{\text{SP}}\mathbf{H}$$

so that

$$\boxed{i\hbar\left(\frac{\partial\rho(t)^{\text{SP}}}{\partial t}\right) = [\mathbf{H}, \rho(t)^{\text{SP}}]} \tag{3.170}$$

that is, the Liouville–Von Neumann equation also called the *Liouville equation* or the *Von Neumann equation.* Note the difference in the sign of the commutator when passing from this equation, which applies to density operator, to that of (3.94) dealing with the observables. The reason is that the density operator is not an observable but is constructed from projectors and thus from kets and bras. The sign difference between Eq. (3.170) governing the time dependence of the density operator and that of (3.95) giving the time dependence of some operators other than the density operator, in the Heisenberg picture, that is,

$$i\hbar\langle\Psi|\left(\frac{\partial\mathbf{A}(t)^{\text{HP}}}{\partial t}\right)|\Psi\rangle = \langle\Psi|[\mathbf{A}^{\text{HP}}(t), \mathbf{H}]|\Psi\rangle$$

3.4.6.1.3 Density operators in statistical equilibrium When an isolated system is not in statistical equilibrium, its total density operator changes with time:

$$\left(\frac{\partial\rho_{\text{Tot}}(t)^{\text{SP}}}{\partial t}\right) \neq 0$$

and will continue to change until the system has attained its statistical equilibrium:

$$\left(\frac{\partial\rho_{\text{Tot}}(t)^{\text{SP}}}{\partial t}\right) = 0$$

In this special situation, it results from Eq. (3.170) that, at equilibrium, it is necessary that

$$[\mathbf{H}, \rho_{\text{Tot}}(t)^{\text{SP}}] = 0$$

3.4.6.1.4 Energy representation of the density operator Owing to the appearance of the Hamiltonian **H** on the right-hand side of the Liouville–Von Neumann

Eq. (3.170), it may be of interest to consider the matrix representation of this equation on the basis of the eigenvectors of the Hamiltonian. Thus, write the eigenvalue equation of the Hamiltonian:

$$\mathbf{H}|\Psi_n\rangle = E_n|\Psi_n\rangle \tag{3.171}$$

Then, on the basis $\{|\Psi_n\rangle\}$, the matrix representation of the Liouville Eq. (3.170) takes the form

$$i\hbar\left(\frac{\partial\langle\Psi_n|\rho(t)^{\text{SP}}|\Psi_m\rangle}{\partial t}\right) = \langle\Psi_n|\mathbf{H}\;\rho(t)^{\text{SP}}|\Psi_m\rangle - \langle\Psi_n|\rho(t)^{\text{SP}}\mathbf{H}|\Psi_m\rangle$$

Due to the eigenvalue equation (3.171), this equation transforms to

$$i\hbar\left(\frac{\partial\langle\Psi_n|\rho(t)^{\text{SP}}|\Psi_m\rangle}{\partial t}\right) = E_n\langle\Psi_n|\rho(t)^{\text{SP}}|\Psi_m\rangle - E_m\langle\Psi_n|\rho(t)^{\text{SP}}|\Psi_m\rangle$$

which may also be written using the notations (3.168) for populations and coherences

$$i\hbar\left(\frac{\partial\rho_{nm}(t)^{\text{SP}}}{\partial t}\right) = (E_n - E_m)\rho_{nm}(t)^{\text{SP}} \tag{3.172}$$

Then, by integration of Eq. (3.172), one obtains

$$\rho_{nm}(t)^{\text{SP}} = \rho_{nm}(0)^{\text{SP}}e^{-i(E_n-E_m)t/\hbar} \tag{3.173}$$

Observe that it appears from Eq. (3.173) that the populations (corresponding to $n=m$) remain constant.

3.4.6.1.5 Canonical transformation on the density operator involving the Schrödinger evolution operator Consider the density operator at initial time $t=0$. Equation (3.167) reads

$$\rho(0)^{\text{SP}} = \sum_i W_i|\Phi_i(0)\rangle\langle\Phi_i(0)| \tag{3.174}$$

At time t, the W_i being constant, the SP density operator becomes

$$\rho(t)^{\text{SP}} = \sum_i |\Phi_i(t)^{\text{SP}}\rangle\langle\Phi_i(t)^{\text{SP}}| \tag{3.175}$$

In the time evolution operator formalism, the time dependence of the kets and of the corresponding bras is given by Eq. (3.77):

$$|\Phi_i(t)^{\text{SP}}\rangle = \mathbf{U}(t)|\Phi_i(0)^{\text{SP}}\rangle \qquad \text{and} \qquad \langle\Phi_i(t)^{\text{SP}}| = \langle\Phi_i(0)^{\text{SP}}|\mathbf{U}(t)^\dagger \tag{3.176}$$

where $\mathbf{U}(t)$ is the time evolution operator (3.82) governed by the Hamiltonian of the system, that is,

$$\mathbf{U}(t) = (e^{-i\mathbf{H}t/\hbar}) \qquad \text{and} \qquad \mathbf{U}(t)^\dagger = \mathbf{U}(t)^{-1} = (e^{i\mathbf{H}t/\hbar})$$

The time-dependent density operator is therefore

$$\rho(t)^{\text{SP}} = \sum_i \mathbf{U}(t)|\Phi_i(0)\rangle\langle\Phi_i(0)|\mathbf{U}(t)^\dagger$$

Hence, in view of Eq. (3.174)

$$\rho(t)^{\text{SP}} = \mathbf{U}(t)\rho(0)\mathbf{U}(t)^\dagger = \mathbf{U}(t)\rho(0)\mathbf{U}(t)^{-1} \tag{3.177}$$

or, writing explicitly the time evolution operator

$$\boxed{\rho(t)^{SP} = (e^{-i\mathbf{H}t/\hbar})\rho(0)^{SP}(e^{+i\mathbf{H}t/\hbar})} \tag{3.178}$$

Note that in the canonical transformation of (3.177) or (3.178), the signs have changed with respect to those appearing in the time dependence of the Heisenberg picture or observables that, according to Eqs. (3.89) and (3.88), are

$$\mathbf{A}(t)^{HP} = (e^{i\mathbf{H}t/\hbar})\mathbf{A}(e^{-i\mathbf{H}t/\hbar}) = \mathbf{U}(t)^{\dagger}\mathbf{AU}(t)$$

3.4.6.2 Interaction picture Liouville equation Suppose that the system that is studied involves a Hamiltonian $\mathbf{H}$ that may be split into an unperturbed part $\mathbf{H}^{\circ}$ and a perturbation $\mathbf{V}$, according to

$$\mathbf{H} = \mathbf{H}^{\circ} + \mathbf{V}$$

Due to the partition of the Hamiltonian, the Liouville–Von Neumann equation (3.170) takes the form

$$i\hbar\left(\frac{\partial\rho(t)^{SP}}{\partial t}\right) = [\mathbf{H}^{\circ}, \rho(t)^{SP}] + [\mathbf{V}, \rho(t)^{SP}] \tag{3.179}$$

Next, keeping in mind that the SP density operator at time t is given by Eq. (3.175),

$$\rho(t)^{SP} = \sum_i W_i|\Phi_i(t)^{SP}\rangle\langle\Phi_i(t)^{SP}|$$

and since the W_i are constant, it is clear that the corresponding IP density operator is given by

$$\rho(t)^{IP} = \sum_i W_i|\Phi_i(t)^{IP}\rangle\langle\Phi_i(t)^{IP}| \tag{3.180}$$

whereas Eq. (3.118) relating the IP and SP kets is

$$|\Psi(t)^{IP}\rangle = \mathbf{U}^{\circ}(t)^{-1}|\Psi(t)^{SP}\rangle \tag{3.181}$$

where

$$\mathbf{U}^{\circ}(t) = (e^{-i\mathbf{H}^{\circ}t/\hbar}) \tag{3.182}$$

Hence, due to Eq. (3.181) and to its Hermitian conjugate, the IP density operator (3.180) reads

$$\boxed{\rho(t)^{IP} \equiv \mathbf{U}^{\circ}(t)^{-1}\rho(t)^{SP}\mathbf{U}^{\circ}(t)} \tag{3.183}$$

Then, premultiplying this equation by $\mathbf{U}^{\circ}(t)$ and postmultiplying it by its inverse, leads to

$$\mathbf{U}^{\circ}(t)\rho(t)^{IP}\mathbf{U}^{\circ}(t)^{-1} = \mathbf{U}^{\circ}(t)\mathbf{U}^{\circ}(t)^{-1}\rho(t)^{SP}\mathbf{U}^{\circ}(t)\mathbf{U}^{\circ}(t)^{-1}$$

so that, on simplification of the right-hand side, one obtains the relation inverse to (3.183), that is,

$$\boxed{\rho(t)^{SP} = \mathbf{U}^{\circ}(t)\rho(t)^{IP}\mathbf{U}^{\circ}(t)^{-1}} \tag{3.184}$$

Moreover, due to Eq. (3.184), Eq. (3.179) yields

$$i\hbar\left(\frac{\partial \boldsymbol{\rho}(t)^{\mathrm{SP}}}{\partial t}\right) = [\mathbf{H}^\circ, \mathbf{U}^\circ(t)\boldsymbol{\rho}(t)^{\mathrm{IP}}\mathbf{U}^\circ(t)^{-1}] + [\mathbf{V}, \mathbf{U}^\circ(t)\boldsymbol{\rho}(t)^{\mathrm{IP}}\mathbf{U}^\circ(t)^{-1}] \tag{3.185}$$

On the other hand, the partial time derivative of Eq. (3.184), reads

$$\left(\frac{\partial \boldsymbol{\rho}(t)^{\mathrm{SP}}}{\partial t}\right) = \left(\frac{\partial \mathbf{U}^\circ(t)}{\partial t}\right)\boldsymbol{\rho}(t)^{\mathrm{IP}}\mathbf{U}^\circ(t)^{-1} + \mathbf{U}^\circ(t)\left(\frac{\partial \boldsymbol{\rho}(t)^{\mathrm{IP}}}{\partial t}\right)\mathbf{U}^\circ(t)^{-1} + \mathbf{U}^\circ(t)\boldsymbol{\rho}(t)^{\mathrm{IP}}\left(\frac{\partial \mathbf{U}^\circ(t)^{-1}}{\partial t}\right) \tag{3.186}$$

Then, by identification of Eqs. (3.185) and (3.186), one obtains

$$\begin{aligned}&[\mathbf{H}^\circ, \mathbf{U}^\circ(t)\boldsymbol{\rho}(t)^{\mathrm{IP}}\mathbf{U}^\circ(t)^{-1}] + [\mathbf{V}, \mathbf{U}^\circ(t)\boldsymbol{\rho}(t)^{\mathrm{IP}}\mathbf{U}^\circ(t)^{-1}] \\ &= i\hbar\left\{\left(\frac{\partial \mathbf{U}^\circ(t)}{\partial t}\right)\boldsymbol{\rho}(t)^{\mathrm{IP}}\mathbf{U}^\circ(t)^{-1} + \mathbf{U}^\circ(t)\left(\frac{\partial \boldsymbol{\rho}(t)^{\mathrm{IP}}}{\partial t}\right)\mathbf{U}^\circ(t)^{-1} \right. \\ &\left. + \mathbf{U}^\circ(t)\boldsymbol{\rho}(t)^{\mathrm{IP}}\left(\frac{\partial \mathbf{U}^\circ(t)^{-1}}{\partial t}\right)\right\}\end{aligned} \tag{3.187}$$

Moreover, observe that, according to Eq. (3.81), and since $\mathbf{H}^\circ$ is Hermitian, the Schrödinger equation governing the dynamics of the unitary evolution operator $\mathbf{U}^\circ(t)$ and its Hermitian conjugate is

$$i\hbar\left(\frac{\partial \mathbf{U}^\circ(t)}{\partial t}\right) = \mathbf{H}^\circ\mathbf{U}^\circ(t) \qquad \text{and} \qquad -i\hbar\left(\frac{\partial \mathbf{U}^\circ(t)^{-1}}{\partial t}\right) = \mathbf{U}^\circ(t)^{-1}\mathbf{H}^\circ$$

These equations allow one to write the first and third right-hand-side terms of Eq. (3.187) according to

$$i\hbar\left(\frac{\partial \mathbf{U}^\circ(t)}{\partial t}\right)\boldsymbol{\rho}(t)^{\mathrm{IP}}\mathbf{U}^\circ(t)^{-1} = \mathbf{H}^\circ\mathbf{U}^\circ(t)\boldsymbol{\rho}(t)^{\mathrm{IP}}\mathbf{U}^{\circ -1}(t)$$

$$i\hbar\mathbf{U}^\circ(t)\boldsymbol{\rho}(t)^{\mathrm{IP}}\left(\frac{\partial \mathbf{U}^\circ(t)^{-1}}{\partial t}\right) = -\mathbf{U}^\circ(t)\boldsymbol{\rho}(t)^{\mathrm{IP}}\mathbf{U}^\circ(t)^{-1}\mathbf{H}^\circ$$

Hence, the sum of these two terms appearing in Eq. (3.187) reads

$$i\hbar\left\{\left(\frac{\partial \mathbf{U}^\circ(t)}{\partial t}\right)\boldsymbol{\rho}(t)^{\mathrm{IP}}\mathbf{U}^\circ(t)^{-1} + \mathbf{U}^\circ(t)\boldsymbol{\rho}(t)^{\mathrm{IP}}\left(\frac{\partial \mathbf{U}^\circ(t)^{-1}}{\partial t}\right)\right\} = [\mathbf{H}^\circ, \mathbf{U}^\circ(t)\boldsymbol{\rho}(t)^{\mathrm{IP}}\mathbf{U}^\circ(t)^{-1}] \tag{3.188}$$

Hence, the left-hand side of Eq. (3.188) is equivalent to the first and third right-hand terms of Eq. (3.187), whereas the right-hand term of Eq. (3.188) is the same as the first commutator appearing on the left-hand side of Eq. (3.187). As a consequence, Eq. (3.187) simplifies to

$$i\hbar\mathbf{U}^\circ(t)\left(\frac{\partial \boldsymbol{\rho}(t)^{\mathrm{IP}}}{\partial t}\right)\mathbf{U}^\circ(t)^{-1} = [\mathbf{V}, \mathbf{U}^\circ(t)\boldsymbol{\rho}(t)^{\mathrm{IP}}\mathbf{U}^\circ(t)^{-1}] \tag{3.189}$$

On the other hand, Eq. (3.179) may be transformed using Eq. (3.184) to

$$i\hbar\left(\frac{\partial \boldsymbol{\rho}(t)^{\mathrm{SP}}}{\partial t}\right) = [\mathbf{H}^\circ, \boldsymbol{\rho}(t)^{\mathrm{SP}}] + [\mathbf{V}, \mathbf{U}^\circ(t)\boldsymbol{\rho}(t)^{\mathrm{IP}}\mathbf{U}^\circ(t)^{-1}]$$

which, owing to Eq. (3.189), yields

$$\left(\frac{\partial \boldsymbol{\rho}(t)^{\mathrm{SP}}}{\partial t}\right) = \frac{1}{i\hbar}[\mathbf{H}^\circ, \boldsymbol{\rho}(t)^{\mathrm{SP}}] + \mathbf{U}^\circ(t)\left(\frac{\partial \boldsymbol{\rho}(t)^{\mathrm{IP}}}{\partial t}\right)\mathbf{U}^\circ(t)^{-1} \tag{3.190}$$

On the other hand, writing explicitly the right-hand commutator of Eq. (3.189) gives

$$i\hbar\mathbf{U}^\circ(t)\left(\frac{\partial \boldsymbol{\rho}(t)^{\mathrm{IP}}}{\partial t}\right)\mathbf{U}^\circ(t)^{-1} = \mathbf{V}\mathbf{U}^\circ(t)\boldsymbol{\rho}(t)^{\mathrm{IP}}\mathbf{U}^\circ(t)^{-1} - \mathbf{U}^\circ(t)\boldsymbol{\rho}(t)^{\mathrm{IP}}\mathbf{U}^\circ(t)^{-1}\mathbf{V}$$

Then, postmultiplying both members of this last equation by $\mathbf{U}^\circ$ and premultiplying them by its inverse, allows us to write

$$i\hbar\mathbf{U}^\circ(t)^{-1}\left(\mathbf{U}^\circ(t)\left(\frac{\partial \boldsymbol{\rho}(t)^{\mathrm{IP}}}{\partial t}\right)\mathbf{U}^\circ(t)^{-1}\right)\mathbf{U}^\circ(t)$$
$$= \mathbf{U}^\circ(t)^{-1}(\mathbf{V}\mathbf{U}^\circ(t)\boldsymbol{\rho}(t)^{\mathrm{IP}}\mathbf{U}^\circ(t)^{-1})\mathbf{U}^\circ(t) - \mathbf{U}^\circ(t)^{-1}(\mathbf{U}^\circ(t)\boldsymbol{\rho}(t)^{\mathrm{IP}}\mathbf{U}^\circ(t)^{-1}\mathbf{V})\,\mathbf{U}^\circ(t)$$

or, on simplification

$$i\hbar\left(\frac{\partial \boldsymbol{\rho}(t)^{\mathrm{IP}}}{\partial t}\right) = \mathbf{U}^\circ(t)^{-1}\mathbf{V}\mathbf{U}^\circ(t)\boldsymbol{\rho}(t)^{\mathrm{IP}} - \boldsymbol{\rho}(t)^{\mathrm{IP}}\mathbf{U}^\circ(t)^{-1}\mathbf{V}\mathbf{U}^\circ(t)$$

a result that may be written

$$i\hbar\left(\frac{\partial \boldsymbol{\rho}(t)^{\mathrm{IP}}}{\partial t}\right) = \mathbf{V}(t)^{\mathrm{IP}}\boldsymbol{\rho}(t)^{\mathrm{IP}} - \boldsymbol{\rho}(t)^{\mathrm{IP}}\mathbf{V}(t)^{\mathrm{IP}} \tag{3.191}$$

where $\mathbf{V}^{\mathrm{IP}}(t)$ is given, in agreement to Eq. (3.88), by

$$\mathbf{V}(t)^{\mathrm{IP}} = \mathbf{U}^\circ(t)^{-1}\mathbf{V}\mathbf{U}^\circ(t) \tag{3.192}$$

Finally, Eq. (3.191) may be expressed in terms of a commutator to give

$$i\hbar\left(\frac{\partial \boldsymbol{\rho}(t)^{\mathrm{IP}}}{\partial t}\right) = [\mathbf{V}(t)^{\mathrm{IP}}, \boldsymbol{\rho}(t)^{\mathrm{IP}}] \tag{3.193}$$

that is, the *IP Liouville–Von Neumann equation* governing the IP density operator, which involves the same sign for the Hamiltonian and density operator commutator as that appearing in the corresponding SP Liouville equation (3.170).

3.4.6.3 Integration of the IP Liouville Equation Formal integration of the IP Liouville–Von Neumann equation from t_0 to t leads to the following integral equation:

$$\boldsymbol{\rho}(t)^{\mathrm{IP}} = \boldsymbol{\rho}(t_0)^{\mathrm{IP}} + \left(\frac{1}{i\hbar}\right)\int_{t_0}^{t} [\mathbf{V}(t'-t_0)^{\mathrm{IP}}, \boldsymbol{\rho}(t'-t_0)^{\mathrm{IP}}]\,dt' \tag{3.194}$$

with, due to Eqs. (3.192) and (3.182),

$$\mathbf{V}(t'-t_0)^{\mathrm{IP}} = e^{i\mathbf{H}^\circ(t'-t_0)/\hbar}\mathbf{V}e^{-i\mathbf{H}^\circ(t'-t_0)/\hbar} \tag{3.195}$$

If the potential $\mathbf{V}$ is small with respect to $\mathbf{H}^\circ$, the integral equation (3.194) may be solved by successive approximations. For this purpose, observe that the time-dependent IP time evolution density operator involved in the commutator appearing on the right-hand side of Eq. (3.194) may be found by using an equation similar to Eq. (3.194), that is,

$$\boldsymbol{\rho}(t'-t_0)^{\mathrm{IP}} = \boldsymbol{\rho}(t_0)^{\mathrm{IP}} + \left(\frac{1}{i\hbar}\right)\int_{t_0}^{t'-t_0} [\mathbf{V}(t''-t_0)^{\mathrm{IP}}, \boldsymbol{\rho}(t''-t_0)^{\mathrm{IP}}]\, dt''$$

so that Eq. (3.194) becomes

$$\begin{aligned}&[\mathbf{V}(t'-t_0)^{\mathrm{IP}}, \boldsymbol{\rho}(t'-t_0)^{\mathrm{IP}}]\\ &= [\mathbf{V}(t'-t_0)^{\mathrm{IP}}, \boldsymbol{\rho}(t_0)^{\mathrm{IP}}] + \left(\frac{1}{i\hbar}\right)\int_0^{t} [\mathbf{V}(t'-t_0)^{\mathrm{IP}}, [\mathbf{V}(t''-t_0)^{\mathrm{IP}}, \boldsymbol{\rho}(t''-t_0)^{\mathrm{IP}}]]\, dt''\end{aligned}$$

Then, inserting this expression into Eq. (3.194) yields

$$\begin{aligned}\boldsymbol{\rho}(t)^{\mathrm{IP}} &= \boldsymbol{\rho}(t_0)^{\mathrm{IP}} + \left(\frac{1}{i\hbar}\right)\int_{t_0}^{t} [\mathbf{V}(t'-t_0)^{\mathrm{IP}}, \boldsymbol{\rho}(t_0)^{\mathrm{IP}}]\, dt'\\ &+ \left(\frac{1}{i\hbar}\right)^2 \int_{t_0}^{t}\int_{t_0}^{t'} [\mathbf{V}(t'-t_0)^{\mathrm{IP}}, [\mathbf{V}(t''-t_0)^{\mathrm{IP}}, \boldsymbol{\rho}(t''-t_0)^{\mathrm{IP}}]]\, dt'\,dt''\end{aligned}$$

Again, by iteration, one obtains

$$\begin{aligned}\boldsymbol{\rho}(t)^{\mathrm{IP}} &= \boldsymbol{\rho}(t_0)^{\mathrm{IP}} + \frac{1}{i\hbar}\int_{t_0}^{t} [\mathbf{V}(t'-t_0)^{\mathrm{IP}}, \boldsymbol{\rho}(t_0)^{\mathrm{IP}}]\, dt'\\ &+ \left(\frac{1}{i\hbar}\right)^2 \int_{t_0}^{t}\int_{t_0}^{t'} [\mathbf{V}(t'-t_0)^{\mathrm{IP}}, [\mathbf{V}(t''-t_0)^{\mathrm{IP}}, \boldsymbol{\rho}(t_0)^{\mathrm{IP}}]]\, dt'\,dt''\\ &+ \left(\frac{1}{i\hbar}\right)^3 \int_{t_0}^{t}\int_{t_0}^{t'}\int_{t_0}^{t''} [\mathbf{V}(t'-t_0)^{\mathrm{IP}}, [\mathbf{V}(t''-t_0)^{\mathrm{IP}}, [\mathbf{V}(t'''-t_0)^{\mathrm{IP}},\\ &\qquad \boldsymbol{\rho}(t'''-t_0)^{\mathrm{IP}}]]]\, dt'\,dt''\,dt'''\end{aligned} \tag{3.196}$$

This operation may be repeated any number of times. However, if the perturbation **V** is weak, the treatment may be limited to the second order in the IP perturbation operator so that Eq. (3.196) becomes truncated at the second order in the perturbation according to

$$\begin{aligned}\boldsymbol{\rho}(t)^{\mathrm{IP}} &\simeq \boldsymbol{\rho}(t_0)^{\mathrm{IP}} + \frac{1}{i\hbar}\int_{t_0}^{t} [\mathbf{V}(t'-t_0)^{\mathrm{IP}}, \boldsymbol{\rho}(t_0)^{\mathrm{IP}}]\, dt'\\ &+ \left(\frac{1}{i\hbar}\right)^2 \int_{t_0}^{t}\int_{t_0}^{t'} [\mathbf{V}(t'-t_0)^{\mathrm{IP}}, [\mathbf{V}(t''-t_0)^{\mathrm{IP}}, \boldsymbol{\rho}(t_0)^{\mathrm{IP}}]]\, dt'\,dt''\end{aligned} \tag{3.197}$$

Recall that the IP perturbation operator is given by Eq. (3.192) and that, when the expression of the IP density operator has been obtained with the help of Eq. (3.197), one may retrieve the time-dependent density operator using Eq. (3.184).

3.5 CONCLUSION

This chapter, which was devoted to different representations of quantum mechanics, has lead to important useful developments. (i) Matrix representation allowing one to replace the eigenvalue equation of an Hermitian operator to be solved by a corresponding matrix eigenvalue equation susceptible to be easily numerically solved, which is of great interest for the study of quantum anharmonic oscillators. (ii) Wave mechanics, which is the representation of quantum mechanics in geometrical space, is in many situations such as atoms or molecules more tractable than matrix or quantum mechanics. Although there is less interest in quantum oscillators than matrix mechanics, it will remain of some interest in visualizing some results dealing with these oscillators. (iii) Density operator approaches are very powerful when studying many-particle systems, particularly for statistical equilibrium situations leading to thermal equilibrium, and will be widely used when studying thermal properties of quantum oscillators. (iv) Time-dependent representations other than the Schrödinger picture where the time dependence resides in the quantum states, which constitute the Heisenberg picture where the time dependence is contained in the Hermitian operators, and the interaction picture, which is a description intermediate between the Schrödinger and Heisenberg pictures and which will be very useful when studying the irreversible dynamics of quantum oscillators coupled to a thermal bath.

The important results concerning the time-dependent Schrödinger, Heisenberg, and interaction pictures are collected into the two following lists:

Schrödinger and Heisenberg pictures
Schrödinger equation and evolution operator:
$i\hbar\frac{\partial}{\partial t}\|\Psi(t)^{SP}\rangle = \mathbf{H}\|\Psi(t)^{SP}\rangle$
Time-dependent ket in the Schrödinger picture:
$\|\Psi(t)^{SP}\rangle = \mathbf{U}(t)^{SP}\|\Psi(0)^{SP}\rangle$
Time-dependent evolution operator in the Schrödinger picture:
$\mathbf{U}(t)^{SP} = e^{-i\mathbf{H}t/\hbar}$
Time-dependent operators in the Heisenberg picture:
$\mathbf{A}(t)^{HP} = \mathbf{U}(t)^{SP-1}\mathbf{A}(0)\mathbf{U}(t)^{SP}$

Dynamic equation governing evolution operators in the Schrödinger picture:
$i\hbar\left(\dfrac{\partial \mathbf{U}(t)^{\mathrm{SP}}}{\partial t}\right) = \mathbf{H}\mathbf{U}(t)^{\mathrm{SP}}$
Dynamic equation governing operators in the Heisenberg picture:
$i\hbar\left(\dfrac{\partial \mathbf{A}(t)^{\mathrm{HP}}}{\partial t}\right) = [\mathbf{A}(t)^{\mathrm{HP}}, \mathbf{H}]$

Interaction picture
Hamiltonian partition and corresponding evolution operators:
$\mathbf{H} = \mathbf{H}^{\circ} + \mathbf{V}$ $\mathbf{U}^{\circ}(t) = e^{-i\mathbf{H}^{\circ}t/\hbar}$ and $\mathbf{U}(t) = e^{-i\mathbf{H}t/\hbar}$
Relation between IP and SP evolution operators:
$\mathbf{U}(t)^{\mathrm{SP}} = \mathbf{U}^{\circ}(t)\mathbf{U}(t)^{\mathrm{IP}}$
Time-dependent operators A in the interaction picture:
$\mathbf{A}(t)^{\mathrm{IP}} = \mathbf{U}^{\circ}(t)^{-1}\mathbf{A}\mathbf{U}^{\circ}(t)$
Dynamic equation governing the interaction picture evolution operator:
$i\hbar\left(\dfrac{\partial \mathbf{U}(t)^{\mathrm{IP}}}{\partial t}\right) = \mathbf{V}(t)^{\mathrm{IP}}\mathbf{U}(t)^{\mathrm{IP}}$
Connection between SP and IP time-dependent kets:
$\lvert\Psi(t)^{\mathrm{SP}}\rangle = \mathbf{U}^{\circ}(t)\lvert\Psi(t)^{\mathrm{IP}}\rangle$

Those dealing with density operators are given as follows:

Density operators
Definition of density operators:
$\rho = \sum_i W_i \lvert\Phi_i\rangle\langle\Phi_i\rvert$
Average values performed over density operators:
$\langle\mathbf{A}\rangle_\rho = \mathbf{tr}\{\rho\mathbf{A}\}$

Liouville equation in the Schrödinger picture:
$i\hbar\left(\dfrac{\partial\rho(t)^{\mathrm{SP}}}{\partial t}\right) = [\mathbf{H}, \rho(t)^{\mathrm{SP}}]$
Statistical entropy:
$S = -k_B \mathbf{tr}\{\rho \ln \rho\}$

BIBLIOGRAPHY

C. Cohen-Tannoudji, B. Diu, and F. Laloe. *Quantum Mechanics*. Wiley-Interscience: Hoboken, NJ, 2006.

A. S. Davydov. *Quantum Mechanics*, 2nd ed. Pergamon Press: Oxford, 1976.

P. A. M. Dirac. *The Principles of Quantum Mechanics*, 4th ed. Oxford University Press: 1982.

W. H. Louisell. *Quantum Statistical Properties of Radiation*. Wiley: New York, 1973.

A. Messiah. *Quantum Mechanics*. Dover Publications: New York, 1999.

L. I. Schiff. *Quantum Mechanics*, 3rd ed. McGraw-Hill: New York, 1968.

CHAPTER 4

SIMPLE MODELS USEFUL FOR QUANTUM OSCILLATOR PHYSICS

INTRODUCTION

Before studing quantum oscillators, which is the principal aim of the present book, it may be useful to apply the information of the previous chapters dealing with the basis of quantum mechanics to three simple models that will be of interest in the future. The first one is the *particle-in-a-box model*, which comprises a single particle enclosed in a box where the potential is zero, this same potential being infinite beyond the box walls. Applying simply the wave mechanics, we shall get quantized energy levels and their associated wavefunctions, the node number of which increases with the energy. It will also be useful to illustrate the quantization of the energy levels and the decrease of the associated wavelength when the energy rises, two concepts we shall meet when discussing the oscillators. The second model to which the present chapter is devoted deals with the interaction between two energy levels, which will be of interest when focusing attention on the local interaction between two excited states of two different oscillators, a situation that occurs in the area of *Fermi resonances*. Finally, the last section treats the probability for a system to pass from one of its stationary energy levels to another if a potential perturbs it. Using a formalism that will later be applied to the interaction of oscillators with the electromagnetic field, it will lead to the important *Fermi golden rule*.

4.1 PARTICLE-IN-A-BOX MODEL

Consider a particle of mass m enclosed in a box of volume $\mathbb{V}$ given by

$$\mathbb{V} = a_x a_y a_z \tag{4.1}$$

dimensions in which the potential is zero while it is infinite outside. Its kinetic energy is

$$\mathbf{T} = \frac{1}{2m}(\mathbf{P}_x^2 + \mathbf{P}_y^2 + \mathbf{P}_z^2)$$

Quantum Oscillators, First Edition. Olivier Henri-Rousseau and Paul Blaise.
© 2011 John Wiley & Sons, Inc. Published 2011 by John Wiley & Sons, Inc.

where the $\mathbf{P}_i^2$ are the components x, y, and z of the momentum. In the wave mechanics representation, the momentum operator obeys Eq. (3.51):

$$P = -i\hbar\frac{\partial}{\partial Q}$$

Thus, the wave mechanics description of the particle-in-a-box kinetic operator reads

$$T = -\frac{\hbar^2}{2m}\left(\frac{\partial^2}{\partial x^2} + \frac{\partial^2}{\partial y^2} + \frac{\partial^2}{\partial z^2}\right)$$

where x, y, and z are the components of Q. Furthermore, since the potential is assumed to be zero inside the box, the wave mechanics description of the potential is simply

$$V = 0$$

The Hamiltonian of the system, which is the sum of the kinetic and potential operators, is in the position representation

$$H = -\frac{\hbar^2}{2m}\left(\frac{\partial^2}{\partial x^2} + \frac{\partial^2}{\partial y^2} + \frac{\partial^2}{\partial z^2}\right)$$

Now, for this three-dimensional (3D) model, the wavefunction of the particle can be written as the product of the wavefunctions along the three independent dimensions, that is,

$$\Psi(x, y, z) = \Psi(x)\Psi(y)\Psi(z) \tag{4.2}$$

Hence, the eigenvalue equation of the Hamiltonian, that is, the time-independent Schrödinger equation, takes the form

$$\begin{aligned} -\frac{\hbar^2}{2m}&\left(\Psi(y)\Psi(z)\frac{\partial^2\Psi(x)}{\partial x^2} + \Psi(x)\Psi(z)\frac{\partial^2\Psi(y)}{\partial y^2} + \Psi(x)\Psi(y)\frac{\partial^2\Psi(z)}{\partial z^2}\right) \\ &= E\Psi(x)\Psi(y)\Psi(z) \end{aligned} \tag{4.3}$$

where E is the Hamiltonian eigenvalue.

4.1.1 Solving the 3D Schrödinger equation

Now, the Hamiltonian eigenvalue E may be written as the sum of the energies along the three dimensions, that is,

$$E = E_x + E_y + E_z \tag{4.4}$$

Thus, the Schrödinger equation (4.3) splits into three independent and equivalent Schrödinger equations corresponding to the three dimensions of the geometrical space, according to

$$\left(\frac{\partial^2\Psi(x)}{\partial x^2}\right) = -k_x^2\Psi(x) \tag{4.5}$$

$$\left(\frac{\partial^2\Psi(y)}{\partial y^2}\right) = -k_y^2\Psi(y) \tag{4.6}$$

$$\left(\frac{\partial^2 \Psi(z)}{\partial z^2}\right) = -k_z^2 \Psi(z) \tag{4.7}$$

with

$$k_x^2 = \frac{2m}{\hbar^2} E_x \tag{4.8}$$

$$k_y^2 = \frac{2m}{\hbar^2} E_y \tag{4.9}$$

$$k_z^2 = \frac{2m}{\hbar^2} E_z \tag{4.10}$$

Observe that since outside the box the potential is infinite, it is impossible to the particle to get out of the box (Fig. 4.1). Therefore, the probability for the particle to be outside the box is zero, and, thus, since this probability is the squared modulus of the wavefunction, the three wavefunctions satisfy the following boundary conditions, for example, leading for the x component to

$$\Psi(x) = 0 \quad \text{if} \quad -\infty < x \leqslant 0 \quad \text{and} \quad a_x \leqslant x < \infty \tag{4.11}$$

$$\Psi(y) = 0 \quad \text{if} \quad -\infty < y \leqslant 0 \quad \text{and} \quad a_y \leqslant y < \infty \tag{4.12}$$

$$\Psi(z) = 0 \quad \text{if} \quad -\infty < z \leqslant 0 \quad \text{and} \quad a_z \leqslant z < \infty \tag{4.13}$$

It appears, therefore, that the partial differential equations (4.5)–(4.7) to be solved are subject to the boundary conditions (4.11)–(4.13). The general solution of Eq. (4.5) is of the form

$$\Psi(x) = A_x \sin(k_x x) + B_x \cos(k_x x)$$

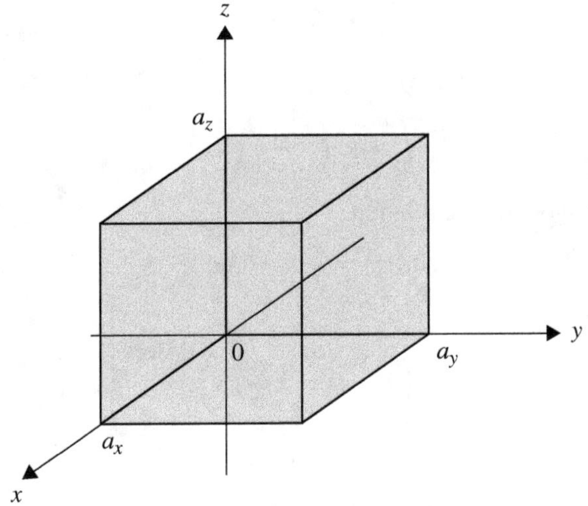

Figure 4.1 Particle-in-a-box model.

where A_x and B_x are two constants. Next, owing to the boundary condition (4.11) implying that, at $x = 0$, $\Psi(0) = 0$, it follows that $B_x = 0$, so that

$$\Psi(x) = A_x \sin(k_x x) \tag{4.14}$$

Now, the same boundary condition (4.11) implying that $x = a$, leads to write $\Psi(a) = 0$ so that, since $A \neq 0$ Eq. (4.14), it is necessary that

$$\sin(k_x a_x) = 0$$

such a condition being verified if

$$k_x a_x = n_x \pi \tag{4.15}$$

where n_x is a number that may take a priori all the integer values between 0 and ∞. That leads one to write the solution (4.14) as

$$\Psi_{n_x}(x) = A_x \sin\left(\frac{n_x \pi}{a_x} x\right) \tag{4.16}$$

In a similar way, one can obtain for the solutions of Eqs. (4.6) and (4.7) subject, respectively, to the boundary conditions (4.12) and (4.13)

$$\Psi_{n_y}(y) = A_y \sin\left(\frac{n_y \pi}{a_y} y\right) \tag{4.17}$$

$$\Psi_{n_z}(z) = A_z \sin\left(\frac{n_z \pi}{a_z} z\right) \tag{4.18}$$

with

$$k_y a_y = n_y \pi \qquad \text{and} \qquad k_z a_z = n_z \pi \tag{4.19}$$

The normalization condition of a wavefunction, which is

$$\int_0^a \Psi_{n_x}(x)^2 \, dx = 1$$

reads for the wavefunction (4.16)

$$A_x^2 \int_0^a \sin^2\left(\frac{n_x \pi}{a_x} x\right) dx = 1 \tag{4.20}$$

Next, using the trigonometric relation

$$\sin^2(z) = \tfrac{1}{2}(1 - \cos 2z)$$

Eq. (4.20) yields

$$A_x^2 \int_0^a \tfrac{1}{2}\left(1 - \cos\left(2\frac{n_x \pi}{a_x} x\right)\right) dx = 1 \tag{4.21}$$

Moreover, due to the fact that

$$\int_0^a \cos\left(2\frac{n_x \pi}{a_x} x\right) dx = 0$$

Eq. (4.21) reduces to

$$\tfrac{1}{2}A_x^2 a_x = 1$$

so that Eq. (4.16) becomes

$$\Psi_{n_x}(x) = \sqrt{\frac{2}{a_x}} \sin\left(\frac{n_x \pi}{a_x} x\right) \tag{4.22}$$

In a similar way, one would obtain for the wavefunctions (4.17) and (4.18),

$$\Psi_{n_y}(y) = \sqrt{\frac{2}{a_y}} \sin\left(\frac{n_y \pi}{a_y} y\right) \tag{4.23}$$

$$\Psi_{n_z}(z) = \sqrt{\frac{2}{a_z}} \sin\left(\frac{n_z \pi}{a_z} z\right) \tag{4.24}$$

4.1.2 3D Wavefunctions and energy levels

Due to Eq. (4.2), and to Eqs. (4.22)–(4.24), the total wavefunction reads

$$\Psi_{n_x,n_y,n_z}(x,y,z) = \frac{2^{3/2}}{\sqrt{\mathbb{V}}} \sin\left(\frac{n_x \pi}{a_x} x\right) \sin\left(\frac{n_y \pi}{a_y} y\right) \sin\left(\frac{n_z \pi}{a_z} z\right) \tag{4.25}$$

where Eq. (4.1) has been used, relating the dimensions a_x, a_y, and a_z of the box to its volume $\mathbb{V}$. Of course, the solutions corresponding to $n_x = 0$, $n_y = 0$, or $n_z = 0$ are without physical meaning since it would imply erroneously that the wavefunction (4.25) and thus the probability of the particle in all the box would be zero. Thus, all the quantum numbers n_x, n_y, and n_z must be integers, starting from unity.

The wavefunction (4.25) appears to be a product of stationary wavefunctions of the form

$$\Psi_{n_x,n_y,n_z}(x,y,z) = \Psi_{n_x}(x)\Psi_{n_y}(y)\Psi_{n_z}(z)$$

with

$$\Psi_{n_x}(x) = \sqrt{\frac{2}{a_x}} \sin\left(\frac{2\pi}{\lambda_{n_x}} x\right)$$

$$\Psi_{n_y}(y) = \sqrt{\frac{2}{a_y}} \sin\left(\frac{2\pi}{\lambda_{n_y}} y\right)$$

$$\Psi_{n_z}(z) = \sqrt{\frac{2}{a_z}} \sin\left(\frac{2\pi}{\lambda_{n_z}} z\right)$$

and where the λ_{n_x}, λ_{n_y}, and λ_{n_z} are wavelengths obeying

$$\lambda_{n_x} = \left(\frac{2a_x}{n_x}\right) \qquad \lambda_{n_y} = \left(\frac{2a_y}{n_y}\right) \qquad \lambda_{n_z} = \left(\frac{2a_z}{n_z}\right) \tag{4.26}$$

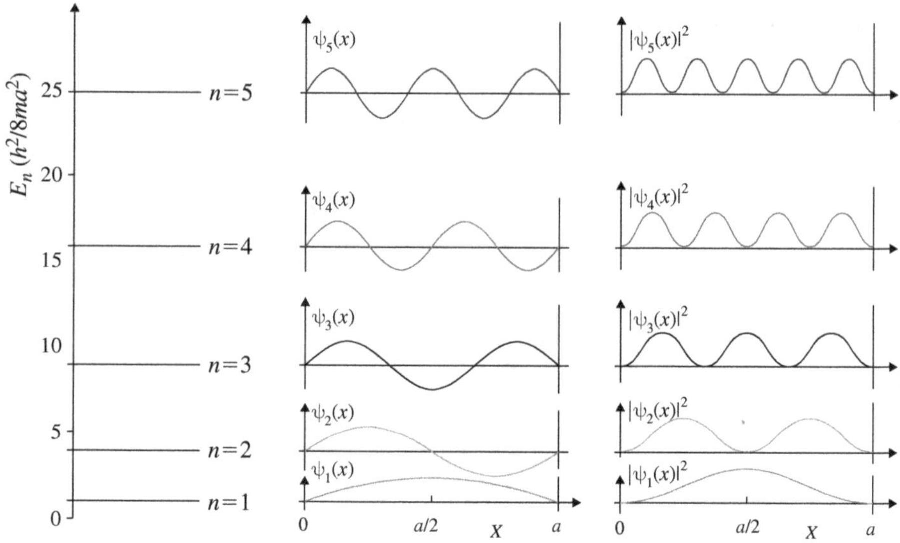

Figure 4.2 One-dimensional particle-in-a-box model. Energy levels and corresponding wavefunctions and probability densities for the four lowest quantum numbers.

In addition, Eqs. (4.8)–(4.10) combined with (4.15)–(4.19) allow us to get following results, with the energies corresponding to the x, y, and z components:

$$E_{n_x} = \left(\frac{\pi^2\hbar^2}{2ma_x^2}\right) n_x^2 \qquad E_{n_y} = \left(\frac{\pi^2\hbar^2}{2ma_y^2}\right) n_y^2 \qquad E_{n_z} = \left(\frac{\pi^2\hbar^2}{2ma_z^2}\right) n_z^2 \tag{4.27}$$

Hence, after passing from $\hbar$ to h, the total energy (4.4) becomes

$$E_{n_x,n_y,n_z} = \frac{h^2}{8m}\left(\frac{n_x^2}{a_x^2} + \frac{n_y^2}{a_y^2} + \frac{n_z^2}{a_z^2}\right) \tag{4.28}$$

It must be emphasized that, since n_x, n_z, and n_z are integers, the energy levels (4.28) are quantized, a result that will be also found later for quantum harmonic oscillators. Figure 4.2, which deals with the x component of the 3D model, gives the dimensionless energy levels and the corresponding wavefunctions for the four lowest quantum number n_x.

Hence, the nodes of the wavefunction are increasing with the quantum number and the corresponding energy level, a situation that will be met later for quantum harmonic oscillators and that is related to the *de Broglie wavelength*, we shall consider some later.

Moreover, when the box is cubic, that is, when

$$a_x = a_y = a_z = a$$

and due to Eq. (4.1), the equation (4.28) giving the energy levels simplifies to

$$E_{n_x,n_y,n_z} = \frac{h^2}{8m}\frac{(n_x^2 + n_y^2 + n_z^2)}{\mathbb{V}^{2/3}} \tag{4.29}$$

Then, in terms of the energy units E°

$$E^\circ = \frac{h^2}{8m\mathbb{V}^{2/3}}$$

and for the two lowest values 1 and 2 of the quantum numbers n_x, n_y, and n_z, the lowest energy levels (4.29) appear to be those given in the tabular expression (4.30):

n_x	n_y	n_z	E_{n_x,n_y,n_z}
1	1	1	1
1	1	2	6
1	2	1	6
2	1	1	6
2	2	1	9
2	1	2	9
1	2	2	9

(4.30)

Inspection of this data shows that different energy levels may have the same energy, that is, they are *degenerate.*

4.1.3 Consequences useful for quantum oscillators

As seen above, the particle-in-a-box model leads to the important result of the energy quantization. However, it leads also to some other interesting consequences, for example, the *de Broglie wavelength* of a quantum particle and a simple understanding of how the energy quantization disappears when the physical dimensions are progressively increasing.

Observe that the energy levels (4.28) are only kinetic in nature since the potential energy is zero inside the box. Thus, they may be written as the sum of the kinetic energies along the three dimensions, that is,

$$E_{n_x,n_y,n_z} = \frac{1}{2m}\left(P_{n_x}^2 + P_{n_y}^2 + P_{n_z}^2\right)$$

Then, by identification of this formal expression with Eq. (4.28), one obtains

$$P_{n_x} = \pm\frac{h}{2a_x}n_x$$

so that, due to Eq. (4.26), it appears that the wavelengths are given by

$$\lambda_{n_x} = \frac{h}{|P_{n_x}|} \tag{4.31}$$

which is the *Louis de Broglie's relation*, which has been experimentally verified for microscopic particles. Note that Fig. 4.2 reveals that the number of nodes of the wavefunctions are increasing with the quantum number n_x, reflecting the fact that in agreement with Eq. (4.31), the modulus of the momentum raises when the de Broglie wavelength decreases, leading, therefore, to an enhancement of the energy since there is no potential.

In the special case of a cubic box, when one of the quantum numbers is increased by a factor of 1, the others remaining constant, Eq. (4.29) reads

$$E_{n_x+1,n_y,n_z} = \left(\frac{h^2}{8m}\right)\frac{(n_x+1)^2+n_y^2+n_z^2}{\mathbb{V}^{2/3}} \tag{4.32}$$

Then, the difference between the two successive energy levels (4.29) and (4.32) yields

$$\Delta E_{n_x,n_x+1} = E_{n_x+1,n_y,n_z} - E_{n_x,n_y,n_z} = \left(\frac{h^2}{8m}\right)\frac{(2n_x+1)}{\mathbb{V}^{2/3}}$$

which, for large quantum numbers, may be approximated as

$$\Delta E_{n_x,n_x+1} = \left(\frac{h^2}{4m}\right)\frac{n_x}{\mathbb{V}^{2/3}} \tag{4.33}$$

This result, which follows from quantum mechanics, holds for microscopic dimensions of atoms or molecules. But there is no reason why it should not also be true for macroscopic systems where the mass (denoted M in place of m) of the particle and the volume in which it is enclosed are those of usual experiences, so that Eq. (4.33) reads

$$\widetilde{\Delta E}_{n_x,n_x+1} = \left(\frac{h^2}{4M}\right)\frac{n_x}{\mathbb{V}^{2/3}} \tag{4.34}$$

As an illustration, when passing from a description of an atomic electron of mass m_e enclosed in a volume $\mathbb{V}$, which is roughly that of the atom of radius a_{at}, to the description of a ball of mass M_B of 1 kg moving in a volume $\mathbb{V}$ around 1 m^3, one has respectively

$$m_e \backsimeq 10^{-30}\,\text{kg} \qquad \text{and} \qquad a_{at} \backsimeq 10^{-10}\,\text{m}$$

$$M_B \backsimeq 1\,\text{kg} \qquad \text{and} \qquad a \backsimeq 1\,\text{m}$$

so that Eq. (4.34) leads to

$$\widetilde{\Delta E}_{n_x,n_x+1} = 10^{-50}\,\Delta E_{n_x,n_x+1} \tag{4.35}$$

In a similar way, when passing from a description of a proton of mass m_p enclosed in a nucleus of volume $\mathbb{V}$, which is roughly that of the third power of the nucleus radius a_{nu}, to that of the Earth of mass M_E moving around the Sun at the distance $a_{\text{Sun}} \backsimeq 10^{11}$ m there are

$$m_p \backsimeq 10^{-27}\,\text{kg} \qquad \text{and} \qquad a_{nu} \backsimeq 10^{-15}\,\text{m}$$

$$M_E \backsimeq 10^{25}\,\text{kg} \qquad \text{and} \qquad a_{\text{Sun}} \backsimeq 10^{11}\,\text{m}$$

so that Eq. (4.34) leads to

$$\widetilde{\Delta E}_{n_x,n_x+1} = 10^{-104}\,\Delta E_{n_x,n_x+1} \tag{4.36}$$

Equations (4.35) and (4.36) illustrate the fact that passing from the microscopic to the macroscopic levels drastically decreases the energy gap between two successive energy levels.

4.2 TWO-ENERGY-LEVEL SYSTEMS

Now, it is time to study the model of two-energy-level systems, which will illustrate the phenomenon called *quantum interference* between kets, which is a simple consequence of the linear properties of quantum mechanics. Such a model will be later applied when studying the interactions dealing with anharmonically coupled oscillators. But, it is suitable to begin the present approach by starting from equations dealing with the more general model of multiple interacting energy levels.

4.2.1 Multiple interacting energy levels

Consider a system the Hamiltonian of which may be split into two parts according to

$$\mathbf{H} = \mathbf{H}^{\circ} + \mathbf{V} \tag{4.37}$$

Now, suppose that the eigenvalue equation of $\mathbf{H}^{\circ}$ is known, that is,

$$\mathbf{H}^{\circ}|\Phi_i\rangle = E_i^{\circ}|\Phi_i\rangle \tag{4.38}$$

Since $\mathbf{H}^{\circ}$ is Hermitian, its eigenvectors are orthogonal so that, if they are normalized, they verify

$$\langle\Phi_i|\Phi_j\rangle = \delta_{ij} \tag{4.39}$$

Owing to Eqs. (4.37)–(4.39), and in the basis $\{|\Phi_i\rangle\}$, the diagonal matrix elements of the full Hamiltonian are

$$\langle\Phi_i|\mathbf{H}|\Phi_i\rangle = E_i^{\circ} + \langle\Phi_i|\mathbf{V}|\Phi_i\rangle \tag{4.40}$$

Now, due to Eq. (4.37), the off-diagonal matrix elements are

$$\langle\Phi_i|\mathbf{H}|\Phi_j\rangle = \langle\Phi_i|(\mathbf{H}^{\circ} + \mathbf{V})|\Phi_j\rangle$$

In addition, owing to the eigenvalue equation (4.38), there is

$$\langle\Phi_i|\mathbf{H}^{\circ}|\Phi_j\rangle = \langle\Phi_i|E_j^{\circ}|\Phi_j\rangle = E_j^{\circ}\langle\Phi_i|\Phi_j\rangle = E_j^{\circ}\delta_{ij}$$

so that only part $\mathbf{V}$ of the Hamiltonian (4.37) couples two different eigenkets of $\mathbf{H}^{\circ}$, according to

$$\langle\Phi_i|\mathbf{H}|\Phi_j\rangle = \langle\Phi_i|\mathbf{V}|\Phi_j\rangle = \beta_{ij}$$

Now, writing

$$\alpha_i = \langle\Phi_i|\mathbf{H}|\Phi_i\rangle \qquad \text{and} \qquad \beta_{ii} = \langle\Phi_i|\mathbf{V}|\Phi_i\rangle \tag{4.41}$$

Eq. (4.40) leads to

$$\alpha_i = E_i^{\circ} + \beta_{ii}$$

Now, the eigenvalue equation of the full Hamiltonian (4.37) is

$$\mathbf{H}|\Psi_\mu\rangle = E_\mu|\Psi_\mu\rangle \quad \text{with} \quad \langle\Psi_\mu|\Psi_\nu\rangle = \delta_{\mu\nu} \tag{4.42}$$

the matrix representation of this eigenvalue equation in the basis of the eigenkets of $\mathbf{H}^\circ$ being

$$\begin{pmatrix} \alpha_1 & \beta_{12} & \dots & \beta_{1N} \\ \beta_{21} & \alpha_2 & \dots & \dots \\ \dots & \dots & \dots & \dots \\ \beta_{N1} & \dots & \dots & \alpha_N \end{pmatrix} \begin{pmatrix} C_{1\mu} \\ C_{2\mu} \\ \dots \\ C_{N\mu} \end{pmatrix} = E_\mu \begin{pmatrix} 1 & 0 & \dots & 0 \\ 0 & 1 & \dots & \dots \\ \dots & \dots & \dots & \dots \\ 0 & \dots & \dots & 1 \end{pmatrix} \begin{pmatrix} C_{1\mu} \\ C_{2\mu} \\ \dots \\ C_{N\mu} \end{pmatrix} \tag{4.43}$$

with

$$C_{\mu i} = \langle \Psi_\mu | \Phi_i \rangle$$

One possibility to obtain the eigenvalues E_μ and the corresponding eigenvectors is to diagonalize the left-hand matrix appearing in Eq. (4.43). However, there is yet another possibility, because this eigenvalue equation may be also written as a system of simultaneous equations:

$$\begin{cases} (\alpha_1 - E_\mu)C_{1\mu} + \beta_{12}C_{2\mu} & + \cdots + \beta_{1N}C_{N\mu} & = 0 \\ \beta_{21}C_{1\mu} \quad + (\alpha_2 - E_\mu)C_{2\mu} & + \cdots + \cdots & = 0 \\ \cdots \quad + \cdots & + \cdots + \cdots & \cdots \\ \beta_{N1}C_{1\mu} \quad + \cdots & + \cdots + (\alpha_N - E_\mu)C_{N\mu} & = 0 \end{cases}$$

Then, since the coefficients $C_{i\mu}$ cannot be zero, this system of equations is satisfied if the corresponding determinant is zero, that is,

$$\begin{vmatrix} (\alpha_1 - E) & \beta_{12} & \dots & \beta_{1N} \\ \beta_{21} & (\alpha_2 - E) & \dots & \dots \\ \dots & \dots & \dots & \dots \\ \beta_{N1} & \dots & \dots & (\alpha_N - E) \end{vmatrix} = 0 \tag{4.44}$$

where we have omitted the subscript μ for the unknown eigenvalues E_μ.

4.2.2 Energies of two interacting levels

In the special situation of two interacting energy levels α_1 and α_2, and where β_{12} is real and thus equal to β_{21}, the matrix representation of the Hamiltonian reduces to

$$\boxed{\mathbf{H}} = \begin{pmatrix} \alpha_1 & \beta \\ \beta & \alpha_2 \end{pmatrix} \tag{4.45}$$

In order to solve the eigenvalue equation

$$\mathbf{H}|\Psi_\pm\rangle = E_\pm|\Psi_\pm\rangle \tag{4.46}$$

consider the secular equation that, according to Eq. (4.44), reads

$$\begin{vmatrix} \alpha_1 - E & \beta \\ \beta & \alpha_2 - E \end{vmatrix} = 0 \quad \text{with} \quad \beta \equiv \beta_{12} \tag{4.47}$$

Then, expanding the determinant according to the usual rule, that is,

$$(\alpha_1 - E)(\alpha_2 - E) - \beta^2 = 0$$

the following second-order equation in E is obtained

$$E^2 + \alpha_1\alpha_2 - E(\alpha_1 + \alpha_2) - \beta^2 = 0$$

the two roots of which are

$$E_\pm = \frac{(\alpha_1 + \alpha_2) \pm \sqrt{(\alpha_1 + \alpha_2)^2 - 4(\alpha_1\alpha_2 - \beta^2)}}{2}$$

or, after simplification

$$E_\pm = \frac{(\alpha_1 + \alpha_2) \pm \sqrt{(\alpha_1 - \alpha_2)^2 + 4\beta^2}}{2} \tag{4.48}$$

Hence, the difference between the two eigenvalues is

$$E_+ - E_- = \sqrt{(\alpha_1 - \alpha_2)^2 + 4\beta^2} \tag{4.49}$$

On the other hand, the eigenvectors of **H** appearing in (4.42), and corresponding to the eigenvalues (4.48), are of the form

$$|\Psi_\pm\rangle = C_{1_\pm}|\Phi_1\rangle + C_{2_\pm}|\Phi_2\rangle \tag{4.50}$$

whereas the orthonormality properties (4.42) of these kets leads to

$$\langle\Psi_-\,|\Psi_+\rangle = 0 \qquad \text{and} \qquad \langle\Psi_+|\,\Psi_+\rangle = \langle\Psi_-|\Psi_-\rangle = 1 \tag{4.51}$$

so that, due to Eq. (4.50),

$$(C_{1_-}\langle\Phi_1| + C_{2_-}\langle\Phi_2|)(C_{1_+}|\Phi_1\rangle + C_{2_+}|\Phi_2\rangle) = 0$$

and thus

$$C_{1_-}C_{1_+}\langle\Phi_1|\Phi_1\rangle + C_{2_-}C_{2_+}\langle\Phi_2|\Phi_2\rangle + C_{1_-}C_{2_+}\langle\Phi_1|\Phi_2\rangle + C_{1_+}C_{2_-}\langle\Phi_2|\Phi_1\rangle = 0$$

Then, owing to the orthonormality conditions appearing in Eq. (4.39), this last expression reduces to

$$C_{1_-}C_{1_+} + C_{2_-}C_{2_+} = 0$$

Likewise, the normality conditions appearing in (4.51) lead to

$$C_{1_-}^2 + C_{2_-}^2 = 1 \qquad \text{and} \qquad C_{1_+}^2 + C_{2_+}^2 = 1 \tag{4.52}$$

When the two interacting levels are degenerate, that is, have the same energy, the two eigenvalues (4.48) of the Hamiltonian **H** reduce to

$$E_\pm = \alpha \pm \beta \quad \text{when} \quad \alpha_1 = \alpha_2 = \alpha \tag{4.53}$$

Then, in order to get the expansion coefficients of the **H** eigenvectors, corresponding to these two eigenvalues, return to Eq. (4.43); however, for the special situation of two interacting levels, that is,

$$\begin{pmatrix} \alpha - E_\pm & \beta \\ \beta & \alpha - E_\pm \end{pmatrix} \begin{pmatrix} C_{1_\pm} \\ C_{2_\pm} \end{pmatrix} = 0 \tag{4.54}$$

which leads to

$$(\alpha - (\alpha \pm \beta))\,C_{1_\pm} = \beta C_{2_\pm}$$

Rearranging, gives, respectively, for the components of the eigenvectors corresponding to the two eigenvalues

$$\frac{C_{1_+}}{C_{2_+}} = 1 \qquad \text{and} \qquad \frac{C_{1_-}}{C_{2_-}} = -1$$

where, of course, the complementary equations (4.52) continue to hold.

4.2.3 Approximate solution far from degeneracy

Now, consider the special situation where

$$|\alpha_1 - \alpha_2| \gg |\beta| \tag{4.55}$$

4.2.3.1 Eigenvalues Before applying this relation, it is convenient to write the eigenvalues (4.48) in the following form:

$$E_{\pm} = \frac{1}{2}\left[(\alpha_1 + \alpha_2) \pm (\alpha_1 - \alpha_2)\sqrt{1 + \frac{4\beta^2}{(\alpha_1 - \alpha_2)^2}}\right] \tag{4.56}$$

where the square root appears to be of the form

$$\sqrt{1+\varepsilon} \quad \text{with} \quad \varepsilon = \frac{4\beta^2}{(\alpha_1 - \alpha_2)^2} \qquad \text{and} \qquad \varepsilon \ll 1$$

Hence, by expansion of the square root up to first order in ε, one has

$$\sqrt{1+\varepsilon} \simeq 1 + \frac{\varepsilon}{2}$$

Thus, when the condition (4.55) is verified, the eigenvalues are

$$E_{\pm} = \frac{1}{2}\left[(\alpha_1 + \alpha_2) \pm (\alpha_1 - \alpha_2)\left(1 + \frac{2\beta^2}{(\alpha_1 - \alpha_2)^2}\right)\right]$$

or

$$\boxed{E_+ = \alpha_1 + \frac{\beta^2}{(\alpha_1 - \alpha_2)} \text{ and } E_- = \alpha_2 - \frac{\beta^2}{(\alpha_1 - \alpha_2)}} \tag{4.57}$$

4.2.3.2 Expansion coefficients of the eigenvectors Generally, the following inequalities hold:

$$\alpha_1 < 0, \qquad \alpha_2 < 0, \qquad \beta < 0$$

In order to get the expansion coefficients of the **H** eigenvectors corresponding to the eigenvalues (4.57), it is convenient to return to Eq. (4.54):

$$\begin{pmatrix} \alpha_1 - E_{\pm} & \beta \\ \beta & \alpha_2 - E_{\pm} \end{pmatrix}\begin{pmatrix} C_{1_\pm} \\ C_{2_\pm} \end{pmatrix} = 0 \tag{4.58}$$

which yields for the eigenvalue E_+ the following equation:

$$(\alpha_1 - E_+)C_{1+} + \beta C_{2+} = 0$$

so that

$$C_{1+} = -C_{2+}\frac{\beta}{(\alpha_1 - E_+)}$$

Then, inserting the expression of E_+ given by (4.57), one obtains

$$C_{1+} = -C_{2+}\frac{\beta}{\left(\alpha_1 - \left\{\alpha_1 + \dfrac{\beta^2}{\alpha_1 - \alpha_2}\right\}\right)}$$

or

$$C_{1+} = C_{2+}\frac{(\alpha_1 - \alpha_2)}{\beta} \tag{4.59}$$

with the normalization of the **H** eigenkets given, of course, by the last equation of (4.52), that is,

$$C_{1+}^2 + C_{2+}^2 = 1$$

In a similar way, Eq. (4.58) yields for the eigenvalue E_-

$$\beta C_{1-} + (\alpha_2 - E_-)C_{2-} = 0$$

while for E_+ to

$$C_{1-} = -C_{2-}\frac{\beta}{(\alpha_1 - \alpha_2)} \tag{4.60}$$

a result that must be combined to the first normalization condition of (4.52), that is,

$$C_{1-}^2 + C_{2-}^2 = 1$$

4.2.3.3 Eigenvectors pictorial representation Next, suppose that

$$\alpha_1 > \alpha_2 \qquad \text{and} \qquad \beta < 0 \tag{4.61}$$

Then, according to the first inequality, and owing to (4.57), the two eigenvalues obey

$$E_+ > E_-$$

Hence, in view of this new inequality combined with the second one appearing in (4.61), Eqs. (4.59) and (4.60) lead to the following results:

$$\frac{|C_{1+}|}{|C_{2+}|} >> 1 \qquad \text{and} \qquad \frac{|C_{1-}|}{|C_{2-}|} << 1 \tag{4.62}$$

$$\frac{C_{1+}}{C_{2+}} < 0 \qquad \text{and} \qquad \frac{C_{1-}}{C_{2-}} > 0 \tag{4.63}$$

The inequalities (4.61)–(4.63) are illustrated by Fig. 4.3 of the two interacting energy level systems with β negative. Figure 4.3 shows that after an interaction induced by **V**, the energy level E_+ is lowered by the amount $|\beta|$ and the other E_- raised by the same amount, the contribution of the two basic interacting levels being the same for the energy level E_+ and opposite for E_-. That may be viewed as considering the ket associated to E_+ as resulting from a *constructive quantum interference* between the interacting levels and that associated to E_- as following from a *destructive one*.

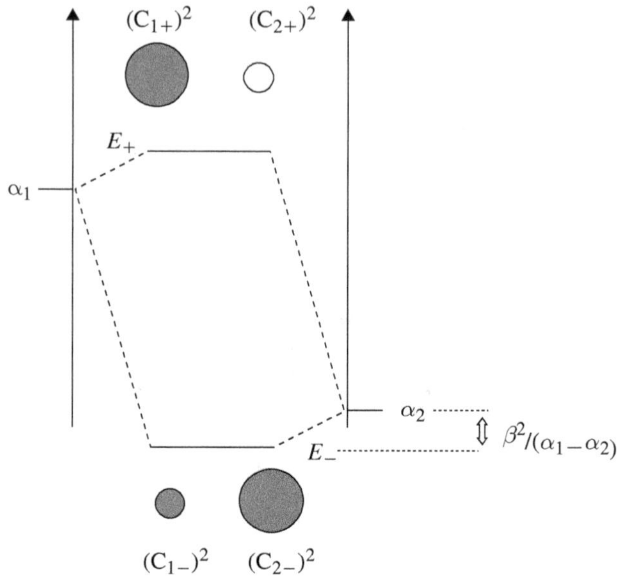

Figure 4.3 Correlation energy levels of two interacting energy levels.

4.2.3.4 Result of second-order perturbation theory In the notation of Eqs. (4.37)–(4.42), the inequality (4.55) and its consequence (4.57) take, respectively, the form

$$|\langle\Phi_i|\mathbf{H}|\Phi_i\rangle - \langle\Phi_j|\mathbf{H}|\Phi_j\rangle| >> \langle\Phi_i|\mathbf{V}|\Phi_j\rangle$$

$$\boxed{\langle\Psi_\pm|\mathbf{H}|\Psi_\pm\rangle \backsimeq \langle\Phi_i|\mathbf{H}|\Phi_i\rangle \pm \frac{\langle\Phi_i|\mathbf{V}|\Phi_j\rangle^2}{\langle\Phi_i|\mathbf{H}|\Phi_i\rangle - \langle\Phi_j|\mathbf{H}|\Phi_j\rangle}} \tag{4.64}$$

with

$$\langle\Psi_\pm|\mathbf{H}|\Psi_\pm\rangle = E_\pm \qquad \langle\Phi_i|\mathbf{H}|\Phi_i\rangle = \alpha_i \qquad \langle\Phi_i|\mathbf{V}|\Phi_j\rangle = \beta$$

Equation (4.64) is the expression for the special case of a two-energy-level system, of *second-order perturbation expansion* of the eigenvalues of the full Hamiltonian **H** in terms of the matrix elements of this Hamiltonian in the basis of the Hamiltonian **H**°.

4.2.4 Dynamics

In order to get the dynamics of the system, it is convenient to write the Hamiltonian matrix (4.45) in the following form:

$$\boxed{\mathbf{H}} = \begin{pmatrix} \dfrac{\alpha_1+\alpha_2}{2} & 0 \\ 0 & \dfrac{\alpha_1+\alpha_2}{2} \end{pmatrix} + \begin{pmatrix} +\dfrac{\alpha_1-\alpha_2}{2} & \beta \\ \beta & -\dfrac{\alpha_1-\alpha_2}{2} \end{pmatrix} \tag{4.65}$$

or

$$\boxed{\mathbf{H}} = \left(\frac{\alpha_1 + \alpha_2}{2}\right)\boxed{\mathbf{1}} + \left(\frac{\alpha_1 - \alpha_2}{2}\right)\boxed{\mathbf{K}} \tag{4.66}$$

where the last right-hand-side matrix is given by

$$\boxed{\mathbf{K}} = \begin{pmatrix} +1 & \dfrac{2\beta}{\alpha_1 - \alpha_2} \\ \dfrac{2\beta}{\alpha_1 - \alpha_2} & -1 \end{pmatrix} \tag{4.67}$$

According to Section 1.3.2, since the two right-hand-side Hermitian matrices of Eq. (4.66) commute, they admit the same eigenvectors, so that the two following eigenvalue equations, both involving the same eigenkets $|\Psi_\pm\rangle$, are satisfied:

$$\boxed{\mathbf{H}}\,|\Psi_\pm\rangle = E_\pm|\Psi_\pm\rangle \qquad \boxed{\mathbf{K}}\,|\Psi_\pm\rangle = K_\pm|\Psi_\pm\rangle$$

where $E_\pm$ are the Hamiltonian eigenvalues we obtained above, whereas $K_\pm$ are the corresponding eigenvalues of the **K** matrix. Hence, due to Eq. (4.66), the eigenvalues of $\boxed{\mathbf{H}}$ read

$$E_\pm = \left(\frac{\alpha_1 + \alpha_2}{2}\right) + \left(\frac{\alpha_1 - \alpha_2}{2}\right)K_\pm \tag{4.68}$$

Next, write the matrix (4.67) as follows:

$$\boxed{\mathbf{K}} = \begin{pmatrix} +1 & \tan\theta \\ \tan\theta & -1 \end{pmatrix}$$

with

$$\tan\theta = \frac{2\beta}{\alpha_1 - \alpha_2} \tag{4.69}$$

Then, since $\boxed{\mathbf{K}}$ and $\boxed{\mathbf{H}}$, which commute, have the same set of eigenvectors $|\Psi_\pm\rangle$, the matricial eigenvalue equation of $\boxed{\mathbf{K}}$ is similar to that for $\boxed{\mathbf{H}}$ given by Eq. (4.58), so that one gets

$$\begin{pmatrix} +1 & \tan\theta \\ \tan\theta & -1 \end{pmatrix}\begin{pmatrix} C_{1\pm} \\ C_{2\pm} \end{pmatrix} = K_\pm \begin{pmatrix} 1 & 0 \\ 0 & 1 \end{pmatrix}\begin{pmatrix} C_{1\pm} \\ C_{2\pm} \end{pmatrix} \tag{4.70}$$

where the $C_{k\pm}$ are the components of the eigenvectors $|\Psi_\pm\rangle$ given by Eq. (4.50). The corresponding secular determinant, which must be zero, that is,

$$\begin{vmatrix} 1 - K_\pm & \tan\theta \\ \tan\theta & -1 - K_\pm \end{vmatrix} = 0$$

leads by expansion to

$$K_\pm^2 - 1 - \tan^2\theta = 0$$

so that

$$K_\pm^2 = 1 + \tan^2\theta = \frac{\cos^2\theta}{\cos^2\theta} + \frac{\sin^2\theta}{\cos^2\theta} = \frac{1}{\cos^2\theta}$$

and thus

$$K_{\pm} = \pm\frac{1}{\cos\theta} \tag{4.71}$$

Hence, the two Hamiltonian eigenvalues (4.68) read

$$E_{\pm} = \left(\frac{\alpha_1 + \alpha_2}{2}\right) \pm \left(\frac{\alpha_1 - \alpha_2}{2\cos\theta}\right)$$

their difference being

$$E_+ - E_- = \frac{\alpha_1 - \alpha_2}{\cos\theta}$$

so that, by inversion, $\cos\theta$ appears to be

$$\cos\theta = \frac{\alpha_1 - \alpha_2}{E_+ - E_-} \tag{4.72}$$

Now, by insertion of Eq. (4.71) into Eq. (4.70), one gets for the situation corresponding to the E_+ eigenvalue

$$\left(1 - \frac{1}{\cos\theta}\right) C_{1+} + \tan\theta\, C_{2+} = 0$$

which reads

$$(\cos\theta - 1)\, C_{1+} + \sin\theta\, C_{2+} = 0 \tag{4.73}$$

Moreover, keeping in mind the trigonometric relation

$$\frac{1 - \cos 2\theta}{2} = \sin^2\theta \tag{4.74}$$

which reads

$$1 - \cos 2\theta = 2\sin\theta\sin\theta$$

the term multiplying C_{1+} in Eq. (4.73) reads

$$\cos\theta - 1 = -2\sin\left(\frac{\theta}{2}\right)\sin\left(\frac{\theta}{2}\right) \tag{4.75}$$

Furthermore, the trigonometric relation

$$\sin 2\theta = 2\sin\theta\cos\theta$$

yields

$$\sin\theta = 2\sin\left(\frac{\theta}{2}\right)\cos\left(\frac{\theta}{2}\right) \tag{4.76}$$

Hence, using Eqs. (4.75) and (4.76), Eq. (4.73) transforms to

$$-\sin\left(\frac{\theta}{2}\right)\sin\left(\frac{\theta}{2}\right) C_{1+} + \sin\left(\frac{\theta}{2}\right)\cos\left(\frac{\theta}{2}\right) C_{2+} = 0$$

so that

$$\frac{C_{1+}}{C_{2+}} = \frac{\cos(\theta/2)}{\sin(\theta/2)}$$

Thus, the two expansion coefficients, which are clearly normalized, read

$$C_{1+} = \cos\left(\frac{\theta}{2}\right) \qquad \text{and} \qquad C_{2+} = \sin\left(\frac{\theta}{2}\right)$$

so that Eq. (4.50) yields

$$|\Psi_+\rangle = \cos\left(\frac{\theta}{2}\right)|\Phi_1\rangle + \sin\left(\frac{\theta}{2}\right)|\Phi_2\rangle \tag{4.77}$$

In a similar way one would obtain

$$|\Psi_-\rangle = -\sin\left(\frac{\theta}{2}\right)|\Phi_1\rangle + \cos\left(\frac{\theta}{2}\right)|\Phi_2\rangle \tag{4.78}$$

which may be verified by observing that the two normalized kets (4.77) and (4.78) are orthogonal. Then, multiplying Eq. (4.77) by $\sin(\theta/2)$, and Eq. (4.78) by $\cos(\theta/2)$, one obtains, after summing these results and simplification,

$$|\Phi_2\rangle = \sin\left(\frac{\theta}{2}\right)|\Psi_+\rangle + \cos\left(\frac{\theta}{2}\right)|\Psi_-\rangle \tag{4.79}$$

or, due to Eqs. (4.77) and (4.78),

$$|\Phi_2\rangle = |\Psi_+\rangle\langle\Psi_+|\Phi_2\rangle + |\Psi_-\rangle\langle\Psi_-|\Phi_2\rangle$$

with

$$\langle\Phi_2|\Psi_+\rangle = \langle\Psi_+|\Phi_2\rangle = \sin\left(\frac{\theta}{2}\right) \qquad \text{and} \qquad \langle\Phi_2|\Psi_-\rangle = \langle\Psi_-|\Phi_2\rangle = \cos\left(\frac{\theta}{2}\right) \tag{4.80}$$

In a similar way, after multiplying Eq. (4.77) by $\cos(\theta/2)$, and Eq. (4.78) by $-\sin(\theta/2)$, and adding the results, one obtains,

$$|\Phi_1\rangle = \cos\left(\frac{\theta}{2}\right)|\Psi_+\rangle - \sin\left(\frac{\theta}{2}\right)|\Psi_-\rangle \tag{4.81}$$

4.2.5 Transition probability from $|\Phi_1\rangle$ to $|\Phi_2\rangle$ due to the V perturbation

Suppose that at an initial time the system is in the state

$$|\Psi(0)\rangle = |\Phi_1\rangle \tag{4.82}$$

At time t, this state will transform into $|\Psi(t)\rangle$ given, according to Eq. (3.85), by

$$|\Psi(t)\rangle = (e^{-i\mathbf{H}t/\hbar})|\Psi(0)\rangle$$

or, owing to the initial condition (4.82), by

$$|\Psi(t)\rangle = (e^{-i\mathbf{H}t/\hbar})|\Phi_1\rangle$$

and thus, according to Eq. (4.81), by

$$|\Psi(t)\rangle = \cos\left(\frac{\theta}{2}\right)(e^{-i\mathbf{H}t/\hbar})|\Psi_+\rangle - \sin\left(\frac{\theta}{2}\right)(e^{-i\mathbf{H}t/\hbar})|\Psi_-\rangle$$

Again, owing to the Hamiltonian eigenvalue equation (4.46), this expression reads

$$|\Psi(t)\rangle = \cos\left(\frac{\theta}{2}\right)(e^{-iE_+t/\hbar})|\Psi_+\rangle - \sin\left(\frac{\theta}{2}\right)(e^{-iE_-t/\hbar})|\Psi_-\rangle$$

Next premultiplying both terms of this last equation by the bra $\langle\Phi_2|$ corresponding to the ket (4.79), that is,

$$\langle\Phi_2|\Psi(t)\rangle = \cos\left(\frac{\theta}{2}\right)(e^{-iE_+t/\hbar})\langle\Phi_2|\Psi_+\rangle - \sin\left(\frac{\theta}{2}\right)(e^{-iE_-t/\hbar})\langle\Phi_2|\Psi_-\rangle$$

then, owing to (4.80), it appears that

$$\langle\Phi_2|\Psi(t)\rangle = \sin\left(\frac{\theta}{2}\right)\cos\left(\frac{\theta}{2}\right)(e^{-iE_+t/\hbar} - e^{-iE_-t/\hbar}) \tag{4.83}$$

Moreover, the probability for the system to jump at time t into the state $|\Phi_2\rangle$ being

$$P_{12}(t) = |\langle\Phi_2|\Psi(t)\rangle|^2$$

becomes with the help of Eq. (4.83)

$$P_{12}(t) = \sin^2\left(\frac{\theta}{2}\right)\cos^2\left(\frac{\theta}{2}\right)(2 - (e^{+i(E_+-E_-)t/\hbar} + e^{-i(E_+-E_-)t/\hbar}))$$

or

$$P_{12}(t) = 2\sin^2\left(\frac{\theta}{2}\right)\cos^2\left(\frac{\theta}{2}\right)\left(1 - \cos\left(\frac{(E_+-E_-)t}{\hbar}\right)\right) \tag{4.84}$$

Furthermore, by aid of the trigonometric relations

$$\sin^2 x = \left(\frac{1-\cos 2x}{2}\right) \qquad \text{and} \qquad \sin x = 2\sin\left(\frac{x}{2}\right)\cos\left(\frac{x}{2}\right)$$

where x is some variable, we have

$$1 - \cos\left(\frac{(E_+-E_-)t}{\hbar}\right) = 2\sin^2\left(\frac{(E_+-E_-)t}{2\hbar}\right)$$

so that Eq. (4.84) may be written

$$P_{12}(t) = \sin^2\theta\,\sin^2\left(\frac{(E_+-E_-)\,t}{2\hbar}\right) \tag{4.85}$$

Now, since we do not know $\sin\theta$, but both $\tan\theta$ and $\cos\theta$, which are, respectively, given by Eqs. (4.69) and (4.72), it is suitable to transform this last equation into

$$P_{12}(t) = \cos^2\theta\tan^2\theta\,\sin^2\left(\frac{(E_+-E_-)t}{2\hbar}\right)$$

so that, due to Eqs. (4.69) and (4.72), the transition probability transforms to

$$P_{12}(t) = 4\frac{\beta^2}{(E_+-E_-)^2}\sin^2\left(\frac{(E_+-E_-)t}{2\hbar}\right)$$

Finally, owing to Eq. (4.49), we have

$$\boxed{P_{12}(t) = \frac{4\beta^2}{(\alpha_1-\alpha_2)^2+4\beta^2}\sin^2\left(\frac{\left(\sqrt{(\alpha_1-\alpha_2)^2+4\beta^2}\right)t}{2\hbar}\right)} \tag{4.86}$$

that is, the *Rabi equation*. Besides, the time-dependent probability $P_{12}(t)$ to jump from $|\Phi_1\rangle$ to $|\Phi_2\rangle$ plus that $P_{11}(t)$ for the system to remain into $|\Phi_1\rangle$ must be unity, one has

$$P_{11}(t) = 1 - P_{12}(t)$$

In the special situation where the two levels are degenerate, Eq. (4.86) reduces to

$$\boxed{P_{12}(t) = \sin^2\left(\frac{\beta t}{\hbar}\right) \quad \text{so that} \quad P_{11}(t) = \cos^2\left(\frac{\beta t}{\hbar}\right) \quad \text{when} \quad \alpha_1 = \alpha_2} \tag{4.87}$$

In the following, many results of this section will be applied to *Fermi resonances*, a physical phenomenon that is met in situations involving anharmonic couplings between molecular oscillators.

4.2.6 Fermi golden rule

In relation with the dynamics of the double energy levels, and to end this chapter, we must now touch on the question of transition probabilities per unit time from one energy level to another one because of a coupling between them, a question that will be of importance when we later study the coupling between molecular oscillators and the electromagnetic field. Thus, consider a system described by a Hamiltonian $\mathbf{H}$ that may be split into two noncommuting parts $\mathbf{H}^\circ$ and $\mathbf{V}$ according to

$$\mathbf{H} = \mathbf{H}^\circ + \mathbf{V} \quad \text{with} \quad [\mathbf{H}^\circ, \mathbf{V}] \neq 0$$

the eigenvalue equation of $\mathbf{H}^\circ$ being

$$\mathbf{H}^\circ|\Psi_k\rangle = E_k|\Psi_k\rangle \tag{4.88}$$

with

$$\langle\Psi_k|\Psi_l\rangle = \delta_{kl} \tag{4.89}$$

We seek the transition probability at time t for the system described by $\mathbf{H}$ to pass from any eigenstate of $\mathbf{H}^\circ$ to another because of the presence of $\mathbf{V}$, that is,

$$|C(l,t|k,0)|^2 = |\langle\Psi_l(t)|\Psi_k(0)\rangle|^2 = \langle\Psi_k(0)|\Psi_l(t)\rangle\langle\Psi_l(t)|\Psi_k(0)\rangle \tag{4.90}$$

Owing to the time-dependent evolution equation, the ket $|\Psi_l(t)\rangle$ evolves with time according to

$$|\Psi_l(t)\rangle = \mathbf{U}(t)|\Psi_l(0)\rangle$$

Now, in the interaction picture, the time evolution operator is given, in terms of the Hamiltonian $\mathbf{H}^\circ$ by Eq. (3.122), that is,

$$\mathbf{U}(t) = (e^{-i\mathbf{H}^\circ t/\hbar})\mathbf{U}(t)^{\mathrm{IP}}$$

Thus, the transition probability (4.90) becomes

$$|C(l,t|k,0)|^2 = \langle\Psi_k(0)|(e^{-i\mathbf{H}^\circ t/\hbar})\mathbf{U}(t)^{\mathrm{IP}}|\Psi_l(0)\rangle\langle\Psi_l(0)|\mathbf{U}(t)^{\mathrm{IP}-1}(e^{i\mathbf{H}^\circ \mathbf{t}/\hbar})|\Psi_l(0)\rangle$$

Again, owing to the eigenvalue equation (4.88), the transition probability transforms to

$$|C(l,t \mid k,0)|^2 = \Psi_k(0)|(e^{-iE_k t/\hbar})\mathbf{U}(t)^{\mathrm{IP}}|\Psi_l(0)\rangle\langle\Psi_l(0)|\mathbf{U}(t)^{\mathrm{IP}-1}(e^{iE_k t/\hbar})|\Psi_k(0)\rangle$$

Or, after simplification

$$|C(l,t|k,0)|^2 = |\langle\Psi_k(0)|\mathbf{U}(t)^{\text{IP}}|\Psi_l(0)\rangle|^2 \tag{4.91}$$

Up to first order, the IP time evolution operator is, according to Eq. (3.125), given by

$$\mathbf{U}(t)^{\text{IP}} = 1 + \left(\frac{1}{i\hbar}\right)\int_0^t \mathbf{V}(t')^{\text{IP}}dt'$$

where

$$\mathbf{V}(t)^{\text{IP}} = (e^{i\mathbf{H}^\circ t/\hbar})\mathbf{V}(e^{-i\mathbf{H}^\circ t/\hbar}) \tag{4.92}$$

Thus, Eq. (4.91) becomes

$$|C(l,t|k,0)|^2 = \left|\langle\Psi_k(0)|\left(1 + \left(\frac{1}{i\hbar}\right)\int_0^t \mathbf{V}(t')^{\text{IP}}dt'\right)|\Psi_l(0)\rangle\right|^2$$

Next, using Eq. (4.92) and simplifying by the orthogonality property (4.89), the transition probability takes the form

$$|C(l,t|k,0)|^2 = \left(\frac{1}{\hbar}\right)^2\left|\int_0^t \langle\Psi_k(0)|(e^{i\mathbf{H}^\circ t'/\hbar})\mathbf{V}(e^{-i\mathbf{H}^\circ t'/\hbar})|\Psi_l(0)\rangle dt'\right|^2 \tag{4.93}$$

Again, the eigenvalue equation allows one to transform this result into

$$|C(l,t|k,0)|^2 = \left(\frac{1}{\hbar}\right)^2\left|\int_0^t \langle\Psi_k(0)|(e^{iE_k t'/\hbar})\mathbf{V}(e^{-iE_l t'/\hbar})|\Psi_l(0)\rangle dt'\right|^2$$

or

$$|C(l,t|k,0)|^2 = \left(\frac{1}{\hbar}\right)^2 |\langle\Psi_k|\mathbf{V}|\Psi_l\rangle|^2\left|\int_0^t (e^{i\omega_{kl}t'})dt'\right|^2 \tag{4.94}$$

with

$$\omega_{kl} = \frac{(E_k - E_l)}{\hbar} \tag{4.95}$$

where the reference to time $t = 0$ has been omitted. By integration, one has

$$\int_0^t (e^{i\omega_{kl}t'})\,dt' = \frac{1}{i}\left(\frac{e^{i\omega_{kl}t} - 1}{\omega_{kl}}\right) \tag{4.96}$$

In addition, the corresponding absolute value is

$$\left|\int_0^t (e^{i\omega_{kl}t'})\,dt'\right|^2 = \frac{2}{\omega_{kl}^2}(1 - \cos\omega_{kl}t)$$

Moreover, by aid of the usual trigonometric relations

$$(1 - \cos\omega_{kl}t) = 2\sin^2\left(\frac{\omega_{kl}t}{2}\right)$$

Eq. (4.94) becomes

$$|C(l,t|k,0)|^2 = 4|\langle\Psi_k|\mathbf{V}|\Psi_l\rangle|^2\left(\frac{\sin^2(\omega_{kl}t/2)}{(\hbar\omega_{kl})^2}\right) \tag{4.97}$$

This last expression holds for any time, up to first order in **V**. Now, consider this expression for large times for which it is convenient to write the second right-hand-side term of Eq. (4.97) in the following way:

$$|C(l,t|k,0)|^2 = 4|\langle\Psi_k|\mathbf{V}|\Psi_l\,\rangle|^2\left(\frac{\sin^2(xt/2\hbar)}{x^2}\right) \tag{4.98}$$

with

$$x = \hbar\omega_{kl} \tag{4.99}$$

Then, for large time t, Eq. (4.98) reads

$$|C(l,t|k,0)|^2 = |\langle\Psi_k|\mathbf{V}|\Psi_l\,\rangle|^2\left(\frac{\sin^2(x/\varepsilon)}{x^2}\right) \tag{4.100}$$

with

$$\frac{t}{2\hbar} = \frac{1}{\varepsilon} \quad \text{with} \quad \varepsilon \to 0 \tag{4.101}$$

Next observe that one of the expressions of the Dirac distribution function is given by Eq. (18.57) of Section 18.6.

$$\delta(x) = \frac{\varepsilon}{\pi}\left(\frac{\sin^2(x/\varepsilon)}{x^2}\right) \quad \text{when} \quad \varepsilon \to 0$$

which, owing to Eqs. (4.99) and (4.100), reads in the present situation

$$\left(\frac{\sin^2(x/\varepsilon)}{x^2}\right) = \frac{\pi t}{2\hbar}\delta(x)$$

so that Eq. (4.100) takes the form

$$|C(l,t|k,0)|^2 = 4|\langle\Psi_k|\mathbf{V}|\Psi_l\,\rangle|^2 t\frac{\pi}{2\hbar}\delta(x)$$

or, in view of Eqs. (4.95) and (4.99),

$$|C(l,t|k,0)|^2 = \frac{2\pi}{\hbar}|\langle\Psi_k|\mathbf{V}|\Psi_l\rangle|^2 t\delta(E_k - E_l) \tag{4.102}$$

Owing to this result, it is now possible to get the first-order transition probability per unit time, which is by definition

$$W(l,t|k,0) = \left(\frac{\partial|C(l,t|k,0)|^2}{\partial t}\right)$$

That gives what is called the *Fermi golden rule*:

$$\boxed{W(l,t|k,0) = \frac{2\pi}{\hbar}|\langle\Psi_k|\mathbf{V}|\Psi_l\,\rangle|^2\delta(E_k - E_l)} \tag{4.103}$$

an equation of the form of (4.103) will be met at the end of this book, dealing with molecular spectroscopy, when studying the interaction of molecular oscillators with electromagnetic field through a potential **V** involving a coupling of their dipolar moments with the electric field.

4.3 CONCLUSION

This chapter, devoted to some quantum models, has lead to the following important results that will be useful in the subsequent studies of quantum oscillators:

Particle-in-a-box energy and de Broglie relation
$E_{n_x n_y n_z} = \left(\frac{h^2}{8ma^2}\right)(n_x^2 + n_y^2 + n_z^2) \qquad \lambda = \frac{h}{p}$
Second-order perturbation energy:
$\langle\Psi_\pm\vert\mathbf{H}\vert\Psi_\pm\rangle \backsimeq \langle\Phi_i\vert\mathbf{H}\vert\Phi_i\rangle \pm \frac{\langle\Phi_i\vert\mathbf{V}\vert\Phi_j\rangle^2}{\langle\Phi_i\vert\mathbf{H}\vert\Phi_i\rangle - \langle\Phi_j\vert\mathbf{H}\vert\Phi_j\rangle}$
Rabi's relation:
$P_{12}(t) = \frac{4\vert\langle\Phi_1\vert\mathbf{H}\vert\Phi_2\rangle\vert^2}{(\langle\Phi_1\vert\mathbf{H}\vert\Phi_1\rangle - \langle\Phi_2\vert\mathbf{H}\vert\Phi_2\rangle)^2 + 4\vert\langle\Phi_1\vert\mathbf{H}\vert\Phi_2\rangle\vert^2} \times \sin^2\left(\frac{\left(\sqrt{(\langle\Phi_1\vert\mathbf{H}\vert\Phi_1\rangle - \langle\Phi_2\vert\mathbf{H}\vert\Phi_2\rangle)^2 + 4\vert\langle\Phi_1\vert\mathbf{H}\vert\Phi_2\rangle\vert^2}\right)t}{2\hbar}\right)$
Fermi's golden rule:
$W(l,t\vert k,0) = \frac{2\pi}{\hbar}\vert\langle\Psi_k\vert\mathbf{V}\vert\Psi_l\rangle\vert^2\delta(E_k - E_l)$

Among them, the result of the particle-in-a-box model showing that waves associated to quantum states obey the de Broglie wavelength law according to which the number of nodes of the stationary waves increases with energy, a property that is to be obeyed by the energy wavefunctions of quantum oscillators. The other is the quantum interference found in the study of two-energy-state systems, which is met in the study of Fermi resonances, a physical phenomenon appearing in anharmonically coupled molecular oscillators. The latter is the time-dependent amplitude probability for a system to pass from one state to another one due to some coupling with the thermal bath, a result that will be widely used when studying coupling of molecular oscillators with the infrared (IR) electromagnetic field.

BIBLIOGRAPHY

C. Cohen-Tannoudji, B. Diu, and F. Laloe. *Quantum Mechanics*. Wiley-Interscience: Hoboken, NJ, 2006.

PART II

SINGLE QUANTUM HARMONIC OSCILLATORS

CHAPTER 5

ENERGY REPRESENTATION FOR QUANTUM HARMONIC OSCILLATORS

INTRODUCTION

The present chapter develops the basis of the quantum approach to harmonic oscillators. The dimensionless creation and annihilation operators are first introduced. Using these operators, which are Hermitian self-conjugate, it is possible to solve the eigenvalue equation of the Hamiltonian and thus to get the values of the energy levels of quantum harmonic oscillators. That also permits one to obtain the corresponding orthonormalized eigenkets, thus providing a basis for the study of quantum oscillators. Moreover, in a subsequent section, the relation governing the action of the raising and lowering operators on the eigenkets of the Hamiltonian are derived, leading to the possibility of finding how the Heisenberg uncertainty relations apply to quantum harmonic oscillators, when they are in some eigenkets of their Hamiltonian. The formalism introduced allows them to verify the validity of the virial theorem. Furthermore, a place is reserved to non-Hermitian operators (*Fermion operators*) playing for two-level systems a role analogous to that of creating annihilation operators (*Boson operators*) for quantum oscillators. Another section is devoted to the wave mechanics representation of the eigenkets of the Hamiltonian, which will permit a pictorial description of these kets in terms of wavefunctions, the corresponding number of nodes increasing with the energy. Finally, the time dependence of the creation and annihilation operators is calculated in the Heisenberg picture and applied to get the time dependence of the basic operators and of their mean values averaged over the eigenkets of the oscillator Hamiltonians.

5.1 HAMILTONIAN EIGENKETS AND EIGENVALUES

The most important result dealing with quantum harmonic oscillators is the knowledge of its quantized energy levels E_n, initially introduced by Planck (1901) in order to explain the spectral density of a black body via

$$E_n = n\hbar\omega \quad \text{with} \quad n = 1, 2, \ldots$$

Quantum Oscillators, First Edition. Olivier Henri-Rousseau and Paul Blaise.
© 2011 John Wiley & Sons, Inc. Published 2011 by John Wiley & Sons, Inc.

Later, this assumed expression of the quantized energy levels was weakly modified by Heisenberg (1925) in its matrix mechanics, which showed that they were given by

$$E_n = \left(n + \tfrac{1}{2}\right)\hbar\omega \quad \text{with} \quad n = 0, 1, 2, \ldots$$

From quantum mechanics, the problem is to solve the eigenvalue equation of the Hamiltonian **H**. One possibility is to solve the second-order partial differential equation, which is the wave mechanics picture of this eigenvalue equation, that is, the time-independent Schrödinger equation. Such an approach supposes to have some knowledge about the theory of partial derivative equations. The other possibility is to pass from the position and momentum Hermitian operators involved in the Hamiltonian to two new dimensionless Hermitian self-conjugated operators, the *ladder operators*, which allow an easy resolution of the eigenvalue equation of the Hamiltonian. It is the latter approach that is chosen in the present section.

5.1.1 Hamiltonian in terms of ladder operators

5.1.1.1 Ladder operators Starting from the Hamiltonian **H** of a quantum harmonic oscillator of angular frequency ω and of reduced mass m coupling two masses, m_1 and m_2, that is

$$\mathbf{H} = \left(\frac{\mathbf{P}^2}{2m} + \frac{1}{2}m\omega^2\mathbf{Q}^2\right) \tag{5.1}$$

where **Q** is the position operator and **P** its conjugate momentum obeying the commutation rule

$$[\mathbf{Q}, \mathbf{P}] = i\hbar$$

where the reduced mass m of the oscillator is given by

$$m = \left(\frac{m_1 m_2}{m_1 + m_2}\right) \tag{5.2}$$

In order to solve the eigenvalue equation of this Hamiltonian, it is convenient to work with the following dimensionless non-Hermitian operators, which are mutually Hermitian conjugates (*the ladder operators*):

$$\mathbf{a} = \sqrt{\frac{m\omega}{2\hbar}}\mathbf{Q} + i\sqrt{\frac{1}{2m\hbar\omega}}\mathbf{P} \tag{5.3}$$

$$\mathbf{a}^\dagger = \sqrt{\frac{m\omega}{2\hbar}}\mathbf{Q} - i\sqrt{\frac{1}{2m\hbar\omega}}\mathbf{P} \tag{5.4}$$

Next, we calculate the commutator of these two conjugate Hermitian operators. From Eqs. (5.3) and (5.4), it reads

$$\mathbf{a}\mathbf{a}^\dagger = (\eta\mathbf{Q} + i\zeta\mathbf{P})(\eta\mathbf{Q} - i\zeta\mathbf{P}) = \eta^2\mathbf{Q}^2 + \zeta^2\mathbf{P}^2 + i\zeta\eta[\mathbf{P}, \mathbf{Q}]$$

$$\mathbf{a}^\dagger\mathbf{a} = (\eta\mathbf{Q} - i\zeta\mathbf{P})(\eta\mathbf{Q} + i\zeta\mathbf{P}) = \eta^2\mathbf{Q}^2 + \zeta^2\mathbf{P}^2 - i\zeta\eta[\mathbf{P}, \mathbf{Q}]$$

with

$$\eta = \sqrt{\frac{m\omega}{2\hbar}} \qquad \text{and} \qquad \zeta = \sqrt{\frac{1}{2m\hbar\omega}}$$

Hence, the commutator of **a** and $\mathbf{a}^\dagger$ reads

$$\mathbf{a}\mathbf{a}^\dagger - \mathbf{a}^\dagger\mathbf{a} = 2i\zeta\eta[\mathbf{P},\mathbf{Q}] = \frac{i}{\hbar}[\mathbf{P},\mathbf{Q}]$$

or, due to the basic commutator (2.3),

$$\boxed{\mathbf{a}\mathbf{a}^\dagger - \mathbf{a}^\dagger\mathbf{a} = [\mathbf{a},\mathbf{a}^\dagger] = \mathbf{1}} \tag{5.5}$$

Next, by inversion of Eqs. (5.3) and (5.4), one obtains the dependence of **Q** and **P** operators with respect to **a** and $\mathbf{a}^\dagger$, respectively,

$$\boxed{\mathbf{Q} = \sqrt{\frac{\hbar}{2m\omega}}(\mathbf{a}^\dagger + \mathbf{a})} \tag{5.6}$$

$$\boxed{\mathbf{P} = i\sqrt{\frac{m\hbar\omega}{2}}(\mathbf{a}^\dagger - \mathbf{a})} \tag{5.7}$$

For reasons that will be clear later, the ladder operators $\mathbf{a}^\dagger$ and **a** are, respectively, often named the *raising* and *lowering* operators or *creation* and *annihilation* operators. Then, the insertion of Eqs. (5.6) and (5.7) into Eq. (5.1) gives

$$\mathbf{H} = \frac{i^2}{2m}\frac{\hbar m\omega}{2}(\mathbf{a}^\dagger - \mathbf{a})^2 + \frac{1}{2}m\omega^2\frac{\hbar}{2m\omega}(\mathbf{a}^\dagger + \mathbf{a})^2$$

or

$$\mathbf{H} = -\frac{\hbar\omega}{4}(\mathbf{a}^\dagger - \mathbf{a})^2 + \frac{\hbar\omega}{4}(\mathbf{a}^\dagger + \mathbf{a})^2$$

Hence,

$$\mathbf{H} = -\frac{\hbar\omega}{4}((\mathbf{a}^\dagger)^2 + (\mathbf{a})^2 - \mathbf{a}^\dagger\mathbf{a} - \mathbf{a}\mathbf{a}^\dagger) + \frac{\hbar\omega}{4}((\mathbf{a}^\dagger)^2 + (\mathbf{a})^2 + \mathbf{a}^\dagger\mathbf{a} + \mathbf{a}\mathbf{a}^\dagger)$$

and, after simplification

$$\mathbf{H} = \frac{\hbar\omega}{2}(\mathbf{a}^\dagger\mathbf{a} + \mathbf{a}\mathbf{a}^\dagger) \tag{5.8}$$

Now, the commutator (5.5) may be written

$$\mathbf{a}\mathbf{a}^\dagger = \mathbf{a}^\dagger\mathbf{a} + 1$$

so that Eq. (5.8) leads to the following fundamental expression for the Hamiltonian of the quantum harmonic oscillator:

$$\boxed{\mathbf{H} = \hbar\omega\left(\mathbf{a}^\dagger\mathbf{a} + \tfrac{1}{2}\right)} \tag{5.9}$$

Observe that this Hamiltonian is Hermitian, as required, since

$$(\mathbf{a}^\dagger\mathbf{a})^\dagger = \mathbf{a}^\dagger\mathbf{a}$$

Now write the Hamiltonian eigenvalue equation to be solved:

$$\mathbf{H}|\{n\}\rangle = E_n|\{n\}\rangle \tag{5.10}$$

where $|\{n\}\rangle$[1] are the eigenvectors and E_n the corresponding eigenvalues, which are real because **H** is Hermitian. Besides, since **H** is Hermitian, its eigenvectors are orthogonal and, if normalized, satisfy

$$\langle\{n\}|\{m\}\rangle = \delta_{nm} \tag{5.11}$$

5.1.2 Resolution of the Hamiltonian eigenvalue

To solve the eigenvalue equation (5.10), define the following Hermitian operator **N**, which commutes with the Hamiltonian **H**, that is,

$$\mathbf{N} = \mathbf{a}^\dagger\mathbf{a} \quad \text{with} \quad [\mathbf{N}, \mathbf{H}] = 0 \qquad \text{and} \qquad \mathbf{N}^\dagger = \mathbf{N} \tag{5.12}$$

5.1.2.1 Commutators [N, a], [N, a†], and eigenvalue equation of N For this purpose, it is necessary to know the commutator $[\mathbf{N}, \mathbf{a}]$ of **N** with the annihilation operator **a**:

$$[\mathbf{N}, \mathbf{a}] = (\mathbf{a}^\dagger\mathbf{a})\mathbf{a} - \mathbf{a}(\mathbf{a}^\dagger\mathbf{a})$$

Now, by changing the position of the second parenthesis, which does not modify anything, the commutator reads

$$[\mathbf{N}, \mathbf{a}] = (\mathbf{a}^\dagger\mathbf{a})\mathbf{a} - (\mathbf{a}\mathbf{a}^\dagger)\mathbf{a} \tag{5.13}$$

In addition, according to Eq. (5.5), that is,

$$\mathbf{a}\mathbf{a}^\dagger = \mathbf{a}^\dagger\mathbf{a} + 1 \tag{5.14}$$

Eq. (5.13) becomes

$$[\mathbf{N}, \mathbf{a}] = \{\mathbf{a}^\dagger\mathbf{a} - (\mathbf{a}^\dagger\mathbf{a} + \mathbf{1})\}\mathbf{a}$$

or

$$\boxed{[\mathbf{N}, \mathbf{a}] = -\mathbf{a} = [\mathbf{a}^\dagger\mathbf{a}, \mathbf{a}]} \tag{5.15}$$

Now, calculate the commutator of **N** with the creation operator $\mathbf{a}^\dagger$. We have

$$[\mathbf{N}, \mathbf{a}^\dagger] = (\mathbf{a}^\dagger\mathbf{a})\mathbf{a}^\dagger - \mathbf{a}^\dagger(\mathbf{a}^\dagger\mathbf{a})$$

which, by changing the first parenthesis position, reads

$$[\mathbf{N}, \mathbf{a}^\dagger] = \mathbf{a}^\dagger(\mathbf{a}\mathbf{a}^\dagger) - \mathbf{a}^\dagger(\mathbf{a}^\dagger\mathbf{a})$$

or, due to Eq. (5.14),

$$[\mathbf{N}, \mathbf{a}^\dagger] = \mathbf{a}^\dagger(\mathbf{a}^\dagger\mathbf{a} + \mathbf{1}) - \mathbf{a}^\dagger(\mathbf{a}^\dagger\mathbf{a})$$

[1]We shall use for the writing of the eigenkets of $\mathbf{a}^\dagger\mathbf{a}$ notations such as $|\{n\}\rangle$, $|(n)\rangle$, and $|[n]\rangle$, which are more complex than the usual ones $|n\rangle$, in order to allow one to distinguish easily different eigenkets belonging to different oscillators characterized by different sets of ladder operators $\mathbf{a}^\dagger\mathbf{a}$, $\mathbf{b}^\dagger\mathbf{b}$, and $\mathbf{c}^\dagger\mathbf{c}$. That will appear to be useful in the following chapters.

so that

$$\boxed{[\mathbf{N}, \mathbf{a}^\dagger] = \mathbf{a}^\dagger = [\mathbf{a}^\dagger\mathbf{a}, \mathbf{a}^\dagger]} \tag{5.16}$$

Next, write the eigenvalue equation of **N**:

$$\boxed{\mathbf{N}|\{n\}\rangle = A_n|\{n\}\rangle} \tag{5.17}$$

where A_n are the eigenvalues of **N**, which are real because **N** is Hermitian, whereas $|\{n\}\rangle$ are the corresponding eigenvectors obeying Eq. (5.11), which must be, therefore, the same as those appearing in Eq. (5.10) because **H** and **N** commute.

5.1.2.2 Action of N on $|a^k\{n\}\rangle$ To solve the eigenvalue equation (5.17) consider the action of the commutator (5.15) on any eigenket of Eq. (5.17), that is,

$$[\mathbf{N}, \mathbf{a}]|\{n\}\rangle = (\mathbf{Na} - \mathbf{aN})|\{n\}\rangle \tag{5.18}$$

which, due to Eq. (5.15), reads

$$(\mathbf{Na} - \mathbf{aN})|\{n\}\rangle = -\mathbf{a}|\{n\}\rangle$$

and which, owing to Eq. (5.17), transforms to

$$(\mathbf{Na} - \mathbf{a}A_n)|\{n\}\rangle = -\mathbf{a}|\{n\}\rangle$$

Then, rearranging, it yields

$$\mathbf{Na}|\{n\}\rangle = \mathbf{a}A_n|\{n\}\rangle - \mathbf{a}|\{n\}\rangle$$

Since A_n is a scalar that commutes with **a**, we have

$$\mathbf{Na}|\{n\}\rangle = (A_n - 1)\mathbf{a}|\{n\}\rangle \tag{5.19}$$

Now, observe that the action of **a** on the eigenstate $|\{n\}\rangle$ yields a new state, which may be written formally as

$$\mathbf{a}|\{n\}\rangle \equiv |\mathbf{a}\{n\}\rangle \tag{5.20}$$

so that Eq. (5.19) reads

$$\mathbf{N}|\mathbf{a}\{n\}\rangle = (A_n - 1)|\mathbf{a}\{n\}\rangle \tag{5.21}$$

Hence, $(A_n - 1)$ is the eigenvalue of **N** corresponding to the ket resulting from the action of **a** on $|\{n\}\rangle$.

Again, consider the action of the commutator (5.15); however, let it now act on the ket defined by Eq. (5.20), that is,

$$[\mathbf{N}, \mathbf{a}]|\mathbf{a}\{n\}\rangle = (\mathbf{Na} - \mathbf{aN})|\mathbf{a}\{n\}\rangle$$

Then, proceeding in the same way as for passing from Eq. (5.18) to (5.19), one finds

$$\mathbf{Na}|\mathbf{a}\{n\}\rangle = (A_n - 2)\mathbf{a}|\mathbf{a}\{n\}\rangle \tag{5.22}$$

Moreover, writing

$$\mathbf{a}|\mathbf{a}\{n\}\rangle = |\mathbf{a}^2\{n\}\rangle$$

Eq. (5.22) takes the form

$$\mathbf{N}|\mathbf{a}^2\{n\}\rangle = (A_n - 2)|\mathbf{a}^2\{n\}\rangle \tag{5.23}$$

Hence, from Eqs. (5.21) and (5.23), one obtains by recurrence

$$\boxed{\mathbf{N}|\mathbf{a}^k\{n\}\rangle = (A_n - k)|\mathbf{a}^k\{n\}\rangle \quad \text{with} \quad |\mathbf{a}^k\{n\}\rangle \equiv \mathbf{a}^k|\{n\}\rangle} \tag{5.24}$$

5.1.2.3 ***Action of N on $|(\mathbf{a}^\dagger)^k\{n\}\rangle$*** Now, consider the action of the commutator (5.16) on any eigenket of Eq. (5.17), that is,

$$[\mathbf{N}, \mathbf{a}^\dagger]|\{n\}\rangle = (\mathbf{N}\mathbf{a}^\dagger - \mathbf{a}^\dagger\mathbf{N})|\{n\}\rangle$$

Then, using Eq. (5.16) to express the left-hand-side member of this last expression, one obtains

$$\mathbf{a}^\dagger|\{n\}\rangle = (\mathbf{N}\mathbf{a}^\dagger - \mathbf{a}^\dagger\,\mathbf{N})|\{n\}\rangle$$

Again, using the eigenvalue equation (5.17), one gets

$$\mathbf{a}^\dagger|\{n\}\rangle = (\mathbf{N}\mathbf{a}^\dagger - \mathbf{a}^\dagger A_n)|\{n\}\rangle$$

and thus, after commuting the scalar A_n with the operator $\mathbf{a}^\dagger$, we have

$$\mathbf{N}\mathbf{a}^\dagger|\{n\}\rangle = (A_n + 1)\mathbf{a}^\dagger|\{n\}\rangle$$

or

$$\mathbf{N}|\mathbf{a}^\dagger\{n\}\rangle = (A_n + 1)|\mathbf{a}^\dagger\{n\}\rangle \tag{5.25}$$

with

$$\mathbf{a}^\dagger|\{n\}\rangle \equiv |\mathbf{a}^\dagger\{n\}\rangle$$

Consider again the action of the commutator (5.16) on $|\mathbf{a}^\dagger\{n\}\rangle$:

$$[\mathbf{N}, \mathbf{a}^\dagger]|\mathbf{a}^\dagger\{n\}\rangle = (\mathbf{N}\mathbf{a}^\dagger - \mathbf{a}^\dagger\mathbf{N})|\mathbf{a}^\dagger\{n\}\rangle$$

Then, proceeding as above, one would obtain

$$\mathbf{N}\mathbf{a}^\dagger|\mathbf{a}^\dagger\{n\}\rangle = (A_n + 2)\mathbf{a}^\dagger|\mathbf{a}^\dagger\{n\}\rangle$$

or, changing the notation,

$$\mathbf{N}|(\mathbf{a}^\dagger)^2\{n\}\rangle = (A_n + 2)|(\mathbf{a}^\dagger)^2\{n\}\rangle \tag{5.26}$$

with

$$\mathbf{a}^\dagger|\mathbf{a}^\dagger\{n\}\rangle \equiv |(\mathbf{a}^\dagger)^2\{n\}\rangle \tag{5.27}$$

Hence, from Eqs. (5.25) and (5.26), one gets by recurrence

$$\boxed{\mathbf{N}|(\mathbf{a}^\dagger)^k\{n\}\rangle = (A_n + k)|(\mathbf{a}^\dagger)^k\{n\}\rangle \quad \text{with} \quad |(\mathbf{a}^\dagger)^k\{n\}\rangle = (\mathbf{a}^\dagger)^k|\{n\}\rangle} \tag{5.28}$$

5.1.2.4 *Discrete character of the eigenvalues A_n*

Starting from the assumed eigenvalue equation (5.17), it has been possible to prove Eqs. (5.24) and (5.28). Rewrite them for comparison:

$$\mathbf{N}|\{n\}\rangle = A_n|\{n\}\rangle$$

$$\mathbf{N}|(\mathbf{a})^k\{n\}\rangle = (A_n - k)|(\mathbf{a})^k\{n\}\rangle$$

$$\mathbf{N}|(\mathbf{a}^\dagger)^k\{n\}\rangle = (A_n + k)|(\mathbf{a}^\dagger)^k\{n\}\rangle$$

with

$$|(\mathbf{a})^k\{n\}\rangle \equiv (\mathbf{a})^k|\{n\}\rangle \qquad \text{and} \qquad |(\mathbf{a}^\dagger)^k\{n\}\rangle \equiv (\mathbf{a}^\dagger)^k|\{n\}\rangle \tag{5.29}$$

By inspection of these equations, it appears that

$|\{n\}\rangle$ is an eigenvector of $\mathbf{N}$ with the corresponding eigenvalue A_n.
$|(\mathbf{a})^k\{n\}\rangle$ is an eigenvector of $\mathbf{N}$ with the corresponding eigenvalue $(A_n - k)$.
$|(\mathbf{a}^\dagger)^k\{n\}\rangle$ is an eigenvector of $\mathbf{N}$ with the corresponding eigenvalue $(A_n + k)$.

Hence, $|\{n\}\rangle$, $(\mathbf{a})^k|\{n\}\rangle$, and $(\mathbf{a}^\dagger)^k|\{n\}\rangle$ are eigenvectors of $\mathbf{N}$ with the eigenvalues A_n $(A_n - k)$ and $(A_n + k)$, respectively.

Thus, it may be inferred that the action of the kth power of the $\mathbf{a}$ operator on an eigenvector of $\mathbf{N}$ lowers by k the eigenvalue A_n of $\mathbf{N}$ corresponding to this eigenvector, whereas the action of the kth power of $\mathbf{a}^\dagger$ on the same eigenvector of $\mathbf{N}$ raises by k the eigenvalue A_n. Hence, the eigenvalues of $\mathbf{N}$ obey the relation

$$\boxed{A_n, A_n \pm 1, A_n \pm 2, \ldots} \tag{5.30}$$

5.1.2.5 *Impossibility for A_n to be negative*

Now let us show that A_n cannot be negative. Thus, consider

$$|\mathbf{a}\{n\}\rangle \equiv \mathbf{a}|\{n\}\rangle \tag{5.31}$$

the Hermitian conjugate of which is

$$\langle\{n\}\mathbf{a}^\dagger| \equiv \langle\{n\}|\mathbf{a}^\dagger \tag{5.32}$$

Then, owing to the property of the norm, requiring

$$\langle\{n\}\mathbf{a}^\dagger|\mathbf{a}\{n\}\rangle \geq 0$$

and according to the notations (5.31) and (5.32), we have

$$\langle\{n\}\mathbf{a}^\dagger|\mathbf{a}\{n\}\rangle = \langle\{n\}|\mathbf{a}^\dagger\mathbf{a}|\{n\}\rangle \tag{5.33}$$

Moreover, due to the definition (5.12) of $\mathbf{N}$, Eq. (5.33) transforms to

$$\langle\{n\}\mathbf{a}^\dagger|\mathbf{a}\{n\}\rangle = \langle\{n\}|\mathbf{N}|\{n\}\rangle$$

which, with the help of the eigenvalue equation (5.17), transforms to

$$\langle\{n\}\mathbf{a}^{\dagger}|\mathbf{a}\{n\}\rangle = \langle\{n\}|A_n|\{n\}\rangle$$

so that A_n being a scalar, is given by

$$A_n = \frac{\langle\{n\}\mathbf{a}^{\dagger}|\mathbf{a}\{n\}\rangle}{\langle\{n\}|\{n\}\rangle}$$

Now, observe that because the norm cannot be negative one has, respectively,

$$\langle\{n\}\mathbf{a}^{\dagger}|\mathbf{a}\{n\}\rangle \geq 0 \qquad \text{and} \qquad \langle\{n\}|\{n\}\rangle \geq 0$$

Hence, the eigenvalues of **N** cannot be negative

$$\boxed{A_n \geq 0} \tag{5.34}$$

5.1.2.6 *Nullity of the lowest eigenvalue* Since the eigenvalue A_n cannot be negative, there exists a lowest eigenvalue A_0 to which is associated an eigenvector denoted $|\{0\}\rangle$, leading to write in this special situation the eigenvalue equation (5.17) according to

$$\mathbf{N}|\{0\}\rangle = A_0|\{0\}\rangle$$

Now, the action of **N** on the ket resulting from the action of **a** on the lowest state $|\{0\}\rangle$ would lead, according to Eq. (5.19), to a new state, eigenvector of **N** with a corresponding eigenvalue $(A_0 - 1)$, which is impossible since A_0 was assumed to be the lowest possible eigenvalue:

$$\mathbf{Na}|\{0\}\rangle = \mathbf{N}|\mathbf{a}\{0\}\rangle = (A_0 - 1)|\mathbf{a}\{0\}\rangle \qquad \text{Impossible}$$

Thereby, owing to this impossibility, $|\{0\}\rangle$ must be the fundamental eigenstate of **N**, leading to write

$$\mathbf{a}|\{0\}\rangle = |\mathbf{a}\{0\}\rangle = 0 \tag{5.35}$$

the Hermitian conjugate of which is

$$\langle\{0\}|\mathbf{a}^{\dagger} = \langle\{0\}\mathbf{a}^{\dagger}| = 0$$

Of course, the norm between the states involved in the two above equations is

$$\langle\{0\}\mathbf{a}^{\dagger}|\mathbf{a}\{0\}\rangle = 0 \tag{5.36}$$

Next, observe that, due to the notations (5.31) and (5.32),

$$\langle\{0\}\mathbf{a}^{\dagger}|\mathbf{a}\{0\}\rangle \equiv \langle\{0\}|\mathbf{a}^{\dagger}\mathbf{a}|\{0\}\rangle \tag{5.37}$$

and, due to Eq. (5.12), that

$$\langle\{0\}|\mathbf{a}^{\dagger}\mathbf{a}|\{0\}\rangle = \langle\{0\}|\mathbf{N}|\{0\}\rangle$$

and, at last, that, owing to Eq. (5.17),

$$\langle\{0\}|\mathbf{N}|\{0\}\rangle = \langle\{0\}|A_0|\{0\}\rangle = A_0\langle\{0\}|\{0\}\rangle \tag{5.38}$$

Again, if $|\{0\}\rangle$ is normalized, that is, if

$$\langle\{0\}|\{0\}\rangle = 1$$

then, in view of Eqs. (5.37) and (5.38), it reads

$$\langle\{0\}\mathbf{a}^{\dagger}|\mathbf{a}\{0\}\rangle = A_0$$

so that, owing to Eq. (5.36), we have

$$\boxed{A_0 = 0} \tag{5.39}$$

5.1.2.7 Solution of the Hamiltonian eigenvalue equation (5.17) In the above section, we studied the eigenvalue equation (5.17), that is,

$$\mathbf{N}|\{n\}\rangle = A_n|\{n\}\rangle$$

for which it was shown that the eigenvalues A_n obey Eqs. (5.30), (5.34), and (5.39), that is,

$$A_n, A_n \pm 1, A_n \pm 2, \ldots \quad \text{with} \quad A_n \geq 0 \qquad \text{and} \qquad A_0 = 0$$

These results show that A_n is of the form

$$A_n = 0, 1, 2, 3, \ldots$$

Hence, writing explicitly the operator $\mathbf{N}$ by aid of Eq. (5.12) leads to writing the following eigenvalue equation:

$$\boxed{(\mathbf{a}^{\dagger}\mathbf{a})|\{n\}\rangle = n|\{n\}\rangle \quad \text{with } n \equiv A_n \qquad \text{and} \qquad n = 0, 1, 2, 3, \ldots} \tag{5.40}$$

Since the Hamiltonian of the quantum harmonic oscillator is given by Eq. (5.9), that is,

$$\mathbf{H} = \left(\mathbf{a}^{\dagger}\mathbf{a} + \tfrac{1}{2}\right)\hbar\omega \tag{5.41}$$

and due to Eq. (5.40), we see that the following eigenvalue equation is satisfied:

$$\boxed{\mathbf{H}|\{n\}\rangle = \hbar\omega\left(n + \tfrac{1}{2}\right)|\{n\}\rangle \quad \text{with} \quad n = 0, 1, 2, 3, \ldots} \tag{5.42}$$

The lowest eigenstate $|\{0\}\rangle$ of the Hamiltonian corresponding to $n=0$ is called *the ground state*, whereas the corresponding residual energy $\hbar\omega/2$ is called *the zero-point energy* of the oscillator. Now, according to Section 1.3.1, since the Hamiltonian operator (5.9) is Hermitian, its eigenvectors are necessarily orthogonal. Thus, if they have been normalized, they form an orthonormal basis obeying

$$\boxed{\langle\{n\}|\{m\}\rangle = \delta_{nm} \qquad \text{and} \qquad \sum_n |\{n\}\rangle\langle\{n\}| = 1} \tag{5.43}$$

5.1.2.8 Zero-point energy as preserving the Heisenberg uncertainty relations It may be of interest to understand the role of the zero-point energy $\hbar\omega/2$ appearing in Eq. (5.42) in the context of the Heisenberg uncertainty relations (2.9) dealing with the momentum and the position operators:

$$\Delta\mathbf{P}\,\Delta\mathbf{Q} \geqslant \frac{\hbar}{2} \tag{5.44}$$

Now, suppose that the energy of the ground state $|\{0\}\rangle$ is zero. Then, since the harmonic potential energy V of the oscillator cannot be negative because quadratic in the position Q, that is,

$$V = \tfrac{1}{2}M\omega^2 Q^2 \tag{5.45}$$

and because the kinetic energy T, which is quadratic in the momentum P, is necessarily positive, that is,

$$T = \frac{P^2}{2M} > 0 \tag{5.46}$$

our supposition would imply that both the kinetic T and potential V energies ought separately to be zero, that is,

$$T = V = 0 \tag{5.47}$$

However, as a consequence of Eqs. (5.45)–(5.47), it would then appear that

$$P = Q = 0 \tag{5.48}$$

Moreover, if Eq. (5.48) was true, that would in turn imply that P and Q would be known without any uncertainty, that is,

$$\Delta P = \Delta Q = 0$$

in contradiction to the Heisenberg uncertainty relations (5.44).

5.1.3 Action of ladder operators on Hamiltonian eigenkets

The solution of the eigenvalue equation of the Hermitian Hamiltonian has not only the merit that it yields energy levels of the oscillators but also the merit that it provides a basis from which it is possible to obtain matrix representations of all operators dealing with quantum oscillators. Since these operators may be written as functions of the position and momentum operators, they may be also expressed as functions of the raising and lowering operators. Therefore, it appears that the knowledge of the action of these operators on the eigenkets of the Hamiltonian will be of much interest from now on. Thus, the aim of this new section will be to treat this point.

5.1.3.1 Action of a Consider the action of $\mathbf{a}$ operator on $|\{n\}\rangle$. Keeping in mind Eq. (5.40) according to which $n \equiv A_n$, Eq. (5.21) reads

$$\mathbf{N}|\mathbf{a}\{n\}\rangle = (n-1)|\mathbf{a}\{n\}\rangle \quad \text{with} \quad n = 0, 1, 2, 3, \ldots \tag{5.49}$$

whereas the eigenvalue equation Eq. (5.17) allows one to write

$$\begin{aligned} \mathbf{N}|\{n\}\rangle &= n|\{n\}\rangle \\ \mathbf{N}|\{n-1\}\rangle &= (n-1)|\{n-1\}\rangle \end{aligned} \tag{5.50}$$

Comparison of the eigenvalue equations (5.49) and (5.50) shows that both equations involve the same operator and the same eigenvalues. Moreover, if the eigenvectors appearing in these eigenvalue equations are not necessarily the same, they must be proportional, so that one may write

$$|\mathbf{a}\{n\}\rangle = \lambda_n|\{n-1\}\rangle$$

where λ_n is a complex scalar. The Hermitian conjugate of this last expression being

$$\langle\{n\}\mathbf{a}^{\dagger}| = \lambda_n^*\langle\{n-1\}|$$

the corresponding norm is thereby

$$\langle\{n\}\mathbf{a}^{\dagger}|\mathbf{a}\{n\}\rangle = |\lambda_n|^2\langle\{n-1\}|\{n-1\}\rangle$$

Next, if the right-hand-side ket is normalized, this last equation reduces to

$$\langle\{n\}\mathbf{a}^{\dagger}|\mathbf{a}\{n\}\rangle = |\lambda_n|^2 \tag{5.51}$$

On the other hand, Eqs. (5.31) and (5.32) allow one to write the left-hand side of Eq. (5.51) as

$$\langle\{n\}\mathbf{a}^{\dagger}|\mathbf{a}\{n\}\rangle = \langle\{n\}|\mathbf{a}^{\dagger}\mathbf{a}|\{n\}\rangle$$

It appears that, due to Eq. (5.40),

$$\langle\{n\}\mathbf{a}^{\dagger}|\mathbf{a}\{n\}\rangle = n\langle\{n\}|\{n\}\rangle = n \tag{5.52}$$

Therefore, by identification of Eqs. (5.51) and (5.52), we have

$$|\lambda_n|^2 = n$$

so that, ignoring the phase factor (if λ_n would be imaginary), which is of no interest,

$$\lambda_n = \sqrt{n}$$

Thus, one obtains the final result of interest:

$$\boxed{\mathbf{a}|\{n\}\rangle = \sqrt{n}|\{n-1\}\rangle} \tag{5.53}$$

As it appears, the action of operator **a** on any eigenstate $|\{n\}\rangle$ of $\mathbf{a}^{\dagger}\mathbf{a}$ corresponding to the eigenvalue n transforms this state into a new eigenstate $|\{n-1\}\rangle$ of $\mathbf{a}^{\dagger}\mathbf{a}$ corresponding to the eigenvalue $(n-1)$. This action may be, therefore, viewed as lowering the eigenvalue of $\mathbf{a}^{\dagger}\mathbf{a}$ by unity and thus the corresponding eigenvector. Hence, **a** is called a *lowering operator*. Observe that the Hermitian conjugate of this equation is

$$\langle\{n\}|\mathbf{a}^{\dagger} = \sqrt{n}\langle\{n-1\}| \tag{5.54}$$

Now, since $|\{0\}\rangle$ is the lowest eigenket of $\mathbf{a}^{\dagger}\mathbf{a}$, Eqs. (5.53) and (5.54) lead to

$$\mathbf{a}|\{0\}\rangle = \langle\{0\}|\mathbf{a}^{\dagger} = 0 \tag{5.55}$$

5.1.3.2 Action of $\mathbf{a}^\dagger$ After finding the action of **a** on the Hamiltonian eigenkets, then pass to that on its Hermitian conjugate $\mathbf{a}^\dagger$. In view of $A_n = n$, Eq. (5.25) reads

$$\mathbf{N}|\mathbf{a}^\dagger\{n\}\rangle = (n+1)|\mathbf{a}^\dagger\{n\}\rangle \quad \text{with} \quad n = 0, 1, 2, 3, \ldots \tag{5.56}$$

whereas the eigenvalue equation (5.17) reads, respectively,

$$\begin{aligned} \mathbf{N}|\{n\}\rangle &= n|\{n\}\rangle \\ \mathbf{N}|\{n+1\}\rangle &= (n+1)|\{n+1\}\rangle \end{aligned} \tag{5.57}$$

Equation (5.57) is analogous to Eq. (5.56) since the **N** operator and the eigenvalues $(n+1)$ are the same in both cases so that the kets appearing in Eqs. (5.56) and (5.57) must be at least proportional to each other, leading one to write

$$|\mathbf{a}^\dagger\{n\}\rangle = \mu_n|\{n+1\}\rangle \tag{5.58}$$

where μ_n is a complex scalar. The corresponding norm is, therefore,

$$\langle\{n\}\mathbf{a}|\mathbf{a}^\dagger\{n\}\rangle = |\mu_n|^2\langle\{n+1\}|\{n+1\}\rangle$$

Again, if the eigenvectors $|\{n+1\}\rangle$ are normalized, this last expression reduces to

$$\langle\{n\}\mathbf{a}|\mathbf{a}^\dagger\{n\}\rangle = |\mu_n|^2 \tag{5.59}$$

In addition, changing the writing with the aid of Eqs. (5.31) and (5.32), the left-hand-side reads

$$\langle\{n\}\mathbf{a}|\mathbf{a}^\dagger\{n\}\rangle = \langle\{n\}|\mathbf{a}\mathbf{a}^\dagger|\{n\}\rangle \tag{5.60}$$

Furthermore, using the commutation rule (5.5), leading to

$$\mathbf{a}\mathbf{a}^\dagger = \mathbf{a}^\dagger\mathbf{a} + 1$$

Eq. (5.60) becomes

$$\langle\{n\}\mathbf{a}|\mathbf{a}^\dagger\{n\}\rangle = \langle\{n\}|(\mathbf{a}^\dagger\mathbf{a} + 1)|\{n\}\rangle \tag{5.61}$$

Moreover, using the eigenvalue equation (5.40), we have

$$(\mathbf{a}^\dagger\mathbf{a} + 1)|\{n\}\rangle = (n+1)|\{n\}\rangle$$

Thus, Eq. (5.61) transforms to

$$\langle\{n\}\mathbf{a}|\mathbf{a}^\dagger\{n\}\rangle = (n+1)\langle\{n\}|\{n\}\rangle$$

or, $|\{n\}\rangle$ being normalized,

$$\langle\{n\}\mathbf{a}|\mathbf{a}^\dagger\{n\}\rangle = n+1 \tag{5.62}$$

Finally, by identification of Eqs. (5.59) and (5.62)

$$|\mu_n|^2 = n+1$$

and ignoring an irrelevant phase factor, we have

$$\mu_n = \sqrt{n+1}$$

Hence, from Eq. (5.58), and due to the notations in (5.31) and (5.32), one obtains the final result:

$$\boxed{\mathbf{a}^{\dagger}|\{n\}\rangle = \sqrt{n+1}|\{n+1\}\rangle} \tag{5.63}$$

the Hermitian conjugate of which is

$$\boxed{\langle\{n\}|\mathbf{a} = \sqrt{n+1}\langle\{n+1\}|} \tag{5.64}$$

Observe that, according to Eq. (5.63), the action of $\mathbf{a}^{\dagger}$ on any ket $|\{n\}\rangle$ is changed into the raised one $|\{n+1\}\rangle$; hence, this operator is called a *raising operator.*

5.1.3.3 Action of different powers of $\mathbf{a}^{\dagger}$ and a Consider the action of different powers of $\mathbf{a}^{\dagger}$. The action of $\mathbf{a}^{\dagger}$ on the lowest state $|\{0\}\rangle$ corresponding to $n=0$, Eq. (5.63) yields

$$\mathbf{a}^{\dagger}|\{0\}\rangle = |\{1\}\rangle \tag{5.65}$$

In addition, due to Eq. (5.63), the second power of $\mathbf{a}^{\dagger}$

$$(\mathbf{a}^{\dagger})^2|\{0\}\rangle = \mathbf{a}^{\dagger}\mathbf{a}^{\dagger}|\{0\}\rangle = \mathbf{a}^{\dagger}|\{1\}\rangle \tag{5.66}$$

or, using again Eq. (5.63),

$$(\mathbf{a}^{\dagger})^2|\{0\}\rangle = \sqrt{1(1+1)}|\{1+1\}\rangle$$

Moreover, the third power of $\mathbf{a}^{\dagger}$ yields, using Eq. (5.63),

$$(\mathbf{a}^{\dagger})^3|\{0\}\rangle = \sqrt{1(1+1)(2+1)}|\{2+1\}\rangle$$

so that, by recurrence, one obtains

$$(\mathbf{a}^{\dagger})^n|\{0\}\rangle = \sqrt{n!}|\{n\}\rangle \tag{5.67}$$

the Hermitian conjugate of which is

$$\langle\{0\}|(\mathbf{a})^n = \sqrt{n!}\langle\{n\}| \tag{5.68}$$

Furthermore, by inversion, Eqs. (5.67) and (5.68) read, respectively,

$$|\{n\}\rangle = \frac{(\mathbf{a}^{\dagger})^n}{\sqrt{n!}}|\{0\}\rangle \tag{5.69}$$

$$\langle\{n\}| = \langle\{0\}|\frac{(\mathbf{a})^n}{\sqrt{n!}} \tag{5.70}$$

Next, passing to the action of different powers of $\mathbf{a}$ on $|\{n\}\rangle$, Eq. (5.53), allows one to write successively

$$(\mathbf{a})|\{n\}\rangle = \sqrt{n}|\{n-1\}\rangle$$

$$(\mathbf{a})^2|\{n\}\rangle = \sqrt{n}(\mathbf{a})|\{n-1\}\rangle$$

$$(\mathbf{a})^2|\{n\}\rangle = \sqrt{n(n-1)}|\{n-2\}\rangle$$

so that, by recurrence, one gets

$$(\mathbf{a})^l|\{n\}\rangle = \sqrt{n(n-1)\cdots(n-l+1)}|\{n-l\}\rangle \tag{5.71}$$

or

$$\boxed{(\mathbf{a})^l|\{n\}\rangle = \frac{\sqrt{n!}}{\sqrt{(n-l)!}}|\{n-l\}\rangle} \tag{5.72}$$

the Hermitian conjugate of which is

$$\langle\{n\}|(\mathbf{a}^\dagger)^l = \frac{\sqrt{n!}}{\sqrt{(n-l)!}}\langle\{n-l\}| \tag{5.73}$$

In a similar way, one would obtain using Eq. (5.63)

$$(\mathbf{a}^\dagger)^l|\{n\}\rangle = \sqrt{n(n+1)\cdots(n+l)}|\{n+l\}\rangle$$

or

$$\boxed{(\mathbf{a}^\dagger)^l|\{n\}\rangle = \frac{\sqrt{(n+l)!}}{\sqrt{(n)!}}|\{n+l\}\rangle} \tag{5.74}$$

for which the Hermitian conjugate is

$$\langle\{n\}|(\mathbf{a})^l = \frac{\sqrt{(n+l)!}}{\sqrt{(n)!}}\langle\{n+l\}|$$

5.1.4 Matrix representation of ladder operators

Knowledge of the action of the ladder operators on the eigenkets and eigenbras of the Hamiltonian allows one to get the matrix representation of these operators. For this purpose, start from the eigenvalue equation

$$\mathbf{N}|(n)\rangle = n|(n)\rangle \quad \text{with} \quad n = 0, 1, 2, \ldots \tag{5.75}$$

keeping in mind that $\mathbf{N}$ is the Hermitian number occupation operator

$$\mathbf{N} = \mathbf{a}^\dagger\mathbf{a} \qquad \text{and} \qquad \mathbf{N} = \mathbf{N}^\dagger \quad \text{since} \quad (\mathbf{a}^\dagger\mathbf{a})^\dagger = \mathbf{a}^\dagger\mathbf{a} \tag{5.76}$$

whereas $\mathbf{a}$ and $\mathbf{a}^\dagger$ are obeying the commutation rules

$$[\mathbf{a}, \mathbf{a}^\dagger]_- \equiv \mathbf{a}\mathbf{a}^\dagger - \mathbf{a}^\dagger\mathbf{a} = 1 \tag{5.77}$$

$$[\mathbf{a}, \mathbf{a}]_- = [\mathbf{a}^\dagger, \mathbf{a}^\dagger]_- = 0 \tag{5.78}$$

and that the kets form an orthogonal basis so that

$$\langle(n)|(m)\rangle = \delta_{mn} \tag{5.79}$$

Note that the subscript $-$ has been introduced in the expressions for commutators (5.77) and (5.78) in order to distinguish them from the anticommutators, which will appear later. At last, recall Eqs. (5.53) and (5.63), that is,

$$\mathbf{a}|(m)\rangle = \sqrt{m}|(m-1)\rangle \qquad \text{and} \qquad \mathbf{a}^\dagger|(m)\rangle = \sqrt{m+1}|(m+1)\rangle \tag{5.80}$$

Thus, in the basis defined by Eq. (5.79) and using Eq. (5.80), the matrix elements of **a** and $\mathbf{a}^\dagger$ read, respectively,

$$\langle (n)|\mathbf{a}|(m)\rangle = \sqrt{m}\langle (n)|(m-1)\rangle = \sqrt{m}\delta_{n,m-1}$$

$$\langle (n)|\mathbf{a}^\dagger|(m)\rangle = \sqrt{m+1}\langle (n)|(m+1)\rangle = \sqrt{m+1}\delta_{n,m+1}$$

As a consequence, the matrix representations of **a** and $\mathbf{a}^\dagger$ read, after arbitrary truncation,

$$\boxed{\mathbf{a}} = \begin{pmatrix} 0 & \sqrt{1} & & & \\ & 0 & \sqrt{2} & & \\ & & 0 & \sqrt{3} & \\ & & & 0 & \sqrt{4} \\ & & & & 0 \end{pmatrix} \quad \text{and} \quad \boxed{\mathbf{a}^\dagger} = \begin{pmatrix} 0 & & & & \\ \sqrt{1} & 0 & & & \\ & \sqrt{2} & 0 & & \\ & & \sqrt{3} & 0 & \\ & & & \sqrt{4} & 0 \end{pmatrix} \tag{5.81}$$

Hence, the matrix representation of the occupation number defined by Eq. (5.76) yields

$$\boxed{\mathbf{N}} = \begin{pmatrix} 0 & & & & \\ \sqrt{1} & 0 & & & \\ & \sqrt{2} & 0 & & \\ & & \sqrt{3} & 0 & \\ & & & \sqrt{4} & 0 \end{pmatrix} \begin{pmatrix} 0 & \sqrt{1} & & & \\ & 0 & \sqrt{2} & & \\ & & 0 & \sqrt{3} & \\ & & & 0 & \sqrt{4} \\ & & & & 0 \end{pmatrix}$$

or, after performing the matrix product,

$$\boxed{\mathbf{N}} = \begin{pmatrix} 0 & & & & \\ & 1 & & & \\ & & 2 & & \\ & & & 3 & \\ & & & & 4 \end{pmatrix}$$

That is in agreement with the result obtained by premultiplying Eq. (5.75) by the bra $\langle (m)|$ to give

$$\langle (m)|\mathbf{N}|(n)\rangle = n\langle (m)|(n)\rangle = n\,\delta_{mn}$$

5.1.5 Heisenberg uncertainty relations

As we have said above, Heisenberg provided the first demonstration of the quantized energy levels of harmonic oscillators and was lead to these results through his anticipation of the uncertainty relations. It is, therefore, important to answer the question of the expression of the Heisenberg uncertainty relation when computed over the eigenstates of the quantum harmonic oscillator Hamiltonian. For this purpose, first consider the required average values of **Q** and $\mathbf{Q}^2$.

Thus, start from the average value of **Q** over the number occupation eigenstates, which, owing to Eq. (5.6), is

$$\langle\{n\}|\mathbf{Q}|\{n\}\rangle = \sqrt{\frac{\hbar}{2m\omega}}\langle\{n\}|(\mathbf{a}^\dagger + \mathbf{a})|\{n\}\rangle \tag{5.82}$$

Again, owing to Eq. (5.53), the average value of **a** appearing in Eq. (5.82) is zero because of the orthogonality of the eigenkets of the Hamiltonian:

$$\langle\{n\}|\mathbf{a}|\{n\}\rangle = \sqrt{n}\langle\{n\}|\{n-1\}\rangle = 0 \tag{5.83}$$

Of course, the Hermitian conjugate must be also zero:

$$\langle\{n\}|\mathbf{a}^\dagger|\{n\}\rangle = \sqrt{n}\langle\{n-1\}|\{n\}\rangle = 0 \tag{5.84}$$

Therefore, it appears that the average value (5.82) of **Q** is zero, that is,

$$\boxed{\langle\{n\}|\mathbf{Q}|\{n\}\rangle = 0} \tag{5.85}$$

Now, owing to Eq. (5.6), the average value of $\mathbf{Q}^2$ reads

$$\langle\{n\}|\mathbf{Q}^2|\{n\}\rangle = \frac{\hbar}{2m\omega}\langle\{n\}|(\mathbf{a}^\dagger + \mathbf{a})^2|\{n\}\rangle \tag{5.86}$$

Moreover, the square appearing on the right-hand-side yields

$$(\mathbf{a}^\dagger + \mathbf{a})^2 = \mathbf{a}^\dagger\mathbf{a}^\dagger + \mathbf{aa} + \mathbf{a}^\dagger\mathbf{a} + \mathbf{aa}^\dagger$$

or, due to the commutation rule (5.77), that gives

$$\boxed{(\mathbf{a}^\dagger + \mathbf{a})^2 = (\mathbf{a}^\dagger)^2 + (\mathbf{a})^2 + 2\mathbf{a}^\dagger\mathbf{a} + 1} \tag{5.87}$$

Furthermore, owing to Eq. (5.53), the two successive actions of **a** on an eigenstate of the Hamiltonian, lead to

$$\mathbf{aa}|\{n\}\rangle = \sqrt{n}\mathbf{a}|\{n-1\}\rangle = \sqrt{n}\sqrt{n-1}|\{n-2\}\rangle$$

The average value of **aa** is, therefore, zero, according to the orthogonality of the eigenkets of the Hamiltonian, that is,

$$\langle\{n\}|(\mathbf{aa})|\{n\}\rangle = \sqrt{n(n-1)}\langle\{n\}|\{n-2\}\rangle = 0 \tag{5.88}$$

Of course, the Hermitian conjugate of this last equation may be obtained by taking for the left-hand-side $\mathbf{a}^\dagger\mathbf{a}^\dagger$ in place of **aa** and permuting, for the right-hand side, the ket and the bra of the scalar product, without changing the real scalar. That is,

$$\langle\{n\}|(\mathbf{a}^\dagger\mathbf{a}^\dagger)|\{n\}\rangle = \sqrt{n(n-1)}\langle\{n-2\}|\{n\}\rangle = 0 \tag{5.89}$$

which is also zero. Finally, in view of Eq. (5.40), the required average value of $\mathbf{a}^\dagger\mathbf{a}$ reads

$$\langle\{n\}|(\mathbf{a}^\dagger\mathbf{a})|\{n\}\rangle = n\langle\{n\}|\{n\}\rangle = n \tag{5.90}$$

Thus, using Eqs. (5.86)–(5.90), one obtains

$$\boxed{\langle\{n\}|\mathbf{Q}^2|\{n\}\rangle = \frac{\hbar}{2m\omega}(2n+1)} \tag{5.91}$$

so that, in view of Eqs. (5.85) and (5.91), the dispersion over $\mathbf{Q}$ appears to be

$$\Delta\mathbf{Q}_{|\{n\}\rangle} = \sqrt{\langle\{n\}|\mathbf{Q}^2|\{n\}\rangle - \langle\{n\}|\mathbf{Q}|\{n\}\rangle^2} = \sqrt{\frac{\hbar}{m\omega}}\sqrt{\left(n+\frac{1}{2}\right)} \tag{5.92}$$

Now, consider the corresponding average value of the momentum, which, according to Eq. (9.35), is

$$\langle\{n\}|\mathbf{P}|\{n\}\rangle = i\sqrt{\frac{\hbar m\omega}{2}}\langle\{n\}|(\mathbf{a}^\dagger - \mathbf{a})|\{n\}\rangle$$

Owing to Eqs. (5.19) and (5.84), it appears to be zero, that is,

$$\boxed{\langle\{n\}|\mathbf{P}|\{n\}\rangle = 0} \tag{5.93}$$

Again, owing to Eq. (9.35), the average value of the square of the momentum is

$$\langle\{n\}|\mathbf{P}^2|\{n\}\rangle = -\frac{\hbar m\omega}{2}\langle\{n\}|(\mathbf{a}^\dagger - \mathbf{a})^2|\{n\}\rangle$$

where, according to the commutation rule, the right-hand-side operator reads

$$(\mathbf{a}^\dagger - \mathbf{a})^2 = \mathbf{a}^\dagger\mathbf{a}^\dagger + \mathbf{a}\mathbf{a} - (2\mathbf{a}^\dagger\mathbf{a} + 1)$$

Hence, by combination of Eqs. (5.88)–(5.90), one gets

$$\boxed{\langle\{n\}|\mathbf{P}^2|\{n\}\rangle = \frac{\hbar m\omega}{2}(2n+1)} \tag{5.94}$$

Thereby, in view of Eqs. (5.93) and (5.94), the dispersion over $\mathbf{P}$ becomes

$$\Delta\mathbf{P}_{|\{n\}\rangle} = \sqrt{\langle\{n\}|\mathbf{P}^2|\{n\}\rangle - \langle\{n\}|\mathbf{P}|\{n\}\rangle^2} = \sqrt{\hbar m\omega}\sqrt{\left(n+\tfrac{1}{2}\right)} \tag{5.95}$$

by combining Eqs. (5.92) and (5.95), one obtains the following expression for the Heisenberg relation as applied to the eigenstates of the Hamiltonian of the harmonic oscillator:

$$\boxed{\Delta\mathbf{Q}_{|\{n\}\rangle}\Delta\mathbf{P}_{|\{n\}\rangle} = \left(n+\tfrac{1}{2}\right)\hbar} \tag{5.96}$$

which is an agreement with the Heisenberg uncertainty relation, which states that

$$\Delta\mathbf{Q}_{|\{n\}\rangle}\Delta\mathbf{P}_{|\{n\}\rangle} \geq \hbar \tag{5.97}$$

5.1.6 Virial theorem

We have seen that the knowledge of the average values of $\mathbf{P}^2$ and $\mathbf{Q}^2$ over the eigenstates of the number occupation operator allowed us to find the uncertainty Heisenberg relation (5.96), which holds for these states. These same average values may also allow us to verify the virial theorem studied in Section 2.4.4. This is the purpose of the present section.

When applied to harmonic oscillators, the virial theorem leads to Eqs. (2.88) and (2.89) from which results the following relation between the average values of the

Hamiltonian and those of the kinetic **T** and potential **V** operators, what may be the stationary state over which the averages are performed:

$$\langle \mathbf{T} \rangle_{|\Psi\rangle} = \langle \mathbf{V} \rangle_{|\Psi\rangle} = \tfrac{1}{2} \langle \mathbf{H} \rangle_{|\Psi\rangle} \tag{5.98}$$

To verify if our above results of the present chapter are in agreement with this theorem, first consider the average value of the harmonic potential over the states $|\{n\}\rangle$, which, being the eigenstates of the harmonic oscillator Hamiltonian, are therefore stationary:

$$\langle \mathbf{V} \rangle_{|\{n\}\rangle} = \tfrac{1}{2} m\omega^2 \langle \{n\}|\mathbf{Q}^2|\{n\}\rangle$$

Owing to Eq. (5.91), this takes the form

$$\langle \mathbf{V} \rangle_{|\{n\}\rangle} = \tfrac{1}{2} \hbar\omega \left(n + \tfrac{1}{2}\right) \tag{5.99}$$

On the other hand, the corresponding average value of the kinetic operator is

$$\langle \mathbf{T} \rangle_{|\{n\}\rangle} = \tfrac{1}{2m} \langle \{n\}|\mathbf{P}^2|\{n\}\rangle$$

or, in view of Eq. (5.94),

$$\langle \mathbf{T} \rangle_{|\{n\}\rangle} = \tfrac{1}{2} \hbar\omega \left(n + \tfrac{1}{2}\right) \tag{5.100}$$

Thus, according to Eq. (5.9), the average value of the harmonic Hamiltonian is

$$\langle \mathbf{H} \rangle_{|\{n\}\rangle} = \hbar\omega \langle \{n\}| \left(\mathbf{a}^\dagger \mathbf{a} + \tfrac{1}{2}\right) |\{n\}\rangle$$

or, due to Eq. (5.90),

$$\langle \mathbf{H} \rangle_{|\{n\}\rangle} = \hbar\omega \left(n + \tfrac{1}{2}\right) \tag{5.101}$$

Hence, as it may be observed, Eqs. (5.99)–(5.101) obey the virial theorem (5.98).

5.1.7 3D Harmonic oscillators

The previous sections dealt with 1D harmonic oscillators. The generalization of 1D results to 3D harmonic oscillators is the aim of the present section.

The kinetic operator **T** of a 3D oscillator of reduced mass m is

$$\mathbf{T} = \left(\frac{\mathbf{P}_x^2 + \mathbf{P}_y^2 + \mathbf{P}_z^2}{2m} \right)$$

where the $\mathbf{P}_x$, $\mathbf{P}_y$, and $\mathbf{P}_z$ are, respectively, the x, y, and z Cartesian components of the momentum operator. On the other hand, the potential operator is

$$\mathbf{V} = \tfrac{1}{2} m \left(\omega_x^2 \mathbf{Q}_x^2 + \omega_y^2 \mathbf{Q}_y^2 + \omega_z^2 \mathbf{Q}_z^2\right)$$

where the $\mathbf{Q}_x$, $\mathbf{Q}_y$, and $\mathbf{Q}_z$ are, respectively, the x, y, and z Cartesian components of the position operator, obeying the commutation rules

$$[\mathbf{Q}_k, \mathbf{P}_l] = i\hbar\, \delta_{kl} \tag{5.102}$$

where k and l run for x, y, and z, whereas the ω_k are the corresponding angular frequencies. Then, the Hamiltonian of the oscillator yields

$$\mathbf{H} = \left(\frac{\mathbf{P}_x^2 + \mathbf{P}_y^2 + \mathbf{P}_z^2}{2m} \right) + \frac{1}{2} m \left(\omega_x^2 \mathbf{Q}_x^2 + \omega_y^2 \mathbf{Q}_y^2 + \omega_z^2 \mathbf{Q}_z^2\right) \tag{5.103}$$

In a similar way as in Eqs. (5.6) and (5.7), one may express the position and momentum operators in terms of dimensionless non-Hermitian operators according to

$$\mathbf{Q}_k = \sqrt{\frac{\hbar}{2m\omega}}(\mathbf{a}_k^\dagger + \mathbf{a}_k) \tag{5.104}$$

$$\mathbf{P}_k = i\sqrt{\frac{m\hbar\omega}{2}}(\mathbf{a}_k^\dagger - \mathbf{a}_k) \tag{5.105}$$

with the following commutation rule between $\mathbf{a}_k$ and $\mathbf{a}_l^\dagger$ resulting from Eq. (5.102):

$$[\mathbf{a}_k, \mathbf{a}_l^\dagger] = \delta_{kl} \tag{5.106}$$

Again, proceeding as at the beginning of this chapter, the Hamiltonian (5.103) takes the form

$$\mathbf{H} = \mathbf{H}_x + \mathbf{H}_y + \mathbf{H}_z \tag{5.107}$$

with

$$\mathbf{H}_k = \hbar\omega_k\left(\mathbf{a}_k^\dagger\mathbf{a}_k + \tfrac{1}{2}\right) \quad \text{with} \quad k = x, y, z$$

Then, since each term $\mathbf{H}_k$ of the Hamiltonian $\mathbf{H}$ has the same structure as that of Eq. (5.41) of the Hamiltonian of 1D harmonic oscillators, one may write for each Hamiltonian $\mathbf{H}_k$ an eigenvalue equation having the same structure as that of (5.42), that is,

$$\hbar\omega_k\left(\mathbf{a}_k^\dagger\mathbf{a}_k + \tfrac{1}{2}\right)|\{n\}_k\rangle = E_{n_k}|\{n\}_k\rangle \tag{5.108}$$

with, for $k = x, y$, and z,

$$E_{n_k} = \hbar\omega_k\left(n_k + \tfrac{1}{2}\right) \qquad \textit{and} \qquad n_k = 0, 1, 2, 3, \ldots$$

In Eq. (5.108), the $|\{n\}_k\rangle$ are the eigenkets of the $\mathbf{H}_k$ Hamiltonians, whereas the E_{n_k} are the corresponding eigenvalues. Of course, since the Hamiltonians $\mathbf{H}_k$ are Hermitian, their eigenkets are orthonormal:

$$\langle\{n\}_k|\{m\}_k\rangle = \delta_{n_k m_k}$$

and, thereby, form a basis allowing us to write for each dimension the closure relation, that is,

$$\sum_{n_k}|\{n\}_k\rangle\langle\{n\}_k| = 1$$

Now, as for the particle-in-a-box model, the full eigenkets of the 3D Hamiltonian (5.107) must be written as the products of the eigenkets of the 1D Hamiltonians $\mathbf{H}_k$, that is

$$|\Psi_{n_x n_y n_z}\rangle = |\{n\}_x\rangle|\{n\}_y\rangle|\{n\}_z\rangle \tag{5.109}$$

whereas the corresponding eigenvalue of the 3D Hamiltonian must be the sum of the corresponding eigenvalues E_{n_k}, that is,

$$E_{n_x n_y n_z} = \hbar\omega_x\left(n_x + \tfrac{1}{2}\right) + \hbar\omega_y\left(n_y + \tfrac{1}{2}\right) + \hbar\omega_z\left(n_z + \tfrac{1}{2}\right) \tag{5.110}$$

Of course, the eigenkets (5.109) and the eigenvalues (5.110) are related through the eigenvalue equation

$$\mathbf{H}|\{n\}_x\rangle|\{n\}_y\rangle|\{n\}_z\rangle$$
$$= \left\{\hbar\omega_x\left(n_x+\tfrac{1}{2}\right)+\hbar\omega_y\left(n_y+\tfrac{1}{2}\right)+\hbar\omega_z\left(n_z+\tfrac{1}{2}\right)\right\}|\{n\}_x\rangle|\{n\}_y\rangle|\{n\}_z\rangle$$

Moreover, when the 3D harmonic oscillator is isotropic, the eigenvalue (5.110) reduces to

$$E_{n_xn_yn_z} = \hbar\omega\left((n_x+n_y+n_z)+\tfrac{3}{2}\right) \tag{5.111}$$

Hence, as for the particle-in-a-box model, it appears that degeneracy occurs for all situations having the same value of $(n_x+n_y+n_z)$ verifying Eq. (5.111).

Owing to the equivalence between the three Hamiltonians $\mathbf{H}_x$, $\mathbf{H}_y$, and $\mathbf{H}_z$ and the Hamiltonian of the 1D harmonic oscillator, all that has been proved for 1D oscillators may be transposed to the 3D ones. Using equations similar to Eqs. (5.53) and (5.63), namely,

$$\mathbf{a}_k|\{n\}_k\rangle = \sqrt{n_k}|\{n_k-1\}\rangle$$

$$\mathbf{a}_k^{\dagger}|\{n\}_k\rangle = \sqrt{n_k+1}|\{n_k+1\}\rangle$$

and by the aid of Eqs. (5.106) and (5.108), it is possible to reproduce for each component of the 3D oscillator the results obtained in the 1D situation, particularly those concerning the Heisenberg uncertainty relations and the virial theorem.

5.2 WAVEFUNCTIONS CORRESPONDING TO HAMILTONIAN EIGENKETS

Although the kets and the corresponding wavefunctions are without direct physical meaning, it may be of interest, for the purpose of physical intuitive investigation, to visualize the forms of the wavefunctions corresponding to the eigenvectors of the Hamiltonian of quantum harmonic oscillators. One of the reasons, which will appear later, is that the number of nodes of these vibrational wavefunctions increases with the corresponding energy in a way that is deeply linked to the de Broglie wavelength rule according to which the kinetic energy increases with the number of nodes of the associated wavelength.

5.2.1 Second-order partial differential equation to be solved

In order to get the expression of the wavefunctions corresponding to the eigenkets of the harmonic Hamiltonian, consider this operator within wave mechanics that reads

$$\hat{H} = \hat{T} + \hat{V}$$

where $\hat{T}$ and $\hat{V}$ are, respectively, the wave mechanical kinetic and potential operators, the first one being given by Eq. (3.51), that is,

$$\hat{T} = -\frac{\hbar^2}{2m}\frac{\partial^2}{\partial Q^2}$$

and the last one simply by

$$\hat{V} = \tfrac{1}{2}m\omega^2 Q^2$$

According to Eq. (3.60), the Hamiltonian of the harmonic oscillator is, therefore,

$$\hat{H} = -\frac{\hbar^2}{2m}\frac{\partial^2}{\partial Q^2} + \frac{1}{2}m\omega^2 Q^2$$

Then, the time-independent Schrödinger equation

$$\hat{H}\Psi_n(Q) = E_n\Psi_n(Q) \tag{5.112}$$

reads

$$-\frac{\hbar^2}{2m}\left(\frac{\partial^2\Psi_n(Q)}{\partial Q^2}\right) + \frac{1}{2}m\omega^2 Q^2\Psi_n(Q) = E_n\Psi_n(Q) \tag{5.113}$$

Of course, since the quantum mechanics and the wave mechanics are equivalent, the eigenvalues of the Hamiltonian appearing in Eq. (5.112) are given by Eq. (5.42), that is,

$$E_n = \hbar\omega\left(n + \tfrac{1}{2}\right) \tag{5.114}$$

As a consequence, the eigenvalue equation (5.113) becomes

$$\frac{\hbar^2}{2m}\left(\frac{\partial^2\Psi_n(Q)}{\partial Q^2}\right) - \left(\frac{m\omega^2}{2}Q^2 - \hbar\omega\left(n + \frac{1}{2}\right)\right)\Psi_n(Q) = 0 \tag{5.115}$$

That is the equation to be solved in order to get the expression of the wavefunction $\Psi_n(Q)$ given by the scalar product

$$\Psi_n(Q) = \langle\{Q\}|\{n\}\rangle$$

its boundary condition being

$$\Psi_n(Q) \to 0 \quad \text{when} \quad Q \to \infty \tag{5.116}$$

Next, perform the following variable change:

$$\xi = \xi^\circ Q \quad \text{with} \quad \xi^\circ = \sqrt{\frac{m\omega}{\hbar}} \tag{5.117}$$

leading to

$$\frac{\partial\xi}{\partial Q} = \xi^\circ \qquad \text{and} \qquad \frac{\partial}{\partial Q} = \frac{\partial}{\partial\xi}\left(\frac{\partial\xi}{\partial Q}\right) = \xi^\circ\frac{\partial}{\partial\xi} \tag{5.118}$$

and thus, using in turn Eq. (5.117), to

$$\frac{\partial^2}{\partial Q^2} = \xi^{\circ 2}\frac{\partial^2}{\partial\xi^2} = \frac{m\omega}{\hbar}\frac{\partial^2}{\partial\xi^2} \tag{5.119}$$

Thereby, using Eqs. (5.117) and (5.119), Eq. (5.115) becomes

$$\frac{\hbar^2}{2m}\frac{m\omega}{\hbar}\left(\frac{\partial^2\Psi_n(\xi)}{\partial\xi^2}\right) - \left(\frac{m\omega^2}{2}\frac{\hbar}{m\omega}\xi^2 - \left(n + \frac{1}{2}\right)\hbar\omega\right)\Psi_n(\xi) = 0$$

or, after simplifying by $\hbar\omega$

$$\left(\frac{\partial^2 \Psi_n(\xi)}{\partial \xi^2}\right) - (\xi^2 - (2n+1))\Psi_n(\xi) = 0 \tag{5.120}$$

with, due to Eq. (5.116), the following boundary condition:

$$\Psi_n(\xi) \to 0 \quad \text{when} \quad \xi \to \infty \tag{5.121}$$

resulting from the fact that the wavefunctions have to be normalized in order to verify that their squared modulus must be a density probability obeying

$$\int_{-\infty}^{+\infty} |\Psi_n(\xi)|^2 d\xi = 1 \tag{5.122}$$

5.2.2 Special solutions of Eq. (5.120)

Now, look at Eq. (5.120) for the lowest state situation corresponding to $n = 0$, that is,

$$\left(\frac{\partial^2 \Psi_0(\xi)}{\partial \xi^2}\right) - (\xi^2 - 1)\Psi_0(\xi) = 0 \tag{5.123}$$

with the same boundary condition (5.121). Search for a solution of the form

$$\Psi_0(\xi) = (e^{\pm\xi^2/2}) \tag{5.124}$$

Then, it yields

$$\left(\frac{\partial e^{\pm\xi^2/2}}{\partial \xi}\right) = \pm\xi(e^{\pm\xi^2/2}) \qquad \left(\frac{\partial^2 e^{\pm\xi^2/2}}{\partial \xi^2}\right) = (\xi^2 \pm 1)(e^{\pm\xi^2/2})$$

so that the second partial derivative of the expression (5.124) reads

$$\left(\frac{\partial^2 \Psi_0(\xi)}{\partial \xi^2}\right) - (\xi^2 \pm 1)\Psi_0(\xi) = 0 \tag{5.125}$$

Thus, it appears that the two solutions of Eq. (5.123) are verified. But, the boundary condition (5.121) being not compatible with the + solution, the physical solution is necessarily the following one:

$$\Psi_0(\xi) = e^{-\xi^2/2}$$

This last equation is the unnormalized ground-state wavefunction of the Hamiltonian of the harmonic oscillator satisfying Eq. (5.115) with the ground-state energy $\hbar\omega/2$, its normalized form being

$$\Psi_0(\xi) = C_0(e^{-\xi^2/2}) \tag{5.126}$$

where C_0 is the normalization constant of the wavefunction.

The normalization constant C_0 must be such that Eq. (5.122) has to be satisfied. Hence, using Eq. (5.117) in order to return from the dimensionless variable ξ to the dimensioned one Q, the normalization of the ground-state wavefunction (5.126) reads

$$(C_0^2)^{-1} = \int_{-\infty}^{+\infty} \exp\left(-\frac{m\omega}{\hbar}Q^2\right) dQ$$

and, thus, after integration

$$(C_0^2)^{-1} = \sqrt{\frac{\pi\hbar}{m\omega}} \qquad \text{or} \qquad C_0 = \left(\frac{m\omega}{\pi\hbar}\right)^{1/4} \tag{5.127}$$

5.2.3 Recurrence relation between wavefunctions

Now, in order to pass from the ground-state wavefunction to the excited wavefunctions, use Eq. (5.63), that is,

$$\mathbf{a}^\dagger |\{n\}\rangle = \sqrt{n+1}|\{n+1\}\rangle$$

Then premultiplying both terms by any bra, Hermitian conjugate of some eigenket of the position operator, one obtains

$$\langle\{Q\}|\{n+1\}\rangle = \frac{1}{\sqrt{n+1}}\langle\{Q\}|\mathbf{a}^\dagger|\{n\}\rangle \tag{5.128}$$

or, due to Eq. (5.4), it reads

$$\langle\{Q\}|\{n+1\}\rangle = \frac{1}{\sqrt{2}}\frac{1}{\sqrt{n+1}}\langle\{Q\}|(\xi^\circ\mathbf{Q} - i\zeta^\circ\mathbf{P})|\{n\}\rangle$$

with

$$\xi^\circ = \sqrt{\frac{m\omega}{\hbar}} \qquad \text{and} \qquad \zeta^\circ = \sqrt{\frac{1}{m\hbar\omega}} \tag{5.129}$$

Again, introduce the closure relation over the eigenstates of the position operator:

$$\langle\{Q\}|\{n+1\}\rangle = \frac{1}{\sqrt{2}}\frac{1}{\sqrt{n+1}}\langle\{Q\}|(\xi^\circ\mathbf{Q} - i\zeta^\circ\mathbf{P})\left\{\int_{-\infty}^{+\infty}|\{Q'\}\rangle\langle\{Q'\}|dQ'\right\}|\{n\}\rangle$$

leading to

$$\langle\{Q\}|\{n+1\}\rangle = \frac{1}{\sqrt{2}}\frac{1}{\sqrt{n+1}}\langle\{Q\}|(\xi^\circ\mathbf{Q} - i\zeta^\circ\mathbf{P})\int_{-\infty}^{+\infty}|\{Q'\}\rangle\langle\{Q'\}|\{n\}\rangle dQ'$$

Hence, Eq. (5.128) becomes

$$\Psi_{n+1}(Q) = \frac{1}{\sqrt{2}}\frac{1}{\sqrt{n+1}}\langle\{Q\}|(\xi^\circ\mathbf{Q} - i\zeta^\circ\mathbf{P})\int_{-\infty}^{+\infty}|\{Q'\}\rangle\Psi_n(Q')\,dQ' \tag{5.130}$$

with

$$\Psi_{n+1}(Q) = \langle\{Q\}|\{n+1\}\rangle \qquad \text{and} \qquad \Psi_n(Q') = \langle\{Q'\}|\{n\}\rangle$$

Next, observe that, according to Eqs. (3.50) and (3.52)

$$\langle\{Q\}|\mathbf{P}|\{Q'\}\rangle = -i\hbar\delta(Q - Q')\frac{\partial}{\partial Q}$$

$$\langle\{Q\}|\mathbf{Q}|\{Q'\}\rangle = \delta(Q - Q')Q'$$

As a consequence, using these expressions and the fact that, in wave mechanics, **Q** acts as a scalar Q, Eq. (5.130) transforms to

$$\Psi_{n+1}(Q) = \frac{1}{\sqrt{2}}\frac{1}{\sqrt{n+1}}\left(\xi^{\circ}Q - \hbar\zeta^{\circ}\frac{\partial}{\partial Q}\right)\Psi_n(Q) \tag{5.131}$$

Again, in view of Eqs. (5.117) and (5.118) and since [cf., Eq. (5.129)] the following relation between ζ° and ξ° exists

$$\hbar\zeta^{\circ} = \frac{1}{\xi^{\circ}}$$

so that Eq. (5.131) takes the final recurrence form

$$\boxed{\Psi_{n+1}(\xi) = \frac{1}{\sqrt{2}}\frac{1}{\sqrt{n+1}}\left(\xi - \frac{\partial}{\partial\xi}\right)\Psi_n(\xi)} \tag{5.132}$$

keeping in mind Eq. (5.117), that is,

$$\xi = \xi^{\circ}Q$$

5.2.4 Obtaining the lowest wavefunctions

5.2.4.1 First excited wavefunction Now, recall that the ground-state wavefunction (5.126), for reasons that will become apparent, may be formally written

$$\Psi_0(\xi) = C_0H_0(\xi)(e^{-\xi^2/2}) \tag{5.133}$$

with

$$H_0(\xi) = 1 \tag{5.134}$$

and where C_0 is the normalization constant of the wavefunction. Apply Eq. (5.132) to the ground-state wavefunction (5.133), that is, for $n=0$

$$\Psi_1(\xi) = \frac{1}{\sqrt{2}}\frac{1}{\sqrt{1}}C_0\left(\xi - \frac{\partial}{\partial\xi}\right)(e^{-\xi^2/2}) \tag{5.135}$$

The partial derivative with respect to ξ being

$$\frac{\partial}{\partial\xi}(e^{-\xi^2/2}) = -(\xi e^{-\xi^2/2})$$

Eq. (5.135) yields

$$\Psi_1(\xi) = C_1 2(\xi e^{-\xi^2/2}) \tag{5.136}$$

where

$$C_1 = \frac{1}{\sqrt{2}}C_0 = \frac{1}{\sqrt{2}}\left(\frac{m\omega}{\pi\hbar}\right)^{1/4} \tag{5.137}$$

Finally, Eq. (5.136) may be written in a form similar to that of Eq. (5.133):

$$\Psi_1(\xi) = C_1H_1(\xi)e^{-\xi^2/2} \quad \text{with} \quad H_1(\xi) = 2\xi \tag{5.138}$$

5.2.4.2 *Second excited wavefunction*

Next, in order to get the second excited-state wavefunction, apply Eq. (5.132) to Eq. (5.136), that is

$$\Psi_2(\xi) = \frac{1}{\sqrt{2}}\frac{1}{\sqrt{2!}}\left(\xi - \frac{\partial}{\partial \xi}\right)\{C_1 2(\xi e^{-\xi^2/2})\}$$

which reads

$$\Psi_2(\xi) = 2C_2\left(\xi(\xi e^{-\xi^2/2}) - \frac{\partial}{\partial \xi}(\xi e^{-\xi^2/2})\right) \tag{5.139}$$

with, in view of Eq. (5.137),

$$C_2 = \frac{1}{\sqrt{2!}}\left(\frac{1}{\sqrt{2}}\right)C_1 = \frac{1}{\sqrt{2!}}\left(\frac{1}{\sqrt{2}}\right)^2 C_0 \tag{5.140}$$

By derivation one obtains

$$\frac{\partial}{\partial \xi}(\xi e^{-\xi^2/2}) = (1-\xi^2)(e^{-\xi^2/2}) \tag{5.141}$$

so that Eq. (5.139) yields

$$\Psi_2(\xi) = C_2(4\xi^2 - 2)(e^{-\xi^2/2}) \tag{5.142}$$

or

$$\Psi_2(\xi) = C_2 H_2(\xi)(e^{-\xi^2/2}) \quad \text{with} \quad H_2(\xi) = 4\xi^2 - 2 \tag{5.143}$$

5.2.4.3 *Third excited wavefunction*

Again, to get the third excited-state wavefunction, apply Eq. (5.132) a third time to Eq. (5.142), leading to

$$\Psi_3(\xi) = C_3\left(\xi - \frac{\partial}{\partial \xi}\right)\{(4\xi^2 - 2)e^{-\xi^2/2}\} \tag{5.144}$$

with

$$C_3 = \frac{1}{\sqrt{2}}\frac{1}{\sqrt{3!}}C_2 = \frac{1}{\sqrt{3!}}\left(\frac{1}{\sqrt{2}}\right)^3 C_0$$

which transforms to

$$\Psi_3(\xi) = C_3\left((4\xi^3 - 2\xi)(e^{-\xi^2/2}) - \frac{\partial}{\partial \xi}\{(4\xi^2 - 2)(e^{-\xi^2/2})\}\right)$$

Then, obtaining by differentiation

$$\frac{\partial}{\partial \xi}(4\xi^2 - 2)(e^{-\xi^2/2}) = (8\xi - (4\xi^2 - 2)\xi)(e^{-\xi^2/2})$$

the wavefunction becomes

$$\Psi_3(\xi) = C_3(8\xi^3 - 12\xi)(e^{-\xi^2/2}) \tag{5.145}$$

or

$$\Psi_3(\xi) = C_3 H_3(\xi)(e^{-\xi^2/2}) \quad \text{with} \quad H_3(\xi) = 8\xi^3 - 12\xi \tag{5.146}$$

5.2.4.4 nth excited wavefunction Note that the functions $H_k(\xi)$ (5.134), (5.138), (5.143), and (5.146) are the first *Hermite polynomials*. Besides, proceeding in a similar way for the higher excited wavefunctions, one would obtain now

$$\Psi_n(\xi) = C_n H_n(\xi)(e^{-\xi^2/2}) \tag{5.147}$$

with

$$C_n = \frac{1}{\sqrt{n!}}\left(\frac{1}{\sqrt{2}}\right)^n C_0$$

or, in view of Eq. (5.127),

$$C_n = \frac{1}{\sqrt{n!}}\left(\frac{1}{\sqrt{2}}\right)^n \left(\frac{m\omega}{\hbar\pi}\right)^{1/4} \tag{5.148}$$

and with, for $n = 4$, 5, and 6,

$$\begin{aligned} H_4(\xi) &= 16\xi^4 - 48\xi^2 + 12 \\ H_5(\xi) &= 32\xi^5 - 160\xi^3 + 120\xi \\ H_6(\xi) &= 64\xi^6 - 480\xi^4 + 720\xi^2 - 120 \end{aligned} \tag{5.149}$$

5.2.4.5 Pictorial representation of the lowest wavefunctions The five lowest wavefunctions and energy levels are pictured in Fig. 5.1a, whereas the corresponding wavefunctions and energy levels of the particle-in-a-box model are shown in Fig. 5.1b.

Observe that, in agreement with Eqs. (5.126), (5.138), (5.143), (5.146), and (5.147), the parity of the wavefunctions $\Psi_n(\xi)$ is alternatively changing, those characterized by *even* quantum numbers n, being *gerade* and the other ones, characterized by *odd* quantum numbers n, being *ungerade*. Observe also that this figure illustrates the node number increase of the wavefunctions when enhancing the quantum number and thus the energy, an evolution that is almost similar to that encountered in the particle-in-a-box model, as may be verified by inspection of Fig. 5.1(b).

5.3 DYNAMICS

In the previous sections we found many important results dealing with static situations of quantum harmonic oscillators. We have now to search the dynamics of these oscillators via the time dependence of the mean values of the basic operators averaged over the eigenkets of the harmonic oscillator Hamiltonian. To get these time dependent average values, it will be suitable to work within the Heisenberg picture (where the operators depend on time whereas the kets remain constant).

5.3.1 Heisenberg equations for oscillator operators

5.3.1.1 Ladder operators Heisenberg equations Look, therefore, at the Heisenberg equation governing the dynamical equation of the lowering operator $\mathbf{a}(t)$, which, according to Eq. (3.94), reads in the present situation

$$i\hbar\left(\frac{d\mathbf{a}(t)}{dt}\right) = [\mathbf{a}(t), \mathbf{H}] \tag{5.150}$$

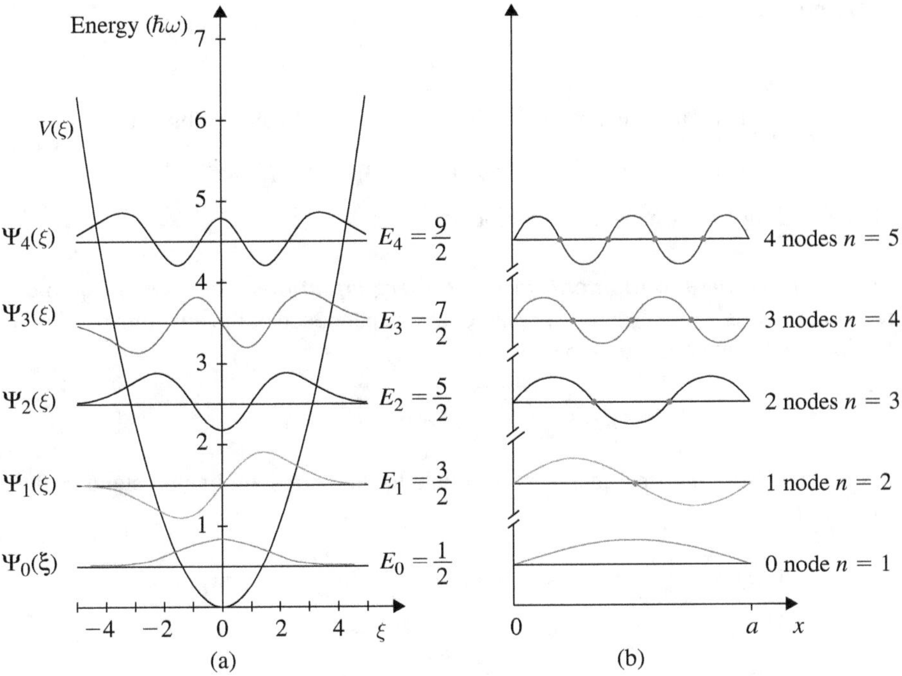

Figure 5.1 Five lowest energy levels and wavefunctions. Comparison between (a) quantum harmonic oscillator and (b) particle-in-a-box model.

which, with the help of Eq. (5.9) defining the Hamiltonian, becomes

$$i\hbar\left(\frac{d\mathbf{a}(t)}{dt}\right) = \hbar\omega\left[\mathbf{a}(t), \left(\mathbf{a}(t)^{\dagger}\mathbf{a}(t) + \frac{1}{2}\right)\right]$$

or

$$i\left(\frac{d\mathbf{a}(t)}{dt}\right) = \omega(\mathbf{a}(t)\mathbf{a}(t)^{\dagger}\mathbf{a}(t) - \mathbf{a}(t)^{\dagger}\mathbf{a}(t)\mathbf{a}(t))$$

and thus

$$\left(\frac{d\mathbf{a}(t)}{dt}\right) = -i\omega[\mathbf{a}(t), \mathbf{a}(t)^{\dagger}]\mathbf{a}(t)$$

Again, the commutation rule (5.5) holds for any time, so that

$$[\mathbf{a}(t), \mathbf{a}(t)^{\dagger}] = 1$$

and thereby

$$\left(\frac{d\mathbf{a}(t)}{dt}\right) = -i\omega\mathbf{a}(t)$$

Hence, after integration, that gives

$$\boxed{\mathbf{a}(t) = \mathbf{a}(0)e^{-i\omega t}} \tag{5.151}$$

the Hermitian conjugate of which being

$$\boxed{\mathbf{a}^\dagger(t) = \mathbf{a}^\dagger(0)e^{i\omega t}} \tag{5.152}$$

Now, observe that when dealing with 3D oscillators, the latter results read

$$\mathbf{a}_k(t) = \mathbf{a}_k(0)e^{-i\omega_k t} \qquad \text{and} \qquad \mathbf{a}_k^\dagger(t) = \mathbf{a}_k^\dagger(0)e^{i\omega_k t}$$

with k standing for the x, y, and z components.

5.3.1.2 Position and momentum Heisenberg equations In view of Eq. (5.6), and owing to Eqs. (5.151) and (5.152), the time dependence of the quantum oscillator position coordinate appears to be

$$\mathbf{Q}(t) = \sqrt{\frac{\hbar}{2m\omega}}(\mathbf{a}^\dagger e^{i\omega t} + \mathbf{a}e^{-i\omega t}) \tag{5.153}$$

Passing then from the imaginary exponentials to the corresponding sine and cosine functions leads to

$$\mathbf{Q}(t) = \sqrt{\frac{\hbar}{2m\omega}}(\mathbf{a}^\dagger \cos\omega t + i\mathbf{a}^\dagger \sin\omega t + \mathbf{a}\cos\omega t - i\mathbf{a}\sin\omega t)$$

or

$$\mathbf{Q}(t) = \sqrt{\frac{\hbar}{2m\omega}}((\mathbf{a}^\dagger + \mathbf{a})\cos\omega t + i(\mathbf{a}^\dagger - \mathbf{a})\sin\omega t) \tag{5.154}$$

so that, due to Eqs. (5.6) and (5.7), we have

$$\boxed{\mathbf{Q}(t) = \mathbf{Q}(0)\cos\omega t + \frac{1}{m\omega}\mathbf{P}(0)\sin\omega t} \tag{5.155}$$

with, respectively,

$$\mathbf{P}(0) = i\sqrt{\frac{\hbar m\omega}{2}}(\mathbf{a}^\dagger - \mathbf{a}) \tag{5.156}$$

$$\mathbf{Q}(0) = \sqrt{\frac{\hbar m\omega}{2}}(\mathbf{a}^\dagger + \mathbf{a}) \tag{5.157}$$

Now, consider the commutator of the position coordinate operators at different times, that is,

$$\begin{aligned}&[\mathbf{Q}(t), \mathbf{Q}(t')]\\ &= \frac{\hbar}{2m\omega}((\mathbf{a}^\dagger e^{i\omega t} + \mathbf{a}e^{-i\omega t})(\mathbf{a}^\dagger e^{i\omega t'} + \mathbf{a}e^{-i\omega t'}) - (\mathbf{a}^\dagger e^{i\omega t'} + \mathbf{a}e^{-i\omega t'})\\ &\quad \times (\mathbf{a}^\dagger e^{i\omega t} + \mathbf{a}e^{-i\omega t}))\end{aligned}$$

which, after performing the products and simplification gets

$$[\mathbf{Q}(t), \mathbf{Q}(t')] = \frac{\hbar}{2m\omega}(\mathbf{a}^\dagger\mathbf{a}(e^{i\omega(t-t')} - e^{-i\omega(t-t')}) + \mathbf{a}\mathbf{a}^\dagger(e^{-i\omega(t-t')} - e^{i\omega(t-t')}))$$

and, thus, after coming back to the sine function, we have

$$[\mathbf{Q}(t), \mathbf{Q}(t')] = \frac{\hbar}{2m\omega}(\mathbf{a}^{\dagger}\mathbf{a}2i \sin(\omega(t-t')) - \mathbf{a}\mathbf{a}^{\dagger}2i \sin(\omega(t-t')))$$

Then, with the help of the commutation rule (5.5), this last result reduces, after simplification, to

$$\boxed{[\mathbf{Q}(t), \mathbf{Q}(t')] = -i\frac{\hbar}{m\omega} \sin(\omega(t-t'))} \tag{5.158}$$

We emphasizes that this commutator differs from zero, that is,

$$[\mathbf{Q}(t), \mathbf{Q}(t')] \neq 0$$

On the other hand, the time dependence of the momentum, which reads in view of Eq. (5.7)

$$\mathbf{P}(t) = i\sqrt{\frac{\hbar m\omega}{2}}(\mathbf{a}^{\dagger}e^{i\omega t} - \mathbf{a}e^{-i\omega t}) \tag{5.159}$$

transforms after passing to the trigonometric functions and using Eq. (5.7)

$$\mathbf{P}(t) = i\sqrt{\frac{\hbar m\omega}{2}}((\mathbf{a}^{\dagger} - \mathbf{a}) \cos \omega t + i(\mathbf{a}^{\dagger} + \mathbf{a}) \sin \omega t)$$

so that

$$\mathbf{P}(t) = i\sqrt{\frac{\hbar m\omega}{2}}(\mathbf{a}^{\dagger} - \mathbf{a}) \cos \omega t - \sqrt{\frac{\hbar m\omega}{2}}(\mathbf{a}^{\dagger} + \mathbf{a}) \sin \omega t \tag{5.160}$$

and, therefore, due to Eqs. (5.156) and (5.157), leads to

$$\boxed{\mathbf{P}(t) = \mathbf{P}(0) \cos \omega t - m\omega\mathbf{Q}(0) \sin \omega t} \tag{5.161}$$

Hence, the commutator of **P** at different times is not zero.

5.3.2 Time dependence of mean values averaged on Hamiltonian eigenkets

Recall that operators, unlike average values, are not directly connected with experience. Thus, it is now necessary to study the dependence of the average values of the operators defined by Eq. (5.155) or (5.154) and by Eq. (5.160) or (5.161).

5.3.2.1 Average Values of Q(t) and P(t) First, consider the average values of the momentum and position coordinates on the eigenkets of the harmonic Hamiltonian. According to Eq. (5.154), that of the position operator reads

$$\langle\{n\}|\mathbf{Q}(t)|\{n\}\rangle = \sqrt{\frac{\hbar}{2m\omega}}(\langle\{n\}|\mathbf{a}^{\dagger}|\{n\}\rangle e^{i\omega t} + \langle\{n\}|\mathbf{a}|\{n\}\rangle e^{-i\omega t})$$

Then, keeping in mind Eq. (5.53) and its Hermitian conjugate, that is,

$$\mathbf{a}|\{n\}\rangle = \sqrt{n}|\{n-1\}\rangle \qquad \text{and} \qquad \langle\{n\}|\mathbf{a}^{\dagger} = \sqrt{n}\langle\{n-1\}|$$

the average value of the position coordinate becomes

$$\langle\{n\}|\mathbf{Q}(t)|\{n\}\rangle = \frac{1}{2}\sqrt{\frac{\hbar}{2m\omega}}(\langle\{n-1\}|\{n\}\rangle e^{i\omega t} + \langle\{n\}|\{n-1\}\rangle e^{-i\omega t})$$

Thus, owing to the orthonormality of the eigenkets of the harmonic oscillator Hamiltonian, we have

$$\langle\{n\}|\mathbf{Q}(t)|\{n\}\rangle = 0 \tag{5.162}$$

Now, proceeding in the same way for the momentum coordinate by the aid of Eq. (5.159), it may be easily shown that

$$\langle\{n\}|\mathbf{P}(t)|\{n\}\rangle = 0 \tag{5.163}$$

Note that the results (5.162) may be also found by the aid of the wave mechanics. In this quantum picture, the left-hand side of Eq. (5.162) reads

$$\langle\{n\}|\mathbf{Q}(t)|\{n\}\rangle = \int_{-\infty}^{+\infty} \Psi_n(Q)Q(t)\Psi_n(Q)\, dQ$$

where

$$\Psi_n(Q) \equiv \langle\{Q\}|\{n\}\rangle$$

Now, since Q commutes with $\Psi_n(Q)$ and irrespective of the time t, which may be omitted, the average value reduces to

$$\langle\{n\}|\mathbf{Q}|\{n\}\rangle = \int_{-\infty}^{+\infty} \Psi_n(Q)^2 Q\, dQ$$

Now, observe that, whatever $\Psi_n(Q)$, the parity of its square is always even, whereas that of Q is odd. Hence, the parity of the integrand, which is the product of that of $\Psi_n(Q)^2$ by that of Q, is always odd, so that the integral involving this integrand must be necessarily zero, the contribution from 0 to $+\infty$ being canceled by that from $-\infty$ to 0, that is,

$$\int_{-\infty}^{+\infty} \Psi_n(Q)^2 Q\, dQ = 0$$

On the other hand, in the Schrödinger picture, Eq. (5.163) takes the form

$$\langle\{n\}(t)|\mathbf{P}|\{n\}(t)\rangle = 0$$

Then, inserting a closure relation over the basis $\{|\{Q\}\rangle\}$ before and after $\mathbf{P}$, it transforms by the aid of Eq. (3.50) into

$$-i\hbar \int_{-\infty}^{+\infty} \Psi_n(Q)^* \frac{\partial}{\partial Q}\Psi_n(Q)\, dQ = 0$$

Of course, Eqs. (5.162) and (5.163) may be immediately extended to the 3D oscillator to give for the three Cartesian components $k = x, y$, and z:

$$\langle\{n\}|\mathbf{Q}_k(t)|\{n\}\rangle = \langle\{n\}|\mathbf{P}_k(t)|\{n\}\rangle = 0$$

5.3.2.2 Average values of V(t) and T(t) Now, consider the time dependence of the average potential energy. In the Heisenberg representation, this average value reads

$$\langle\{n\}|\mathbf{V}(t)|\{n\}\rangle = \tfrac{1}{2}m\omega^2\langle\{n\}|\mathbf{Q}(t)^2|\{n\}\rangle \tag{5.164}$$

Next, the right-hand-side average value may be expressed in terms of the raising and lowering operators by the aid of Eq. (5.153):

$$\langle\{n\}|\mathbf{Q}(t)^2|\{n\}\rangle = \frac{\hbar}{2m\omega}\langle\{n\}|(\mathbf{a}^\dagger e^{i\omega t} + \mathbf{a}e^{-i\omega t})^2|\{n\}\rangle \tag{5.165}$$

By expansion of the square involved on the right-hand side, one finds

$$(\mathbf{a}^\dagger e^{i\omega t} + \mathbf{a}e^{-i\omega t})^2 = (\mathbf{a}^\dagger)^2 e^{2i\omega t} + (\mathbf{a})^2 e^{-2i\omega t} + \mathbf{a}^\dagger\mathbf{a} + \mathbf{a}\mathbf{a}^\dagger \tag{5.166}$$

or, after using the usual commutation rule (5.5) between the raising and lowering operators, Eq. (5.165) becomes

$$\langle\{n\}|\mathbf{Q}(t)^2|\{n\}\rangle = \frac{\hbar}{2m\omega}\langle\{n\}|((\mathbf{a}^\dagger)^2 e^{2i\omega t} + (\mathbf{a})^2 e^{-2i\omega t} + 2\mathbf{a}^\dagger\mathbf{a} + 1)|\{n\}\rangle$$

By inserting this result into Eq. (5.164), the average value of the potential energy is

$$\langle\{n\}|\mathbf{V}(t)|\{n\}\rangle = \tfrac{1}{4}\hbar\omega\langle\{n\}|((\mathbf{a}^\dagger)^2 e^{2i\omega t} + (\mathbf{a})^2 e^{-2i\omega t} + 2\mathbf{a}^\dagger\mathbf{a} + 1)|\{n\}\rangle$$

so that, owing to Eqs. (5.53) and (5.63),

$$\langle\{n\}|(\mathbf{a})^2|\{n\}\rangle = \sqrt{n}\sqrt{n-1}\langle\{n\}|\{n-2\}\rangle = 0$$

$$\langle\{n\}|(\mathbf{a}^\dagger)^2|\{n\}\rangle = \sqrt{n+2}\sqrt{n+1}\langle\{n\}|\{n+2\}\rangle = 0$$

Now, due to Eq. (5.40)

$$\langle\{n\}|(\mathbf{a}^\dagger\mathbf{a})|\{n\}\rangle = n\langle\{n\}|\{n\}\rangle = n$$

Hence, using these equations, the average value of the potential energy becomes

$$\langle\{n\}|\mathbf{V}(t)|\{n\}\rangle = \tfrac{1}{2}\hbar\omega\left(n + \tfrac{1}{2}\right) = const. \tag{5.167}$$

Hence, the average potential energy remains constant throughout time and equal to half the energy when the oscillator is in any eigenstate $|\{n\}\rangle$ of its Hamiltonian as can be directly obtained from the virial theorem.

In a similar way, one would find for the mean kinetic energy averaged over an Hamiltonian eigenket

$$\langle\{n\}|\left(\frac{\mathbf{P}(t)^2}{2m}\right)|\{n\}\rangle = \frac{1}{2}\hbar\omega\left(n + \frac{1}{2}\right) = const. \tag{5.168}$$

with

$$\mathbf{P}(t)^2 = -\frac{\hbar m\omega}{2}(\mathbf{a}^\dagger e^{i\omega t} - \mathbf{a}e^{-i\omega t})^2$$

Of course, results (5.167) and (5.168) may be generalized to the x, y, and z components of 3D harmonic oscillators, leading one to write for each component

$$\langle\{n\}_k|\mathbf{V}_k(t)|\{n\}_k\rangle = \langle\{n\}_k|\left(\frac{\mathbf{P}_k(t)^2}{2m}\right)|\{n\}_k\rangle = \frac{1}{2}\hbar\omega_k\left(n_k + \frac{1}{2}\right)$$

5.4 BOSON AND FERMION OPERATORS

As observed above, the non-Hermitian annihilation and creation operators $\mathbf{a}$ and $\mathbf{a}^\dagger$ are very important in the quantum approach of harmonic oscillators. They are often called *Boson operators* because they are related to the *Bose–Einstein statistics* where the number of particles inside a nondegenerate energy level is arbitrary. On the other hand, in the study of double-energy-level systems one meets non-Hermitian operators $\mathbf{a}_f$ and $\mathbf{a}_f^\dagger$, which play for these simple systems a role presenting analogies with those of $\mathbf{a}$ and $\mathbf{a}^\dagger$ in the quantum oscillators theory. (In all this section, the subscript f refers to Fermions and the corresponding double-energy-level systems.) These new operators are called *Fermion operators* because they are related to the *Fermi–Dirac statistics* where the number of particles inside a nondegenerate energy level can be only either zero or unity. Owing to the importance of Fermion operators description in many double-energy-level system studies, and of their deep analogy with the Boson operators, it is convenient to treat here the Fermion operators.

Consider a two-energy-level system, the Hamiltonian eigenvalue equation of which is

$$\mathbf{H}|(k)_f\rangle = E_k|(k)_f\rangle \quad \text{with} \quad k = 0, 1$$

where E_k are the two eigenvalues, that is, E_0 and E_1, with $E_1 > E_0$, whereas $|(k)_f\rangle$ are the corresponding eigenkets $|(0)_f\rangle$ and $|(1)_f\rangle$ of $\mathbf{H}$, the first one being the ground state and the last one the excited state as illustrated in Fig. 5.2.

Figure 5.2 Fermion energy levels and corresponding eigenkets.

When normalized and orthogonal, these states read

$$\langle (1)_f | (1)_f \rangle = \langle (0)_f | (0)_f \rangle = 1 \tag{5.169}$$

$$\langle (1)_f | (0)_f \rangle = 0 \tag{5.170}$$

Now, by analogy with the Boson operators $\mathbf{a}^\dagger$ and $\mathbf{a}$, introduce two kinds of non-Hermitian operators $\mathbf{a}_f^\dagger$ and $\mathbf{a}_f$, the *Fermion operators*, obeying the following anticommutation rules:

$$[\mathbf{a}_f, \mathbf{a}_f]_+ = [\mathbf{a}_f^\dagger, \mathbf{a}_f^\dagger]_+ = 0 \tag{5.171}$$

$$[\mathbf{a}_f, \mathbf{a}_f^\dagger]_+ = 1 \tag{5.172}$$

where the anticommutator is defined by

$$[\mathbf{A}, \mathbf{B}]_+ \equiv \mathbf{AB} + \mathbf{BA}$$

The two Fermion operators are assumed to act on the kets $|(1)_f\rangle$ and $|(0)_f\rangle$ according to

$$\mathbf{a}_f |(1)_f\rangle = |(0)_f\rangle \tag{5.173}$$

$$\mathbf{a}_f^\dagger |(0)_f\rangle = |(1)_f\rangle \tag{5.174}$$

$$\mathbf{a}_f |(0)_f\rangle = 0 \tag{5.175}$$

and

$$\mathbf{a}_f^\dagger |(1)_f\rangle = 0 \tag{5.176}$$

Now, by analogy with the Hermitian number occupation operator (5.12) of Boson operators, introduce here the Hermitian operators defined by

$$\mathbf{N}_f = \mathbf{a}_f^\dagger \mathbf{a}_f \quad \text{with} \quad \mathbf{N}_f = \mathbf{N}_f^\dagger \tag{5.177}$$

since

$$(\mathbf{a}_f^\dagger \mathbf{a}_f)^\dagger = \mathbf{a}_f^\dagger \mathbf{a}_f$$

Then, owing to Eq. (5.177), the action of $\mathbf{N}_f$ on the excited state reads

$$\mathbf{N}_f |(1)_f\rangle = \mathbf{a}_f^\dagger \mathbf{a}_f |(1)_f\rangle$$

or, owing to Eq. (5.173),

$$\mathbf{N}_f |(1)_f\rangle = \mathbf{a}_f^\dagger |(0)_f\rangle$$

and thus, due to Eq. (5.174),

$$\mathbf{N}_f |(1)_f\rangle = 1 |(1)_f\rangle \tag{5.178}$$

On the other hand, the action of $\mathbf{N}_f$ on the ground state reads

$$\mathbf{N}_f |(0)_f\rangle = \mathbf{a}_f^\dagger \mathbf{a}_f |(0)_f\rangle$$

which, according to Eq. (5.175), reads

$$\mathbf{N}_f|(0)_f\rangle = 0 \tag{5.179}$$

Equations (5.178) and (5.179) may be viewed as corresponding to the eigenvalue equation (5.75), whereas Eqs. (5.173) and (5.174) may be put into correspondence with Eq. (5.80). Moreover, the anticommutation rules (5.172) and (5.171) dealing with the Fermion operators are in correspondence with the commutation rule of the Boson operators (5.77) and (5.78). Finally, the Hermitian operator $\mathbf{N}_f$ defined by Eq. (5.177) is the *Fermion operator analog* of the number occupation operator $\mathbf{N}$ (5.76) involving Boson operators.

It is now of interest to give matrix representations of the Fermion operators. For this purpose, it is convenient to represent the two orthogonal kets $|(0)_f\rangle$ and $|(1)_f\rangle$, by two orthonormalized column vectors of dimension 2 according to

$$|(0)_f\rangle = \begin{pmatrix} 0 \\ 1 \end{pmatrix} \qquad \text{and} \qquad |(1)_f\rangle = \begin{pmatrix} 1 \\ 0 \end{pmatrix} \tag{5.180}$$

which satisfy the orthonormality properties, since one obtains, respectively,

$$\langle(1)_f|(1)_f\rangle = (1 \quad 0)\begin{pmatrix} 1 \\ 0 \end{pmatrix} = 1$$

$$\langle(0)_f|(0)_f\rangle = (0 \quad 1)\begin{pmatrix} 0 \\ 1 \end{pmatrix} = 1$$

$$\langle(1)_f|(0)_f\rangle = (1 \quad 0)\begin{pmatrix} 0 \\ 1 \end{pmatrix} = 0$$

Then, in order to be compatible with Eqs. (5.171)–(5.179), the matrix representations of the two Fermion operators $\mathbf{a}_f$ and $\mathbf{a}_f^\dagger$ of the Hermitian operator $\mathbf{N}_f$ have to be chosen in such a way as

$$\boxed{\mathbf{a}_f} = \begin{pmatrix} 0 & 0 \\ 1 & 0 \end{pmatrix} \qquad \text{and} \qquad \boxed{\mathbf{a}_f^\dagger} = \begin{pmatrix} 0 & 1 \\ 0 & 0 \end{pmatrix} \tag{5.181}$$

These matrix representations, which may be compared to those of the Boson operators $\mathbf{a}$ and $\mathbf{a}^\dagger$ given by Eq. (5.81), satisfy, as required, the anticommutation relation (5.172) because

$$[\mathbf{a}_f, \mathbf{a}_f^\dagger]_+ = \begin{pmatrix} 0 & 0 \\ 1 & 0 \end{pmatrix}\begin{pmatrix} 0 & 1 \\ 0 & 0 \end{pmatrix} + \begin{pmatrix} 0 & 1 \\ 0 & 0 \end{pmatrix}\begin{pmatrix} 0 & 0 \\ 1 & 0 \end{pmatrix}$$

leading after matrix multiplication to

$$[\mathbf{a}_f, \mathbf{a}_f^\dagger]_+ = \begin{pmatrix} 0 & 0 \\ 0 & 1 \end{pmatrix} + \begin{pmatrix} 1 & 0 \\ 0 & 0 \end{pmatrix} = \begin{pmatrix} 1 & 0 \\ 0 & 1 \end{pmatrix} = \boxed{\mathbf{1}}$$

In the same way, the matrix representation of the anticommutator of $\mathbf{a}_f$ with itself, as required by (5.171), reads,

$$[\mathbf{a}_f, \mathbf{a}_f]_+ = \begin{pmatrix} 0 & 0 \\ 1 & 0 \end{pmatrix}\begin{pmatrix} 0 & 0 \\ 1 & 0 \end{pmatrix} + \begin{pmatrix} 0 & 0 \\ 1 & 0 \end{pmatrix}\begin{pmatrix} 0 & 0 \\ 1 & 0 \end{pmatrix} = \boxed{\mathbf{0}}$$

One would find in a similar way that the anticommutator of $\mathbf{a}_f^\dagger$ with itself is also zero as required by (5.171). Moreover, using Eqs. (5.181), the matrix representation of the Hermitian operator $\boxed{\mathbf{N}_f}$ defined by Eq. (5.177) is

$$\boxed{\mathbf{N}_f} = \begin{pmatrix} 0 & 1 \\ 0 & 0 \end{pmatrix} \begin{pmatrix} 0 & 0 \\ 1 & 0 \end{pmatrix} = \begin{pmatrix} 1 & 0 \\ 0 & 0 \end{pmatrix} \tag{5.182}$$

Moreover, using Eqs. (5.180) and (5.182), the matrix representation of the action of the operator $\boxed{\mathbf{N}_f}$ on $|(1)_f\rangle$ and $|(0)_f\rangle$, reads

$$\boxed{\mathbf{N}_f}\,|(1)_f\rangle = \begin{pmatrix} 1 & 0 \\ 0 & 0 \end{pmatrix} \begin{pmatrix} 1 \\ 0 \end{pmatrix} \quad \text{and} \quad \boxed{\mathbf{N}_f}\,|(0)_f\rangle = \begin{pmatrix} 1 & 0 \\ 0 & 0 \end{pmatrix} \begin{pmatrix} 0 \\ 1 \end{pmatrix}$$

so it appears that Eqs. (5.178) and (5.179) are satisfied, since the two expressions above yield, respectively,

$$\boxed{\mathbf{N}_f}\,|(1)_f\rangle = \begin{pmatrix} 1 \\ 0 \end{pmatrix} = |(1)_f\rangle$$

and

$$\boxed{\mathbf{N}_f}\,|(0)_f\rangle = \begin{pmatrix} 0 \\ 0 \end{pmatrix} = 0$$

On the other hand, the matrix representation of the actions of the two Fermion operators $\mathbf{a}_f$ and $\mathbf{a}_f^\dagger$ on the two states, lead, respectively, as required by Eqs. (5.173)–(5.176), to

$$\mathbf{a}_f|(1)_f\rangle = \begin{pmatrix} 0 & 0 \\ 1 & 0 \end{pmatrix} \begin{pmatrix} 1 \\ 0 \end{pmatrix} = \begin{pmatrix} 0 \\ 1 \end{pmatrix} = |(0)_f\rangle$$

$$\mathbf{a}_f|(0)_f\rangle = \begin{pmatrix} 0 & 0 \\ 1 & 0 \end{pmatrix} \begin{pmatrix} 0 \\ 1 \end{pmatrix} = \begin{pmatrix} 0 \\ 0 \end{pmatrix} = 0$$

$$\mathbf{a}_f^\dagger|(1)_f\rangle = \begin{pmatrix} 0 & 1 \\ 0 & 0 \end{pmatrix} \begin{pmatrix} 1 \\ 0 \end{pmatrix} = \begin{pmatrix} 0 \\ 0 \end{pmatrix} = 0$$

$$\mathbf{a}_f^\dagger|(0)_f\rangle = \begin{pmatrix} 0 & 1 \\ 0 & 0 \end{pmatrix} \begin{pmatrix} 0 \\ 1 \end{pmatrix} = \begin{pmatrix} 1 \\ 0 \end{pmatrix} = |(1)_f\rangle$$

5.5 CONCLUSION

In the present chapter devoted to the study of single isolated quantum harmonic oscillators, we have obtained many important results. Among them, there are the eigenvalues of the Hamiltonian and the action of the raising and lowering operators on the orthonormalized eigenvectors of the Hamiltonian, which constitute a basis in the state space. We have also considered the time evolution of the average values performed over these kets, of different observables dealing with the oscillators. All these important results, which are convenient to know, are collected in the following list:

Relations between ladder operators and position and momentum operators
$\mathbf{Q} = \sqrt{\frac{\hbar}{2m\omega}}(\mathbf{a}^\dagger + \mathbf{a})$ and $\mathbf{P} = i\sqrt{\frac{m\hbar\omega}{2}}(\mathbf{a}^\dagger - \mathbf{a})$ $[\mathbf{a}, \mathbf{a}^\dagger] = 1$
Eigenvalue equation of the harmonic Hamiltonian:
$\mathbf{H} = \hbar\omega(\mathbf{a}^\dagger\mathbf{a} + \frac{1}{2})$ $\hbar\omega(\mathbf{a}^\dagger\mathbf{a} + \frac{1}{2})\,|\{n\}\rangle = \hbar\omega(n + \frac{1}{2})\,|\{n\}\rangle$ with $n = 0, 1, 2, 3, \ldots$ $\langle\{n\}|\{m\}\rangle = \delta_{nm}$
Action of the ladder operators on the harmonic Hamiltonian eigenkets:
$\mathbf{a}|\{n\}\rangle = \sqrt{n}|\{n-1\}\rangle$ and $\mathbf{a}^\dagger|\{n\}\rangle = \sqrt{n+1}|\{n+1\}\rangle$
Time dependence of the Boson operators:
$\mathbf{a}(t) = \mathbf{a}(0)e^{-i\omega t}$ and $\mathbf{a}^\dagger(t) = \mathbf{a}^\dagger(0)e^{i\omega t}$

Finally, the analytical expressions for the vibrational wavefunctions associated with the quantized energy levels exist, which yield some knowledge concerning the corresponding somewhat "esoteric" kets.

BIBLIOGRAPHY

C. Cohen-Tannoudji, B. Diu, and F. Laloe. *Quantum Mechanics*. Wiley-Interscience: Hoboken, NJ, 2006.

H. Eyring, J. Walter, and G. E. Kimball. *Quantum Chemistry*. Wiley: New York, 1944.

W. H. Louisell. *Quantum Statistical Properties of Radiation*. Wiley: New York, 1973.

CHAPTER 6

COHERENT STATES AND TRANSLATION OPERATORS

INTRODUCTION

The previous chapter focused essentially on the stationary orthonormal eigenstates of the harmonic oscillator Hamiltonian, which form a basis of the state space. There exist other states dealing with harmonic oscillators that also play a very fundamental role in the area of quantum harmonic oscillators. They are the *coherent states* $|\{\alpha\}\rangle$, which are, by definition, the eigenkets of the lowering operator $\mathbf{a}$. They are of great importance for numerous reasons. The first one is that the coherent states, whatever they are, minimize the Heisenberg uncertainty relations. Another one, which is deeply connected to the first one, is that the harmonic oscillator operators, when averaged on it, lead to behaviors that are more and more classical when the eigenvalue α corresponding to the eigenket $|\{\alpha\}\rangle$ is increasing. In addition, they are good simple examples of how the formalism of quantum mechanics operates. Moreover, since they are the eigenkets of a non-Hermitian operator, they illustrate that, unlike the number occupation operator, they do not necessarily admit real eigenvalues and furthermore are continuous and nonorthogonal. Moreover, they play a fundamental role in the area of the quantum theory of light, the average values of the electric field operators performed on them, being reached via the corresponding classical fields. For all these reasons, coherent states now merit study.

Thus, the present chapter will begin by deducing its definition, the expansion of a coherent state on the eigenkets of the number occupation operator. Then, the scalar product between two coherent states will be calculated. The chapter will continue by proving that the Heisenberg uncertainty relation is always minimal for coherent states. Then, it will be shown that coherent states may be generated by the action of the translation operator. One section concerns the time dependence of coherent states. Thus, it will be possible to obtain the wave representation of coherent states and of their time dependence. In another section, it will be also possible to calculate by the aid of the translation operator, the Franck–Condon factors, that is, the scalar products between any eigenfunction of the harmonic oscillator Hamiltonian and another one that has been translated with respect to the first one. The chapter ends with the quest for the energy levels of driven harmonic oscillators, which are deeply connected to the properties of coherent states and of translation operators. This is the opportunity

Quantum Oscillators, First Edition. Olivier Henri-Rousseau and Paul Blaise.
© 2011 John Wiley & Sons, Inc. Published 2011 by John Wiley & Sons, Inc.

to test numerically approximate approaches of these levels through truncated matrix representations of the driven oscillator Hamiltonian in the basis of the eigenkets of the harmonic oscillator.

6.1 COHERENT-STATE PROPERTIES

6.1.1 Definition and expansion within Hamiltonian eigenkets basis

By definition, coherent states $|\{\alpha\}\rangle$ are the eigenkets of the lowering **a** operator, obeying therefore

$$\mathbf{a}|\{\alpha\}\rangle = \alpha|\{\alpha\}\rangle \tag{6.1}$$

where α are the corresponding eigenvalues, the Hermitian conjugate of which is

$$\langle\{\alpha\}|\mathbf{a}^{\dagger} = \alpha^{*}\langle\{\alpha\}| \tag{6.2}$$

Observe that, since **a** is not Hermitian, its eigenvalues are not necessarily real and different coherent states are not necessarily orthogonal. An important information dealing with coherent states is contained in their expansion over the eigenkets of the occupation number operator, that is, of the harmonic Hamiltonian. Keeping in mind the eigenvalue equation (5.40), that is,

$$\mathbf{a}^{\dagger}\mathbf{a}|\{n\}\rangle = n|\{n\}\rangle$$

with, of course, since $\mathbf{a}^{\dagger}\mathbf{a}$ is Hermitian,

$$\mathbf{1} = \sum_{n=0}^{\infty} |\{n\}\rangle\langle\{n\}| \qquad \text{and} \qquad \langle\{m\}|\{n\}\rangle = \delta_{mn} \tag{6.3}$$

In order to get the expansion of a coherent state on this basis $\{|\{n\}\rangle\}$, introduce the unity operator resulting from the closure relation as follows:

$$|\{\alpha\}\rangle = \sum_{n=0}^{\infty} |\{n\}\rangle\langle\{n\}|\{\alpha\}\rangle$$

so that

$$|\{\alpha\}\rangle = \sum_{n=0}^{\infty} C_n(\alpha)|\{n\}\rangle \tag{6.4}$$

where $C_n(\alpha)$ is the scalar product given by

$$C_n(\alpha) = \langle\{n\}|\{\alpha\}\rangle$$

On the other hand, observe that, by action on the left of **a** on both sides of Eq. (6.4), one gets

$$\mathbf{a}|\{\alpha\}\rangle = \sum_{n=0}^{\infty} C_n(\alpha)\mathbf{a}|\{n\}\rangle$$

Next, using Eq. (5.53) in order to find the expression of the right-hand side of this last equation, one finds

$$\mathbf{a}|\{\alpha\}\rangle = \sum_{n=0}^{\infty} C_n(\alpha))\sqrt{n}|\{n-1\}\rangle$$

Again, using eigenvalue equation (6.1) we have

$$\alpha|\{\alpha\}\rangle = \sum_{n=0}^{\infty} C_n(\alpha)\sqrt{n}|\{n-1\}\rangle$$

and after using Eq. (6.4) for the left-hand side of this last equation, one gets

$$\alpha \sum_{n=0}^{\infty} C_n(\alpha)|\{n\}\rangle = \sum_{n=0}^{\infty} C_n(\alpha)\sqrt{n}|\{n-1\}\rangle \tag{6.5}$$

Now, observe that the first term involved on the right-hand side of Eq. (6.5) corresponding to $n=0$ is zero since no eigenkets of the harmonic oscillator Hamiltonian under $|\{0\}\rangle$ exist. Thus, performing the following variable change

$$n \rightarrow n+1$$

Eq. (6.5) reads

$$\alpha \sum_{n=0}^{\infty} C_n(\alpha)|\{n\}\rangle = \sum_{n=0}^{\infty} C_{n+1}(\alpha)\sqrt{n+1}|\{n\}\rangle$$

Since this last expression must be true for each term of the sum, the following relation must be verified:

$$C_{n+1}(\alpha)\sqrt{n+1} = \alpha C_n(\alpha) \tag{6.6}$$

which yields for $n=0$

$$\sqrt{1}C_1(\alpha) = \alpha C_0(\alpha) \tag{6.7}$$

and for $n=1$

$$\sqrt{2}C_2(\alpha) = \alpha C_1(\alpha)$$

Then, inserting in this last result Eq. (6.7), one obtains

$$C_2(\alpha) = \frac{\alpha^2}{\sqrt{2}}C_0(\alpha) \tag{6.8}$$

Moreover, for $n=2$, Eq. (6.6) gives

$$\sqrt{3}C_3(\alpha) = \alpha C_2(\alpha)$$

which, using Eq. (6.8) leads to

$$C_3(\alpha) = \frac{\alpha^3}{\sqrt{3}\sqrt{2}}C_0(\alpha) \tag{6.9}$$

Hence, one obtains by recurrence from Eqs. (6.7)–(6.9)

$$C_n(\alpha) = \frac{\alpha^n}{\sqrt{n!}} C_0(\alpha) \tag{6.10}$$

Allowing to transform the expansion (6.4) of the coherent state into

$$|\{\alpha\}\rangle = C_0(\alpha) \sum_{n=0}^{\infty} \frac{\alpha^n}{\sqrt{n!}} |\{n\}\rangle \tag{6.11}$$

Furthermore, in order to find the expression of the unknown coefficient $C_0(\alpha)$, use the Hermitian conjugate of Eq. (6.11), that is,

$$\langle\{\alpha\}| = C_0(\alpha)^* \sum_{m=0}^{\infty} \frac{(\alpha^*)^m}{\sqrt{m!}} \langle\{m\}| \tag{6.12}$$

allowing, with the help of Eq. (6.11), to get the norm of the coherent state

$$\langle\{\alpha\}|\{\alpha\}\rangle = |C_0(\alpha)|^2 \sum_m \sum_n \frac{(\alpha^*)^m (\alpha)^n}{\sqrt{m!}\sqrt{n!}} \langle\{m\}|\{n\}\rangle$$

which, in view of the orthonormality property appearing in (6.3), reduces to

$$\langle\{\alpha\}|\{\alpha\}\rangle = |C_0(\alpha)|^2 \sum_n \frac{|\alpha|^{2n}}{n!} \langle\{n\}|\{n\}\rangle \tag{6.13}$$

Again, owing to Eq. (6.3), and after imposing the coherent state to be normalized, it reads

$$\langle\{\alpha\}|\{\alpha\}\rangle = 1 \qquad \text{and} \qquad \langle\{n\}|\{n\}\rangle = 1$$

so that Eq. (6.13) reduces to

$$|C_0(\alpha)|^2 \sum_n \frac{|\alpha|^{2n}}{n!} = 1$$

which, due to the expansion properties of the exponential, yields

$$|C_0(\alpha)|^2 = e^{-|\alpha|^2} \tag{6.14}$$

At last, passing from the squared absolute value $|C_0(\alpha)|^2$ to the corresponding $C_0(\alpha)$, and after neglecting a phase factor $e^{i\varphi}$ without interest, one obtains

$$C_0(\alpha) = e^{-|\alpha|^2/2} \tag{6.15}$$

so that the recurrence equation (6.10) becomes

$$C_n(\alpha) = e^{-|\alpha|^2/2} \frac{\alpha^n}{\sqrt{n!}}$$

which allows us to transform Eq. (6.4) into

$$\boxed{|\{\alpha\}\rangle = e^{-|\alpha|^2/2} \sum_n \left(\frac{\alpha^n}{\sqrt{n!}} \right) |\{n\}\rangle} \tag{6.16}$$

This 1D result may be generalized to three dimensions, leading, for each x, y, and z components, to coherent states of the form

$$|\{\alpha\}_k\rangle = e^{-|\alpha_k|^2/2} \sum_{n_k} \left(\frac{\alpha_k^{n_k}}{\sqrt{n_k!}} \right) |\{n\}_k\rangle$$

obeying the eigenvalue equations

$$\mathbf{a}_k |\{\alpha\}_k\rangle = \alpha_k |\{\alpha\}_k\rangle$$

6.1.2 Scalar products and closure relations

Since coherent states are the eigenkets of **a**, which is non-Hermitian, different coherent states being different eigenkets of **a**, have no reason to be orthogonal, since this property is specific to eigenkets of Hermitian operators. Hence, because of the absence of orthogonality between two coherent states characterized by two different eigenvalues of α, it is necessary to determine their scalar product.

6.1.2.1 Scalar products For this purpose, consider a coherent state $|\{\beta\}\rangle$ obeying an expression of the same form as Eq. (6.1), which reads

$$\mathbf{a}|\{\beta\}\rangle = \beta\, |\{\beta\}\rangle \tag{6.17}$$

where β is the corresponding eigenvalue. The expansion of this new coherent state is of course given by an expression similar to Eq. (6.16), so that its Hermitian conjugate reads as the bra (6.12), that is,

$$\langle\{\beta\}| = e^{-|\beta|^2/2} \sum_m \frac{\beta^{*m}}{\sqrt{m!}} \langle\{m\}|$$

Thereby, the scalar product of $|\{\alpha\}\rangle$, defined by Eq. (6.16) and of $|\{\beta\}\rangle$ given by Eq. (6.17), yields

$$\langle\{\beta\}|\{\alpha\}\rangle = e^{-|\beta|^2/2} e^{-|\alpha|^2/2} \sum_m \sum_n \frac{\alpha^n}{\sqrt{n!}} \frac{\beta^{*m}}{\sqrt{m!}} \langle\{m\}|\{n\}\rangle$$

Next, using the orthonormality properties

$$\langle\{m\}|\{n\}\rangle = \delta_{mn}$$

the above scalar product reduces to

$$\langle\{\beta\}|\{\alpha\}\rangle = e^{-|\beta|^2/2} e^{-|\alpha|^2/2} \sum_n \frac{(\alpha\beta^*)^n}{n!}$$

or, after passing to exponentials, to

$$\langle\{\beta\}|\{\alpha\}\rangle = \exp\left(-\frac{|\beta|^2 + |\alpha|^2}{2} \right) e^{\alpha\beta^*}$$

and thus

$$\boxed{\langle\{\beta\}|\{\alpha\}\rangle = e^{-|\alpha-\beta|^2/2}} \tag{6.18}$$

6.1.2.2 Closure relations Since coherent states are not orthogonal, they cannot be used to generate a standard discrete closure relation. However, despite this difficulty, it is possible to obtain a continuous closure relation given by

$$I = \frac{1}{\pi} \int_{-\infty}^{+\infty} \int_{-\infty}^{+\infty} |\{\alpha\}\rangle \, \langle\{\alpha\}| d\,\mathrm{Re}\,(\alpha)\, d\,\mathrm{Im}\,(\alpha) \tag{6.19}$$

In order to prove that Eq. (6.19) is unity, first start from the α complex eigenvalue written as

$$\alpha = \rho e^{i\varphi} \tag{6.20}$$

where ρ and φ are both real, the differentiation of which yields

$$d\alpha = e^{i\varphi}(d\rho + i\rho\, d\varphi)$$

or, after passing to the trigonometric expression,

$$d\alpha = (\cos\varphi + i\sin\varphi)\,(d\rho + i\rho\, d\varphi)$$

and thus

$$d\alpha = d\{\mathrm{Re}\,(\alpha)\} + id\{\mathrm{Im}\,(\alpha)\}$$

where $\mathrm{Re}\,(\alpha)$ and $\mathrm{Im}\,(\alpha)$ are, respectively, the real and imaginary parts of α obeying

$$\begin{pmatrix} d\,\mathrm{Re}\,(\alpha) \\ d\,\mathrm{Im}\,(\alpha) \end{pmatrix} = \begin{pmatrix} \cos\varphi & -\rho\sin\varphi \\ \sin\varphi & +\rho\cos\varphi \end{pmatrix} \begin{pmatrix} d\rho \\ d\varphi \end{pmatrix} \tag{6.21}$$

Next, consider the product of $d\,\mathrm{Re}(\alpha)$ and $d\,\mathrm{Im}(\alpha)$, namely

$$d\,\mathrm{Re}(\alpha)\, d\,\mathrm{Im}(\alpha) = \det\,(\mathbf{J})\, d\rho\, d\varphi \tag{6.22}$$

where $\mathbf{J}$ is the Jacobian, that is, the determinant corresponding to the matrix involved in Eq. (6.21), that is,

$$\mathbf{J} = \begin{vmatrix} \cos\varphi & -\rho\sin\varphi \\ \sin\varphi & +\rho\cos\varphi \end{vmatrix}$$

Hence, the product (6.22) reads

$$d\,\mathrm{Re}\,(\alpha)\, d\,\mathrm{Im}\,(\alpha) = \rho(\cos^2\varphi + \sin^2\varphi) d\rho\, d\varphi$$

or, after simplifications

$$d\,\mathrm{Re}\,(\alpha)\, d\,\mathrm{Im}\,(\alpha) = \rho\, d\rho\, d\varphi \tag{6.23}$$

Now, in view of Eq. (6.16) and of its Hermitian conjugate, one may write

$$|\{\alpha\}\rangle \, \langle\{\alpha\}| = e^{-|\alpha|^2} \sum_m \sum_n \frac{\alpha^n}{\sqrt{n!}} \frac{\alpha^{*m}}{\sqrt{m!}} |\{n\}\rangle \, \langle\{m\}|$$

which, using Eq. (6.20), yields

$$|\{\alpha\}\rangle \, \langle\{\alpha\}| = e^{-\rho^2} \sum_m \sum_n \frac{\rho^{n+m}}{\sqrt{n!}\sqrt{m!}} e^{i(n-m)\varphi} |\{n\}\rangle\langle\{m\}| \tag{6.24}$$

As a consequence of Eqs. (6.23) and (6.24), the integral I appearing in Eq. (6.19) becomes

$$I = \frac{1}{\pi}\sum_m\sum_n \frac{|\{n\}\rangle\,\langle\{m\}|}{\sqrt{n!}\sqrt{m!}}\int_0^{+\infty} e^{-\rho^2}\rho^{n+m}\rho\,d\rho\int_0^{2\pi} e^{i(n-m)\varphi}d\varphi \tag{6.25}$$

The last right-hand-side integral has the following solutions:

$$\int_0^{2\pi} e^{i(n-m)\varphi}d\varphi = 2\pi \quad \text{if} \quad n = m$$

$$= \frac{1}{i\,(n-m)}[e^{i(n-m)\varphi}]_0^{2\pi} = 0 \quad \text{if the integers } n \neq m$$

so that

$$\int_0^{2\pi} e^{i(n-m)\varphi}d\varphi = 2\pi\delta_{nm} \tag{6.26}$$

Then, Eq. (6.25) reduces to

$$I = \sum_n I_n \frac{|\{n\}\rangle\langle\{n\}|}{n!} \tag{6.27}$$

with

$$I_n = 2\int_{-\infty}^{+\infty} e^{-\rho^2}\rho^{2n}\rho\,d\rho \tag{6.28}$$

Again, performing the variable change

$$u = \rho^2$$

the integrals (6.22) yield

$$I_n = \int_0^{+\infty} e^{-u}u^n du = n!$$

Owing to this result, and according to the closure relation appearing in (5.43), the integral (6.27) reduces to

$$I = \pi\sum_n |\{n\}\rangle\langle\{n\}| = \pi$$

Consequently, keeping in mind that, in view of Eqs. (6.20) and (6.23),

$$\rho e^{i\varphi} = \alpha \qquad \text{and} \qquad \rho\,d\rho\,d\varphi = d\text{Re}\,(\alpha)\,d\text{Im}\,(\alpha)$$

it appears that the closure relation over the coherent sates (6.19) is

$$\boxed{\int_{-\infty}^{+\infty}\int_{-\infty}^{+\infty} |\{\alpha\}\rangle\langle\{\alpha\}|\,d\text{Re}\,(\alpha)\,d\text{Im}\,(\alpha) = 1} \tag{6.29}$$

6.2 POISSON DENSITY OPERATOR

Consider a pure density operator of oscillators described by coherent states, which according to Eq. (3.139), reads

$$\rho_\alpha = |\{\alpha\}\rangle\langle\{\alpha\}| \tag{6.30}$$

with

$$\mathbf{a}|\{\alpha\}\rangle = \alpha|\{\alpha\}\rangle \qquad \text{and} \qquad \langle\{\alpha\}|\mathbf{a}^\dagger = \langle\{\alpha\}|\alpha^*$$

and

$$[\mathbf{a}, \mathbf{a}^\dagger] = 1 \qquad \text{and} \qquad \alpha = \alpha^\circ e^{i\varphi} \tag{6.31}$$

Then, owing to Eq. (6.16), Eq. (6.30) becomes

$$\rho_\alpha = e^{-\alpha^{\circ 2}} \sum_m \sum_n \left(\frac{(\alpha)^m}{\sqrt{m!}}\right) |\{m\}\rangle\langle\{n\}| \left(\frac{(\alpha^*)^n}{\sqrt{n!}}\right)$$

or, in view of Eq. (6.31),

$$\rho_\alpha = e^{-\alpha^{\circ 2}} \sum_m \sum_n \left(\frac{(\alpha^\circ)^{m+n}}{\sqrt{m!}\sqrt{n!}}\right) |\{m\}\rangle\langle\{n\}| e^{i(m-n)\varphi} \tag{6.32}$$

Next, performing the average of this density operator over the phase φ according to

$$\bar{\rho}_\alpha = \frac{1}{2\pi} \int_0^{2\pi} \rho_\alpha \, d\varphi$$

Eq. (6.32) transforms to

$$\bar{\rho}_\alpha = e^{-\alpha^{\circ 2}} \sum_m \sum_n \left(\frac{(\alpha^\circ)^{m+n}}{\sqrt{m!}\sqrt{n!}}\right) |\{m\}\rangle\langle\{n\}| \frac{1}{2\pi} \int_0^{2\pi} e^{i(m-n)\varphi} d\varphi$$

which after integration using Eq. (6.26) yields

$$\bar{\boldsymbol{\rho}}_\alpha = e^{-\alpha^{\circ 2}} \sum_m \sum_n \left(\frac{(\alpha^\circ)^{m+n}}{\sqrt{m!}\sqrt{n!}}\right) |\{m\}\rangle\langle\{n\}| \delta_{mn}$$

or

$$\bar{\boldsymbol{\rho}}_\alpha = e^{-\alpha^{\circ 2}} \sum_n \left(\frac{(\alpha^\circ)^{2n}}{n!}\right) |\{n\}\rangle\langle\{n\}| \tag{6.33}$$

Now, observe that this density operator is diagonal in the basis $\{|\{n\}\rangle\}$ and that its diagonal matrix elements are given by the following Poisson distribution:

$$\langle\{n\}|\bar{\boldsymbol{\rho}}_\alpha|\{n\}\rangle = e^{-\alpha^{\circ 2}} \left(\frac{(\alpha^\circ)^{2n}}{n!}\right) \tag{6.34}$$

On the other hand, if the unaveraged density operator (6.32) is not diagonal, its matrix elements are given by

$$\langle\{m\}|\rho_\alpha|\{n\}\rangle = e^{-\alpha^{\circ 2}} \left(\frac{(\alpha^\circ)^{m+n}}{\sqrt{m!}\sqrt{n!}} \right) e^{i(m-n)\varphi} \tag{6.35}$$

Thus, the diagonal matrix elements (6.35) are the same as the diagonal ones (6.34) of the averaged density operator (6.33).

6.3 AVERAGE AND FLUCTUATION OF ENERGY

An important property of coherent states is that these states allow us to easily obtain the mean values of operators averaged on them, whereas another interest is that these mean values exhibit physical properties that are close to those of classical harmonic oscillators. The following sections will thus be devoted to calculate mean values of operators averaged over coherent states, the present one dealing with the average value of the Hamiltonian and of its square, allowing one to find the energy fluctuations within coherent states.

6.3.1 Average Hamiltonian

First, consider the mean value of the harmonic oscillator Hamiltonian averaged over coherent states, that is,

$$\langle \mathbf{H} \rangle_\alpha = \langle\{\alpha\}|\mathbf{H}\,|\{\alpha\}\rangle$$

which becomes in view of Eq. (5.9)

$$\langle \mathbf{H} \rangle_\alpha = \hbar\omega \langle\{\alpha\}|\left(\mathbf{a}^\dagger \mathbf{a} + \tfrac{1}{2}\right)|\{\alpha\}\rangle$$

or

$$\langle \mathbf{H} \rangle_\alpha = \hbar\omega \langle\{\alpha\}|\mathbf{a}^\dagger \mathbf{a}|\{\alpha\}\rangle + \tfrac{1}{2}\hbar\omega\,\langle\{\alpha\}|\{\alpha\}\rangle \tag{6.36}$$

Moreover, in view of Eqs. (6.1) and (6.2), the right-hand-side average value appearing in Eq. (6.36) reads

$$\langle\{\alpha\}|\mathbf{a}^\dagger \mathbf{a}|\{\alpha\}\rangle = \langle\{\alpha\}|\alpha^*\alpha|\{\alpha\}\rangle = |\alpha|^2 \langle\{\alpha\}|\{\alpha\}\rangle \tag{6.37}$$

so that, if the coherent states are normalized, the average value (6.36) of the energy takes the form

$$\boxed{\langle \mathbf{H} \rangle_\alpha = \hbar\omega\left(|\alpha|^2 + \tfrac{1}{2}\right)} \tag{6.38}$$

which may be generalized to 3D oscillators:

$$\langle \mathbf{H} \rangle_\alpha = \hbar\omega_x\left(|\alpha_x|^2 + \tfrac{1}{2}\right) + \hbar\omega_y\left(|\alpha_y|^2 + \tfrac{1}{2}\right) + \hbar\omega_z\left(|\alpha_z|^2 + \tfrac{1}{2}\right)$$

6.3.2 Average squared Hamiltonian

Next, consider the corresponding average value of the squared Hamiltonian, that is,

$$\langle \mathbf{H}^2 \rangle_\alpha = \langle \{\alpha\} | \mathbf{H}^2 | \{\alpha\} \rangle$$

which, by comparing Eq. (5.9), reads

$$\langle \mathbf{H}^2 \rangle_\alpha = (\hbar\omega)^2 \langle \{\alpha\} | \left(\mathbf{a}^\dagger \mathbf{a} + \tfrac{1}{2} \right)^2 | \{\alpha\} \rangle \tag{6.39}$$

or

$$\langle \mathbf{H}^2 \rangle_\alpha = (\hbar\omega)^2 \langle \{\alpha\} | \left(\mathbf{a}^\dagger \mathbf{a} \mathbf{a}^\dagger \mathbf{a} + \mathbf{a}^\dagger \mathbf{a} + \tfrac{1}{4} \right) | \{\alpha\} \rangle \tag{6.40}$$

Again, in view of Eqs. (6.1) and (6.2), the average value of the quadruple product of operators over coherent states takes the form

$$\langle \{\alpha\} | \mathbf{a}^\dagger \mathbf{a} \mathbf{a}^\dagger \mathbf{a} | \{\alpha\} \rangle = \langle \{\alpha\} | \alpha^* \mathbf{a} \mathbf{a}^\dagger \alpha | \{\alpha\} \rangle$$

and thus α being a scalar,

$$\langle \{\alpha\} | \mathbf{a}^\dagger \mathbf{a} \mathbf{a}^\dagger \mathbf{a} | \{\alpha\} \rangle = |\alpha|^2 \langle \{\alpha\} | \mathbf{a} \mathbf{a}^\dagger | \{\alpha\} \rangle \tag{6.41}$$

Furthermore, because of the basic commutator (5.5), the right-hand-side matrix element of Eq. (6.41) becomes

$$\langle \{\alpha\} | \mathbf{a} \mathbf{a}^\dagger | \{\alpha\} \rangle = \langle \{\alpha\} | \left(\mathbf{a}^\dagger \mathbf{a} + \mathbf{1} \right) | \{\alpha\} \rangle$$

and thus using in turn Eqs. (6.1) and (6.2),

$$\langle \{\alpha\} | \mathbf{a} \mathbf{a}^\dagger | \{\alpha\} \rangle = \left(|\alpha|^2 + 1 \right) \langle \{\alpha\} | \{\alpha\} \rangle = |\alpha|^2 + 1$$

so that Eq. (6.41) becomes

$$\langle \{\alpha\} | \mathbf{a}^\dagger \mathbf{a} \mathbf{a}^\dagger \mathbf{a} | \{\alpha\} \rangle = |\alpha|^2 \left(|\alpha|^2 + 1 \right)$$

As a consequence and according to Eqs. (6.40) and (6.37), it yields

$$\langle \{\alpha\} | \left(\mathbf{a}^\dagger \mathbf{a} + \tfrac{1}{2} \right)^2 | \{\alpha\} \rangle = |\alpha|^2 \left(|\alpha|^2 + 1 \right) + |\alpha|^2 + \tfrac{1}{4}$$

so that the average value of the squared Hamiltonian given by Eq. (6.39) becomes

$$\boxed{\langle \mathbf{H}^2 \rangle_\alpha = (\hbar\omega)^2 \left(|\alpha|^4 + 2|\alpha|^2 + \tfrac{1}{4} \right)} \tag{6.42}$$

which for 3D oscillators gives

$$\begin{aligned} \langle \mathbf{H}^2 \rangle_\alpha = {} & (\hbar\omega_x)^2 \left(|\alpha_x|^4 + 2|\alpha_x|^2 + \tfrac{1}{4} \right) + (\hbar\omega_y)^2 \left(|\alpha_y|^4 + 2|\alpha_y|^2 + \tfrac{1}{4} \right) \\ & + (\hbar\omega_z)^2 \left(|\alpha_z|^4 + 2|\alpha_z|^2 + \tfrac{1}{4} \right) \end{aligned}$$

6.3.3 Energy fluctuations

It is now possible to obtain an expression for the relative energy fluctuation of harmonic oscillators within coherent states. The energy fluctuation $\Delta \mathbf{H}_\alpha$ is formally given by

$$\Delta \mathbf{H}_\alpha = \sqrt{\langle \mathbf{H}^2 \rangle_\alpha - \langle \mathbf{H} \rangle_\alpha^2}$$

and, comparing Eqs. (6.38) and (6.42), we have

$$\Delta \mathbf{H}_\alpha = \hbar\omega\sqrt{\left(|\alpha|^4 + 2|\alpha|^2 + \tfrac{1}{4}\right) - \left(|\alpha|^4 + |\alpha|^2 + \tfrac{1}{4}\right)}$$

which reduces to

$$\Delta \mathbf{H}_\alpha = \hbar\omega|\alpha| \tag{6.43}$$

Clearly, the energy fluctuation is not zero because the average value of the Hamiltonian has been performed over a state that is not an eigenstate of this operator.

Now, the relative energy fluctuation reads, with the help of Eqs. (6.38) and (6.43),

$$\left(\frac{\Delta \mathbf{H}_\alpha}{\langle \mathbf{H} \rangle_\alpha}\right) = \frac{|\alpha|}{|\alpha|^2 + \frac{1}{2}}$$

which, when $|\alpha|$ becomes important, simplifies to

$$\left(\frac{\Delta \mathbf{H}_\alpha}{\langle \mathbf{H} \rangle_\alpha}\right) \backsimeq \frac{1}{|\alpha|} \quad \text{when} \quad |\alpha| >> 1$$

which, in turn, vanishes when $|\alpha|$ becomes very large:

$$\boxed{\left(\frac{\Delta \mathbf{H}_\alpha}{\langle \mathbf{H} \rangle_\alpha}\right) \to 0 \quad \text{when} \quad |\alpha| \to \infty}$$

This last result, which holds also for 3D oscillators, narrows the behavior of a classical harmonic oscillator for which the energy is always exact, according to classical mechanics.

6.4 COHERENT STATES AS MINIMIZING HEISENBERG UNCERTAINTY RELATIONS

Coherent states that present such classical asymptotic behavior also minimize the Heisenberg uncertainty relations, which we shall now prove.

6.4.1 Average values of the first and second moments of Q and P

For this purpose, it is necessary to obtain the mean values of the **Q** and **P** operators and of their squares averaged over coherent states.

6.4.1.1 Q and P average values First start from the position operator **Q** averaged over coherent states:

$$\langle \mathbf{Q} \rangle_\alpha = \langle \{\alpha\}| \; \mathbf{Q} \; |\{\alpha\}\rangle$$

which, in view of the expression (5.6) of **Q** in terms of the ladder operators, becomes

$$\langle \mathbf{Q} \rangle_\alpha = \sqrt{\frac{\hbar}{2m\omega}} \langle \{\alpha\}|(\mathbf{a}^\dagger + \mathbf{a})|\{\alpha\}\rangle$$

and due to the eigenvalue equations, (6.1) and (6.2), transforms to

$$\langle \mathbf{Q} \rangle_\alpha = \sqrt{\frac{\hbar}{2m\omega}} \langle \{\alpha\}|(\alpha^* + \alpha)|\{\alpha\}\rangle$$

Moreover, since α and α^* are scalars and when coherent states are normalized, this result reduces to

$$\boxed{\langle \mathbf{Q} \rangle_\alpha = \sqrt{\frac{\hbar}{2m\omega}} \{\alpha^* + \alpha\}} \tag{6.44}$$

Observe that, while the **Q** mean value averaged over Hamiltonian eigenstates is zero, those averaged over the coherent state are not so.

Now, Eq. (5.7) allows us to write the **P** average value over a coherent state, according to

$$\langle \mathbf{P} \rangle_\alpha = \langle \{\alpha\}|\ \mathbf{P}\ |\{\alpha\}\rangle = i\sqrt{\frac{m\hbar\omega}{2}} \langle \{\alpha\}|(\mathbf{a}^\dagger - \mathbf{a})|\{\alpha\}\rangle$$

which, proceeding in the same way as above using Eqs. (6.1) and (6.2), reads

$$\boxed{\langle \mathbf{P} \rangle_\alpha = i\sqrt{\frac{m\hbar\omega}{2}} \{\alpha^* - \alpha\}} \tag{6.45}$$

6.4.1.2 Q^2 and P^2 average values Now, in order to find the dispersion of **P** and **Q** within a coherent state, one has first to get the mean values of $\mathbf{P}^2$ and $\mathbf{Q}^2$ within these states. Then, with the help of Eq. (5.6) it may be written

$$\langle \mathbf{Q}^2 \rangle_\alpha = \langle \{\alpha\}|\mathbf{Q}^2|\{\alpha\}\rangle = \frac{\hbar}{2m\omega} \langle \{\alpha\}|(\mathbf{a}^\dagger + \mathbf{a})^2|\{\alpha\}\rangle$$

or

$$\langle \mathbf{Q}^2 \rangle_\alpha = \frac{\hbar}{2m\omega} \langle \{\alpha\}|\left((\mathbf{a}^\dagger)^2 + (\mathbf{a})^2 + \mathbf{a}^\dagger\mathbf{a} + \mathbf{a}\mathbf{a}^\dagger\right)|\{\alpha\}\rangle$$

which, in view of the commutation rule (5.5), transforms to

$$\langle \mathbf{Q}^2 \rangle_\alpha = \frac{\hbar}{2m\omega} \langle \{\alpha\}|\left(\mathbf{a}^\dagger\mathbf{a}^\dagger + \mathbf{a}\mathbf{a} + 2\mathbf{a}^\dagger\mathbf{a} + 1\right)|\{\alpha\}\rangle$$

or

$$\langle \mathbf{Q}^2 \rangle_\alpha = \frac{\hbar}{2m\omega}[\langle \{\alpha\}|\mathbf{a}^\dagger\mathbf{a}^\dagger|\{\alpha\}\rangle + \langle \{\alpha\}|\mathbf{a}\mathbf{a}|\{\alpha\}\rangle + 2\langle \{\alpha\}|\mathbf{a}^\dagger\mathbf{a}|\{\alpha\}\rangle + \langle \{\alpha\}|\{\alpha\}\rangle] \tag{6.46}$$

Next, due to Eq. (6.1) we have

$$\mathbf{a}\mathbf{a}\ |\{\alpha\}\rangle = \mathbf{a}\ \alpha|\{\alpha\}\rangle$$

where

$$\mathbf{a}\mathbf{a}|\{\alpha\}\rangle = \alpha\mathbf{a}\ |\{\alpha\}\rangle = (\alpha)^2\ |\{\alpha\}\rangle \tag{6.47}$$

the Hermitian conjugate of which is

$$\langle \{\alpha\}|\mathbf{a}^\dagger\mathbf{a}^\dagger = (\alpha^*)^2 \langle \{\alpha\}| \tag{6.48}$$

Hence, comparing Eqs. (6.47) and (6.48), the average value (6.46) becomes

$$\langle \mathbf{Q}^2 \rangle_\alpha = \frac{\hbar}{2m\omega} \left((\alpha^*)^2 + (\alpha)^2 + 2\alpha^*\alpha + 1 \right) \langle \{\alpha\} | \{\alpha\} \rangle$$

Finally, rearranging and assuming normalized coherent states, one obtains

$$\boxed{\langle \mathbf{Q}^2 \rangle_\alpha = \frac{\hbar}{2m\omega}((\alpha + \alpha^*)^2 + 1)} \tag{6.49}$$

Now, the average value $\langle \mathbf{P}^2 \rangle_\alpha$ reads, in view of Eq. (5.7),

$$\langle \mathbf{P}^2 \rangle_\alpha = \langle \{\alpha\} | \mathbf{P}^2 | \{\alpha\} \rangle = -\frac{m\hbar\omega}{2} \langle \{\alpha\} | (\mathbf{a}^\dagger - \mathbf{a})^2 | \{\alpha\} \rangle$$

which transforms after expanding the right-hand-side term into

$$\langle \mathbf{P}^2 \rangle_\alpha = -\frac{m\hbar\omega}{2} \langle \{\alpha\} | ((\mathbf{a}^\dagger)^2 + (\mathbf{a})^2 - 2\mathbf{a}^\dagger\mathbf{a} - 1) | \{\alpha\} \rangle$$

so that, proceeding in the same way as for $\mathbf{Q}^2$, we have

$$\boxed{\langle \mathbf{P}^2 \rangle_\alpha = -\frac{m\hbar\omega}{2}((\alpha^* - \alpha)^2 - 1)} \tag{6.50}$$

6.4.2 Heisenberg uncertainty relations

It is now possible to get the dispersion over $\mathbf{Q}$

$$\Delta \mathbf{Q}_\alpha = \sqrt{\langle \mathbf{Q}^2 \rangle_\alpha - \langle \mathbf{Q} \rangle_\alpha^2}$$

which, comparing Eqs. (6.44) and (6.49), reads

$$\Delta \mathbf{Q}_\alpha = \sqrt{\frac{\hbar}{2m\omega}} \tag{6.51}$$

Now, the dispersion over $\mathbf{P}$ is

$$\Delta \mathbf{P}_\alpha = \sqrt{\langle \mathbf{P}^2 \rangle_\alpha - \langle \mathbf{P} \rangle_\alpha^2}$$

which, owing to Eqs. (6.45) and (6.50), becomes

$$\Delta \mathbf{P}_\alpha = \sqrt{\frac{m\hbar\omega}{2}} \tag{6.52}$$

Thus, the product of the uncertainties (6.51) and (6.52) yields, for arbitrary α,

$$\boxed{\Delta \mathbf{Q}_\alpha \Delta \mathbf{P}_\alpha = \frac{\hbar}{2}} \tag{6.53}$$

Thus, this 1D uncertainty relation, which may be generalized to three dimensions, is the minimum compatible with the Heisenberg uncertainty relations.

6.5 DYNAMICS

Owing to the semiclassical properties of the mean values of operators averaged over coherent states, it would be interesting to find if the dynamics of these mean values behave also semiclassically. Since the time dependence of Boson operators is known, it is convenient to perform the dynamic investigations dealing with coherent states within the time-dependent Heisenberg representation instead of the time-dependent Schrödinger one.

6.5.1 Position and momentum time-dependent average values

6.5.1.1 One-dimensional oscillators First, consider the Heisenberg time dependence of the mean value of $\mathbf{Q}(t)$ average over coherent states, which, owing to Eq. (5.153), reads

$$\langle\{\alpha\}|\mathbf{Q}(t)|\{\alpha\}\rangle = \sqrt{\frac{\hbar}{2m\omega}}\left(\langle\{\alpha\}|\mathbf{a}^{\dagger}(t)|\{\alpha\}\rangle + \langle\{\alpha\}|\mathbf{a}(t)|\{\alpha\}\rangle\right) \tag{6.54}$$

Now, in view of Eqs. (5.151) and (5.152), we have

$$\langle\{\alpha\}|\mathbf{a}(t)|\{\alpha\}\rangle = \langle\{\alpha\}|\mathbf{a}(0)|\{\alpha\}\rangle e^{-i\omega t}$$

$$\langle\{\alpha\}|\mathbf{a}^{\dagger}(t)|\{\alpha\}\rangle = \langle\{\alpha\}|\mathbf{a}^{\dagger}(0)|\{\alpha\}\rangle e^{i\omega t}$$

so that due to (6.1)

$$\langle\{\alpha\}|\mathbf{a}(t)|\{\alpha\}\rangle = \alpha e^{-i\omega t}\langle\{\alpha\}|\{\alpha\}\rangle = \alpha(t)$$

$$\langle\{\alpha\}|\mathbf{a}^{\dagger}(t)|\{\alpha\}\rangle = \alpha e^{i\omega t}\langle\{\alpha\}|\{\alpha\}\rangle = \alpha^{*}(t)$$

with

$$\alpha(t) = \alpha e^{-i\omega t} \tag{6.55}$$

As a consequence, Eq. (6.54) becomes

$$\langle\{\alpha\}|\ \mathbf{Q}(t)|\{\alpha\}\rangle = \sqrt{\frac{\hbar}{2m\omega}}(\alpha^{*}e^{i\omega t} + \alpha e^{-i\omega t})$$

which, if α is real, reduces to

$$\boxed{\langle\{\alpha\}|\mathbf{Q}(t)|\{\alpha\}\rangle = 2\alpha\sqrt{\frac{\hbar}{2m\omega}}\cos\omega t} \tag{6.56}$$

6.5.1.2 Two-dimensional oscillators Now pass from 1D to 2D oscillators for which the time-dependent $\mathbf{Q}(t)$ operator reads

$$\mathbf{Q}(t) = \mathbf{Q}_x(t) + \mathbf{Q}_y(t)$$

Moreover, defining the coherent states dealing with the x and y components using

$$\mathbf{a}_k(t)|\{\alpha\}_k\rangle = \alpha_k(t)|\{\alpha\}_k\rangle \quad \text{with} \quad \mathbf{Q}_k(t) = \sqrt{\frac{\hbar}{2m\omega}}(\mathbf{a}_k^{\dagger}(t) + \mathbf{a}_k(t))$$

where k stands for x or y, the mean values of $\mathbf{Q}_x(t)$ and $\mathbf{Q}_y(t)$ averaged on their corresponding coherent states must be given by expressions similar to that of (6.54), that is,

$$\langle\{\alpha\}_k|\mathbf{Q}_k(t)|\{\alpha\}_k\rangle = \sqrt{\frac{\hbar}{2m\omega}}(\langle\{\alpha\}_k|\mathbf{a}_k^\dagger(t)|\{\alpha\}_k\rangle + \langle\{\alpha\}_k|\mathbf{a}_k(t)|\{\alpha\}_k\rangle)$$

so that, using the eigenvalue equation defining the x and y coherent states, it is obtained

$$\langle\{\alpha\}_k|\mathbf{Q}_k(t)|\{\alpha\}_k\rangle = \sqrt{\frac{\hbar}{2m\omega}}\left(\langle\{\alpha\}_k|\alpha_k^*(t)|\{\alpha\}_k\rangle + \langle\{\alpha\}_k|\alpha_k(t)|\{\alpha\}_k\rangle\right) \tag{6.57}$$

Next, if, for some reason, dephasing between $\alpha_x(t)$ and $\alpha_y(t)$ exists so that

$$\alpha_x(t) = \alpha e^{-i\omega t} \qquad \text{and} \qquad \alpha_y(t) = \alpha e^{\pm i\pi/2} e^{-i\omega t} \tag{6.58}$$

the two average values (6.57) read, respectively,

$$\langle\{\alpha\}_x|\mathbf{Q}_x(t)|\{\alpha\}_x\rangle = 2\alpha\sqrt{\frac{\hbar}{2m\omega}}\cos(\omega t) \tag{6.59}$$

$$\langle\{\alpha\}_y|\mathbf{Q}_y^\pm(t)|\{\alpha\}_y\rangle = 2\alpha\sqrt{\frac{\hbar}{2m\omega}}\cos\left(\omega t \mp \frac{\pi}{2}\right) \tag{6.60}$$

Hence, for 2D oscillators obeying Eq. (6.58), the mean value of $\mathbf{Q}$ averaged over coherent states, yields

$$\begin{aligned}\langle\{\alpha\}_y|\langle\{\alpha\}_x|\mathbf{Q}^\pm(t)|\{\alpha\}_x\rangle|\{\alpha\}_y\rangle &= \langle\{\alpha\}_y|\langle\{\alpha\}_x|\mathbf{Q}_x(t)|\{\alpha\}_x\rangle|\{\alpha\}_y\rangle \\ &\quad + \langle\{\alpha\}_x|\langle\{\alpha\}_y|\mathbf{Q}_y^\pm(t)|\{\alpha\}_y\rangle|\{\alpha\}_x\rangle\end{aligned}$$

or, due to Eqs. (6.59) and (6.60) and after simplification,

$$\langle\{\alpha\}_y|\langle\{\alpha\}_x|\mathbf{Q}^\pm(t)|\{\alpha\}_x\rangle|\{\alpha\}_y\rangle = 2\alpha\sqrt{\frac{\hbar}{2m\omega}}(\cos(\omega t) \mp \sin(\omega t)) \tag{6.61}$$

Hence, the two $\pm$ equations (6.61) constitute two inverse polarized circular motions.

Next, using Eq. (5.159), for the averaged momentum, which may be in correspondence with the average value of $\mathbf{Q}(t)$ given by Eq. (6.56), one would obtain

$$\boxed{\langle\{\alpha\}_y|\langle\{\alpha\}_x|\mathbf{P}(t)|\{\alpha\}_x\rangle|\{\alpha\}_y\rangle = -2\alpha\sqrt{\frac{\hbar m\omega}{2}}\sin\omega t} \tag{6.62}$$

6.5.2 Kinetic and potential time-dependent average values

Now, using Eqs. (6.56) and (6.62), giving the average values of $\mathbf{Q}(t)$ and $\mathbf{P}(t)$, it is possible to get the time dependence of the average potential and kinetic energy operators $\mathbf{V}(t)$ and $\mathbf{T}(t)$. For the first one, which is

$$\langle\{\alpha\}|\mathbf{V}(t)|\{\alpha\}\rangle = \tfrac{1}{2}m\omega^2\langle\{\alpha\}|\mathbf{Q}(t)^2|\{\alpha\}\rangle \tag{6.63}$$

the right-hand-side average value may be expressed in terms of the raising and lowering operators from Eq. (5.153), that is,

$$\langle\{\alpha\}|\mathbf{Q}(t)^2|\{\alpha\}\rangle = \frac{\hbar}{2m\omega}\langle\{\alpha\}|\left(\mathbf{a}^\dagger e^{i\omega t} + \mathbf{a}e^{-i\omega t}\right)^2|\{\alpha\}\rangle \tag{6.64}$$

which using Eq. (5.166) and the commutation rule (5.5), Eq. (6.64) reads

$$\langle\{\alpha\}|\mathbf{Q}(t)^2|\{\alpha\}\rangle = \frac{\hbar}{2m\omega}\langle\{\alpha\}|\left(\left(\mathbf{a}^\dagger\right)^2 e^{2i\omega t} + (\mathbf{a})^2 e^{-2i\omega t} + 2\mathbf{a}^\dagger\mathbf{a}+1\right)|\{\alpha\}\rangle$$

so that Eq. (6.63) becomes

$$\langle\{\alpha\}|\mathbf{V}(t)|\{\alpha\}\rangle = \tfrac{1}{4}\hbar\omega\langle\{\alpha\}|(\left(\mathbf{a}^\dagger\right)^2 e^{2i\omega t} + (\mathbf{a})^2 e^{-2i\omega t} + 2\mathbf{a}^\dagger\mathbf{a}+1)|\{\alpha\}\rangle$$

Again, owing to Eq. (6.1) and to the Hermitian conjugate, one obtains, respectively, after a double action of **a** on the right and of $\mathbf{a}^\dagger$ on the left,

$$\langle\{\alpha\}|\,(\mathbf{a})^2\,|\{\alpha\}\rangle = \alpha^2\langle\{\alpha\}|\{\alpha\}\rangle = \alpha^2 \tag{6.65}$$

$$\langle\{\alpha\}|\left(\mathbf{a}^\dagger\right)^2|\{\alpha\}\rangle = \alpha^{*2}\langle\{\alpha\}|\{\alpha\}\rangle = \alpha^{*2}$$

Besides, owing to Eq. (6.2) and its Hermitian conjugate, one gets

$$\langle\{\alpha\}|\left(\mathbf{a}^\dagger\mathbf{a}\right)^2|\{\alpha\}\rangle = |\alpha|^2\langle\{\alpha\}|\{\alpha\}\rangle = |\alpha|^2 \tag{6.66}$$

so that, by the aid of these last three equations, the average value of the potential energy becomes

$$\langle\{\alpha\}|\mathbf{V}(t)|\{\alpha\}\rangle = \tfrac{1}{4}\hbar\omega(\alpha^{*2}e^{2i\omega t} + \alpha^2 e^{-2i\omega t} + 2|\alpha|^2 + 1)$$

which, passing to sine and cosine functions, transforms to

$$\langle\{\alpha\}|\mathbf{V}(t)|\{\alpha\}\rangle = \tfrac{1}{4}\hbar\omega((\alpha^2+\alpha^{*2})\cos 2\omega t\ - i(\alpha^2-\alpha^{*2})\sin 2\omega t + 2|\alpha|^2+1) \tag{6.67}$$

which, in turn, if α is real, reduces to

$$\langle\{\alpha\}|\mathbf{V}(t)|\{\alpha\}\rangle = \tfrac{1}{2}\hbar\omega\left(\alpha^2(\cos 2\omega t + 1) + \tfrac{1}{2}\right) \tag{6.68}$$

On the other hand, the corresponding average value of the kinetic energy, which reads

$$\langle\{\alpha\}|\mathbf{T}(t)|\{\alpha\}\rangle = \frac{1}{2m}\langle\{\alpha\}|\mathbf{P}(t)^2|\{\alpha\}\rangle \tag{6.69}$$

may be found, proceeding in a similar way as for the potential energy by the aid of Eqs. (5.159) and (5.166):

$$\langle\{\alpha\}|\mathbf{T}(t)|\{\alpha\}\rangle = \tfrac{1}{4}\hbar\omega\langle\{\alpha\}|((2\mathbf{a}^\dagger\mathbf{a}+1) - ((\mathbf{a}^\dagger)^2 e^{2i\omega t} + (\mathbf{a})^2 e^{-2i\omega t}))|\{\alpha\}\rangle$$

Thus, in view of Eqs. (6.65) and (6.66), we have

$$\langle\{\alpha\}|\mathbf{T}(t)|\{\alpha\}\rangle = \tfrac{1}{2}\hbar\omega\left(\alpha^2(1 - \cos 2\omega t) + \tfrac{1}{2}\right) \tag{6.70}$$

Observe that, owing to Eqs. (6.67) and (6.70), the average value of the Hamiltonian is a constant given by

$$\langle\{\alpha\}|\mathbf{H}|\{\alpha\}\rangle = \langle\{\alpha\}|\mathbf{T}(t)|\{\alpha\}\rangle + \langle\{\alpha\}|\mathbf{V}(t)|\{\alpha\}\rangle = \hbar\omega(\alpha^2 + \tfrac{1}{2})$$

that is, as required, in agreement with Eq. (6.38). Observe also the difference in the behavior of time dependence of the average values of the kinetic and potential operators when passing from the eigenstates of the Hamiltonian to the coherent states. Whereas they are constant when the quantum average is performed on the Hamiltonian eigenstates, they become time dependent when passing to coherent states, although coming back and forth in such a way as the average energy remains constant.

6.6 TRANSLATION OPERATORS

The coherent states are deeply connected to the translation operators. As we shall see in this section, the translation operators generate coherent states.

6.6.1 Action of translation operators on ladder operators

To prove this, first seek the action of the translation operator

$$\mathbf{A}(Q^\circ) = \exp\left(-\frac{iQ^\circ \mathbf{P}}{\hbar}\right) \tag{6.71}$$

on the raising and lowering operators. For this, consider the unitary operator given by Eq. (2.95) where $\mathbf{P}$ is the momentum operator and Q° a scalar having the dimension of a length. According to Eq. (2.102) the following canonical transformation holds:

$$\mathbf{A}(Q^\circ)^{-1}\mathbf{Q}\,\mathbf{A}(Q^\circ) = \mathbf{Q} + Q^\circ \tag{6.72}$$

Now, when, expressed in terms of the raising and lowering operators using Eq. (5.7), the translation operator takes the form

$$\mathbf{A}(Q^\circ) = \mathbf{A}(\alpha^\circ) = \exp\left\{-i\frac{Q^\circ}{\hbar}i\sqrt{\frac{\hbar m\omega}{2}}(\mathbf{a}^\dagger - \mathbf{a})\right\} \tag{6.73}$$

or

$$\mathbf{A}(\alpha^\circ) = e^{\alpha^\circ \mathbf{a}^\dagger - \alpha^\circ \mathbf{a}} = e^{-\alpha^\circ \mathbf{a} + \alpha^\circ \mathbf{a}^\dagger} \tag{6.74}$$

with

$$\alpha^\circ = Q^\circ\sqrt{\frac{m\omega}{2\hbar}} \tag{6.75}$$

The second right-hand-side expression in (6.74) has been written to underline the fact that the order of the operators involved in the exponential is irrelevant. Besides, observe that if α° is changed into $-\alpha^\circ$ into Eq. (6.74), this equation transforms to

$$\mathbf{A}(-\alpha^\circ) = e^{-\alpha^\circ \mathbf{a}^\dagger + \alpha^\circ \mathbf{a}} = e^{\alpha^\circ \mathbf{a} - \alpha^\circ \mathbf{a}^\dagger} \tag{6.76}$$

the right-hand side, which is simply the inverse of the translation operator $\mathbf{A}(\alpha^\circ)$ given by Eq. (6.74), so that

$$\mathbf{A}(-\alpha^\circ) = \mathbf{A}(\alpha^\circ)^{-1} \tag{6.77}$$

Now, use Glauber's theorem (1.78) in order to transform the translation operator (6.74) and its inverse (6.76), into products of exponential operators involving only $\mathbf{a}^\dagger$ or $\mathbf{a}$, according to

$$\mathbf{A}(\alpha^\circ) = (e^{\alpha^\circ \mathbf{a}^\dagger})(e^{-\alpha^\circ \mathbf{a}})e^{[\alpha^\circ \mathbf{a}^\dagger, \alpha^\circ \mathbf{a}]/2} = (e^{-\alpha^\circ \mathbf{a}})(e^{\alpha^\circ \mathbf{a}^\dagger})e^{[\alpha^\circ \mathbf{a}, \alpha^\circ \mathbf{a}^\dagger]/2}$$

$$\mathbf{A}(\alpha^\circ)^{-1} = (e^{-\alpha^\circ \mathbf{a}^\dagger})(e^{\alpha^\circ \mathbf{a}})e^{-[\alpha^\circ \mathbf{a}^\dagger, \alpha^\circ \mathbf{a}]/2} = (e^{\alpha^\circ \mathbf{a}})(e^{-\alpha^\circ \mathbf{a}^\dagger})e^{-[\alpha^\circ \mathbf{a}, \alpha^\circ \mathbf{a}^\dagger]/2}$$

or, since $[\mathbf{a}^\dagger, \mathbf{a}] = -1$,

$$\boxed{\mathbf{A}(\alpha^\circ) = (e^{\alpha^\circ \mathbf{a}^\dagger})(e^{-\alpha^\circ \mathbf{a}})e^{-\alpha^{\circ 2}/2} = (e^{-\alpha^\circ \mathbf{a}})(e^{\alpha^\circ \mathbf{a}^\dagger})e^{\alpha^{\circ 2}/2}} \tag{6.78}$$

$$\mathbf{A}(\alpha^\circ)^{-1} = (e^{-\alpha^\circ \mathbf{a}^\dagger})(e^{\alpha^\circ \mathbf{a}})e^{\alpha^{\circ 2}/2} = (e^{\alpha^\circ \mathbf{a}})(e^{-\alpha^\circ \mathbf{a}^\dagger})e^{-\alpha^{\circ 2}/2} \tag{6.79}$$

Next, return to the situation where α is real and thus equal to α°. Then, owing to Eqs. (6.75), (6.74) and (6.77), Eq. (6.72) reads

$$\sqrt{\frac{\hbar}{2m\omega}}(e^{-(\alpha^\circ \mathbf{a}^\dagger - \alpha^\circ \mathbf{a})})(\mathbf{a}^\dagger + \mathbf{a})(e^{(\alpha^\circ \mathbf{a}^\dagger - \alpha^\circ \mathbf{a})}) = \sqrt{\frac{\hbar}{2m\omega}}((\mathbf{a}^\dagger + \mathbf{a}) + (\alpha^\circ + \alpha^\circ)) \tag{6.80}$$

or, after simplification and use of Eq. (6.74),

$$\boxed{\mathbf{A}(\alpha^\circ)^{-1}(\mathbf{a}^\dagger + \mathbf{a})\,\mathbf{A}(\alpha^\circ) = (\mathbf{a}^\dagger + \alpha^\circ) + (\mathbf{a} + \alpha^\circ)} \tag{6.81}$$

which, owing to the first expression of (6.78) and (6.79), appears after simplification to be given by

$$(e^{\alpha^\circ \mathbf{a}})(e^{-\alpha^\circ \mathbf{a}^\dagger})(\mathbf{a}^\dagger + \mathbf{a})(e^{\alpha^\circ \mathbf{a}^\dagger})(e^{-\alpha^\circ \mathbf{a}}) = \mathbf{a}^\dagger + \alpha^\circ + \mathbf{a} + \alpha^\circ \tag{6.82}$$

However, the right-hand side of the latter equation may be expressed as

$$\begin{aligned}(e^{\alpha^\circ \mathbf{a}})(e^{-\alpha^\circ \mathbf{a}^\dagger})(\mathbf{a}^\dagger + \mathbf{a})(e^{\alpha^\circ \mathbf{a}^\dagger})(e^{-\alpha^\circ \mathbf{a}}) &= (e^{\alpha^\circ \mathbf{a}})(e^{-\alpha^\circ \mathbf{a}^\dagger}\mathbf{a}^\dagger e^{\alpha^\circ \mathbf{a}^\dagger})(e^{-\alpha^\circ \mathbf{a}}) \\ &\quad + (e^{-\alpha^\circ \mathbf{a}^\dagger})(e^{\alpha^\circ \mathbf{a}}\mathbf{a}e^{-\alpha^\circ \mathbf{a}})(e^{\alpha^\circ \mathbf{a}^\dagger})\end{aligned}$$

Hence, after simplifications, because the function of an operator commutes with this operator, Eq. (6.82) yields

$$(e^{\alpha^\circ \mathbf{a}})(e^{-\alpha^\circ \mathbf{a}^\dagger})(\mathbf{a}^\dagger + \mathbf{a})(e^{\alpha^\circ \mathbf{a}^\dagger})(e^{-\alpha^\circ \mathbf{a}}) = (e^{\alpha^\circ \mathbf{a}})\mathbf{a}^\dagger(e^{-\alpha^\circ \mathbf{a}}) + (e^{-\alpha^\circ \mathbf{a}^\dagger})\mathbf{a}(e^{\alpha^\circ \mathbf{a}^\dagger}) \tag{6.83}$$

As a consequence, due to Eqs. (6.82) and (6.83), it appears that

$$\boxed{(e^{\alpha^\circ \mathbf{a}})\mathbf{a}^\dagger(e^{-\alpha^\circ \mathbf{a}}) = \mathbf{a}^\dagger + \alpha^\circ} \tag{6.84}$$

$$\boxed{(e^{-\alpha^\circ \mathbf{a}^\dagger})\mathbf{a}(e^{\alpha^\circ \mathbf{a}^\dagger}) = \mathbf{a} + \alpha^\circ} \tag{6.85}$$

6.6.2 Action of translation operators on Hamiltonian ground states

Now, study the action of the translation operators given by Eq. (6.78), on the ground state of the Hamiltonian of the quantum harmonic oscillator, which reads

$$\mathbf{A}(\alpha)|\{0\}\rangle = \left\{e^{\frac{-|\alpha|^2}{2}}\right\}(e^{\alpha \mathbf{a}^\dagger})(e^{-\alpha^* \mathbf{a}})|\{0\}\rangle$$

In order to get the action of the exponential operators on $|\{0\}\rangle$, expand the exponential operators as

$$\mathbf{A}(\alpha)|\{0\}\rangle = \left\{e^{\frac{-|\alpha|^2}{2}}\right\}\sum_n\left\{\frac{(\alpha \mathbf{a}^\dagger)^n}{n!}\right\}\sum_m\left\{\frac{(-\alpha^* \mathbf{a})^m}{m!}\right\}|\{0\}\rangle \tag{6.86}$$

Furthermore, due to Eqs. (5.53), that is,

$$\mathbf{a}|\{n\}\rangle = \sqrt{n}|\{n-1\}\rangle \quad \text{leading to} \quad \mathbf{a}|\{0\}\rangle = 0 \tag{6.87}$$

so that

$$\mathbf{a}^n|\{0\}\rangle = \delta_{o\,n}|\{0\}\rangle$$

Eq. (6.86) reduces to

$$\mathbf{A}(\alpha)|\{0\}\rangle = \left\{e^{\frac{-|\alpha|^2}{2}}\right\}\sum_n\left\{\frac{\alpha^n}{n!}(\mathbf{a}^\dagger)^n\right\}|\{0\}\rangle \tag{6.88}$$

or, keeping in mind Eq. (5.67), that is,

$$(\mathbf{a}^\dagger)^n|\{0\}\rangle = \sqrt{n!}|\{n\}\rangle \tag{6.89}$$

Eq. (6.88) transforms to

$$\mathbf{A}(\alpha)|\{0\}\rangle = \left\{e^{\frac{-|\alpha|^2}{2}}\right\}\sum_n\left\{\frac{\alpha^n}{n!}\sqrt{n!}\right\}|\{n\}\rangle$$

or, after simplification,

$$\mathbf{A}(\alpha)|\{0\}\rangle = \left\{e^{\frac{-|\alpha|^2}{2}}\right\}\sum_n\left\{\frac{\alpha^n}{\sqrt{n!}}\right\}|\{n\}\rangle \tag{6.90}$$

Now, since the expansion of coherent states is given by Eq. (6.16), that is,

$$|\{\alpha\}\rangle = \left\{e^{\frac{-|\alpha|^2}{2}}\right\}\sum_n\left\{\frac{\alpha^n}{\sqrt{n!}}\right\}|\{n\}\rangle \tag{6.91}$$

Then, by identification of Eqs. (6.90) and (6.91), it follows that

$$\boxed{\mathbf{A}(\alpha)|\{0\}\rangle = |\{\alpha\}\rangle} \tag{6.92}$$

or, according to Eq. (6.74)

$$\boxed{e^{\alpha\mathbf{a}^\dagger-\alpha^*\mathbf{a}}|\{0\}\rangle = |\{\alpha\}\rangle \quad \text{with} \quad \mathbf{a}|\{\alpha\}\rangle = \alpha|\{\alpha\}\rangle} \tag{6.93}$$

and thus, after using Glauber's theorem (1.79)

$$\boxed{\left\{e^{\frac{-|\alpha|^2}{2}}\right\}(e^{\alpha\mathbf{a}^\dagger})(e^{-\alpha^*\mathbf{a}})|\{0\}\rangle = |\{\alpha\}\rangle} \tag{6.94}$$

Moreover, observe that due to the last equation of (6.87), leading to

$$(e^{-\alpha^*\mathbf{a}})|\{0\}\rangle = \sum_{n=0}^{\infty}\frac{\alpha^{*n}}{n!}(\mathbf{a})^n|\{0\}\rangle = |\{0\}\rangle$$

Eq. (6.94) simplifies to

$$\boxed{\left\{e^{\frac{-|\alpha|^2}{2}}\right\}(e^{\alpha\mathbf{a}^\dagger})|\{0\}\rangle = |\{\alpha\}\rangle} \tag{6.95}$$

6.6.3 Product of translation operators

Now, consider the two following translation operators where ξ and ζ are c-numbers:

$$\mathbf{A}(\xi) = (e^{\xi\mathbf{a}^\dagger - \xi^*\mathbf{a}}) \qquad \mathbf{A}(\zeta) = (e^{\zeta\mathbf{a}^\dagger - \zeta^*\mathbf{a}})$$

Their product

$$\mathbf{A}(\zeta)\mathbf{A}(\xi) = (e^{\zeta\mathbf{a}^\dagger - \zeta^*\mathbf{a}})(e^{\xi\mathbf{a}^\dagger - \xi^*\mathbf{a}})$$

due to the Glauber theorem (1.78), becomes

$$(e^{\zeta\mathbf{a}^\dagger - \zeta^*\mathbf{a}})(e^{\xi\mathbf{a}^\dagger - \xi^*\mathbf{a}}) = (e^{(\zeta\mathbf{a}^\dagger - \zeta^*\mathbf{a}) + (\xi\mathbf{a}^\dagger - \xi^*\mathbf{a})})(e^{[(\xi\mathbf{a}^\dagger - \xi^*\mathbf{a}),\ (\zeta\mathbf{a}^\dagger - \zeta^*\mathbf{a})]/2}) \tag{6.96}$$

Now, since $[\mathbf{a}, \mathbf{a}^\dagger] = 1$, the commutator appearing on the last right-hand-side term is

$$[(\xi\mathbf{a}^\dagger - \xi^*\mathbf{a}), (\zeta\mathbf{a}^\dagger - \zeta^*\mathbf{a})] = -(\xi\zeta^* - \xi^*\zeta)$$

so that Eq. (6.96) yields

$$\boxed{(e^{\zeta\mathbf{a}^\dagger - \zeta^*\mathbf{a}})(e^{\xi\mathbf{a}^\dagger - \xi^*\mathbf{a}}) = (e^{(\xi+\zeta)\mathbf{a}^\dagger - (\xi^*+\zeta^*)\mathbf{a}})(e^{-(\xi\zeta^* - \xi^*\zeta)/2})} \tag{6.97}$$

6.7 COHERENT-STATE WAVEFUNCTIONS

Owing to the quasi-classical behavior of coherent states, it may be of interest to visualize them through their wave mechanics representation, which is the purpose of the present section.

6.7.1 Wavefunctions

According to Eq. (3.43), the wavefunction corresponding to the coherent state is the scalar product

$$\Phi_\alpha(Q) = \langle\{Q\}|\{\alpha\}\rangle \tag{6.98}$$

which, in view of (6.92), reads

$$\Phi_\alpha(Q) = \langle\{Q\}|\mathbf{A}(\alpha)|\{0\}\rangle \tag{6.99}$$

with, according to Eq. (6.74),

$$\mathbf{A}(\alpha) = e^{\alpha\mathbf{a}^\dagger - \alpha^*\mathbf{a}} \tag{6.100}$$

Again, in view of Eqs. (5.3) and (5.4), the argument of the translation operator reads

$$\alpha\mathbf{a}^\dagger - \alpha^*\mathbf{a} = \left(\alpha\sqrt{\frac{m\omega}{2\hbar}}\mathbf{Q} - i\alpha\sqrt{\frac{1}{2m\hbar\omega}}\mathbf{P}\right) - \left(\alpha^*\sqrt{\frac{m\omega}{2\hbar}}\mathbf{Q} + i\alpha^*\sqrt{\frac{1}{2m\hbar\omega}}\mathbf{P}\right)$$

or, after rearranging,

$$\alpha\mathbf{a}^\dagger - \alpha^*\mathbf{a} = (\alpha - \alpha^*)\sqrt{\frac{m\omega}{2\hbar}}\mathbf{Q} - i(\alpha + \alpha^*)\sqrt{\frac{1}{2m\hbar\omega}}\mathbf{P} \tag{6.101}$$

Furthermore, one obtains, respectively, by inversion of Eqs. (6.44) and (6.45)

$$(\alpha - \alpha^*)\sqrt{\frac{m\omega}{2\hbar}} = \frac{i}{\hbar}\langle\{\alpha\}|\mathbf{P}|\{\alpha\}\rangle = \frac{i}{\hbar}\langle\mathbf{P}\rangle_\alpha \tag{6.102}$$

$$(\alpha + \alpha^*)\sqrt{\frac{1}{2m\hbar\omega}} = \frac{1}{\hbar}\langle\{\alpha\}|\mathbf{Q}|\{\alpha\}\rangle = \frac{1}{\hbar}\langle\mathbf{Q}\rangle_\alpha \tag{6.103}$$

Therefore, owing to Eqs. (6.101)–(6.103), the translation operator (6.100) takes the form

$$\mathbf{A}(\alpha) = e^{i(\langle\mathbf{P}\rangle_\alpha\mathbf{Q} - \langle\mathbf{Q}\rangle_\alpha\mathbf{P})/\hbar}$$

which by the aid of Glauber's theorem (1.78) transforms to

$$\mathbf{A}(\alpha) = (\mathbf{e}^{-i\langle\mathbf{Q}\rangle_\alpha\mathbf{P}/\hbar})(e^{i\langle\mathbf{P}\rangle_\alpha\mathbf{Q}/\hbar})(e^{-\zeta}) \tag{6.104}$$

In the preceding equation, ζ is given by

$$\zeta = \frac{1}{2\hbar^2}\langle\mathbf{P}\rangle_\alpha\langle\mathbf{Q}\rangle_\alpha[\mathbf{Q},\mathbf{P}]$$

which due to $[\mathbf{Q},\mathbf{P}] = i\hbar$ reads

$$\zeta = \frac{i}{2\hbar}\langle\mathbf{P}\rangle_\alpha\langle\mathbf{Q}\rangle_\alpha$$

Or, because of Eqs. (6.102) and (6.103) leads by inversion to

$$\langle\mathbf{Q}\rangle_\alpha = \sqrt{\frac{\hbar}{2m\omega}}\{\alpha^* + \alpha\} \qquad \text{and} \qquad \langle\mathbf{P}\rangle_\alpha = i\sqrt{\frac{\hbar m\omega}{2}}\{\alpha^* - \alpha\}$$

so that we have

$$\zeta = -\tfrac{1}{4}\{\alpha^{*2} - \alpha^2\} \tag{6.105}$$

Here, it is possible to get an explicit expression for the coherent state wavefunction (6.99) corresponding to the coherent state, that is,

$$\Phi_\alpha(Q) = \langle\{Q\}|\{\alpha\}\rangle = \langle\{Q\}|\mathbf{A}(\alpha)|\{0\}\rangle$$

which due to Eqs. (6.104) and (6.105) takes the form

$$\Phi_\alpha(Q) = \langle\{Q\}|(e^{-i\langle\mathbf{Q}\rangle_\alpha\mathbf{P}/\hbar})(e^{i\langle\mathbf{P}\rangle_\alpha\mathbf{Q}/\hbar})|\{0\}\rangle(e^{1/4\{\alpha^{*2}-\alpha^2\}}) \tag{6.106}$$

Now, observe that Eq. (2.119) allows us to write

$$(e^{i\langle\mathbf{Q}\rangle_\alpha\mathbf{P}/\hbar})|\{Q\}\rangle = |\{Q - \langle\mathbf{Q}\rangle_\alpha\}\rangle$$

the Hermitian conjugate of this last equation of which is

$$\langle\{Q\}|(e^{-i\langle\mathbf{Q}\rangle_\alpha\mathbf{P}/\hbar}) = \langle\{Q - \langle\mathbf{Q}\rangle_\alpha\}| \tag{6.107}$$

so that the coherent state wavefunction (6.106) takes the form

$$\Phi_\alpha(Q) = \langle\{Q - \langle\mathbf{Q}\rangle_\alpha\}|(e^{i\langle\mathbf{P}\rangle_\alpha\mathbf{Q}/\hbar})|\{0\}\rangle(e^{1/4\{\alpha^{*2}-\alpha^2\}}) \tag{6.108}$$

In Eq. (6.108), the exponential operator is acting on the left on an eigenbra of the position operator $\mathbf{Q}$ and of all operators functions of $\mathbf{Q}$. Hence, the following eigenvalue equation is verified

$$\langle\{Q - \langle\mathbf{Q}\rangle_\alpha\}|(e^{i\langle\mathbf{P}\rangle_\alpha\mathbf{Q}/\hbar}) = \langle\{Q - \langle\mathbf{Q}\rangle_\alpha\}|(e^{i\langle\mathbf{P}\rangle_\alpha(Q-\langle\mathbf{Q}\rangle_\alpha)/\hbar})$$

so that Eq. (6.108) transforms to

$$\Phi_\alpha(Q) = \langle\{Q - \langle\mathbf{Q}\rangle_\alpha\}|\{0\}\rangle(e^{i\langle\mathbf{P}\rangle_\alpha(Q-\langle\mathbf{Q}\rangle_\alpha)/\hbar})(e^1/4\{\alpha^{*2} - \alpha^2\}) \tag{6.109}$$

On the other hand, since the ground-state wavefunction of the quantum harmonic oscillator Hamiltonian is

$$\Psi_0(Q) = \langle\{Q\}|\{0\}\rangle$$

the scalar product involved in Eq. (6.109) is nothing but the displaced ground-state wavefunction, the origin of which has been displaced by the amount $\langle\mathbf{Q}\rangle_\alpha$, that is,

$$\langle\{Q - \langle\mathbf{Q}\rangle_\alpha\}|\{0\}\rangle = \Psi_0(Q - \langle\mathbf{Q}\rangle_\alpha) \tag{6.110}$$

Now, keeping in mind that the ground-state wavefunction may be obtained using Eqs. (5.126) and (5.127), leading to

$$\Psi_0(Q) = \left(\frac{m\omega}{\hbar\pi}\right)^{1/4} \exp\left(-\frac{m\omega}{2\hbar}Q^2\right)$$

the translated wavefunction (6.110) becomes

$$\Psi_0(Q - \langle\mathbf{Q}\rangle_\alpha) = \left(\frac{m\omega}{\hbar\pi}\right)^{1/4} \exp\left[-\frac{m\omega}{2\hbar}(Q - \langle\mathbf{Q}\rangle_\alpha)^2\right] \tag{6.111}$$

Finally, in view of Eq. (6.111), and using the expression (6.51) of the uncertainty $\Delta\mathbf{Q}$ performed over a coherent state, Eq. (6.109) becomes

$$\boxed{\Phi_\alpha(Q) = \left(\frac{m\omega}{\hbar\pi}\right)^{1/4} \exp\left(-\left(\frac{Q - \langle\mathbf{Q}\rangle_\alpha}{2\Delta Q_\alpha}\right)^2\right) \exp\left(\tfrac{i}{\hbar}\langle\mathbf{P}\rangle_\alpha(Q - \langle\mathbf{Q}\rangle_\alpha)\right) \exp\left(\tfrac{1}{4}\{\alpha^{*2} - \alpha^2\}\right)} \tag{6.112}$$

with

$$\langle\mathbf{P}\rangle_\alpha = \langle\{\alpha\}|\mathbf{P}|\{\alpha\}\rangle] \qquad \text{and} \qquad \Delta Q_\alpha = \sqrt{\frac{\hbar}{2m\omega}}$$

Note that Eq. (6.112) may be shortly written in a narrowing form encountered in *wavelet* theory, that is,

$$\Phi_\alpha(Q) = Ke^{-(Q/Q^\circ)^2}e^{i\lambda Q}$$

where K, Q°, and λ are constants that may be obtained when passing from Eq. (6.112) to this expression.

6.7.2 Time-dependent coherent-state wavefunctions

It has been shown above that the wave mechanics representation of the coherent state is given by Eq. (6.112). Now, we require its time dependence, and for this purpose we transform Eq. (6.112) involving the average values of $\langle \mathbf{Q} \rangle_\alpha$ and $\langle \mathbf{P} \rangle_\alpha$ by taking in place of them the corresponding time-dependent average values $\langle \mathbf{Q}(t) \rangle_\alpha$ and $\langle \mathbf{P}(t) \rangle_\alpha$, leading us to write

$$\Phi_\alpha(Q,t) = \left(\frac{m\omega}{\hbar\pi}\right)^{1/4} \exp\left[-\left(\frac{Q - \langle \mathbf{Q}(t) \rangle_\alpha}{2\Delta Q_\alpha}\right)^2\right] \exp\left[\frac{i}{\hbar}\langle \mathbf{P}(t) \rangle_\alpha (Q - \langle \mathbf{Q}(t) \rangle_\alpha)\right]$$
$$\times\,(\exp 1/4\{\alpha(t)^{*2} - \alpha(t)^2\}) \tag{6.113}$$

where $\alpha(t)$ is now given by Eq. (6.55). Next, in the Schrödinger picture, and due to Eqs. (6.56) and (6.62), the time-dependent average values involved in Eq. (6.113) are given by

$$\langle \mathbf{Q}(t) \rangle_\alpha = \langle \{\alpha\} | \mathbf{Q}(t) | \{\alpha\} \rangle = \alpha\sqrt{\frac{2\hbar}{m\omega}} \cos\omega t$$

$$\langle \mathbf{P}(t) \rangle_\alpha = \langle \{\alpha\} | \mathbf{P}(t) | \{\alpha\} \rangle = -\alpha\sqrt{2\hbar m\omega} \sin\omega t$$

so that the coherent-state wavefunction (6.113) reads

$$\Phi_\alpha(Q,t) = \left(\frac{m\omega}{\hbar\pi}\right)^{1/4} \exp\left[-\frac{m\omega}{2\hbar}(Q - \alpha\sqrt{\tfrac{2\hbar}{m\omega}} \cos\omega t)^2\right]$$
$$\times \exp\left[-\frac{i}{\hbar}\alpha\sqrt{2\hbar m\omega} \sin\omega t_\alpha \left(Q - \alpha\sqrt{\tfrac{2\hbar}{m\omega}} \cos\omega t\right)\right] \left(\exp\left(i\frac{\alpha^2}{2} \sin 2\omega t\right)\right) \tag{6.114}$$

where α stands for the eigenvalue of the coherent state at initial time. Figure 6.1 reports the time dependence of the corresponding modulus $|\Phi_\alpha(Q,t)|^2$:

$$|\Phi_\alpha(Q,t)|^2 = \left(\frac{m\omega}{\hbar\pi}\right)^{1/2} \exp\left[-\frac{m\omega}{\hbar}(Q - \alpha\sqrt{\tfrac{2\hbar}{m\omega}} \cos\omega t)^2\right] \tag{6.115}$$

Inspection of this figure shows that the coherent state initially localized on the right-hand side of the equilibrium position moves back and forth around this position without spreading.

6.8 FRANCK–CONDON FACTORS

One has sometimes to compute the overlap integral (*Franck–Condon factors*) between the eigenfunctions of two oscillator Hamiltonians, the harmonic potentials of which are displaced, and thus not orthogonal, as illustrated in Fig. 6.2. Franck–Condon factors are met, for instance, in the area of electronic molecular spectroscopy where the subbands of the electronic line shapes correspond to transitions between vibrational states of the ground and first excited electronic states, the latter being displaced. They are also found in theories dealing with IR line shapes of weak H-bonded species.

As we have seen above, the energy wavefunctions of the quantum harmonic oscillator are given by the following scalar product:

$$\Psi_n(Q) = \langle \{Q\} | \{n\} \rangle \tag{6.116}$$

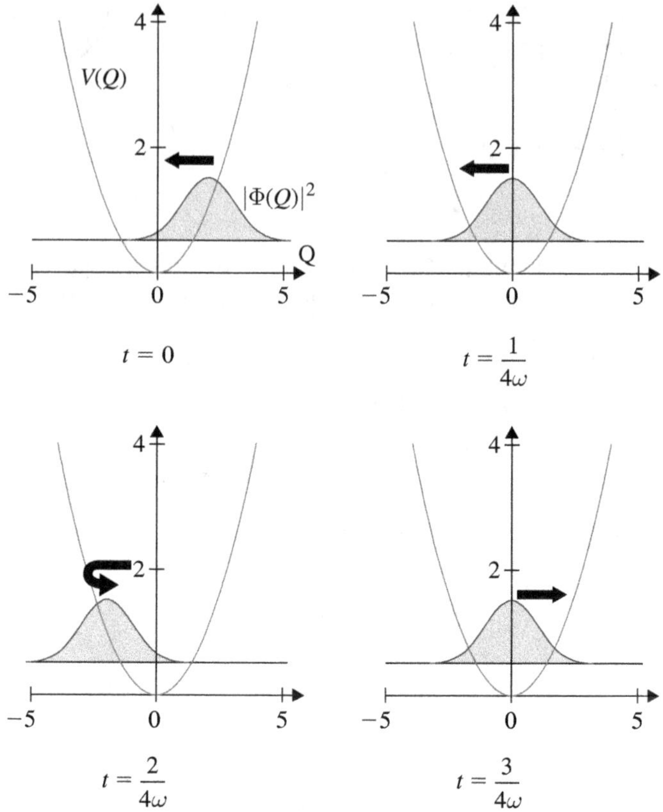

Figure 6.1 Time evolution of the probability density (6.115) of a coherent-state wavefunction, with Q expressed in $\sqrt{\frac{\hbar}{2m\omega}}$ small units, t small in ω^{-1} small units, and $\alpha = 1$,

where $|\{n\}\rangle$ is an eigenket of the number occupation operator. Now, we have shown that the action of the translation operator on an eigenket of the position operator is given by Eq. (2.118), that is,

$$\mathbf{A}(Q^\circ)|\{Q\}\rangle = |\{Q + Q^\circ\}\rangle \tag{6.117}$$

Since the translation operator is unitary, so that its inverse is its Hermitian conjugate, the Hermitian conjugate of Eq. (6.117) is

$$\langle\{Q\}|\mathbf{A}(Q^\circ)^{-1} = \langle\{Q + Q^\circ\}| \tag{6.118}$$

On the other hand, the wavefunction $\{\Psi_m(Q + Q^\circ)\}$ displaced by the amount Q° with respect to that $\Psi_n(Q)$ defined by Eq. (6.116) is given by the scalar product

$$\{\Psi_m(Q + Q^\circ)\} = \langle\{Q + Q^\circ\}|\{m\}\rangle$$

or, in view of Eq. (6.118)

$$\{\Psi_m(Q + Q^\circ)\} = \langle\{Q\}|\mathbf{A}(Q^\circ)^{-1}|\{m\}\rangle \tag{6.119}$$

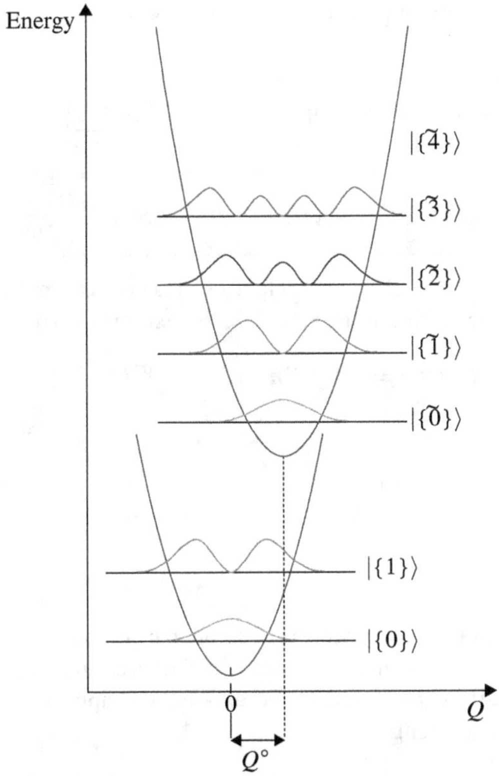

Figure 6.2 Displaced oscillator wavefunctions generating Franck–Condon factors.

Now, look at the Franck–Condon factors, that is, the following overlap integrals

$$\boxed{S_{nm}(Q^\circ) = \int_{-\infty}^{\infty} \{\Psi_n(Q)^*\}\{\Psi_m(Q + Q^\circ)\}dQ} \tag{6.120}$$

which, in view of Eqs. (6.116) and (6.119), take the form

$$S_{nm}(Q^\circ) = \int_{-\infty}^{\infty} \langle\{n\}|\{Q\}\rangle\langle\{Q\}|\mathbf{A}(Q^\circ)^{-1}|\{m\}\rangle \, dQ$$

a result that may be simplified using the closure relation involving the eigenstates of the position operator, that is,

$$\int_{-\infty}^{\infty} |\{Q\}\rangle\langle\{Q\}| \, dQ = 1$$

Thus

$$S_{nm}(Q^\circ) = \langle\{n\}|\mathbf{A}(Q^\circ)^{-1}|\{m\}\rangle \tag{6.121}$$

Next, pass to Boson operators for the translation operators appearing in Eq. (6.121). Then, in view of Eq. (6.74), we have

$$\boxed{S_{nm}(\alpha^\circ) = \langle\{n\}|\,(e^{-\alpha^\circ(\mathbf{a}^\dagger - \mathbf{a})})\,|\{m\}\rangle} \tag{6.122}$$

$$\boxed{S_{nm}(\alpha^\circ) \equiv S_{nm}(Q^\circ) \quad \text{with} \quad \alpha^\circ = Q^\circ\sqrt{\frac{m\omega}{2\hbar}}} \tag{6.123}$$

In order to calculate the Franck–Condon factors, it is convenient to use for the inverse translation operator appearing in Eq. (6.121), the expression (6.79) leading to

$$S_{nm}(\alpha^\circ) = e^{\alpha^{\circ 2}/2}\langle\{n\}|(e^{\alpha^\circ \mathbf{a}})(e^{-\alpha^\circ \mathbf{a}^\dagger})\,|\{m\}\rangle \tag{6.124}$$

or

$$S_{nm}(\alpha^\circ) = e^{\alpha^{\circ 2}/2}\langle\{A\}_n|\{B\}_m\rangle \tag{6.125}$$

with

$$|\{B\}_m\rangle = (e^{-\alpha^\circ \mathbf{a}^\dagger})\,|\{m\}\rangle \qquad \text{and} \qquad \langle\{A\}_n| = \langle\{n\}|(e^{\alpha^\circ \mathbf{a}}) \tag{6.126}$$

We must now compute the scalar product appearing on the right-hand side of Eq. (6.125), and then find in a first place the expression of the ket defined by Eq. (6.126). To obtain it, first expand the exponential appearing on the right-hand side of Eq. (6.126), according to

$$|\{B\}_m\rangle = \sum_k \left(\frac{(-1)^k \alpha^{\circ k} (\mathbf{a}^\dagger)^k}{k!}\right)|\{m\}\rangle \tag{6.127}$$

which, due to Eq. (6.89), is

$$|\{m\}\rangle = \left(\frac{(\mathbf{a}^\dagger)^m}{\sqrt{m!}}\right)|\{0\}\rangle \tag{6.128}$$

Hence, Eq. (6.127) transforms to

$$|\{B\}_m\rangle = \sum_k \left(\frac{(-1)^k \alpha^{\circ k} (\mathbf{a}^\dagger)^{k+m}}{k!\sqrt{m!}}\right)|\{0\}\rangle \tag{6.129}$$

Again, using Eq. (6.128), we have

$$(\mathbf{a}^\dagger)^{k+m}|\{0\}\rangle = \sqrt{(k+m)!}|\{k+m\}\rangle$$

so that Eq. (6.129) becomes

$$|\{B\}_m\rangle = \sum_k \left(\frac{(-1)^k \alpha^{\circ k}\sqrt{(k+m)!}}{k!\sqrt{m!}}\right)|\{k+m\}\rangle \tag{6.130}$$

Now, save that the minus sign is changed into a plus sign, the bra appearing on the right-hand side of Eq. (6.125) is the Hermitian conjugate of Eq. (6.126), so that it is given by an expression similar to that of (6.130), except for the presence of the power of (-1). Hence, α° being real, this bra appears to be

$$\langle\{A\}_n| = \sum_l \left(\frac{\alpha^{\circ l}\sqrt{(l+n)!}}{l!\sqrt{n!}}\right)\langle\{l+n\}| \tag{6.131}$$

Thus, as a consequence of Eqs. (6.130) and (6.131), the Franck–Condon factors (6.125) take the form

$$S_{nm}(\alpha^\circ) = e^{\alpha^{\circ 2}/2}\sum_k\sum_l\left(\frac{(-1)^k\alpha^{\circ k+l}\sqrt{(l+n)!}\sqrt{(k+m)!}}{l!\sqrt{n!}k!\sqrt{m!}}\right)\langle\{l+n\}|\{k+m\}\rangle \tag{6.132}$$

Finally, due to

$$\langle\{l+n\}|\{k+m\}\rangle = \delta_{l,k+m-n}$$

Eq. (6.132) reduces to

$$\boxed{S_{nm}(\alpha^\circ) = e^{\alpha^{\circ 2}/2}\sum_{k=n-m}\left(\frac{(-1)^k\alpha^{\circ 2k+m-n}\,(k+m)!}{(k+m-n)!\sqrt{n!}k!\sqrt{m!}}\right) \quad \text{with} \quad n \geq m} \tag{6.133}$$

with a similar expression for the situation where $m > n$ in which m is changed into n and vice versa.

6.9 DRIVEN HARMONIC OSCILLATORS

Using the work in the present chapter, it is now possible to find the energy levels of driven harmonic oscillators, the Hamiltonian of which is

$$\mathbf{H}_{\mathrm{Dr}} = \left(\frac{\mathbf{P}^2}{2M} + \frac{1}{2}M\omega^2\mathbf{Q}^2\right) + b\mathbf{Q}$$

Passing to Boson operators by the aid of Eqs. (5.6), (5.7), and (5.9), this Hamiltonian becomes

$$\mathbf{H}_{\mathrm{Dr}} = \hbar\omega\left(\mathbf{a}^\dagger\mathbf{a} + \tfrac{1}{2}\right) + \alpha^\circ\hbar\omega(\mathbf{a}^\dagger + \mathbf{a}) \tag{6.134}$$

with

$$\alpha^\circ = \frac{b}{\omega}\sqrt{\frac{1}{2M\hbar\omega}}$$

6.9.1 Diagonalization of driven Hamiltonians by aid of translation operators

6.9.1.1 Canonical transformations involving translation operators In order to diagonalize the Hamiltonian operator (6.134), consider the matrix elements of this operator in the basis of the eigenstates of the quantum harmonic oscillator:

$$\langle\{n\}|\mathbf{H}_{\mathrm{Dr}}|\{m\}\rangle = \langle\{n\}|\{(\mathbf{a}^\dagger\mathbf{a} + \tfrac{1}{2}) + \alpha^\circ(\mathbf{a}^\dagger + \mathbf{a})\}|\{m\}\rangle\hbar\omega \tag{6.135}$$

Next, insert the unity operator built up from the translation operator through

$$\mathbf{1} = \mathbf{A}(\alpha^\circ)^{-1}\mathbf{A}(\alpha^\circ)$$

with

$$\mathbf{A}(\alpha^\circ) = e^{\alpha^\circ \mathbf{a}^\dagger - \alpha^\circ \mathbf{a}} \tag{6.136}$$

in such a way as to write

$$\begin{aligned}
&\langle\{n\}|\mathbf{A}(\alpha^\circ)^{-1}\mathbf{A}(\alpha^\circ)\mathbf{H}_{\mathrm{Dr}}\mathbf{A}(\alpha^\circ)^{-1}\mathbf{A}(\alpha^\circ)|\{m\}\\
&\quad = \langle\{n\}|\mathbf{A}(\alpha^\circ)^{-1}\mathbf{A}(\alpha^\circ)\\
&\qquad \times \left\{\left(\mathbf{a}^\dagger\mathbf{a} + \tfrac{1}{2}\right) + \alpha^\circ(\mathbf{a}^\dagger + \mathbf{a})\right\}\mathbf{A}(\alpha^\circ)^{-1}\mathbf{A}(\alpha^\circ)|\{m\}\rangle\hbar\omega
\end{aligned} \tag{6.137}$$

Now, observe that the action of the translation operator transforms the eigenstates of the harmonic Hamiltonian into new displaced ones according to

$$\mathbf{A}(\alpha^\circ)|\{n\}\rangle = |\{\tilde{n}\}\rangle \tag{6.138}$$

$$\langle\{n\}|\mathbf{A}(\alpha^\circ)^{-1} = \langle\{\tilde{n}\}| \tag{6.139}$$

In order to get the expression of the real oscillator wavefunction corresponding to the transformed ket (6.138) observe that, due to Eq. (6.116), the wavefunction corresponding to the states $|\{n\}\rangle$ is given by

$$\Psi_n(Q) = \langle\{Q\}|\{n\}\rangle = \langle\{n\}|\{Q\}\rangle$$

whereas the wavefunction corresponding to the bra $\langle\{\tilde{n}\}|$ appearing in Eq. (6.139), and resulting from the action of the translation operator (involving a real α°), is

$$\Psi_{\tilde{n}}(Q) = \langle\{Q\}|\{\tilde{n}\}\rangle$$

or, due to Eq. (6.139),

$$\Psi_{\tilde{n}}(Q) = \langle\{Q\}|\mathbf{A}(\alpha^\circ)^{-1}|\{n\}\rangle$$

and thus, owing to Eqs. (6.75) and (6.77),

$$\Psi_{\tilde{n}}(Q) = \langle\{Q\}|\mathbf{A}(-Q^\circ)|\{n\}\rangle \tag{6.140}$$

with

$$\alpha^\circ = Q^\circ\sqrt{\frac{m\omega}{2\hbar}}$$

Next, due to Eq. (6.117) leading to

$$\mathbf{A}(-Q^\circ)|\{Q\}\rangle = |\{Q - Q^\circ\}\rangle$$

Eq. (6.140) reads

$$\Psi_{\tilde{n}}(Q) = \langle\{Q - Q^\circ\}|\{n\}\rangle$$

or

$$\Psi_{\tilde{n}}(Q) = \Psi_n(Q - Q^\circ)$$

with

$$\Psi_n(Q - Q^\circ) = \langle\{Q - Q^\circ\}|\{n\}\rangle$$

In a similar way, one would obtain

$$\Psi_{\tilde{m}}(Q) = \Psi_m(Q - Q^\circ) = \langle\{Q - Q^\circ\}|\{m\}\rangle$$

Now, in the context of the transformed states (6.138) and (6.139) corresponding to the wavefunction $\Psi_n(Q - Q^\circ)$, which is displaced by the amount $-Q^\circ$, let us introduce the following transformed Hamiltonian:

$$\tilde{\mathbf{H}}_{\mathrm{Dr}} = \mathbf{A}(\alpha^\circ)\mathbf{H}_{\mathrm{Dr}}\mathbf{A}(\alpha^\circ)^{-1} \tag{6.141}$$

6.9.1.2 *Hamiltonian diagonalization by the canonical transformation*

Then, owing to Eqs. (6.137) and (6.141), the matrix elements of the transformed Hamiltonian $\tilde{\mathbf{H}}_{\mathrm{Dr}}$ take the form

$$\begin{aligned}\langle\{\tilde{n}\}|\tilde{\mathbf{H}}_{\mathrm{Dr}}|\{\tilde{m}\}\rangle &= \tfrac{1}{2}\hbar\omega + \langle\{\tilde{n}\}|\mathbf{A}(\alpha^\circ)\mathbf{a}^\dagger\mathbf{a}\ \mathbf{A}(\alpha^\circ)^{-1}|\{\tilde{m}\}\rangle\hbar\omega \\ &+ \alpha^\circ\langle\{\tilde{n}\}|\mathbf{A}(\alpha^\circ)(\mathbf{a}^\dagger + \mathbf{a})\mathbf{A}(\alpha^\circ)^{-1}|\{\tilde{m}\}\rangle\hbar\omega\end{aligned} \tag{6.142}$$

Moreover, observe that, according to Eq. (6.81), one has

$$\mathbf{A}(\alpha^\circ)\mathbf{a}^\dagger\mathbf{A}(\alpha^\circ)^{-1} = \mathbf{a}^\dagger - \alpha^\circ \tag{6.143}$$

$$\mathbf{A}(\alpha^\circ)\mathbf{a}\mathbf{A}(\alpha^\circ)^{-1} = \mathbf{a} - \alpha^\circ \tag{6.144}$$

Now, in order to get the result of the canonical transformation on $\mathbf{a}^\dagger\mathbf{a}$ appearing in Eq. (6.142), insert between $\mathbf{a}^\dagger$ and $\mathbf{a}$ the unity operator built up from the unitary translation operator, as follows:

$$\mathbf{A}(\alpha^\circ)\mathbf{a}^\dagger\mathbf{a}\mathbf{A}(\alpha^\circ)^{-1} = \mathbf{A}(\alpha^\circ)\mathbf{a}^\dagger\mathbf{A}(\alpha^\circ)^{-1}\mathbf{A}(\alpha^\circ)\ \mathbf{a}\ \mathbf{A}(\alpha^\circ)^{-1}$$

Then, in view of Eqs. (6.143) and (6.144), we have

$$\mathbf{A}(\alpha^\circ)\mathbf{a}^\dagger\mathbf{a}\mathbf{A}(\alpha^\circ)^{-1} = (\mathbf{a}^\dagger - \alpha^\circ)(\mathbf{a} - \alpha^\circ)$$

Hence, owing to this result and to Eqs. (6.143) and (6.144), the sum of the transformed operators appearing on the right-hand side of Eq. (6.142) yields

$$\begin{aligned}&\mathbf{A}(\alpha^\circ)\mathbf{a}^\dagger\mathbf{a}\ \mathbf{A}(\alpha^\circ)^{-1} + \mathbf{A}(\alpha^\circ)(\mathbf{a}^\dagger + \mathbf{a})\mathbf{A}(\alpha^\circ)^{-1} \\ &= \mathbf{a}^\dagger\mathbf{a} - \alpha^\circ(\mathbf{a}^\dagger + \mathbf{a}) + \alpha^{\circ 2} + \alpha^\circ(\mathbf{a}^\dagger + \mathbf{a}) - 2\alpha^{\circ 2}\end{aligned}$$

or, after simplification

$$\mathbf{A}(\alpha^\circ)(\mathbf{a}^\dagger\mathbf{a} + \alpha^\circ(\mathbf{a}^\dagger + \mathbf{a}))\mathbf{A}(\alpha^\circ)^{-1} = \mathbf{a}^\dagger\mathbf{a} - \alpha^{\circ 2}$$

Therefore, according to these results and to Eqs. (6.138) and (6.139), Eq. (6.142) reduces to

$$\langle\{\tilde{n}\}|\tilde{\mathbf{H}}_{\mathrm{Dr}}|\{\tilde{m}\}\rangle = \langle\{\tilde{n}\}|\left(\mathbf{a}^\dagger\mathbf{a} + \tfrac{1}{2} - \alpha^{\circ 2}\right)|\{\tilde{m}\}\rangle\hbar\omega$$

or, due to Eq. (5.40),

$$\langle\{\tilde{n}\}|\tilde{\mathbf{H}}_{\mathrm{Dr}}|\{\tilde{m}\}\rangle = \left\{\left(\tilde{n} + \tfrac{1}{2}\right) - \alpha^{\circ 2}\right\}\hbar\omega\delta_{\tilde{m}\tilde{n}} \tag{6.145}$$

Therefore, since, according to Eq. (6.145), the Hamiltonian $\tilde{\mathbf{H}}_{\mathrm{Dr}}$ is diagonal in the basis $\{|\{\tilde{n}\}\rangle\}$ obtained from that $\{|\{n\}\rangle\}$ through the canonical transformation (6.138), it appears that the following eigenvalue equation has been solved:

$$\tilde{\mathbf{H}}_{\mathrm{Dr}}|\{\tilde{n}\}\rangle = E_{\tilde{n}}\ |\{\tilde{n}\}\rangle$$

with the eigenvalues

$$E_{\tilde{n}} = \left\{\left(\tilde{n}+\tfrac{1}{2}\right) - \alpha^{\circ 2}\right\}\hbar\omega$$

6.9.2 Diagonalization of the Hamiltonian matrix representation

Besides the above canonical diagonalization of the driven harmonic oscillator Hamiltonian (6.134), it is also possible to diagonalize the matrix representation (6.135) of this Hamiltonian.

6.9.2.1 Matrix elements of the driven Hamiltonian in the basis of the harmonic Hamiltonian Consider Eq. (6.135):

$$\langle\{n\}|\mathbf{H}_{\mathrm{Dr}}|\{m\}\rangle = \langle\{n\}|\left(\mathbf{a}^{\dagger}\mathbf{a} + \alpha^{\circ}(\mathbf{a}^{\dagger}+\mathbf{a}) + \tfrac{1}{2}\right)|\{m\}\rangle\hbar\omega$$

Next, in view of Eq. (5.40), we have

$$\langle\{n\}|\mathbf{H}_{\mathrm{Dr}}|\{\ m\}\rangle = \alpha^{\circ}\langle\{n\}|(\mathbf{a}^{\dagger}+\mathbf{a})|\{m\}\rangle\hbar\omega + \left(m+\tfrac{1}{2}\right)\langle\{n\}|\{m\}\rangle\hbar\omega \tag{6.146}$$

Now, keeping in mind Eq. (5.40), allowing one to write

$$\mathbf{a}|\{m\}\rangle = \sqrt{m}|\{m-1\}\rangle \qquad \text{and} \qquad \langle\{n\}|\mathbf{a}^{\dagger} = \sqrt{n}\langle\{n-1\}|$$

the matrix elements (6.146) read

$$\begin{aligned}\langle\{n\}|\mathbf{H}_{\mathrm{Dr}}|\{m\}\rangle &= \alpha^{\circ}(\sqrt{n}\ \langle\{n-1\}|\{m\}\rangle\hbar\omega + \sqrt{m}\ \langle\{n\}|\{m-1\}\rangle) \\ &\quad + (m+\tfrac{1}{2})\langle\{n\}|\{m\}\rangle\hbar\omega\end{aligned} \tag{6.147}$$

which are zero because of the orthonormality properties of the eigenstates of the quantum harmonic Hamiltonian, except the following cases:

$$\langle\{n\}|\mathbf{H}_{\mathrm{Dr}}|\{n\}\rangle = \hbar\omega\left(n+\tfrac{1}{2}\right) \tag{6.148}$$

$$\langle\{n\}|\mathbf{H}_{\mathrm{Dr}}|\{n-1\}\rangle = \alpha^{\circ}\hbar\omega(\sqrt{n} + \sqrt{n-1}) \tag{6.149}$$

with, since the matrix is Hermitian,

$$\langle\{n-1\}|\mathbf{H}_{\mathrm{Dr}}|\{n\}\rangle = \langle\{n\}|\mathbf{H}_{\mathrm{Dr}}|\{n-1\}\rangle \tag{6.150}$$

6.9.2.2 Truncation and diagonalization of the matrix representation The matrix elements involved in the matrix representation of the Hamiltonian (6.134) may be computed using Eqs. (6.148)–(6.150). This Hamiltonian matrix may be built up by starting from the ground-state $|\{0\}\rangle$ and then increasing progressively the quantum number n associated with the kets $|\{n\}\rangle$ and with the bras $\langle\{n\}|$. Now, since the kets and bras appearing in Eq. (6.147) belong to a basis that is infinite, the matrix representation must also be infinite. Thus, in order to be numerically diagonalized, the Hamiltonian matrix (6.135) of the Hamiltonian (6.134) must be truncated beyond some value n° of

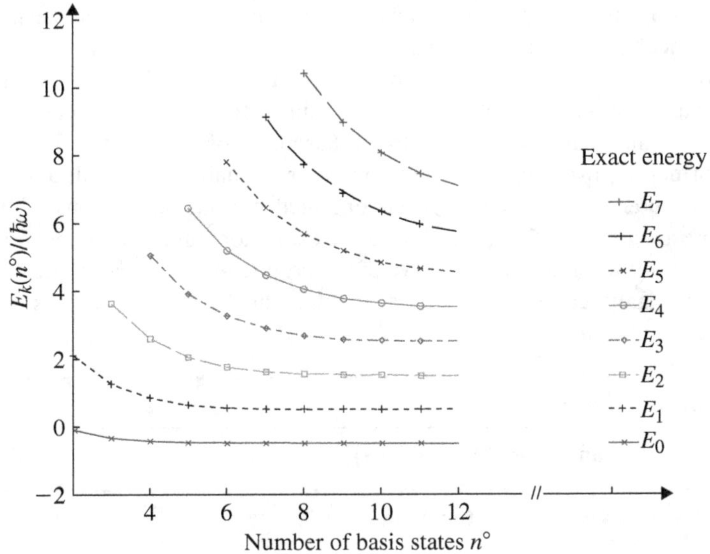

Figure 6.3 Stabilization of the energy of the eight lowest eigenvalues $E_k(n^\circ)/\hbar\omega^\circ$ with respect to n°. (See color insert.)

the quantum number n, leading, therefore, to a finite square $(n^\circ + 1) \times (n^\circ + 1)$ matrix involving the parameters $\hbar\omega$ and α°. The diagonalization of this truncated matrix leads to approximate solutions $E_k(n^\circ)$ of the exact eigenvalue equation

$$\mathbf{H}|\Psi_k(n^\circ)\rangle = E_k|\Psi_k(n^\circ)\rangle$$

Figure 6.3 shows the dependence of the eight lowest eigenvalues $E_k(n^\circ)$ on n° when $\alpha^\circ = 1$.

Inspection of the figure shows that when n° is progressively increased, the lowest eigenvalues $E_k(n^\circ)$ decrease progressively and then stabilize toward their exact values obtained by the aid of Eq. (6.145). Such a result manifests the ability to satisfactorily obtain the eigenvalues of a Hamiltonian by diagonalizing its truncated matrix representations by increasing progressively its dimensions until energy level stabilization occurs. That will be later applied to get the energy levels of anharmonic oscillators for which the direct diagonalization of the Hamiltonian is very difficult or impossible to perform. Now, it may be of interest to observe that, as required from the variation theorem (2.25), the energy of the ground state is lowered when improving the accuracy of the corresponding eigenfunction by increasing the dimension of the truncated basis.

6.10 CONCLUSION

In this chapter, devoted to coherent states assumed to be eigenstates of the lowering operator, the following results have been obtained: (i) the expansion of the coherent states over the eigenstates of the harmonic oscillator Hamiltonian, (ii) the fact that they minimize the Heisenberg uncertainty relations, (iii) the fact that they may be generated by the action of the translation operator on these eigenkets, and (iv) their wave mechanics representation. In addition, using the translation operator, it has been possible to get the overlap (Franck–Condon factors) between two mutually translated eigenstates of the harmonic Hamiltonian. Moreover, it has been shown how to diagonalize the Hamiltonian of a driven harmonic oscillator by aid of a canonical transformation involving translation operators. Finally, this result allows one to verify the accuracy of the energy levels obtained by diagonalizations of truncated matrix representations of the driven harmonic oscillator Hamiltonian, opening therefore a possibility to obtain numerically the energy levels of anharmonic oscillators for which no analytical expression is available. The most important results of this chapter are listed as follows:

Definition of coherent states
$\mathbf{a}\|\{\alpha\}\rangle = \alpha\|\{\alpha\}\rangle$ and $\langle\{\alpha\}\|\mathbf{a}^\dagger = \alpha^*\langle\{\alpha\}\|$
Coherent-state expansion in terms of the $\mathbf{a}^\dagger\mathbf{a}$ eigenkets:
$\|\{\alpha\}\rangle = e^{-\|\alpha\|^2/2}\sum_n\left(\frac{\alpha^n}{\sqrt{n!}}\right)\|\{n\}\rangle$
Scalar product between two coherent states:
$\langle\{\beta\}\|\{\alpha\}\rangle = e^{-\|\alpha-\beta\|^2/2}$
Closure relations over coherent states:
$\int_{-\infty}^{+\infty}\int_{-\infty}^{+\infty}\|\{\alpha\}\rangle\langle\{\alpha\}\|d\,\mathrm{Re}(\alpha)d\,\mathrm{Im}(\alpha) = 1$
Translation operators:
$\mathbf{A}(\alpha^\circ) = e^{\alpha^\circ\mathbf{a}^\dagger-\alpha^\circ\mathbf{a}} = e^{-\alpha^\circ\mathbf{a}+\alpha^\circ\mathbf{a}^\dagger}$
Generation of coherent states by action of the translation operator:
$\|\{\alpha\}\rangle = \mathbf{A}(\alpha^\circ)\|\{0\}\rangle$

BIBLIOGRAPHY

P. Carruthers and M. Nieto. *Am. J. Phys.*, **33** (1965): 537.

C. Cohen-Tannoudji, B. Diu and F. Laloe, *Quantum Mechanics*. Wiley-Interscience: Hoboken, NJ, 2006.

P. A. M. Dirac. *The Principles of Quantum Mechanics*, 4th ed. Oxford University Press: Oxford, 1982.

S. Koide. *Z. Naturforschg.*, **15a** (1960): 123–128.

W. H. Louisell. *Quantum Statistical Properties of Radiation*. Wiley: New York, 1973.

CHAPTER 7

BOSON OPERATOR THEOREMS

INTRODUCTION

In Chapter 5, some important properties of the ladder operators were found, particularly their action on the eigenkets of the number occupation operator. However, many other theorems dealing with the Boson operators, which are also important when working for not only a single quantum harmonic oscillator but in the context of anharmonic oscillators or sets of coupled harmonic oscillators, exist. The aim of the present chapter is to treat these theorems. This chapter deals with by canonical transformations involving the ladder operators. Then, we consider the normal and antinormal ordering formalism that allows one to pass from equations dealing with noncommuting ladder operators to equations involving only scalars, which are easier to solve and then, after solution, to return to the operator equations that are themselves the solutions of the starting operator equations. A final section illustrates this formalism by applying the procedure to the calculation of time evolution operators of driven quantum harmonic oscillators.

7.1 CANONICAL TRANSFORMATIONS

Here, we shall prove theorems dealing with different canonical transformations on functions of Boson operators, involving operators that are also functions of these Boson operators.

7.1.1 Transformations involving translation operators

Start from the Baker–Campbell–Hausdorff relation given by Eq. (1.77):

$$(e^{\xi \mathbf{A}})\{\mathbf{f}(\mathbf{B})\}(e^{-\xi \mathbf{A}}) = \{\mathbf{f}(e^{\xi \mathbf{A}}\mathbf{B}e^{-\xi \mathbf{A}})\} \tag{7.1}$$

where ξ is a c-number, whereas $\mathbf{f}$, $\mathbf{A}$, and $\mathbf{B}$ are linear operators. Now, apply this relation to the situation where $\mathbf{A}$ is the Boson operator $\mathbf{a}$ and where $\mathbf{f}(\mathbf{B})$ is a function of both $\mathbf{a}$ and its Hermitian conjugate $\mathbf{a}^\dagger$, that is,

$$\mathbf{A} = \mathbf{a} \qquad \text{and} \qquad \mathbf{f}(\mathbf{B}) = \mathbf{f}(\mathbf{a}, \mathbf{a}^\dagger)$$

Then, Eq. (7.1) takes the form

$$(e^{\xi \mathbf{a}})\{\mathbf{f}(\mathbf{a}, \mathbf{a}^\dagger)\}(e^{-\xi \mathbf{a}}) = \{\mathbf{f}((e^{\xi \mathbf{a}}\mathbf{a}e^{-\xi \mathbf{a}}), (e^{\xi \mathbf{a}}\mathbf{a}^\dagger e^{-\xi \mathbf{a}}))\} \tag{7.2}$$

Quantum Oscillators, First Edition. Olivier Henri-Rousseau and Paul Blaise.
© 2011 John Wiley & Sons, Inc. Published 2011 by John Wiley & Sons, Inc.

Now, since any function of a single operator commutes with this operator, the following relation is verified:

$$(e^{\xi\mathbf{a}})\,\mathbf{a}\,(e^{-\xi\mathbf{a}}) = \mathbf{a} \tag{7.3}$$

whereas, due to Eq. (6.84)

$$(e^{\xi\mathbf{a}})\,\mathbf{a}^{\dagger}(e^{-\xi\mathbf{a}}) = \mathbf{a}^{\dagger} + \xi \tag{7.4}$$

Thus, owing to Eqs. (7.3) and (7.4), Eq. (7.2) transforms to

$$\boxed{(e^{\xi\mathbf{a}})\{\mathbf{f}(\mathbf{a},\mathbf{a}^{\dagger})\}(e^{-\xi\mathbf{a}}) = \{\mathbf{f}(\mathbf{a},\mathbf{a}^{\dagger} + \xi)\}} \tag{7.5}$$

Now, apply Eq. (7.1) to another special situation by changing ξ into $-\xi^*$ and taking

$$\mathbf{A} = \mathbf{a}^{\dagger} \qquad \text{and} \qquad \{\mathbf{f}(\mathbf{B})\} = \{\mathbf{f}(\mathbf{a},\mathbf{a}^{\dagger})\}$$

Then, Eq. (7.1) reads

$$(e^{-\xi^*\mathbf{a}^{\dagger}})\{\mathbf{f}(\mathbf{a},\mathbf{a}^{\dagger})\}(e^{\xi^*\mathbf{a}^{\dagger}}) = \{\mathbf{f}(e^{-\xi^*\mathbf{a}^{\dagger}}\mathbf{a}e^{\xi^*\mathbf{a}^{\dagger}}, e^{-\xi^*\mathbf{a}^{\dagger}}a^{\dagger}e^{\xi^*\mathbf{a}^{\dagger}})\} \tag{7.6}$$

Of course, for the same reasons as those used to obtain Eq. (7.3), one has

$$(e^{-\xi^*\mathbf{a}^{\dagger}})\,\mathbf{a}^{\dagger}(e^{\xi^*\mathbf{a}^{\dagger}}) = \mathbf{a}^{\dagger}$$

Besides, according to Eq. (6.85), we have

$$(e^{-\xi^*\mathbf{a}^{\dagger}})\,\mathbf{a}(e^{\xi^*\mathbf{a}^{\dagger}}) = \mathbf{a} + \xi^*$$

Thus, the canonical transformation (7.6) reads

$$\boxed{(e^{-\xi^*\mathbf{a}^{\dagger}})\{\mathbf{f}(\mathbf{a},\mathbf{a}^{\dagger})\}(e^{\xi^*\mathbf{a}^{\dagger}}) = \{\mathbf{f}(\mathbf{a} + \xi^*,\mathbf{a}^{\dagger})\}} \tag{7.7}$$

Next, consider the general transformation

$$(e^{-\xi^*\mathbf{a}^{\dagger}})(e^{\xi\mathbf{a}})\{\mathbf{f}(\mathbf{a},\mathbf{a}^{\dagger})\}(e^{-\xi\mathbf{a}})(e^{\xi^*\mathbf{a}^{\dagger}}) \equiv (e^{-\xi^*\mathbf{a}^{\dagger}})\{(e^{\xi\mathbf{a}})\{\mathbf{f}(\mathbf{a},\mathbf{a}^{\dagger})\}(e^{-\xi\mathbf{a}})\}(e^{\xi^*\mathbf{a}^{\dagger}})$$

which, due to Eq. (7.5), reads

$$(e^{-\xi^*\mathbf{a}^{\dagger}})(e^{\xi\mathbf{a}})\{\mathbf{f}(\mathbf{a},\mathbf{a}^{\dagger})\}(e^{-\xi\mathbf{a}})(e^{\xi^*\mathbf{a}^{\dagger}}) = (e^{-\xi^*\mathbf{a}^{\dagger}})\{\mathbf{f}(\mathbf{a},\mathbf{a}^{\dagger} + \xi)\}(e^{\xi^*\mathbf{a}^{\dagger}})$$

and owing to Eq. (7.7) transforms to

$$\boxed{(e^{-\xi^*\mathbf{a}^{\dagger}})(e^{\xi\mathbf{a}})\{\mathbf{f}(\mathbf{a},\mathbf{a}^{\dagger})\}(e^{-\xi\mathbf{a}})(e^{\xi^*\mathbf{a}^{\dagger}}) = \{\mathbf{f}(\mathbf{a} + \xi^*,\mathbf{a}^{\dagger} + \xi)\}} \tag{7.8}$$

Hence, using the Glauber theorem (1.79), we have

$$(e^{[\mathbf{a}^{\dagger},\mathbf{a}]|\xi|^2/2})(e^{-\xi^*\mathbf{a}^{\dagger}+\xi\mathbf{a}})\{\mathbf{f}(\mathbf{a},\mathbf{a}^{\dagger})\}(e^{\xi^*\mathbf{a}^{\dagger}-\xi\mathbf{a}})(e^{-[\mathbf{a}^{\dagger},\mathbf{a}]|\xi|^2/2}) = \{\mathbf{f}(\mathbf{a} + \xi^*,\mathbf{a}^{\dagger} + \xi)\}$$

or, after simplification,

$$(e^{-\xi^*\mathbf{a}^{\dagger}+\xi\mathbf{a}})\{\mathbf{f}(\mathbf{a},\mathbf{a}^{\dagger})\}(e^{\xi^*\mathbf{a}^{\dagger}-\xi\mathbf{a}}) = \{\mathbf{f}(\mathbf{a} + \xi^*,\mathbf{a}^{\dagger} + \xi)\}$$

so that, noting the definition of the translation operator (6.74),

$$\boxed{\mathbf{A}(\xi)^{-1}\{\mathbf{f}(\mathbf{a},\mathbf{a}^{\dagger})\}\mathbf{A}(\xi) = \{\mathbf{f}(\mathbf{a} + \xi^*,\mathbf{a}^{\dagger} + \xi)\}} \tag{7.9}$$

with

$$\mathbf{A}(\xi) = (e^{\xi^*\mathbf{a}^{\dagger}-\xi\mathbf{a}}) \tag{7.10}$$

7.1.2 Transformations involving number occupation operator exponentials

7.1.2.1 Transformations of the ladder operators Now, consider the following canonical transformation in which ξ is a c-number:

$$\mathbf{g}(\xi) = (e^{\xi\mathbf{a}^\dagger\mathbf{a}})\mathbf{a}(e^{-\xi\mathbf{a}^\dagger\mathbf{a}}) \tag{7.11}$$

which for $\xi = 0$ reads

$$\mathbf{g}(0) = \mathbf{a} \tag{7.12}$$

The derivative of (7.11) with respect to ξ yields

$$\left(\frac{d\mathbf{g}(\xi)}{d\xi}\right) = \left(\frac{d}{d\xi}e^{\xi\mathbf{a}^\dagger\mathbf{a}}\right)\mathbf{a}(e^{-\xi\mathbf{a}^\dagger\mathbf{a}}) + (e^{\xi\mathbf{a}^\dagger\mathbf{a}})\mathbf{a}\left(\frac{d}{d\xi}e^{-\xi\mathbf{a}^\dagger\mathbf{a}}\right)$$

or

$$\left(\frac{d\mathbf{g}(\xi)}{d\xi}\right) = (\mathbf{a}^\dagger\mathbf{a}\, e^{\xi\mathbf{a}^\dagger\mathbf{a}})\,\mathbf{a}\,(e^{-\xi\mathbf{a}^\dagger\mathbf{a}}) - (e^{\xi\mathbf{a}^\dagger\mathbf{a}})\,\mathbf{a}\,(\mathbf{a}^\dagger\mathbf{a}\, e^{-\xi\mathbf{a}^\dagger\mathbf{a}}) \tag{7.13}$$

Again, since $\mathbf{a}^\dagger\mathbf{a}$ commutes with all functions of the product $\mathbf{a}^\dagger\mathbf{a}$, the first right-hand-side term of Eq. (7.13) becomes

$$(\mathbf{a}^\dagger\mathbf{a}\, e^{\xi\mathbf{a}^\dagger\mathbf{a}})\mathbf{a}(e^{-\xi\mathbf{a}^\dagger\mathbf{a}}) = (e^{\xi\mathbf{a}^\dagger\mathbf{a}}\mathbf{a}^\dagger\mathbf{a})\,\mathbf{a}\,(e^{-\xi\mathbf{a}^\dagger\mathbf{a}})$$

so that eq. (7.13) transforms to

$$\left(\frac{d\mathbf{g}(\xi)}{d\xi}\right) = (e^{\xi\mathbf{a}^\dagger\mathbf{a}})(\mathbf{a}^\dagger\mathbf{a}\mathbf{a} - \mathbf{a}\mathbf{a}^\dagger\mathbf{a})(e^{-\xi\mathbf{a}^\dagger\mathbf{a}})$$

or

$$\left(\frac{d\mathbf{g}(\xi)}{d\xi}\right) = (e^{\xi\mathbf{a}^\dagger\mathbf{a}})[\mathbf{a}^\dagger, \mathbf{a}]\,\mathbf{a}\,(e^{-\xi\mathbf{a}^\dagger\mathbf{a}}) \tag{7.14}$$

Again, using $[\mathbf{a}, \mathbf{a}^\dagger] = 1$, Eq. (7.14) transforms to

$$\left(\frac{d\mathbf{g}(\xi)}{d\xi}\right) = -(e^{\xi\mathbf{a}^\dagger\mathbf{a}})\,\mathbf{a}\,(e^{-\xi\mathbf{a}^\dagger\mathbf{a}})$$

and, in view of Eq. (7.11), into

$$\left(\frac{d\mathbf{g}(\xi)}{d\xi}\right) = -\mathbf{g}(\xi) \tag{7.15}$$

Next, by derivation of both terms of Eq. (7.15) with respect to ξ, that is,

$$\left(\frac{d^2\mathbf{g}(\xi)}{d\xi^2}\right) = -\left(\frac{d\,\mathbf{g}(\xi)}{d\xi}\right)$$

and, due to Eq. (7.15), it reads

$$\left(\frac{d^2\mathbf{g}(\xi)}{d\xi^2}\right) = \mathbf{g}(\xi)$$

Again, by recurrence, one obtains

$$\left(\frac{d^n\mathbf{g}(\xi)}{d\xi^n}\right) = (-1)^n\mathbf{g}(\xi) \tag{7.16}$$

Now, at $\xi = 0$, and in view of Eq. (7.12), the nth derivative given by Eq. (7.16) reduces to

$$\left(\frac{d^n\mathbf{g}(\xi)}{d\xi^n}\right)_{\xi=0} = (-1)^n\mathbf{a} \tag{7.17}$$

Next, write the Taylor expansion of the function (7.11) around $\xi = 0$, that is,

$$\mathbf{g}(\xi) = \mathbf{g}(0) + \left(\frac{d\mathbf{g}(\xi)}{d\xi}\right)_{\xi=0}\xi + \frac{1}{2}\left(\frac{d^2\mathbf{g}(\xi)}{d\xi^2}\right)_{\xi=0}\xi^2 + \frac{1}{3!}\left(\frac{d^3\mathbf{g}(\xi)}{d\xi^3}\right)_{\xi=0}\xi^3\ldots$$

Hence, comparing Eqs. (7.12) and (7.17), this expansion takes the form

$$\mathbf{g}(\xi) = \mathbf{a}\left(1 - \xi + \frac{1}{2}\xi^2 - \frac{1}{3!}\xi^3\ldots\ldots\right)$$

or, passing from the expansion to its corresponding exponential expression,

$$\mathbf{g}(\xi) = \mathbf{a}(e^{-\xi})$$

so that, due to Eq. (7.11), we have

$$\boxed{(e^{\xi\mathbf{a}^\dagger\mathbf{a}})\,\mathbf{a}\,(e^{-\xi\mathbf{a}^\dagger\mathbf{a}}) = \mathbf{a}\,(e^{-\xi})} \tag{7.18}$$

Moreover, by a similar inference as that allowing to pass from Eq. (7.11) to Eq. (7.18), we have

$$\boxed{(e^{\xi\mathbf{a}^\dagger\mathbf{a}})\,\mathbf{a}^\dagger(e^{-\xi\mathbf{a}^\dagger\mathbf{a}}) = \mathbf{a}^\dagger(e^{\xi})} \tag{7.19}$$

Apply Eqs. (7.18) and (7.19) to reproduce the results of the integration of the Heisenberg equation governing the dynamics of the ladder operators, keeping in mind Eq. (3.88) governing the time dependence of an operator $\mathbf{A}$ in the Heisenberg picture, that is,

$$\mathbf{A}(t)^{\mathrm{HP}} = (e^{i\mathbf{H}t/\hbar})\mathbf{A}(e^{-i\mathbf{H}t/\hbar})$$

which reads for the Boson operator $\mathbf{a}$, and when the Hamiltonian $\mathbf{H}$ is that of a harmonic oscillator

$$\mathbf{a}(t)^{\mathrm{HP}} = (e^{i\omega t\mathbf{a}^\dagger\mathbf{a}})\,\mathbf{a}\,(e^{-i\omega t\mathbf{a}^\dagger\mathbf{a}}) \tag{7.20}$$

Then, applying Eqs. (7.18) and (7.19) to the situation where $\xi = i\omega t$, it yields

$$(e^{i\omega t\mathbf{a}^\dagger\mathbf{a}})\,\mathbf{a}\,(e^{-i\omega t\mathbf{a}^\dagger\mathbf{a}}) = \mathbf{a}\,(e^{-i\omega t}) \tag{7.21}$$

$$(e^{i\omega t\mathbf{a}^\dagger\mathbf{a}})\,\mathbf{a}^\dagger(e^{i\omega t\mathbf{a}^\dagger\mathbf{a}}) = \mathbf{a}^\dagger(e^{i\omega t}) \tag{7.22}$$

so that Eq. (7.20) reads

$$\mathbf{a}(t)^{\mathrm{HP}} = \mathbf{a}e^{-i\omega t}$$

the Hermitian conjugate of which is

$$\mathbf{a}^\dagger(t)^{\mathrm{HP}} = \mathbf{a}^\dagger e^{i\omega t}$$

One may verify that these results are equivalent to those of (5.151) and (5.152) obtained by integration of the Heisenberg equation (3.94).

7.1.2.2 Transformations on functions of the ladder operators Now, consider the following transformation:

$$(e^{\xi \mathbf{a}^\dagger \mathbf{a}})\{\mathbf{f}(\mathbf{a})\}(e^{-\xi \mathbf{a}^\dagger \mathbf{a}}) \tag{7.23}$$

where $\mathbf{f}(\mathbf{a})$ is a function of $\mathbf{a}$ that may be expanded according to

$$\{\mathbf{f}(\mathbf{a})\} = \sum_n \{C_n\}(\mathbf{a})^n \tag{7.24}$$

where the $\{C_n\}$ are the scalar coefficients of the expansion. The latter expansion may be transformed using the following unity operator:

$$\mathbf{1} = (e^{-\xi \mathbf{a}^\dagger \mathbf{a}})(e^{\xi \mathbf{a}^\dagger \mathbf{a}}) \tag{7.25}$$

according to

$$\{\mathbf{f}(\mathbf{a})\} = \sum_n \{C_n\}\{\mathbf{a}\mathbf{1}\mathbf{a} \cdots \mathbf{1}\mathbf{a}\}_n$$

so that Eq. (7.24) reads

$$\{\mathbf{f}(\mathbf{a})\} = \sum_n \{C_n\}\{\mathbf{a}(e^{-\xi \mathbf{a}^\dagger \mathbf{a}})(e^{\xi \mathbf{a}^\dagger \mathbf{a}})\mathbf{a} \cdots (e^{-\xi \mathbf{a}^\dagger \mathbf{a}})(e^{\xi \mathbf{a}^\dagger \mathbf{a}})\mathbf{a}\}_n$$

Then, the transformation (7.23) becomes

$$(e^{\xi \mathbf{a}^\dagger \mathbf{a}})\{\mathbf{f}(\mathbf{a})\}e^{(-\xi \mathbf{a}^\dagger \mathbf{a})} = \sum_n \{C_n\}\{(e^{\xi \mathbf{a}^\dagger \mathbf{a}}\mathbf{a}e^{-\xi \mathbf{a}^\dagger \mathbf{a}})(e^{\xi \mathbf{a}^\dagger \mathbf{a}}\mathbf{a}e^{-\xi \mathbf{a}^\dagger \mathbf{a}}) \cdots (e^{\xi \mathbf{a}^\dagger \mathbf{a}}\mathbf{a}e^{-\xi \mathbf{a}^\dagger \mathbf{a}})\}_n$$

Again, using Eq. (7.18), we have

$$(e^{\xi \mathbf{a}^\dagger \mathbf{a}})\{\mathbf{f}(\mathbf{a})\}(e^{-\xi \mathbf{a}^\dagger \mathbf{a}}) = \sum_n \{C_n\}(\mathbf{a}e^{-\xi})^n$$

Thus, comparing (7.24), one obtains

$$(e^{\xi \mathbf{a}^\dagger \mathbf{a}})\{\mathbf{f}(\mathbf{a})\}(e^{-\xi \mathbf{a}^\dagger \mathbf{a}}) = \{\mathbf{f}(\mathbf{a}e^{-\xi})\} \tag{7.26}$$

In a similar way, one would find

$$(e^{\xi \mathbf{a}^\dagger \mathbf{a}})\{\mathbf{f}(\mathbf{a}^\dagger)\}(e^{-\xi \mathbf{a}^\dagger \mathbf{a}}) = \{\mathbf{f}(\mathbf{a}^\dagger e^{\xi})\} \tag{7.27}$$

Now, consider a function of both $\mathbf{a}^\dagger$ and $\mathbf{a}$, defined by the following expansion:

$$\mathbf{f}\{(\mathbf{a}^\dagger, \mathbf{a})\} = \sum_{l,m,\ldots,r,\ldots,s,\ldots,u} \{C_{l,m,\ldots,r,\ldots,s,\ldots,u}\}\{(\mathbf{a}^\dagger)^l(\mathbf{a})^m \cdots (\mathbf{a})^r \cdots (\mathbf{a}^\dagger)^s \cdots (\mathbf{a}^\dagger)^u\} \tag{7.28}$$

where the $\{C_{l,m,\ldots,r\ldots,s\ldots,u}\}$ are the scalar coefficients of the expansion. Then, consider the following transformation over this function:

$$\{\mathbf{F}(\mathbf{a}^\dagger, \mathbf{a})\} = (e^{\xi \mathbf{a}^\dagger \mathbf{a}})\{\mathbf{f}(\mathbf{a}^\dagger, \mathbf{a})\}(e^{-\xi \mathbf{a}^\dagger \mathbf{a}}) \tag{7.29}$$

Next, insert in the following way into the right-hand side of Eq. (7.29), the unity operator Eq. (7.25):

$$\{\mathbf{F}(\mathbf{a}^\dagger, \mathbf{a})\} = \sum_{l,m,\ldots,r,\ldots,s,\ldots,u} \{C_{l,m,\ldots,r,\ldots,s,\ldots,u}\}$$
$$\times (e^{\xi\mathbf{a}^\dagger\mathbf{a}})(\mathbf{a}^\dagger)^l(e^{-\xi\mathbf{a}^\dagger\mathbf{a}}e^{\xi\mathbf{a}^\dagger\mathbf{a}})(\mathbf{a})^m(e^{-\xi\mathbf{a}^\dagger\mathbf{a}}e^{\xi\mathbf{a}^\dagger\mathbf{a}})$$
$$\cdots(\mathbf{a})^r\cdots(\mathbf{a}^\dagger)^s\cdots(e^{-\xi\mathbf{a}^\dagger\mathbf{a}}e^{\xi\mathbf{a}^\dagger\mathbf{a}})(\mathbf{a}^\dagger)^u(e^{-\xi\mathbf{a}^\dagger\mathbf{a}})$$

Then, using Eqs. (7.26) and (7.27), we have

$$\{\mathbf{F}(\mathbf{a}^\dagger, \mathbf{a})\} = \sum_{l,m,\ldots,r\ldots,s\ldots,u} \{C_{l,m,\ldots,r\ldots,s\ldots,u}\}$$

Hence, comparing Eqs. (7.29) and Eq.(7.28), we have

$$(e^{\xi\mathbf{a}^\dagger\mathbf{a}})\{\mathbf{f}(\mathbf{a}^\dagger, \mathbf{a})\}(e^{-\xi\mathbf{a}^\dagger\mathbf{a}}) = \{\mathbf{f}(\mathbf{a}^\dagger e^{\xi}, \mathbf{a}e^{-\xi})\} \tag{7.30}$$

so that, when $\xi = i\omega t$, one obtains

$$\boxed{(e^{i\omega t\mathbf{a}^\dagger\mathbf{a}})\{\mathbf{f}(\mathbf{a}^\dagger, \mathbf{a})\}(e^{-i\omega t\mathbf{a}^\dagger\mathbf{a}}) = \{\mathbf{f}(\mathbf{a}^\dagger e^{i\omega t}, \mathbf{a}e^{-i\omega t})\}} \tag{7.31}$$

7.2 NORMAL AND ANTINORMAL ORDERING FORMALISM

We shall now deal with a formalism that allows us to transform an equation involving the noncommuting Boson operators into new scalar ones involving partial derivatives, which may be solved, the obtained solutions being inversely converted into expressions involving the ladder operators, which are the solutions of the above operator equations we want to solve. This formalism concerns what is called, the *normal and antinormal ordering*.

7.2.1 Normal and antinormal ordering

To introduce this formalism, start from the very simple operator

$$\{\mathbf{f}(\mathbf{a}, \mathbf{a}^\dagger)\} = \mathbf{a}\mathbf{a}^\dagger \tag{7.32}$$

which, due to the commutation rule $[\mathbf{a}, \mathbf{a}^\dagger] = 1$, and thus to $\mathbf{a}\mathbf{a}^\dagger = \mathbf{a}^\dagger\mathbf{a} + 1$, reads

$$\{\mathbf{f}(\mathbf{a}, \mathbf{a}^\dagger)\} = \mathbf{a}^\dagger\mathbf{a} + 1 \tag{7.33}$$

In the latter equation, the raising operator $\mathbf{a}^\dagger$ is before the lowering $\mathbf{a}$, at the opposite of the situation given by Eq. (7.32). In case (7.33), the operators $\mathbf{a}^\dagger$ and $\mathbf{a}$ are said to be in *normal form*, whereas in case (7.32), they are said to be in *antinormal* form. The following notations are used, respectively, for the two equivalent normal $\{n\}$ and antinormal $\{a\}$ expressions:

$$\{\mathbf{f}^{\{n\}}(\mathbf{a}, \mathbf{a}^\dagger)\} = \mathbf{a}^\dagger\mathbf{a} + 1 \qquad \text{and} \qquad \{\mathbf{f}^{\{a\}}(\mathbf{a}, \mathbf{a}^\dagger)\} = \mathbf{a}\mathbf{a}^\dagger$$

Of course, since the two expressions are equivalent, one must write

$$\{\mathbf{f}^{\{n\}}(\mathbf{a},\mathbf{a}^\dagger)\} = \{\mathbf{f}^{\{a\}}(\mathbf{a},\mathbf{a}^\dagger)\}$$

Now consider a very general operator $\mathbf{f}(\mathbf{a},\mathbf{a}^\dagger)$, which is some function of the operators $\mathbf{a}$ and $\mathbf{a}^\dagger$, susceptible to be expanded according to

$$\{\mathbf{f}(\mathbf{a},\mathbf{a}^\dagger)\} = \sum_{l,m,\ldots,r,\ldots,s,\ldots,u} \{C_{l,m,\ldots,r,\ldots,s,\ldots,u}\}(\mathbf{a}^\dagger)^l(\mathbf{a})^m \cdots (\mathbf{a})^r \cdots (\mathbf{a}^\dagger)^s \cdots (\mathbf{a}^\dagger)^u \quad (7.34)$$

where $\{C_{l,m,\ldots,r,\ldots,s,\ldots,u}\}$ are the expansion coefficients. Then, it is possible by systematic aid of $\mathbf{a}\mathbf{a}^\dagger = \mathbf{a}^\dagger\mathbf{a} + 1$ resulting from the commutation rule $[\mathbf{a},\mathbf{a}^\dagger] = 1$ to write this operator (7.34) either in normal or in antinormal form, according to

$$\{\mathbf{f}^{\{n\}}(\mathbf{a},\mathbf{a}^\dagger)\} = \sum_{rs}\{f^n_{rs}\}(\mathbf{a}^\dagger)^r(\mathbf{a})^s \quad (7.35)$$

$$\{\mathbf{f}^{\{a\}}(\mathbf{a},\mathbf{a}^\dagger)\} = \sum_{sr}\{f^a_{sr}\}(\mathbf{a})^s(\mathbf{a}^\dagger)^r \quad (7.36)$$

Here, f^n_{rs} and f^a_{sr} are, respectively, the expansion coefficients of the normal and antinormal form of the operator (7.34). Of course, as above, the three expressions (7.34)–(7.36) being equivalent, one may write

$$\{\mathbf{f}(\mathbf{a},\mathbf{a}^\dagger)\} = \{\mathbf{f}^{\{a\}}(\mathbf{a},\mathbf{a}^\dagger)\} = \{\mathbf{f}^{\{n\}}(\mathbf{a},\mathbf{a}^\dagger\}) \quad (7.37)$$

7.2.2 Normal and antinormal ordering operators

Now, consider two linear operators $\hat{\mathbf{N}}$ and $\hat{\mathbf{A}}$, the inverses of which are $\hat{\mathbf{N}}^{-1}$ and $\hat{\mathbf{A}}^{-1}$.

$$\hat{\mathbf{N}}\hat{\mathbf{N}}^{-1} = \hat{\mathbf{N}}^{-1}\hat{\mathbf{N}} = 1 \quad (7.38)$$

$$\hat{\mathbf{A}}\hat{\mathbf{A}}^{-1} = \hat{\mathbf{A}}^{-1}\hat{\mathbf{A}} = 1 \quad (7.39)$$

Then, assume that $\hat{\mathbf{N}}^{-1}$ and $\hat{\mathbf{A}}^{-1}$ allow one to transform, respectively, the normal and antinormal series expansion (7.35) and (7.36) of Boson operators, to the corresponding series expansion of scalars in which the $\mathbf{a}$ and $\mathbf{a}^\dagger$ operators have been transformed, respectively, into the scalars α and α^*:

$$\hat{\mathbf{N}}^{-1}\{\mathbf{f}^{\{n\}}(\mathbf{a},\mathbf{a}^\dagger)\} = \{f^{\{n\}}(\alpha,\alpha^*)\} \quad (7.40)$$

$$\hat{\mathbf{A}}^{-1}\{\mathbf{f}^{\{a\}}(\mathbf{a},\mathbf{a}^\dagger)\} = \{f^{\{a\}}(\alpha,\alpha^*)\} \quad (7.41)$$

with, respectively,

$$\{f^{\{n\}}(\alpha^*,\alpha)\} = \sum_{rs}\{\{f^n_{rs}\}(\alpha^*)^r(\alpha)^s\} \quad (7.42)$$

$$\{f^{\{a\}}(\alpha^*,\alpha)\} = \sum_{sr}\{\{f^a_{sr}\}(\alpha)^s(\alpha^*)^r\} \quad (7.43)$$

Premultiply Eqs. (7.40) and (7.41) by $\hat{\mathbf{N}}$ and $\hat{\mathbf{A}}$, respectively,

$$\hat{\mathbf{N}}\hat{\mathbf{N}}^{-1}\{\mathbf{f}^{\{n\}}(\mathbf{a},\mathbf{a}^\dagger)\} = \hat{\mathbf{N}}\{f^{\{n\}}(\alpha,\alpha^*)\}$$

$$\hat{\mathbf{A}}\hat{\mathbf{A}}^{-1}\{\mathbf{f}^{\{a\}}(\mathbf{a},\mathbf{a}^\dagger)\} = \hat{\mathbf{A}}\{f^{\{a\}}(\alpha,\alpha^*)\}$$

Then, using Eqs. (7.38) and (7.39), one obtains

$$\boxed{\{\mathbf{f}^{\{n\}}(\mathbf{a}, \mathbf{a}^\dagger)\} = \hat{\mathbf{N}}\{f^{\{n\}}(\alpha, \alpha^*)\}} \qquad \text{and} \qquad \boxed{\{\mathbf{f}^{\{a\}}(\mathbf{a}, \mathbf{a}^\dagger)\} = \hat{\mathbf{A}}\{f^{\{a\}}(\alpha, \alpha^*)\}} \tag{7.44}$$

showing that the linear operators $\hat{\mathbf{N}}$ and $\hat{\mathbf{A}}$ transform the scalar functions (7.42) and (7.43) into the corresponding normal and antinormal operators (7.35) and (7.36).

7.2.3 Commutators of Boson operators with functions of Boson operators

Now, before continuing the study of normal and antinormal formalism, it is necessary to get some commutators of Boson operators with functions of them, and for this purpose consider the following expression:

$$[(\mathbf{a}^\dagger)^2, \mathbf{a}] = \mathbf{a}^\dagger\mathbf{a}^\dagger\mathbf{a} - \mathbf{a}\mathbf{a}^\dagger\mathbf{a}^\dagger \tag{7.45}$$

Then, in view of the commutation rule $\mathbf{a}\mathbf{a}^\dagger - \mathbf{a}^\dagger\mathbf{a} = 1$, the last right-hand side of Eq. (7.45) transforms to

$$(\mathbf{a}\mathbf{a}^\dagger)\mathbf{a}^\dagger = (\mathbf{a}^\dagger\mathbf{a} + 1)\mathbf{a}^\dagger$$

so that Eq. (7.45) reads

$$[(\mathbf{a}^\dagger)^2, \mathbf{a}] = \mathbf{a}^\dagger\mathbf{a}^\dagger\mathbf{a} - \mathbf{a}^\dagger\mathbf{a}\mathbf{a}^\dagger - \mathbf{a}^\dagger \tag{7.46}$$

or, factorizing,

$$[(\mathbf{a}^\dagger)^2, \mathbf{a}] = \mathbf{a}^\dagger(\mathbf{a}^\dagger\mathbf{a} - \mathbf{a}\mathbf{a}^\dagger) - \mathbf{a}^\dagger$$

Hence, using in turn the commutation rule of Bosons, leads to

$$[(\mathbf{a}^\dagger)^2, \mathbf{a}] = -2\mathbf{a}^\dagger \tag{7.47}$$

Hence, one obtains by derivation

$$2\mathbf{a}^\dagger = \left(\frac{\partial(\mathbf{a}^\dagger)^2}{\partial\mathbf{a}^\dagger}\right)$$

so that Eq. (7.47) reads

$$\boxed{[(\mathbf{a}^\dagger)^2, \mathbf{a}] = -\left(\frac{\partial(\mathbf{a}^\dagger)^2}{\partial\mathbf{a}^\dagger}\right)} \tag{7.48}$$

Next, consider the commutator of **a** with the third power of its Hermitian conjugate:

$$[(\mathbf{a}^\dagger)^3, \mathbf{a}] = [\mathbf{a}^\dagger(\mathbf{a}^\dagger)^2, \mathbf{a}] \tag{7.49}$$

Again, recall that according to Eq. (1.75), the following relation holds between commutators of operators **A**, **B**, and **C**:

$$[\mathbf{BC}, \mathbf{A}] = [\mathbf{B}, \mathbf{A}]\mathbf{C} + \mathbf{B}[\mathbf{C}, \mathbf{A}] \tag{7.50}$$

Then, in order to apply this theorem to Eq. (7.49), take

$$\mathbf{B} = \mathbf{a}^\dagger, \qquad \mathbf{C} = (\mathbf{a}^\dagger)^2 \qquad \mathbf{A} = \mathbf{a}$$

so that, by application of theorem (7.50), we have

$$[\mathbf{B},\mathbf{A}] = [\mathbf{a}^\dagger,\mathbf{a}] = -1$$

and

$$[\mathbf{C},\mathbf{A}] = [(\mathbf{a}^\dagger)^2,\mathbf{a}]$$

which, due to (7.47), yields

$$[\mathbf{C},\mathbf{A}] = -2\mathbf{a}^\dagger$$

Thus, Eq. (7.49), reads

$$[(\mathbf{a}^\dagger)^3,\mathbf{a}] = -(\mathbf{a}^\dagger)^2 \quad - 2(\mathbf{a}^\dagger)^2 \quad = -3(\mathbf{a}^\dagger)^2$$

The latter result may be also expressed in terms of the derivative of the third power of $\mathbf{a}^\dagger$ with respect to $\mathbf{a}^\dagger$:

$$[(\mathbf{a}^\dagger)^3,\mathbf{a}] = -\left(\frac{\partial(\mathbf{a}^\dagger)^3}{\partial\mathbf{a}^\dagger}\right) \tag{7.51}$$

Moreover, one obtains by recurrence of Eqs. (7.48) and (7.51)

$$\boxed{[(\mathbf{a}^\dagger)^n,\mathbf{a}] = -\left(\frac{\partial(\mathbf{a}^\dagger)^n}{\partial\mathbf{a}^\dagger}\right)} \tag{7.52}$$

In order to obtain the Hermitian conjugate of this expression, first write it explicitly according to

$$(\mathbf{a}^\dagger)^n\mathbf{a} - \mathbf{a}(\mathbf{a}^\dagger)^n \quad = -\left(\frac{\partial(\mathbf{a}^\dagger)^n}{\partial\mathbf{a}^\dagger}\right)$$

Next, to get the full Hermitian conjugate of the latter expression, take the Hermitian conjugate of each term and then invert the result, so that

$$\mathbf{a}^\dagger(\mathbf{a})^n - (\mathbf{a})^n\mathbf{a}^\dagger \quad = -\left(\frac{\partial(\mathbf{a})^n}{\partial\mathbf{a}}\right) \tag{7.53}$$

and thus

$$\boxed{[(\mathbf{a})^n,\mathbf{a}^\dagger] = \left(\frac{\partial(\mathbf{a})^n}{\partial\mathbf{a}}\right)} \tag{7.54}$$

Furthermore, consider the following equation involving commutators:

$$[\mathbf{a}^\dagger,\{\mathbf{f}(\mathbf{a},\mathbf{a}^\dagger)\}] = [\mathbf{a}^\dagger,\{\mathbf{f}^{\{a\}}(\mathbf{a},\mathbf{a}^\dagger)\}]$$

which holds because of Eq. (7.37) expressing the equivalence between any function of the Boson operators and its antinormal order form. Owing to Eq. (7.36), this commutator takes the form

$$[\mathbf{a}^\dagger,\{\mathbf{f}(\mathbf{a},\mathbf{a}^\dagger)\}] = \sum_{rs}\{f^a_{rs}\}[\mathbf{a}^\dagger,(\mathbf{a})^r(\mathbf{a}^\dagger)^s] \tag{7.55}$$

which, using Eqs. (7.50) and (7.55), becomes

$$[\mathbf{a}^\dagger,\{\mathbf{f}(\mathbf{a},\mathbf{a}^\dagger)\}] = \sum_{rs}\{f^a_{rs}\}([\mathbf{a}^\dagger,(\mathbf{a})^r](\mathbf{a}^\dagger)^s + (\mathbf{a})^r[\mathbf{a}^\dagger,(\mathbf{a}^\dagger)^s]) \tag{7.56}$$

Now, we remark that in the latter expression, the second right-hand-side commutator is zero, so that this equation reduces to

$$[\mathbf{a}^\dagger, \{\mathbf{f}(\mathbf{a}, \mathbf{a}^\dagger)\}] = \sum_{rs} \{f_{rs}^{a}\}[\mathbf{a}^\dagger, (\mathbf{a})^r](\mathbf{a}^\dagger)^s \tag{7.57}$$

Moreover, writing r in place of n, Eq. (7.53), we have

$$[\mathbf{a}^\dagger, (\mathbf{a})^r] = -r(\mathbf{a})^{r-1}$$

so that Eq. (7.57) simplifies to

$$[\mathbf{a}^\dagger, \{\mathbf{f}(\mathbf{a}, \mathbf{a}^\dagger)\}] = -\sum_{rs} \{f_{rs}^{a}\} r(\mathbf{a})^{r-1}(\mathbf{a}^\dagger)^s \tag{7.58}$$

Moreover, observe that, owing to Eq. (7.36) the right-hand side of this last equation is just the partial derivative with respect to **a** of the antinormal ordered expression of the function of Boson operators. Hence, the commutator (7.58) becomes

$$[\mathbf{a}^\dagger, \{\mathbf{f}(\mathbf{a}, \mathbf{a}^\dagger)\}] = -\left(\frac{\partial \mathbf{f}^{\{a\}}(\mathbf{a}, \mathbf{a}^\dagger)}{\partial \mathbf{a}}\right)$$

and, thus, owing to Eq. (7.37),

$$[\mathbf{a}^\dagger, \{\mathbf{f}(\mathbf{a}, \mathbf{a}^\dagger)\}] = -\left(\frac{\partial \mathbf{f}(\mathbf{a}, \mathbf{a}^\dagger)}{\partial \mathbf{a}}\right) \tag{7.59}$$

In a similar way, one would obtain

$$[\mathbf{a}, \{\mathbf{f}(\mathbf{a}, \mathbf{a}^\dagger)\}] = \left(\frac{\partial \mathbf{f}(\mathbf{a}, \mathbf{a}^\dagger)}{\partial \mathbf{a}^\dagger}\right) \tag{7.60}$$

7.2.4 Average values over coherent states

Now, consider the following operator written in order form:

$$\{\mathbf{f}^{\{n\}}(\mathbf{a}, \mathbf{a}^\dagger)\} = \sum_{rs} \{f_{rs}^{n}\}(\mathbf{a}^\dagger)^r(\mathbf{a})^s$$

Then, according to Eq. (7.35), its average value over a coherent state is

$$\langle\{\alpha\}|\{\mathbf{f}^{\{n\}}(\mathbf{a}, \mathbf{a}^\dagger)\}|\{\alpha\}\rangle = \sum_{rs} \{f_{rs}^{n}\}\langle\{\alpha\}|(\mathbf{a}^\dagger)^r(\mathbf{a})^s|\{\alpha\}\rangle$$

Next, keeping in mind the definitions (6.1) and (6.2) of coherent states

$$\mathbf{a}|\{\alpha\}\rangle = \alpha|\{\alpha\}\rangle \qquad \text{and} \qquad \langle\{\alpha\}|\mathbf{a}^\dagger = \langle\{\alpha\}|\alpha^* \tag{7.61}$$

and applying them to the above average value, one finds

$$\langle\{\alpha\}|\{\mathbf{f}^{\{n\}}(\mathbf{a}, \mathbf{a}^\dagger)\}|\{\alpha\}\rangle = \sum_{rs} \{f_{rs}^{n}\}\langle\{\alpha\}|(\alpha^*)^r(\alpha)^s|\{\alpha\}\rangle$$

or

$$\langle\{\alpha\}|\mathbf{f}^{\{n\}}(\mathbf{a}, \mathbf{a}^\dagger)|\{\alpha\}\rangle = \sum_{rs} \{f_{rs}^{n}\}(\alpha^*)^r(\alpha)^s\langle\{\alpha\}|\{\alpha\}\rangle$$

so that if the coherent state is normalized,

$$\langle\{\alpha\}|\mathbf{f}^{\{n\}}(\mathbf{a},\mathbf{a}^\dagger)|\{\alpha\}\rangle = \sum_{rs}\{f_{rs}^n\}(\alpha^*)^r(\alpha)^s$$

Therefore, owing to Eq. (7.42), we have

$$\boxed{\langle\{\alpha\}|\mathbf{f}^{\{n\}}(\mathbf{a},\mathbf{a}^\dagger)|\{\alpha\}\rangle = f^{\{n\}}(\alpha^*,\alpha)} \tag{7.62}$$

which, due to Eq. (7.37), reads also

$$\boxed{\langle\{\alpha\}|\mathbf{f}(\mathbf{a},\mathbf{a}^\dagger)|\{\alpha\}\rangle = f^{\{n\}}(\alpha^*,\alpha)} \tag{7.63}$$

The latter equation shows that the average value of an arbitrary operator function of Boson operators performed on a coherent state is the scalar function defined by Eq. (7.42).

7.2.5 Expression of $|\{\alpha\}\rangle\langle\{\alpha\}|\mathbf{a}$ and of its Hermitian conjugate

Start from the first equation of Eq. (7.61), that is,

$$\mathbf{a}|\{\alpha\}\rangle = \alpha|\{\alpha\}\rangle \tag{7.64}$$

Then, postmultiply both sides of this equation by the bra $\langle\{\alpha\}|$, and one obtains

$$\mathbf{a}|\{\alpha\}\rangle\langle\{\alpha\}| = \alpha|\{\alpha\}\rangle\langle\{\alpha\}| \tag{7.65}$$

Now, the question must be posed: What is the result of the following expression?

$$|\{\alpha\}\rangle\langle\{\alpha\}|\mathbf{a} = ?$$

To answer, write the coherent state in terms of the action of the translation operator on the ground state of the Hamiltonian of the harmonic oscillator by the aid of Eq. (6.94), that is,

$$|\{\alpha\}\rangle = (e^{-|\alpha|^2/2})(e^{\alpha\mathbf{a}^\dagger})(e^{-\alpha^*\mathbf{a}})|\{0\}\rangle \tag{7.66}$$

the Hermitian conjugate of which is

$$\langle\{\alpha\}| = \langle\{0\}|(e^{-\alpha\mathbf{a}^\dagger})(e^{\alpha^*\mathbf{a}})(e^{-|\alpha|^2/2}) \tag{7.67}$$

Then, using the two above equations, we have

$$|\{\alpha\}\rangle\langle\{\alpha\}| = (e^{-|\alpha|^2/2})(e^{\alpha\mathbf{a}^\dagger})(e^{-\alpha^*\mathbf{a}})|\{0\}\rangle\langle\{0\}|(e^{-\alpha\mathbf{a}^\dagger})(e^{\alpha^*\mathbf{a}})(e^{-|\alpha|^2/2}) \tag{7.68}$$

an expression which may be simplified in the following way. First, observe that, by expansion of $\exp\{-\alpha^*\mathbf{a}\}$, it is possible to write

$$(e^{-\alpha^*\mathbf{a}})|\{0\}\rangle = \left\{1+\alpha^*\mathbf{a}+\frac{(\alpha^*\mathbf{a})^2}{2!}+\frac{(\alpha^*\mathbf{a})^3}{3!}+\cdots\right\}|\{0\}\rangle \tag{7.69}$$

Then, due to Eq. (5.35), that is,

$$\mathbf{a}|\{0\}\rangle = 0$$

and thus

$$\mathbf{a}^n|\{0\}\rangle = 0 \quad \text{except if} \quad n = 0$$

Eq. (7.69) simplifies to

$$(e^{-\alpha^*\mathbf{a}})|\{0\}\rangle = |\{0\}\rangle \tag{7.70}$$

the Hermitian conjugate of which is

$$\langle\{0\}|(e^{-\alpha\mathbf{a}^\dagger}) = \langle\{0\}| \tag{7.71}$$

As a consequence of Eqs. (7.70) and (7.71), Eq. (7.68) simplifies to

$$|\{\alpha\}\rangle\langle\{\alpha\}| = (e^{-|\alpha|^2/2})(e^{\alpha\mathbf{a}^\dagger})|\{0\}\rangle\langle\{0\}|(e^{\alpha^*\mathbf{a}})(e^{-|\alpha|^2/2})$$

Next, postmultiplying both right- and left-hand-side terms of this last equation by **a** and after rearranging, using

$$e^{-|\alpha|^2/2}e^{-|\alpha|^2/2} = e^{-\alpha\alpha^*} \tag{7.72}$$

one obtains

$$|\{\alpha\}\rangle\langle\{\alpha\}|\mathbf{a} = (e^{-\alpha\alpha^*})(e^{\alpha\mathbf{a}^\dagger})|\{0\}\rangle\langle\{0\}|(e^{\alpha^*\mathbf{a}})\mathbf{a} \tag{7.73}$$

Moreover, due to the expression of the following partial derivative

$$\left(\frac{\partial e^{\alpha^*\mathbf{a}}}{\partial\alpha^*}\right) = (e^{\alpha^*\mathbf{a}})\mathbf{a}$$

Eq. (7.73) transforms to

$$|\{\alpha\}\rangle\langle\{\alpha\}|\mathbf{a} = (e^{-\alpha\alpha^*})(e^{\alpha\mathbf{a}^\dagger})|\{0\}\rangle\langle\{0\}|\left(\frac{\partial e^{\alpha^*\mathbf{a}}}{\partial\alpha^*}\right) \tag{7.74}$$

Again, the partial derivative of the exponential operator with respect to α^* commutes with the bra, the ket, and the operator $e^{\alpha\mathbf{a}^\dagger}$, which do not depend on α, thus allowing one to transform Eq. (7.74) into

$$|\{\alpha\}\rangle\langle\{\alpha\}|\mathbf{a} = (e^{-\alpha\alpha^*})\left(\frac{\partial}{\partial\alpha^*}\{(e^{\alpha\mathbf{a}^\dagger})|\{0\}\rangle\langle\{0\}|(e^{\alpha^*\mathbf{a}})\}\right) \tag{7.75}$$

Furthermore, denoting

$$\{\mathbf{f}(\alpha^*\mathbf{a}, \alpha\mathbf{a}^\dagger)\} = (e^{\alpha\mathbf{a}^\dagger})|\{0\}\rangle\langle\{0\}|(e^{\alpha^*\mathbf{a}}) \tag{7.76}$$

Eq. (7.75) reads

$$|\{\alpha\}\rangle\langle\{\alpha\}|\mathbf{a} = e^{-\alpha\alpha^*}\left(\frac{\partial\mathbf{f}(\alpha^*\mathbf{a}, \alpha\mathbf{a}^\dagger)}{\partial\alpha^*}\right) \tag{7.77}$$

Now, observe that the following relation is verified:

$$\left(\frac{\partial e^{-\alpha\alpha^*}\mathbf{f}(\alpha^*\mathbf{a}, \alpha\mathbf{a}^\dagger)}{\partial\alpha^*}\right) = -\alpha(e^{-\alpha\alpha^*})\{\mathbf{f}(\alpha^*\mathbf{a}, \alpha\mathbf{a}^\dagger)\} + (e^{-\alpha\alpha^*})\left(\frac{\partial\mathbf{f}(\alpha^*\mathbf{a}, \alpha\mathbf{a}^\dagger)}{\partial\alpha^*}\right)$$

Then, rearranging gives

$$(e^{-\alpha\alpha^*})\left(\frac{\partial\mathbf{f}(\alpha^*\mathbf{a}, \alpha\mathbf{a}^\dagger)}{\partial\alpha^*}\right) = \left(\frac{\partial}{\partial\alpha^*} + \alpha\right)\{(e^{-\alpha\alpha^*})\{\mathbf{f}(\alpha^*\mathbf{a}, \alpha\mathbf{a}^\dagger)\}\} \tag{7.78}$$

As a consequence of Eq. (7.78), Eq. (7.77) transforms with the help of Eq. (7.76) into

$$|\{\alpha\}\rangle\langle\{\alpha\}|\mathbf{a} = \left(\frac{\partial}{\partial\alpha^*} + \alpha\right)\{(e^{-\alpha\alpha^*})(e^{\alpha\mathbf{a}^\dagger})|\{0\}\rangle\ \langle\{0\}|(e^{\alpha^*\mathbf{a}})\}$$

or, owing to Eq. (7.72)

$$|\{\alpha\}\rangle\langle\{\alpha\}|\mathbf{a} = \left(\frac{\partial}{\partial\alpha^*} + \alpha\right)\{(e^{-|\alpha|^2})(e^{\alpha\mathbf{a}^\dagger})|\{0\}\rangle\ \langle\{0\}|(e^{\alpha^*\mathbf{a}})\}$$

Then after rearranging the exponential involving the scalar $|\alpha|^2$, we have

$$|\{\alpha\}\rangle\langle\{\alpha\}|\mathbf{a} = \left(\frac{\partial}{\partial\alpha^*} + \alpha\right)\{(e^{-|\alpha|^2/2})(e^{\alpha\mathbf{a}^\dagger})|\{0\}\rangle\ \langle\{0\}|(e^{\alpha^*\mathbf{a}})(e^{-|\alpha|^2/2})\}$$

Again, in view of Eq. (7.71), it may be written in the more complex form

$$|\{\alpha\}\rangle\langle\{\alpha\}|\mathbf{a} = \left(\frac{\partial}{\partial\alpha^*} + \alpha\right)\{(e^{-|\alpha|^2/2})(e^{\alpha\mathbf{a}^\dagger})(e^{-\alpha^*\mathbf{a}})|\{0\}\rangle\ \langle\{0\}|(e^{-\alpha\mathbf{a}^\dagger})(e^{\alpha^*\mathbf{a}})(e^{-|\alpha|^2/2})\}$$

Finally, using Eqs. (7.66) and (7.67), one obtains the result

$$\boxed{|\{\alpha\}\rangle\langle\{\alpha\}|\mathbf{a} = \left(\frac{\partial}{\partial\alpha^*} + \alpha\right)\{|\{\alpha\}\rangle\langle\{\alpha\}|\}} \tag{7.79}$$

the Hermitian conjugate of which is

$$\boxed{\mathbf{a}^\dagger|\{\alpha\}\rangle\langle\{\alpha\}| = \left(\frac{\partial}{\partial\alpha} + \alpha^*\right)\{|\{\alpha\}\rangle\langle\{\alpha\}|\}}$$

7.2.6 Theorems dealing with normal and antinormal ordering

7.2.6.1 Normal ordering Consider the following normal ordered expansion of any operator function of Boson operators

$$\mathbf{f}(\mathbf{a},\mathbf{a}^\dagger) = \sum_{l,m,\ldots,r,\ldots,s,\ldots,u} \{C_{l,m,\ldots,r,\ldots,s,\ldots,u}\}(\mathbf{a}^\dagger)^l(\mathbf{a})^m\cdots(\mathbf{a})^r\cdots(\mathbf{a}^\dagger)^s\cdots(\mathbf{a}^\dagger)^u$$

$$\boxed{|\{\alpha\}\rangle\langle\{\alpha\}|\mathbf{a} = \left(\frac{\partial}{\partial\alpha^*} + \alpha\right)\{|\{\alpha\}\rangle\langle\{\alpha\}|\}} \tag{7.80}$$

Now, consider the average value of this operator over a coherent state allowing one to write via Eqs. (7.62), that is,

$$\langle\{\alpha\}|\mathbf{f}(\mathbf{a},\mathbf{a}^\dagger)|\{\alpha\}\rangle = \{f(\alpha,\alpha^*)\}$$

Moreover, introduce after the operator of the left-hand side of Eq. (7.63), a closure relation over some basis $\{|\Psi_k\rangle\}$

$$f^{\{n\}}(\alpha,\alpha^*) = \sum_k \langle\{\alpha\}|\mathbf{f}(\mathbf{a},\mathbf{a}^\dagger)|\Psi_k\rangle\langle\Psi_k|\{\alpha\}\rangle$$

which, on commuting the scalar products with the matrix elements, transforms to

$$\{f^{\{n\}}(\alpha,\alpha^*)\} = \sum_k \langle\Psi_k|\{\alpha\}\rangle\langle\{\alpha\}|\mathbf{f}(a,\mathbf{a}^\dagger)|\Psi_k\rangle$$

so that the right-hand side of this last equation may be written formally as the trace over an arbitrary basis $\{|\Psi_k\rangle\}$ according to

$$\{f^{\{n\}}(\alpha,\alpha^*)\} = \mathbf{tr}\{|\{\alpha\}\rangle\langle\{\alpha\}|\mathbf{f}(\mathbf{a},\mathbf{a}^\dagger)\}$$

Next, owing to Eq. (7.80), this average value reads

$$\{f^{\{n\}}(\alpha,\alpha^*)\} = \sum_{l,m,\ldots,r,\ldots,s,\ldots,u} \{C_{l,m,\ldots,r,\ldots,s,\ldots,u}\}\mathbf{tr}\{|\{\alpha\}\rangle\langle\{\alpha\}|(\mathbf{a}^\dagger)^l(\mathbf{a})^m\cdots(\mathbf{a})^r\cdots(\mathbf{a}^\dagger)^s\cdots(\mathbf{a}^\dagger)^u\} \tag{7.81}$$

Now, the Hermitian conjugate of Eq. (7.65) is

$$|\{\alpha\}\rangle\langle\{\alpha\}|\mathbf{a}^\dagger = \alpha^*|\{\alpha\}\rangle\langle\{\alpha\}| \tag{7.82}$$

whereas Eq. (7.79) reads

$$|\{\alpha\}\rangle\langle\{\alpha\}|\mathbf{a} = \left(\alpha + \frac{\partial}{\partial\alpha^*}\right)|\{\alpha\}\rangle\langle\{\alpha\}| \tag{7.83}$$

so that, by iteration of Eqs. (7.82) and (7.83), we have

$$|\{\alpha\}\rangle\langle\{\alpha\}|(\mathbf{a}^\dagger)^r = (\alpha^*)^r|\{\alpha\}\rangle\langle\{\alpha\}| \tag{7.84}$$

$$|\{\alpha\}\rangle\langle\{\alpha\}|(\mathbf{a})^s = \left(\alpha + \frac{\partial}{\partial\alpha^*}\right)^s|\{\alpha\}\rangle\langle\{\alpha\}| \tag{7.85}$$

Hence, Eq. (7.81) transforms to

$$f\{^{\{n\}}(\alpha,\alpha^*)\} = \sum_{l,m,\ldots,r,\ldots,s,\ldots,u} \{C_{l,m,\ldots,r,\ldots,s,\ldots,u}\} \times \mathbf{tr}\left\{(\alpha^*)^l\left(\alpha + \frac{\partial}{\partial\alpha^*}\right)^m\cdots\left(\alpha + \frac{\partial}{\partial\alpha^*}\right)^r\cdots(\alpha^*)^s\cdots(\alpha^*)^u|\{\alpha\}\rangle\langle\{\alpha\}|\right\}$$

Now, writing explicitly the trace, and using the fact that the bras $\langle\Psi_k|$ commute with the α and α^* and the partial derivative with respect to α^*, that gives

$$\{f^{\{n\}}(\alpha,\alpha^*)\} = \sum_{l,m,\ldots,r,\ldots,s,\ldots,u} \{C_{l,m,\ldots,r,\ldots,s,\ldots,u}\}\sum_k \times\left\{(\alpha^*)^l\left(\alpha + \frac{\partial}{\partial\alpha^*}\right)^m\cdots\left(\alpha + \frac{\partial}{\partial\alpha^*}\right)^r\cdots(\alpha^*)^s\cdots(\alpha^*)^u\right\} \times\langle\Psi_k|\{\alpha\}\rangle\langle\{\alpha\}|\Psi_k\rangle \tag{7.86}$$

Moreover, since

$$\sum_k\langle\Psi_k|\{\alpha\}\rangle\langle\{\alpha\}|\Psi_k\rangle = \sum_k|\langle\Psi_k|\{\alpha\}\rangle|^2 = 1$$

Eq. (7.86) reduces to

$$\{f^{\{n\}}(\alpha,\alpha^*)\} = f\left(\alpha + \frac{\partial}{\partial\alpha^*},\alpha^*\right) \tag{7.87}$$

with

$$\left\{f\left(\alpha+\frac{\partial}{\partial\alpha^*},\alpha^*\right)\right\} = \sum_{l,m,\ldots,r,\ldots,s,\ldots,u} \{C_{l,m,\ldots,r,\ldots,s,\ldots,u}\}$$

$$\times\left\{(\alpha^*)^l\left(\alpha+\frac{\partial}{\partial\alpha^*}\right)^m\cdots\left(\alpha+\frac{\partial}{\partial\alpha^*}\right)^r\cdots(\alpha^*)^s\cdots(\alpha^*)^u\right\}$$

Moreover, premultiply both terms of Eq. (7.87) by the ordering operator $\hat{\mathbf{N}}$, that is,

$$\hat{\mathbf{N}}\{f^{\{n\}}(\alpha^*,\alpha)\} = \hat{\mathbf{N}}\left\{f\left(\alpha+\frac{\partial}{\partial\alpha^*}\alpha^*\right)\right\}$$

Then, owing to the first equation of (7.44), one obtains the important result

$$\boxed{\{\mathbf{f}^{\{n\}}(\mathbf{a},\mathbf{a}^\dagger)\} = \hat{\mathbf{N}}\left\{f\left(\alpha+\frac{\partial}{\partial\alpha^*},\alpha^*\right)\right\}} \tag{7.88}$$

7.2.6.2 Theorem dealing with antinormal ordering Consider the following antinormal ordered expansion of a function of Boson operators:

$$\{\mathbf{f}^{\{a\}}(\mathbf{a},\mathbf{a}^\dagger)\} = \sum_{s,r}\{f^a_{sr}\}(\mathbf{a})^s(\mathbf{a}^\dagger)^r \tag{7.89}$$

Then, using the closure relation (6.19) on coherent states, that is,

$$\frac{1}{\pi}\int_{-\infty}^{+\infty}\int_{-\infty}^{+\infty}|\{\alpha\}\rangle\langle\{\alpha\}|d\{\text{Re}(\alpha)\}d\{\text{Im}(\alpha)\} = 1$$

Eq. (7.89) reads

$$\{\mathbf{f}^{\{a\}}(\mathbf{a},\mathbf{a}^\dagger)\} = \frac{1}{\pi}\int_{-\infty}^{+\infty}\int_{-\infty}^{+\infty}\sum_{s,r}\{f^a_{sr}\}(\mathbf{a})^s|\{\alpha\}\rangle\langle\{\alpha\}|\,(\mathbf{a}^\dagger)^r d\{\text{Re}(\alpha)\}d\{\text{Im}(\alpha)\}$$

Again, keeping in mind the properties of the coherent state, that is,

$$(\mathbf{a})|\{\alpha\}\rangle = (\alpha)|\{\alpha\}\rangle \qquad \text{and} \qquad \langle\{\alpha\}\,|(\mathbf{a}^\dagger) = \langle\{\alpha\}|\,(\alpha^*)$$

and leading by iteration

$$(\mathbf{a})^s|\{\alpha\}\rangle = (\alpha)^s|\{\alpha\}\rangle \qquad \text{and} \qquad \langle\{\alpha\}|(\mathbf{a}^\dagger)^r = \langle\{\alpha\}|(\alpha^*)^r$$

the previous equation becomes

$$\{\mathbf{f}^{\{a\}}(\mathbf{a},\mathbf{a}^\dagger)\} = \frac{1}{\pi}\int_{-\infty}^{+\infty}\int_{-\infty}^{+\infty}\sum_{s,r}\{f^a_{sr}\}(\alpha)^s(\alpha^*)^r|\{\alpha\}\rangle\langle\{\alpha\}|\,d\{\text{Re}(\alpha)\}d\{\text{Im}(\alpha)\} \tag{7.90}$$

Finally, due to Eq. (7.43), that is,

$$\sum_{s,r}\{f^a_{sr}\}(\alpha)^s(\alpha^*)^r = f^{\{a\}}(\alpha^*,\alpha)$$

Eq. (7.90) yields

$$\{\mathbf{f}^{\{a\}}(\mathbf{a},\mathbf{a}^\dagger)\} = \frac{1}{\pi}\int_{-\infty}^{+\infty}\int_{-\infty}^{+\infty} f^{\{a\}}(\alpha^*,\alpha)|\{\alpha\}\rangle\langle\{\alpha\}|\,d\{\mathrm{Re}(\alpha)\}d\{\mathrm{Im}(\alpha)\}$$

7.2.7 Generalization of theorems dealing with normal ordering

We start from the partial derivative with respect to $\mathbf{a}^\dagger$ of the normal ordered expansion (7.35), that is,

$$\left(\frac{\partial \mathbf{f}^{\{n\}}(\mathbf{a},\mathbf{a}^\dagger)}{\partial \mathbf{a}^\dagger}\right) = \frac{\partial}{\partial \mathbf{a}^\dagger}\left(\sum_{r,s}\{f_{rs}^n\}(\mathbf{a}^\dagger)^r(\mathbf{a})^s\right)$$

which reads

$$\left(\frac{\partial \mathbf{f}^{\{n\}}(\mathbf{a},\mathbf{a}^\dagger)}{\partial \mathbf{a}^\dagger}\right) = \sum_{r,s}\{f_{rs}^n\}r(\mathbf{a}^\dagger)^{r-1}(\mathbf{a})^s \tag{7.91}$$

then, owing to the first equation of (7.44), the right-hand side of Eq. (7.91) becomes

$$\sum_{rs}\{f_{rs}^n\}r(\mathbf{a}^\dagger)^{r-1}(\mathbf{a})^s = \hat{\mathbf{N}}\left\{\sum_{r,s}\{f_{rs}^n\}r(\alpha^*)^{r-1}(\alpha)^s\right\}$$

so that Eq. (7.91) yields

$$\left(\frac{\partial \mathbf{f}^{\{n\}}(\mathbf{a},\mathbf{a}^\dagger)}{\partial \mathbf{a}^\dagger}\right) = \hat{\mathbf{N}}\left\{\sum_{r,s}\{f_{rs}^n\}r(\alpha^*)^{r-1}(\alpha)^s\right\}$$

a result that may also be expressed as

$$\left(\frac{\partial \mathbf{f}^{\{n\}}(\mathbf{a},\mathbf{a}^\dagger)}{\partial \mathbf{a}^\dagger}\right) = \hat{\mathbf{N}}\left\{\frac{\partial}{\partial \alpha^*}\sum_{r,s}\{f_{rs}^n\}(\alpha^*)^r(\alpha)^s\right\}$$

or, owing to Eq. (7.42), as

$$\left(\frac{\partial \mathbf{f}^{\{n\}}(\mathbf{a},\mathbf{a}^\dagger)}{\partial \mathbf{a}^\dagger}\right) = \hat{\mathbf{N}}\left\{\left(\frac{\partial}{\partial \alpha^*}f^n(\alpha^*,\alpha)\right)\right\} \tag{7.92}$$

Recall the commutator given by Eq. (7.60), that is,

$$\{\mathbf{a},\mathbf{f}^{\{n\}}(\mathbf{a},\mathbf{a}^\dagger)\} = \left(\frac{\partial \mathbf{f}^{\{n\}}(\mathbf{a},\mathbf{a}^\dagger)}{\partial \mathbf{a}^\dagger}\right)$$

allows one to write

$$\{\mathbf{a}\mathbf{f}^{\{n\}}(\mathbf{a},\mathbf{a}^\dagger)\} = \{\mathbf{f}^{\{n\}}(\mathbf{a},\mathbf{a}^\dagger)\mathbf{a}\} + \left(\frac{\partial \mathbf{f}^{\{n\}}(\mathbf{a},\mathbf{a}^\dagger)}{\partial \mathbf{a}^\dagger}\right) \tag{7.93}$$

Now, observe that the first right-hand-side term of this last equation, which appears to be in normal order, may be viewed as the result of the action of the normal order operator according to

$$\{\mathbf{f}^{\{n\}}(\mathbf{a},\mathbf{a}^\dagger)\mathbf{a}\} = \hat{\mathbf{N}}\{f^{\{n\}}(\alpha^*,\alpha)\alpha\} = \hat{\mathbf{N}}\{\alpha f^{\{n\}}(\alpha^*,\alpha)\}$$

Now, the last right-hand-side term of Eq. (7.93) is given by Eq. (7.92), so that Eq. (7.93) may be written

$$\{\mathbf{a}\mathbf{f}^{\{n\}}(\mathbf{a},\mathbf{a}^\dagger)\} = \hat{\mathbf{N}}\{\alpha f^{\{n\}}(\alpha^*,\alpha)\} + \hat{\mathbf{N}}\left\{\frac{\partial f^{\{n\}}(\alpha^*,\alpha)}{\partial\alpha^*}\right\}$$

or

$$\{\mathbf{a}\mathbf{f}^{\{n\}}(\mathbf{a},\mathbf{a}^\dagger)\} = \hat{\mathbf{N}}\left\{\left(\alpha + \frac{\partial}{\partial\alpha^*}\right) f^{\{n\}}(\alpha^*,\alpha)\right\} \tag{7.94}$$

By generalization of Eq. (7.94), one now obtains

$$(\mathbf{a})^m\{\mathbf{f}^{\{n\}}(\mathbf{a},\mathbf{a}^\dagger)\} = \hat{\mathbf{N}}\left\{\left(\alpha + \frac{\partial}{\partial\alpha^*}\right)^m \{f^{\{n\}}(\alpha^*,\alpha)\}\right\} \tag{7.95}$$

which, due to Eq. (7.37), reads

$$(\mathbf{a})^m\{\mathbf{f}(\mathbf{a},\mathbf{a}^\dagger)\} = \hat{\mathbf{N}}\left\{\left(\alpha + \frac{\partial}{\partial\alpha^*}\right)^m \{f^{\{n\}}(\alpha^*,\alpha)\}\right\} \tag{7.96}$$

Next, multiplying both the right- and left-hand sides of Eq. (7.96) by $\hat{\mathbf{N}}^{-1}$, that is,

$$\hat{\mathbf{N}}^{-1}(\mathbf{a})^m\{\mathbf{f}(\mathbf{a},\mathbf{a}^\dagger)\} = \hat{\mathbf{N}}^{-1}\hat{\mathbf{N}}\left\{\left(\alpha + \frac{\partial}{\partial\alpha^*}\right)^m \{f^{\{n\}}(\alpha^*,\alpha)\}\right\}$$

yields after simplification, with the help of Eq. (7.38)

$$\boxed{\hat{\mathbf{N}}^{-1}\,(\mathbf{a})^m\,\{\mathbf{f}(\mathbf{a},\mathbf{a}^\dagger)\} = \left(\alpha + \frac{\partial}{\partial\alpha^*}\right)^m \{f^{\{n\}}(\alpha^*,\alpha)\}} \tag{7.97}$$

On the other hand, it is clear that the following relation is satisfied since the Boson operator $\mathbf{a}^\dagger$ is in front of the function of the normal ordered expression $\mathbf{f}^{\{n\}}(\mathbf{a},\mathbf{a}^\dagger)$ of the Boson operators:

$$\boxed{(\mathbf{a}^\dagger)^k\{\mathbf{f}^{\{n\}}(\mathbf{a},\mathbf{a}^\dagger)\} = \hat{\mathbf{N}}\{(\alpha^*)^k\{f^{\{n\}}(\alpha^*,\alpha)\}\}} \tag{7.98}$$

Then, proceeding in the same way as for passing from Eq. (7.96) to Eq. (7.97), one obtains

$$\boxed{\hat{\mathbf{N}}^{-1}\{(\mathbf{a}^\dagger)^k\{\mathbf{f}(\mathbf{a},\mathbf{a}^\dagger)\}\} = (\alpha^*)^k\{f^{\{n\}}(\alpha^*,\alpha)\}} \tag{7.99}$$

Now, consider an operator equation dealing with Boson operators of the form

$$\{\mathbf{F}(\mathbf{a},\mathbf{a}^\dagger)\} = \{(\mathbf{a}^\dagger)^k(\mathbf{a})^m\mathbf{f}(\mathbf{a},\mathbf{a}^\dagger)\}$$

The question now is what may be in the scalar language an expression of the form

$$\hat{\mathbf{N}}^{-1}\{\mathbf{F}(\mathbf{a},\mathbf{a}^\dagger)\} = \hat{\mathbf{N}}^{-1}\{(\mathbf{a}^\dagger)^k(\mathbf{a})^m\mathbf{f}(\mathbf{a},\mathbf{a}^\dagger)\} \tag{7.100}$$

Owing to Eq. (7.95), it takes the form

$$\hat{\mathbf{N}}^{-1}\{\mathbf{F}(\mathbf{a},\mathbf{a}^\dagger)\} = \hat{\mathbf{N}}^{-1}\left\{(\mathbf{a}^\dagger)^k\hat{\mathbf{N}}\left\{\left(\alpha + \frac{\partial}{\partial\alpha^*}\right)^m \{f^{\{n\}}(\alpha,\alpha^*)\}\right\}\right\} \tag{7.101}$$

However, due to Eq. (7.40) and to the equivalence between $\mathbf{F}(\mathbf{a},\mathbf{a}^\dagger)$ and $\mathbf{F}^{\{n\}}(\mathbf{a},\mathbf{a}^\dagger)$, we may write

$$\hat{\mathbf{N}}^{-1}\{\mathbf{F}(\mathbf{a},\mathbf{a}^\dagger)\} = \hat{\mathbf{N}}^{-1}\{\mathbf{F}^{\{n\}}(\mathbf{a},\mathbf{a}^\dagger)\} = \{F^{\{n\}}(\alpha,\alpha^*)\}$$

it yields

$$\hat{\mathbf{N}}^{-1}\{(\mathbf{a}^\dagger)^k\{\ldots\}\} = (\alpha^*)^k\hat{\mathbf{N}}^{-1}\{\ldots\}$$

Hence, since $(\mathbf{a}^\dagger)^k$ is in front of $\{\ldots\}$, it appears that Eq. (7.101) leads to

$$\{F^{\{n\}}(\alpha,\alpha^*)\} = (\alpha^*)^k\hat{\mathbf{N}}^{-1}\hat{\mathbf{N}}\left\{\left(\alpha+\frac{\partial}{\partial\alpha^*}\right)^m\{f^{\{n\}}(\alpha,\alpha^*)\}\right\} \tag{7.102}$$

so that, due to Eq. (7.38), Eq. (7.102) simplifies to

$$\boxed{\{F^{\{n\}}(\alpha,\alpha^*)\} = (\alpha^*)^k\left(\alpha+\frac{\partial}{\partial\alpha^*}\right)^m\{f^{\{n\}}(\alpha,\alpha^*)\}} \tag{7.103}$$

7.2.8 Another theorem of interest

Consider the following linear transformation on the ground state of $\mathbf{a}^\dagger\mathbf{a}$:

$$(e^{x\mathbf{a}^\dagger\mathbf{a}})(e^{y\mathbf{a}^\dagger})|\{0\}\rangle$$

where x and y are complex scalars. Insert between the last operator and the ket the unity operator built up from the first operator, that is,

$$(e^{-x\mathbf{a}^\dagger\mathbf{a}})(e^{x\mathbf{a}^\dagger\mathbf{a}}) = 1$$

Hence

$$(e^{x\mathbf{a}^\dagger\mathbf{a}})(e^{y\mathbf{a}^\dagger})|\{0\}\rangle = (e^{x\mathbf{a}^\dagger\mathbf{a}})(e^{y\mathbf{a}^\dagger})(e^{-x\mathbf{a}^\dagger\mathbf{a}})(e^{x\mathbf{a}^\dagger\mathbf{a}})|\{0\}\rangle \tag{7.104}$$

Next, in view of Eq. (7.27) and taking

$$\mathbf{f}(\mathbf{a}^\dagger) = (e^{y\mathbf{a}^\dagger})$$

Eq. (7.104) reads

$$(e^{x\mathbf{a}^\dagger\mathbf{a}})(e^{y\mathbf{a}^\dagger})(e^{-x\mathbf{a}^\dagger\mathbf{a}}) = (e^{y\mathbf{a}^\dagger e^x})$$

Thus, Eq. (7.104) becomes

$$(e^{x\mathbf{a}^\dagger\mathbf{a}})(e^{y\mathbf{a}^\dagger})|\{0\}\rangle = (e^{y\mathbf{a}^\dagger e^x})(e^{x\mathbf{a}^\dagger\mathbf{a}})|\{0\}\rangle$$

However, since $|\{0\}\rangle$ is the ground state of $\mathbf{a}^\dagger\mathbf{a}$, with corresponding zero eigenvalue, the series expansion of the exponential of $\mathbf{a}^\dagger\mathbf{a}$ on the ground state is zero except for the first term of the expansion, i.e.

$$(e^{x\mathbf{a}^\dagger\mathbf{a}})|\{0\}\rangle = \left(1+x\mathbf{a}^\dagger\mathbf{a}+\frac{x^2(\mathbf{a}^\dagger\mathbf{a})^2}{2!}+\cdots\right)|\{0\}\rangle = |\{0\}\rangle \tag{7.105}$$

Hence, Eq. (7.105) leads to the final result

$$\boxed{(e^{x\mathbf{a}^\dagger\mathbf{a}})(e^{y\mathbf{a}^\dagger})|\{0\}\rangle = (e^{y\mathbf{a}^\dagger e^x})|\{0\}\rangle} \tag{7.106}$$

7.3 TIME EVOLUTION OPERATOR OF DRIVEN HARMONIC OSCILLATORS

With the help of the theorems proved above, it is now possible to study the dynamics of driven quantum harmonic oscillators. For this purpose start from their Hamiltonian, which reads

$$\mathbf{H} = \left(\frac{\mathbf{P}^2}{2M} + \frac{1}{2}M\Omega^2\mathbf{Q}^2\right) + b\mathbf{Q}$$

Then, the dynamics of this system is governed by the time evolution operator, a solution of the Schrödinger equation

$$i\hbar\left(\frac{\partial \mathbf{U}(t)}{\partial t}\right) = \mathbf{H}\,\mathbf{U}(t) \quad \text{with} \quad \mathbf{U}(0) = 1$$

Next, in order to solve this equation, it is suitable to work within the interaction picture and thus to make the following partition:

$$\mathbf{H} = \mathbf{H}^\circ + b\mathbf{Q}$$

with

$$\mathbf{H}^\circ = \left(\frac{\mathbf{P}^2}{2M} + \frac{1}{2}M\Omega^2\mathbf{Q}^2\right) \tag{7.107}$$

Recall that the time evolution operator $\mathbf{U}(t)$ is related to the IP time evolution operator $\mathbf{U}(t)^{\mathrm{IP}}$ through Eq. (3.122)

$$\mathbf{U}(t) = \mathbf{U}^\circ(t)\mathbf{U}(t)^{\mathrm{IP}} \tag{7.108}$$

with

$$\mathbf{U}^\circ(t) = (e^{-i\mathbf{H}^\circ t/\hbar}) \tag{7.109}$$

Hence, according to Eq. (3.114), the IP time evolution operator obeys the IP Schrödinger equation

$$i\hbar\left(\frac{\partial \mathbf{U}(t)^{\mathrm{IP}}}{\partial t}\right) = b\mathbf{Q}(t)^{\mathrm{IP}}\mathbf{U}(t)^{\mathrm{IP}} \tag{7.110}$$

with the boundary condition

$$\mathbf{U}(0)^{\mathrm{IP}} = 1 \tag{7.111}$$

Next, due to Eq. (3.108), the interaction picture coordinate $\mathbf{Q}(t)^{\mathrm{IP}}$ appearing in Eq. (7.110) is given by

$$\mathbf{Q}(t)^{\mathrm{IP}} = \mathbf{U}^\circ(t)^{-1}\mathbf{Q}(0)\mathbf{U}^\circ(t) \tag{7.112}$$

Moreover, the iteration solution of the integral equation Eq. (7.110) is of the form of (3.124), however, up to infinite order

$$\mathbf{U}(t)^{\mathrm{IP}} \equiv 1 + \left(\frac{b}{i\hbar}\right)\int_0^t \mathbf{Q}(t')^{\mathrm{IP}}dt' + \left(\frac{b}{i\hbar}\right)^2\int_0^t \mathbf{Q}(t')^{\mathrm{IP}}dt'\int_0^{t'} \mathbf{Q}(t'')^{\mathrm{IP}}dt'' + \cdots$$

a solution that may be written formally as

$$\mathbf{U}(t)^{\mathrm{IP}} = \widehat{\mathbf{P}}\left\{\exp\left(-ib\int_0^t \mathbf{Q}(t')^{\mathrm{IP}}dt'\right)\right\} \tag{7.113}$$

where $\widehat{\mathbf{P}}$ is the Dyson time-ordering operator met in Eq. (3.87). Recall that, owing to Eq. (5.158), the position operators $\mathbf{Q}(t)^{\mathrm{IP}}$ and $\mathbf{Q}(t')^{\mathrm{IP}}$ at different times given by Eq. (7.112) do not commute. Next, passing to Boson operators by the aid of Eqs. (5.6) and (5.7), using Eqs. (5.9) and (7.107), and finally, after neglecting the zero-point energy which is irrelevant for the present purpose, the unperturbed time evolution operator (7.109) reads

$$\mathbf{U}^{\circ}(t) = (e^{-i\mathbf{a}^{\dagger}\mathbf{a}\Omega t}) \tag{7.114}$$

whereas the IP time evolution operator (7.110) becomes

$$i\left(\frac{\partial\mathbf{U}(\mathbf{a},\mathbf{a}^{\dagger},t)^{\mathrm{IP}}}{\partial t}\right) = \alpha^{\circ}\Omega(\mathbf{a}^{\dagger}(t)^{\mathrm{IP}} + \mathbf{a}(t)^{\mathrm{IP}})\mathbf{U}(\mathbf{a},\mathbf{a}^{\dagger},t)^{\mathrm{IP}} \tag{7.115}$$

with

$$\alpha^{\circ} = \frac{b}{\Omega}\sqrt{\frac{\hbar}{2M\Omega}} \tag{7.116}$$

In Eq. (7.115), the IP time-dependent Boson operator is given by

$$\mathbf{a}(t)^{\mathrm{IP}} = \mathbf{U}^{\circ}(t)^{-1}\mathbf{a}\mathbf{U}^{\circ}(t)$$

or in view of Eq. (7.114)

$$\mathbf{a}(t)^{\mathrm{IP}} = (e^{i\mathbf{a}^{\dagger}\mathbf{a}\Omega t})\mathbf{a}(e^{-i\mathbf{a}^{\dagger}\mathbf{a}\Omega t})$$

and thus, according to Eq. (7.21)

$$\mathbf{a}(t)^{\mathrm{IP}} = \mathbf{a}e^{-i\Omega t} \tag{7.117}$$

so that Eq. (7.112) reads

$$\mathbf{Q}(t)^{\mathrm{IP}} = \sqrt{\frac{\hbar}{2M\Omega}}(\mathbf{a}^{\dagger}e^{i\Omega t} + \mathbf{a}e^{-i\Omega t}) \tag{7.118}$$

Therefore, Eq. (7.115) becomes

$$i\left(\frac{\partial\mathbf{U}(\mathbf{a},\mathbf{a}^{\dagger},t)^{\mathrm{IP}}}{\partial t}\right) = \alpha^{\circ}\Omega\,(\mathbf{a}^{\dagger}e^{i\Omega t} + \mathbf{a}e^{-i\Omega t})\{\mathbf{U}(\mathbf{a},\mathbf{a}^{\dagger},t)^{\mathrm{IP}}\} \tag{7.119}$$

Now, solve the differential equation (7.119) by the aid of the normal ordering procedure according to which it is possible to pass from operators that are functions of the noncommutative Boson operators to scalars. That is possible using the inverse of the $\hat{\mathbf{N}}$ operator and of Eqs. (7.97) and (7.99) to get

$$\hat{\mathbf{N}}^{-1}\{\mathbf{a}^{\dagger}\{\mathbf{U}(\mathbf{a},\mathbf{a}^{\dagger},t)^{\mathrm{IP}}\}\} = \alpha^{*}\{U^{\{n\}}(\alpha,\alpha^{*},t)\} \tag{7.120}$$

$$\hat{\mathbf{N}}^{-1}\{\mathbf{a}\{\mathbf{U}(\mathbf{a},\mathbf{a}^{\dagger},t)^{\mathrm{IP}}\}\} = \left(\alpha + \frac{\partial}{\partial\alpha^{*}}\right)\{U^{\{n\}}(\alpha,\alpha^{*},t)\} \tag{7.121}$$

$$\hat{\mathbf{N}}^{-1}\left\{\frac{\partial \mathbf{U}(\mathbf{a},\mathbf{a}^{\dagger},t)^{\mathrm{IP}}}{\partial t}\right\} = \left(\frac{\partial U^{\{n\}}(\alpha,\alpha^*,t)}{\partial t}\right) \tag{7.122}$$

Thus, it is possible to pass from the partial derivative Eq. (7.119) to

$$i\left(\frac{\partial U^{\{n\}}(\alpha,\alpha^*,t)}{\partial t}\right) = \alpha^\circ\,\Omega\left(\alpha^* e^{i\Omega t} + \left(\alpha + \frac{\partial}{\partial \alpha^*}\right)e^{-i\Omega t}\right)\{U^{\{n\}}(\alpha,\alpha^*,t)\} \tag{7.123}$$

with, corresponding to Eq. (7.111), the boundary condition

$$\{U^{\{n\}}(\alpha,\alpha^*,0)\} = 1 \tag{7.124}$$

Now, in order to solve the partial derivative equation (7.123), let us write

$$\{U^{\{n\}}(\alpha,\alpha^*,t)\} = e^{G(t)} \tag{7.125}$$

Next, in terms of the new scalar function $G(t)$, the partial derivatives of $U^{\{n\}}(\alpha,\alpha^*,t)$ with respect to the scalars t and α^* are, respectively,

$$\left(\frac{\partial U^{\{n\}}(\alpha,\alpha^*,t)}{\partial t}\right) = \left(\frac{\partial G(t)}{\partial t}\right)U^{\{n\}}(\alpha,\alpha^*,t)$$

$$\left(\frac{\partial U^{\{n\}}(\alpha,\alpha^*,t)}{\partial \alpha^*}\right) = \left(\frac{\partial G(t)}{\partial \alpha^*}\right)U^{\{n\}}(\alpha,\alpha^*,t)$$

Thus, owing to these equations, and after simplification by $U^{\{n\}}(\alpha,\alpha^*,t)$, Eq. (7.123) becomes

$$i\left(\frac{\partial G(t)}{\partial t}\right) = \alpha^\circ\Omega\left\{e^{i\Omega t}\alpha^* + e^{-i\Omega t}\alpha + e^{-i\Omega t}\left(\frac{\partial G(t)}{\partial \alpha^*}\right)\right\} \tag{7.126}$$

Again, assume for the intermediate function $G(t)$ appearing in Eq. (7.125), an expression of the form

$$G(t) = A(t) + B(t)\alpha + C(t)\alpha^* \tag{7.127}$$

Here, $A(t)$, $B(t)$, and $C(t)$ are unknown functions to be found, which, due to Eqs. (7.124) and (7.125), must satisfy at the initial time

$$A(0) = B(0) = C(0) = 1$$

Then, in terms of these new functions, the partial derivatives involved in (7.126) are, respectively,

$$\left(\frac{\partial G(t)}{\partial \alpha^*}\right) = C(t)$$

$$\left(\frac{\partial G(t)}{\partial t}\right) = \left(\frac{\partial A(t)}{\partial t}\right) + \left(\frac{\partial B(t)}{\partial t}\right)\alpha + \left(\frac{\partial C(t)}{\partial t}\right)\alpha^*$$

so that Eq. (7.126) becomes

$$i\left(\frac{\partial A(t)}{\partial t} + \frac{\partial B(t)}{\partial t}\alpha + \frac{\partial C(t)}{\partial t}\alpha^*\right) = \alpha^\circ\Omega\{e^{i\Omega t}\alpha^* + e^{-i\Omega t}\alpha + e^{-i\Omega t}C(t)\}$$

which, by identification, leads to

$$i\left(\frac{\partial A(t)}{\partial t}\right) = \alpha^\circ \Omega e^{-i\Omega t} C(t)$$

$$i\left(\frac{\partial B(t)}{\partial t}\right) = \alpha^\circ \Omega e^{-i\Omega t}$$

$$i\left(\frac{\partial C(t)}{\partial t}\right) = \alpha^\circ \Omega e^{i\Omega t}$$

Solving these equations yields, respectively, to

$$C(t) = -\alpha^\circ(e^{i\Omega t} - 1) \tag{7.128}$$

$$B(t) = \alpha^\circ(e^{-i\Omega t} - 1) \tag{7.129}$$

$$A(t) = i\alpha^{\circ 2}\Omega t + \alpha^\circ B(t) \tag{7.130}$$

Thus, in view of Eqs. (7.128)–(7.130), Eq. (7.127) becomes

$$G(t) = i\alpha^{\circ 2}\Omega t + \alpha^{\circ 2}(e^{-i\Omega t} - 1) + \alpha^\circ(e^{-i\Omega t} - 1)\alpha - \alpha^\circ(e^{i\Omega t} - 1)\alpha^*$$

so that Eq. (7.125) reads

$$\{U^{(n)}(\alpha, \alpha^*, t)\} = (e^{i\alpha^{\circ 2}\Omega t})e^{\Phi^\circ(t)\alpha^{\circ 2}} e^{-\Phi^\circ(t)^*\alpha^*} e^{\Phi^\circ(t)\alpha} \tag{7.131}$$

with

$$\Phi^\circ(t) \equiv \alpha^\circ(e^{-i\Omega t} - 1) \tag{7.132}$$

Now, by the aid of Eq. (7.131), it is possible to return to the time evolution operator, using the normal ordering operator $\hat{\mathbf{N}}$ prompting one to write

$$\hat{\mathbf{N}}\{U^{(n)}(\alpha, \alpha^*, t)\} = (e^{i\alpha^{\circ 2}\Omega t})e^{\Phi^\circ(t)\alpha^{\circ 2}}\hat{\mathbf{N}}\{e^{-\Phi^\circ(t)^*\alpha^*} e^{\Phi^\circ(t)\alpha}\} \tag{7.133}$$

Then, according to the first equation of (7.44), we have

$$\hat{\mathbf{N}}\{U^{\{n\}}(\alpha, \alpha^*, t)\} = \{\mathbf{U}(\mathbf{a}^\dagger, \mathbf{a}, t)^{\text{IP}}\} \tag{7.134}$$

$$\hat{\mathbf{N}}\{e^{-\Phi^\circ(t)^*\alpha^*} e^{\Phi^\circ(t)\alpha}\} = (e^{-\Phi^\circ(t)^*\mathbf{a}^\dagger})(e^{\Phi^\circ(t)\mathbf{a}}) \tag{7.135}$$

Hence, from Eqs. (7.133)–(7.135), Eq. (7.131) allows us to obtain the IP time evolution operator in the form

$$\{\mathbf{U}(\mathbf{a}, \mathbf{a}^\dagger, t)^{\text{IP}}\} = (e^{i\alpha^{\circ 2}\Omega t})e^{\alpha^\circ\Phi^\circ(t)}(e^{-\Phi^\circ(t)^*\mathbf{a}^\dagger})(e^{\Phi^\circ(t)\mathbf{a}}) \tag{7.136}$$

Next, use the Glauber–Weyl theorem (1.79) to transform the right-hand-side product of exponential operators

$$(e^{-\Phi^\circ(t)^*\mathbf{a}^\dagger})(e^{\Phi^\circ(t)\mathbf{a}}) = e^{-[\Phi^\circ(t)^*\mathbf{a}^\dagger, \Phi^\circ(t)\mathbf{a}]/2}(e^{-\Phi^\circ(t)^*\mathbf{a}^\dagger + \Phi^\circ(t)\mathbf{a}}) \tag{7.137}$$

The commutator appearing on the right-hand side is

$$[\Phi^\circ(t)^*\mathbf{a}^\dagger, \Phi^\circ(t)\mathbf{a}] = |\Phi^\circ(t)|^2[\mathbf{a}^\dagger, \mathbf{a}] = -|\Phi^\circ(t)|^2$$

Consequently, Eq. (7.137) becomes

$$(e^{-\Phi^\circ(t)^*\mathbf{a}^\dagger})(e^{\Phi^\circ(t)\mathbf{a}}) = e^{|\Phi^\circ(t)|^2/2}(e^{-\Phi^\circ(t)^*\mathbf{a}^\dagger + \Phi^\circ(t)\mathbf{a}})$$

Hence, Eq. (7.136) takes the form

$$\mathbf{U}(\mathbf{a}, \mathbf{a}^\dagger, t)^{\mathrm{IP}} = (e^{i\alpha^{\circ 2}\Omega t})e^{\alpha^\circ\Phi^\circ(t) + |\Phi^\circ(t)|^2/2}(e^{-\Phi^\circ(t)^*\mathbf{a}^\dagger + \Phi^\circ(t)\mathbf{a}}) \tag{7.138}$$

Again, owing to Eq. (7.132), it appears that

$$\left\{\alpha^\circ\Phi^\circ(t) + \tfrac{1}{2}|\Phi^\circ(t)|^2\right\} = \alpha^{\circ 2}((e^{-i\Omega t} - 1) + \tfrac{1}{2}(2 - e^{i\Omega t} - e^{-i\Omega t}))$$

or

$$\left\{\alpha^\circ\Phi^\circ(t) + \tfrac{1}{2}|\Phi^\circ(t)|^2\right\} = \alpha^{\circ 2}\left(e^{-i\Omega t} - \tfrac{1}{2}e^{i\Omega t} - \tfrac{1}{2}e^{-i\Omega t}\right)$$

and

$$\left\{\alpha^\circ\Phi^\circ(t) + \tfrac{1}{2}|\Phi^\circ(t)|^2\right\} = -i\alpha^{\circ 2}\sin\Omega t$$

so that Eq. (7.138) transforms to

$$\mathbf{U}(\mathbf{a}, \mathbf{a}^\dagger, t)^{\mathrm{IP}} = (e^{i\alpha^{\circ 2}\Omega t})(e^{-i\alpha^{\circ 2}\sin\Omega t})(e^{-\Phi^\circ(t)^*\mathbf{a}^\dagger + \Phi^\circ(t)\mathbf{a}}) \tag{7.139}$$

As a consequence, owing to Eqs. (7.114) and (7.139), the full time evolution operator (7.108) takes the form

$$\boxed{\mathbf{U}(t) = (e^{i\alpha^{\circ 2}\Omega t})(e^{-i\alpha^{\circ 2}\sin\Omega t})(e^{-i\mathbf{a}^\dagger\mathbf{a}\Omega t})(e^{-\Phi^\circ(t)^*\mathbf{a}^\dagger + \Phi^\circ(t)\mathbf{a}})} \tag{7.140}$$

Now, in view of Eqs. (7.108), (7.113), and (7.114), it appears that this equation may be also written, after simplification, as

$$\boxed{\widehat{\mathbf{P}}\left\{\exp\left(-ib\int_0^t \mathbf{Q}(t')^{\mathrm{IP}}dt'\right)\right\} = (e^{i\alpha^{\circ 2}\Omega t})(e^{-i\alpha^{\circ 2}\sin\Omega t})(e^{-\Phi^\circ(t)^*\mathbf{a}^\dagger + \Phi^\circ(t)\mathbf{a}})} \tag{7.141}$$

or, due to Eq. (7.118), for the inverse of Eq. (7.141)

$$\widehat{\mathbf{P}}\left\{\exp\left(i\alpha^\circ\int_0^t s[\mathbf{a}^\dagger e^{i\Omega t'} + \mathbf{a}e^{-i\Omega t'}]dt'\right)\right\} = (e^{-i\alpha^{\circ 2}\Omega t})(e^{i\alpha^{\circ 2}\sin\Omega t})(e^{\Phi^\circ(t)^*\mathbf{a}^\dagger - \Phi^\circ(t)\mathbf{a}}) \tag{7.142}$$

7.4 CONCLUSION

This chapter dealt with the theoretical properties of the ladder operators, more elaborate than those found in Chapter 5, and has lead to theorems allowing us to make canonical transformations concerning these operators, particularly those involving translation operators and the other time evolution operators. It has also given the most important results concerning normal and antinormal ordering formalism, allowing one to transform quantum equations dealing with noncommuting ladder operators, to scalar equivalent ones having the form of partial differential equations. All the results gained in this chapter will be widely used, particularly when studying the reversible or irreversible dynamics of quantum oscillators, more specially the following ones:

Canonical transformations on ladder operators
Involving the translation operator:
$(e^{-\xi^*\mathbf{a}^\dagger+\xi\mathbf{a}})\{\mathbf{f}(\mathbf{a},\mathbf{a}^\dagger)\}(e^{\xi^*\mathbf{a}^\dagger-\xi\mathbf{a}}) = \{\mathbf{f}(\mathbf{a}+\xi^*,\mathbf{a}^\dagger+\xi)\}$
Involving the Hamiltonian:
$(e^{\xi\mathbf{a}^\dagger\mathbf{a}})\{\mathbf{f}(\mathbf{a}^\dagger,\mathbf{a})\}(e^{-\xi\mathbf{a}^\dagger\mathbf{a}}) = \{\mathbf{f}(\mathbf{a}^\dagger e^{\xi},\mathbf{a}e^{-\xi})\}$ $\boxed{(e^{i\omega t\mathbf{a}^\dagger\mathbf{a}})\{\mathbf{f}(\mathbf{a}^\dagger,\mathbf{a})\}(e^{-i\omega t\mathbf{a}^\dagger\mathbf{a}}) = \{\mathbf{f}(\mathbf{a}^\dagger e^{i\omega t},\mathbf{a}e^{-i\omega t})\}}$
Normal ordering formalism
Operator to be transformed:
$\{\mathbf{F}(\mathbf{a},\mathbf{a}^\dagger)\} = \{(\mathbf{a}^\dagger)^k(\mathbf{a})^m\mathbf{f}(\mathbf{a},\mathbf{a}^\dagger)\}$
Passage from an operator to its corresponding scalar:
$\hat{\mathbf{N}}^{-1}\{\mathbf{F}(\mathbf{a},\mathbf{a}^\dagger)\} = \{F^{\{n\}}(\alpha,\alpha^*)\}$
The corresponding scalar expression:
$\boxed{F^{\{n\}}(\alpha,\alpha^*)\} = (\alpha^*)^k\left(\alpha+\frac{\partial}{\partial\alpha^*}\right)^m\{f^{\{n\}}(\alpha,\alpha^*)\}}$

BIBLIOGRAPHY

W. H. Louisell. *Quantum Statistical Properties of Radiation*. Wiley: New York, 1973.

CHAPTER 8

PHASE OPERATORS AND SQUEEZED STATES

INTRODUCTION

In the present chapter, we study new operators, states, and theorems concerning harmonic oscillators that are less usual than those previously studied. Up to now, in the study of quantum oscillators, we have not yet encountered the concept of *phase*, which is usual in the area of classical oscillators. The corresponding quantum phase may be treated using the *phase operator*, which is the object of the first section of this chapter, where it will be applied to situations in which the quantum harmonic oscillator is described by coherent states. Now, other quantum states exist that resemble coherent states, although they are more complex. They are the *squeezed states*. One of their characteristic properties leads to uncertainty relations, which evoke phase properties because they involve time-dependent oscillatory momentum and position uncertainties coming back and forth. These squeezed states will be comprehensively treated in the second section of this chapter, which ends with a study of the *Bogolioubov–Valatin* transformation allowing one to diagonalize some Hamiltonians involving the product of Boson operators of the same forms as those appearing in squeezed states.

8.1 PHASE OPERATORS

We begin with *phase operators* evoking the phases appearing in the physics of classical oscillators.

8.1.1 Phase operators in the basis of harmonic Hamiltonian eigenkets

For this purpose, define a new operator Φ through the raising and lowering operators according to the following equations:

$$\mathbf{a} = \sqrt{\mathbf{a}^\dagger \mathbf{a} + 1}(e^{i\Phi})$$

$$\mathbf{a}^\dagger = (e^{-i\Phi})\sqrt{\mathbf{a}^\dagger \mathbf{a} + 1}$$

Quantum Oscillators, First Edition. Olivier Henri-Rousseau and Paul Blaise.
© 2011 John Wiley & Sons, Inc. Published 2011 by John Wiley & Sons, Inc.

which read by inversion

$$(e^{i\Phi}) = \left(\frac{1}{\sqrt{\mathbf{a}^\dagger\mathbf{a}+1}}\right)\mathbf{a} \tag{8.1}$$

$$(e^{-i\Phi}) = \mathbf{a}^\dagger\left(\frac{1}{\sqrt{\mathbf{a}^\dagger\mathbf{a}+1}}\right) \tag{8.2}$$

Now, we prove that the operators appearing on the left-hand side of these two last equations are unitary. Thus, take the product of Eq. (8.1) by Eq. (8.2):

$$(e^{i\Phi})(e^{-i\Phi}) = \left(\frac{1}{\sqrt{\mathbf{a}^\dagger\mathbf{a}+1}}\right)\mathbf{a}\mathbf{a}^\dagger\left(\frac{1}{\sqrt{\mathbf{a}^\dagger\mathbf{a}+1}}\right)$$

which, owing to the commutation rule (5.5), yields

$$(e^{i\Phi})(e^{-i\Phi}) = \left(\frac{1}{\sqrt{\mathbf{a}^\dagger\mathbf{a}+1}}\right)(\mathbf{a}^\dagger\mathbf{a}+1)\left(\frac{1}{\sqrt{\mathbf{a}^\dagger\mathbf{a}+1}}\right) \tag{8.3}$$

Now, one may write the last two terms of the right-hand side of this equation as

$$(\mathbf{a}^\dagger\mathbf{a}+1)\left(\frac{1}{\sqrt{\mathbf{a}^\dagger\mathbf{a}+1}}\right) = \sqrt{\mathbf{a}^\dagger\mathbf{a}+1}\sqrt{\mathbf{a}^\dagger\mathbf{a}+1}\left(\frac{1}{\sqrt{\mathbf{a}^\dagger\mathbf{a}+1}}\right)$$

or

$$(\mathbf{a}^\dagger\mathbf{a}+1)\left(\frac{1}{\sqrt{\mathbf{a}^\dagger\mathbf{a}+1}}\right) = \sqrt{\mathbf{a}^\dagger\mathbf{a}+1}$$

so that Eq. (8.3) becomes

$$(e^{i\Phi})(e^{-i\Phi}) = \left(\frac{1}{\sqrt{\mathbf{a}^\dagger\mathbf{a}+1}}\right)\sqrt{\mathbf{a}^\dagger\mathbf{a}+1}$$

which reduces to

$$(e^{i\Phi})(e^{-i\Phi}) = 1 \tag{8.4}$$

Moreover, the action of the operator (8.1) on an eigenstate of the harmonic oscillator Hamiltonian yields

$$(e^{i\Phi})|\{n\}\rangle = \frac{1}{\sqrt{\mathbf{a}^\dagger\mathbf{a}+1}}\mathbf{a}\;|\{n\}\rangle$$

and, owing to Eq. (5.53),

$$(e^{i\Phi})|\{n\}\rangle = \frac{1}{\sqrt{\mathbf{a}^\dagger\mathbf{a}+1}}\sqrt{n}|\{n-1\}\rangle \tag{8.5}$$

Again, write the following formal expansion of the square root

$$\frac{1}{\sqrt{y}} = \sum_k C_k y^k \tag{8.6}$$

where an explicit expression of the expansion coefficients C_k need not be given. Applying this expansion to $y = \mathbf{a}^\dagger\mathbf{a} + 1$ reads

$$\frac{1}{\sqrt{\mathbf{a}^\dagger\mathbf{a}+1}} = \sum_k C_k(\mathbf{a}^\dagger\mathbf{a}+1)^k \quad \text{with} \quad k = 0, 1, \ldots$$

or, after postmultiplying by $|\{n\}\rangle$

$$\frac{1}{\sqrt{\mathbf{a}^\dagger \mathbf{a}+1}}|\{n\}\rangle = \sum_k C_k(\mathbf{a}^\dagger \mathbf{a}+1)^k|\{n\}\rangle \tag{8.7}$$

Then, due to Eq. (5.42), this last expression transforms to

$$\frac{1}{\sqrt{\mathbf{a}^\dagger \mathbf{a}+1}}|\{n\}\rangle = \sum_k C_k(n+1)^k|\{n\}\rangle \tag{8.8}$$

Furthermore, applying in turn Eq. (8.6) with $y = n+1$, that is,

$$\sum_k C_k(n+1)^k = \frac{1}{\sqrt{n+1}}$$

Eq. (8.8) becomes

$$\frac{1}{\sqrt{\mathbf{a}^\dagger \mathbf{a}+1}}|\{n\}\rangle = \frac{1}{\sqrt{n+1}}|\{n\}\rangle$$

or, changing $n+1$ into n, and thus n into $n-1$

$$\frac{1}{\sqrt{\mathbf{a}^\dagger \mathbf{a}+1}}|\{n-1\}\rangle = \frac{1}{\sqrt{n}}|\{n-1\}\rangle \tag{8.9}$$

Now, recall Eq. (8.5), that is,

$$(e^{i\Phi})|\{n\}\rangle = \frac{1}{\sqrt{\mathbf{a}^\dagger \mathbf{a}+1}}\sqrt{n}|\{n-1\}\rangle \tag{8.10}$$

which, in view of Eq. (8.9) transforms after simplification into

$$(e^{i\Phi})|\{n\}\rangle = |\{n-1\}\rangle \tag{8.11}$$

In a similar way, one obtains for the corresponding operator defined by Eq. (8.2)

$$(e^{-i\Phi})|\{n\}\rangle = |\{n+1\}\rangle \tag{8.12}$$

Hence, as a consequence of Eqs. (8.11) and (8.12), and owing to the orthonormality of the eigenstates of the harmonic oscillator, the matrix elements of the operators (8.1) and (8.2) satisfy

$$\langle\{m\}|e^{i\Phi}|\{n\}\rangle = \langle\{m\}|\{n-1\}\rangle = \delta_{m,n-1} \tag{8.13}$$

$$\langle\{m\}|e^{-i\Phi}|\{n\}\rangle = \langle\{m\}|\{n+1\}\rangle = \delta_{m,n+1} \tag{8.14}$$

Now, introduce the following operators

$$\cos\Phi = \tfrac{1}{2}(e^{i\Phi} + e^{-i\Phi}) \tag{8.15}$$

$$\sin\Phi = \tfrac{1}{2i}(e^{i\Phi} - e^{-i\Phi}) \tag{8.16}$$

Then, in the basis of the eigenstates of the harmonic oscillator, the matrix elements of the first operator read

$$\langle\{m\}|\cos\Phi|\{n\}\rangle = \tfrac{1}{2}\{\langle\{m\}|e^{i\Phi}|\{n\}\rangle + \langle\{m\}|e^{-i\Phi}|\{n\}\rangle\}$$

or, owing to Eqs. (8.13) and (8.14), we have

$$\langle\{m\}|\cos\Phi|\{n\}\rangle = \tfrac{1}{2}(\delta_{m,n-1} + \delta_{m,n+1})$$

In a similar way, we have

$$\langle\{m\}|\sin\Phi|\{n\}\rangle = \tfrac{1}{2i}(\delta_{m,n-1} - \delta_{m,n+1})$$

As a consequence of these two last equations, it appears that the diagonal elements of these operators are zero, that is,

$$\langle\{n\}|\cos\Phi|\{n\}\rangle = \langle\{n\}|\sin\Phi|\{n\}\rangle = 0 \tag{8.17}$$

Moreover, consider the matrix elements of $\cos^2\Phi$, which, due to Eq. (8.15), read

$$\langle\{m\}|\cos^2\Phi|\{n\}\rangle = \tfrac{1}{4}\{\langle\{m\}|(e^{2i\Phi} + e^{-2i\Phi})|\{n\}\rangle + \langle\{m\}|2|\{n\}\rangle\} \tag{8.18}$$

However, since

$$(e^{2i\Phi})|\{n\}\rangle = (e^{i\Phi})(e^{i\Phi})|\{n\}\rangle$$

and, in view of Eq. (8.11), it appears that

$$(e^{2i\Phi})|\{n\}\rangle = (e^{i\Phi})|\{n-1\}\rangle$$

or, using in turn Eq. (8.11)

$$(e^{2i\Phi})|\{n\}\rangle = |\{n-2\}\rangle \tag{8.19}$$

In a similar way, using Eq. (8.12), we have

$$(e^{-2i\Phi})|\{n\}\rangle = |\{n+2\}\rangle \tag{8.20}$$

Then, Eqs. (8.19) and (8.20) allow to transform Eq. (8.18) into

$$\langle\{m\}|\cos^2\Phi|\{n\}\rangle = \tfrac{1}{4}\langle\{m\}|(|\{n-2\}\rangle + |\{n+2\}\rangle) + \tfrac{1}{2}\delta_{mn}$$

or, using the orthonormality properties of the states involved in this expression, it reduces to

$$\langle\{m\}|\cos^2\Phi|\{n\}\rangle = \tfrac{1}{4}(\delta_{m,n-2} + \delta_{m,n+2}) + \tfrac{1}{2}\delta_{mn} \tag{8.21}$$

In like manner

$$\langle\{m\}|\sin^2\Phi|\{n\}\rangle = -\tfrac{1}{4}(\delta_{m,n-2} + \delta_{m,n+2}) - \tfrac{1}{2}\delta_{mn} \tag{8.22}$$

Hence, the diagonal elements of Eqs. (8.21) and (8.22), reduce simply to

$$\langle\{n\}|\cos^2\Phi|\{n\}\rangle = \langle\{n\}|\sin^2\Phi|\{n\}\rangle = \tfrac{1}{2} \tag{8.23}$$

Moreover, the dispersions of $\cos\Phi$ and $\sin\Phi$ in the states $|\{n\}\rangle$, which are, respectively, given by

$$(\Delta\cos\Phi)_{|n\rangle} = \sqrt{\langle\{n\}|\cos^2\Phi|\{n\}\rangle - \langle\{n\}|\cos\Phi|\{n\}\rangle^2}$$

$$(\Delta\sin\Phi)_{|n\rangle} = \sqrt{\langle\{n\}|\sin^2\Phi|\{n\}\rangle - \langle\{n\}|\sin\Phi|\{n\}\rangle^2}$$

read, due to Eqs. (8.17) and (8.23),

$$(\Delta\cos\Phi)_{|n\rangle} = (\Delta\sin\Phi)_{|n\rangle} = \sqrt{\tfrac{1}{2}}$$

which indicates a random dispersion of the phase.

8.1.2 Commutation rule involving the phase operators

Now, seek the commutator of $\mathbf{a}^\dagger\mathbf{a}$ with the operator defined by Eq. (8.1), that is,

$$[\mathbf{a}^\dagger\mathbf{a},(e^{i\Phi})] = \mathbf{a}^\dagger\mathbf{a}\frac{1}{\sqrt{\mathbf{a}^\dagger\mathbf{a}+1}}\mathbf{a} - \frac{1}{\sqrt{\mathbf{a}^\dagger\mathbf{a}+1}}\mathbf{a}\mathbf{a}^\dagger\mathbf{a} \tag{8.24}$$

Because $\mathbf{a}^\dagger\mathbf{a}$ commutes with all its powers, and in view of the expression of the commutator (5.5) of $\mathbf{a}^\dagger$ and $\mathbf{a}$, Eq. (8.24) reduces to

$$[\mathbf{a}^\dagger\mathbf{a},(e^{i\Phi})] = \frac{1}{\sqrt{\mathbf{a}^\dagger\mathbf{a}+1}}\mathbf{a}^\dagger\mathbf{a}\mathbf{a} - \frac{1}{\sqrt{\mathbf{a}^\dagger\mathbf{a}+1}}(\mathbf{a}^\dagger\mathbf{a}+1)\mathbf{a}$$

or, factorizing and rearranging,

$$[\mathbf{a}^\dagger\mathbf{a},(e^{i\Phi})] = -\frac{1}{\sqrt{\mathbf{a}^\dagger\mathbf{a}+1}}\mathbf{a}$$

and after simplification and use of Eq. (8.1),

$$[\mathbf{a}^\dagger\mathbf{a},(e^{i\Phi})] = -(e^{i\Phi}) \tag{8.25}$$

Similarly

$$[\mathbf{a}^\dagger\mathbf{a},(e^{-i\Phi})] = (e^{-i\Phi}) \tag{8.26}$$

Hence, due to Eqs. (8.15) and (8.16), Eqs. (8.25) and (8.26) lead to

$$[\mathbf{a}^\dagger\mathbf{a}, \cos\Phi] = -i\sin\Phi \tag{8.27}$$

$$[\mathbf{a}^\dagger\mathbf{a}, \sin\Phi] = i\cos\Phi \tag{8.28}$$

It is now possible to get the product of the uncertainties over $\mathbf{a}^\dagger\mathbf{a}$ and $\cos\Phi$ or $\sin\Phi$. Keeping in mind that the product of uncertainties of two operators $\mathbf{A}$ and $\mathbf{B}$ calculated over kets $|\Psi\rangle$ is given by Eq. (2.49), that is,

$$(\Delta A_\Psi)^2(\Delta B_\Psi)^2 \geq -\tfrac{1}{4}\langle\Psi|[\mathbf{A},\mathbf{B}]|\Psi\rangle^2$$

and taking $\mathbf{A} = \mathbf{a}^\dagger\mathbf{a}$ and $\mathbf{B} = \cos\Phi$ or $\sin\Phi$, Eqs. (8.27) and (8.28) allow us to get the uncertainty relations, which read in the present situation

$$(\Delta\mathbf{a}^\dagger\mathbf{a})^2_\Psi(\Delta\cos\Phi)^2_\Psi \geq \tfrac{1}{4}\langle\Psi|\sin\Phi|\Psi\rangle^2$$

$$(\Delta\mathbf{a}^\dagger\mathbf{a})^2_\Psi(\Delta\sin\Phi)^2_\Psi \geq \tfrac{1}{4}\langle\Psi|\cos\Phi|\Psi\rangle^2$$

or

$$(\Delta\mathbf{a}^\dagger\mathbf{a})_\Psi(\Delta\cos\Phi)_\Psi \geq \tfrac{1}{2}|\langle\Psi|\sin\Phi|\Psi\rangle|$$

$$(\Delta\mathbf{a}^\dagger\mathbf{a})_\Psi(\Delta\sin\Phi)_\Psi \geq \tfrac{1}{2}|\langle\Psi|\cos\Phi|\Psi\rangle|$$

8.1.3 Phase operators within coherent-state picture

8.1.3.1 Diagonal matrix elements of cos Φ Consider now the average value of the operator $\cos\Phi$ performed over a coherent-state picture:

$$\langle\{\alpha\}|\cos\Phi|\{\alpha\}\rangle = \tfrac{1}{2}(\langle\{\alpha\}|e^{i\Phi}|\{\alpha\}\rangle + \langle\{\alpha\}|e^{-i\Phi}|\{\alpha\}\rangle) \qquad (8.29)$$

Keeping in mind the expansion of a coherent state on the eigenstates of the harmonic oscillator Hamiltonian given by Eq. (6.16), that is,

$$|\{\alpha\}\rangle = e^{-|\alpha|^2/2}\sum_n \left(\frac{(\alpha)^n}{\sqrt{n!}}\right)|\{n\}\rangle \qquad (8.30)$$

$$\langle\{\alpha\}| = e^{-|\alpha|^2/2}\sum_m \left(\frac{(\alpha^*)^m}{\sqrt{m!}}\right)\langle\{m\}| \qquad (8.31)$$

Eq. (8.29) reads

$$\langle\{\alpha\}|\cos\Phi|\{\alpha\}\rangle = \frac{1}{2}e^{-|\alpha|^2}\sum_n\sum_m \left(\frac{(\alpha^*)^m}{\sqrt{m!}}\frac{(\alpha)^n}{\sqrt{n!}}\right)(\langle\{m\}|e^{i\Phi}|\{n\}\rangle + \langle\{m\}|e^{-i\Phi}|\{n\}\rangle)$$

so that, passing from the imaginary exponentials to the corresponding cosine function, and due to Eqs. (8.13) and (8.14), we have

$$\langle\{\alpha\}|\cos\Phi|\{\alpha\}\rangle = \frac{1}{2}e^{-|\alpha|^2}\sum_n\sum_m \left(\frac{(\alpha^*)^m}{\sqrt{m!}}\frac{(\alpha)^n}{\sqrt{n!}}\right)(\delta_{m,n-1} + \delta_{m,n+1})$$

or

$$\langle\{\alpha\}|\cos\Phi|\{\alpha\}\rangle = \frac{1}{2}e^{-|\alpha|^2}\sum_n \left(\frac{(\alpha^*)^{n-1}}{\sqrt{(n-1)!}}\frac{(\alpha)^n}{\sqrt{n!}} + \frac{(\alpha^*)^{n+1}}{\sqrt{(n+1)!}}\frac{(\alpha)^n}{\sqrt{n!}}\right)$$

Since $(n-1)!$ cannot start from $n = 0$, the summation must be redefined in the terms where $(n-1)!$ appears, by changing n into $n+1$, leading to

$$\langle\{\alpha\}|\cos\Phi|\{\alpha\}\rangle = \frac{1}{2}e^{-|\alpha|^2}\sum_n \left(\frac{(\alpha^*)^n}{\sqrt{n!}}\frac{(\alpha)^{n+1}}{\sqrt{(n+1)!}} + \frac{(\alpha^*)^{n+1}}{\sqrt{(n+1)!}}\frac{(\alpha)^n}{\sqrt{n!}}\right)$$

or

$$\langle\{\alpha\}|\cos\Phi|\{\alpha\}\rangle = \frac{1}{2}e^{-|\alpha|^2}\sum_n \left(\frac{|\alpha|^{2n}}{\sqrt{n!}}\frac{(\alpha+\alpha^*)}{\sqrt{(n+1)!}}\right) \qquad (8.32)$$

Next, writing

$$\alpha = |\alpha|e^{i\theta} \qquad (8.33)$$

Eq. (8.32) yields

$$\langle\{\alpha\}|\cos\Phi|\{\alpha\}\rangle = e^{-|\alpha|^2}\cos\theta\sum_n \left(\frac{|\alpha|^{2n+1}}{\sqrt{n!}\sqrt{(n+1)!}}\right)$$

or, using $(n+1)! = (n+1)n!$

$$\langle\{\alpha\}|\cos\Phi|\{\alpha\}\rangle = e^{-|\alpha|^2}\cos\theta\sum_n \left(\frac{|\alpha|^{2n+1}}{n!\sqrt{n+1}}\right)$$

8.1.3.2 Diagonal matrix element of $\cos^2 \Phi$ Now, consider the average value of $\cos^2 \Phi$ over a coherent state that, owing to Eq. (8.15), reads

$$\langle\{\alpha\}|\cos^2 \Phi|\{\alpha\}\rangle = \tfrac{1}{4}\langle\{\alpha\}|(e^{i\Phi} + e^{-i\Phi})^2|\{\alpha\}\rangle$$

or, after expansion of the right-hand-side squared term, and using Eq. (8.4), we have

$$\langle\{\alpha\}|\cos^2 \Phi|\{\alpha\}\rangle = \tfrac{1}{4}[\langle\{\alpha\}|(e^{2i\Phi} + e^{-2i\Phi})|\{\alpha\}\rangle + 2\langle\{\alpha\}|\{\alpha\}\rangle]$$

Next, assuming that the coherent states are normalized, and after using Eqs. (8.30) and (8.31), these matrix elements become

$$\langle\{\alpha\}|\cos^2 \Phi|\{\alpha\}\rangle = \frac{1}{2} + \frac{1}{4}e^{-|\alpha|^2}\sum_n\sum_m\left(\frac{(\alpha^*)^m}{\sqrt{m!}}\frac{(\alpha)^n}{\sqrt{n!}}\right)[\langle\{m\}|(e^{2i\Phi} + e^{-2i\Phi})|\{n\}\rangle] \tag{8.34}$$

or, due to Eqs. (8.19) and (8.20), the equation above becomes

$$\langle\{\alpha\}|\cos^2 \Phi|\{\alpha\}\rangle = \frac{1}{2} + \frac{1}{4}e^{-|\alpha|^2}\sum_n\sum_m\left(\frac{(\alpha^*)^m}{\sqrt{m!}}\frac{(\alpha)^n}{\sqrt{n!}}\right)\langle\{m\}|(|\{n-2\}\rangle + |\{n+2\}\rangle)$$

and, thus, after using the orthonormality properties,

$$\langle\{\alpha\}|\cos^2 \Phi|\{\alpha\}\rangle = \frac{1}{2} + \frac{1}{4}e^{-|\alpha|^2}\sum_n\sum_m\left(\frac{(\alpha^*)^m}{\sqrt{m!}}\frac{(\alpha)^n}{\sqrt{n!}}\right)(\delta_{m,n-2} + \delta_{m,n+2})$$

or

$$\langle\{\alpha\}|\cos^2 \Phi|\{\alpha\}\rangle = \frac{1}{2} + \frac{1}{4}e^{-|\alpha|^2}\sum_n\left(\frac{(\alpha^*)^{n-2}}{\sqrt{(n-2)!}}\frac{(\alpha)^n}{\sqrt{n!}} + \frac{(\alpha^*)^{n+2}}{\sqrt{(n+2)!}}\frac{(\alpha)^n}{\sqrt{n!}}\right) \tag{8.35}$$

Moreover, as above, shift the index in the sum containing $(n-2)!$ by changing n into $n+2$, so that Eq. (8.35) reads

$$\langle\{\alpha\}|\cos^2 \Phi|\{\alpha\}\rangle = \frac{1}{2} + \frac{1}{4}e^{-|\alpha|^2}\sum_n\frac{(\alpha^*)^n(\alpha)^{n+2} + (\alpha^*)^{n+2}(\alpha)^n}{\sqrt{n!}\sqrt{(n+2)!}}$$

or, using Eq. (8.33),

$$\langle\{\alpha\}|\cos^2 \Phi|\{\alpha\}\rangle = \frac{1}{2} + \frac{1}{2}e^{-|\alpha|^2}|\alpha|^2\cos(2\theta)\sum_n\left(\frac{|\alpha|^{2n}}{\sqrt{n!}\sqrt{(n+2)!}}\right)$$

so that, using finally $(n+2)! = (n+2)(n+1)n!$, we have

$$\langle\{\alpha\}|\cos^2 \Phi|\{\alpha\}\rangle = \frac{1}{2} + e^{-|\alpha|^2}|\alpha|^2\left(\cos^2\theta - \frac{1}{2}\right)\sum_n\left(\frac{|\alpha|^{2n}}{n!\sqrt{(n+2)(n+1)}}\right)$$

8.2 SQUEEZED STATES

We have often encountered coherent states that may be viewed as the result of the action of the translation operator

$$\mathbf{A}(\alpha) = \exp\{\alpha\mathbf{a}^\dagger - \alpha\mathbf{a}\}$$

on any eigenstate $|\{n\}\rangle$ of the harmonic oscillator Hamiltonian, and we have found that, whatever they may be, these states minimize the Heisenberg uncertainty relations.

8.2.1 Canonical transformations on ladder operators using squeezing operators

However, other interesting states exist that may be considered as generalization of coherent states, the *squeezed states*, which may be obtained by the action on the kets $|\{n\}\rangle$ of the following operator:

$$\mathbf{S}(z) = \exp\left\{\tfrac{1}{2}(z\mathbf{a}^{\dagger 2} - z^*\mathbf{a}^2)\right\} \tag{8.36}$$

the Hermitian conjugate of which is

$$\mathbf{S}(z)^\dagger = \exp\left\{\tfrac{1}{2}(z^*\mathbf{a}^2 - z\mathbf{a}^{\dagger 2})\right\} = \exp\left\{-\tfrac{1}{2}(z\mathbf{a}^{\dagger 2} - z^*\mathbf{a}^2)\right\}$$

whereas its inverse is

$$\mathbf{S}(z)^{-1} = \exp\left\{-\tfrac{1}{2}(z\mathbf{a}^{\dagger\,2} - z^*\mathbf{a}^2)\right\} = \mathbf{S}(-z) \tag{8.37}$$

that implies that the operator (8.36) is unitary since obeying

$$\mathbf{S}(z)^\dagger = \mathbf{S}(z)^{-1} \tag{8.38}$$

Next, consider the following canonical transformation:

$$\mathbf{S}(z)\mathbf{a}\mathbf{S}(z)^{-1} = \exp\left\{\tfrac{1}{2}(z\mathbf{a}^{\dagger 2} - z^*\mathbf{a}^2)\right\}\mathbf{a}\exp\left\{-\tfrac{1}{2}(z\mathbf{a}^{\dagger 2} - z^*\mathbf{a}^2)\right\} \tag{8.39}$$

To perform this transformation, one may use the Baker–Campbell–Hausdorff formula (1.76):

$$e^{\xi\mathbf{A}}\mathbf{B}e^{-\xi\mathbf{A}} = \mathbf{B} + [\mathbf{A},\mathbf{B}]\xi + \frac{1}{2}[\mathbf{A},[\mathbf{A},\mathbf{B}]]\xi^2 + \frac{1}{3!}[\mathbf{A},[\mathbf{A},[\mathbf{A},\mathbf{B}]]]\xi^3 + \cdots$$

where $\mathbf{A}$ and $\mathbf{B}$ are two linear operators and ξ a scalar. If one defines the operator $\mathbf{D}$ as

$$\mathbf{D} = \xi\mathbf{A}$$

the Baker–Campbell–Hausdorff relation reads

$$\boxed{e^{\mathbf{D}}\mathbf{B}e^{-\mathbf{D}} = \mathbf{B} + [\mathbf{D},\mathbf{B}] + \frac{1}{2}[\mathbf{D},[\mathbf{D},\mathbf{B}]] + \frac{1}{3!}[\mathbf{D},[\mathbf{D},[\mathbf{D},\mathbf{B}]]] + \cdots} \tag{8.40}$$

Hence, setting in Eq. (8.40)

$$\mathbf{B} = \mathbf{a} \qquad \text{and} \qquad \mathbf{D} = \tfrac{1}{2}(z\mathbf{a}^{\dagger 2} - z^*\mathbf{a}^2)$$

Eq. (8.39) takes the form

$$\begin{aligned}\mathbf{S}(z)\mathbf{a}\mathbf{S}(z)^{-1} &= \mathbf{a} + \frac{1}{2}[(z\mathbf{a}^{\dagger 2} - z^*\mathbf{a}),\mathbf{a}] \\ &+ \frac{1}{2!}\frac{1}{4}[(z\mathbf{a}^{\dagger 2} - z^*\mathbf{a}),[(z\mathbf{a}^{\dagger 2} - z^*\mathbf{a}),\mathbf{a}]] \\ &+ \frac{1}{3!}\frac{1}{8}[(z\mathbf{a}^{\dagger 2} - z^*\mathbf{a})[(z\mathbf{a}^{\dagger 2} - z^*\mathbf{a}),[(z\mathbf{a}^{\dagger 2} - z^*\mathbf{a}),\mathbf{a}]] + \cdots\end{aligned} \tag{8.41}$$

The first commutator appearing in this equation simplifies to

$$[(z\mathbf{a}^{\dagger 2} - z^*\mathbf{a}^2), \mathbf{a}] = z[\mathbf{a}^{\dagger 2}, \mathbf{a}] \tag{8.42}$$

Now, keeping in mind that the right-hand-side commutator is given by Eq. (7.47), that is,

$$[\mathbf{a}^{\dagger 2}, \mathbf{a}] = -2\mathbf{a}^{\dagger} \tag{8.43}$$

the left-hand-side commutator of Eq. (8.42) yields

$$\boxed{[(z\mathbf{a}^{\dagger 2} - z^*\mathbf{a}), \mathbf{a}] = -2z\mathbf{a}^{\dagger}} \tag{8.44}$$

Therefore, the double commutator appearing in Eq. (8.41) reads

$$\begin{aligned}[(z\mathbf{a}^{\dagger 2} - z^*\mathbf{a}^2), [(z\mathbf{a}^{\dagger 2} - z^*\mathbf{a}^2), \mathbf{a}]] &= [(z\mathbf{a}^{\dagger 2} - z^*\mathbf{a}^2), -2z\mathbf{a}^{\dagger}] \\ &= [(-z^*\mathbf{a}^2), -2z\mathbf{a}^{\dagger}] = 2|z|^2[\mathbf{a}^2, \mathbf{a}^{\dagger}] \end{aligned} \tag{8.45}$$

Now, the right-hand-side commutator of Eq. (8.45) is

$$[\mathbf{a}^2, \mathbf{a}^{\dagger}] = \mathbf{aaa}^{\dagger} - \mathbf{a}^{\dagger}\mathbf{aa} = \mathbf{a}(\mathbf{a}^{\dagger}\mathbf{a} + 1) - \mathbf{a}^{\dagger}\mathbf{aa}$$

or

$$[\mathbf{a}^2, \mathbf{a}^{\dagger}] = (\mathbf{aa}^{\dagger} - \mathbf{a}^{\dagger}\mathbf{a})\mathbf{a} + \mathbf{a} = 2\mathbf{a} \tag{8.46}$$

so that, the double commutator (8.45) becomes

$$\boxed{[(z\mathbf{a}^{\dagger 2} - z^*\mathbf{a}^2), [(z\mathbf{a}^{\dagger 2} - z^*\mathbf{a}^2), \mathbf{a}]] = 4|z|^2\mathbf{a}} \tag{8.47}$$

Again, due to Eq. (8.47), the triple commutator appearing in Eq. (8.41) reads

$$\begin{aligned}&[(z\mathbf{a}^{\dagger 2} - z^*\mathbf{a}), [(z\mathbf{a}^{\dagger 2} - z^*\mathbf{a}^2), [(z\mathbf{a}^{\dagger 2} - z^*\mathbf{a}), \mathbf{a}]]] \\ &= [(z\mathbf{a}^{\dagger 2} - z^*\mathbf{a}^2), 4|z|^2\mathbf{a}] = 4|z|^2 z[\mathbf{a}^{\dagger 2}, \mathbf{a}]\end{aligned}$$

or, using in turn Eq. (8.43)

$$[(z\mathbf{a}^{\dagger 2} - z^*\mathbf{a}), [(z\mathbf{a}^{\dagger 2} - z^*\mathbf{a}^2), [(z\mathbf{a}^{\dagger 2} - z^*\mathbf{a}), \mathbf{a}]]] = -8z|z|^2\mathbf{a}^{\dagger} \tag{8.48}$$

Moreover, according to Eq. (8.48), the quadruple commutator of Eq. (8.41) takes the form

$$\begin{aligned}&[(z\mathbf{a}^{\dagger 2} - z^*\mathbf{a}^2), [(z\mathbf{a}^{\dagger 2} - z^*\mathbf{a}^2), [(z\mathbf{a}^{\dagger 2} - z^*\mathbf{a}^2), [(z\mathbf{a}^{\dagger 2} - z^*\mathbf{a}^2), \mathbf{a}]]]] \\ &= [(z\mathbf{a}^{\dagger 2} - z^*\mathbf{a}^2), -8z|z|^2\mathbf{a}^{\dagger}] = 8|z|^4[\mathbf{a}^2, \mathbf{a}^{\dagger}]\end{aligned}$$

so that, owing to Eq. (8.46), it becomes

$$\boxed{[(z\mathbf{a}^{\dagger 2} - z^*\mathbf{a}^2), [(z\mathbf{a}^{\dagger 2} - z^*\mathbf{a}^2), [(z\mathbf{a}^{\dagger 2} - z^*\mathbf{a}^2), [(z\mathbf{a}^{\dagger 2} - z^*\mathbf{a}^2), \mathbf{a}]]]] = 16|z|^4\mathbf{a}} \tag{8.49}$$

At last, collecting the results from (8.44), (8.47), (8.48), and (8.49), the canonical transformation (8.41) appears to be

$$\mathbf{S}(z)\mathbf{a}\mathbf{S}(z)^{-1} = \mathbf{a}\left(1+\frac{1}{2!}|z|^2+\frac{1}{4!}|z|^4\right) - \mathbf{a}^{\dagger}\left(z + \frac{1}{3!}z|z|^2 + \frac{1}{5!}z|z|^4\right)$$

or

$$\mathbf{S}(z)\mathbf{a}\mathbf{S}(z)^{-1} = \mathbf{a}\left(1+\frac{1}{2!}|z|^2+\frac{1}{4!}|z|^4+\cdots\right) - \mathbf{a}^\dagger\frac{z}{|z|}\left(|z|+\frac{1}{3!}|z|^3+\frac{1}{5!}|z|^5+\cdots\right) \tag{8.50}$$

Next, due to the following expansions of the hyperbolic sine and cosine functions,

$$\cosh z = 1 + \frac{z^2}{2!} + \frac{z^4}{4!} + \cdots$$

$$\sinh z = z + \frac{z^3}{3!} + \frac{z^5}{5!} + \cdots$$

where

$$\cosh z = \left(\frac{e^z + e^{-z}}{2}\right) \qquad \text{and} \qquad \sinh z = \left(\frac{e^z - e^{-z}}{2}\right) \tag{8.51}$$

it appears that the canonical transformation (8.50) reduces to the compact expression

$$\boxed{\mathbf{S}(z)\mathbf{a}\mathbf{S}(z)^{-1} = \mathbf{a}\cosh|z| - \frac{z}{|z|}\mathbf{a}^\dagger\sinh|z|} \tag{8.52}$$

Again, changing z to $-z$, and using Eqs. (8.36) and (8.38), Eq. (8.52) yields

$$\boxed{\mathbf{S}(z)^{-1}\mathbf{a}\mathbf{S}(z) = \mathbf{a}\cosh|z| + \frac{z}{|z|}\mathbf{a}^\dagger\sinh|z|} \tag{8.53}$$

Observe that the Hermitian conjugate of this equation is

$$(\mathbf{S}(z)^{-1}\mathbf{a}\mathbf{S}(z))^\dagger = \mathbf{a}^\dagger\cosh|z| + \mathbf{a}\frac{z}{|z|}|z| \tag{8.54}$$

its left-hand side being

$$(\mathbf{S}(z)^{-1}\mathbf{a}\mathbf{S}(z))^\dagger = (\mathbf{S}(z))^\dagger\mathbf{a}^\dagger(\mathbf{S}(z)^{-1})^\dagger$$

or, since the operator $\mathbf{S}(z)$ is unitary,

$$(\mathbf{S}(z)^{-1}\mathbf{a}\mathbf{S}(z))^\dagger = \mathbf{S}(z)^{-1}\mathbf{a}^\dagger\mathbf{S}(z) \tag{8.55}$$

so that identifying Eqs. (8.54) and (8.55) leads to

$$\boxed{\mathbf{S}(z)^{-1}\mathbf{a}^\dagger\mathbf{S}(z) = \mathbf{a}^\dagger\cosh|z| + \frac{z}{|z|}\mathbf{a}\sinh|z|} \tag{8.56}$$

8.2.2 Uncertainty relations for squeezed state

8.2.2.1 Squeezed state Introduce the squeezed states according to

$$|\{z(t),\alpha(t)\}\rangle = \mathbf{U}^\circ(t)^{-1}\mathbf{A}(\alpha^\circ)\mathbf{S}(|z|)|\{0\}\rangle \tag{8.57}$$

Here $|\{0\}\rangle$ is the ground state of the harmonic Hamiltonian, whereas $\mathbf{S}(|z|)$ is the squeezing operator defined by Eq. (8.36), and $\mathbf{A}(\alpha^\circ)$ the translation operator defined by Eq. (6.74), $\mathbf{U}^\circ(t)$ being the time evolution operator constructed from $\mathbf{a}^\dagger\mathbf{a}$, that is,

$$\mathbf{S}(|z|) = (e^{(|z|\mathbf{a}^{\dagger 2} - |z|\mathbf{a}^2)/2})$$

$$\mathbf{A}(\alpha^\circ) = (e^{\alpha^\circ \mathbf{a}^\dagger - \alpha^\circ \mathbf{a}})$$

$$\mathbf{U}^\circ(t) = e^{-i\omega t \mathbf{a}^\dagger \mathbf{a}}$$

Hence, Eq. (8.57) becomes

$$|\{z(t), \alpha(t)\}\rangle = (e^{i\omega t \mathbf{a}^\dagger \mathbf{a}})\,(e^{\alpha^\circ \mathbf{a}^\dagger - \alpha^\circ \mathbf{a}})\,(e^{(|z|\mathbf{a}^{\dagger 2} - |z|\mathbf{a}^2/2)})|\{0\}\rangle$$

Now, insert between the translation and the squeezing operators and between the squeezing operator and the ground-state eigenket, the unity operator built up from the time evolution operator, that is,

$$(e^{-i\omega t \mathbf{a}^\dagger \mathbf{a}})\,(e^{i\omega t \mathbf{a}^\dagger \mathbf{a}}) = 1$$

We have

$$\begin{aligned} |\{z(t), \alpha(t)\}\rangle &= (e^{i\omega t \mathbf{a}^\dagger \mathbf{a}})\, e^{\alpha^\circ \mathbf{a}^\dagger - \alpha^\circ \mathbf{a}}\,(e^{-i\omega t \mathbf{a}^\dagger \mathbf{a}})\,(e^{i\omega t \mathbf{a}^\dagger \mathbf{a}}) \\ &\quad \times e^{(|z|\mathbf{a}^{\dagger 2} - |z|\mathbf{a}^2)/2}\,(e^{-i\omega t \mathbf{a}^\dagger \mathbf{a}})\,(e^{i\omega t \mathbf{a}^\dagger \mathbf{a}})|\{0\}\rangle \end{aligned} \tag{8.58}$$

Next, keeping in mind Eq. (7.31),

$$(e^{i\omega t \mathbf{a}^\dagger \mathbf{a}})\,\mathbf{f}(\mathbf{a}^\dagger, \mathbf{a})\,(e^{-i\omega t \mathbf{a}^\dagger \mathbf{a}}) = \{\mathbf{f}(\mathbf{a}^\dagger e^{i\omega t}, \mathbf{a}\, e^{-i\omega t})\}$$

and applying this expression, we have

$$(e^{i\omega t \mathbf{a}^\dagger \mathbf{a}})\,(e^{\alpha^\circ \mathbf{a}^\dagger - \alpha^\circ \mathbf{a}})\,(e^{-i\omega t \mathbf{a}^\dagger \mathbf{a}}) = e^{\alpha^*(t)\mathbf{a}^\dagger - \alpha(t)\mathbf{a}} = \{\mathbf{A}(\alpha(t))\} \tag{8.59}$$

$$(e^{i\omega t \mathbf{a}^\dagger \mathbf{a}})\{e^{(|z|\mathbf{a}^{\dagger 2} - |z|\mathbf{a}^2)/2}\}\,(e^{-i\omega t \mathbf{a}^\dagger \mathbf{a}}) = \{e^{(z^*(t)\mathbf{a}^{\dagger 2} - z(t)\mathbf{a}^2)/2}\} = \{\mathbf{S}(z(t))\} \tag{8.60}$$

where

$$\alpha(t) = \alpha^\circ e^{-i\omega t} \qquad \text{and} \qquad z(t) = |z| e^{-2i\omega t} \tag{8.61}$$

Now, expansion of the exponential operator appearing in Eq. (8.58) suggests writing the two last terms of the right-hand side of this equation as

$$(e^{i\omega t \mathbf{a}^\dagger \mathbf{a}})|\{0\}\rangle = \left(1 + i\omega t \mathbf{a}^\dagger \mathbf{a} + \frac{(i\omega t \mathbf{a}^\dagger \mathbf{a})^2}{2!} + \cdots\right)|\{0\}\rangle \tag{8.62}$$

so that keeping in mind that the action of $\mathbf{a}^\dagger \mathbf{a}$ on its ground state $|\{0\}\rangle$ is zero, that is,

$$\mathbf{a}^\dagger \mathbf{a}|\{0\}\rangle = 0|\{0\}\rangle$$

Then, it appears that all terms involved in the sum of Eq. (8.62) are zero, except that corresponding to $n = 0$, which acts on the ground state as the unity operator, so that

$$(e^{i\omega t \mathbf{a}^\dagger \mathbf{a}})|\{0\}\rangle = |\{0\}\rangle \tag{8.63}$$

Hence, owing to Eqs. (8.59), (8.60), and (8.63), Eq. (8.58) becomes

$$\boxed{|\{z(t), \alpha(t)\}\rangle = \mathbf{A}(\alpha(t))\mathbf{S}(z(t))|\{0\}\rangle} \tag{8.64}$$

8.2.3 Ladder operator functions averaged over squeezed states

8.2.3.1 Average values of $a^\dagger$ and a Now, consider the time-dependent mean value of $\mathbf{a}^\dagger$ averaged over the squeezed state (8.57), that is,

$$\langle \mathbf{a}(t)^\dagger \rangle_{z,\alpha} = \langle \{z(t), \alpha(t)\} | \mathbf{a}^\dagger | \{z(t), \alpha(t)\} \rangle \tag{8.65}$$

which, due to Eq. (8.64), takes the form

$$\langle \mathbf{a}(t)^\dagger \rangle_{z,\alpha} = \langle \{0\} | \mathbf{S}(z(t))^{-1} \mathbf{A}(\alpha(t))^{-1} \, \mathbf{a}^\dagger \mathbf{A}(\alpha(t)) \mathbf{S}(z(t)) | \{0\} \rangle \tag{8.66}$$

Next, keeping in mind Eq. (6.81), which implies that the action of the time-dependent translation operator on the Boson operators gives

$$\mathbf{A}(\alpha(t))^{-1} \mathbf{a}^\dagger \mathbf{A}(\alpha(t)) = \mathbf{a}^\dagger + \alpha(t)^*$$

$$\mathbf{A}(\alpha(t))^{-1} \mathbf{a} \mathbf{A}(\alpha(t)) = \mathbf{a} + \alpha(t) \tag{8.67}$$

Using also the fact,

$$\mathbf{S}(z(t))^{-1} (\mathbf{a}^\dagger + \alpha(t)^*) \mathbf{S}(z(t)) = \alpha(t)^* + \mathbf{S}(z(t))^{-1} \mathbf{a}^\dagger \mathbf{S}(z(t))$$

and taking into account Eq. (8.56), it appears that

$$\mathbf{S}(z(t))^{-1} \mathbf{A}(\alpha(t))^{-1} \mathbf{a}^\dagger \mathbf{A}(\alpha(t)) \mathbf{S}(z(t)) = \mathbf{a}^\dagger \cosh |z| + \mathbf{a} \frac{z(t)}{|z|} \sinh |z| + \alpha(t)^* \tag{8.68}$$

so that Eq. (8.66) becomes

$$\langle \mathbf{a}(t)^\dagger \rangle_{z,\alpha} = \langle \{0\} | \mathbf{a}^\dagger \cosh |z| | \{0\} \rangle + \langle \{0\} | \, \mathbf{a} \, (e^{-2i\omega t}) \sinh |z| | \{0\} \rangle + \alpha(t)^*$$

Finally, in view of Eq. (8.61), and using

$$\mathbf{a} | \{0\} \rangle = 0 \qquad \text{and} \qquad \langle \{0\} | \mathbf{a}^\dagger = 0$$

we have

$$\boxed{\langle \mathbf{a}(t)^\dagger \rangle_{z,\alpha} = \alpha(t)^* = \alpha^\circ e^{i\omega t}} \tag{8.69}$$

the Hermitian conjugate of which is

$$\langle \mathbf{a}(t) \rangle_{z,\alpha} = \alpha(t) = \alpha^\circ e^{-i\omega t} \tag{8.70}$$

8.2.3.2 Average values of $(a^\dagger)^2$ and $(a)^2$ Moreover, the average value of the square of $\mathbf{a}(t)^\dagger$ defined by

$$\langle (\mathbf{a}(t)^\dagger)^2 \rangle_{z,\alpha} = \langle \{z(t), \alpha(t)\} | (\mathbf{a}^\dagger)^2 | \{z(t), \alpha(t)\} \rangle \tag{8.71}$$

yields, with the help of Eq. (8.64),

$$\langle (\mathbf{a}(t)^\dagger)^2 \rangle_{z,\alpha} = \langle \{0\} | \mathbf{S}(z(t))^{-1} \mathbf{A}(\alpha(t))^{-1} \mathbf{a}^\dagger \mathbf{a}^\dagger \mathbf{A}(\alpha(t)) \mathbf{S}(z(t)) | \{0\} \rangle$$

Again, insert the following unity operator:

$$1 = (\mathbf{A}(\alpha(t)) \mathbf{S}(z(t)) \mathbf{S}(z(t))^{-1} \mathbf{A}(\alpha(t))^{-1}) \tag{8.72}$$

between the two raising operators, as follows:

$$\begin{aligned}\langle(\mathbf{a}(t)^\dagger)^2\rangle_{z,\alpha} &= \langle\{0\}|\mathbf{S}(z(t))^{-1}\mathbf{A}(\alpha(t))^{-1}\mathbf{a}^\dagger \\ &\quad\times(\mathbf{A}(\alpha(t))\mathbf{S}(z(t))\mathbf{S}(z(t))^{-1}\mathbf{A}(\alpha(t))^{-1})\mathbf{a}^\dagger\mathbf{A}(\alpha(t))\mathbf{S}(z(t))|\{0\}\rangle\end{aligned}$$

leading to

$$\langle(\mathbf{a}(t)^\dagger)^2\rangle_{z,\alpha} = \langle\{0\}|(\mathbf{S}(z(t))^{-1}\mathbf{A}(\alpha(t))^{-1}\mathbf{a}^\dagger\mathbf{A}(\alpha(t))\mathbf{S}(z(t)))^2|\{0\}\rangle$$

and thus, using Eq. (8.68),

$$\langle(\mathbf{a}(t)^\dagger)^2\rangle_{z,\alpha} = \langle\{0\}|(\mathbf{a}^\dagger\cosh|z| + (e^{-2i\omega t})\mathbf{a}\sinh|z| + \alpha(t)^*)^2|\{0\}\rangle$$

Again, performing the square involved on the right-hand-side average value yields

$$\begin{aligned}\langle(\mathbf{a}(t)^\dagger)^2\rangle_{z,\alpha} &= \langle\{0\}|\{(\mathbf{a}^\dagger)^2\cosh^2|z| + (e^{4i\omega t})(\mathbf{a})^2\sinh^2|z| \\ &\quad + (2\mathbf{a}^\dagger\mathbf{a}+1)\sinh|z|e^{2i\omega t}\cosh|z|\}|\{0\}\rangle \\ &\quad + 2\alpha(t)^*\langle\{0\}|(\cosh|z|\mathbf{a}^\dagger + 2e^{2i\omega t}\mathbf{a}\sinh|z|)|\{0\}\rangle + \alpha(t)^{*2}\end{aligned}$$

or, after simplifications, due to

$$\mathbf{a}^\dagger\mathbf{a}|\{0\}\rangle = 0$$

$$\langle\{0\}|(\mathbf{a}^\dagger)^2|\{0\}\rangle = \langle\{0\}|(\mathbf{a})^2|\{0\}\rangle = 0$$

$$\boxed{\langle(\mathbf{a}(t)^\dagger)^2\rangle_{z,\alpha} = (e^{-2i\omega t})\sinh|z|\cosh|z| + \alpha^{\circ 2}(e^{2i\omega t})} \tag{8.73}$$

the Hermitian conjugate of which reads

$$\boxed{\langle(\mathbf{a}(t))^2\rangle_{z,\alpha} = (e^{2i\omega t})\sinh|z|\cosh|z| + \alpha^{\circ 2}(e^{-2i\omega t})} \tag{8.74}$$

8.2.3.3 Average value of $\mathbf{a}^\dagger\mathbf{a}$ Finally, consider the average value of the occupation number $\mathbf{a}^\dagger\mathbf{a}$:

$$\langle(\mathbf{a}^\dagger\mathbf{a})\rangle_{z,\alpha} = \langle\{z(t),\alpha(t)\}|\mathbf{a}^\dagger\mathbf{a}|\{z(t),\alpha(t)\}\rangle \tag{8.75}$$

Then, insert in the following way, between the two Boson operators, the unity operator (8.72):

$$\begin{aligned}\langle(\mathbf{a}^\dagger\mathbf{a})\rangle_{z,\alpha} &= \langle\{0\}|(\mathbf{S}(z(t))^{-1}\mathbf{A}(\alpha(t))^{-1}\mathbf{a}^\dagger\mathbf{A}(\alpha(t))\mathbf{S}(z(t))) \\ &\quad\times(\mathbf{S}(z(t))^{-1}\mathbf{A}(\alpha(t))^{-1}\mathbf{a}\mathbf{A}(\alpha(t))\mathbf{S}(z(t)))|\{0\}\rangle\end{aligned} \tag{8.76}$$

Hence, using Eq. (8.68), Eq. (8.76) yields

$$\langle(\mathbf{a}^\dagger\mathbf{a})\rangle_{z,\alpha} = \langle\{0\}|(\mathbf{a}^\dagger\cosh|z| + \mathbf{a}(e^{-2i\omega t})\sinh|z| + \alpha^\circ(e^{i\omega t}))\times \mathrm{hc}|\{0\}\rangle$$

where hc is the Hermitian conjugate

$$\mathrm{hc} = \mathbf{a}\cosh|z| + \mathbf{a}^\dagger(e^{2i\omega t})\sinh|z| + \alpha^\circ(e^{-i\omega t})$$

The product involved in the right-hand-side term reads

$$\begin{aligned}(\mathbf{a}^\dagger \cosh|z| + \mathbf{a}e^{-2i\omega t}\sinh|z| + \alpha^\circ(e^{i\omega t})) \times \text{hc} \\ = |\alpha|^2 + \alpha^\circ(e^{-i\omega t})(\mathbf{a}^\dagger \cosh|z| + \mathbf{a}(e^{-2i\omega t})\sinh|z|) \\ + \mathbf{a}^\dagger\mathbf{a}\cosh^2|z| + (2\mathbf{a}^\dagger\mathbf{a}+1)\sinh^2|z| \\ + ((e^{-2i\omega t})(\mathbf{a})^2 + (e^{2i\omega t})(\mathbf{a}^\dagger)^2)\cosh|z|\sinh|z| \\ + \alpha^\circ(e^{i\omega t})(\mathbf{a}\cosh(|z|) + \mathbf{a}^\dagger(e^{2i\omega t})\sinh|z|)\end{aligned}$$

so that Eq. (8.76) appears to be simply

$$\boxed{\langle(\mathbf{a}^\dagger\mathbf{a})\rangle_{z,\alpha} = |\alpha|^2 + \sinh^2|z|} \tag{8.77}$$

Therefore, the mean value of the oscillator Hamiltonian (5.9) averaged over the squeezed states is given by

$$\langle\mathbf{H}\rangle_{z,\alpha} = \hbar\omega\langle\{z(t),\alpha(t)\}|\left(\mathbf{a}^\dagger\mathbf{a} + \tfrac{1}{2}\right)|\{z(t),\alpha(t)\}\rangle$$

which reads, after using Eq. (8.77),

$$\langle\mathbf{H}\rangle_{z,\alpha} = \hbar\omega\left(|\alpha|^2 + \sinh^2|z| + \tfrac{1}{2}\right)$$

8.2.4 Uncertainty relations for squeezed states

8.2.4.1 Average values of Q and P operators It is now possible, using Eqs. (8.69) and (8.70), to get the expression for the corresponding average value of the position operator **Q** and of its conjugate momentum **P**. First, begin with **Q** leading to

$$\langle\mathbf{Q}(t)\rangle_{z,\alpha} = \langle\{z(t),\alpha(t)\}|\mathbf{Q}|\{z(t),\alpha(t)\}\rangle$$

which, according to Eq. (5.6), reads

$$\langle\mathbf{Q}(t)\rangle_{z,\alpha} = \sqrt{\frac{\hbar}{2m\omega}}(\langle\{z(t),\alpha(t)\}|\mathbf{a}^\dagger|\{z(t),\alpha(t)\}\rangle + \langle\{z(t),\alpha(t)\}|\mathbf{a}|\{z(t),\alpha(t)\}\rangle) \tag{8.78}$$

or, due to the definition equation (8.65) and to its Hermitian conjugate,

$$\langle\mathbf{Q}(t)\rangle_{z,\alpha} = \sqrt{\frac{\hbar}{2m\omega}}(\langle\mathbf{a}(t)^\dagger\rangle_{z,\alpha} + \langle\mathbf{a}(t)\rangle_{z,\alpha})$$

and in view of Eqs. (8.69) and (8.70), it yields

$$\boxed{\langle\mathbf{Q}(t)\rangle_{z,\alpha} = 2\alpha^\circ\sqrt{\frac{\hbar}{2m\omega}}\cos\omega t} \tag{8.79}$$

On the other hand, the corresponding average value of the momentum **P**

$$\langle\mathbf{P}(t)\rangle_{z,\alpha} = \langle\{z(t),\alpha(t)\}|\mathbf{P}|\{z(t),\alpha(t)\}\rangle$$

reads, according to Eqs. (5.7) and (8.65),

$$\langle\mathbf{P}(t)\rangle_{z,\alpha} = i\sqrt{\frac{m\hbar\omega}{2}}(\langle\mathbf{a}(t)^\dagger\rangle_{z,\alpha} - \langle\mathbf{a}(t)\rangle_{z,\alpha})$$

so that, in view of Eqs. (8.69) and (8.70), we have

$$\boxed{\langle \mathbf{P}(t)\rangle_{z,\alpha} = 2\alpha^{\circ}\sqrt{\frac{m\hbar\omega}{2}}\sin\omega t} \tag{8.80}$$

8.2.4.2 *Mean value of Q^2 and corresponding fluctuation* Now consider the following average value of the squared position operator

$$\langle \mathbf{Q}(t)^2\rangle_{z,\alpha} = \langle\{z(t),\alpha(t)\}|\mathbf{Q}^2|\{z(t),\alpha(t)\}\rangle$$

Then, using Eq. (8.57), and the fact that the squeezing and translation operators are unitary, it becomes

$$\langle \mathbf{Q}(t)^2\rangle_{z,\alpha} = \frac{\hbar}{2m\omega}\langle\{0\}|\mathbf{S}(z(t))^{-1}\mathbf{A}(\alpha(t))^{-1}((\mathbf{a}^{\dagger})^2+(\mathbf{a})^2+2\mathbf{a}^{\dagger}\mathbf{a}+1)\mathbf{A}(\alpha(t))\mathbf{S}(z(t))|\{0\}\rangle$$

or, due to Eq. (8.71) and its Hermitian conjugate, and also to Eq. (8.75)

$$\langle \mathbf{Q}(t)^2\rangle_{z,\alpha} = \frac{\hbar}{2m\omega}(\langle(\mathbf{a}^{\dagger}(t))^2\rangle_{z,\alpha} + \langle(\mathbf{a}(t))^2\rangle_{z,\alpha} + 2\langle(\mathbf{a}^{\dagger}\mathbf{a}(t))\rangle_{z,\alpha} + 1)$$

so that, with Eqs. (8.73), (8.74), and (8.77), we have

$$\begin{aligned}\langle \mathbf{Q}(t)^2\rangle_{z,\alpha} &= \frac{\hbar}{2m\omega}((\alpha(t)^{*2} + \alpha(t)^2) + 2\cos 2\omega t\sinh|z|\cosh|z| \\ &\quad + 2|\alpha(t)|^2 + 2\sinh^2|z| + 1)\end{aligned} \tag{8.81}$$

Next, using the trigonometric formulas

$$\cos 2\omega t = \cos^2\omega t - \sin^2\omega t \tag{8.82}$$

the product of hyperbolic sine and cosine functions leads to

$$\sinh|z|\cosh|z| = \frac{e^{|z|}-e^{-|z|}}{2}\frac{e^{|z|}+e^{-|z|}}{2} = \frac{e^{2|z|}-e^{-2|z|}}{4}$$

$$2\sinh(|z|)\cosh(|z|) = \sinh(2|z|) \tag{8.83}$$

$$2\sinh^2(|z|) = 2\left(\frac{e^{|z|}-e^{-|z|}}{2}\right)^2 = \frac{e^{2|z|}+e^{-2|z|}-2}{2}$$

$$2\sinh^2|z| = \cosh 2|z| - 1 \tag{8.84}$$

Then, Eq. (8.81) reduces to

$$\langle \mathbf{Q}(t)^2\rangle_{z,\alpha} = \frac{\hbar}{2m\omega}\{(\alpha^*(t)+\alpha(t))^2 + (\cos^2\omega t - \sin^2\omega t)\sinh 2|z| + \cosh 2|z|\}$$

which may be also written as

$$\begin{aligned}\langle \mathbf{Q}(t)^2\rangle_{z,\alpha} &= \frac{\hbar}{2m\omega}\{(\alpha^*(t)+\alpha(t))^2 + (\cos^2\omega t - \sin^2\omega t)\sinh 2|z| \\ &\quad + (\cos^2\omega t + \sin^2\omega t)\cosh 2|z|\}\end{aligned}$$

or

$$\langle \mathbf{Q}(t)^2 \rangle_{z,\alpha} = \frac{\hbar}{2m\omega} \left\{ (\alpha^*(t) + \alpha(t))^2 + (\cos^2 \omega t - \sin^2 \omega t) \frac{e^{2|z|} - e^{-2|z|}}{2} \right.$$
$$\left. + \ (\cos^2 \omega t + \sin^2 \omega t) \frac{e^{2|z|} + e^{-2|z|}}{2} \right\}$$

and thus

$$\boxed{\langle \mathbf{Q}(t)^2 \rangle_{z,\alpha} = \frac{\hbar}{2m\omega} \{ (\alpha^*(t) + \alpha(t))^2 + e^{2|z|} \cos^2 \omega t + e^{-2|z|} \sin^2 \omega t \}} \tag{8.85}$$

Hence, according to Eqs. (8.79) and (8.85), the fluctuation of the coordinate operator defined by

$$\Delta \mathbf{Q}_{z,\alpha}(t) = \sqrt{\langle \mathbf{Q}(t)^2 \rangle_{z,\alpha} - \langle \mathbf{Q}(t) \rangle_{z,\alpha}^2}$$

reads

$$\boxed{\Delta \mathbf{Q}_{z,\alpha}(t) = \sqrt{\frac{\hbar}{2m\omega}} \sqrt{e^{2|z|} \cos^2 \omega t + e^{-2|z|} \sin^2 \omega t}} \tag{8.86}$$

On the other hand, the following average value of the squared momentum operator

$$\langle \mathbf{P}(t)^2 \rangle_{z,\alpha} = \langle \{z(t), \alpha(t)\} | \mathbf{P}^2 | \{z(t), \alpha(t)\} \rangle \tag{8.87}$$

reads, in view of Eqs. (5.7), (8.57), (8.71), and (8.75),

$$\langle \mathbf{P}(t)^2 \rangle_{z,\alpha} = -\frac{\hbar m\omega}{2} \{ \langle (\mathbf{a}^\dagger(t))^2 \rangle_{z,\alpha} + \langle (\mathbf{a}(t))^2 \rangle_{z,\alpha} - (2\langle \mathbf{a}^\dagger \mathbf{a} \rangle_{z,\alpha} + 1) \}$$

or, using Eqs. (8.73), (8.74), and (8.77),

$$\langle \mathbf{P}(t)^2 \rangle_{z,\alpha} = -\frac{\hbar m\omega}{2} \{ (\alpha(t)^{*2} + \alpha(t)^2) + 2 \cos 2\omega t \sinh |z| \cosh |z|$$
$$- (2|\alpha(t)|^2 + 2 \sinh^2 |z| + 1) \}$$

which can be rearranged by the aid of Eqs. (8.82)–(8.84), according to

$$\boxed{\langle \mathbf{P}(t)^2 \rangle_{z,\alpha} = \frac{\hbar m\omega}{2} \{ (\alpha^*(t) - \alpha(t))^2 + e^{-2|z|} \cos^2 \omega t - e^{2|z|} \sin^2 \omega t \}} \tag{8.88}$$

Moreover, owing to Eqs. (8.80) and (8.88), the fluctuation of the momentum

$$\Delta \mathbf{P}_{z,\alpha}(t) = \sqrt{\langle \mathbf{P}(t)^2 \rangle_{z,\alpha} - \langle \mathbf{P}(t) \rangle_{z,\alpha}^2}$$

takes the form

$$\boxed{\Delta \mathbf{P}_{z,\alpha}(t) = \sqrt{\frac{\hbar m\omega}{2}} \sqrt{e^{-2|z|} \cos^2 \omega t + e^{2|z|} \sin^2 \omega t}} \tag{8.89}$$

so that multiplying this result by Eq. (8.86) leads to the following uncertainty relation:

$$\boxed{\Delta \mathbf{P}_{z,\alpha}(t) \Delta \mathbf{Q}_{z,\alpha}(t) = \frac{\hbar}{2} (e^{-2|z|} \cos^2 \omega t + e^{2|z|} \sin^2 \omega t)} \tag{8.90}$$

8.3 BOGOLIUBOV–VALATIN TRANSFORMATION

We shall end the present chapter with the products of Boson operators of the form $\mathbf{a}^\dagger\mathbf{a}^\dagger$ and $\mathbf{a}\,\mathbf{a}$, with the *Bogolioubov–Valatin* transformation allowing one to diagonalize the following Hamiltonian:

$$\mathbf{H} = \hbar\omega_1\mathbf{a}_1^\dagger\mathbf{a}_1 + \hbar\omega_2\mathbf{a}_2^\dagger\mathbf{a}_2 + \hbar\omega_{12}(\mathbf{a}_1^\dagger\mathbf{a}_2^\dagger + \mathbf{a}_1\mathbf{a}_2) \tag{8.91}$$

where ω_1 and ω_2 are the angular frequencies of the two oscillators, $\hbar\omega_{12}$ is the energetic coupling parameter, where $\mathbf{a}_1$, $\mathbf{a}_1^\dagger$ and $\mathbf{a}_2$, $\mathbf{a}_2^\dagger$ are ladder operators satisfying the commutation rule

$$[\mathbf{a}_i, \mathbf{a}_j^\dagger] = \delta_{ij} \tag{8.92}$$

We attempt to diagonalize this Hamiltonian so that it will read

$$\mathbf{H} = E^\circ + \hbar\Omega_1\mathbf{b}_1^\dagger\mathbf{b}_1 + \hbar\Omega_2\mathbf{b}_2^\dagger\mathbf{b}_2 \tag{8.93}$$

where $\mathbf{b}_i$ and $\mathbf{b}_i^\dagger$ are new Boson operators satisfying the commutation relation

$$[\mathbf{b}_i, \mathbf{b}_j^\dagger] = \delta_{ij} \tag{8.94}$$

and where the Ω_i are the angular frequencies of the decoupled oscillators, whereas E° is some reference energy.

The diagonalization of the Hamiltonian (8.91) into (8.93) may be performed through the linear Bogolioubov–Valatin transformation

$$\mathbf{b}_1 = \mathbf{a}_1\cosh\varphi + \mathbf{a}_2^\dagger\sinh\varphi \qquad \mathbf{b}_1^\dagger = \mathbf{a}_1^\dagger\cosh\varphi + \mathbf{a}_2\sinh\varphi \tag{8.95}$$

$$\mathbf{b}_2 = \mathbf{a}_1^\dagger\sinh\varphi - \mathbf{a}_2\cosh\varphi \qquad \mathbf{b}_2^\dagger = \mathbf{a}_1\sinh\varphi - \mathbf{a}_2^\dagger\cosh\varphi \tag{8.96}$$

In order to determine the transformation parameter φ, suppose that the commutator of $\mathbf{b}_1$ with the Hamiltonian (8.91) is equal to that of $\mathbf{b}_1^\dagger\mathbf{b}_1$ with the Hamiltonian (8.93). In this context, observe that, owing to Eq. (8.94), the commutator of $\mathbf{b}_1$ with the Hamiltonian (8.93) is simply

$$[\mathbf{b}_1, \mathbf{H}] = \hbar\Omega_1[\mathbf{b}_1, \mathbf{b}_1^\dagger\mathbf{b}_1] = \hbar\Omega_1\mathbf{b}_1$$

or, using the first equation appearing in (8.95), we have

$$\boxed{[\mathbf{b}_1, \mathbf{H}] = (\mathbf{a}_1\cosh\varphi + \mathbf{a}_2^\dagger\sinh\varphi)\hbar\Omega_1} \tag{8.97}$$

On the other hand, according to the first equation of (8.95), the commutator of $\mathbf{b}_1$ with the Hamiltonian (8.91), reads

$$[\mathbf{b}_1, \mathbf{H}] = [(\mathbf{a}_1\cosh\varphi + \mathbf{a}_2^\dagger\sinh\varphi), (\hbar\omega_1\mathbf{a}_1^\dagger\mathbf{a}_1 + \hbar\omega_2\mathbf{a}_2^\dagger\mathbf{a}_2 + \hbar\omega_{12}(\mathbf{a}_1^\dagger\mathbf{a}_2^\dagger + \mathbf{a}_1\mathbf{a}_2))]$$

or

$$\begin{aligned}[\mathbf{b}_1, \mathbf{H}] &= [\mathbf{a}_1, (\hbar\omega_1\mathbf{a}_1^\dagger\mathbf{a}_1 + \hbar\omega_2\mathbf{a}_2^\dagger\mathbf{a}_2 + \hbar\omega_{12}(\mathbf{a}_1^\dagger\mathbf{a}_2^\dagger + \mathbf{a}_1\mathbf{a}_2))]\cosh\varphi \\ &\quad + [\mathbf{a}_2^\dagger, (\hbar\omega_1\mathbf{a}_1^\dagger\mathbf{a}_1 + \hbar\omega_2\mathbf{a}_2^\dagger\mathbf{a}_2) + \hbar\omega_{12}(\mathbf{a}_1^\dagger\mathbf{a}_2^\dagger + \mathbf{a}_1\mathbf{a}_2)]\sinh\varphi\end{aligned} \tag{8.98}$$

Again, owing to the commutation rules (8.92), and applying Eqs. (5.15) and (5.16), it appears that

$$[\mathbf{a}_1, \mathbf{a}_1^\dagger \mathbf{a}_1] = \mathbf{a}_1 \qquad \text{and} \qquad [\mathbf{a}_1, \mathbf{a}_1^\dagger \mathbf{a}_2^\dagger] = \mathbf{a}_2^\dagger$$

$$[\mathbf{a}_2^\dagger, \mathbf{a}_2^\dagger \mathbf{a}_2] = \mathbf{a}_2^\dagger \qquad \text{and} \qquad [\mathbf{a}_2^\dagger, \mathbf{a}_1 \mathbf{a}_2] = -\mathbf{a}_1$$

Thereby, Eq. (8.98) transforms to

$$\boxed{[\mathbf{b}_1, \mathbf{H}] = \mathbf{a}_1(\hbar\omega_1 \cosh\varphi - \hbar\omega_{12}\sinh\varphi) + \mathbf{a}_2^\dagger(\hbar\omega_{12}\cosh\varphi - \hbar\omega_2 \sinh\varphi)} \tag{8.99}$$

Then, equating the two commutators (8.97) and (8.99)yields

$$\hbar\Omega_1(\mathbf{a}_1 \cosh\varphi + \mathbf{a}_2^\dagger \sinh\varphi) = \mathbf{a}_1(\hbar\omega_1 \cosh\varphi - \hbar\omega_{12}\sinh\varphi)$$
$$+ \mathbf{a}_2^\dagger(\hbar\omega_{12}\cosh\varphi - \hbar\omega_2 \sinh\varphi)$$

or, after identification, since the Boson operators cannot be zero

$$(\Omega_1 - \omega_1)\cosh\varphi + \omega_{12}\sinh\varphi = 0 \tag{8.100}$$

$$-\omega_{12}\cosh\varphi + (\Omega_1 + \omega_2)\sinh\varphi = 0 \tag{8.101}$$

Now, since the coefficients $\cosh\varphi$ and $\sinh\varphi$ are different from zero, the set of equations (8.100) and (8.101) is satisfied if

$$\begin{vmatrix} (\Omega_1 - \omega_1) & \omega_{12} \\ -\omega_{12} & (\Omega_1 + \omega_2) \end{vmatrix} = 0$$

or, after expansion of the determinant,

$$\Omega_1^2 - \Omega_1(\omega_1 - \omega_2) + (\omega_{12}^2 - \omega_1\omega_2) = 0$$

In the two solutions of this equation

$$\Omega_1 = \tfrac{1}{2}((\omega_1 - \omega_2) \pm \sqrt{(\omega_1 + \omega_2)^2 - 4\omega_{12}^2})$$

the one that must be selected is that allowing Ω_1 in Eq. (8.91) to be equal to ω_1 when the coupling ω_{12} is zero, that is,

$$\boxed{\Omega_1 = \tfrac{1}{2}((\omega_1 - \omega_2) + \sqrt{(\omega_1 + \omega_2)^2 - 4\omega_{12}^2})} \tag{8.102}$$

Moreover, according to Eq. (8.100), the ratio of the coefficients $\sinh\varphi$ and $\cosh\varphi$ reads

$$\frac{\sinh\varphi}{\cosh\varphi} = \tanh\varphi = \frac{\omega_1 - \Omega_1}{\omega_{12}}$$

In like manner for the commutator of $\mathbf{b}_2$ with the Hamiltonian $\mathbf{H}$ given, using Eq. (8.91) or (8.93), and with the help of the first equation of (8.96), one obtains for the second angular frequency appearing in Eq. (8.93)

$$\boxed{\Omega_2 = \tfrac{1}{2}((\omega_2 - \omega_1) + \sqrt{(\omega_1 + \omega_2)^2 - 4\omega_{12}^2})} \tag{8.103}$$

Finally, it is possible to get the expression of E° appearing in Eq. (8.93) by transforming this equation using Eqs. (8.95) and (8.96), that is,

$$\mathbf{H} = E^\circ + \hbar\Omega_1((\mathbf{a}_1^\dagger \cosh\varphi + \mathbf{a}_2 \sinh\varphi)(\mathbf{a}_1 \cosh\varphi + \mathbf{a}_2^\dagger \sinh\varphi))$$
$$+ \hbar\Omega_2((\mathbf{a}_1 \sinh\varphi - \mathbf{a}_2^\dagger \cosh\varphi)(\mathbf{a}_1^\dagger \sinh\varphi - \mathbf{a}_2 \cosh\varphi))$$

or, due to the commutation rules (8.92), one obtains

$$\mathbf{H} = E^\circ + \hbar\Omega_1(\mathbf{a}_1^\dagger\mathbf{a}_1 \cosh^2\varphi + (\mathbf{a}_2^\dagger\mathbf{a}_2 + 1)\sinh^2\varphi + (\mathbf{a}_1^\dagger\mathbf{a}_2^\dagger + \mathbf{a}_1\mathbf{a}_2)\sinh\varphi\cosh\varphi)$$
$$+ \hbar\Omega_2((\mathbf{a}_1^\dagger\mathbf{a}_1 + 1)\sinh^2\varphi + \mathbf{a}_2^\dagger\mathbf{a}_2 \cosh^2\varphi - (\mathbf{a}_1^\dagger\mathbf{a}_2^\dagger + \mathbf{a}_1\mathbf{a}_2)\sinh\varphi\cosh\varphi)$$

a result that simplifies to

$$\mathbf{H} = E^\circ + (\hbar\Omega_1 + \hbar\Omega_2)\sinh^2\varphi + \hbar\Omega_1(\cosh^2\varphi + \sinh^2\varphi)\mathbf{a}_1^\dagger\mathbf{a}_1$$
$$+ \hbar\Omega_2(\cosh^2\varphi + \sinh^2\varphi)\mathbf{a}_2^\dagger\mathbf{a}_2$$
$$+ (\hbar\Omega_1 - \hbar\Omega_2)\sinh\varphi\cosh\varphi(\mathbf{a}_1^\dagger\mathbf{a}_2^\dagger + \mathbf{a}_1\mathbf{a}_2)$$

Finally, equating this last expression to that of $\mathbf{H}$ given by Eq. (8.91), leads for the coefficients of $\mathbf{a}_1^\dagger\mathbf{a}_1$, $\mathbf{a}_2^\dagger\mathbf{a}_2$, $\mathbf{a}_1^\dagger\mathbf{a}_2^\dagger$, and $\mathbf{a}_1\mathbf{a}_2$ the following results:

$$\hbar\omega_k = \hbar\Omega_k(\cosh^2\varphi + \sinh^2\varphi) \quad \text{with} \quad k = 1, 2$$

$$\hbar\omega_{12} = (\hbar\Omega_1 - \hbar\Omega_2)\sinh\varphi\cosh\varphi$$

and also to the conclusion that

$$\boxed{E^\circ + (\hbar\Omega_1 + \hbar\Omega_2)\sinh^2\varphi = 0} \tag{8.104}$$

a result allowing one to get E° appearing in the diagonal Hamiltonian (8.93).

8.4 CONCLUSION

This chapter was devoted to various questions related to the notion of phase for quantum oscillators, which is not without connection with the squeezed states susceptible to be obtained through the action of squeezing operators having the structure of translation operators in which the ladder operators have been replaced by their squared expression. It ended by the Bogoliubov–Valatin transformation allowing one to diagonalize the Hamiltonian of coupled oscillators via terms quadratic in the raising and lowering operators. Even though many results will not be used later, they are nevertheless important in many studies lying beyond the scope of the present work particularly in quantum optics.

BIBLIOGRAPHY

A. S. Davydov. *Quantum Mechanics*, 2nd ed. Pergamon Press: Oxford, New York, 1976.
J. R. Klauder and B. Skagerstam. *Coherent States*. World Scientific: Singapore, 1985.
R. Loudon. *The Quantum Theory of Light*, 3rd ed. Oxford University Press: Oxford, 2000.
W. H. Louisell. *Quantum Statistical Properties of Radiation*. Wiley: New York, 1973.

PART III

ANHARMONICITY

CHAPTER 9

ANHARMONIC OSCILLATORS

INTRODUCTION

In a previous chapter, we studied the general properties of single-degree-of-freedom quantum harmonic oscillators for which it was possible to solve the Hamiltonian eigenvalue equation and thus to get the energy levels. The calculation was also made for the driven harmonic oscillator by diagonalization of its Hamiltonian through canonical transformations, using the translation operators. Then, it was seen that it is possible to reproduce numerically the exact Hamiltonian eigenvalues obtained by this last procedure, after diagonalization of the truncated matrix representation of the driven Hamiltonian in the basis of the eigenkets of the harmonic oscillator Hamiltonian.

The aim of the present chapter is to find the energy levels of various anharmonic oscillators of interest for which it is not possible to diagonalize the Hamiltonian, using this numerical procedure for the driven harmonic oscillator. We shall first study the energy levels of oscillators for which the harmonic potential is perturbed by a cubic term. Second, we shall consider oscillators in a Morse potential, a physical model that applies to the vibrational behavior of diatomic molecules. Finally, we shall consider particles in a double-well potential, a model that applies to the inversion of ammonia for which tunneling may proceed through the barrier potential between the two wells. However, before commencing these studies, it may be of interest to find how quantum mechanics predicts the form of the anharmonic potentials in which the nuclei of diatomic molecules move.

9.1 MODEL FOR DIATOMIC MOLECULE POTENTIALS

For this purpose, we shall introduce in this section a very crude model applying to the H_2^+ molecular ion, which is the most simple diatomic molecule, since it involves only one electron and two protons. We shall attempt to simplify the notations, using the centimeter–gram–second (cgs) system.

The average kinetic energy $\langle T \rangle$ of the electron belonging to the molecular ion H_2^+ may be approximated by that of the *1D particle-in-a-box* model for which,

Quantum Oscillators, First Edition. Olivier Henri-Rousseau and Paul Blaise.
© 2011 John Wiley & Sons, Inc. Published 2011 by John Wiley & Sons, Inc.

according to Eq. (4.28), the 1D energy levels are given by

$$E_{n_x} = n_x^2 \frac{h^2}{8ma_x^2} \quad \text{with} \quad n_x = 1, 2, \ldots$$

where m is the mass of the particle and a_x the dimension of the 1D box. We shall assume that this kinetic energy is given by the 1D ground state of this model, that is,

$$\langle T \rangle = \frac{h^2}{8ma_x^2}$$

Moreover, in the present situation, the dimension a_x of the box may be viewed roughly as the distance R between the two protons.

For the electronic ground state, this leads us to write the average kinetic energy $\langle T \rangle$:

$$\langle T \rangle = \frac{A_T}{R^2} \quad \text{with} \quad A_T = \frac{h^2}{8m} \tag{9.1}$$

where m is now the mass of the single electron.

On the other hand, the average Coulombic potential energy $\langle V \rangle$ of the ion is the sum of the average attraction energies $\langle V_1 \rangle$ and $\langle V_2 \rangle$ between the single electron and the two protons and of the repulsion energy between the two protons, which is inversely proportional to R, that is,

$$\langle V \rangle = \langle V_1 \rangle + \langle V_2 \rangle + \frac{e^2}{R} \tag{9.2}$$

where e is the elementary electrical charge. The two average attraction energies $\langle V_1 \rangle$ and $\langle V_2 \rangle$ must be the same for symmetry reasons. For each of them, one may roughly assume that they are proportional to the inverse of the average distance between the electron and the nuclei, and more crudely that this average distance is proportional to half the distance R, leading us to write

$$\langle V_1 \rangle = \langle V_2 \rangle = -\frac{e^2}{R/2}$$

Hence, the average potential energy (9.2) becomes

$$\langle V \rangle = -4\frac{e^2}{R} + \frac{e^2}{R}$$

or

$$\langle V \rangle = -\frac{A_V}{R} \quad \text{with} \quad A_V = 3e^2 \tag{9.3}$$

Then, for this crude linear model, the electronic energy $\langle E \rangle$ of the molecular ion is simply the sum of $\langle T \rangle$ and $\langle V \rangle$, given, respectively, by Eqs. (9.1) and (9.3):

$$\langle E \rangle = \left\{ \frac{A_T}{R^2} - \frac{A_V}{R} \right\} \tag{9.4}$$

The evolution of E with respect to R given by Eq. (9.4) is reproduced in Fig. 9.1.

Inspection of this figure shows, as expected, a minimum of the energy function (9.4), which appears to be the result of a compromise between the positive

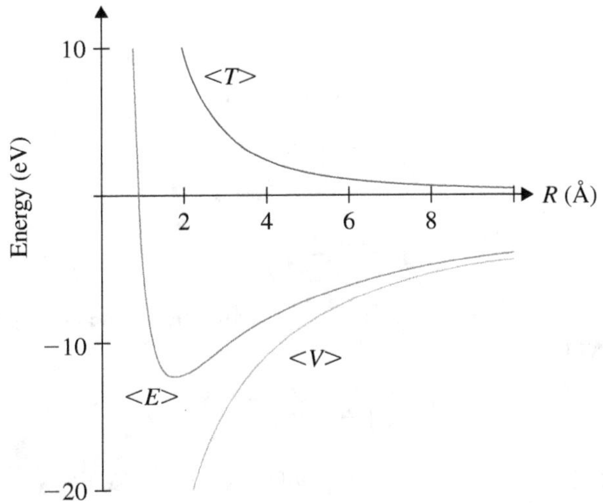

Figure 9.1 Total energy of the molecular ion H_2^+ as a compromise between a repulsive electronic kinetic energy and an attractive potential energy. Energies are in electron volts and distances in Ångström.

kinetic energy, which is decreasing in R, and the negative potential energy, which is correlatively increasing.

By Taylor expansion of the energy $\langle E\rangle$, denoted $E(R)$ near its minimum, corresponding to the internuclear distance $R = R^\circ$ of the energy curve yields

$$E(R) = (E)_{R^\circ} + \left(\frac{\partial E}{\partial R}\right)_{R=R^\circ}(R - R^\circ) + \frac{1}{2!}\left(\frac{\partial^2 E}{\partial R^2}\right)_{R=R^\circ}(R - R^\circ)^2$$
$$+ \frac{1}{3!}\left(\frac{\partial^3 E}{\partial R^3}\right)_{R=R^\circ}(R - R^\circ)^3 + \frac{1}{4!}\left(\frac{\partial^4 E}{\partial R^4}\right)_{R=R^\circ}(R - R^\circ)^4 + \cdots \quad (9.5)$$

Of course, at the minimum of the energy function, the first derivative is zero, that is,

$$\left(\frac{\partial E}{\partial R}\right)_{R_e} = 0 \quad (9.6)$$

Hence, taking as energy reference

$$(E)_{R_e} = 0 \quad \text{at} \quad R = R_e$$

and near the minimum, the Taylor expansion (9.5) reads

$$E(R) = \frac{1}{2!}k_e(R - R_e)^2 + \frac{1}{3!}g_e(R - R_e)^3 + \frac{1}{4!}j_e(R - R_e)^4 + \cdots \quad (9.7)$$

with, respectively,

$$k_e = \left(\frac{\partial^2 E}{\partial R^2}\right)_{R=R_e} \tag{9.8}$$

$$g_e = \left(\frac{\partial^3 E}{\partial R^3}\right)_{R=R_e} \tag{9.9}$$

$$j_e = \left(\frac{\partial^4 E}{\partial R^4}\right)_{R=R_e} \tag{9.10}$$

In order to get the equilibrium distance R_e, we differentiate the energy function $E(R)$ (9.4) with respect to R:

$$\left(\frac{\partial E}{\partial R}\right) = -\frac{2A_T}{R^3} + \frac{A_V}{R^2} \tag{9.11}$$

Thus, for the equilibrium distance R_e for which Eq. (9.6) is verified, Eq. (9.11) yields

$$-\frac{2A_T}{R_e^3} + \frac{A_V}{R_e^2} = 0$$

so that the equilibrium distance appears to be given by

$$R_e = 2\frac{A_T}{A_V} \tag{9.12}$$

or, owing to Eqs. (9.1) and (9.3),

$$R_e = \left(\frac{\hbar^2}{12me^2}\right) \tag{9.13}$$

and so, inserting numerical values,

$$R_e = 1.73 \times 10^{-8}\,\text{cm} = 1.73\,\text{Å}$$

Now, due to Eq. (9.13) and owing to Eqs. (9.8) and (9.11), the constant k_e reads

$$k_e = \frac{6A_T}{R_e^4} - \frac{2A_V}{R_e^3}$$

or, using Eq. (9.12) for R_e,

$$k_e = 6\left\{A_T\left(\frac{A_V}{2A_T}\right)^4 - A_V\left(\frac{A_V}{2A_T}\right)^3\right\}$$

and, after rearranging and simplifying

$$k_e = \frac{1}{8}\left(\frac{(A_V)^4}{(A_T)^3}\right) \tag{9.14}$$

In a similar way, one may obtain for the constants g_e and j_e defined by Eq. (9.9) and (9.10) the following result:

$$g_e = -\frac{3}{8}\left(\frac{(A_V)^5}{(A_T)^4}\right) \tag{9.15}$$

$$j_e = \frac{9}{8}\left(\frac{(A_V)^6}{(A_T)^5}\right) \tag{9.16}$$

Now, introduce the well-known dimensionless *fine structure constant* α and the *Compton wavelength* λ_c defined, respectively, in the CGS system by

$$\alpha = \frac{e^2}{\hbar c} = 2\pi \frac{e^2}{hc} = \frac{1}{137} \tag{9.17}$$

$$\lambda_c = \left(\frac{h}{mc}\right) = 2.43 \times 10^{-10}\,\text{cm} \quad \text{with } c\text{, the light velocity} \tag{9.18}$$

so that

$$\left(\frac{\lambda_c}{\alpha}\right) = \frac{h}{mc}\frac{\hbar c}{2\pi e^2} = \frac{h^2}{2\pi m e^2} \tag{9.19}$$

Then, due to Eq. (9.19), the equilibrium distance (9.13) may be approximated by

$$R_e = \frac{\pi}{6}\frac{\lambda_c}{\alpha} \backsimeq 0.5\left(\frac{\lambda_c}{\alpha}\right) \tag{9.20}$$

Proceeding in a similar way for k_e, g_e, and j_e, defined by Eqs. (9.14)–(9.16), we have

$$k_e = 3.33\alpha^2(mc^2)\left(\frac{\alpha}{\lambda_c}\right)^2 \backsimeq 3\alpha^2(mc^2)\left(\frac{\alpha}{\lambda_c}\right)^2 > 0 \tag{9.21}$$

$$g_e = 38.11\alpha^2(mc^2)\left(\frac{\alpha}{\lambda_c}\right)^3 \backsimeq -3 \times 12\alpha^2(mc^2)\left(\frac{\alpha}{\lambda_c}\right)^3 < 0 \tag{9.22}$$

$$j_e = 436.78\alpha^2(mc^2)\left(\frac{\alpha}{\lambda_c}\right)^4 \backsimeq 3 \times (12)^2\alpha^2(mc^2)\left(\frac{\alpha}{\lambda_c}\right)^4 < 0 \tag{9.23}$$

Hence, the Taylor expansion of the energy (9.7) yields

$$E(R) \backsimeq 3\alpha^2(mc^2)\left\{\frac{1}{2}\left(\frac{\alpha}{\lambda_c}\right)^2(R-R_e)^2 - \frac{12}{3!}\left(\frac{\alpha}{\lambda_c}\right)^3(R-R_e)^3 + \frac{(12)^2}{4!}\left(\frac{\alpha}{\lambda_c}\right)^4(R-R_e)^4\right\} \tag{9.24}$$

Again, owing to Eq. (9.20), which reads,

$$\frac{\alpha}{\lambda_c} = \frac{1}{2R_e}$$

and after making the approximation $\frac{5}{4} \backsimeq 1$, the expansion (9.24) may be approached by

$$E(R) \backsimeq 3\alpha^2(mc^2) \times \left(\frac{1}{2}\right)^3\left\{\left(\frac{(R-R_e)}{R_e}\right)^2 - 2\left(\frac{(R-R_e)}{R_e}\right)^3 + 3\left(\frac{(R-R_e)}{R_e}\right)^4\right\} \tag{9.25}$$

Next, near the equilibrium distance R_e, where

$$\frac{(R-R_e)}{R_e} << 1$$

then it reads

$$3\left(\frac{(R-R_e)}{R_e}\right)^4 << 2\left(\frac{(R-R_e)}{R_e}\right)^3 << \left(\frac{(R-R_e)}{R_e}\right)^2$$

so that the Taylor expansion (9.7) may be approximated by the first quadratic term, that is, by

$$E(R) = \tfrac{1}{2}k_e(R-R_e)^2 \tag{9.26}$$

from which the following force may be obtained through:

$$F(R) = -\left(\frac{\partial E(R)}{\partial R}\right) \backsimeq -k_e(R-R_e) \tag{9.27}$$

Equation (9.27) now allows one to write the classical dynamics equation

$$M\left(\frac{d^2(R-R_e)}{dt^2}\right) = -k_e(R-R_e)$$

where M is the reduced mass of the two protons. The solution of the latter equation is, of course,

$$(R(t)-R_e) = (R(0)-R_e)\cos(\omega t+\varphi)$$

where φ is some phase, and ω is an angular frequency given by

$$\boxed{\omega = \sqrt{\frac{k_e}{M}}} \tag{9.28}$$

Hence, due to this expression of ω, the electronic energy (9.26) of the molecular ion may be written

$$E(R) = \tfrac{1}{2}M\omega^2(R-R_e)^2 \tag{9.29}$$

This electronic energy $E(R)$ may be viewed as a potential in which the two protons of the molecular ion are allowed to oscillate with angular frequency ω. Then, according to quantum mechanics, one has to write in place of $E(R)$ given by Eq. (9.29) the following potential operator $\mathbf{V}(\mathbf{Q})$:

$$\mathbf{V}(\mathbf{Q}) = \tfrac{1}{2}M\omega^2\mathbf{Q}^2$$

where $\mathbf{Q}$ is the quantum position operator corresponding to the classical elongation Q given by

$$Q = R - R_e$$

Thus, in view of Eq. (9.21), the angular frequency (9.28) takes the form

$$\omega \backsimeq \alpha^2\left(\frac{c}{\lambda_c}\right)\sqrt{3\frac{m}{M}} \tag{9.30}$$

Moreover, observe that the ratio of the electron and proton masses m/M is about 10^3 so that, using Eqs. (9.17) and (9.18) for α and λ_c and for c its numerical value $3 \times 10^{10}\,\text{cm·s}^{-1}$, this angular frequency appears to be

$$\omega \backsimeq 3.6 \times 10^{14}\,\text{rad·s}^{-1}$$

Again, passing to vibrational frequency ν, one obtains a value that is in the infrared area:

$$\nu = \frac{\omega}{2\pi} \backsimeq 5 \times 10^{13}\ \text{Hz} \tag{9.31}$$

Now, return to the expansion of the potential energy (9.25). If one limits oneself to the quadratic term, the potential becomes harmonic and that used in the previous chapters dealing with quantum harmonic oscillator. Now, if one likes to go further in the potential energy expansion, the largest correction to incorporate is cubic in the elongation and thus in the **Q** operator. The next section is devoted to treating quantum anharmonic oscillators involving such potentials.

9.2 HARMONIC OSCILLATOR PERTURBED BY A Q^3 POTENTIAL

For this purpose, consider the following Hamiltonian:

$$\mathbf{H} = \frac{\mathbf{P}^2}{2M} + \mathbf{V}_{\text{Cub}}(\mathbf{Q}) \tag{9.32}$$

Here **Q**, **P**, and M are, respectively, the position and momentum operators and the reduced mass of the oscillator, whereas $\mathbf{V}_{\text{Cub}}(\mathbf{Q})$ is the cubic anharmonic potential

$$\mathbf{V}_{\text{Cub}}(\mathbf{Q}) = \tfrac{1}{2}M\omega^2\mathbf{Q}^2 + \xi\mathbf{Q}^3 \quad \text{with} \quad \xi = \frac{g_e}{3!} < 0 \tag{9.33}$$

where ξ is the dimensionless anharmonic cubic parameter related to the g_e coefficient of the expansion (9.7), which is negative, according to Eq. (9.22). Now consider the raising and lowering operators using Eqs. (5.6) and (5.7), that is,

$$\mathbf{Q} = \sqrt{\frac{\hbar}{2M\omega}}(\mathbf{a}^\dagger + \mathbf{a}) \tag{9.34}$$

$$\mathbf{P} = i\sqrt{\frac{M\hbar\omega}{2}}(\mathbf{a}^\dagger - \mathbf{a}) \tag{9.35}$$

Then, the Hamiltonian (9.32) becomes

$$\mathbf{H} = \hbar\omega\left\{\left(\mathbf{a}^\dagger\,\mathbf{a} + \tfrac{1}{2}\right) + \eta(\mathbf{a}^\dagger + \mathbf{a})^3\right\} \tag{9.36}$$

with

$$\eta = \xi\left(\frac{\hbar}{2M\omega}\right)^{3/2} < 0 \tag{9.37}$$

9.2.1 Matrix representation of the anharmonic Hamiltonian in the basis of the harmonic Hamiltonian eigenkets

Now, consider the basis in which $\mathbf{a}^\dagger\mathbf{a}$ is diagonal, that is,

$$\mathbf{a}^\dagger\mathbf{a}|\{n\}\rangle = n|\{n\}\rangle \quad \text{with} \quad \langle\{m\}|\{n\}\rangle = \delta_{mn} \tag{9.38}$$

In this basis, the matrix representation of the full Hamiltonian is

$$\langle\{m\}|\mathbf{H}|\{n\}\rangle = \hbar\omega\langle\{m\}|\left\{\left(\mathbf{a}^\dagger\,\mathbf{a} + \tfrac{1}{2}\right) + \eta(\mathbf{a}^\dagger + \mathbf{a})^3\right\}|\{n\}\rangle$$

Then, in view of the orthonormality properties appearing in (9.38), it reduces to

$$\langle\{m\}|\mathbf{H}|\{n\}\rangle = \hbar\omega\left\{\left(n + \tfrac{1}{2}\right)\delta_{mn} + \eta\langle\{m\}|(\mathbf{a}^\dagger + \mathbf{a})^3|\{n\}\rangle\right\} \tag{9.39}$$

Again, the matrix elements of the cubic perturbation terms appearing in Eq. (9.39) are

$$\begin{aligned}
&\langle\{m\}|(\mathbf{a}^\dagger + \mathbf{a})^3|\{n\}\rangle \\
&\quad = \langle\{m\}|\mathbf{a}^\dagger\mathbf{a}^\dagger\mathbf{a}^\dagger|\{n\}\rangle + \langle\{m\}|\mathbf{a}^\dagger\mathbf{a}\mathbf{a}|\{n\}\rangle + \langle\{m\}|\mathbf{a}^\dagger\mathbf{a}^\dagger\mathbf{a}|\{n\}\rangle \\
&\qquad + \langle\{m\}|\mathbf{a}^\dagger\mathbf{a}\mathbf{a}^\dagger|\{n\}\rangle + \langle\{m\}|\mathbf{a}\mathbf{a}^\dagger\mathbf{a}^\dagger|\{n\}\rangle \\
&\qquad + \langle\{m\}|\mathbf{a}\mathbf{a}\mathbf{a}|\{n\}\rangle + \langle\{m\}|\mathbf{a}\mathbf{a}^\dagger\mathbf{a}|\{n\}\rangle + \langle\{m\}|\mathbf{a}\mathbf{a}\mathbf{a}^\dagger|\{n\}\rangle
\end{aligned} \tag{9.40}$$

Next, keeping in mind Eqs. (5.53) and (5.63), that is,

$$\mathbf{a}^\dagger|\{n\}\rangle = \sqrt{n+1}|\{n+1\}\rangle \qquad \text{and} \qquad \mathbf{a}|\{n\}\rangle = \sqrt{n}|\{n-1\}\rangle$$

and after using the orthornormality properties (9.38), the following expressions are obtained for the various matrix elements (9.40):

$$\langle\{m\}|\mathbf{a}^\dagger\mathbf{a}^\dagger\mathbf{a}^\dagger|\{n\}\rangle = \sqrt{(n+1)(n+2)(n+3)}\delta_{m,n+3} \tag{9.41}$$

$$\langle\{m\}|\mathbf{a}^\dagger\mathbf{a}\mathbf{a}|\{n\}\rangle = (n-1)\sqrt{n}\delta_{m,n-1} \tag{9.42}$$

$$\langle\{m\}|\mathbf{a}^\dagger\mathbf{a}^\dagger\mathbf{a}|\{n\}\rangle = n\sqrt{n+1}\delta_{m,n+1} \tag{9.43}$$

$$\langle\{m\}|\mathbf{a}^\dagger\mathbf{a}\mathbf{a}^\dagger|\{n\}\rangle = (n+1)\sqrt{n+1}\delta_{m,n+1} \tag{9.44}$$

$$\langle\{m\}|\mathbf{a}\mathbf{a}^\dagger\mathbf{a}^\dagger|\{n\}\rangle = \sqrt{n+1}(n+2)\delta_{m,n+1} \tag{9.45}$$

$$\langle\{m\}|\mathbf{a}\mathbf{a}\mathbf{a}|\{n\}\rangle = \sqrt{n}\sqrt{(n-1)(n-2)}\delta_{m,n-3} \tag{9.46}$$

$$\langle\{m\}|\mathbf{a}\mathbf{a}^\dagger\mathbf{a}|\{n\}\rangle = n\sqrt{n}\delta_{m,n-1} \tag{9.47}$$

$$\langle\{m\}|\mathbf{a}\mathbf{a}\mathbf{a}^\dagger|\{n\}\rangle = (n+1)\sqrt{n}\delta_{m,n-1} \tag{9.48}$$

9.2.2 Diagonalization of the truncated matrix representations of the Hamiltonian

The matrix elements involved in the matrix representation (9.39) of the Hamiltonian may be computed using Eq. (9.40) by the aid of Eqs. (9.41)–(9.48). This may be accomplished by starting from the ground state $|\{0\}\rangle$ and increasing progressively the quantum number associated to the ket $|\{n\}\rangle$. Since the basis appearing in Eq. (9.38) is infinite, the matrix representation must be also infinite. Thus, in order to be numerically diagonalized, the matrix representation (9.39) must be truncated after some value n° of the quantum number n characterizing $|\{n\}\rangle$. Then, we get a finite square $(n^\circ + 1) \times (n^\circ + 1)$ Hamiltonian matrix in the basis $\{|\{n\}\rangle\}$ expressed in terms of $\hbar\omega$

and η. For example, when $\eta = 0.017$ and for a truncation corresponding to $n° = 9$, the matrix representation reads as follows:

$$\boxed{\frac{\mathbf{H}}{\hbar\omega}} =$$

0.500	−0.051		−0.042						
−0.051	**1.500**	−0.144		−0.083					
	−0.144	**2.500**	−0.265		−0.132				
−0.042		−0.265	**3.500**	−0.408		−0.186			
	−0.083		−0.408	**4.500**	−0.570		−0.246		
		−0.132		−0.570	**5.500**	−0.750		−0.312	
			−0.186		−0.750	**6.500**	−0.945		−0.382
				−0.246		−0.945	**7.500**	−1.154	
					−10.312		−1.154	**8.500**	−1.377
						−0.382		−1.377	**9.500**

(9.49)

A truncated matrix such as (9.49) may be diagonalized by standard procedures leading to approximate numerical solutions of the eigenvalue equation

$$\mathbf{H}|\Psi_k(n°)\rangle = E_k(n°)|\Psi_k(n°)\rangle \tag{9.50}$$

Here the $E_k(n°)$ are the $n°$-dependent approximate eigenvalues of (9.49), whereas the $|\Psi_k(n°)\rangle$ are the corresponding $n°$-dependent approximate eigenvectors, given by

$$|\Psi_k(n°)\rangle = \sum_n C_{nk}(n°)|\{n\}\rangle \tag{9.51}$$

where the $C_{nk}(n°)$ are components of the orthogonal matrix allowing one to diagonalize the matrix (9.49) of dimension $n°$. Of course, the eigenvalues $E_k(n°)$ and the eigenvectors $|\Psi_k(n°)\rangle$, which are, respectively, the approximate eigenvalues and eigenkets of the Hamiltonian (9.36), may be assumed to change with the dimension $n°$ of the truncated matrix representation of this Hamiltonian in such a way as to stabilize themselves as for the driven harmonic oscillator.

Recall that we have previously studied this driven harmonic oscillator by two different methods. In the first one the Hamiltonian operator was diagonalized using a canonical transformation so that its eigenvalues, that is, the energy levels and the corresponding eigenkets were obtained exactly. In the last one, a matrix representation of the Hamiltonian was performed in the basis of the eigenkets of the harmonic Hamiltonian, and then the truncated matrices of increasing dimension $n°$ were diagonalized, leading to approximate energy levels that progressively decrease to converge toward the exact eigenvalues obtained by the first method, when $n°$ was progressively increased.

Figure 9.2 shows that, as for the driven harmonic oscillator, the approximate energy levels $E_k(n°)$ obtained by diagonalization of the matrix representation of the full Hamiltonian (for $\eta = -0.017$) stabilize indeed when the dimension $n°$ of the truncated matrix is progressively increased.

Hence, the stabilized energy levels may be considered as satisfactorily fitting the exact energy levels.

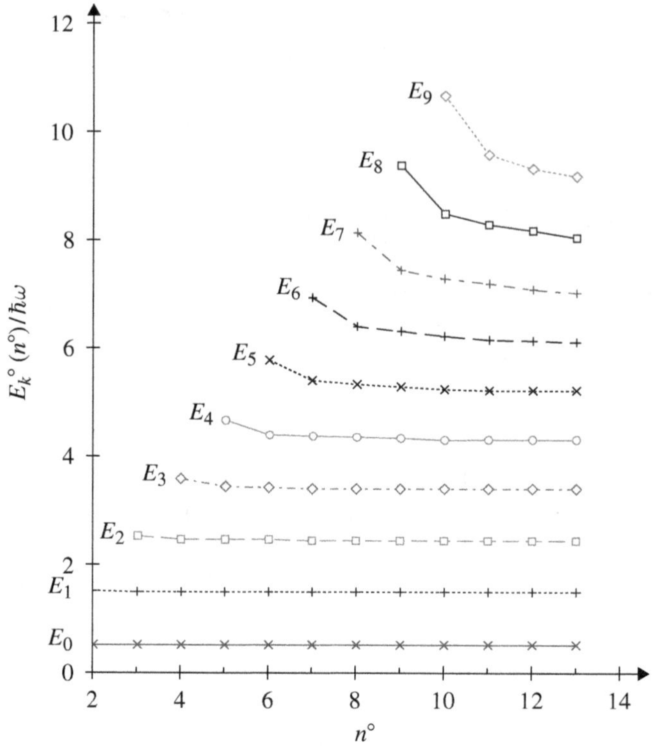

Figure 9.2 Progressive stabilization of the eigenvalues appearing in Eq. (9.50) with the dimension n° of the truncated matrix representation ($\eta = -0.017$). (See color insert.)

9.2.3 Accuracy criterion of the approximate method using the virial theorem

It has been shown above that the virial theorem allows one to write the following expression for the average value of the kinetic operator $\mathbf{T}$:

$$2\langle\Psi_k(n^{\circ})|\mathbf{T}|\Psi_k(n^{\circ})\rangle = \langle\Psi_k(n^{\circ})|\mathbf{Q}\left(\frac{\partial\mathbf{V}(\mathbf{Q})}{\partial\mathbf{Q}}\right)|\Psi_k(n^{\circ})\rangle \tag{9.52}$$

where $\mathbf{Q}$ is the coordinate operator and $\mathbf{V}(\mathbf{Q})$ the potential operator. This expression holds if the averages of the operators are performed over a stationary state $|\Psi_k\rangle$. In our study of the present anharmonic oscillator, we are interested in the eigenvalues and the corresponding eigenvectors of the Hamiltonian. Actually, these eigenvectors are stationary states, so that the average values performed on them ought to satisfy Eq. (9.52), if they were exact. But they are only approximate so that it may be of interest to show how Eq. (9.52) is approximately satisfied.

Keeping in mind Eq. (9.33), that is,

$$\mathbf{V}_{\mathsf{Cub}}(\mathbf{Q}) = \tfrac{1}{2}M\omega^2\mathbf{Q}^2 + \xi\mathbf{Q}^3 \tag{9.53}$$

the partial derivative of the potential appearing in Eq. (9.52) yields

$$\left(\frac{\partial \mathbf{V}_{\text{Cub}}(\mathbf{Q})}{\partial \mathbf{Q}}\right) = M\omega^2\mathbf{Q} + 3\xi\mathbf{Q}^2$$

and thus

$$\mathbf{Q}\left(\frac{\partial \mathbf{V}_{\text{Cub}}(\mathbf{Q})}{\partial \mathbf{Q}}\right) = M\omega^2\mathbf{Q}^2 + 2\xi\mathbf{Q}^3 + \xi\mathbf{Q}^3$$

or

$$\mathbf{Q}\left(\frac{\partial \mathbf{V}_{\text{Cub}}(\mathbf{Q})}{\partial \mathbf{Q}}\right) = 2\mathbf{V}_{\text{Cub}}(\mathbf{Q}) + \xi\mathbf{Q}^3$$

Hence, Eq. (9.52) reads

$$\langle\Psi_k(n^\circ)|\mathbf{T}|\Psi_k(n^\circ)\rangle = \langle\Psi_k(n^\circ)|\mathbf{V}_{\text{Cub}}(\mathbf{Q})|\Psi_k(n^\circ)\rangle + \tfrac{1}{2}\xi\langle\Psi_k(n^\circ)|\mathbf{Q}^3|\Psi_k(n^\circ)\rangle \quad (9.54)$$

Moreover, the corresponding average value of the Hamiltonian (9.32), that is,

$$\langle\Psi_k(n^\circ)|\mathbf{H}|\Psi_k(n^\circ)\rangle = \langle\mathbf{T}\rangle_{\Psi_k} + \langle\Psi_k(n^\circ)|\mathbf{V}_{\text{Cub}}(\mathbf{Q})|\Psi_k(n^\circ)\rangle$$

appears, due to Eq. (9.54), to be given by

$$\langle\Psi_k(n^\circ)|\mathbf{H}|\Psi_k(n^\circ)\rangle = 2\langle\Psi_k(n^\circ)|\mathbf{V}_{\text{Cub}}(\mathbf{Q})|\Psi_k(n^\circ)\rangle + \tfrac{1}{2}\xi\langle\Psi_k(n^\circ)|\mathbf{Q}^3|\Psi_k(n^\circ)\rangle \quad (9.55)$$

On the other hand, in view of Eq. (9.50), it may be also given by

$$\langle\Psi_k(n^\circ)|\mathbf{H}|\Psi_k(n^\circ)\rangle = E_k(n^\circ) \quad (9.56)$$

Hence, a good test for the accuracy of the approximate kets $|\Psi_k(n^\circ)\rangle$ may be the differences $\Delta E_k(n^\circ)$ between Eqs. (9.56) and (9.55), which ought to be zero if the kets were the exact eigenvectors of the Hamiltonian.

$$\Delta E_k(n^\circ) = E_k(n^\circ) - \{2\langle\Psi_k(n^\circ)|\mathbf{V}_{\text{Cub}}(\mathbf{Q})|\Psi_k(n^\circ)\rangle + \tfrac{1}{2}\xi\langle\Psi_k(n^\circ)|\mathbf{Q}^3|\Psi_k(n^\circ)\rangle\}$$

Then using Eqs. (9.51) and (9.53), the virial theorem leads to

$$\Delta E_k(n^\circ) = E_k(n^\circ) - \sum_n\sum_m C_{nk}(n^\circ)C_{km}(n^\circ) \times \left\{M\omega^2\langle\{m\}|(\mathbf{Q}^2)|\{n\}\rangle + \tfrac{5}{2}\xi\langle\{m\}|(\mathbf{Q}^3)|\{n\}\rangle\right\}$$

Passing to Boson operators for the cubic term yields

$$\Delta E_k(n^\circ) = E_k(n^\circ) - \sum_n\sum_m C_{nk}(n^\circ)C_{km}(n^\circ) \times \left\{M\omega^2\langle\{m\}|\mathbf{Q}^2|\{n\}\rangle + \frac{5}{2}\xi\left(\frac{\hbar}{2M\omega}\right)^{3/2}\langle\{m\}|(\mathbf{a}^\dagger + \mathbf{a})^3|\{n\}\rangle\right\} \quad (9.57)$$

or in dimensionless units, given by (9.37), that is,

$$\xi = \eta\hbar\omega\left(\frac{2M\omega}{\hbar}\right)^{3/2}$$

$$\Delta E_k(n^\circ) = E_k(n^\circ) - \sum_n\sum_m C_{nk}(n^\circ)C_{km}(n^\circ)\hbar\omega$$

$$\times\left\{\frac{1}{2}\langle\{m\}|(\mathbf{a}^\dagger + \mathbf{a})^2|\{n\}\rangle + \frac{5}{2}\eta\langle\{m\}|(\mathbf{a}^\dagger + \mathbf{a})^3|\{n\}\rangle\right\}$$

Of course, the matrix elements that are cubic with respect to the Boson operators may also be obtained by the aid of Eqs. (9.40) and (9.48). On the other hand, according to Eq. (5.87), those that are quadratic in $(\mathbf{a}^\dagger + \mathbf{a})$ are given by

$$\langle\{m\}|(\mathbf{a}^\dagger + \mathbf{a})^2|\{n\}\rangle = \langle\{m\}|((\mathbf{a}^\dagger)^2 + (\mathbf{a})^2 + 2\mathbf{a}^\dagger\mathbf{a} + 1)|\{n\}\rangle \tag{9.58}$$

and thus, due to Eqs. (5.53) and (5.63),

$$\langle\{m\}|(\mathbf{a}^\dagger + \mathbf{a})^2|\{n\}\rangle = (2n+1)\delta_{mn} + \sqrt{n+1}\sqrt{n+2}\,\delta_{m,n+2} + \sqrt{n}\sqrt{n-2}\,\delta_{m,n-2} \tag{9.59}$$

Hence, the verification that the virial theorem is satisfied may be performed by considering the energy difference between $\Delta E_k(n^\circ)/E_k(n^\circ)$ and zero, the smaller being this term and the best accurate being the kets $|\Psi_k(n^\circ)\rangle$ and the corresponding energy levels $E_k(n^\circ)$. In Fig. 9.3, we show the evolution of the relative dispersion $\Delta E_k(n^\circ)/E_k(n^\circ)$ as a function of n°, for the six lowest energy levels.

This criteria for accuracy of the absolute dispersion appears to be more drastic than stabilization of the energy as n° is raised. The graph in Fig. 9.3 shows that even if the energy is stabilized with respect to an enhancement of the number of basis terms, the virial theorem is less and less satisfied.

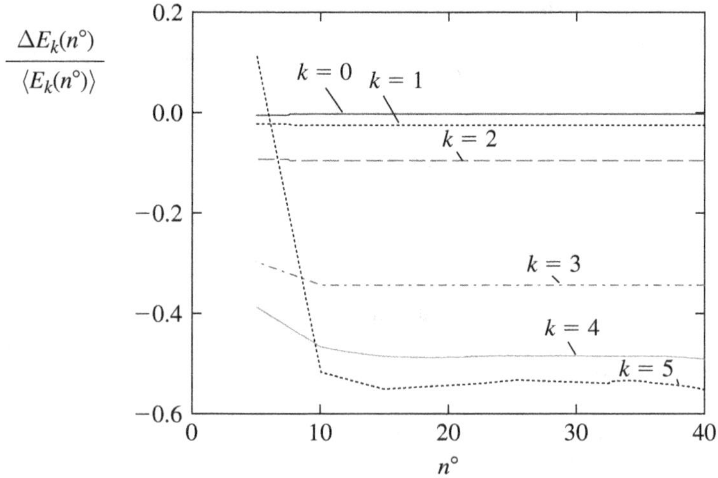

Figure 9.3 Relative dispersion of the difference between the energy levels and the virial theorem. (See color insert.)

9.3 MORSE OSCILLATOR

In the previous section, we approximated the Morse potential for diatomic nuclei by introducing in the harmonic potential the greatest correction, which is cubic with respect to the elongation. However, to be more rigorous it is necessary to take explicitly into account the Morse potential as a whole. For this purpose, one may use the following expression, which is *stricto sensu* what is called a *Morse potential*, to describe the Morse potential operator $\mathbf{V}_{\text{Morse}}$

$$\mathbf{V}_{\text{Morse}} = D_e(1 - e^{-\beta\mathbf{Q}})^2 \tag{9.60}$$

Here, $\mathbf{Q}$ is the position coordinate operator, D_e is the dissociation energy of the diatomic molecule, whereas β is a parameter that is function of the force constant k_e of the oscillator near the equilibrium geometry and of D_e

$$\beta = \left(\frac{k_e}{2D_e}\right)^{1/2}$$

The Hamiltonian governing the vibration of this diatomic molecule is described by

$$\mathbf{H} = \frac{\mathbf{P}^2}{2m} + \mathbf{V}_{\text{Morse}} \tag{9.61}$$

9.3.1 Analytical solution of the Hamiltonian eigenvalue equation

In the wave mechanics picture, the time-independent Schrödinger equation obtained from the Hamiltonian (9.61) is

$$-\frac{\hbar^2}{2M}\left(\frac{\partial^2\Psi_n(Q)}{\partial Q^2}\right) + D_e(1 - e^{-\beta\mathbf{Q}})^2\Psi_n(Q) = E_n\Psi_n(Q) \tag{9.62}$$

where $\Psi_n(Q)$ are the eigenfunctions and E_n the corresponding eigenvalues. Without solving the Schrödinger equation (9.62), the wavefunctions $\Psi_n(Q)$ are found to be[1]

$$\Psi_n(Q) = \left(\frac{2d\beta}{N_n}\right)^{1/2} \exp\{-de^{-\beta Q}\}(2de^{-\beta Q})^{(2d-2n-1)/2}\mathbb{L}_{2d-n-1}^{2d-2n-1}(2de^{-\beta Q}) \tag{9.63}$$

where $\mathbb{L}_{2d-n-1}^{2d-2n-1}(2de^{-\beta Q})$ are the associated Laguerre polynomials given by

$$\mathbb{L}_{2d-n-1}^{2d-2n-1}\left(2de^{-\beta Q}\right) = \sum_{k=0}^{n}(-1)^{k+1}\frac{\{(2d-n-1)!\}^2}{(n-k)!(2d+k-2n-1)!k!}(2de^{-\beta Q})^k$$

$$d = \frac{(2MD_e)^{1/2}}{\beta\hbar}$$

whereas N_n are the normalization constant

$$N_n = ((2d-n-1)!)^2\sum_{s=0}^{n}\frac{(2d-2n+s-2)!}{s!}$$

[1]See, for instance, Mu Sang Lee, L. A. Carreira, and D. A. Berkowitz, *Bull. Korean. Chem. Soc.*, **7** (1986): 6–7.

Of course, since the Hamiltonian (9.62) is Hermitian, the eigenfunctions form an orthonormal basis and thus obey

$$\int \Psi_m(Q)\Psi_n(Q)\,dQ = \delta_{mn}$$

Also, the corresponding energy levels E_n are[2]

$$E_n = \hbar\omega^\circ \left(n + \frac{1}{2}\right) - \frac{(\hbar\omega^\circ)^2}{4D_e}\left(n + \frac{1}{2}\right)^2 \tag{9.64}$$

Now, the analytical solutions (9.63) and (9.64) are exact with respect to the model to which they apply, they are unfortunately not easy to handle as a basis for matrix representations of Hamiltonians of more complex systems in which the Morse potential plays a role. This is the reason why an analytical solution of the Schrödinger equation (9.62) is not given here, and why we shall consider the numerical diagonalization of Hamiltonians involving the Morse potential, susceptible to be generalized to other anharmonic Hamiltonians.

9.3.2 Matrix elements of the Morse potential within the harmonic Hamiltonian eigenkets

9.3.2.1 Morse potential in terms of the ladder operators First, find the expression relating β to the force constant k_e and to the dissociation energy D_e. For this purpose, calculate the second derivative of the potential (9.60) in equilibrium geometry, that is, at the minimum of the Morse curve taken as the origin for the elongation operator $\mathbf{Q}$:

$$\left(\frac{\partial^2 \mathbf{V}_{\text{Morse}}}{\partial \mathbf{Q}^2}\right)_{\mathbf{Q}=0} = 2\beta^2 D_e \tag{9.65}$$

Now, the force constant k_e is by definition the second derivative of the potential at its minimum:

$$k_e = \left(\frac{\partial^2 \mathbf{V}_{\text{Morse}}}{\partial \mathbf{Q}^2}\right)_{\mathbf{Q}=0} \tag{9.66}$$

Thus, after identification of Eqs. (9.65) and (9.66), the β parameter appears to be given by

$$\beta = \sqrt{\frac{k_e}{2D_e}} \tag{9.67}$$

On the other hand, the force constant of a harmonic potential characterized by the angular frequency ω and the reduced mass M is

$$k_e = M\omega^2 \tag{9.68}$$

[2] P. M. Morse, *Phys. Rev.*, **34** (1929): 57–64.

Hence, if this harmonic potential is to be that which reduces the Morse potential at its minimum, Eq. (9.67) transforms using Eq. (9.60) into

$$\beta = \omega\sqrt{\frac{M}{2D_e}}$$

Then, in order to make explicit the Hamiltonian of the harmonic oscillator in the expression (9.61) of the Morse Hamiltonian, it is convenient to write Eq. (9.61) as

$$\mathbf{H} = \frac{\mathbf{P}^2}{2m} + \{\mathbf{V}_{\text{Harm}} - \mathbf{V}_{\text{Harm}}\} + \mathbf{V}_{\text{Morse}} \tag{9.69}$$

with

$$\{\mathbf{V}_{\text{Harm}}\} = \tfrac{1}{2}M\omega^2\mathbf{Q}^2 \tag{9.70}$$

leading to

$$\mathbf{H} = \left(\frac{\mathbf{P}^2}{2M} + \frac{1}{2}M\omega^2\mathbf{Q}^2\right) + (\mathbf{V}_{\text{Morse}} - \mathbf{V}_{\text{Harm}})$$

Next, pass to Boson operators by the aid of Eqs. (9.34) and (9.35). Then, the Morse potential operator (9.60) takes the form

$$\mathbf{V}_{\text{Morse}} = D_e(1 - 2e^{-\beta^\circ(\mathbf{a}^\dagger+\mathbf{a})} + e^{-2\beta^\circ(\mathbf{a}^\dagger+\mathbf{a})}) \tag{9.71}$$

where

$$\beta^\circ = \beta\sqrt{\frac{\hbar}{2M\omega}} = \frac{1}{2}\sqrt{\frac{\hbar\omega}{D_e}} \tag{9.72}$$

whereas the harmonic potential operator (9.70) reads

$$\mathbf{V}_{\text{Harm}} = \frac{\hbar\omega}{4}(\mathbf{a}^\dagger + \mathbf{a})^2 \tag{9.73}$$

Moreover, expansions of the square on the right-hand side give, using the commutation rule $[\mathbf{a}, \mathbf{a}^\dagger] = 1$,

$$\mathbf{V}_{\text{Harm}} = \frac{\hbar\omega}{4}((\mathbf{a}^\dagger)^2 + (\mathbf{a})^2) + \frac{\hbar\omega}{2}\left(\mathbf{a}^\dagger\mathbf{a} + \frac{1}{2}\right) \tag{9.74}$$

Finally, the sum of the potential (9.74) and of the kinetic energy operator, that is, the Hamiltonian of a harmonic oscillator yields, according to Eqs. (5.6) and (5.7),

$$\left(\frac{\mathbf{P}^2}{2M} + \frac{1}{2}M\omega^2\mathbf{Q}^2\right) = \hbar\omega\left(\mathbf{a}^\dagger\mathbf{a} + \frac{1}{2}\right) \tag{9.75}$$

As a consequence of Eqs. (9.71), (9.74), and (9.75), the full Hamiltonian (9.69) transforms to

$$\mathbf{H} = \hbar\omega\left\{\tfrac{1}{2}\left(\mathbf{a}^\dagger\mathbf{a} + \tfrac{1}{2}\right) - \tfrac{1}{4}((\mathbf{a}^\dagger)^2 + (\mathbf{a})^2) + \zeta(1 - 2e^{-\beta^\circ(\mathbf{a}^\dagger+\mathbf{a})} + e^{-2\beta^\circ(\mathbf{a}^\dagger+\mathbf{a})})\right\}$$

with

$$\zeta = \frac{D_e}{\hbar\omega}$$

This may be also written in the condensed form:

$$\mathbf{H} = \hbar\omega\left\{\tfrac{1}{2}(\mathbf{a}^{\dagger}\mathbf{a} + \tfrac{1}{2}) - \tfrac{1}{4}((\mathbf{a}^{\dagger})^2 + (\mathbf{a})^2) + \zeta(1 - 2\mathbf{F}(\beta^{\circ}) + \mathbf{F}(2\beta^{\circ}))\right\} \tag{9.76}$$

with

$$\mathbf{F}(\beta^{\circ}) = e^{-\beta^{\circ}(\mathbf{a}^{\dagger}+\mathbf{a})} \qquad \text{and} \qquad \mathbf{F}(2\beta^{\circ}) = e^{-2\beta^{\circ}(\mathbf{a}^{\dagger}+\mathbf{a})}$$

Moreover, using the Glauber–Weyl theorem (1.79), the latter operator reads

$$\mathbf{F}(\beta^{\circ}) = e^{-\beta^{\circ}\mathbf{a}^{\dagger}} e^{-\beta^{\circ}\mathbf{a}}(e^{-\beta^{\circ 2}[\mathbf{a}^{\dagger},\mathbf{a}]/2})$$

which, owing to the commutation rule $\left[\mathbf{a}, \mathbf{a}^{\dagger}\right] = 1$, it transforms to

$$\mathbf{F}(\beta^{\circ}) = e^{-\beta^{\circ}\mathbf{a}^{\dagger}} e^{-\beta^{\circ}\mathbf{a}}(e^{\beta^{\circ 2}/2})$$

Finally, expanding the exponentials involving the Boson operators according to

$$\mathbf{F}(\beta^{\circ}) = (e^{\beta^{\circ 2}/2})\sum_{k=0}^{\infty}\frac{(-1)^k\beta^{\circ k}(\mathbf{a}^{\dagger})^k}{k!}\sum_{l=0}^{\infty}\frac{(-1)^l\beta^{\circ l}(\mathbf{a})^l}{l!}$$

leads to

$$\mathbf{F}(\beta^{\circ}) = (e^{\beta^{\circ 2}/2})\sum_{k=0}^{\infty}\sum_{l=0}^{\infty}\frac{(-1)^{k+l}\beta^{\circ k+l}}{k!l!}(\mathbf{a}^{\dagger})^k(\mathbf{a})^l \tag{9.77}$$

9.3.2.2 *Matrix elements of the Morse Hamiltonian*

Next, consider the matrix elements of the full Hamiltonian (9.76) in the basis $\{|\{n\}\rangle\}$ of the eigenkets of $\mathbf{a}^{\dagger}\mathbf{a}$:

$$\mathbf{a}^{\dagger}\mathbf{a}|\{n\}\rangle = n|\{n\}\rangle \quad \text{with} \quad \langle\{n\}|\{n\}\rangle = \delta_{mn} \tag{9.78}$$

They are

$$\begin{aligned}\hbar\omega\langle\{m\}|\mathbf{H}|\{n\}\rangle &= \tfrac{1}{4} + \tfrac{1}{2}\langle\{m\}|(\mathbf{a}^{\dagger}\mathbf{a})|\{n\}\rangle - \tfrac{1}{4}\langle\{m\}|(\mathbf{a}^{\dagger})^2|\{n\}\rangle - \tfrac{1}{4}\langle\{m\}|(\mathbf{a})^2|\{n\}\rangle \\ &\quad + \zeta(1 - 2\langle\{m\}|\mathbf{F}(\beta^{\circ})|\{n\}\rangle + \langle\{m\}|\mathbf{F}(2\beta^{\circ})|\{n\}\rangle)\end{aligned} \tag{9.79}$$

Owing to Eq. (5.42), the first-right-hand-side matrix elements of Eq. (9.79) are

$$\langle\{m\}|(\mathbf{a}^{\dagger}\mathbf{a})|\{n\}\rangle = n\,\delta_{mn} \tag{9.80}$$

Then, one obtains, respectively, by the aid of Eqs. (5.53) and (5.63) for the two matrix elements of the squared Boson operators appearing in Eq. (9.79)

$$\langle\{m\}|(\mathbf{a}^{\dagger})^2|\{n\}\rangle = \sqrt{(n+1)(n+2)}\delta_{m,n+2} \tag{9.81}$$

$$\langle\{m\}|(\mathbf{a})^2|\{n\}\rangle = \sqrt{(m+1)(m+2)}\delta_{m+2,n} \tag{9.82}$$

Now, consider the matrix elements of the operator $\mathbf{F}(\beta^{\circ})$ (9.77), that is,

$$\langle\{m\}|\mathbf{F}(\beta^{\circ})|\{n\}\rangle = (e^{\beta^{\circ 2}/2})\sum_{k=0}^{\infty}\sum_{l=0}^{\infty}\frac{(-1)^{k+l}(\beta^{\circ})^{k+l}}{k!l!}\langle\{m\}|(\mathbf{a}^{\dagger})^k(\mathbf{a})^l|\{n\}\rangle \tag{9.83}$$

Owing to Eqs. (5.72) and (5.73), the matrix elements involved on the right-hand side of (9.83) are given by

$$\langle\{m\}|(\mathbf{a}^{\dagger})^k(\mathbf{a})^l|\{n\}\rangle = \frac{\sqrt{m!}}{\sqrt{(m-k)!}}\frac{\sqrt{n!}}{\sqrt{(n-l)!}}\langle\{m-k\}|\{n-l\}\rangle$$

or, using the orthornormality properties appearing in Eq. (9.78),

$$\langle\{m\}|\mathbf{a}^{\dagger k}\mathbf{a}^{l}|\{n\}\rangle = \frac{\sqrt{m!}}{\sqrt{(m-k)!}}\frac{\sqrt{n!}}{\sqrt{(n-l)!}}\{\delta(n-l,m-k)\}$$

so that the matrix element (9.83) takes the form

$$\langle\{m\}|\mathbf{F}(\beta^\circ)|\{n\}\rangle = e^{\beta^{\circ 2}/2}\sum_{k=0}^{\infty}\sum_{l=0}^{\infty}\frac{(-1)^{k+l}\beta^{\circ k+l}}{k!l!}\frac{\sqrt{m!}}{\sqrt{(m-k)!}}\frac{\sqrt{n!}}{\sqrt{(n-l)!}}\{\delta(n-l,m-k)\} \tag{9.84}$$

Now, because

$$\{\delta(n-l,m-k)\} = 1 \quad \text{if} \quad l = n-m+k$$

and

$$\{\delta(n-l,m-k)\} = 0 \quad \text{otherwise}$$

Eq. (9.84) leads, respectively, according to whether $m \geq n$ or $m < n$, to

$$\langle\{m\}|\mathbf{F}(\beta^\circ)|\{n\}\rangle = e^{1/2\beta^{\circ 2}}\sum_{k=n-m}^{m}\frac{(-1)^{(n-m+2k)}\beta^{\circ(n-m+2k)}}{k!(n-m+k)!}\frac{\sqrt{m!}\sqrt{n!}}{(m-k)!} \quad \text{if} \quad m \geq n \tag{9.85}$$

and a similar expression for $n > m$ in which n and m have to be permuted.

Of course, the matrix elements $\langle\{m\}|\mathbf{F}(2\beta^\circ)|\{n\}\rangle$ appearing in Eq. (9.79) may be obtained by the aid of Eq. (9.85) in which the argument β° has been replaced by $2\beta^\circ$.

9.3.3 Diagonalization of the truncated Hamiltonian matrix representation

It is now possible to construct various truncated matrix representations (9.79) of the full Hamiltonian (9.61) by the aid of Eqs. (9.76), (9.80), (9.81), (9.82), and (9.85), the dimension n° of which being progressively increased, and then to numerically diagonalize them. This procedure leads to numerical approximations of the eigenvalue equation of the full Hamiltonian:

$$\begin{aligned}\mathbf{H}|\Psi_k(n^\circ)\rangle &= E_k(n^\circ)|\Psi_k(n^\circ)\rangle \\ |\Psi_k(n^\circ)\rangle &= \sum_{n=0}^{N-1} C_{nk}(n^\circ)|\{n\}\rangle\end{aligned} \tag{9.86}$$

where $E_k(n^\circ)$ are the approximate eigenvalue functions of n°, whereas $|\Psi_k(n^\circ)\rangle$ are the corresponding eigenvectors, the components of which in the basis $\{|\{n\}\rangle\}$ are $C_{nk}(n^\circ)$. As previously for the driven harmonic oscillator and for the anharmonic oscillator involving a cubic potential, the approximate eigenvalues E_k of the Hamiltonian matrix are progressively stabilized.

As an illustration we reproduce the result of a numerical calculation where the dissociation energy of the Morse potential is $D_e = 50\,\hbar\omega$. The 10 lowest energy levels

$E_k(n^\circ)$ expressed in units $\hbar\omega$ are given in (9.87) and compared to the eigenvalues of the harmonic Hamiltonians:

	E_0	E_1	E_2	E_3	E_4	E_5	E_6	E_7	E_8	E_9
Morse	0.499	1.489	2.469	3.489	4.309	5.349	6.289	7.219	8.139	9.049
Harm	0.500	1.500	2.500	3.500	4.500	5.500	6.500	7.500	8.500	9.500

(9.87)

Inspection of (9.87) shows that the energy levels of the Morse Hamiltonian are more and more lowered with respect to those of the harmonic Hamiltonian when the energy levels are raising, the explanation lying in the fact that there is more place for the atoms to move inside the potential when passing from the harmonic to the Morse potential, which induces a lowering of the average kinetic energy according to the particle-in-a-box model.

On the other hand the 10 first components $C_{nk}\,(n^\circ)$ of the eigenvectors $|\Psi_k(n^\circ)\rangle$ corresponding to the 10 lowest energy levels E_k are given in (9.88):

	$\|\Psi_0\rangle$	$\|\Psi_1\rangle$	$\|\Psi_2\rangle$	$\|\Psi_3\rangle$	$\|\Psi_4\rangle$	$\|\Psi_5\rangle$	$\|\Psi_6\rangle$	$\|\Psi_7\rangle$	$\|\Psi_8\rangle$	$\|\Psi_9\rangle$
C_{0k}	−0.998	−0.053	0.001	0.013	0.005	0.001	0.001	−0.000	0.000	−0.000
C_{1k}	−0.053	0.987	−0.149	−0.012	−0.026	−0.015	−0.005	0.002	−0.001	0.001
C_{2k}	−0.003	0.148	0.950	0.268	0.039	0.041	0.029	−0.013	0.006	−0.003
C_{3k}	−0.015	0.016	0.266	−0.872	−0.393	−0.086	−0.059	0.048	−0.026	0.012
C_{4k}	−0.002	0.030	0.046	−0.388	0.742	0.507	0.155	−0.084	0.071	−0.045
C_{5k}	−0.000	0.007	0.051	−0.095	0.497	−0.555	−0.585	0.241	−0.119	0.097
C_{6k}	−0.000	0.001	0.019	−0.080	0.162	−0.570	0.320	−0.606	0.333	−0.165
C_{7k}	−0.000	0.001	0.005	−0.039	0.117	−0.239	0.586	0.058	−0.549	0.410
C_{8k}	0.000	0.000	0.003	−0.012	0.069	−0.164	0.314	0.529	−0.199	−0.409
C_{9k}	0.000	0.000	0.001	−0.007	0.027	−0.109	0.216	0.369	0.396	−0.406

(9.88)

Inspection of (9.88) shows that the 5 lowest eigenvectors have only components on the 10 lowest eigenstates $|\{n\}\rangle$ of the harmonic oscillator Hamiltonian. This result allows one to get numerical representations of the 5 lowest wavefunctions $\Psi_k(Q)$ corresponding to these 5 lowest eigenvectors $|\Psi_k\rangle$ defined by

$$\Psi_k(Q,n^\circ) = \langle\{Q\}|\Psi_k(n^\circ)\rangle \quad \text{with} \quad \mathbf{Q}|\{Q\}\rangle = Q|\{Q\}\rangle$$

Owing to Eq. (9.86) these wavefunctions are given by

$$\Psi_k(Q,n^\circ) = \langle\{Q\}|\sum_{n=0}^{4} C_{nk}(n^\circ)|\{n\}\rangle$$

or

$$\Psi_k(Q,n^\circ) = \sum_{n=0}^{4} C_{nk}(n^\circ)\Phi_n(Q) \tag{9.89}$$

where the $\Phi_n(Q)$ are the wavefunctions of the harmonic oscillators defined by

$$\Phi_n(Q,n^\circ) = \langle\{Q\}|\{n\}\rangle$$

Now, it has been seen above that, in dimensionless position, coordinate ξ is given by Eq. (5.117), that is,

$$\xi = \sqrt{\frac{m\omega}{\hbar}} Q$$

the wavefunctions of the harmonic oscillator Hamiltonian are given by Eq. (5.147):

$$\Phi_n(\xi) = K_n H_n(\xi) e^{-\xi^2/2} \tag{9.90}$$

The coefficients K_n are given by Eq. (5.148):

$$K_n = \frac{1}{\sqrt{n!}} \left(\frac{1}{\sqrt{2}}\right)^n \left(\frac{m\omega}{\hbar\pi}\right)^{1/4} \tag{9.91}$$

whereas $H_n(\xi)$ are the Hermite polynomials, the five lowest of which being, respectively, given by Eqs. (5.134), (5.138), (5.143), (5.146), and (5.149)

$$H_0(\xi) = 1 \qquad H_1(\xi) = 2\xi \qquad H_2(\xi) = 4\xi^2 - 2 \tag{9.92}$$

$$H_3(\xi) = 8\xi^3 - 12\xi \qquad H_4(\xi) = 16\xi^4 - 48\xi^2 + 12 \tag{9.93}$$

Equations (9.89) and (9.90)–(9.93) allow one to get pictorial representations of the wavefunctions of the Morse oscillator using the values of the expansion coefficients C_{nk} appearing in (9.88), as shown in Fig. 9.4.

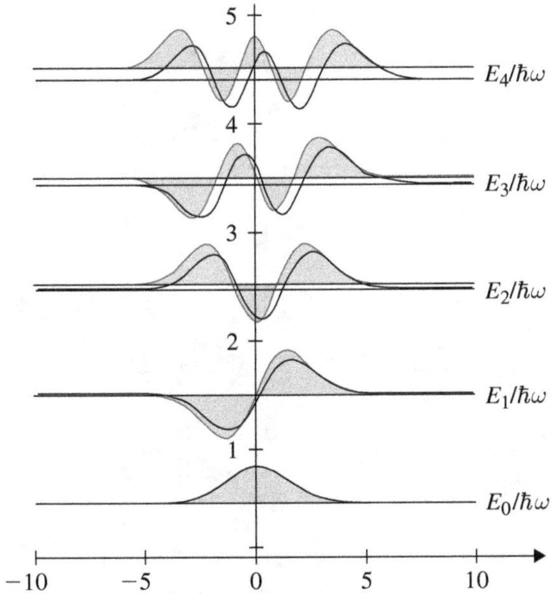

Figure 9.4 Five lowest wavefunctions $\Psi_k(\xi)$ of the Morse Hamiltonian compared to the five symmetric or antisymmetric lowest wavefunctions $\Phi_n(\xi)$ of the harmonic Hamiltonian. The length unit is $Q^{\circ\circ} = \sqrt{h/2m\omega}$. (See color insert.)

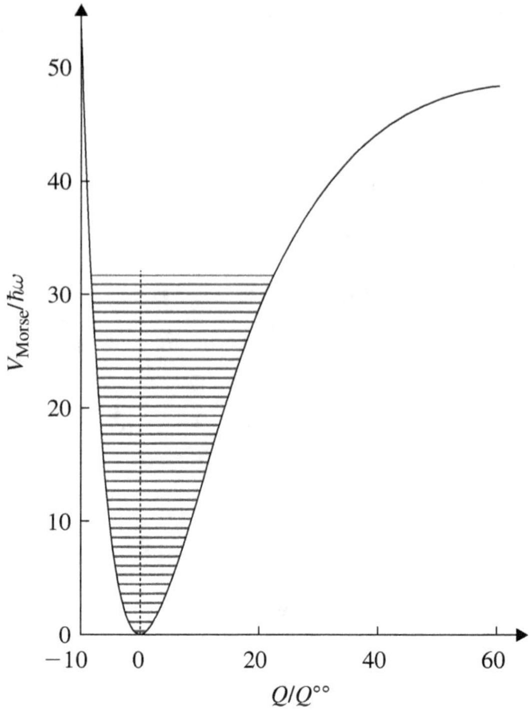

Figure 9.5 The 40 lowest energy levels of the Morse oscillator. The length unit is $Q^{\circ\circ} = \sqrt{\hbar/2m\omega}$.

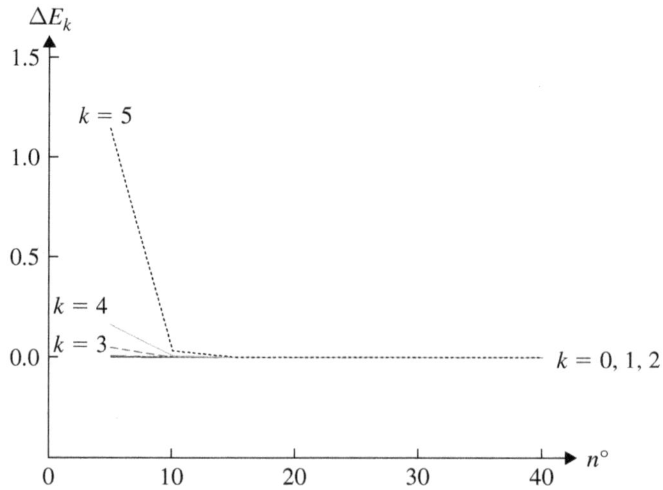

Figure 9.6 Energy gap between the numerical and exact eigenvalues for a Morse oscillator.

Finally, it may be convenient to reproduce the 40 lowest energy levels. That is illustrated in Fig. 9.5, which illustrates a situation that would roughly apply, for instance, to the deuterium molecule D_2.

Also, it would be of interest to compare these numerical eigenvalues $E_k(n°)$ obtained by diagonalization of the Hamiltonian (9.76) with the analytical E_k° obtained with Eq. (9.64). In Fig. 9.6 we show, as a function of the number of basis states $n°$ of the numerical procedure, the gap $\Delta E_k(n°)$ between the numerical eigenvalues and the exact analytical ones, defined by

$$\Delta E_k(n°) = E_k(n°) - E_k^\circ$$

9.4 QUADRATIC POTENTIALS PERTURBED BY COSINE FUNCTIONS

In situations dealing with the biophysics of proteins, modulated harmonic potentials are encountered that may be modeled by potentials that, translated in quantum mechanics, read

$$\mathbf{V}(\mathbf{Q}) = \frac{1}{2}m\omega^2\mathbf{Q}^2 + A\cos\left(\frac{k\mathbf{Q}}{\hbar}\right)$$

where A and k are parameters characterizing the system. The Hamiltonian $\mathbf{H}$ corresponding to such a potential is then

$$\mathbf{H} = \frac{\mathbf{p}^2}{2m} + \mathbf{V}(\mathbf{Q}) \tag{9.94}$$

Passing to Boson operators, the Hamiltonian (9.94) reads

$$\mathbf{H} = \hbar\omega\left\{\left(\mathbf{a}^\dagger\mathbf{a}+\tfrac{1}{2}\right) + \alpha\cos(\beta(\mathbf{a}^\dagger + \mathbf{a}))\right\}$$

or

$$\mathbf{H} = \hbar\omega\left\{\left(\mathbf{a}^\dagger\mathbf{a}+\tfrac{1}{2}\right) + \frac{\alpha}{2}(e^{i\beta(\mathbf{a}^\dagger+\mathbf{a})} + e^{-i\beta(\mathbf{a}^\dagger+\mathbf{a})})\right\} \tag{9.95}$$

with

$$\alpha = \frac{A}{\hbar\omega} \qquad \text{and} \qquad \beta = \frac{k}{\hbar}\sqrt{\frac{\hbar}{2m\omega}}$$

Now, to get the energy levels of the system, we have to diagonalize the matrix representation of the Hamiltonian (9.95) in the basis where $\mathbf{a}^\dagger\mathbf{a}$ is diagonal and calculate the matrix elements $\langle\{m\}|\mathbf{H}|\{n\}\rangle$ of this expression. The matrix elements of the Hamiltonian (9.95) are of the form

$$\begin{aligned}\langle\{m\}|\mathbf{H}|\{n\}\rangle &= \hbar\omega\Big(\langle\{m\}|\left(\mathbf{a}^\dagger\mathbf{a}+\tfrac{1}{2}\right)|\{n\}\rangle \\ &\quad + \frac{\alpha}{2}\{\langle\{m\}|(e^{i\beta(\mathbf{a}^\dagger+\mathbf{a})})|\{n\}\rangle + \langle\{m\}|(e^{-i\beta(\mathbf{a}^\dagger+\mathbf{a})})|\{n\}\rangle\}\Big)\end{aligned} \tag{9.96}$$

Again, using the Glauber theorem (1.79), the exponential operators of the right-hand side of the latter equation become

$$\langle e^{\pm i\beta(\mathbf{a}^\dagger+\mathbf{a})}\rangle = \langle e^{-(\pm i\beta)^2[\mathbf{a}^\dagger,\mathbf{a}]/2}\rangle\langle e^{\pm i\beta\mathbf{a}^\dagger}\rangle\langle e^{\pm i\beta\mathbf{a}}\rangle$$

and thus, due to $[\mathbf{a}^\dagger, \mathbf{a}] = -1$

$$\langle e^{\pm i\beta(\mathbf{a}^\dagger+\mathbf{a})}\rangle = \langle e^{-\beta^2/2}\rangle\langle e^{\pm i\beta\mathbf{a}^\dagger}\rangle\langle e^{\pm i\beta\mathbf{a}}\rangle$$

Again, after expanding the two last exponentials, the corresponding matrix elements read

$$\langle\{m\}|\langle e^{\pm i\beta(\mathbf{a}^\dagger+\mathbf{a})}\rangle|\{n\}\rangle = \langle e^{-\beta^2/2}\rangle\sum_{k=0}\sum_{l}\frac{(\pm i)^{k+l}\beta^{k+l}}{k!l!}\langle\{m\}|(\mathbf{a}^\dagger)^k(\mathbf{a})^l|\{n\}\rangle \quad (9.97)$$

Then, using Eq. (5.72), and its Hermitian conjugate (5.73), that is,

$$(\mathbf{a})^l|\{n\}\rangle = \frac{\sqrt{n!}}{\sqrt{(n-l)!}}|\{n-l\}\rangle \qquad \text{and} \qquad \langle\{m\}|(\mathbf{a}^\dagger)^k = \langle\{m-k\}|\frac{\sqrt{m!}}{\sqrt{(m-k)!}}$$

Eq. (9.97) yields

$$\begin{aligned}&\langle\{m\}|\langle e^{\pm i\beta(\mathbf{a}^\dagger+\mathbf{a})}\rangle|\{n\}\rangle\\ &= \langle e^{-\beta^2/2}\rangle\sum_{k}\sum_{l}\frac{(\pm i)^{k+l}\beta^{k+l}}{k!l!}\frac{\sqrt{m!}}{\sqrt{(m-k)!}}\frac{\sqrt{n!}}{\sqrt{(n-l)!}}\langle\{m-k\}|\{n-l\}\rangle\end{aligned}$$

or due to the orthogonality properties

$$\begin{aligned}&\langle\{m\}|\langle e^{\pm i\beta(\mathbf{a}^\dagger+\mathbf{a})}\rangle|\{n\}\rangle\\ &= \langle e^{-\beta^2/2}\rangle\sum_{k}\sum_{l}\frac{(\pm i)^{k+l}\beta^{k+l}}{k!l!}\frac{\sqrt{m!}}{\sqrt{(m-k)!}}\frac{\sqrt{n!}}{\sqrt{(n-l)!}}\{\delta_{m-k,n-l}\}\end{aligned} \quad (9.98)$$

and thus after simplification because

$$\{\delta_{m-k,n-l}\} = 0 \quad \text{except if} \quad l = n - m + k$$

$$\langle\{m\}|\langle e^{\pm i\beta(\mathbf{a}^\dagger+\mathbf{a})}\rangle|\{n\}\rangle = \langle e^{-\beta^2/2}\rangle\sum_{k}\frac{(\pm i)^{2k+n-m}\beta^{2k+n-m}}{k!(k+n-m)!}\frac{\sqrt{m!}\sqrt{n!}}{(m-k)!} \quad (9.99)$$

Take care that in Eq. (9.99) the forms of the factorials imply that in the sum over k, this variable runs from $k = (m-n)$ to m if $m \geqslant n$, and from $k = (n-m)$ to n if $m < n$.

Then, numerical diagonalizations of the truncated Hamiltonian matrix representation (9.96), the dimension n° of which is progressively increased, have to be performed until the required stabilization of the energy levels $E_k(n^\circ)$ with respect to the dimension of the basis has been attained. Of course, the stabilized energy levels $E_k(n^\circ)$ and the corresponding eigenkets $|\Psi_k(n^\circ)\rangle$ obey the formal eigenvalue equation

$$\mathbf{H}|\Psi_k(n^\circ)\rangle = E_k(n^\circ)|\Psi_k(n^\circ)\rangle \quad \text{with} \quad |\Psi_k(n^\circ)\rangle = \sum_{n}\{C_{nk}(n^\circ)\}|\{n\}\rangle \quad (9.100)$$

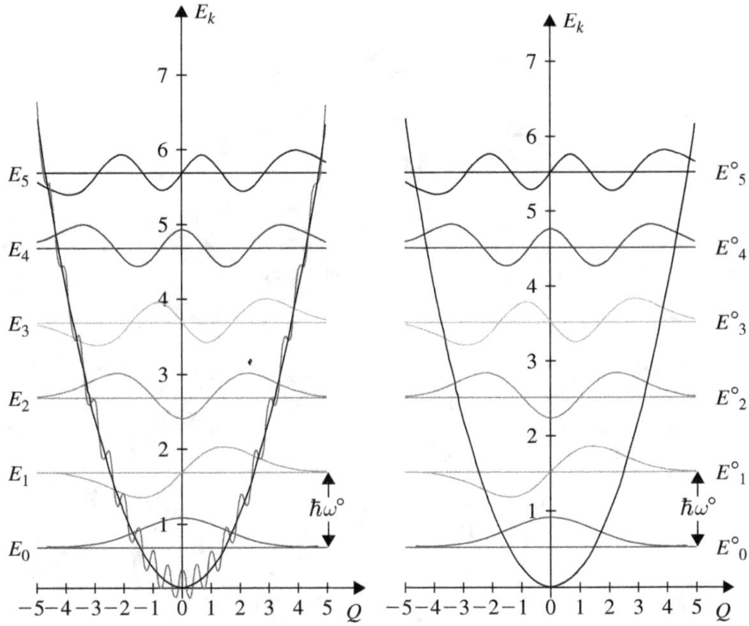

Figure 9.7 Comparison between the energy levels calculated by Eq. (9.100) and the wavefunctions obtained by Eq. (9.101) and the energy levels and the wavefunctions of the harmonic oscillator. (See color insert.)

the wavefunctions $\Psi_k(Q, n^\circ)$ corresponding to the kets $|\Psi_k(n^\circ)\rangle$ being given by

$$\{\Psi_k(Q, n^\circ)\} = \langle\{Q\}|\Psi_k(n^\circ)\rangle = \sum_n \{C_{nk}(n^\circ)\}\langle\{Q\}|\{n\}\rangle = \sum_n \{C_{nk}(n^\circ)\}\Phi_n(Q) \tag{9.101}$$

Figure 9.7 gives a modulated potential and the corresponding energy levels obtained.

As may be noted, the form of the potential practically does not affect the energy spacing of the levels, which remains close to that of the harmonic oscillator and does not sensitively modify the corresponding wavefunctions.

9.5 DOUBLE-WELL POTENTIAL AND TUNNELING EFFECT

Many situations dealing with molecules involve double-well potentials in which the nuclei of atoms are moving. For example, consider the gaseous NH_3 ammonia molecule. This tetrahedral molecule is pyramidal, shaped with the three hydrogen atoms forming the base and the nitrogen atom at the top, as shown in Fig. 9.8.

The nitrogen atom sees a double-well potential with one well on either side of the plane defined by the hydrogen atoms, as shown in Fig. 9.9.

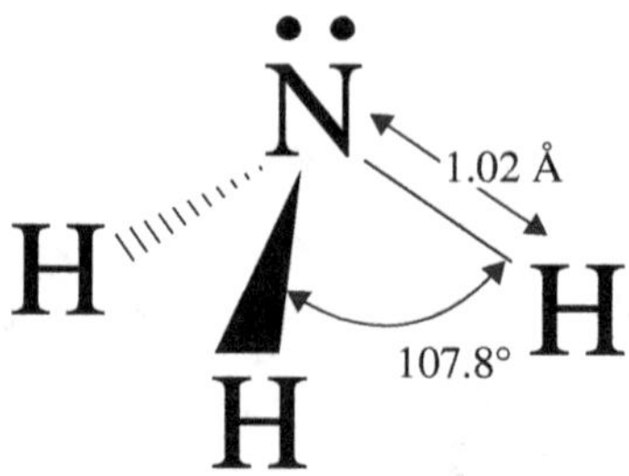

Figure 9.8 Ammonia molecule.

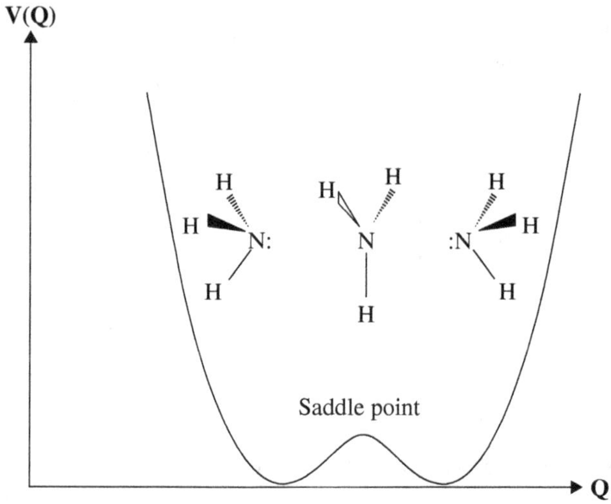

Figure 9.9 Double-well ammonia potential.

9.5.1 Hamiltonians in terms of the ladder operators

For generality, we shall consider an asymmetric double-well potential. The Hamiltonian $\mathbf{H}$ of a factitious particle of mass M moving in this potential is the sum of its kinetic operator and of the corresponding potential one $\mathbf{V}_{\text{wells}}(\mathbf{Q})$:

$$\mathbf{H} = \frac{\mathbf{P}^2}{2M} + \mathbf{V}_{\text{wells}}(\mathbf{Q}) \tag{9.102}$$

In order to have a suitable expression for this double-well asymmetric potential, one may choose to describe it by a quartic potential $\mathbf{Q}^4$ perturbed by a quadratic potential $\mathbf{Q}^2$ such as

$$\mathbf{V}_{\text{wells}}(\mathbf{Q}) = A\mathbf{Q}^4 - B\,(\mathbf{Q}-C)^2 \tag{9.103}$$

Here A, B, and C are parameters characterizing the asymmetric potential function, A having the dimension of an energy per the fourth power of a length, B of an energy per squared length, and C that of a length.

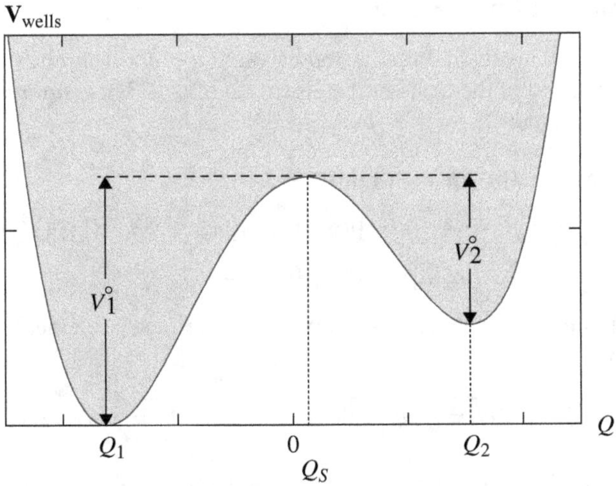

Figure 9.10 Example of double-well potential $V(Q)$ defined by Eq. (9.103) in terms of the geometric parameters V_1°, V_2°, Q_S, Q_1 and Q_2 defined in the text.

Figure 9.10 illustrates an asymmetric double-well potential (9.103) in which are shown the coordinates of the two minima Q_1 and Q_2 and the Q_S of the saddle point and also the energy barriers V_1° and V_2° separating the two energy minima from the energy maximum.

Owing to Eq. (9.103), the Hamiltonian (9.102) takes the form

$$\mathbf{H} = \frac{\mathbf{P}^2}{2M} + A\mathbf{Q}^4 - B(\mathbf{Q}-C)^2 \tag{9.104}$$

where M is the reduced mass of interest. Adding and substracting the harmonic potential $\frac{1}{2}M\omega^{\circ 2}\mathbf{Q}^2$ where ω° is some reference angular frequency yields

$$\mathbf{H} = \left(\frac{\mathbf{P}^2}{2M} + \frac{1}{2}M\omega^{\circ 2}\mathbf{Q}^2\right) - \frac{1}{2}M\omega^{\circ 2}\mathbf{Q}^2 + A\mathbf{Q}^4 - B(\mathbf{Q}-C)^2 \tag{9.105}$$

Now, from A, B, and C, we define the following dimensionless parameters:

$$\xi = \frac{A}{\hbar\omega^\circ}\left(\frac{\hbar}{2M\omega^\circ}\right)^2 \qquad \beta = \frac{B}{\hbar\omega^\circ}\left(\frac{\hbar}{2M\omega^\circ}\right) \qquad \eta = C\sqrt{\frac{2M\omega^\circ}{\hbar}}$$

hence Eq. (9.105) becomes

$$\mathbf{H} = \hbar\omega^\circ\left\{\left(\mathbf{a}^\dagger\mathbf{a}+\tfrac{1}{2}\right) - \tfrac{1}{4}(\mathbf{a}^\dagger + \mathbf{a})^2 + \xi(\mathbf{a}^\dagger + \mathbf{a})^4 - \beta((\mathbf{a}^\dagger + \mathbf{a}) - \eta)^2\right\}$$

or

$$\mathbf{H} = \hbar\omega^\circ\left\{\left(\mathbf{a}^\dagger\mathbf{a}+\tfrac{1}{2}\right) - \left(\beta - \tfrac{1}{4}\right)(\mathbf{a}^\dagger + \mathbf{a})^2 + \xi(\mathbf{a}^\dagger + \mathbf{a})^4 + 2\beta\eta(\mathbf{a}^\dagger + \mathbf{a}) - \beta\eta^2\right\} \tag{9.106}$$

9.5.2 Hamiltonian matrix elements

To obtain the energy levels, the matrix representation of the Hamiltonian **H** (9.106) has to be diagonalized in the basis of the eigenkets of $\mathbf{a}^\dagger\mathbf{a}$. This requires expressions for the matrix elements:

$$\begin{aligned}\langle\{n\}|\mathbf{H}|\{m\}\rangle &= \hbar\omega^\circ\langle\{n\}|\left(\mathbf{a}^\dagger\mathbf{a}+\tfrac{1}{2}\right)|\{m\}\rangle \\ &\quad -\hbar\omega^\circ\left(\beta-\tfrac{1}{4}\right)\langle\{n\}|(\mathbf{a}^\dagger+\mathbf{a})^2|\{m\}\rangle + \hbar\omega^\circ\xi\langle\{n\}|(\mathbf{a}^\dagger+\mathbf{a})^4|\{m\}\rangle \\ &\quad + 2\hbar\omega^\circ\beta\eta\langle\{n\}|(\mathbf{a}^\dagger+\mathbf{a})|\{m\}\rangle - \hbar\omega^\circ\beta\eta^2 \end{aligned} \tag{9.107}$$

Due to the commutation rule of Boson operators, the matrix elements of $(\mathbf{a}^\dagger+\mathbf{a})^2$ are by Eq. (9.59)

$$\langle\{n\}|(\mathbf{a}^\dagger+\mathbf{a})^2|\{m\}\rangle = \sqrt{m+1}\sqrt{m+2}\,\delta_{n,m+2} + \sqrt{m}\sqrt{m-1}\,\delta_{n,m-2} + (2m+1)\delta_{nm} \tag{9.108}$$

Now, from the usual commutation rules $[\mathbf{a},\mathbf{a}^\dagger]=1$, development of $(\mathbf{a}^\dagger+\mathbf{a})^4$ yields

$$\begin{aligned}(\mathbf{a}^\dagger+\mathbf{a})^4 &= (4\mathbf{a}^\dagger\mathbf{a}+1) + 4(\mathbf{a}^\dagger\mathbf{a})^2 + (\mathbf{a}^\dagger)^4 + (\mathbf{a})^4 \\ &\quad + ((\mathbf{a}^\dagger)^2(\mathbf{a})^2 + (\mathbf{a})^2(\mathbf{a}^\dagger)^2) + \mathbf{2}((\mathbf{a}^\dagger)^2+(\mathbf{a})^2)(\mathbf{a}^\dagger\mathbf{a}) \\ &\quad + \mathbf{2}(\mathbf{a}^\dagger\mathbf{a}+1)((\mathbf{a}^\dagger)^2+(\mathbf{a})^2)\end{aligned} \tag{9.109}$$

Then, Eq. (5.71) and its Hermitian conjugate allow one to find the following results:

$$\langle\{n\}|(\mathbf{a}^\dagger)^4|\{m\}\rangle = \sqrt{(m+1)(m+2)(m+3)(m+4)}\,\delta_{n,m+4}$$

Moreover, due to Eqs. (5.53) and (5.54)

$$(\mathbf{a})^2|\{m\}\rangle = \sqrt{m(m-1)}|\{m-2\}\rangle$$

the Hermitian conjugate of which is, after taking n in place of m,

$$\langle\{n\}|(\mathbf{a}^\dagger)^2 = \sqrt{n(n-1)}\langle\{n-2\}|$$

Hence, it follows that

$$\langle\{n\}|(\mathbf{a}^\dagger)^2(\mathbf{a})^2|\{m\}\rangle = \sqrt{n(n-1)}\sqrt{m(m-1)}\langle\{n-2\}|\{m-2\}\rangle$$

so that, after simplification, it reads

$$\langle\{n\}|(\mathbf{a}^\dagger)^2(\mathbf{a})^2|\{m\}\rangle = \sqrt{n(n-1)}\sqrt{m(m-1)}\,\delta_{nm} \tag{9.110}$$

On the other hand, one has, respectively,

$$\begin{aligned}\mathbf{a}|\{m\}\rangle &= \sqrt{m}|\{m-1\}\rangle \\ \langle\{n\}|(\mathbf{a}^\dagger)^3 &= \sqrt{(n)(n-1)(n-2)}\langle\{n-3\}|\end{aligned}$$

so that

$$\langle\{n\}|(\mathbf{a}^\dagger)^3\mathbf{a}|\{m\}\rangle = \sqrt{(m)(n)(n-1)(n-2)}\langle\{n-3\}|\{m-1\}\rangle = \delta_{n-3,m-1}$$

and thus

$$\langle\{n\}|(\mathbf{a}^\dagger)^3\mathbf{a}|\{m\}\rangle = \sqrt{(m)(n)(n-1)(n-2)}\,\delta_{n-2,m} \tag{9.111}$$

Again, keeping in mind that the matrix elements (1.30) of a real non-Hermitian operator **B** obey

$$\langle\{n\}|\mathbf{B}|\{m\}\rangle = \langle\{m\}|\mathbf{B}^\dagger|\{n\}\rangle^* = \langle\{m\}|\mathbf{B}^\dagger|\{n\}\rangle \tag{9.112}$$

it follows, respectively, from Eqs. (9.110) and (9.111) that

$$\begin{aligned}
\langle\{m\}|(\mathbf{a})^2(\mathbf{a}^\dagger)^2|\{n\}\rangle &= \sqrt{n(n-1)}\sqrt{m(m-1)}\,\delta_{nm} \\
\langle\{n\}|\mathbf{a}(\mathbf{a}^\dagger)^3|\{m\}\rangle &= \sqrt{(m)(n)(n-1)(n-2)}\,\delta_{n-2,m}
\end{aligned}$$

Hence, the matrix elements (9.109) read

$$\begin{aligned}
\langle\{n\}|(\mathbf{a}^\dagger+\mathbf{a})^4|\{m\}\rangle &= 3(2m^2+2m+1)\,\delta_{nm} + (2(2m+3)\sqrt{m+1}\sqrt{m+2})\,\delta_{n,m+2} \\
&\quad + (\sqrt{(m+1)(m+2)(m+3)(m+4)})\,\delta_{n,m+4} \\
&\quad + \sqrt{(m)(m-1)(m-2)(m-3)}\,\delta_{n,m-4} \\
&\quad + (2(2m-1)\sqrt{(m)(m-1)})\,\delta_{n,m-2}
\end{aligned} \tag{9.113}$$

Thus, it is possible, using Eqs. (9.107)–(9.113) to build up the matrix representation of the Hamiltonian (9.106) and then to diagonalize different truncated matrix representations of this Hamiltonian, the dimensions of which have to be progressively increased until stabilization of the lowest eigenvalues occurs.

9.5.3 Tunneling in symmetric double-well potentials

It is now possible to consider the tunneling effect, which may occur through the potential barrier of a double-wells potential. For this purpose, consider the two lowest eigenstates of the Hamiltonian (9.102) obtained by diagonalization of the matrix representation (9.107) of dimension n°, which obey the eigenvalue equations

$$\mathbf{H}|\Psi_k(n^\circ)\rangle = E_k(n^\circ)|\Psi_k(n^\circ)\rangle \quad \text{with} \quad k = 0, 1 \tag{9.114}$$

The diagonalization of the matrix representation of this Hamiltonian gives, respectively, for the two lowest quasi-degenerate eigenvalues (in $\hbar\omega^\circ$ units with three decimals)

$$E_0(n^\circ) = 1.357 \qquad \text{and} \qquad E_1(n^\circ) = 1.357$$

whereas it gives for the two corresponding eigenvectors

$$|\Psi_k(n^\circ)\rangle = \sum_n C_{nk}(n^\circ)|\{n\}\rangle \quad \text{with} \quad C_{nk}(n^\circ) = \langle\{n\}|\Psi_k(n^\circ)\rangle \tag{9.115}$$

and for the expansion coefficients $C_{nk}(n^\circ)$ the following tabular data in (9.116) in which we restrict the number of component eigenstates to 20:

n	$\langle\{n\}\vert\Psi_0\rangle$	$\langle\{n\}\vert\Psi_1\rangle$	n	$\langle\{n\}\vert\Psi_0\rangle$	$\langle\{n\}\vert\Psi_1\rangle$
0	0.305	0.000	10	0.035	0.000
1	0.000	0.598	11	0.000	0.038
2	0.786	0.000	12	0.011	0.000
3	0.000	0.760	13	0.000	−0.011
4	0.526	0.000	14	−0.014	0.000
5	0.000	0.212	15	0.000	−0.004
6	−0.037	0.000	16	0.004	0.000
7	0.000	−0.135	17	0.000	0.005
8	−0.102	0.000	18	0.000	0.000
9	0.000	−0.020	19	0.000	−0.002

(9.116)

Next, introduce the wavefunctions corresponding to the kets (9.115) by premultiplying Eq. (9.115) by the bra $\langle\{Q\}|$ to give

$$\Psi_k(Q, n^\circ) = \sum_n C_{nk}(n^\circ)\Phi_n(Q, n^\circ) \tag{9.117}$$

with

$$\Psi_k(Q, n^\circ) = \langle\{Q\}|\Psi_k(n^\circ)\rangle \qquad \text{and} \qquad \Phi_n(Q, n^\circ) = \langle\{Q\}|\{n\}\rangle \tag{9.118}$$

In addition, recall that the wavefunctions $\Phi_n(Q)$ corresponding to the eigenkets of the harmonic oscillator Hamiltonian may be expressed in terms of $H_n(\alpha Q)$ and given by Eq. (5.147), that is,

$$\Phi_n(Q, n^\circ) = K_n(e^{-\alpha^2 Q^2/2})H_n(\alpha Q) \tag{9.119}$$

where $H_n(\alpha Q)$ are the Hermite polynomials and where α and K_n are, respectively, given by

$$\alpha = \sqrt{\frac{M\omega^\circ}{\hbar}} \qquad \text{and} \qquad K_n = \left(\frac{\alpha}{\sqrt{\pi}2^n n!}\right)^{1/2}$$

Then, using the expansion coefficients $C_{nk}(n^\circ)$ given in (9.116), and resulting from the diagonalization of the matrix representation (9.107), Eqs. (9.117) and (9.119) allow one to obtain the wavefunctions corresponding to the eigenstates of the Hamiltonian $\mathbf{H}(\mathbf{a}, \mathbf{a}^\dagger)$. Figure 9.11 gives the picture of the six lowest wavefunctions and their corresponding energy levels for any symmetric double-well potential. Inspection of the figure shows that the four energy levels lying below the barrier are split, E_0 and E_1 being quasi-degenerate, whereas the two others $E_2(n^\circ)$ and $E_3(n^\circ)$ involve a large splitting. It shows also that the wavefunctions $\Psi_2(Q, n^\circ)$ and $\Psi_3(Q, n^\circ)$ corresponding to the energy levels $E_2(n^\circ)$ and $E_3(n^\circ)$ represent residual amplitudes inside the potential barrier, a region that would be forbidden for the particle from the viewpoint of classical mechanics. This penetration of the particle through the energy barrier is a manifestation of what is called the *tunneling effect*.

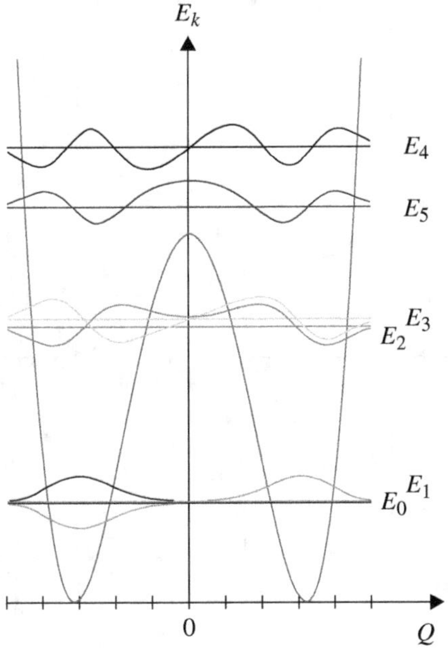

Figure 9.11 Representation of the six lowest wavefunctions and the corresponding energy levels for symmetrical double-well potential. (See color insert.)

In addition, Fig. 9.12 deals with the influence on tunneling of the asymmetry parameter η of the double-well potential. Inspection of this figure shows that increasing the values of the dimensionless asymmetry parameter η of the potential induces drastic changes in the wavefunctions corresponding to the energy levels lying below the barrier, since their delocalizations are strongly vanishing as soon as the asymmetry is slightly increasing.

In may be of interest to study the precise behavior of the particle when tunneling is occurring. The best way would be to deal with that involving $\Psi_2(Q)$ and $\Psi_3(Q)$ appearing in Fig. 9.11 and in which $n°$ has been omitted. But, although the tunneling effect is weaker for the levels E_0 and E_1 than for those E_2 and E_3, it is, however, simpler to use $\Psi_0(Q)$ and $\Psi_1(Q)$ because of their great simplicity, allowing a better visual picture. Thus, to get such a picture, assume that, at an initial time, the system is described by either $\Theta_+(Q, 0)$ or $\Theta_-(Q, 0)$, which are combinations of the wavefunctions $\Psi_0(Q, 0)$ and $\Psi_1(Q, 0)$ at initial time $t = 0$:

$$\Theta_\pm(Q, 0) = \frac{1}{\sqrt{2}}(\Psi_0(Q, 0) \pm \Psi_1(Q, 0)) \tag{9.120}$$

Figure 9.13 gives a schematic representation of the two wavefunctions involved in the expansions (9.120), which according to (9.116) are given by

$$\begin{aligned}\Psi_0(Q, 0) = {} & 0.30\Phi_0(Q) + 0.79\Phi_2(Q) + 0.53\Phi_4(Q) - 0.04\Phi_6(Q) - 0.10\Phi_8(Q) \\ & + 0.03\Phi_{10}(Q)\end{aligned}$$

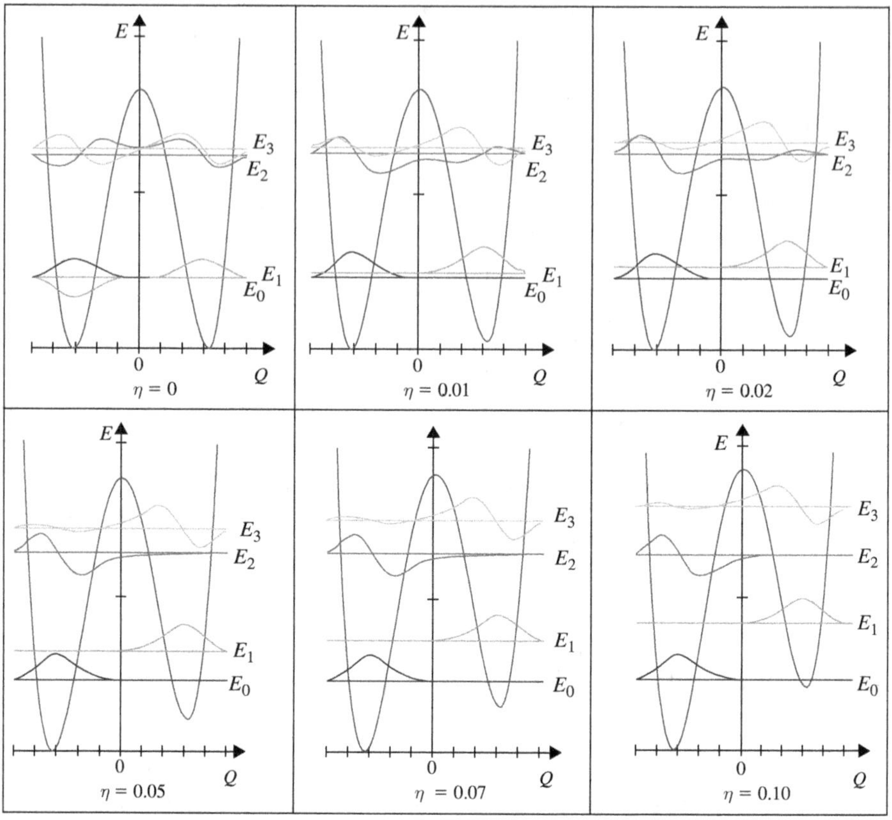

Figure 9.12 Influence of the double-well potential asymmetry on the eigenstates of the double-well potential Hamiltonian.

$$\Psi_1(Q,0) = 0.60\Phi_1(Q) + 0.76\Phi_3(Q) + 0.21\Phi_5(Q) - 0.013\Phi_7(Q) - 0.02\Phi_9(Q) + 0.04\Phi_{11}(Q)$$

Thus, $\Theta_+(Q,0)$ and $\Theta_-(Q,0)$ are each localized on one of the two potential minima.

Next, at time t, the linear combination (9.120) reads

$$\{\Theta_\pm(Q,t)\} = \frac{1}{\sqrt{2}}(\{\Psi_0(Q,t)\} \pm \{\Psi_1(Q,t)\}) \tag{9.121}$$

where the two time-dependent wavefunctions $\Psi_0(Q,t)$ and $\Psi_1(Q,t)$ are solutions of the Schrödinger equations:

$$i\hbar\left(\frac{\partial\Psi_k(Q,t)}{\partial Q}\right) = \mathbf{H}\{\Psi_k(Q,t)\} \quad \text{with} \quad k = 0, 1$$

Since, according to Eq. (9.114), the Hamiltonian $\mathbf{H}$ here acts on some of its eigenstates, the corresponding eigenvalues are the energy levels E_K with $K = 0, 1$, so that

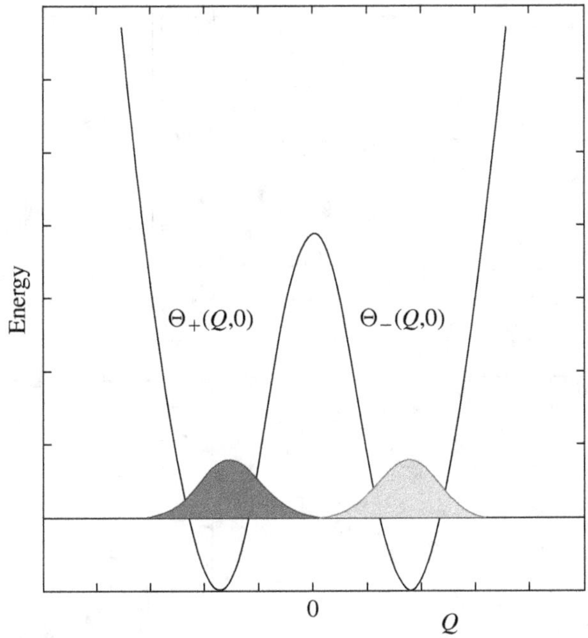

Figure 9.13 Schematic representation of the two wavefunctions (9.120).

the two time-dependent Schrödinger equations become

$$i\hbar\left(\frac{\partial\Psi_k(Q,t)}{\partial Q}\right) = E_k\{\Psi_k(Q,t)\} \quad \text{with} \quad k = 0, 1$$

the integration of which leads to

$$\{\Psi_k(Q,t)\} = \{\Psi_k(Q,0)\}e^{-iE_kt/\hbar} \quad \text{with} \quad k = 0, 1 \qquad (9.122)$$

Hence, at time t the linear combinations of the two time-dependent wavefunctions (9.121) take the form

$$\{\Theta_\pm(Q,t)\} = \frac{1}{\sqrt{2}}(\{\Psi_0(Q,0)\}e^{-iE_0t/\hbar} \pm \{\Psi_1(Q,0)\}e^{-iE_1t/\hbar})$$

so that the corresponding squared modulus read

$$|\Theta_+(Q,t)|^2 = \{\Psi_0(Q,0)\}^2 + \{\Psi_1(Q,0)\}^2 + \{\Psi_0(Q,0)\}\{\Psi_1(Q,0)\}\cos\omega_{01}t \quad (9.123)$$

$$|\Theta_-(Q,t)|^2 = \{\Psi_0(Q,0)\}^2 + \{\Psi_1(Q,0)\}^2 - \{\Psi_0(Q,0)\}\{\Psi_1(Q,0)\}\sin\omega_{01}t \quad (9.124)$$

with

$$\omega_{01} = \hbar/(E_0 - E_1)$$

Figure 9.14 gives some changes with time of Eq. (9.124), that is, of the probability to find the system at any position Q within the double-well potential.

Inspection shows that this density probability oscillates from one side to the other of the double well at the angular frequency ω_{01}. Hence, the system tunnels back

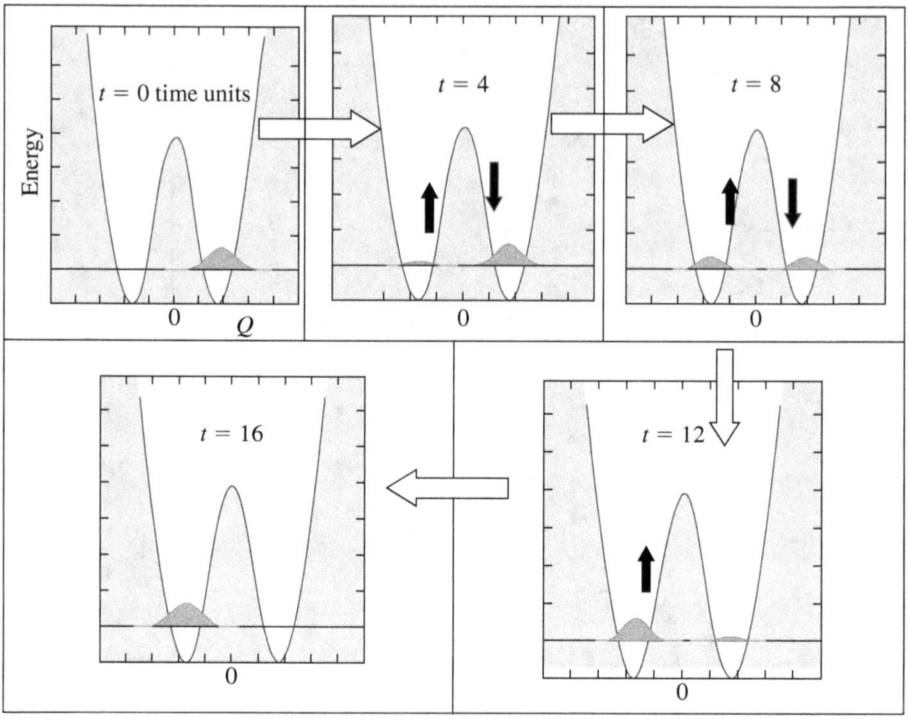

Figure 9.14 Probability density (9.124) for different times t expressed in units ω^{-1}.

and forth through the barrier potential. In a similar way, one would obtain, for the two *gerade* g and *ungerade* u levels corresponding to the energy levels E_2 and E_3, the following results:

$$|\Theta_g(Q,t)|^2 = \{\Psi_2(Q,0)\}^2 + \{\Psi_3(Q,0)\}^2 + \{\Psi_2(Q,0)\}\{\Psi_3(Q,0)\}\cos\omega_{23}t$$

$$|\Theta_u(Q,t)|^2 = \{\Psi_2(Q,0)\}^2 + \{\Psi_3(Q,0)\}^2 - \{\Psi_2(Q,0)\}\{\Psi_3(Q,0)\}\sin\omega_{23}t$$

with

$$\omega_{23} = \hbar/(E_2 - E_3)$$

In such a situation, the pictorial representation of the time evolution of the wavepacket would be more complex, the tunneling effect occurring with an angular frequency larger than for Eq. (9.124) since it appears from Fig. 9.11 that

$$\omega_{23} > \omega_{01}$$

the oscillations would of course remain.

9.6 CONCLUSION

This chapter has shown how to treat the energy levels of quantum anharmonic oscillators for which it is not generally possible to solve the Hamiltonian eigenvalue equation. That was performed by numerically diagonalizing the matrix representations of the different Hamiltonians in the basis of the eigenkets of the harmonic oscillator Hamiltonian, a method the accuracy of which had been satisfactorily tested in a previous chapter on the driven harmonic oscillator and which was tested in the present chapter on anharmonic oscillators involving a cubic term perturbing the harmonic potential with the help of the virial theorem, and on anharmonic oscillators involving Morse potentials, by comparison with the analytical solutions, which exist for this model. The numerical calculations involving anharmonic potentials, which did not modify sensitively the harmonic potential, lead one to conclude that the anharmonic perturbation does not modify drastically the energy levels, whereas those dealing with double-well potentials reveal the possibility of tunneling through the potential barrier separating the two wells, when the potential is symmetric.

BIBLIOGRAPHY

P. M. Morse. *Phys. Rev.*, **34** (1929): 57–64.

CHAPTER 10

OSCILLATORS INVOLVING ANHARMONIC COUPLINGS

INTRODUCTION

This chapter is devoted to some realistic models of oscillators that interact through anharmonic couplings and that are important in vibrational spectroscopy. First, we shall study Fermi resonances, which are considered as the splitting of vibrational energy levels because of residual anharmonic coupling. Then, we treat the strong anharmonic coupling model used in theoretical approaches to the IR line shapes of weak H-bonded species and according to which there exists a special kind of anharmonic coupling between a high- and a low-frequency mode summing from a dependence of the angular frequency of the fast mode on the coordinate of the slow one. In a subsequent section, we show that for weak H-bond species, an adiabatic separation may be performed between the motions of the fast and slow oscillators, leading to effective Hamiltonians describing a driven slow mode, the strength of the driven term increasing according to the degree of excitation of the fast oscillator. Finally, Fermi resonances and Davydov coupling are incorporated in the strong anharmonic coupling model in the context of adiabatic approximation, for the special case of H-bonded centrosymmetric cyclic dimers in which strong exchange may occur between two degenerate adiabatic excited states.

10.1 FERMI RESONANCES

First, we consider *Fermi resonances*, a phenomenon that occurs in vibrational spectroscopy when the residual anharmonic coupling between some normal modes cannot be neglected.

Thus, consider the model of two harmonic oscillators of angular frequencies ω° and Ω, which are anharmonically coupled, their Hamiltonians being

$$\mathbf{H} = \mathbf{H}^\circ + \mathbf{H}_{\text{Int}} \tag{10.1}$$

with, respectively,

$$\mathbf{H}^\circ = \left(\frac{\mathbf{p}^2}{2m} + \frac{1}{2}m\omega^{\circ 2}\,\mathbf{q}^2\right) + \left(\frac{\mathbf{P}^2}{2M} + \frac{1}{2}M\Omega^2\,\mathbf{Q}^2\right) \tag{10.2}$$

Quantum Oscillators, First Edition. Olivier Henri-Rousseau and Paul Blaise.
© 2011 John Wiley & Sons, Inc. Published 2011 by John Wiley & Sons, Inc.

$$\mathbf{H}_{\text{Int}} = \lambda\, \mathbf{Q}^2 \mathbf{q} \tag{10.3}$$

Here, m and M are, respectively, the reduced masses of the oscillators, whereas $\mathbf{q}$ and $\mathbf{Q}$ and $\mathbf{p}$ and $\mathbf{P}$ are the corresponding position and momentum operators, λ being the coefficient of the anharmonic coupling between the two oscillators. Of course, the position and momentum operators of the two oscillators obey

$$[\mathbf{Q}, \mathbf{P}] = [\mathbf{q}, \mathbf{p}] = i\hbar$$

$$[\mathbf{Q}, \mathbf{p}] = [\mathbf{P}, \mathbf{q}] = 0$$

Passing to Boson operators using form (5.6) and (5.7), the Hamiltonian (10.2) transforms to

$$\mathbf{H}^\circ = \hbar\omega^\circ \left(\mathbf{b}^\dagger \mathbf{b} + \tfrac{1}{2}\right) + \hbar\Omega \left(\mathbf{a}^\dagger \mathbf{a} + \tfrac{1}{2}\right)$$

with

$$[\mathbf{a}, \mathbf{a}^\dagger] = [\mathbf{b}, \mathbf{b}^\dagger] = 1$$

$$[\mathbf{b}, \mathbf{a}^\dagger] = [\mathbf{a}, \mathbf{b}^\dagger] = [\mathbf{b}^\dagger, \mathbf{a}] = [\mathbf{a}^\dagger, \mathbf{b}] = 0$$

while that (10.3) yields

$$\mathbf{H}_{\text{Int}} = \xi\hbar\Omega(\mathbf{b}^\dagger + \mathbf{b})(\mathbf{a}^\dagger + \mathbf{a})^2 \tag{10.4}$$

with

$$\xi = \lambda \left(\frac{\hbar}{2m\omega^\circ}\right)\sqrt{\left(\frac{\hbar}{2M\Omega}\right)} \tag{10.5}$$

Then, in order to diagonalize the Hamiltonian (10.1) in the basis defined by Eqs. (10.38)–(10.40), the three eigenvalue equations corresponding to the three lowest energy levels read

$$\mathbf{H}^\circ|\{0\}\rangle|[0]\rangle = \{E^{\{0\}[0]}\}|\{0\}\rangle|[0]\rangle$$

$$\mathbf{H}^\circ|\{1\}\rangle|[0]\rangle = \{E^{\{1\}[0]}\}|\{1\}\rangle|[0]\rangle \tag{10.6}$$

$$\mathbf{H}^\circ|\{1\}\rangle|[2]\rangle = \{E^{\{0\}[2]}\}|\{1\}\rangle|[2]\rangle \tag{10.7}$$

where the three lowest eigenvalues are given by

$$\{E^{\{0\}[0]}\} = \tfrac{1}{2}\hbar\Omega + \tfrac{1}{2}\hbar\omega^\circ \qquad \{E^{\{1\}[0]}\} = \tfrac{3}{2}\hbar\omega^\circ + \tfrac{1}{2}\hbar\Omega \qquad \{E^{\{0\}[2]}\} = \tfrac{1}{2}\hbar\omega^\circ + \tfrac{5}{2}\hbar\Omega \tag{10.8}$$

In many situations in the spectroscopy of vibrational states, the two excited states defined by Eqs. (10.6) and (10.7) play an important role, when Ω is around half ω° and when the coupling Hamiltonian of the form (10.4) is very weak yet cannot be completely neglected. Then, the coupling Hamiltonian induces an interaction between the two levels $\{E^{\{1\}[0]}\}$ and $\{E^{\{0\}[2]}\}$, which is called a *Fermi resonance*.

To treat them, consider the two Hamiltonian matrix elements built up from the two excited states appearing in Eqs. (10.6) and (10.7), that is, owing to Eq. (10.4)

$$\langle\{1\}|\langle[0]|\mathbf{H}_{\text{Int}}|[2]\rangle|\{0\}\rangle = \xi\hbar\Omega\langle\{1\}|\langle[0]|(\mathbf{b}^\dagger + \mathbf{b})(\mathbf{a}^\dagger + \mathbf{a})^2|[2]\rangle|\{0\}\rangle \tag{10.9}$$

The squared operator involving the Boson operators $\mathbf{a}^{\dagger}$ and $\mathbf{a}$ is given by Eq. (5.87), that is,

$$(\mathbf{a}^{\dagger}+\mathbf{a})^2 = (2\mathbf{a}^{\dagger}\mathbf{a}+1+(\mathbf{a}^{\dagger})^2+(\mathbf{a})^2)$$

Moreover, observe that owing to Eqs. (5.40) and (5.63), and to the orthogonality of the involved kets, we have

$$\langle[0]|(2\mathbf{a}^{\dagger}\mathbf{a}+1)|[2]\rangle = 5\langle[0]|[2]\rangle = 0$$

$$\langle[0]|(\mathbf{a}^{\dagger})^2|[2]\rangle = \sqrt{4\times 3}\langle[0]|[4]\rangle = 0$$

Thus, the matrix element (10.9) becomes

$$\langle\{1\}|\langle[0]|\mathbf{H}_{\text{Int}}|[2]\rangle|\{0\}\rangle = \xi\hbar\Omega\langle\{1\}|\mathbf{b}^{\dagger}|\{0\}\rangle\langle[0]|(\mathbf{a})^2|[2]\rangle$$

Now after action of the operators $\mathbf{b}^{\dagger}$ and $\mathbf{a}^2$, one gets, with the help of Eqs. (5.40) and (5.66)

$$\langle\{1\}|\langle[0]|\mathbf{H}_{\text{Int}}|\,[2]\rangle|\{0\}\rangle = \xi\hbar\Omega\langle\{1\}|1|\{1\}\rangle\langle[0]|\sqrt{2}|[0]\rangle = \sqrt{2}\xi\hbar\Omega \tag{10.10}$$

the Hermitian conjugate of which is

$$\langle\{0\}|\langle[2]|\mathbf{H}_{\text{Int}}|[0]\rangle|\{1\}\rangle = \langle\{0\}|\mathbf{b}|\{1\}\rangle\langle[2]|(\mathbf{a}^{\dagger})^2|[0]\rangle = \sqrt{2}\xi\hbar\Omega \tag{10.11}$$

When one limits the present approach to the representation of the full Hamiltonian (10.1) in the two-level subspaces defined by the eigenvalue equations (10.6) and (10.7), the following 2×2 matrix representation is obtained:

	$\mid[0]\rangle\mid\{1\}\rangle$	$\mid[2]\rangle\mid\{0\}\rangle$
$\langle\{1\}\mid\langle[0]\mid$	$\hbar\omega^{\circ}+\left(\frac{1}{2}\hbar\omega^{\circ}+\frac{1}{2}\hbar\Omega\right)$	$\sqrt{2}\xi\hbar\Omega$
$\langle\{0\}\mid\langle[2]\mid$	$\sqrt{2}\xi\hbar\Omega$	$2\hbar\Omega+\left(\frac{1}{2}\hbar\omega^{\circ}+\frac{1}{2}\hbar\Omega\right)$

(10.12)

Now, this matrix representation has the same structure as that (4.45) met in the study of two-level systems. Letting

$$\alpha_1 = \tfrac{3}{2}\hbar\omega^{\circ}+\tfrac{1}{2}\hbar\Omega \qquad \alpha_2 = \tfrac{1}{2}\hbar\omega^{\circ}+\tfrac{5}{2}\hbar\Omega \qquad \beta = \sqrt{2}\xi\hbar\Omega \tag{10.13}$$

one has to solve an equation of the form of (4.54)

$$\begin{pmatrix}\alpha_1 - E_{\pm} & \beta \\ \beta & \alpha_2 - E_{\pm}\end{pmatrix}\begin{pmatrix}\{C_{1_{\pm}}\} \\ \{C_{2_{\pm}}\}\end{pmatrix} = 0$$

where $E_{\pm}$ are the eigenvalues of the matrix (10.12), whereas the $C_{k_{\pm}}$, are the expansion coefficients of the corresponding eigenvectors $|\Psi_{\pm}\rangle$, that is,

$$|\Psi_{\pm}\rangle = \{C_{1_{\pm}}\}|\{1\}\rangle|\,[0]\rangle + \{C_{2_{\pm}}\}|\{0\}\rangle|\,[2]\rangle \tag{10.14}$$

Owing to Eqs. (4.56) and (10.13), these energy levels are

$$E_{\pm} = \frac{1}{2}\left\{\hbar(\omega^{\circ}+2\Omega) \pm \hbar(\omega^{\circ}-2\Omega)\sqrt{1+\frac{4(\sqrt{2}\xi\hbar\Omega)^2}{(\hbar\omega^{\circ}-2\hbar\Omega)^2}}\right\} \tag{10.15}$$

Moreover, if the coupling parameter is small with respect to the energy gap between the two interacting vibrational levels, that is,

$$\beta^2 << (\alpha_1-\alpha_2)^2$$

or

$$4(\sqrt{2}\xi\Omega)^2 << (\omega^\circ - 2\Omega)^2 \tag{10.16}$$

then, according to Eq. (4.57), the energy levels may be approximated by

$$E_+ = \hbar\omega^\circ + \Delta E \qquad \text{and} \qquad E_- = 2\hbar\Omega - \Delta E \tag{10.17}$$

with

$$\Delta E = \frac{(2\sqrt{2}\xi\hbar\Omega)^2}{(\hbar\omega^\circ - 2\hbar\Omega)}$$

Thus, ultimately, when the inequality (10.16) holds, Eqs. (4.59) and (4.60) lead, respectively, to

$$\{C_{1+}\} = \frac{1}{\varepsilon}\{C_{2+}\} \tag{10.18}$$

$$\{C_{1-}\} = -\varepsilon\{C_{2-}\} \tag{10.19}$$

with

$$\varepsilon = \frac{2\sqrt{2}\xi\Omega}{(\omega^\circ - 2\Omega)} < 1 \tag{10.20}$$

$$\{C_{1\pm}\}^2 = 1 - \{C_{2\pm}\}^2$$

Hence, owing to Eqs. (10.18) and (10.19), the unnormalized eigenvectors (10.14) corresponding to the $\pm$ situations read

$$|\Psi_+\rangle = \{C_{2+}\}\left\{\frac{1}{\varepsilon}|\{1\}\rangle|[0]\rangle + |\{0\}\rangle|[2]\rangle\right\} \simeq C_{2+}|\{1\}\rangle|\,[0]\rangle \tag{10.21}$$

$$|\Psi_-\rangle = -\{C_{2-}\}\{\varepsilon|\{1\}\rangle|[0]\rangle - |\{0\}\rangle|[2]\rangle\} \simeq C_{2-}|\{0\}\rangle|\,[2]\rangle \tag{10.22}$$

From Eqs. (10.17), (10.21), and (10.22), it appears that the vibrational state $|\Psi_+\rangle$ roughly looks like the basic state $|\{1\}\rangle|[0]\rangle$ and has an energy $\hbar\omega^\circ$ that is weakly increased by the small amount ΔE, whereas the other vibrational state $|\Psi_-\rangle$, which roughly looks like the basic state $|\{0\}\rangle|[2]\rangle$, has an energy $2\hbar\Omega$ that is weakly stabilized by the same small amount ΔE.

Thus, one may say that $|\Psi_+\rangle$ is roughly the first excited state $|\{1\}\rangle$ of the fast oscillator of angular frequency ω° with an energy $\hbar\omega^\circ + \Delta E$ above that $\hbar\omega^\circ/2$ of its ground state $|\{0\}\rangle$, whereas $|\Psi_-\rangle$ is the second excited state $|[2]\rangle$, of the fast oscillator of angular frequency Ω with an energy $2\hbar\Omega - \Delta E$ above that $\hbar\Omega/2$ of its ground state $|[0]\rangle$.

10.2 STRONG ANHARMONIC COUPLING THEORY

Now, we shall study strong anharmonic coupling theory, which is used in the theory of weak H-bonded species.

10.2.1 Model for bare H-bonded species

For this purpose, consider a system of two oscillators that are anharmonically coupled, the full Hamiltonian of which is given by

$$\mathbf{H}_{\text{Tot}} = \mathbf{H}_{\text{Fast}} + \mathbf{H}_{\text{Slow}} \tag{10.23}$$

Here, $\mathbf{H}_{\text{Slow}}$ is the harmonic Hamiltonian of an oscillator of low angular frequency Ω given by

$$\mathbf{H}_{\text{Slow}} = \left(\frac{\mathbf{P}^2}{2M} + \frac{1}{2}M\Omega^2\,\mathbf{Q}^2\right) \quad \text{with} \quad [\mathbf{Q},\mathbf{P}] = i\hbar \tag{10.24}$$

where $\mathbf{Q}$ and $\mathbf{P}$ are the conjugate position and momentum coordinates, whereas M is the corresponding reduced mass. On the other hand, $\mathbf{H}_{\text{Fast}}$ is the Hamiltonian of a high angular frequency oscillator, the conjugate position and momentum coordinates of which are $\mathbf{q}$ and $\mathbf{p}$, and is given by

$$\mathbf{H}_{\text{Fast}} = \frac{\mathbf{p}^2}{2m} + \frac{1}{2}m\{\omega(\mathbf{Q})\}^2\,\mathbf{q}^2 \quad \text{with} \quad [\mathbf{q},\mathbf{p}] = i\hbar \tag{10.25}$$

where m is the corresponding reduced mass, whereas $\omega(\mathbf{Q})$ is its angular frequency, which is assumed to depend linearly on the coordinate $\mathbf{Q}$ of the low-frequency oscillator according to

$$\omega(\mathbf{Q}) = \omega^\circ + b^\circ\mathbf{Q} \tag{10.26}$$

where ω° is the angular frequency of the high-frequency oscillator when $\mathbf{Q}=0$, that is, at the minimum of the harmonic potential of this low-frequency mode, whereas b° is a constant. Owing to this linear dependence on $\mathbf{Q}$, the Hamiltonian (10.25) of the high-frequency mode is

$$\mathbf{H}_{\text{Fast}} = \left(\frac{\mathbf{p}^2}{2m} + \frac{1}{2}m\omega^{\circ 2}\,\mathbf{q}^2\right) + b^\circ m\omega^\circ\,\mathbf{q}^2\mathbf{Q} + \frac{1}{2}b^{\circ 2}m\mathbf{q}^2\mathbf{Q}^2 \tag{10.27}$$

Owing to Eqs. (10.24) and (10.27), the full Hamiltonian (10.23) may be written

$$\mathbf{H}_{\text{Tot}} = \mathbf{H}^\circ + \mathbf{H}_{\text{Int}} \tag{10.28}$$

with, respectively,

$$\mathbf{H}^\circ = \left(\frac{\mathbf{p}^2}{2m} + \frac{1}{2}m\omega^{\circ 2}\,\mathbf{q}^2\right) + \left(\frac{\mathbf{P}^2}{2M} + \frac{1}{2}M\Omega^2\,\mathbf{Q}^2\right) \tag{10.29}$$

$$\mathbf{H}_{\text{Int}} = b^\circ m\omega^\circ\,\mathbf{q}^2\mathbf{Q} + \tfrac{1}{2}b^{\circ 2}m\,\mathbf{q}^2\mathbf{Q}^2 \tag{10.30}$$

Now, passing to Boson operators according to Eqs. (5.6) and (5.7), that is, in the present situation to

$$\mathbf{Q} = \sqrt{\frac{\hbar}{2M\Omega}}(\mathbf{a}^\dagger + \mathbf{a}) \quad \text{and} \quad \mathbf{P} = i\sqrt{\frac{M\hbar\Omega}{2}}(\mathbf{a}^\dagger - \mathbf{a}) \quad \text{with} \quad [\mathbf{a},\mathbf{a}^\dagger] = 1 \tag{10.31}$$

$$\mathbf{q} = \sqrt{\frac{\hbar}{2m\omega^\circ}}(\mathbf{b}^\dagger + \mathbf{b}) \qquad \text{and} \qquad \mathbf{p} = i\sqrt{\frac{m\hbar\omega^\circ}{2}}(\mathbf{b}^\dagger - \mathbf{b}) \quad \text{with} \quad [\mathbf{b}, \mathbf{b}^\dagger] = 1 \tag{10.32}$$

and defining the dimensionless parameter

$$\alpha^\circ = \frac{b^\circ}{\Omega}\sqrt{\frac{\hbar}{2M\Omega}} \tag{10.33}$$

the Hamiltonians (10.29) and (10.30) take the respective forms

$$\mathbf{H}^\circ = \hbar\omega^\circ\left(\mathbf{b}^\dagger\mathbf{b} + \tfrac{1}{2}\right) + \hbar\Omega\left(\mathbf{a}^\dagger\mathbf{a} + \tfrac{1}{2}\right) \tag{10.34}$$

$$\mathbf{H}_{\text{Int}} = \hbar\Omega\{2\alpha^\circ(\mathbf{b}^\dagger + \mathbf{b})^2(\mathbf{a}^\dagger + \mathbf{a}) + \alpha^{\circ 2}(\mathbf{b}^\dagger + \mathbf{b})^2(\mathbf{a}^\dagger + \mathbf{a})^2\}$$

Again, using the commutation rules

$$[\mathbf{a}, \mathbf{a}^\dagger] = 1 \quad [\mathbf{b}, \mathbf{b}^\dagger] = 1$$

$$[\mathbf{b}, \mathbf{a}^\dagger] = 0 \quad [\mathbf{a}, \mathbf{b}^\dagger] = 0$$

the latter Hamiltonian becomes

$$\mathbf{H}_{\text{Int}} = \alpha^\circ\hbar\Omega\left\{2\{\mathbf{H}_{\text{Int},1}\}(\mathbf{a}^\dagger + \mathbf{a}) + \alpha^\circ\frac{\Omega}{\omega^\circ}\{\mathbf{H}_{\text{Int},1}\}\{\mathbf{H}_{\text{Int},2}\}\right\} \tag{10.35}$$

with

$$\{\mathbf{H}_{\text{Int},1}\} = (\mathbf{b}^\dagger)^2 + (\mathbf{b})^2 + 2\mathbf{b}^\dagger\mathbf{b} + \mathbf{1} \tag{10.36}$$

$$\{\mathbf{H}_{\text{Int},2}\} = (\mathbf{a}^\dagger)^2 + (\mathbf{a})^2 + 2\mathbf{a}^\dagger\mathbf{a} + \mathbf{1} \tag{10.37}$$

respectively.

10.2.2 Hamiltonian matrix representation

In order to diagonalize the Hamiltonian (10.28), consider the two bases $\{|\{k\}\rangle\}$ and $\{|(m)\rangle\}$ defined by the eigenvalue equations

$$\mathbf{b}^\dagger\mathbf{b}|\{k\}\rangle = k|\{k\}\rangle \quad \text{with} \quad \langle\{k\}|\{l\}\rangle = \delta_{kl} \tag{10.38}$$

$$\mathbf{a}^\dagger\mathbf{a}|(m)\rangle = m|(m)\rangle \quad \text{with} \quad \langle(m)|(n)\rangle = \delta_{mn} \tag{10.39}$$

From them, one may build up the following tensor product of basis $\{|\{k\},(m)\rangle\}$ according to

$$|\{k\},(m)\rangle = |\{k\}\rangle|(m)\rangle \quad \text{with} \quad \langle\{k\},(m)|(n),\{l\}\rangle = \delta_{kl}\,\delta_{mn} \tag{10.40}$$

Then, in this basis, it is possible to obtain the matrix elements of the full Hamiltonian, those pertaining to the Hamiltonian (10.34) being given by

$$\langle(m),\{k\}|\mathbf{H}^\circ|\{l\},(n)\rangle = \left(\left(l + \tfrac{1}{2}\right)\hbar\omega^\circ + \left(n + \tfrac{1}{2}\right)\hbar\Omega\right)\delta_{kl}\delta_{mn} \tag{10.41}$$

Now, in order to obtain the matrix elements of the interaction Hamiltonian (10.35), recall Eq. (5.53), that is,

$$\mathbf{b}|\{k\}\rangle = \sqrt{k}|\{k-1\}\rangle \qquad \text{and} \qquad \mathbf{a}|(n)\rangle = \sqrt{n}|(n-1)\rangle$$

which allows one to obtain the following results for the matrix elements of the components (10.36) and (10.37) of the Hamiltonian (10.35):

$$\langle\{k\}|\mathbf{H}_{\text{Int},1}|\{l\}\rangle = ((2l+1)\delta_{kl} + \sqrt{l(l-1)}\delta_{k,l-2} + \sqrt{k(k-1)}\delta_{k-2,l})$$

and

$$\langle(m)|\mathbf{H}_{\text{Int},2}|(n)\rangle = ((2n+1)\delta_{mn} + \sqrt{n(n-1)}\delta_{m,n-2} + \sqrt{m(m-1)}\delta_{m-2,n})$$

Thus, the matrix elements of the two operators (10.36) and (10.37) involved in Eq. (10.35) take on the form

$$\begin{aligned}
&\langle(m),\{k\}|\{\mathbf{H}_{\text{Int},1}\}(\mathbf{a}^{\dagger}+\mathbf{a})|\{l\},(n)\rangle \\
&= ((2l+1)\delta_{kl} + \sqrt{l(l-1)}\delta_{k,l-2} + \sqrt{k(k-1)}\delta_{k-2,l}) \\
&\times ((2l+1)\delta_{kl} + \sqrt{l(l-1)}\delta_{k,l-2} + \sqrt{k(k-1)}\delta_{k-2,l}) \\
&\times (\sqrt{n}\delta_{m,n-1} + \sqrt{m}\delta_{m-1,n})
\end{aligned} \tag{10.42}$$

$$\begin{aligned}
&\langle(m),\{k\}|\mathbf{H}_{\text{Int},1}\mathbf{H}_{\text{Int},2}|\{l\},(n)\rangle \\
&= ((2l+1)\delta_{kl} + \sqrt{l(l-1)}\delta_{k,l-2} + \sqrt{k(k-1)}\delta_{k-2,l}) \\
&\times ((2n+1)\delta_{mn} + \sqrt{n(n-1)}\delta_{m,n-2} + \sqrt{m(m-1)}\delta_{m-2,n})
\end{aligned} \tag{10.43}$$

Hence, using Eqs. (10.41)–(10.43), it is possible to get a matrix representation of the full Hamiltonian (10.28) in terms of $\hbar\Omega$, α°, and ε. Then, if one truncates the basis corresponding to the high and low angular frequency oscillators, respectively, to l° and n°, one obtains a square matrix of dimension $\{(l^{\circ}+1)\times(n^{\circ}+1)\}^2$.

10.3 STRONG ANHARMONIC COUPLING WITHIN THE ADIABATIC APPROXIMATION

Now, we shall show that within the strong anharmonic coupling theory, it is possible to make a so-called *adiabatic separation* between the motions of the high- and low-frequency modes anharmonically coupled because the angular frequency of the high-frequency mode is greater than that of the low-frequency mode by more than one order of magnitude.

10.3.1 Starting equations

Start from the fast mode Hamiltonian (10.25):

$$\mathbf{H}_{\text{Fast}} = \left(\frac{\mathbf{p}^2}{2m} + \frac{1}{2}m(\omega(\mathbf{Q}))^2\,\mathbf{q}^2\right) \tag{10.44}$$

the eigenvalue equation of which reads

$$\mathbf{H}_{\text{Fast}}|\Phi_k(\mathbf{Q})\rangle = \hbar\omega(\mathbf{Q})\left(k + \tfrac{1}{2}\right)|\Phi_k(\mathbf{Q})\rangle \tag{10.45}$$

where $|\Phi_k(\mathbf{Q})\rangle$ are its eigenstates, which depend parametrically on $\mathbf{Q}$, whereas the wavefunctions $\Phi_k(q, \mathbf{Q})$ are corresponding to the kets $|\Phi_k(\mathbf{Q})\rangle$ through the scalar products

$$\Phi_k(q, \mathbf{Q}) = \langle \mathbf{q}|\Phi_k(\mathbf{Q})\rangle$$

where $|\mathbf{q}\rangle$ is an eigenket of the coordinate operator $\mathbf{q}$ of the fast mode, corresponding to the eigenvalue q. Now, the wavefunction $\Phi_k(q, \mathbf{Q})$ of the fast mode is a function of q, which depends parametrically on the coordinate $\mathbf{Q}$ of the H-bond bridge. Note that Eqs. (10.44) and (10.45) hold, whatever the value of $\mathbf{Q}$ may be, so that they are true in the special situation $\mathbf{Q} = 0$ for which the following notation will be used. Here, one may write

$$|\Phi_k(\mathbf{Q} = 0)\rangle \equiv |\{k\}\rangle \tag{10.46}$$

Now, suppose that the Hamiltonian of the H-bond bridge changes with the degree of excitation of the fast mode. Hence, one may consider for the bridge as many effective Hamiltonians $\mathbf{H}^{\{k\}}$ as there are values for the quantum number k appearing in Eqs. (10.44) and (10.45). For each of these effective Hamiltonians, write their eigenvalue equations as

$$\{\mathbf{H}^{\{k\}}\}\left|\chi_n^{\{k\}}\right\rangle = \{E_n^{\{k\}}\}\left|\chi_n^{\{k\}}\right\rangle \tag{10.47}$$

keeping in mind that when $k = 0$, the effective Hamiltonian, which is then $\mathbf{H}^{\{0\}}$, reduces to that of a free harmonic oscillator, that is,

$$\{\mathbf{H}^{\{0\}}\} = \left(\frac{\mathbf{P}^2}{2M} + \frac{1}{2}M\Omega^2\mathbf{Q}^2\right) \tag{10.48}$$

so that the following equivalence holds between the eigenkets $\left|\chi_n^{\{0\}}\right\rangle$ of $\{\mathbf{H}^{\{0\}}\}$ and those $\left|(n)\right\rangle$ of the harmonic oscillator Hamiltonian, that is,

$$\left|\chi_n^{\{0\}}\right\rangle = \left|(n)\right\rangle$$

The eigenkets $\left|\Phi_k(\mathbf{Q})\right\rangle$ of the Hamiltonian (10.44), and those defined by the equations (10.47) and (10.48) $\left|\chi_n^{\{k\}}\right\rangle$ form an orthonormal basis characterized by the equations

$$\left\langle\Phi_k(\mathbf{Q})\middle|\Phi_l(\mathbf{Q})\right\rangle = \delta_{kl} \quad \text{and} \quad \sum_k \left|\Phi_k(\mathbf{Q})\right\rangle\left\langle\Phi_k(\mathbf{Q})\right| = 1 \tag{10.49}$$

$$\left\langle\chi_m^{\{k\}}\middle|\chi_n^{\{k\}}\right\rangle = \delta_{mn} \quad \text{and} \quad \sum_n \left|\chi_n^{\{k\}}\right\rangle\left\langle\chi_n^{\{k\}}\right| = 1 \tag{10.50}$$

Again, from the basis (10.49) and (10.50), it is possible to construct as many tensor product bases as there are k values of the form

$$\left\{ \left|\Phi_k(\mathbf{Q})\right\rangle \left|\chi_n^{\{k\}}\right\rangle \right\} \quad \text{with} \quad k = 0, 1, 2, \ldots$$

All these bases obey the orthonormality properties and the corresponding closure relations given by

$$\left\langle \chi_m^{\{l\}} \right| \left\langle \Phi_l(\mathbf{Q}) \middle| \Phi_k(\mathbf{Q}) \right\rangle \left| \chi_n^{\{k\}} \right\rangle = \delta_{kl}\, \delta_{mn} \tag{10.51}$$

$$\sum_k \sum_n \left|\Phi_k(\mathbf{Q})\right\rangle \left|\chi_n^{\{k\}}\right\rangle \left\langle \chi_n^{\{k\}}\right| \left\langle \Phi_k(\mathbf{Q})\right| = 1 \tag{10.52}$$

10.3.2 Diabatic and adiabatic partition

Now, keeping in mind that, according to Eqs. (10.24) and (10.25), the total Hamiltonian of the two anharmonically coupled oscillators is

$$\mathbf{H}_{\mathrm{Tot}} = \left(\frac{\mathbf{P}^2}{2M} + \frac{1}{2} M \Omega^2 \mathbf{Q}^2 \right) + \left(\frac{\mathbf{p}^2}{2m} + \frac{1}{2} m (\omega(\mathbf{Q}))^2\, \mathbf{q}^2 \right) \tag{10.53}$$

premultiply and postmultiply it by the closure relations (10.52) in the following way:

$$\mathbf{H}_{\mathrm{Tot}} = \sum_k \sum_n \times \left|\Phi_k(\mathbf{Q})\right\rangle \left|\chi_n^{\{k\}}\right\rangle \left\langle \chi_n^{\{k\}}\right| \left\langle \Phi_k(\mathbf{Q})\right| \mathbf{H}_{\mathrm{Tot}} \sum_l \sum_m \left|\Phi_l(\mathbf{Q})\right\rangle \left|\chi_m^{\{l\}}\right\rangle \left\langle \chi_m^{\{l\}}\right| \left\langle \Phi_l(\mathbf{Q})\right|$$

Again, make the following partition:

$$\mathbf{H}_{\mathrm{Tot}} = \mathbf{H}_{\mathrm{Adiab}} + \mathbf{H}_{\mathrm{Diab}} \tag{10.54}$$

with, respectively,

$$\mathbf{H}_{\mathrm{Adiab}} = \sum_k \sum_n \times \left|\Phi_k(\mathbf{Q})\right\rangle \left|\chi_n^{\{k\}}\right\rangle \left\langle \chi_n^{\{k\}}\right| \left\langle \Phi_k(\mathbf{Q})\right| \mathbf{H}_{\mathrm{Tot}} \left|\Phi_k(\mathbf{Q})\right\rangle \left|\chi_n^{\{k\}}\right\rangle \left\langle \chi_n^{\{k\}}\right| \left\langle \Phi_k(\mathbf{Q})\right| \tag{10.55}$$

$$\mathbf{H}_{\mathrm{Diab}} = \sum_k \sum_n \sum_{l \neq k} \sum_{m \neq n} \times \left|\Phi_k(\mathbf{Q})\right\rangle \left|\chi_n^{\{k\}}\right\rangle \left\langle \chi_n^{\{k\}}\right| \left\langle \Phi_k(\mathbf{Q})\right| \mathbf{H}_{\mathrm{Tot}} \left|\Phi_l(\mathbf{Q})\right\rangle \left|\chi_m^{\{k\}}\right\rangle \left\langle \chi_m^{\{k\}}\right| \left\langle \Phi_l(\mathbf{Q})\right| \tag{10.56}$$

The first Hamiltonian $\mathbf{H}_{\mathrm{Adiab}}$ is the adiabatic part, whereas the latter $\mathbf{H}_{\mathrm{Diab}}$ is the diabatic one.

10.3.3 Weakness of the diabatic part of the Hamiltonian

Consider at first the matrix elements involved in the diabatic Hamiltonian (10.56), which in view of Eq. (10.53) reads

$$\begin{aligned}&\langle\chi_n^{\{k\}}|\langle\Phi_k(\mathbf{Q})|\mathbf{H}_{\mathrm{Tot}}|\Phi_l(\mathbf{Q})\rangle|\chi_m^{\{l\}}\rangle\\&=\langle\chi_n^{\{k\}}|\langle\Phi_k(\mathbf{Q})|\left\{\left(\frac{\mathbf{P}^2}{2M}+\frac{1}{2}M\Omega^2\mathbf{Q}^2\right)+\left(\frac{\mathbf{p}^2}{2m}+\frac{1}{2}m(\omega(\mathbf{Q}))^2\,\mathbf{q}^2\right)\right\}|\Phi_l(\mathbf{Q})\rangle|\chi_m^{\{l\}}\rangle\end{aligned} \tag{10.57}$$

Then, owing to the eigenvalue equations (10.45) and (10.44), Eq. (10.57) becomes

$$\begin{aligned}&\langle\chi_n^{\{k\}}|\langle\Phi_k(\mathbf{Q})|\mathbf{H}_{\mathrm{Tot}}|\Phi_l(\mathbf{Q})\rangle|\chi_m^{\{l\}}\rangle\\&=\langle\chi_n^{\{k\}}|\langle\Phi_k(\mathbf{Q})|\left\{\left(\frac{\mathbf{P}^2}{2M}+\frac{1}{2}M\Omega^2\mathbf{Q}^2\right)+\left(l+\frac{1}{2}\right)\hbar\,\omega(\mathbf{Q})\right\}|\Phi_l(\mathbf{Q})\rangle|\chi_m^{\{l\}}\rangle\end{aligned} \tag{10.58}$$

Next, because of the orthonormality properties appearing in Eq. (10.49) and since $k \neq l$, the following matrix elements involved in Eq. (10.58) are zero:

$$\begin{aligned}&\langle\chi_n^{\{k\}}|\langle\Phi_k(\mathbf{Q})|\left(\left(l+\frac{1}{2}\right)\hbar\,\omega(\mathbf{Q})\right)|\Phi_l(\mathbf{Q})\rangle|\chi_m^{\{l\}}\rangle\\&=\langle\chi_n^{\{k\}}|\langle\Phi_k(\mathbf{Q})||\Phi_l(\mathbf{Q})\rangle\left(\left(l+\tfrac{1}{2}\right)\hbar\,\omega(\mathbf{Q})\right)|\chi_m^{\{l\}}\rangle=0\end{aligned} \tag{10.59}$$

Now, the dependence of the ket $|\Phi_l(\mathbf{Q})\rangle$ on the $\mathbf{Q}$ coordinate is parametric, Hence, $\mathbf{Q}$ does not act on this ket as an operator but as a scalar; thus the following matrix elements involved in Eq. (10.58) are also zero:

$$\begin{aligned}&\langle\chi_n^{\{k\}}|\langle\Phi_k(\mathbf{Q})|\mathbf{Q}^2|\Phi_l(\mathbf{Q})\rangle|\chi_m^{\{l\}}\rangle\\&=\langle\chi_n^{\{k\}}|\mathbf{Q}^2|\chi_m^{\{l\}}\rangle\langle\Phi_k(\mathbf{Q})|\Phi_l(\mathbf{Q})\rangle=0\quad\text{since}\quad k\neq l\end{aligned} \tag{10.60}$$

Moreover, the transition matrix elements of the kinetic energy operator involved in Eq. (10.58) read

$$\begin{aligned}&\langle\chi_n^{\{k\}}|\langle\Phi_k(\mathbf{Q})|\left(\frac{\mathbf{P}^2}{2M}\right)|\Phi_l(\mathbf{Q})\rangle|\chi_m^{\{l\}}\rangle\\&=\frac{1}{2M}\langle\chi_n^{\{k\}}|\langle\Phi_k(\mathbf{Q})|\mathbf{P}|\Phi_l(\mathbf{Q})\rangle\mathbf{P}|\chi_m^{\{l\}}\rangle\\&\quad+\langle\Phi_k(\mathbf{Q})|\left(\frac{\mathbf{P}^2}{2M}\right)|\Phi_l(\mathbf{Q})\rangle\langle\chi_n^{\{k\}}|\chi_m^{\{l\}}\rangle\\&\quad+\langle\Phi_k(\mathbf{Q})|\Phi_l(\mathbf{Q})\rangle\langle\chi_n^{\{k\}}|\left(\frac{\mathbf{P}^2}{2M}\right)|\chi_m^{\{l\}}\rangle\end{aligned} \tag{10.61}$$

Hence, via the orthogonality of the kets of the fast mode, that is,

$$\langle\Phi_k(\mathbf{Q})|\Phi_l(\mathbf{Q})\rangle=0$$

Eq. (10.61) simplifies to

$$\langle \chi_n^{\{k\}} | \langle \Phi_k(\mathbf{Q}) | \left(\frac{\mathbf{P}^2}{2M} \right) | \Phi_l(\mathbf{Q}) \rangle | \chi_m^{\{l\}} \rangle$$

$$= \frac{1}{2M} \langle \chi_n^{\{k\}} | \langle \Phi_k(\mathbf{Q}) | \mathbf{P} | \Phi_l(\mathbf{Q}) \rangle \mathbf{P} | \chi_m^{\{l\}} \rangle + \langle \Phi_k(\mathbf{Q}) | \left(\frac{\mathbf{P}^2}{2M} \right) | \Phi_l(\mathbf{Q}) \rangle \langle \chi_n^{\{k\}} | \chi_m^{\{l\}} \rangle$$

Finally, the **P** operator may be expressed in the **Q** representation using of Eqs. (3.50) and (3.51) according to

$$\mathbf{P} = -i\hbar \left(\frac{\partial}{\partial \mathbf{Q}} \right) \tag{10.62}$$

where in the present situation the partial derivative with respect to the position involves an operator **Q** instead of a scalar Q because of the parametric dependence of the kets of the fast mode $|\Phi_l(\mathbf{Q})\rangle$ on **Q**. Then, because of Eqs. (10.59)–(10.62), the matrix elements (10.58) take the form

$$\langle \chi_n^{\{k\}} | \langle \Phi_k(\mathbf{Q}) | \mathbf{H}_{Tot} | \Phi_l(\mathbf{Q}) \rangle | \chi_m^{\{l\}} \rangle$$

$$= -\frac{\hbar^2}{2M} \langle \Phi_k(\mathbf{Q}) | \left(\frac{\partial}{\partial \mathbf{Q}} \right) | \Phi_l(\mathbf{Q}) \rangle \langle \chi_n^{\{k\}} | \left(\frac{\partial}{\partial \mathbf{Q}} \right) | \chi_m^{\{l\}} \rangle$$

$$-\frac{\hbar^2}{2M} \langle \Phi_k(\mathbf{Q}) | \left(\frac{\partial^2}{\partial \mathbf{Q}^2} \right) | \Phi_l(\mathbf{Q}) \rangle \langle \chi_n^{\{k\}} | \chi_m^{\{l\}} \rangle \tag{10.63}$$

Next, in order to evaluate the different transition matrix elements appearing on the right-hand side of Eq. (10.63), consider the commutator of $(\partial/\partial\mathbf{Q})$ with the Hamiltonian (10.25), that is,

$$\langle \Phi_k(\mathbf{Q}) | \left[\mathbf{H}_{\text{Fast}}, \left(\frac{\partial}{\partial \mathbf{Q}} \right) \right] | \Phi_l(\mathbf{Q}) \rangle$$

$$= \langle \Phi_k(\mathbf{Q}) | \mathbf{H}_{\text{Fast}} \left(\frac{\partial}{\partial \mathbf{Q}} \right) | \Phi_l(\mathbf{Q}) \rangle - \langle \Phi_k(\mathbf{Q}) | \left(\frac{\partial}{\partial \mathbf{Q}} \right) \mathbf{H}_{\text{Fast}} | \Phi_l(\mathbf{Q}) \rangle \tag{10.64}$$

So, according to Eqs. (10.44) and (10.45), we have

$$\langle \Phi_k(\mathbf{Q}) | \mathbf{H}_{\text{Fast}} = k\hbar\omega(\mathbf{Q}) \langle \Phi_k(\mathbf{Q}) |$$

$$\mathbf{H}_{\text{Fast}} | \Phi_l(\mathbf{Q}) \rangle = l\hbar\omega(\mathbf{Q}) | \Phi_l(\mathbf{Q}) \rangle$$

so that Eq. (10.64) transforms to

$$\langle \Phi_k(\mathbf{Q}) | \mathbf{H}_{\text{Fast}} \left(\frac{\partial}{\partial \mathbf{Q}} \right) | \Phi_l(\mathbf{Q}) \rangle - \langle \Phi_k(\mathbf{Q}) | \left(\frac{\partial}{\partial \mathbf{Q}} \right) \mathbf{H}_{\text{Fast}} | \Phi_l(\mathbf{Q}) \rangle$$

$$= \hbar\omega(\mathbf{Q})(k - l) \langle \Phi_k(\mathbf{Q}) | \left(\frac{\partial}{\partial \mathbf{Q}} \right) | \Phi_l(\mathbf{Q}) \rangle \tag{10.65}$$

Now, the Hamiltonian (10.25) may be split into its kinetic and potential parts according to

$$\mathbf{H}_{\text{Fast}} = \mathbf{T}_{\text{Fast}} + \mathbf{V}^{\circ}(\mathbf{q}, \mathbf{Q})$$

where $\mathbf{T}_{\text{Fast}}$ and $\mathbf{V}^{\circ}(\mathbf{q},\mathbf{Q})$ are, respectively, the kinetic and potential operators of the high-frequency mode, that is,

$$\mathbf{T}_{\text{Fast}} = \frac{\mathbf{p}^2}{2m}$$

$$\mathbf{V}^{\circ}(\mathbf{q},\mathbf{Q}) = \tfrac{1}{2}m\{\omega(\mathbf{Q})\}^2\,\mathbf{q}^2$$

the potential operator, due to Eq. (10.27) reading

$$\mathbf{V}^{\circ}(\mathbf{q},\mathbf{Q}) = \tfrac{1}{2}m\omega^{\circ 2}\,\mathbf{q}^2 + b^{\circ}m\omega^{\circ}\,\mathbf{q}^2\mathbf{Q} + \tfrac{1}{2}b^{\circ 2}m\mathbf{q}^2\mathbf{Q}^2 \tag{10.66}$$

Furthermore, the kinetic operator $\mathbf{T}_{\text{Fast}}$ of the fast mode commutes with the partial derivative of the H-bond bridge coordinate $\mathbf{Q}$ since it belongs to another space, that is,

$$\left[\mathbf{T}_{\text{Fast}},\left(\frac{\partial}{\partial\mathbf{Q}}\right)\right] = 0 \tag{10.67}$$

Thus, comparing Eq. (10.67), Eq. (10.64) reads

$$\begin{aligned}&\langle\Phi_k(\mathbf{Q})|\left[\mathbf{H}_{\text{Fast}},\left(\frac{\partial}{\partial\mathbf{Q}}\right)\right]|\Phi_l(\mathbf{Q})\rangle\\&= \langle\Phi_k(\mathbf{Q})|\mathbf{V}^{\circ}(\mathbf{q},\mathbf{Q})\left(\frac{\partial}{\partial\mathbf{Q}}\right)|\Phi_l(\mathbf{Q})\rangle - \langle\Phi_k(\mathbf{Q})|\left(\frac{\partial}{\partial\mathbf{Q}}\right)\mathbf{V}^{\circ}(\mathbf{q},\mathbf{Q})|\Phi_l(\mathbf{Q})\rangle\end{aligned} \tag{10.68}$$

or

$$\begin{aligned}\langle\Phi_k(\mathbf{Q})|\left[\mathbf{H}_{\text{Fast}},\left(\frac{\partial}{\partial\mathbf{Q}}\right)\right]|\Phi_l(\mathbf{Q})\rangle &= \langle\Phi_k(\mathbf{Q})|\mathbf{V}^{\circ}(\mathbf{q},\mathbf{Q})\left(\frac{\partial}{\partial\mathbf{Q}}\right)|\Phi_l(\mathbf{Q})\rangle\\&-\langle\Phi_k(\mathbf{Q})|\left\{\left(\frac{\partial\mathbf{V}^{\circ}(\mathbf{q},\mathbf{Q})}{\partial\mathbf{Q}}\right) + \mathbf{V}^{\circ}(\mathbf{q},\mathbf{Q})\left(\frac{\partial}{\partial\mathbf{Q}}\right)\right\}|\Phi_l(\mathbf{Q})\rangle\end{aligned}$$

hence

$$\langle\Phi_k(\mathbf{Q})|\left[\mathbf{H}_{\text{Fast}},\left(\frac{\partial}{\partial\mathbf{Q}}\right)\right]|\Phi_l(\mathbf{Q})\rangle = -\langle\Phi_k(\mathbf{Q})|\left(\frac{\partial\mathbf{V}^{\circ}(\mathbf{q},\mathbf{Q})}{\partial\mathbf{Q}}\right)|\Phi_l(\mathbf{Q})\rangle \tag{10.69}$$

Then, identifying (10.65) and (10.69), we have

$$\langle\Phi_k(\mathbf{Q})|\left(\frac{\partial}{\partial\mathbf{Q}}\right)|\Phi_l(\mathbf{Q})\rangle(k-l)\hbar\omega(\mathbf{Q}) = -\langle\Phi_k(\mathbf{Q})|\left(\frac{\partial\mathbf{V}^{\circ}(\mathbf{q},\mathbf{Q})}{\partial\mathbf{Q}}\right)|\Phi_l(\mathbf{Q})\rangle \tag{10.70}$$

so that, due to Eq. (10.26),

$$\langle\Phi_k(\mathbf{Q})|\left(\frac{\partial}{\partial\mathbf{Q}}\right)|\Phi_l(\mathbf{Q})\rangle = -\frac{\langle\Phi_k(\mathbf{Q})|\left(\dfrac{\partial\mathbf{V}^{\circ}(\mathbf{q},\mathbf{Q})}{\partial\mathbf{Q}}\right)|\Phi_l(\mathbf{Q})\rangle}{(k-l)(\hbar\omega^{\circ}+\hbar b^{\circ}\mathbf{Q})} \tag{10.71}$$

Again, in view of Eqs. (10.26) and of the last equation appearing in (10.66), the operator appearing on the numerator of the right-hand side of Eq. (10.71) becomes

$$\left(\frac{\partial\mathbf{V}^{\circ}(\mathbf{q},\mathbf{Q})}{\partial\mathbf{Q}}\right) = mb^{\circ}(\omega^{\circ}+b^{\circ}\mathbf{Q})\mathbf{q}^2 \tag{10.72}$$

so that Eq. (10.71) reduces to

$$\langle\Phi_k(\mathbf{Q})|\left(\frac{\partial}{\partial\mathbf{Q}}\right)|\Phi_l(\mathbf{Q})\rangle = -\left(\frac{mb^\circ}{\hbar}\right)\frac{\langle\Phi_k(\mathbf{Q})|\,\mathbf{q}^2|\Phi_l(\mathbf{Q})\rangle}{(k-l)} \tag{10.73}$$

Now, according to Eq. (10.33), the b° anharmonic parameter is related to the corresponding dimensionless one α° by

$$b^\circ = \alpha^\circ\Omega\sqrt{\frac{2M\Omega}{\hbar}}$$

Hence, passing from $\mathbf{q}$ to the corresponding Boson operators $\mathbf{b}^\dagger$ and $\mathbf{b}$ defined by Eqs. (10.32), Eq. (10.73) becomes

$$\langle\Phi_k(\mathbf{Q})|\left(\frac{\partial}{\partial\mathbf{Q}}\right)|\Phi_l(\mathbf{Q})\rangle = -\left(\frac{\alpha^\circ}{2}\right)\left(\frac{\Omega}{\omega^\circ}\right)\left(\frac{1}{Q^{\circ\circ}}\right)\frac{\langle\Phi_k(\mathbf{Q})|(\mathbf{b}^\dagger+\mathbf{b})^2|\Phi_l(\mathbf{Q})\rangle}{(k-l)} \tag{10.74}$$

Furthermore, using the commutation rule of Boson operators $[\mathbf{b},\mathbf{b}^\dagger]=1$, the right-hand-side matrix elements take the form

$$\langle\Phi_k(\mathbf{Q})|(\mathbf{b}^\dagger+\mathbf{b})^2|\Phi_l(\mathbf{Q})\rangle = \langle\Phi_k(\mathbf{Q})|((\mathbf{b}^\dagger)^2+(\mathbf{b})^2+2\mathbf{b}^\dagger\mathbf{b}+1)|\Phi_l(\mathbf{Q})\rangle$$

Moreover, since the kets $|\Phi_l(\mathbf{Q})\rangle$ are the eigenkets of the Hamiltonian (10.44) involved in the eigenvalue equation (10.45), and on applying Eqs. (5.53) and (5.63), it appears that

$$\mathbf{b}|\Phi_l(\mathbf{Q})\rangle = \sqrt{l}|\Phi_{l-1}(\mathbf{Q})\rangle \qquad \text{and} \qquad \mathbf{b}^\dagger|\Phi_l(\mathbf{Q})\rangle = \sqrt{l+1}|\Phi_{l+1}(\mathbf{Q})\rangle$$

Hence, after evaluating the right-hand-side matrix elements, Eq. (10.74) transforms to

$$\langle\Phi_k(\mathbf{Q})\,|\left(\frac{\partial}{\partial\mathbf{Q}}\right)|\Phi_l(\mathbf{Q})\rangle = -\left(\frac{\alpha^\circ}{2}\right)\left(\frac{\Omega}{\omega^\circ}\right)\sqrt{\frac{2M\Omega}{\hbar}}C_{kl} \tag{10.75}$$

with

$$C_{kl} = \frac{\sqrt{(l+1)(l+2)}\delta_{k,l+2}+\sqrt{(l)(l-1)}\delta_{k,\,l-2}+(2l+1)\delta_{kl}}{(k-l)} \tag{10.76}$$

Then, the matrix elements appearing in Eq. (10.63), which are of the form

$$-i\hbar\langle\chi_n^{\{k\}}|\left(\frac{\partial}{\partial\mathbf{Q}}\right)|\chi_m^{\{l\}}\rangle = \langle\chi_n^{\{k\}}|\mathbf{P}|\chi_m^{\{l\}}\rangle$$

read, after passing to the corresponding Boson operators by the aid of Eqs. (10.31),

$$-\hbar\langle\chi_n^{\{k\}}|\left(\frac{\partial}{\partial\mathbf{Q}}\right)|\chi_m^{\{l\}}\rangle = \sqrt{\frac{M\hbar\Omega}{2}}\langle\chi_n^{\{k\}}|(\mathbf{a}^\dagger-\mathbf{a})|\chi_m^{\{l\}}\rangle$$

so that

$$-\frac{\hbar^2}{2M}\langle\chi_n^{\{k\}}|\left(\frac{\partial}{\partial\mathbf{Q}}\right)|\chi_m^{\{l\}}\rangle = \hbar\Omega\frac{1}{2}\sqrt{\frac{\hbar}{2M\Omega}}\langle\chi_n^{\{k\}}|(\mathbf{a}^\dagger-\mathbf{a})|\chi_m^{\{l\}}\rangle \tag{10.77}$$

Then, applying in turn Eqs. (5.53) and (5.63) to the Boson operators $\mathbf{a}$ and $\mathbf{a}^\dagger$ yields

$$\mathbf{a}\left|\chi_m^{\{l\}}\right\rangle = \sqrt{m}\left|\chi_{m-1}^{\{l\}}\right\rangle$$

$$\mathbf{a}^\dagger\left|\chi_m^{\{l\}}\right\rangle = \sqrt{m+1}\left|\chi_{m+1}^{\{l\}}\right\rangle$$

so that the right-hand-side matrix elements of Eq. (10.77) are

$$\left\langle\chi_n^{\{k\}}\right|(\mathbf{a}^\dagger - \mathbf{a})\left|\chi_m^{\{l\}}\right\rangle = \sqrt{m+1}\,\delta_{n,m+1} + \sqrt{m}\,\delta_{n,m-1}$$

allowing one to transform Eq. (10.77) as

$$\frac{\hbar^2}{2M}\left\langle\chi_n^{\{k\}}\right|\left(\frac{\partial}{\partial\mathbf{Q}}\right)\left|\chi_m^{\{l\}}\right\rangle = \hbar\Omega\frac{1}{2}\sqrt{\frac{\hbar}{2M\Omega}}(\sqrt{m+1}\,\delta_{n,m+1} + \sqrt{m}\,\delta_{n,m-1}) \tag{10.78}$$

Therefore, in view of Eqs. (10.75), (10.76), and (10.78), the first right-hand-side term of Eq. (10.63) becomes

$$\begin{aligned}&-\frac{\hbar^2}{2M}\langle\Phi_k(\mathbf{Q})|\left(\frac{\partial}{\partial\mathbf{Q}}\right)|\Phi_l(\mathbf{Q})\rangle\left\langle\chi_n^{\{k\}}\right|\left(\frac{\partial}{\partial\mathbf{Q}}\right)\left|\chi_m^{\{l\}}\right\rangle\\&=\hbar\Omega\left(\frac{\alpha^\circ}{4}\right)\left(\frac{\Omega}{\omega^\circ}\right)\frac{\sqrt{(l+1)(l+2)}\,\delta_{k,l+2} + \sqrt{l(l-1)}\,\delta_{k,l-2} + (2l+1)\delta_{k,l})}{(k-l)}\\&\times(\sqrt{m+1}\,\delta_{n,m+1} + \sqrt{m}\,\delta_{n,m-1})\end{aligned} \tag{10.79}$$

Observe at this step that the quantum numbers k and l and also m and n must be small if the dimensionless parameter α° is near unity or smaller, just as weak or intermediate H bonds for which the ratio of the slow and fast mode angular frequencies obeys roughly

$$\frac{\Omega}{\omega^\circ} \backsim \frac{1}{20} \tag{10.80}$$

Hence, it appears that the matrix elements (10.79) cannot exceed a few of the energy $\hbar\Omega$ of the fast mode, which in turn is around $\frac{1}{20}$ of the energy of the fast mode so that one may use the approximation

$$-\frac{\hbar^2}{2M}\langle\Phi_k(\mathbf{Q})|\left(\frac{\partial}{\partial\mathbf{Q}}\right)|\Phi_l(\mathbf{Q})\rangle\left\langle\chi_n^{\{k\}}\right|\left(\frac{\partial}{\partial\mathbf{Q}}\right)\left|\chi_m^{\{l\}}\right\rangle \simeq 0 \tag{10.81}$$

Now, inserting the closure relation appearing in Eq. (10.49) in the matrix elements of $\partial^2/\partial\mathbf{Q}^2$ involved in Eq. (10.63),

$$\langle\Phi_k(\mathbf{Q})|\left(\frac{\partial^2}{\partial\mathbf{Q}^2}\right)|\Phi_l(\mathbf{Q})\rangle = \sum_j\langle\Phi_k(\mathbf{Q})|\left(\frac{\partial}{\partial\mathbf{Q}}\right)|\Phi_j(\mathbf{Q})\rangle\langle\Phi_j(\mathbf{Q})|\left(\frac{\partial}{\partial\mathbf{Q}}\right)|\Phi_l(\mathbf{Q})\rangle$$

multiplying by $-\hbar^2/2M$ and using Eq. (10.75), we have

$$-\frac{\hbar^2}{2M}\langle\Phi_k(\mathbf{Q})|\left(\frac{\partial^2}{\partial\mathbf{Q}^2}\right)|\Phi_l(\mathbf{Q})\rangle = \hbar\Omega\left(\frac{\alpha^\circ}{2}\right)^2\left(\frac{\Omega}{\omega^\circ}\right)^2\sum_j C_{kj}C_{jl} \tag{10.82}$$

Moreover, in view of (10.80), the square of the ratio Ω/ω°, which is around 1/400 so that the matrix element (10.82) is vanishing, so that the following approximation is quite justified

$$-\frac{\hbar^2}{2M}\langle\Phi_k(\mathbf{Q})|\left(\frac{\partial^2}{\partial\mathbf{Q}^2}\right)|\Phi_l(\mathbf{Q})\rangle \simeq 0 \tag{10.83}$$

As a consequence, Eqs. (10.81) and (10.83) allow us to conclude that, for weak to medium H bonds, the diabatic Hamiltonian (10.56), otherwise (10.63), may be neglected so that the full Hamiltonian (10.54) reduces to its adiabatic part (10.55), that is,

$$\mathbf{H}_{\text{Tot}} = \mathbf{H}_{\text{Adiab}} \tag{10.84}$$

10.3.3.1 H-bond bridge effective Hamiltonians Consider the adiabatic Hamiltonian (10.55), which due to Eq. (10.53), reads

$$\begin{aligned}\mathbf{H}_{\text{Adiab}} = &\sum_k\sum_n |\Phi_k(\mathbf{Q})\rangle|\chi_n^{\{k\}}\rangle\langle\chi_n^{\{k\}}|\langle\Phi_k(\mathbf{Q})| \\ &\times\left\{\left(\frac{\mathbf{P}^2}{2M}+\frac{1}{2}M\Omega^2\mathbf{Q}^2\right)+\frac{\mathbf{p}^2}{2m}+\frac{1}{2}m(\omega(\mathbf{Q}))^2\,\mathbf{q}^2\right\} \\ &\times|\Phi_k(\mathbf{Q})\rangle|\chi_n^{\{k\}}\rangle\langle\chi_n^{\{k\}}|\langle\Phi_k(\mathbf{Q})|\end{aligned} \tag{10.85}$$

Then, in view of Eqs. (10.26), (10.44), and (10.45), the matrix elements involved on the right-hand side of Eq. (10.85) become

$$\begin{aligned}&\langle\chi_n^{\{k\}}|\langle\Phi_k(\mathbf{Q})|\left\{\left(\frac{\mathbf{P}^2}{2M}+\frac{1}{2}M\Omega^2\mathbf{Q}^2\right)+\frac{\mathbf{p}^2}{2m}+\frac{1}{2}m(\omega(\mathbf{Q}))^2\,\mathbf{q}^2\right\}|\Phi_k(\mathbf{Q})\rangle|\chi_n^{\{k\}}\rangle \\ &=\langle\chi_n^{\{k\}}|\langle\Phi_k(\mathbf{Q})|\left\{\left(\frac{\mathbf{P}^2}{2M}+\frac{1}{2}M\Omega^2\mathbf{Q}^2\right)+\left(k+\frac{1}{2}\right)\hbar\,(\omega^\circ+b^\circ\mathbf{Q})\right\}|\Phi_k(\mathbf{Q})\rangle|\chi_n^{\{k\}}\rangle\end{aligned}$$

Moreover, since $\left(k+\frac{1}{2}\right)\hbar\,\omega^\circ$ is a scalar, the right-hand side of the latter equation yields

$$\begin{aligned}&\langle\chi_n^{\{k\}}|\langle\Phi_k(\mathbf{Q})|\left\{\left(\frac{\mathbf{P}^2}{2M}+\frac{1}{2}M\Omega^2\mathbf{Q}^2\right)+\left(k+\frac{1}{2}\right)\,\hbar\,(\omega^\circ+b^\circ\mathbf{Q})\right\}|\Phi_k(\mathbf{Q})\rangle|\chi_n^{\{k\}}\rangle \\ &=\left(k+\frac{1}{2}\right)\hbar\,\omega^\circ+\langle\chi_n^{\{k\}}|\left\{\left(\frac{\mathbf{P}^2}{2M}+\frac{1}{2}M\Omega^2\mathbf{Q}^2\right)+\left(k+\frac{1}{2}\right)\,\hbar\,b^\circ\mathbf{Q}\right\}|\chi_n^{\{k\}}\rangle\end{aligned} \tag{10.86}$$

Hence, comparing Eq. (10.86), the adiabatic Hamiltonian (10.85) becomes

$$\begin{aligned}\mathbf{H}_{\text{Adiab}} = &\sum_k\sum_n |\Phi_k(\mathbf{Q})\rangle|\chi_n^{\{k\}}\rangle \\ &\times\left(\left(k+\frac{1}{2}\right)\hbar\,\omega^\circ+\langle\chi_n^{\{k\}}|\left\{\left(\frac{\mathbf{P}^2}{2M}+\frac{1}{2}M\Omega^2\mathbf{Q}^2\right)\right.\right. \\ &\left.\left.+\left(k+\frac{1}{2}\right)\,\hbar\,b^\circ\mathbf{Q}\right\}|\chi_n^{\{k\}}\rangle\right)\langle\chi_n^{\{k\}}|\langle\Phi_k(\mathbf{Q})|\end{aligned}$$

or

$$\mathbf{H}_{\text{Adiab}} = \sum_k \sum_n |\Phi_k(\mathbf{Q})\rangle\langle\Phi_k(\mathbf{Q})|$$
$$\times \left\{ \sum_n |\chi_n^{\{k\}}\rangle\langle\chi_n^{\{k\}}| \left\{ \left(k+\frac{1}{2}\right)\hbar\omega^\circ + \left(\frac{\mathbf{P}^2}{2M} + \frac{1}{2}M\Omega^2\mathbf{Q}^2\right) \right.\right.$$
$$\left.\left. + \left(k+\frac{1}{2}\right)\hbar b^\circ \mathbf{Q} \right\} |\chi_n^{\{k\}}\rangle\langle\chi_n^{\{k\}}| \right\}$$

Finally, using the closure relation (10.52), the adiabatic Hamiltonian reduces to

$$\mathbf{H}_{\text{Adiab}} = \sum_k |\Phi_k(\mathbf{Q})\rangle\langle\Phi_k(\mathbf{Q})|$$
$$\times \left\{ \left(k+\frac{1}{2}\right)\hbar\omega^\circ + \left(\frac{\mathbf{P}^2}{2M} + \frac{1}{2}M\Omega^2\mathbf{Q}^2\right) + \left(k+\frac{1}{2}\right)\hbar b^\circ \mathbf{Q} \right\} \tag{10.87}$$

Now, observe that this adiabatic Hamiltonian is the sum of effective Hamiltonians of the H-bond bridge oscillators corresponding to the different degrees of excitation of the fast mode via the projectors $|\Phi_k(\mathbf{Q})\rangle\langle\Phi_k(\mathbf{Q})|$. Since the parametric dependence on $\mathbf{Q}$ does not modify the structure of Eq. (10.87), we may simply write the adiabatic Hamiltonian (10.87), using (10.46), according to

$$\mathbf{H}_{\text{Adiab}} = \sum_k \{\mathbf{H}^{\{k\}}\}|\{k\}\rangle\langle\{k\}| \tag{10.88}$$

with the effective Hamiltonians given by

$$\{\mathbf{H}^{\{k\}}\} = \left(\frac{\mathbf{P}^2}{2M} + \frac{1}{2}M\Omega^2\mathbf{Q}^2\right) + \left(k+\frac{1}{2}\right)\hbar b^\circ\mathbf{Q} + \left(k+\frac{1}{2}\right)\hbar\omega^\circ \tag{10.89}$$

Next, passing to Boson operators using Eq. (10.31), the effective Hamiltonians transform into

$$\{\mathbf{H}_{\text{I}}^{\{k\}}\} = \left\{\left(\mathbf{a}^\dagger\mathbf{a}+\tfrac{1}{2}\right) + \alpha^\circ\left(k+\tfrac{1}{2}\right)(\mathbf{a}^\dagger+\mathbf{a})\right\}\hbar\Omega + \left(k+\tfrac{1}{2}\right)\hbar\omega^\circ \tag{10.90}$$

where α° is given by Eq. (10.33). Now, in order to remove the driven term $\alpha^\circ(\mathbf{a}^\dagger+\mathbf{a})\hbar\Omega/2$, using Eq. (7.9), that is, taking in this equation the real scalar α° in place of the complex one ξ, namely

$$\mathbf{A}(\alpha^\circ)^{-1}\{\mathbf{f}(\mathbf{a},\mathbf{a}^\dagger)\}\mathbf{A}(\alpha^\circ) = \{\mathbf{f}(\mathbf{a}+\alpha^\circ, \mathbf{a}^\dagger+\alpha^\circ)\} \tag{10.91}$$

we make the following canonical transformation of the Hamiltonian (10.90):

$$\{\widetilde{\mathbf{H}}_{\text{II}}^{\{k\}}\} = \mathbf{A}\left(\frac{\alpha^\circ}{2}\right)\{\mathbf{H}_{\text{I}}^{\{k\}}\}\mathbf{A}\left(\frac{\alpha^\circ}{2}\right)^{-1} \qquad \text{with} \quad \mathbf{A}\left(\frac{\alpha^\circ}{2}\right) = e^{\alpha^\circ(\mathbf{a}^\dagger-\mathbf{a})/2}$$

leading to

$$\{\widetilde{\mathbf{H}}_{\text{II}}^{\{k\}}\} = \left\{\left(\mathbf{a}^\dagger\mathbf{a}+\tfrac{1}{2}\right) + k\alpha^\circ(\mathbf{a}^\dagger+\mathbf{a}) - \left(k+\tfrac{1}{4}\right)\alpha^{\circ 2}\right\}\hbar\Omega + \left(k+\tfrac{1}{2}\right)\hbar\omega^\circ \tag{10.92}$$

Changing the energy reference by subtracting from the effective Hamiltonians (10.92) the same term $(\alpha^{\circ 2}\hbar\Omega/4 - \hbar\omega^{\circ}/2)$ yields new effective Hamiltonians defined by

$$\{\mathbf{H}_{\text{II}}^{\{k\}}\} = \{\widetilde{\mathbf{H}}_{\text{II}}^{\{k\}}\} + \tfrac{1}{4}\alpha^{\circ 2}\hbar\Omega - \tfrac{1}{2}\hbar\omega^{\circ}$$

which read, respectively, for $k = 0$ and $k = 1$

$$\boxed{\{\mathbf{H}_{\text{II}}^{\{0\}}\} = \hbar\Omega\left(\mathbf{a}^{\dagger}\mathbf{a} + \tfrac{1}{2}\right)} \tag{10.93}$$

$$\boxed{\{\mathbf{H}_{\text{II}}^{\{1\}}\} = \hbar\Omega\left\{\left(\mathbf{a}^{\dagger}\mathbf{a} + \tfrac{1}{2}\right) + \alpha^{\circ}(\mathbf{a}^{\dagger} + \mathbf{a}) - \alpha^{\circ 2}\right\} + \hbar\omega^{\circ}} \tag{10.94}$$

10.3.4 New representation for effective Hamiltonians

In order to remove the driven term appearing in the effective Hamiltonian (10.94), without affecting the diagonal effective Hamiltonian (10.93), make the following selective canonical transformations on the different effective Hamiltonians, which are functions of k, that is,

$$\{\mathbf{H}_{\text{III}}^{\{k\}}\} = \mathbf{A}(k\alpha^{\circ})\{\mathbf{H}_{\text{II}}^{\{k\}}\}\mathbf{A}(k\alpha^{\circ})^{-1} \tag{10.95}$$

where $\mathbf{A}(k\alpha^{\circ})$ is the translation operator defined by

$$[\mathbf{A}(k\alpha^{\circ})] = (e^{k\alpha^{\circ}(\mathbf{a}^{\dagger} - \mathbf{a})}) \tag{10.96}$$

Then, one obtains with the help of Eq. (10.91), taking $k\alpha^{\circ}$ in place of α°.

In this new representation denoted {III} resulting from the canonical transformation (10.95), the effective Hamiltonians of the slow mode (10.93) corresponding to the ground state $|\{0\}\rangle$ of the fast mode is unmodified, whereas that (10.94), related to the situation where the fast mode has jumped into its first excited state $|\{1\}\rangle$, is diagonalized, allowing us to write

$$\{\mathbf{H}_{\text{III}}^{\{0\}}\} = \hbar\Omega\left(\mathbf{a}^{\dagger}\mathbf{a} + \tfrac{1}{2}\right)$$

$$\{\mathbf{H}_{\text{III}}^{\{1\}}\} = \hbar\Omega\left\{\left(\mathbf{a}^{\dagger}\mathbf{a} + \tfrac{1}{2}\right) - 2\alpha^{\circ 2}\right\} + \hbar\omega^{\circ}$$

The passage from representation {II} to {III} does not affect the eigenstates of the slow mode harmonic Hamiltonian when the fast mode is in its ground state $|\{0\}\rangle$ but affects them when this mode has jumped into its first excited state $|\{1\}\rangle$. In the latter situation we have

$$|(n)_{\text{III}}^{\{k\}}\rangle = \mathbf{A}(k\alpha^{\circ})|(n)\rangle \tag{10.97}$$

Now, pass to the wavefunction corresponding to the kets (10.97). Those corresponding to $k = 0$ are simply the wavefunctions of the harmonic oscillator, given by the scalar product

$$\langle\{Q\}|(n)_{\text{III}}^{\{0\}}\rangle = \chi_n(Q) \tag{10.98}$$

where the $\langle\{Q\}|$ are the eigenbras of the position operator $\mathbf{Q}$. On the other hand, those corresponding to $k = 1$ are given by

$$\langle\{Q\}|(n)^{\{1\}}_{\text{III}}\rangle = \langle\{Q\}|\mathbf{A}(k\alpha^\circ)|(n)\rangle$$

Now, observe that after passing from Boson operators to the $\mathbf{P}$ operator, the translation operator (10.96) becomes

$$\mathbf{A}(k\alpha^\circ) = e^{-ik\alpha^\circ Q^{\circ\circ}\mathbf{P}} \quad \text{with} \quad Q^{\circ\circ} = \sqrt{\frac{\hbar}{2M\Omega}} \tag{10.99}$$

so that

$$\langle\{Q\}|(n)^{\{1\}}_{\text{III}}\rangle = \langle\{Q\}|(e^{-i\alpha^\circ Q^{\circ\circ}\mathbf{P}})|(n)\rangle \tag{10.100}$$

Then, using Eq. (2.120), it yields

$$\langle\{Q\}|(n)^{\{1\}}_{\text{III}}\rangle = \langle\{Q - \alpha^\circ Q^{\circ\circ}\}|(n)\rangle = \chi_n(Q{-}\alpha^\circ Q^{\circ\circ}) \tag{10.101}$$

Examination of Eqs. (10.98) and (10.101) shows that, in quantum representation {III}, the excitation of the fast mode moves the origin of the slow mode wavefunctions toward shorter lengths. This may be viewed as a translation of the slow mode potential that is induced by the excitation of the fast mode.

As a consequence of the translation of the origin of the slow mode potential induced by the excitation of the fast mode, there is an overlap between the wavefunctions of the H-bond bridge corresponding, respectively, to the ground state of the fast mode and to its first excited state, that is,

$$\langle(m)^{\{0\}}_{\text{III}}|(n)^{\{1\}}_{\text{III}}\rangle = \int_{-\infty}^{\infty} \chi_m(\mathbf{Q})\chi_n(\mathbf{Q}{-}\alpha^\circ Q^{\circ\circ})d\mathbf{Q} = \{A_{mn}(\alpha^\circ)\}$$

These overlaps, which are matrix elements of the translation operator, are the well-known Franck–Condon factors:

$$\{A_{mn}(\alpha^\circ)\} = \langle(m)|(e^{\alpha^\circ(\mathbf{a}^\dagger - \mathbf{a})})|(n)\rangle \tag{10.102}$$

Now, since, when $k = 1$, Eq. (10.97) reads for the ground state $|(0)\rangle$

$$|(0)^{\{1\}}_{\text{III}}\rangle = (e^{\alpha^\circ(\mathbf{a}^\dagger - \mathbf{a})})|(0)\rangle \tag{10.103}$$

then, since according to Eq. (6.95) the action of the translation operator on the ground state of the harmonic oscillator leads to a coherent state $|\{\alpha^\circ\}\rangle$, Eq. (10.103) becomes

$$|(0)^{\{1\}}_{\text{III}}\rangle = |\{\alpha\}\rangle \quad \text{with here} \quad \mathbf{a}|\{\alpha\}\rangle = \alpha^\circ|\{\alpha\}\rangle \tag{10.104}$$

Moreover, due to the expansion (6.16) of a coherent state on the eigenstates of the harmonic oscillator, this ket becomes

$$|(0)^{\{1\}}_{\text{III}}\rangle = \left(\exp\left(-\frac{\alpha^{\circ 2}}{2}\right)\right)\sum_n \frac{\alpha^{\circ n}}{\sqrt{n!}}|(n)\rangle \tag{10.105}$$

where it must be kept in mind that, owing to Eqs. (10.96) and (10.97), there is the equivalence

$$|(n)\rangle \equiv |(0)^{\{0\}}_{\text{III}}\rangle$$

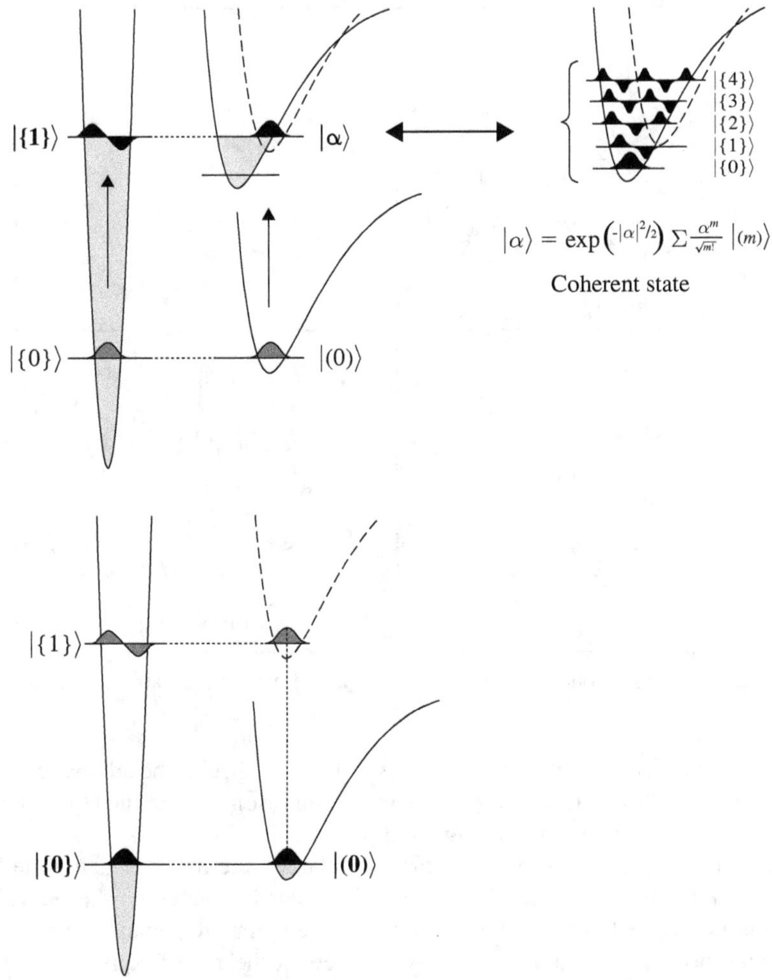

Figure 10.1 Excitation of the fast mode changing the ground state of the H-bond bridge oscillator into a coherent state.

The advantage in passing from quantum representations $\{\mathrm{II}\}$ to $\{\mathrm{III}\}$, is that the Hamiltonian of the H-bond bridge, which is driven in representation $\{\mathrm{II}\}$ when the fast mode is in the state $|\{1\}\rangle$, loses its driven property when passing to representation $\{\mathrm{II}\}$. According to Eq. (10.105), the ground-state of the H-bond bridge corresponding to the ground-state situation of the fast mode $|\{0\}\rangle$ becomes in representation $\{\mathrm{III}\}$ a coherent state, after excitation of the high-frequency mode to the state $|\{1\}\rangle$. This is illustrated in Fig. 10.1.

10.4 FERMI RESONANCES AND STRONG ANHARMONIC COUPLING WITHIN ADIABATIC APPROXIMATION

Fermi resonances, which are well known to play an important role in the area of vibrational processes of H-bonded species, have to be incorporated to the strong

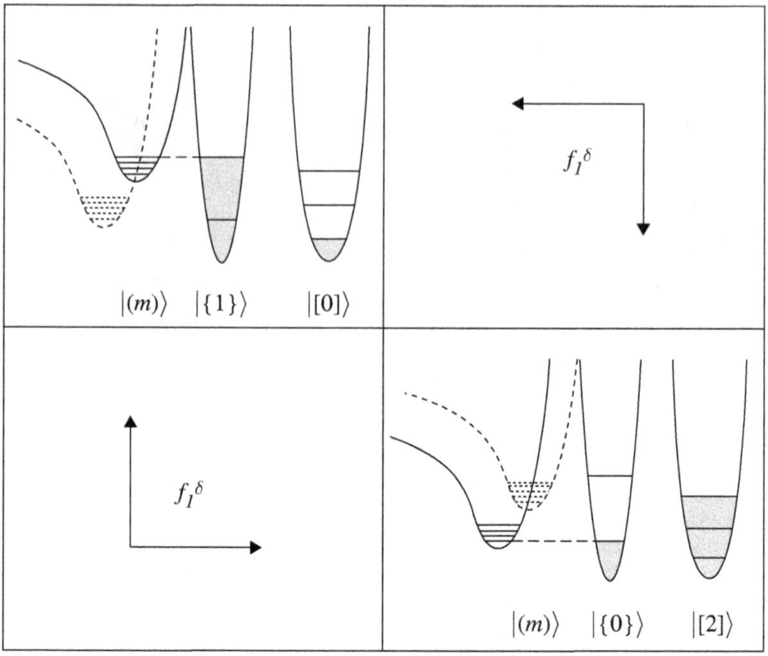

Figure 10.2 Fermi resonance in H-bonded species within the adiabatic approximation.

anharmonic coupling model. Since for weak H-bonded species the adiabatic approximation holds, it allows one to incorporate the Fermi resonances in the H-bond model involving this approximation. See Fig. 10.2.

Consider the model formed by three oscillators, the first one being the high-frequency mode of the H-bonded species, of angular frequency ω, the second one corresponding to the H-bond bridge mode of low angular frequency Ω, and the last one, a vibrational mode anharmonically coupled to the high-frequency mode, its angular frequency ω_δ being around half that of ω. Hence, the full Hamiltonian reads

$$\mathbf{H}_{\text{Tot}} = \mathbf{H}_{\text{Fast}} + \mathbf{H}_{\text{Slow}} + \mathbf{H}_\delta + \mathbf{H}_{\text{Int}} \tag{10.106}$$

Here, $\mathbf{H}_{\text{Slow}}$ is the harmonic Hamiltonian of low angular frequency Ω given by

$$\mathbf{H}_{\text{Slow}} = \left(\frac{\mathbf{P}^2}{2M} + \frac{1}{2} M\Omega^2 \mathbf{Q}^2 \right) \tag{10.107}$$

$\mathbf{H}_\delta$ is the harmonic Hamiltonian of the oscillator of angular frequency ω_δ given by

$$\mathbf{H}_\delta = \left(\frac{\mathbf{p}_\delta^2}{2m_\delta} + \frac{1}{2} m_\delta \omega_\delta^2 \, \mathbf{q}_\delta^2 \right) \tag{10.108}$$

In Eqs. (10.107) and (10.108), $\mathbf{Q}$ and $\mathbf{q}_\delta$ are the position operators of these oscillators, whereas $\mathbf{P}$ and $\mathbf{p}_\delta$ are their conjugate momentum coordinates, whereas M and m_δ are the corresponding reduced masses. On the other hand, the operator $\mathbf{H}_{\text{Fast}}$ appearing in Eq. (10.106) is the Hamiltonian of the high angular frequency mode given by Eq. (10.25):

$$\mathbf{H}_{\text{Fast}} = \frac{\mathbf{p}^2}{2m} + \frac{1}{2}m\{\omega(\mathbf{Q})\}^2\,\mathbf{q}^2 \quad \text{with} \quad [\mathbf{q},\mathbf{p}] = i\hbar$$

where $\mathbf{q}$ and $\mathbf{p}$ are the conjugate position and momentum coordinates, m the reduced mass, whereas $\omega(\mathbf{Q})$ is the angular frequency, which is assumed to depend linearly on the coordinate $\mathbf{Q}$ of the low-frequency oscillator according to

$$\omega(\mathbf{Q}) = \omega^\circ + b^\circ \mathbf{Q}$$

where ω° is the angular frequency of the high-frequency oscillator when $\mathbf{Q} = 0$, that is, at the minimum of the harmonic potential of this low-frequency mode, whereas b° is a constant. Finally, $\mathbf{H}_{\text{Int}}$ represents an anharmonic coupling Hamiltonian of the form

$$\mathbf{H}_{\text{Int}} = \lambda \mathbf{Q}^2 \mathbf{q}$$

Now, the various momentum and position operators obey the commutation rules:

$$[\mathbf{q},\mathbf{p}] = [\mathbf{Q},\mathbf{P}] = [\mathbf{q}_\delta,\mathbf{p}_\delta] = i\hbar$$

$$[\mathbf{q},\mathbf{P}] = [\mathbf{Q},\mathbf{P}] = [\mathbf{q}_\delta,\mathbf{P}] = [\mathbf{q}_\delta,\mathbf{p}] = 0$$

Consider the oscillators of high angular frequency and those of the H-bond bridge of slow angular frequency Ω, the difference between the fast and slow angular frequencies allowing us to perform the adiabatic approximation. Moreover, suppose that there is a residual anharmonic coupling between the high-frequency mode and that of the ω_δ angular frequency and also that the angular frequency ω_δ is near that of $\omega^\circ/2$. That leads us to write

$$\omega^\circ >> \Omega \qquad \text{and} \qquad \omega^\circ \simeq 2\omega_\delta$$

We are therefore concerned with a situation where it is necessary to combine the effective Hamiltonian representation of the Ω slow mode in which its driven character changes with the excitation degree of the ω° fast mode to that of the Fermi resonance representation of the anharmonic coupling of the ω° and ω_δ modes.

Next, passing to Boson operators yields, respectively,

$$\left(\frac{\mathbf{p}^2}{2m} + \frac{1}{2}m\omega^{\circ 2}\,\mathbf{q}^2\right) = \left(\mathbf{b}^\dagger\mathbf{b} + \frac{1}{2}\right)\hbar\omega^\circ$$

$$\left(\frac{\mathbf{P}^2}{2M} + \frac{1}{2}M\Omega^2\,\mathbf{Q}^2\right) = \left(\mathbf{a}^\dagger\mathbf{a} + \frac{1}{2}\right)\hbar\Omega$$

$$\left(\frac{\mathbf{p}_\delta^2}{2m_\delta} + \frac{1}{2}m_\delta\omega_\delta^2\,\mathbf{q}_\delta^2\right) = \left(\mathbf{c}^\dagger\mathbf{c} + \frac{1}{2}\right)\hbar\omega_\delta$$

with

$$[\mathbf{a},\mathbf{a}^\dagger] = 1 \qquad [\mathbf{b},\mathbf{b}^\dagger] = 1 \qquad [\mathbf{c},\mathbf{c}^\dagger] = 1$$

$$[\mathbf{a},\mathbf{b}^\dagger] = 0 \qquad [\mathbf{b},\mathbf{c}^\dagger] = 0 \qquad [\mathbf{a},\mathbf{c}^\dagger] = 0$$

Moreover, the eigenvalue equations of the occupation numbers corresponding to these harmonic Hamiltonians read, respectively,

$$\mathbf{b}^\dagger\mathbf{b}|\{k\}\rangle = k|\{k\}\rangle$$

$$\mathbf{a}^\dagger\mathbf{a}|(n)\rangle = n|(n)\rangle$$

$$\mathbf{c}^\dagger\mathbf{c}|[l]\rangle = l|[l]\rangle$$

Next, the diagonal Hamiltonian built up from the three above harmonic Hamiltonians, that is,

$$\mathbf{H}^\circ = \left(\mathbf{b}^\dagger\mathbf{b} + \tfrac{1}{2}\right)\hbar\omega^\circ + \left(\mathbf{a}^\dagger\mathbf{a} + \tfrac{1}{2}\right)\hbar\Omega + \left(\mathbf{c}^\dagger\mathbf{c} + \tfrac{1}{2}\right)\hbar\omega_\delta$$

admits the eigenvalue equation

$$\mathbf{H}^\circ|\{k\}\rangle|(n)\rangle|[l]\rangle = \left(\left(k+\tfrac{1}{2}\right)\hbar\omega^\circ + \left(n+\tfrac{1}{2}\right)\hbar\Omega + \left(l+\tfrac{1}{2}\right)\hbar\omega_\delta\right)|\{k\}\rangle|(n)\rangle|[l]\rangle$$

The Fermi resonance implies a coupling between the situation where the fast mode ω° is in its ground state $|\{0\}\rangle$ and the ω_δ one is in its second excited state $|[2]\rangle$, and the quasi-resonant situation where the fast mode ω° is in its first excited state $|\{1\}\rangle$ and the ω_δ is in its ground state $|[0]\rangle$. Now, in the adiabatic approximation, the strong anharmonic coupling leads to an effective Hamiltonian describing the Ω slow mode given, respectively, by Eq. (10.93) for $\mathbf{H}^{\{0\}}$ and by Eq. (10.94) for $\mathbf{H}^{\{1\}}$, according to the fact that the ω° high-frequency mode is in its ground state $|\{0\}\rangle$ or in its first excited state $|\{1\}\rangle$. As a consequence, one must focus attention on the Hamiltonian of the coupled oscillators in the subspace spanned by

$$|\{1\}\rangle|(n)\rangle|[0]\rangle \qquad \text{and} \qquad |\{0\}\rangle|(m)\rangle|[2]\rangle \quad \text{with} \quad m,\ n = 0, 1, \ldots$$

Hence, keeping in mind Eq. (10.13), the quantum description leads to the following matrix representation of the Hamiltonian:

	$\vert\{1\}\rangle\vert(n)\rangle\vert[0]\rangle$	$\vert\{0\}\rangle\vert(n)\rangle\vert[2]\rangle$
$\langle[0]\vert\langle(m)\vert\langle\{1\}\vert$	$\left\{\mathbf{M}_{\{1\}[0]}^{\{1\}[0]}\right\}$	$\left\{\mathbf{M}_{\{0\}[2]}^{\{1\}[0]}\right\}$
$\langle[2]\vert\langle(m)\vert\langle\{0\}\vert$	$\left\{\mathbf{M}_{\{0\}[2]}^{\{1\}[0]}\right\}$	$\left\{\mathbf{M}_{\{0\}[2]}^{\{0\}[2]}\right\}$

(10.109)

with, respectively, after neglecting the zero-point energies of the ω_δ bending and of the ω° high-frequency modes,

$$\left\{\mathbf{M}_{\{1\}[0]}^{\{1\}[0]}\right\} = \left\{\left(\mathbf{a}^\dagger\mathbf{a}+\tfrac{1}{2}\right) + \alpha^\circ(\mathbf{a}^\dagger + \mathbf{a}) - \alpha^{\circ 2}\right\}\hbar\Omega + \hbar\omega^\circ \tag{10.110}$$

$$\left\{\mathbf{M}_{\{0\}[2]}^{\{0\}[2]}\right\} = \left(\mathbf{a}^\dagger\mathbf{a}+\tfrac{1}{2}\right)\hbar\Omega + 2\hbar\omega_\delta \tag{10.111}$$

$$\left\{\mathbf{M}_{\{0\}[2]}^{\{1\}[0]}\right\} = \left\{\mathbf{M}_{\{1\}[0]}^{\{0\}[2]}\right\} = \sqrt{2}\xi\hbar\omega_\delta \tag{10.112}$$

with, according to Eq. (10.5),

$$\xi = \lambda\frac{\hbar}{2m\omega^\circ}\sqrt{\frac{\hbar}{2M\Omega}}$$

Thus, the block matrices (10.111) and (10.112) are diagonal, with elements

$$\langle (m)| \left\{ \mathbf{M}_{\{0\}[2]}^{\{0\}[2]} \right\} |(n)\rangle = \left\{ \left(n+\tfrac{1}{2}\right) \hbar\Omega + 2\hbar\omega_\delta \right\} \delta_{mn} \tag{10.113}$$

$$\langle (m)| \left\{ \mathbf{M}_{\{0\}[2]}^{\{1\}[0]} \right\} |(n)\rangle = \sqrt{2}\xi\hbar\omega_\delta \delta_{mn}$$

while the block matrix (10.110), which is nondiagonal, involves elements that are given, respectively, by Eqs. (6.148)–(6.150), the diagonal ones being

$$\langle (n)| \left\{ \mathbf{M}_{\{1\}[0]}^{\{1\}[0]} \right\} |(n)\rangle = \hbar\Omega \left(n+\tfrac{1}{2}\right) + \hbar\omega^\circ \; - \alpha^{\circ 2}\hbar\Omega$$

and the off-diagonal ones being

$$\langle (m)| \left\{ \mathbf{M}_{\{1\}[0]}^{\{1\}[0]} \right\} |(n)\rangle = \mathbf{2}\alpha^\circ \hbar\Omega(\sqrt{n}\delta_{m,n-1} + \sqrt{m}\delta_{m-1,n}) \tag{10.114}$$

It is, therefore, possible, using Eqs. (10.113)–(10.114), to obtain all the matrix elements of the representation (10.109). In order to make the diagonalization numerically tractable, it is necessary to truncate the matrix representations (10.110) and (10.111). Then, if the truncatures conserve the n° lowest energy levels $|(n)\rangle$ used for the matrix representations, they yield a $2n^\circ \times 2n^\circ$ Hamiltonian matrix to be diagonalized. Of course, n° has to be chosen so as the lowest eigenvalues and eigenvectors of interest remain stable with respect to an increase in n°.

10.5 DAVYDOV AND STRONG ANHARMONIC COUPLINGS

When two oscillators a and b have the same angular frequency, an interaction may occur between two degenerate situations, one in which the b oscillator is in its first excited state $|\{1\}_b\rangle$ and the a oscillator is in its ground state $|\{0\}_a\rangle$ and the other corresponding to the inverse situation in which the b oscillator is in its ground state state $|\{0\}_b\rangle$ and the a oscillator is in its first excited state $|\{1\}_a\rangle$. This interaction, called *Davydov coupling,* which induces a splitting of the degenerate energy levels corresponding to the ket products $|\{1\}_b\rangle|\{0\}_a\rangle$ and $|\{0\}_b\rangle|\{1\}_a\rangle$, is summarized in Fig. 10.3.

Such an interaction occurs, for instance, in centrosymmetric cyclic dimers of H-bonded species. However, it has to be taken into account together with the strong anharmonic coupling involved in each H-bonded entity of the dimer and that has been previously studied in this chapter.

10.5.1 H-bonded cyclic dimer of carboxylic acid

Consider a cyclic dimer of carboxylic acid with two H-bond bridges. The two moieties of the dimer are labeled a and b. For cyclic symmetric dimers of H bonds, there are two degenerate high-frequency modes and two degenerate low-frequency H-bond vibrations, as shown in Fig. 10.4.

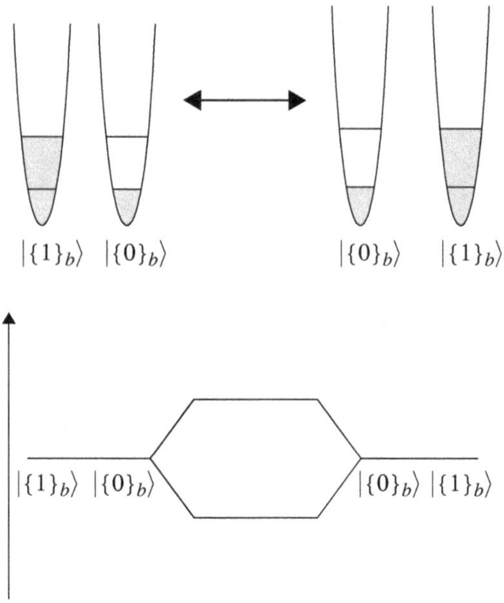

Figure 10.3 Davydov coupling.

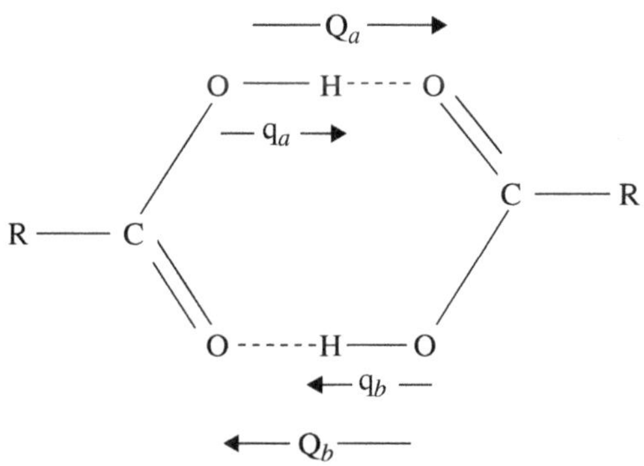

Figure 10.4 Degenerate modes of a centrosymmetric H-bonded dimer.

For each moiety of the H-bonded cyclic dimer, the adiabatic separation between the high- and low-frequency modes leads, for the slow H-bond bridge oscillators, to effective Hamiltonians that differ whether the high-frequency mode is either in its ground state or in its first excited state, the oscillator of the bridge becoming driven when the fast mode passes from its ground state to its first excited state. Moreover, when one of the two identical fast modes is excited, then, because of the symmetry of the cyclic dimer a nonadiabatic Davydov interaction V° may occur, leading to an energy exchange between this excited state and that of the other identical fast mode of the dimer. This underlying physics is the aim of the present section.

In order to visualize this physics, it must be kept in mind that the description of a driven oscillator is equivalent to another one where the potential of this oscillator is displaced. Hence, the above adiabatic description of the H-bond bridge of the two moieties, is equivalent to the new picture where the potential of the H-bond bridge is displaced when the high-frequency mode is passing from its ground state to its first excited state. Then, within this new description, the nonadiabatic Davydov coupling between the two adiabatic representations of the H-bond bridges may be viewed as coupling two equivalent physical situations: In the first situation, the high-frequency oscillator b is in its first excited state $|\{1\}_b\rangle$ and the potential of the H-bond bridge to which it is coupled is displaced, whereas the other fast oscillator a is in its ground state $|\{0\}_a\rangle$, and the potential of the corresponding H-bond bridge is undisplaced. In the other situation, inversely, the high-frequency oscillator b oscillators being in its ground state $|\{0\}_b\rangle$, the potential of the corresponding H-bond bridge to which it is coupled being undisplaced, whereas the other fast oscillator a is in its first excited state $|\{1\}_a\rangle$, the potential of the corresponding H-bond bridge being displaced. This is summarized in Fig. 10.5, where, in order to distinguish clearly the potentials of the high- and low-frequency modes, those of the slow H-bond bridge have been depicted by Morse curves, although in the following these potentials will be assumed to be harmonic.

In Fig. 10.5, the kets $|(m)_a\rangle$ and $|(m)_b\rangle$ are the eigenkets of the Hamiltonians of the H-bond bridges belonging to the two moieties a and b.

Now, return to the initial description of the system, working in terms of effective Hamiltonians and for which it will be assumed that the H-bond bridge may be viewed as harmonic. First, it may be observed that because of the symmetry of the dimer,

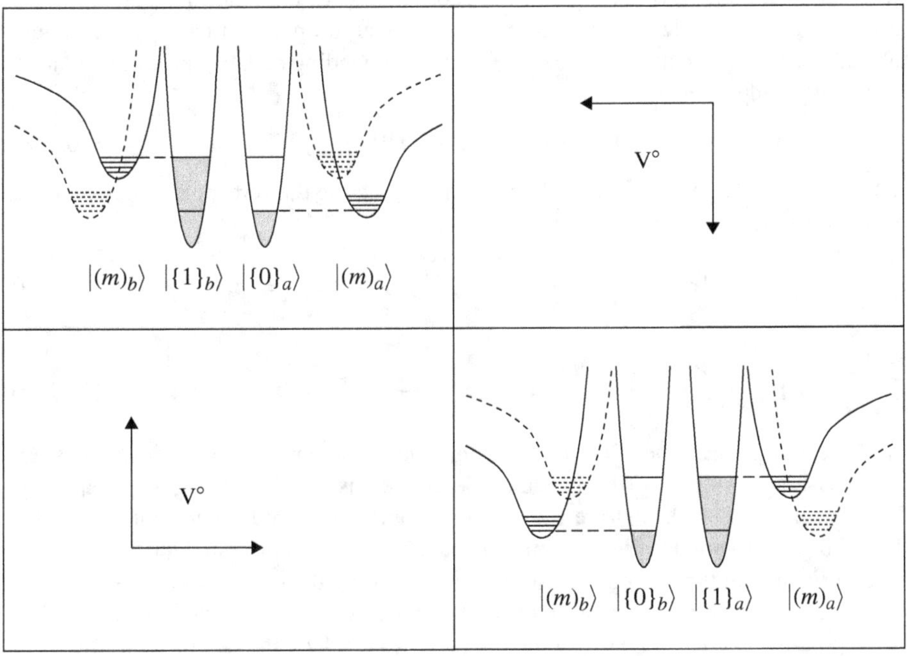

Figure 10.5 Davydov coupling in H-bonded centrosymmetric cyclic dimers.

there exist a $\widehat{\mathbf{C}}_2$ operator (with $\widehat{\mathbf{C}}_2^2 = 1$), which exchanges the coordinates $\mathbf{Q}_i$ of the two H-bond bridges of the cyclic dimer according to

$$\widehat{\mathbf{C}}_2\mathbf{Q}_a = \mathbf{Q}_b \qquad \widehat{\mathbf{C}}_2\mathbf{Q}_b = \mathbf{Q}_a \tag{10.115}$$

In the strong anharmonic coupling theory, the Hamiltonians of the high-frequency modes are

$$\{\mathbf{H}_{\text{Fast}}\}_i = \left(\frac{\mathbf{p}_i^2}{2m} + \frac{m(\omega(\mathbf{Q}_i))^2\mathbf{q}_i^2}{2}\right) \qquad i = a, b$$

where $\mathbf{p}_i$ and $\mathbf{q}_i$ are the coordinates and the conjugate momenta of the two degenerate high-frequency modes, the angular frequencies $\omega(\mathbf{Q}_i)$ of which are the same and supposed to depend on the coordinate of the H-bond bridge. Expansion to first order of the angular frequency $\omega(\mathbf{Q}_i)$ of the fast mode with respect to the coordinate $\mathbf{Q}_i$ of the H-bond bridge yields

$$\omega(\mathbf{Q}_i) = \omega^\circ + b^\circ\mathbf{Q}_i \tag{10.116}$$

where ω° is the angular frequency of the two degenerate fast modes when the corresponding H-bond bridge coordinates are at equilibrium, whereas b° is a constant.

Again, write the eigenvalue equation of the two high-frequency modes when the H-bond bridge modes are at equilibrium, that is, when $\mathbf{Q}_i = 0$:

$$\left(\frac{\mathbf{p}_i^2}{2m} + \frac{m\omega^{\circ 2}\mathbf{q}_i^2}{2}\right)|\{k\}_i\rangle = \left(k_i + \frac{1}{2}\right)\hbar\omega^\circ|\{k\}_i\rangle \tag{10.117}$$

In the adiabatic approximation, and in accordance with the conditions encountered in the study of a single H-bond bridge, the full Hamiltonians of each dimer moiety take the form of a sum of effective Hamiltonians depending on the degree of excitation of the fast modes:

$$\{\mathbf{H}_{\text{Adiab}}\}_i = \{\mathbf{H}^{\{0\}}\}_i|\{0\}_i\rangle\langle\{0\}_i| + \{\mathbf{H}^{\{1\}}\}_i|\{1\}_i\rangle\langle\{1\}_i| \tag{10.118}$$

with, respectively, in view of Eq. (10.89), and neglecting the zero-point energy $\hbar\omega^\circ/2$ of the fast mode

$$\{\mathbf{H}^{\{0\}}\}_i = \left(\frac{\mathbf{P}_i^2}{2M} + \frac{M\Omega^2\mathbf{Q}_i^2}{2}\right) \quad \text{with} \quad i = a, b \tag{10.119}$$

$$\{\mathbf{H}^{\{1\}}\}_i = \left(\frac{\mathbf{P}_i^2}{2M} + \frac{M\Omega^2\mathbf{Q}_i^2}{2}\right) + \hbar b^\circ\mathbf{Q}_i + \hbar\omega^\circ \quad \text{with} \quad i = a, b \tag{10.120}$$

In these equations, the $\mathbf{P}_i$ are the conjugate momenta of the coordinates $\mathbf{Q}_i$ of the H-bond bridges of the two moieties, whereas Ω is their angular frequency. The Hamiltonians (10.119) are those of the undriven quantum harmonic oscillator describing the H-bond bridge moieties a and b, whereas Hamiltonian (10.120) is that of the driven quantum harmonic oscillators describing the a H-bond bridge moiety.

Next, consider an excitation of the fast mode of one moiety of the dimer. The corresponding excited state is resonant with the state where the fast mode of the other moiety is excited. Thus, some Davydov coupling may occur when one of the fast mode has been excited. The Hamiltonian of the cyclic dimer involving Davydov coupling between the first excited state of the high-frequency oscillator a of one moiety, and the excited state of the oscillator b of the other moiety and vice versa is given by

$$\{\mathbf{H}_{\mathrm{Dav}}\} = \{\mathbf{H}_{\mathrm{Adiab}}\}_a + \{\mathbf{H}_{\mathrm{Adiab}}\}_b + \mathbf{V}_{ab} \tag{10.121}$$

Here, $\mathbf{V}_{ab}$ is the Davydov coupling Hamiltonian between the first excited state of the two high-frequency oscillators:

$$\mathbf{V}_{ab} = \hbar V^{\circ}(|\{1\}_a\rangle\langle\{0\}_b| + |\{0\}_a\rangle\langle\{1\}_b|)$$

$$\widehat{\mathbf{C}}_2\mathbf{V}_{ab} = \mathbf{V}_{ab}$$

The eigenvalue equations of the two harmonic H-bond bridge Hamiltonians are

$$\left(\frac{\mathbf{P}_i^2}{2M} + \frac{M\Omega^2\mathbf{Q}_i^2}{2}\right)|(m)_i\rangle = \left(m_i + \frac{1}{2}\right)\hbar\Omega|(m)_i\rangle \tag{10.122}$$

Again, from these states and those given by Eq. (10.117), we can construct the following tensor product of states with m_i $(i = a, b)$ running from 0 to ∞ by

$$\begin{aligned}
\left|\Psi_{m_a,m_b}^{\{0,0\}}\right\rangle &= |\{0\}_a\rangle|(m)_a\rangle|\{0\}_b\rangle|(m)_b\rangle \\
\left|\Psi_{m_a,m_b}^{\{1,0\}}\right\rangle &= |\{1\}_a\rangle|(m)_a\rangle|\{0\}_b\rangle|(m)_b\rangle \\
\left|\Psi_{m_a,m_b}^{\{0,1\}}\right\rangle &= |\{0\}_a\rangle|(m)_a\rangle|\{1\}_b\rangle|(m)_b\rangle
\end{aligned} \tag{10.123}$$

Next, define the following vectors of kets according to

$$\begin{pmatrix} |\Phi_{\{a,b\}}^{\{0,0\}}\rangle \\ |\Phi_{\{a,b\}}^{\{1,0\}}\rangle \\ |\Phi_{\{a,b\}}^{\{0,1\}}\rangle \end{pmatrix} \equiv \begin{pmatrix} |\{0\}_a\rangle|\{0\}_b\rangle \\ |\{1\}_a\rangle|\{0\}_b\rangle \\ |\{0\}_a\rangle\ |\{1\}_b\rangle \end{pmatrix} \tag{10.124}$$

Then, in the basis (10.124), the Hamiltonian (10.121) is

$$\{\mathbf{H}_{\mathrm{Dav}}\} = \begin{pmatrix} \{\mathbf{H}^{\{0,0\}}\} & 0 & 0 \\ 0 & \{\mathbf{H}^{\{1,0\}}\} & \mathbf{V}^{\circ} \\ 0 & \mathbf{V}^{\circ} & \{\mathbf{H}^{\{0,1\}}\} \end{pmatrix} \tag{10.125}$$

with, respectively,

$$\{\mathbf{H}^{\{0,0\}}\} = \{\mathbf{H}^{\{0\}}\}_a + \{\mathbf{H}^{\{0\}}\}_b \tag{10.126}$$

$$\{\mathbf{H}^{\{1,0\}}\} = \{\mathbf{H}^{\circ\{1\}}\}_a + \{\mathbf{H}^{\circ\{0\}}\}_b \tag{10.127}$$

Now, observe that the action of the parity operator $\widehat{\mathbf{C}}_2$ on the Hamiltonian $\{\mathbf{H}^{\{1,0\}}\}$ transforms it into$\{\mathbf{H}^{\{0,1\}}\}$ and vice versa,

$$\widehat{\mathbf{C}}_2\{\mathbf{H}^{\{i,j\}}\} = \{\mathbf{H}^{\{j,i\}}\} \tag{10.128}$$

whereas it does not affect the Hamiltonian $\mathbf{V}^{\circ}$:

$$\widehat{\mathbf{C}}_2\mathbf{V}^{\circ} = \mathbf{V}^{\circ} \tag{10.129}$$

Note also that the following explicit expressions hold for any quantity **A** such as operator or ket:

$$\widehat{\mathbf{C}}_2\{\mathbf{H}^{\{i,j\}}\}\mathbf{A} = \{\mathbf{H}^{\{j,i\}}\}\widehat{\mathbf{C}}_2\mathbf{A} \qquad \widehat{\mathbf{C}}_2\mathbf{V}^{\circ}\mathbf{A} = \mathbf{V}^{\circ}\widehat{\mathbf{C}}_2\mathbf{A} \tag{10.130}$$

Moreover, in the following, we shall use the fact that the square of the parity operator is unity, that is,

$$(\widehat{\mathbf{C}}_2)^2 = \widehat{\mathbf{1}} \tag{10.131}$$

10.5.2 A 2 × 2 matrices commutator

Now, we prove that the following commutator is zero:

$$\left[\begin{pmatrix} 0 & \widehat{\mathbf{C}}_2 \\ \widehat{\mathbf{C}}_2 & 0 \end{pmatrix}, \begin{pmatrix} \{\mathbf{H}^{\{1,0\}}\} & \mathbf{V}^{\circ} \\ \mathbf{V}^{\circ} & \{\mathbf{H}^{\{0,1\}}\} \end{pmatrix}\right] = 0 \tag{10.132}$$

For this purpose, observe that

$$\begin{pmatrix} 0 & \widehat{\mathbf{C}}_2 \\ \widehat{\mathbf{C}}_2 & 0 \end{pmatrix}\begin{pmatrix} \{\mathbf{H}^{\{1,0\}}\} & \mathbf{V}^{\circ} \\ \mathbf{V}^{\circ} & \{\mathbf{H}^{\{0,1\}}\} \end{pmatrix} = \begin{pmatrix} (\widehat{\mathbf{C}}_2\mathbf{V}^{\circ}) & (\widehat{\mathbf{C}}_2\{\mathbf{H}^{\{0,1\}}\}) \\ (\widehat{\mathbf{C}}_2\{\mathbf{H}^{\{1,0\}}\}) & (\widehat{\mathbf{C}}_2\mathbf{V}^{\circ}) \end{pmatrix}$$

or, using Eq. (10.130),

$$\begin{pmatrix} 0 & \widehat{\mathbf{C}}_2 \\ \widehat{\mathbf{C}}_2 & 0 \end{pmatrix}\begin{pmatrix} \{\mathbf{H}^{\{1,0\}}\} & \mathbf{V}^{\circ} \\ \mathbf{V}^{\circ} & \{\mathbf{H}^{\{0,1\}}\} \end{pmatrix} = \begin{pmatrix} (\mathbf{V}^{\circ}\widehat{\mathbf{C}}_2) & (\{\mathbf{H}^{\{1,0\}}\}\widehat{\mathbf{C}}_2) \\ (\{\mathbf{H}^{\{0,1\}}\}\widehat{\mathbf{C}}_2^{\circ}) & (\mathbf{V}^{\circ}\widehat{\mathbf{C}}_2) \end{pmatrix} \tag{10.133}$$

On the other hand, the inverse product of the matrices yields

$$\begin{pmatrix} \{\mathbf{H}^{\{1,0\}}\} & \mathbf{V}^{\circ} \\ \mathbf{V}^{\circ} & \{\mathbf{H}^{\{0,1\}}\} \end{pmatrix}\begin{pmatrix} 0 & \widehat{\mathbf{C}}_2 \\ \widehat{\mathbf{C}}_2 & 0 \end{pmatrix} = \begin{pmatrix} (\mathbf{V}^{\circ}\widehat{\mathbf{C}}_2) & (\{\mathbf{H}^{\{1,0\}}\}\widehat{\mathbf{C}}_2) \\ (\{\mathbf{H}^{\{0,1\}}\}\widehat{\mathbf{C}}_2) & (\mathbf{V}^{\circ}\widehat{\mathbf{C}}_2) \end{pmatrix} \tag{10.134}$$

a result that is identical to that in (10.133) so that Eq. (10.132) is verified.

10.5.3 Eigenvectors of the matrix built up from $\widehat{\mathbf{C}}_2$

Now, consider the action of the matrix constructed from the $\widehat{\mathbf{C}}_2$ operator, which satisfies the commutator (10.132) on the following spinor:

$$\{\tilde{\beta}^{(+)}\} = \begin{pmatrix} \{\beta^{(+)}\} \\ \widehat{\mathbf{C}}_2\{\beta^{(+)}\} \end{pmatrix} \tag{10.135}$$

It reads

$$\begin{pmatrix} 0 & \widehat{\mathbf{C}}_2 \\ \widehat{\mathbf{C}}_2 & 0 \end{pmatrix}\begin{pmatrix} \{\beta^{(+)}\} \\ \widehat{\mathbf{C}}_2\{\beta^{(+)}\} \end{pmatrix} = \begin{pmatrix} \widehat{\mathbf{C}}_2\widehat{\mathbf{C}}_2\{\beta^{(+)}\} \\ \widehat{\mathbf{C}}_2\{\beta^{(+)}\} \end{pmatrix}$$

or, due to

$$\widehat{\mathbf{C}}_2^2 = 1 \tag{10.136}$$

$$\begin{pmatrix} 0 & \widehat{\mathbf{C}}_2 \\ \widehat{\mathbf{C}}_2 & 0 \end{pmatrix}\begin{pmatrix} \{\beta^{(+)}\} \\ \widehat{\mathbf{C}}_2\{\beta^{(+)}\} \end{pmatrix} = +1\begin{pmatrix} \{\beta^{(+)}\} \\ \widehat{\mathbf{C}}_2\{\beta^{(+)}\} \end{pmatrix} \tag{10.137}$$

Again, repeat the action of this same matrix constructed from the $\widehat{\mathbf{C}}_2$ operator, on the other spinors

$$\{\tilde{\beta}^{(-)}\} = \begin{pmatrix} \{\beta^{(-)}\} \\ -\widehat{\mathbf{C}}_2\{\beta^{(-)}\} \end{pmatrix} \tag{10.138}$$

to give

$$\begin{pmatrix} 0 & \widehat{\mathbf{C}}_2 \\ \widehat{\mathbf{C}}_2 & 0 \end{pmatrix} \begin{pmatrix} \{\beta^{(-)}\} \\ -\widehat{\mathbf{C}}_2\{\beta^{(-)}\} \end{pmatrix} = \begin{pmatrix} -\widehat{\mathbf{C}}_2\widehat{\mathbf{C}}_2\{\beta^{(-)}\} \\ \widehat{\mathbf{C}}_2\{\beta^{(+)}\} \end{pmatrix}$$

or, due to Eq. (10.136),

$$\begin{pmatrix} 0 & \widehat{\mathbf{C}}_2 \\ \widehat{\mathbf{C}}_2 & 0 \end{pmatrix} \begin{pmatrix} \{\beta^{(-)}\} \\ -\widehat{\mathbf{C}}_2\{\beta^{(-)}\} \end{pmatrix} = -1 \begin{pmatrix} \{\beta^{(-)}\} \\ -\widehat{\mathbf{C}}_2\{\beta^{(-)}\} \end{pmatrix} \tag{10.139}$$

Equations (10.137) and (10.138) show that the spinors (10.135) and (10.138) are eigenvectors of the matrix built up from the $\widehat{\mathbf{C}}_2$ operator, which verifies the commutator (10.132), so that they are also eigenvectors of the operators matrix involved in this commutator.

10.5.4 Diagonalization of the 2 x 2 matrix involving coupled effective Hamiltonians

Thus, these spinors may be used to diagonalize the operator matrix involved in the commutator (10.132). For this purpose, premultiply these spinors by this operator matrix:

$$\begin{pmatrix} \{\mathbf{H}^{\{1,0\}}\} & \mathbf{V}^{\circ} \\ \mathbf{V}^{\circ} & \{\mathbf{H}^{\{0,1\}}\} \end{pmatrix} \begin{pmatrix} \{\beta^{(\pm)}\} \\ \pm\widehat{\mathbf{C}}_2\{\beta^{(\pm)}\} \end{pmatrix} = \begin{pmatrix} (\{\mathbf{H}^{\{1,0\}}\} \pm \mathbf{V}^{\circ}\widehat{\mathbf{C}}_2)\{\beta^{(\pm)}\} \\ (\mathbf{V}^{\circ} \pm \{\mathbf{H}^{\{0,1\}}\}\widehat{\mathbf{C}}_2)\{\beta^{(\pm)}\} \end{pmatrix} \tag{10.140}$$

Now, insert the unity operator resulting from Eq. (10.136) in the following way:

$$\begin{pmatrix} \{\mathbf{H}^{\{1,0\}}\} & \mathbf{V}^{\circ} \\ \mathbf{V}^{\circ} & \{\mathbf{H}^{\{0,1\}}\} \end{pmatrix} \begin{pmatrix} \{\beta^{(\pm)}\} \\ \pm\widehat{\mathbf{C}}_2\{\beta^{(\pm)}\} \end{pmatrix} = \begin{pmatrix} (\{\mathbf{H}^{\{1,0\}}\} \pm \mathbf{V}^{\circ}\widehat{\mathbf{C}}_2)\{\beta^{(\pm)}\} \\ (\mathbf{V}^{\circ} \pm \{\mathbf{H}^{\{0,1\}}\}\widehat{\mathbf{C}}_2)\widehat{\mathbf{C}}_2^2\{\beta^{(\pm)}\} \end{pmatrix}$$

It reads

$$\begin{pmatrix} \{\mathbf{H}^{\{1,0\}}\} & \mathbf{V}^{\circ} \\ \mathbf{V}^{\circ} & \{\mathbf{H}^{\{0,1\}}\} \end{pmatrix} \begin{pmatrix} \{\beta^{(\pm)}\} \\ \pm\widehat{\mathbf{C}}_2\{\beta^{(\pm)}\} \end{pmatrix} = \begin{pmatrix} (\{\mathbf{H}^{\{1,0\}}\} \pm \mathbf{V}^{\circ}\widehat{\mathbf{C}}_2)\{\beta^{(\pm)}\} \\ (\mathbf{V}^{\circ}\widehat{\mathbf{C}}_2 \pm \{\mathbf{H}^{\{0,1\}}\}\widehat{\mathbf{C}}_2^2)\widehat{\mathbf{C}}_2\{\beta^{(\pm)}\} \end{pmatrix}$$

and thus, after simplification using (10.136)

$$\begin{pmatrix} \{\mathbf{H}^{\{1,0\}}\} & \mathbf{V}^{\circ} \\ \mathbf{V}^{\circ} & \{\mathbf{H}^{\{0,1\}}\} \end{pmatrix} \begin{pmatrix} \{\beta^{(\pm)}\} \\ \pm\widehat{\mathbf{C}}_2\{\beta^{(\pm)}\} \end{pmatrix}$$
$$= \begin{pmatrix} \{\mathbf{H}^{\{1,0\}}\} \pm \mathbf{V}^{\circ}\widehat{\mathbf{C}}_2 & 0 \\ 0 & \{\mathbf{H}^{\{0,1\}}\} \pm \mathbf{V}^{\circ}\widehat{\mathbf{C}}_2 \end{pmatrix} \begin{pmatrix} \{\beta^{(\pm)}\} \\ (\pm\widehat{\mathbf{C}}_2)\{\beta^{(\pm)}\} \end{pmatrix} \tag{10.141}$$

Hence, due to Eq. (10.141), the nondiagonal block matrix (10.125) built up from effective Hamiltonians, may be put in the following diagonal block form:

$$\{\mathbf{H}_{\text{Dav}}\} = \begin{pmatrix} \{\mathbf{H}^{\{0,0\}}\} & 0 & 0 \\ 0 & \{\mathbf{H}^{\{1,0\}}\} + \mathbf{V}^{\circ}\widehat{\mathbf{C}}_2 & 0 \\ 0 & 0 & \{\mathbf{H}^{\{0,1\}}\} - \mathbf{V}^{\circ}\widehat{\mathbf{C}}_2 \end{pmatrix} \tag{10.142}$$

10.5.5 Passing to symmetrical coordinates

Note that Eq. (10.142) may be also written

$$\{\mathbf{H}_{\text{Dav}}\} = \{\mathbf{H}^{\{0,0\}}\} + \{\mathbf{H}^{\{1,1\}}_{\{+\}}\}\{\beta^{(+)}\} + \{\mathbf{H}^{\{1,1\}}_{\{-\}}\}\{\beta^{(-)}\} \tag{10.143}$$

with

$$\{\mathbf{H}^{\{1,1\}}_{\{\pm\}}\} \equiv \{\mathbf{H}^{\{1,0\}}\} \pm \mathbf{V}^{\circ}\widehat{\mathbf{C}}_2 \tag{10.144}$$

In addition, owing to Eq. (10.119), the Hamiltonian (10.126) becomes

$$\{\mathbf{H}^{\{0,0\}}\} = \left(\frac{\mathbf{P}_a^2}{2M} + \frac{M\Omega^2\mathbf{Q}_a^2}{2}\right) + \left(\frac{\mathbf{P}_b^2}{2M} + \frac{M\Omega^2\mathbf{Q}_b^2}{2}\right)$$

whereas owing to Eqs. (10.119), (10.120), and (10.127), Eq. (10.144) reads

$$\begin{aligned}\{\mathbf{H}^{\{1,1\}}_{\{\pm\}}\} &= \left(\frac{\mathbf{P}_a^2}{2M} + \frac{M\Omega^2\mathbf{Q}_a^2}{2}\right) + (\hbar b^{\circ}\mathbf{Q}_a + \hbar\omega^{\circ} - \alpha^{\circ 2}\hbar\Omega) \\ &+ \left(\frac{\mathbf{P}_b^2}{2M} + \frac{M\Omega^2\mathbf{Q}_b^2}{2}\right) \pm \mathbf{V}^{\circ}\widehat{\mathbf{C}}_2\end{aligned} \tag{10.145}$$

Now, recall that the action of the parity operator transforms one coordinate of the H-bond bridge into an other one:

$$\widehat{\mathbf{C}}_2\mathbf{Q}_a = \mathbf{Q}_b \qquad \widehat{\mathbf{C}}_2\mathbf{Q}_b = \mathbf{Q}_a$$

Then, in order to use the symmetry properties of the system, consider the symmetrical coordinates according to

$$\mathbf{Q}_g = \left(\frac{\mathbf{Q}_a + \mathbf{Q}_b}{\sqrt{2}}\right) \quad \text{and} \quad \mathbf{Q}_u = \frac{\mathbf{Q}_a - \mathbf{Q}_b}{\sqrt{2}} \tag{10.146}$$

$$\mathbf{P}_g = \left(\frac{\mathbf{P}_a + \mathbf{P}_b}{\sqrt{2}}\right) \quad \text{and} \quad \mathbf{P}_u = \frac{\mathbf{P}_a - \mathbf{P}_b}{\sqrt{2}} \tag{10.147}$$

with, owing to Eq. (10.115),

$$\widehat{\mathbf{C}}_2\mathbf{Q}_g = \mathbf{Q}_g \quad \text{and} \quad \widehat{\mathbf{C}}_2\mathbf{Q}_u = -\mathbf{Q}_u \tag{10.148}$$

In the symmetrical coordinates, the following sums remain unchanged:

$$\mathbf{P}_a^2 + \mathbf{P}_b^2 = \mathbf{P}_g^2 + \mathbf{P}_u^2$$

$$\mathbf{Q}_a^2 + \mathbf{Q}_b^2 = \mathbf{Q}_g^2 + \mathbf{Q}_u^2$$

Hence, in view of Eq. (10.119), the Hamiltonian (10.126) becomes

$$\{\mathbf{H}^{\{0,0\}}\} = \{\mathbf{H}_g^{\{0\}}\} + \{\mathbf{H}_u^{\{0\}}\}$$

with

$$\{\mathbf{H}_g^{\{0\}}\} = \left(\frac{\mathbf{P}_g^2}{2M} + \frac{M\Omega^2\mathbf{Q}_g^2}{2}\right) \quad \text{and} \quad \{\mathbf{H}_u^{\{0\}}\} = \left(\frac{\mathbf{P}_u^2}{2M} + \frac{M\Omega^2\mathbf{Q}_u^2}{2}\right) \tag{10.149}$$

Now, in view of Eq. (10.120), the Hamiltonian (10.127) transforms to

$$\{\mathbf{H}^{\{1,0\}}\} = \{\mathbf{H}_g^{\{1\}}\} + \{\mathbf{H}_u^{\{1\}}\} + \hbar\omega^\circ - \alpha^{\circ 2}\hbar\Omega \tag{10.150}$$

with, respectively,

$$\{\mathbf{H}_g^{\{1\}}\} = \left(\frac{\mathbf{P}_g^2}{2M} + \frac{M\Omega^2\mathbf{Q}_g^2}{2}\right) + \hbar b^\circ \frac{\mathbf{Q}_g}{\sqrt{2}} \tag{10.151}$$

$$\{\mathbf{H}_u^{\{1\}}\} = \left(\frac{\mathbf{P}_u^2}{2M} + \frac{M\Omega^2\mathbf{Q}_u^2}{2}\right) + \hbar b^\circ \frac{\mathbf{Q}_u}{\sqrt{2}} \tag{10.152}$$

Next, examine carefully how the parity operator $\widehat{\mathbf{C}}_2$ acts on the tensor product of space states in which the gerade (g) and ungerade (u) **P** and **Q** operators work. Since the $\widehat{\mathbf{C}}_2$ operator cannot modify either kets or operators of the g symmetry, we can infer that it only works on kets and operators belonging to the u space states, which may be denoted

$$\widehat{\mathbf{C}}_2 = \widehat{\mathbf{C}}_{2u}\mathbf{1}_g = \widehat{\mathbf{C}}_{2u}$$

Thus, the last right-hand-side operator appearing in Eq. (10.145) reads

$$\mathbf{V}^\circ\widehat{\mathbf{C}}_2 = \mathbf{V}^\circ\widehat{\mathbf{C}}_{2u}$$

so that the Hamiltonian (10.145) becomes

$$\begin{aligned}\{\mathbf{H}_{\{\pm\}}^{\{1,1\}}\} &= \hbar\omega^\circ - \alpha^{\circ 2}\hbar\Omega + \left\{\left(\frac{\mathbf{P}_g^2}{2M} + \frac{M\Omega^2\mathbf{Q}_g^2}{2}\right) + \hbar b^\circ\frac{\mathbf{Q}_g}{\sqrt{2}}\right\} \\ &+ \left\{\left(\frac{\mathbf{P}_u^2}{2M} + \frac{M\Omega^2\mathbf{Q}_u^2}{2}\right) + \hbar b^\circ\frac{\mathbf{Q}_u}{\sqrt{2}} \pm \mathbf{V}^\circ\widehat{\mathbf{C}}_{2u}\right\}\end{aligned} \tag{10.153}$$

10.5.6 Symmetry properties of the eigenstates of the Hamiltonians (10.149)

In the following, it will be of interest to know the symmetry properties of the eigenvectors of the g and u ground states effective Hamiltonians appearing in Eq. (10.149), which verify therefore

$$\{\mathbf{H}_g^{\{0\}}\}|(n)_{\text{ger}}\rangle = \left(n_g + \tfrac{1}{2}\right)\hbar\Omega|(n)_{\text{ger}}\rangle$$

$$\{\mathbf{H}_u^{\{0\}}\}|(n)_{\text{ung}}\rangle = \left(n_u + \tfrac{1}{2}\right)\hbar\Omega|(n)_{\text{ung}}\rangle$$

For this purpose, look at the corresponding wavefunction defined by the scalar products

$$\{\Psi_{n_g}(\mathbf{Q}_g)\} = \langle\{\mathbf{Q}_g\}|(n)_{\text{ger}}\rangle \qquad \text{and} \qquad \{\Psi_{n_u}(\mathbf{Q}_u)\} = \langle\{\mathbf{Q}_u\}|(n)_{\text{ung}}\rangle \tag{10.154}$$

Due to Eq. (5.147), their corresponding dimensionless expressions read

$$\{\Psi_{n_g}(\xi_g)\} = C_n\{H_{n_g}(\xi_g)\}e^{-\xi_g^2/2} \qquad \text{and} \qquad \{\Psi_{n_u}(\xi_u)\} = C_n\{H_{n_u}(\xi_u)\}e^{-\xi_u^2/2} \tag{10.155}$$

where $H_{n_g}(\xi_g)$ and $H_{n_u}(\xi_u)$ are Hermite polynomials of the same kind as those appearing in Section 5.2.3, whereas C_n are the normalization constant, and ξ_g and ξ_u the dimensionless coordinates defined by

$$\xi_g = \sqrt{\frac{M\Omega}{\hbar}}\mathbf{Q}_g \qquad \text{and} \qquad \xi_u = \sqrt{\frac{M\Omega}{\hbar}}\mathbf{Q}_u$$

Then, due to Eq. (10.148), the action of the parity operator on these dimensionless coordinates are

$$\widehat{\mathbf{C}}_2\xi_g = \xi_g \qquad \text{and} \qquad \widehat{\mathbf{C}}_2\xi_u = -\xi_u$$

so that

$$\widehat{\mathbf{C}}_2\xi_g^n = \xi_g^n \qquad \text{and} \qquad \widehat{\mathbf{C}}_2\xi_u^n = (-1)^n\xi_u^n \tag{10.156}$$

and, owing to the Taylor expansion of the exponentials of ξ_g^2 and ξ_u^2

$$\widehat{\mathbf{C}}_2e^{-\xi_g^2/2} = e^{-\xi_g^2/2} \qquad \text{and} \qquad \widehat{\mathbf{C}}_2e^{-\xi_u^2/2} = e^{-\xi_u^2/2}$$

Next, owing to Eqs. (5.134), (5.138), (5.143), and (5.146), the first Hermite polynomials involved in (10.155) read

$$\{H_{o_g}(\xi_g)\} = 1 \qquad \{H_{1_g}(\xi_g)\} = 2\xi_g \qquad \{H_{2_g}(\xi_g)\} = 4\xi_g^2 - 2$$
$$\{H_{3_g}(\xi_g)\} = 8\xi_g^3 - 12\xi_g$$

$$\{H_{o_u}(\xi_u)\} = 1 \qquad \{H_{1_u}(\xi_u)\} = 2\xi_u \qquad \{H_{2_u}(\xi_u)\} = 4\xi_u^2 - 2$$
$$\{H_{3_u}(\xi_u)\} = 8\xi_u^3 - 12\xi_u$$

Hence, since the powers of ξ_g or ξ_u appearing in these Hermite polynomials are alternatively even or odd, and owing to the symmetry properties (10.156), it appears that

$$\widehat{\mathbf{C}}_2\{H_{n_g}(\xi_g)\} = \{H_{n_g}(\xi_g)\} \qquad \text{and} \qquad \widehat{\mathbf{C}}_2\{H_{n_u}(\xi_u)\} = (-1)^{n_u}\{H_{n_u}(\xi_u)\}$$

As a consequence, the action of the parity operator on the dimensionless wavefunctions (10.155) yields

$$\widehat{\mathbf{C}}_2\{\Psi_{n_g}(\xi_g)\} = \{\Psi_{n_g}(\xi_g)\} \qquad \text{and} \qquad \widehat{\mathbf{C}}_2\{\Psi_{n_u}(\xi_u)\} = (-1)^{n_u}\{\Psi_{n_u}(\xi_u)\}$$

so that since the dimensioned wavefunctions (10.154) must have the same symmetry as the dimensionless ones

$$\widehat{\mathbf{C}}_2|(n)_{\text{ger}}\rangle = |(n)_{\text{ger}}\rangle \qquad \text{and} \qquad \widehat{\mathbf{C}}_2|(n)_{\text{ung}}\rangle = (-1)^{n_u}|(n)_{\text{ung}}\rangle \tag{10.157}$$

10.5.7 Final diagonal form of the Davydov Hamiltonian

Again, after separation of the different actions within the different g and u subspaces, the Hamiltonian (10.153) becomes

$$\{\mathbf{H}_{\{\pm\}}^{\{1,1\}}\} = \{\mathbf{H}_g^{\{1\}}\} + \{\mathbf{H}_{\{u^+\}}^{\{1\}}\} + \{\mathbf{H}_{\{u^-\}}^{\{1\}}\}$$

with, respectively,

$$\{\mathbf{H}_g^{\{1\}}\} = \left\{\left(\frac{\mathbf{P}_g^2}{2M} + \frac{M\Omega^2\mathbf{Q}_g^2}{2}\right) + \hbar b^\circ \frac{\mathbf{Q}_g}{\sqrt{2}}\right\} + \hbar\omega^\circ - \alpha^{\circ 2}\hbar\Omega$$

$$\{\mathbf{H}_{\{u^\pm\}}^{\{1\}}\} = \left\{\left(\frac{\mathbf{P}_u^2}{2M} + \frac{M\Omega^2\mathbf{Q}_u^2}{2}\right) + \hbar b^\circ \frac{\mathbf{Q}_u}{\sqrt{2}} \pm \mathbf{V}^\circ\widehat{\mathbf{C}}_2\right\} \tag{10.158}$$

Hence, the Hamiltonian (10.142) has the block form

$$\{\mathbf{H}_{\mathrm{Dav}}\} = \begin{pmatrix} \{\mathbf{H}_g^{\{0\}}\} & 0 & 0 & 0 & 0 \\ 0 & \{\mathbf{H}_g^{\{1\}}\} & 0 & 0 & 0 \\ 0 & 0 & \{\mathbf{H}_u^{\{0\}}\} & 0 & 0 \\ 0 & 0 & 0 & \{\mathbf{H}_{\{u^+\}}^{\{1\}}\} & 0 \\ 0 & 0 & 0 & 0 & \{\mathbf{H}_{\{u^-\}}^{\{1\}}\} \end{pmatrix} \tag{10.159}$$

Moreover, after passing to Boson operators, and in the basis of the eigenkets of the u harmonic Hamiltonian $\{\mathbf{H}_u^{\{0\}}\}$ defined in (10.149), the matrix elements of the Hamiltonians (10.158) are

$$\langle (m)_{\mathrm{ung}} | \{\mathbf{H}_{\{u^\pm\}}^{\{1\}}\} | (n)_{\mathrm{ung}} \rangle = \hbar\Omega \langle (m)_{\mathrm{ung}} | \left(\mathbf{a}_u^\dagger \mathbf{a}_u + \frac{1}{2}\right) | (n)_{\mathrm{ung}} \rangle$$
$$+ \frac{\alpha^\circ}{\sqrt{2}}\hbar\Omega \langle (m)_{\mathrm{ung}} | (\mathbf{a}_u^\dagger + \mathbf{a}_u) | (n)_{\mathrm{ung}} \rangle \pm \mathbf{V}^\circ \langle (m)_{\mathrm{ung}} | \widehat{\mathbf{C}}_2 | (n)_{\mathrm{ung}} \rangle$$

Of course, the two first kinds of matrix elements involved on the right-hand side of this last equation are given by

$$\langle (m)_{\mathrm{ung}} | \left(\mathbf{a}_u^\dagger \mathbf{a}_u + \tfrac{1}{2}\right) | (n)_{\mathrm{ung}} \rangle = \left(n + \tfrac{1}{2}\right)\delta_{mn}$$

and

$$\langle (m)_{\mathrm{ung}} | (\mathbf{a}_u^\dagger + \mathbf{a}_u) | (n)_{\mathrm{ung}} \rangle = \sqrt{n+1}\,\delta_{m,n+1} + \sqrt{n}\,\delta_{m,n-1}$$

At last, due to the last equation of (10.157) the matrix elements involving $\pm\mathbf{V}^\circ$ times the parity operator read

$$\pm\mathbf{V}^\circ \langle (m)_{\mathrm{ung}} | \widehat{\mathbf{C}}_2 | (n)_{\mathrm{ung}} \rangle = \pm\mathbf{V}^\circ (-1)^{n_u} \delta_{m_g n_g} \tag{10.160}$$

See Fig. 10.6.

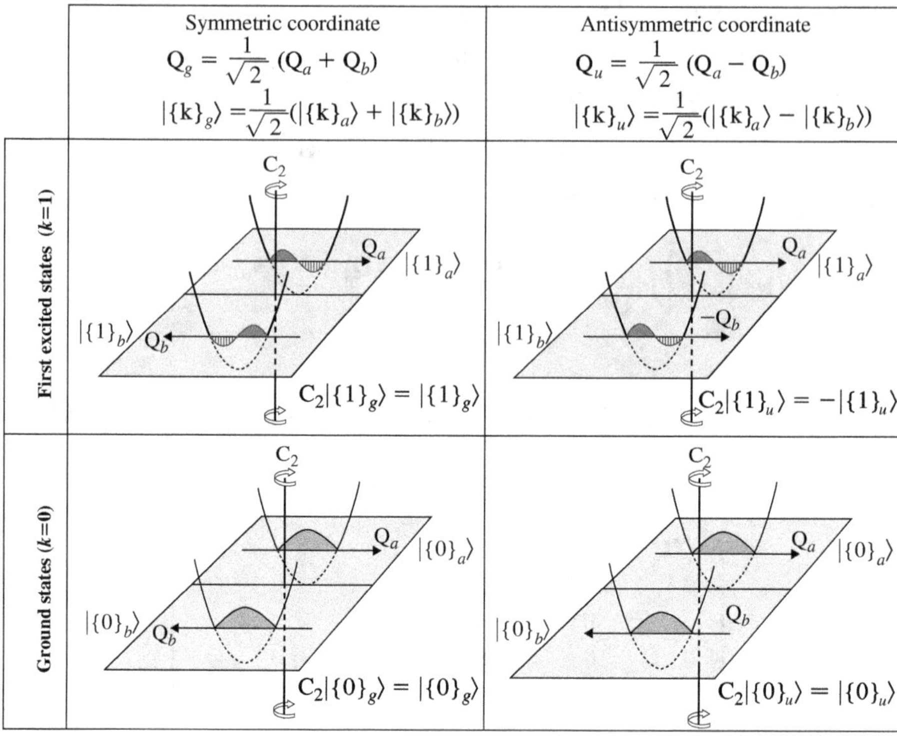

Figure 10.6 Effects of the parity operator C_2 on the ground and the first excited states of the symmetrized g and u eigenfunctions of the g and u quantum harmonic oscillators involved in the centrosymmetric cyclic dimer.

10.6 CONCLUSION

In this chapter we have studied various kinds of anharmonic coupling between oscillators. (i) The first section dealt with Fermi resonances involving two oscillators the frequencies of which are roughly half that of the other, which are coupled through an anharmonic coupling that is linear in the high-frequency mode coordinate and quadratic in the low-frequency one. It lead one to conclude a quantum interference between the first excited state of the fast mode and the second excited state of the slow one. (ii) The second section concerned the strong anharmonic coupling theory encountered in the quantum approach of the IR spectra of weak H-bonded species. According to this theory, the high angular frequency of the molecular oscillator involving a proton depends on the elongation of the very low frequency H-bond bridge, leading to some complex anharmonic coupling involving two kinds of terms, the first one quadratic in the elongation of the two molecular oscillators and the last one quadratic in the elongation of the fast mode and linear in that of the H-bond bridge. For this kind of anharmonic coupling, it was shown that it is possible to make an adiabatic separation between the slow and fast motions leading to effective Hamiltonians describing the H-bond bridge that depend on the excitation degree of the high-frequency mode. (iii) The subsequent section devoted to the study of the combination of Fermi resonance and of the strong anharmonic coupling theory showed that the strong anharmonic coupling enhances the sensitivity to Fermi resonances. (iv) The last section treated the system of four oscillators appearing in centrosymmetric cyclic H-bonded dimers.

BIBLIOGRAPHY

O. Henri-Rousseau and P. Blaise. *Advances in Chemical Physics*, Vol. 139. Wiley: New York, 2008, pp. 245–496.

R. Fulton and M. Gouterman. *J. Chem. Phys.*, **35** (1961): 1059.

Y. Maréchal, Thesis, Grenoble, 1968.

A. Witkowski and M. J. Wojcik. *Chem. Phys.*, **21** (1977): 385.

Y. Maréchal and A. Witkowski. *J. Chem. Phys.*, **48** (1968): 3637.

PART IV

OSCILLATOR POPULATIONS IN THERMAL EQUILIBRIUM

CHAPTER 11

DYNAMICS OF A LARGE SET OF COUPLED OSCILLATORS

INTRODUCTION

In the last two chapters we studied situations involving either the anharmonicity of the potential of a single oscillator or the anharmonicity occurring in the interactions between a limited number of coupled oscillators. However, these studies were focused on only the stationary states, and, therefore, they ignored dynamical aspects. Now let us consider not a single or a small number of oscillators in their stationary states but the dynamics of a large set of coupled oscillators. However, owing to difficulties, we shall limit the present study to coupling, which is quadratic with respect to the Boson operators. An important feature of this approach is the time evolution of the average values of the Hamiltonians of each oscillator, a result that will be used later to reveal an evolution that appears as irreversible when viewed through a *coarse-grained* analysis. Hence, the present chapter is important in relating the reversible behavior of quantum oscillators on the scale of atoms and molecules to the irreversible behavior of a very large set of oscillators used to model quantities on the macroscopic scale.

11.1 DYNAMIC EQUATIONS IN THE NORMAL ORDERING FORMALISM

11.1.1 Schrödinger equation for an infinite set of coupled oscillators

We consider the full Hamiltonian of a coupled chain of oscillators:

$$\mathbf{H}_{\text{Full}} = \hbar \sum_i \omega_{ii} \mathbf{a}_i^\dagger \mathbf{a}_i + \hbar \sum_i \sum_{j \neq i} \omega_{ij} \mathbf{a}_i^\dagger \mathbf{a}_j \tag{11.1}$$

Here, $\mathbf{a}_i$ is the Boson operator describing the ith oscillator, $\mathbf{a}_i^\dagger$ is its Hermitian conjugate, ω_{ii} is the angular frequency of the ith oscillator, and ω_{ij} $(i \neq j)$ the coupling between the ith and jth oscillators.

We shall assume that at initial time $t = 0$, all the oscillators are in their ground state except one (labeled 1), which is excited and described by a coherent state. The

Quantum Oscillators, First Edition. Olivier Henri-Rousseau and Paul Blaise.
© 2011 John Wiley & Sons, Inc. Published 2011 by John Wiley & Sons, Inc.

non-Hermitian eigenvalue equation describing the coherent state $|\{\alpha^\circ\}_1\rangle$ of the excited oscillator is

$$\mathbf{a}_1|\{\alpha^\circ\}_1\rangle = \alpha_1^\circ|\{\alpha^\circ\}_1\rangle$$

where α_1° is the corresponding eigenvalue. Now, the ground states of all the other oscillators obey the eigenvalue equations

$$\mathbf{a}_i^\dagger \mathbf{a}_i|\{0\}_j\rangle = 0|\{0\}_j\rangle = 0 \quad \text{with} \quad j \neq 1$$

and the full ket describing the set of oscillators at some initial time is

$$|\Psi_{\text{Full}}(0)\rangle = |\{\alpha^\circ\}_1\rangle \prod_{j\neq 1} |\{0\}_j\rangle \tag{11.2}$$

Our aim is to obtain the time evolution operator $\mathbf{U}(t)$ of this system of oscillators, thus allowing one to pass from the ket $|\Psi_{\text{Full}}(0)\rangle$ to the ket $|\Psi_{\text{Full}}(t)\rangle$ at time t via

$$\mathbf{U}(t)|\Psi_{\text{Full}}(0)\rangle = |\Psi_{\text{Full}}(t)\rangle$$

The time evolution operator $\mathbf{U}(t)$ of the system obeys the Schrödinger equation

$$i\hbar\left(\frac{\partial \mathbf{U}(t)}{\partial t}\right) = \mathbf{H}_{\text{Full}}\{\mathbf{U}(t)\} \tag{11.3}$$

with the boundary condition

$$\{\mathbf{U}(0)\} = 1 \tag{11.4}$$

Thus, in view of Eq. (11.1)

$$i\left(\frac{\partial \mathbf{U}(t)}{\partial t}\right) = \sum_i \omega_{ii}\mathbf{a}_i^\dagger \mathbf{a}_i\{\mathbf{U}(t)\} + \sum_i\sum_{j\neq i}\omega_{ij}\mathbf{a}_i^\dagger \mathbf{a}_j\{\mathbf{U}(t)\} \tag{11.5}$$

This equation shows that $\mathbf{U}(t)$ is a function of all the Boson operators that do not commute for a given oscillator. In order to solve it, we use the normal ordering procedure:

$$i\hat{\mathbf{N}}^{-1}\left(\frac{\partial \mathbf{U}(t)}{\partial t}\right) = \sum_i \omega_{ii}\hat{\mathbf{N}}^{-1}\{\mathbf{a}_i^\dagger \mathbf{a}_i\{\mathbf{U}(t)\}\} + \sum_i\sum_{j\neq i}\omega_{ij}\hat{\mathbf{N}}^{-1}\{\mathbf{a}_i^\dagger \mathbf{a}_j\{\mathbf{U}(t)\}\}$$

Then, applying Eqs. (7.40), (7.101), and (7.99) we have

$$\hat{\mathbf{N}}^{-1}\left\{\left(\frac{\partial \mathbf{U}(t)}{\partial t}\right)\right\} = \left(\frac{\partial U^{(n)}(t)}{\partial t}\right)$$

$$\hat{\mathbf{N}}^{-1}\{\mathbf{a}_i^\dagger \mathbf{a}_i\{\mathbf{U}(t)\}\} = \{\alpha_i^*\}\left(\{\alpha_i\} + \frac{\partial}{\partial \alpha_i^*}\right)\{U^{(n)}(t)\}$$

As a consequence, one may pass from the partial differential equation (11.5) dealing with noncommuting Boson operators to the one involving scalars only, namely

$$i\left(\frac{\partial U^{(n)}(t)}{\partial t}\right) = \sum_i \omega_{ii}\{\alpha_i^*\}\left(\{\alpha_i\} + \frac{\partial}{\partial \alpha_i^*}\right)$$
$$\times \{U^{(n)}(t)\} + \sum_i\sum_{j\neq i}\omega_{ij}\{\alpha_i^*\}\left(\{\alpha_j\} + \frac{\partial}{\partial \alpha_j^*}\right)\{U^{(n)}(t)\}$$

which leads after factorization to

$$i\left(\frac{\partial U^{(n)}(t)}{\partial t}\right) = \sum_i \left(\{\alpha_i^*\alpha_i\}\omega_{ii}\{U^{(n)}(t)\} + \{\alpha_i^*\}\omega_{ii}\left(\frac{\partial U^{(n)}(t)}{\partial \alpha_i^*}\right)\right)$$
$$+\sum_i\sum_{j\neq i}\left(\{\alpha_i^*\alpha_j\}\omega_{ij}\{U^{(n)}(t)\} + \{\alpha_i^*\}\omega_{ij}\left(\frac{\partial U^{(n)}(t)}{\partial \alpha_j^*}\right)\right) \tag{11.6}$$

with, resulting from Eq. (11.4), the following boundary condition

$$\{U^{(n)}(0)\} = 1 \tag{11.7}$$

11.1.2 Change of variable from $U^{(n)}(t)$ to $G^{(n)}(t)$

In order to solve Eq. (11.6), define the following new time-dependent scalar variable

$$\boxed{\{U^{(n)}(t)\} = \exp\{G^{(n)}(t)\}} \tag{11.8}$$

with, according to Eq. (11.7)

$$\{G^{(n)}(0)\} = 0 \tag{11.9}$$

Then, the partial derivatives of Eq. (11.8) with respect to t and α_i^* yield, respectively,

$$\left(\frac{\partial U^{(n)}(t)}{\partial t}\right) = \exp\{G^{(n)}(t)\}\left(\frac{\partial G^{(n)}(t)}{\partial t}\right) = \{U^{(n)}(t)\}\left(\frac{\partial G^{(n)}(t)}{\partial t}\right)$$

$$\left(\frac{\partial U^{(n)}(t)}{\partial \alpha_i^*}\right) = \exp\{G^{(n)}(t)\}\left(\frac{\partial G^{(n)}(t)}{\partial \alpha_i^*}\right) = \{U^{(n)}(t)\}\left(\frac{\partial G^{(n)}(t)}{\partial \alpha_i^*}\right)$$

so that Eq. (11.6) becomes

$$\{iU^{(n)}(t)\}\left(\frac{\partial G^{(n)}(t)}{\partial t}\right) = \sum_i \left(\{\alpha_i^*\alpha_i\}\omega_{ii}\{U^{(n)}(t)\} + \{\alpha_i^*\}\omega_{ii}\{U^{(n)}(t)\}\left(\frac{\partial G^{(n)}(t)}{\partial \alpha_i^*}\right)\right)$$
$$+\sum_i\sum_{j\neq i}\left(\{\alpha_i^*\alpha_j\}\omega_{ij}\{U^{(n)}(t)\} + \{\alpha_i^*\}\omega_{ij}\{U^{(n)}(t)\}\left(\frac{\partial G^{(n)}(t)}{\partial \alpha_j^*}\right)\right)$$

or, after simplifying via the scalar $U^{(n)}(t)$

$$i\left(\frac{\partial G^{(n)}(t)}{\partial t}\right) = \left(\sum_i\{\alpha_i^*\alpha_i\}\omega_{ii} + \sum_i\{\alpha_i^*\}\omega_{ii}\left(\frac{\partial G^{(n)}(t)}{\partial \alpha_i^*}\right)\right)$$
$$+\left(\sum_i\sum_{j\neq i}\{\alpha_i^*\alpha_j\}\omega_{ij} + \sum_i\sum_{j\neq i}\{\alpha_i^*\}\omega_{ij}\left(\frac{\partial G^{(n)}(t)}{\partial \alpha_j^*}\right)\right) \tag{11.10}$$

11.1.3 Expansion of $G^{(n)}(t)$ in terms of time-dependent scalar functions

Now, assume that $G^{(n)}(t)$ may be expanded in the following way:

$$G^{(n)}(t) = \sum_i \{\alpha_i^* \alpha_i\} A_{ii}(t) + \sum_i \sum_{j \neq i} \{\alpha_i^* \alpha_j\} A_{ij}(t) \tag{11.11}$$

where the $A_{ij}(t)$ are time-dependent quantities to be determined that, due to Eq. (11.9), must obey the boundary conditions

$$A_{ii}(0) = A_{ij}(0) = 0 \tag{11.12}$$

By partial differentiation of Eq. (11.11) with respect to α_i^* and t, we have

$$\left(\frac{\partial G^{(n)}(t)}{\partial \alpha_i^*}\right) = \left(\{\alpha_i\} A_{ii}(t) + \sum_{j \neq i} \{\alpha_j\} A_{ij}(t)\right) \tag{11.13}$$

$$\boxed{\left(\frac{\partial G^{(n)}(t)}{\partial t}\right) = \sum_i \{\alpha_i^* \alpha_i\} \left(\frac{\partial A_{ii}(t)}{\partial t}\right) + \sum_i \sum_{j \neq i} \{\alpha_i^* \alpha_j\} \left(\frac{\partial A_{ij}(t)}{\partial t}\right)} \tag{11.14}$$

Hence, due to Eq. (11.13), Eq. (11.10) becomes

$$i\left(\frac{\partial G^{(n)}(t)}{\partial t}\right) = \sum_i \{\alpha_i^* \alpha_i\} \omega_{ii} + \sum_i \{\alpha_i^*\} \omega_{ii} \left(\{\alpha_i\} A_{ii}(t) + \sum_{j \neq i} \{\alpha_j\} A_{ij}(t)\right)$$
$$+ \sum_i \sum_{j \neq i} \{\alpha_i^* \alpha_j\} \omega_{ij} + \sum_i \sum_{j \neq i} \{\alpha_i^*\} \omega_{ij} \left(\{\alpha_j\} A_{jj}(t) + \sum_{l \neq j} \{\alpha_l\} A_{jl}(t)\right)$$

or

$$i\left(\frac{\partial G^{(n)}(t)}{\partial t}\right) = \sum_i \{\alpha_i^* \alpha_i\} \omega_{ii} + \sum_i \{\alpha_i^* \alpha_i\} \omega_{ii} A_{ii}(t) + \sum_i \sum_{j \neq i} \{\alpha_i^* \alpha_j\} \omega_{ii} A_{ij}(t)$$
$$+ \sum_i \sum_{j \neq i} \{\alpha_i^* \alpha_j\} \omega_{ij} + \sum_i \sum_{j \neq i} \{\alpha_i^* \alpha_j\} \omega_{ij} A_{jj}(t)$$
$$+ \sum_i \sum_{j \neq i} \sum_{l \neq j} \{\alpha_i^* \alpha_l\} \omega_{ij} A_{jl}(t)$$

so that

$$i\left(\frac{\partial G^{(n)}(t)}{\partial t}\right) = \sum_i \{\alpha_i^* \alpha_i\} (\omega_{ii}(A_{ii}(t) + 1)) + \sum_i \sum_{j \neq i} \{\alpha_i^* \alpha_j\} \omega_{ii} A_{ij}(t)$$
$$+ \sum_i \sum_{j \neq i} \{\alpha_i^* \alpha_j\} \omega_{ij} (A_{jj}(t) + 1)$$
$$+ \left\{\sum_i \sum_{j \neq i} \sum_{l \neq j} \{\alpha_i^* \alpha_l\} \omega_{ij} A_{jl}(t)\right\} \tag{11.15}$$

Next, the last term of Eq. (11.15) may be written in such a way so as to distinguish for l the situation where $l = i$ those with $l \neq i$ to yield

$$\left\{\sum_{i}\sum_{j\neq i}\sum_{l\neq j}\{\alpha_i^*\alpha_l\}\omega_{ij}A_{jl}(t)\right\} = \sum_{i}\sum_{j\neq i}\{\alpha_i^*\alpha_i\}\omega_{ij}A_{ji}(t) + \sum_{i}\sum_{j\neq i}\sum_{l\neq j\neq i}\{\alpha_i^*\alpha_l\}\omega_{ij}A_{jl}(t) \quad (11.16)$$

Moreover, since nothing is changed in the sums appearing on the last right-hand-side term of Eq. (11.16) if a permutation of the l and j subscripts is made, this latter expression becomes

$$\left\{\sum_{i}\sum_{j\neq i}\sum_{l\neq j}\{\alpha_i^*\alpha_l\}\omega_{ij}A_{jl}(t)\right\} = \left\{\sum_{i}\sum_{j\neq i}\{\alpha_i^*\alpha_i\}\omega_{ij}A_{ji}(t)\right\} + \sum_{i}\sum_{l\neq i}\sum_{j\neq l\neq i}\{\alpha_i^*\alpha_j\}\omega_{il}A_{lj}(t) \quad (11.17)$$

Hence, using Eq. (11.17), Eq. (11.15) becomes

$$\begin{aligned} i\left(\frac{\partial G^{(n)}(t)}{\partial t}\right) &= \sum_{i}\{\alpha_i^*\alpha_i\}\omega_{ii}(A_{ii}(t)+1) + \left\{\sum_{i}\sum_{j\neq i}\{\alpha_i^*\alpha_i\}\omega_{ij}A_{ji}(t)\right\} \\ &\quad + \sum_{i}\sum_{j\neq i}\{\alpha_i^*\alpha_j\}(\omega_{ii}A_{ij}(t) + \omega_{ij}(A_{jj}(t)+1)) \\ &\quad + \left\{\sum_{i}\sum_{j\neq i}\sum_{l\neq i}\{\alpha_i^*\alpha_j\}\omega_{il}A_{lj}(t)\right\} \end{aligned}$$

or, after rearranging,

$$\begin{aligned} i\left(\frac{\partial G^{(n)}(t)}{\partial t}\right) &= \sum_{i}\{\alpha_i^*\alpha_i\}\left(\omega_{ii}(A_{ii}(t)+1) + \sum_{j\neq i}\omega_{ij}A_{ji}(t)\right) \\ &\quad + \sum_{i}\sum_{j\neq i}\{\alpha_i^*\alpha_j\}\left(\omega_{ii}A_{ij}(t) + \omega_{ij}(A_{jj}(t)+1) + \sum_{l\neq i}\omega_{il}A_{lj}(t)\right) \end{aligned} \quad (11.18)$$

11.1.4 Dynamical equations for $A_{ij}(t)$

Identification of Eqs. (11.14) and (11.18) leads to

$$i\{\alpha_i^*\alpha_i\}\left(\frac{\partial A_{ii}(t)}{\partial t}\right) = \{\alpha_i^*\alpha_i\}\left(\omega_{ii}(A_{ii}(t)+1) + \sum_{j\neq i}\omega_{ij}A_{jj}(t)\right)$$

$$i\{\alpha_i^*\alpha_j\}\left(\frac{\partial A_{ij}(t)}{\partial t}\right) = \{\alpha_i^*\alpha_j\}\left(\omega_{ii}A_{ij}(t) + \omega_{ij}(A_{jj}(t)+1) + \sum_{l\neq j, l\neq i}\omega_{il}A_{lj}(t)\right)$$

which after simplification lead, respectively, to

$$i\left(\frac{\partial A_{ii}(t)}{\partial t}\right) = \omega_{ii}(A_{ii}(t)+1) + \sum_{j\neq i}\omega_{ij}A_{ij}(t) \tag{11.19}$$

$$i\left(\frac{\partial A_{ij}(t)}{\partial t}\right) = \omega_{ii}A_{ij}(t) + \omega_{ij}(A_{jj}(t)+1) + \sum_{l\neq j, l\neq i}\omega_{il}A_{lj}(t) \tag{11.20}$$

Now, make changes of variable

$$\mathbb{A}_{ij}(t) = (A_{ij}(t) + \delta_{ij}) \tag{11.21}$$

for which the boundary condition (11.12) reads

$$\mathbb{A}_{ij}(0) = \delta_{ij} \tag{11.22}$$

Then, observing that due to Eq. (11.21)

$$\left(\frac{\partial A_{ij}(t)}{\partial t}\right) = \left(\frac{\partial \mathbb{A}_{ij}(t)}{\partial t}\right)$$

Eqs. (11.19) and (11.20) transform, respectively, into the following set of coupled first-order time-dependent equations:

$$\boxed{i\left(\frac{\partial \mathbb{A}_{ii}(t)}{\partial t}\right) = \omega_{ii}\mathbb{A}_{ii}(t) + \sum_{j\neq i}\omega_{ij}\mathbb{A}_{ij}(t)} \tag{11.23}$$

$$\boxed{i\left(\frac{\partial \mathbb{A}_{ij}(t)}{\partial t}\right) = \omega_{ii}\mathbb{A}_{ij}(t) + \omega_{ij}\mathbb{A}_{jj}(t) + \sum_{l\neq j, l\neq i}\omega_{il}\mathbb{A}_{lj}(t)} \tag{11.24}$$

11.1.5 Set of equations governing the $\mathbb{A}_{ij}(t)$

Now, to emphasis that the linear set of coupled first-order equations (11.23) and (11.24) takes on block matrix form, write them for the special situation of three coupled oscillators:

$$i\frac{\partial}{\partial t}\begin{pmatrix}\mathbb{A}_{11}(t)\\ \mathbb{A}_{21}(t)\\ \mathbb{A}_{31}(t)\\ \mathbb{A}_{12}(t)\\ \mathbb{A}_{22}(t)\\ \mathbb{A}_{32}(t)\\ \mathbb{A}_{13}(t)\\ \mathbb{A}_{23}(t)\\ \mathbb{A}_{33}(t)\end{pmatrix} = \begin{pmatrix}\omega_{11} & \omega_{12} & \omega_{13} & 0 & 0 & 0 & 0 & 0 & 0\\ \omega_{21} & \omega_{22} & \omega_{23} & 0 & 0 & 0 & 0 & 0 & 0\\ \omega_{31} & \omega_{32} & \omega_{33} & 0 & 0 & 0 & 0 & 0 & 0\\ 0 & 0 & 0 & \omega_{11} & \omega_{12} & \omega_{13} & 0 & 0 & 0\\ 0 & 0 & 0 & \omega_{21} & \omega_{22} & \omega_{23} & 0 & 0 & 0\\ 0 & 0 & 0 & \omega_{31} & \omega_{32} & \omega_{33} & 0 & 0 & 0\\ 0 & 0 & 0 & 0 & 0 & 0 & \omega_{11} & \omega_{12} & \omega_{13}\\ 0 & 0 & 0 & 0 & 0 & 0 & \omega_{21} & \omega_{22} & \omega_{23}\\ 0 & 0 & 0 & 0 & 0 & 0 & \omega_{31} & \omega_{32} & \omega_{33}\end{pmatrix}\begin{pmatrix}\mathbb{A}_{11}(t)\\ \mathbb{A}_{21}(t)\\ \mathbb{A}_{31}(t)\\ \mathbb{A}_{12}(t)\\ \mathbb{A}_{22}(t)\\ \mathbb{A}_{32}(t)\\ \mathbb{A}_{13}(t)\\ \mathbb{A}_{23}(t)\\ \mathbb{A}_{33}(t)\end{pmatrix}$$

Then, it appears that Eqs. (11.23) and (11.24) take the form of three identical sets of linear differential equations, the first of them being

$$i\frac{\partial}{\partial t}\begin{pmatrix}\mathbb{A}_{11}(t)\\ \mathbb{A}_{21}(t)\\ \mathbb{A}_{31}(t)\end{pmatrix} = \begin{pmatrix}\omega_{11} & \omega_{12} & \omega_{13}\\ \omega_{21} & \omega_{22} & \omega_{23}\\ \omega_{31} & \omega_{32} & \omega_{33}\end{pmatrix}\begin{pmatrix}\mathbb{A}_{11}(t)\\ \mathbb{A}_{21}(t)\\ \mathbb{A}_{31}(t)\end{pmatrix}$$

Hence, for N oscillators, N identical sets of differential equations are obtained, the first being

$$i\frac{\partial}{\partial t}\begin{pmatrix} \mathbb{A}_{1k}(t) \\ \mathbb{A}_{2k}(t) \\ \cdots \\ \cdots \\ \cdots \\ \mathbb{A}_{Nk}(t) \end{pmatrix} = \begin{pmatrix} \omega_{11} & \omega_{12} & \cdots & \omega_{1k} & \cdots & \omega_{1N} \\ \omega_{21} & \omega_{22} & \cdots & & & \\ \cdots & \cdots & \cdots & & & \\ \omega_{k1} & & & \omega_{kk} & \cdots & \omega_{kN} \\ \cdots & & & \cdots & \cdots & \cdots \\ \omega_{N1} & & & \omega_{Nk} & \cdots & \omega_{NN} \end{pmatrix}\begin{pmatrix} \mathbb{A}_{1k}(t) \\ \mathbb{A}_{2k}(t) \\ \cdots \\ \cdots \\ \cdots \\ \mathbb{A}_{Nk}(t) \end{pmatrix} \tag{11.25}$$

which may be written formally

$$\left(\frac{\partial[\mathbf{A}_k(t)]}{\partial t}\right) = -i\boxed{\boldsymbol{\Omega}}\,[\mathbf{A}_k(t)] \tag{11.26}$$

where the time-dependent vector $[\mathbf{A}_k(t)]$ and the time-independent $\boxed{\Omega}$ matrix are, respectively, given by

$$[\mathbf{A}_k(t)] = \begin{pmatrix} \mathbb{A}_{1k}(t) \\ \mathbb{A}_{2k}(t) \\ \cdots \\ \mathbb{A}_{kk}(t) \\ \cdots \\ \mathbb{A}_{Nk}(t) \end{pmatrix} \quad \text{and} \quad \boxed{\Omega} = \begin{pmatrix} \omega_{11} & \omega_{12} & \cdots & \omega_{1k} & \cdots & \omega_{1N} \\ \omega_{21} & \omega_{22} & \cdots & & & \\ \cdots & \cdots & \cdots & & & \\ \omega_{k1} & & & \omega_{kk} & \cdots & \omega_{kN} \\ \cdots & & & \cdots & \cdots & \cdots \\ \omega_{N1} & & & \omega_{Nk} & \cdots & \omega_{NN} \end{pmatrix} \tag{11.27}$$

11.2 SOLVING THE LINEAR SET OF DIFFERENTIAL EQUATIONS (11.27)

Now, to solve Eq. (11.26), we use the unitary transformation that diagonalizes the $\boxed{\Omega}$ matrix according to

$$\boxed{\mathbf{P}}^{-1}\boxed{\boldsymbol{\Omega}}\boxed{\mathbf{P}} = \boxed{\boldsymbol{\Lambda}} \quad \text{with} \quad \boxed{\mathbf{P}}\boxed{\mathbf{P}}^{-1} = \boxed{\mathbf{1}} \tag{11.28}$$

where $\boxed{\mathbf{P}}$ is the eigenvector matrix of $\boxed{\boldsymbol{\Omega}}$, $\boxed{\mathbf{P}}^{-1}$ the inverse of $\boxed{\mathbf{P}}$, and $\boxed{\boldsymbol{\Lambda}}$ the diagonal matrix involving the eigenvalues of $\boxed{\boldsymbol{\Omega}}$. Then, premultiply both members of Eq. (11.26) by $\boxed{\mathbf{P}}^{-1}$ and insert the unity operator $\boxed{\mathbf{P}}\boxed{\mathbf{P}}^{-1}$ between $\boxed{\boldsymbol{\Omega}}$ and $[\mathbf{A}_k(t)]$, to get

$$\boxed{\mathbf{P}}^{-1}\left(\frac{\partial[\mathbf{A}_k(t)]}{\partial t}\right) = -i\boxed{\mathbf{P}}^{-1}\boxed{\boldsymbol{\Omega}}\boxed{\mathbf{P}}\boxed{\mathbf{P}}^{-1}[\mathbf{A}_k(t)]$$

or, using (11.28)

$$\left(\frac{\partial[\mathbf{W}_k(t)]}{\partial t}\right) = -i\boxed{\boldsymbol{\Lambda}}[\mathbf{W}_k(t)] \tag{11.29}$$

with

$$[\mathbf{W}_k(t)] = \boxed{\mathbf{P}}^{-1}[\mathbf{A}_k(t)] \tag{11.30}$$

After integration, Eq. (11.29) gives

$$[\mathbf{W}_k(t)] = [\mathbf{W}_k(0)](e^{-i\boxed{\mathbf{\Lambda}}t}) \tag{11.31}$$

with

$$[\mathbf{W}_k(t)] = \begin{pmatrix} W_{1k}(t) \\ W_{2k}(t) \\ \cdots \\ W_{kk}(t) \\ \cdots \\ W_{Nk}(t) \end{pmatrix} \quad \text{and} \quad \boxed{\mathbf{\Lambda}} = \begin{pmatrix} \lambda_1 & 0 & 0 & & & \\ 0 & \lambda_2 & 0 & & & \\ 0 & 0 & \cdots & \cdots & 0 & \\ & \cdots & \cdots & \lambda_k & \cdots & \\ \cdots & \cdots & 0 & \cdots & \cdots & 0 \\ \cdots & \cdots & \cdots & \cdots & 0 & \lambda_N \end{pmatrix} \tag{11.32}$$

where $[\mathbf{W}_k(0)]$ is the value of $[\mathbf{W}_k(t)]$ at the initial time while λ_k are the eigenvalues of the matrix $\boxed{\mathbf{\Lambda}}$. The lth component $W_{lk}(t)$ of the vector $[\mathbf{W}_k(t)]$ obeys, therefore,

$$W_{lk}(t) = W_{lk}(0)e^{-i\lambda_l t} \tag{11.33}$$

Moreover, owing to Eq. (11.30), the vector $[\mathbf{W}_k(0)]$ at initial time yields

$$[\mathbf{W}_k(0)] = \boxed{\mathbf{P}}^{-\mathbf{1}}[\mathbf{A}_k(0)] \tag{11.34}$$

the lth component $W_{lk}(0)$ of which reads

$$W_{lk}(0) = \sum_j P_{lj}^{-1}\mathbb{A}_{jk}(0) \tag{11.35}$$

where P_{lj}^{-1} is the lth component of the jth column of the unitary matrix involved in the linear transformation (11.34). Hence, owing to Eq. (11.22), this expression transforms to

$$W_{lk}(0) = \sum_j P_{lj}^{-1}\delta_{jk} = P_{lk}^{-1}$$

Thus, Eq. (11.33) becomes

$$W_{lk}(t) = P_{lk}^{-1}(e^{-i\lambda_l t}) \tag{11.36}$$

In addition, premultiplying each member of the canonical transformation (11.30) by $\boxed{\mathbf{P}}$ and simplifying using (11.28) yields

$$[\mathbf{A}_k(t)] = \boxed{\mathbf{P}}[\mathbf{W}_k(t)]$$

so that, due to Eq. (11.27), the jth component of the vector $[\mathbf{A}_k(t)]$ appears to be

$$\mathbb{A}_{jk}(t) = \sum_l P_{jl}W_{lk}(t)$$

or, owing to Eq. (11.36), it transforms to

$$\mathbb{A}_{jk}(t) = \sum_l P_{jl}P_{lk}^{-1}(e^{-i\lambda_l t})$$

Hence, in view of Eq. (11.21)

$$\delta_{jk} + A_{jk}(t) = \sum_l P_{jl}P_{lk}^{-1}(e^{-i\lambda_l t}) \tag{11.37}$$

so that

$$\boxed{A_{jk}(t) = f_{jk}(t) - \delta_{jk}} \tag{11.38}$$

with

$$f_{jk}(t) = \sum_l P_{jl} P_{lk}^{-1} e(^{-i\lambda_l t}) \tag{11.39}$$

Since the P_{rs} are the components of the orthogonal matrix $\boxed{\mathbf{P}}$, which diagonalizes according to Eq. (11.28) the matrix $\boxed{\mathbf{\Omega}}$, they satisfy

$$P_{lk}^{-1} = P_{kl}$$

so that Eq. (11.39) yields

$$\boxed{f_{jk}(t) = \sum_l P_{jl} P_{kl} (e^{-i\lambda_l t})} \tag{11.40}$$

11.3 OBTAINMENT OF THE DYNAMICS

11.3.1 Time evolution operator

Now, keeping in mind that, when the distinction between the situations $k = j$ and those $k \neq j$ are removed, Eq. (11.11) takes on the simplified form

$$\{G^{(n)}(t)\} = \sum_k \sum_j A_{kj}(t) \alpha_k^* \alpha_j$$

and seeing that the $A_{jk}(t)$ are given by Eq. (11.38), the expression (11.8) of $U^{(n)}(t)$ is

$$\{U^{(n)}(t)\} = \exp\left\{ \sum_k \sum_j A_{kj}(t) \alpha_k^* \alpha_j \right\} \tag{11.41}$$

or, after using the properties of exponentials,

$$U^{(n)}(t) = \prod_j \exp\left\{ \sum_k A_{kj}(t) \alpha_k^* \alpha_j \right\} \tag{11.42}$$

Next, premultiplying both members of this last equation by the normal ordering operators, from (7.44), one obtains, respectively, for each member

$$\hat{\mathbf{N}}\{U^{(n)}(t)\} = \mathbf{U}(t)$$

$$\hat{\mathbf{N}}\left\{ \exp\left\{ \sum_k \sum_j A_{kj}(t) \alpha_k^* \alpha_j \right\} \right\} = \exp\left\{ \sum_k \sum_j A_{kj}(t) \mathbf{a}_k^\dagger \mathbf{a}_j \right\}$$

so that Eq. (11.42) leads to the following expression for the time evolution operator governed by the Schrödinger equation (11.3):

$$\{\mathbf{U}(t)\} = \exp\left\{ \sum_k \sum_j A_{kj}(t) \mathbf{a}_k^\dagger \mathbf{a}_j \right\} \tag{11.43}$$

11.3.2 Time-dependent state

Due to Eq. (3.77), this time evolution operator (11.43) allows one to determine the time dependence of the state describing the set of the linear chain of oscillators, which at initial time was given by Eq. (11.2) through

$$|\Psi_{\text{Full}}(t)\rangle = \{\mathbf{U}(t)\}|\Psi_{\text{Full}}(0)\rangle$$

or, due to Eq. (11.2),

$$|\Psi_{\text{Full}}(t)\rangle = \mathbf{U}(t)|\{\alpha^\circ\}_1\rangle \prod_{j\neq 1} |\{0\}_j\rangle$$

and thus, owing to Eq. (11.43),

$$|\Psi_{\text{Full}}(t)\rangle = \exp\left\{\sum_k A_{k1}(t)\mathbf{a}_k^\dagger \mathbf{a}_1\right\} |\{\alpha^\circ\}_1\rangle \prod_{j\neq 1} \exp\left\{\sum_k A_{kj}(t)\mathbf{a}_k^\dagger \mathbf{a}_j\right\} |\{0\}_j\rangle \quad (11.44)$$

We note that the exponential under the product over $j \neq 1$ involving a sum over k in its argument may be rewritten

$$\exp\left\{\sum_k A_{kj}(t)\mathbf{a}_k^\dagger \mathbf{a}_j\right\} = \prod_k \exp\{A_{kj}(t)\mathbf{a}_k^\dagger \mathbf{a}_j\} \quad (11.45)$$

in which the exponential may be expanded, that is,

$$\exp\{A_{kj}(t)\mathbf{a}_k^\dagger \mathbf{a}_j\} = \sum_n \left(\frac{A_{kj}(t)^n(\mathbf{a}_k^\dagger)^n}{n!}\right)(\mathbf{a}_j)^n$$

the action of such an operator on the ground state $|\{0\}_j\rangle$ of $\mathbf{a}_j^\dagger \mathbf{a}_j$ being

$$\exp\{A_{kj}(t)\mathbf{a}_k^\dagger \mathbf{a}_j\}|\{0\}_j\rangle = \sum_n \left(\frac{A_{kj}(t)^n(\mathbf{a}_k^\dagger)^n}{n!}\right)(\mathbf{a}_j)^n|\{0\}_j\rangle$$

Then, due to Eq. (5.53), that is,

$$\mathbf{a}|\{m\}\rangle = \sqrt{m}|\{m-1\}\rangle$$

which in the special situation of the ket $|\{0\}\rangle$ yields

$$\mathbf{a}|\{0\}\rangle = 0$$

so that

$$(\mathbf{a}_j)^n|\{0\}_j\rangle = 0 \quad \text{if} \quad n \neq 0 \qquad \text{and} \qquad (\mathbf{a}_j)^n|\{0\}_j\rangle = 1|\{0\}_j\rangle \quad \text{if} \quad n = 0 \quad (11.46)$$

Hence

$$\exp\{A_{kj}(t)\mathbf{a}_k^\dagger \mathbf{a}_j\}|\{0\}_j\rangle = 1|\{0\}_j\rangle$$

Now, since this result holds for all the terms that are summed in Eq. (11.44), it follows that

$$\prod_{j\neq 1} \exp\left(\sum_k A_{kj}(t)\mathbf{a}_k^\dagger \mathbf{a}_j\right)|\{0\}_j\rangle = 1\prod_{j\neq 1}|\{0\}_j\rangle$$

allows one to simplify Eq. (11.44) into

$$|\Psi_{\text{Full}}(t)\rangle = \exp\left\{\sum_k A_{k1}(t)\mathbf{a}_k^\dagger \mathbf{a}_1\right\} |\{\alpha^\circ\}_1\rangle \prod_{j\neq 1} |\{0\}_j\rangle$$

or

$$|\Psi_{\text{Full}}(t)\rangle = \prod_k \exp\{A_{k1}(t)\mathbf{a}_k^\dagger \mathbf{a}_1\} |\{\alpha^\circ\}_1\rangle \prod_{j\neq 1} |\{0\}_j\rangle \tag{11.47}$$

Now, the eigenvalue equation characterizing the coherent state describing at initial time the oscillator labeled 1 is

$$\mathbf{a}_1 |\{\alpha^\circ\}_1\rangle = \alpha_1^\circ |\{\alpha^\circ\}_1\rangle$$

so that Eq. (11.47) transforms to

$$|\Psi_{\text{Full}}(t)\rangle = \prod_k \exp\{A_{k1}(t)\mathbf{a}_k^\dagger \alpha_1^\circ\} |\{\alpha^\circ\}_1\rangle \prod_{j\neq 1} |\{0\}_j\rangle$$

Moreover, allowing each operator to act on the ket belonging to its specific space, one obtains

$$|\Psi_{\text{Full}}(t)\rangle = \exp\{A_{11}(t)\alpha_1^\circ \mathbf{a}_1^\dagger\} |\{\alpha^\circ\}_1\rangle \prod_{k\neq 1} \exp\left(A_{k1}(t)\alpha_1^\circ \mathbf{a}_k^\dagger\right) |\{0\}_k\rangle$$

Next, comparing Eq. (11.38),

$$A_{11}(t) = f_{11}(t) - 1$$

$$A_{k1}(t) = f_{k1}(t) \quad \text{if} \quad k \neq 1$$

$$|\Psi_{\text{Full}}(t)\rangle = \exp\{(f_{11}(t) - 1)\alpha_1^\circ \mathbf{a}_1^\dagger\} |\{\alpha^\circ\}_1\rangle \prod_{k\neq 1} \exp\{f_{k1}(t)\alpha_1^\circ \mathbf{a}_k^\dagger\} |\{0\}_k\rangle$$

so that

$$\boxed{|\Psi_{\text{Full}}(t)\rangle = \exp\{f_{11}(t)\alpha_1^\circ \mathbf{a}_1^\dagger\}(e^{-\alpha_1^\circ \mathbf{a}_1^\dagger}) |\{\alpha^\circ\}_1\rangle \prod_{k\neq 1} \exp\{f_{k1}(t)\alpha_1^\circ \mathbf{a}_k^\dagger\} |\{0\}_k\rangle} \tag{11.48}$$

Now, observe that, according to Eq. (7.66), the coherent state $|\{\alpha^\circ\}_1\rangle$ involved in Eq. (11.48) may be viewed as the result of

$$|\{\alpha^\circ\}_1\rangle = \exp\left\{\frac{|\alpha_1^\circ|^2}{2}\right\} (e^{\alpha^\circ{}_1 \mathbf{a}_1^\dagger})(e^{-\alpha^\circ{}_1^* \mathbf{a}_1^\dagger}) |\{0\}_1\rangle \tag{11.49}$$

which, after a Taylor expansion of $\exp(-\alpha_1^{\circ *}\mathbf{a}_1)$ and with Eq. (11.46), reduces to

$$|\{\alpha^\circ\}_1\rangle = \exp\left\{\frac{|\alpha_1^\circ|^2}{2}\right\} (e^{\alpha^\circ{}_1 \mathbf{a}_1^\dagger}) |\{0\}_1\rangle \tag{11.50}$$

so that the action of $\exp(-\alpha_1^\circ \mathbf{a}_1^\dagger)$ on $|\{\alpha^\circ\}_1\rangle$ appearing in Eq. (11.48) simplifies to

$$(e^{-\alpha^\circ{}_1 \mathbf{a}_1^\dagger}) |\{\alpha^\circ\}_1\rangle = \exp\left\{\frac{|\alpha_1^\circ|^2}{2}\right\} (e^{-\alpha^\circ{}_1 \mathbf{a}_1^\dagger})(e^{\alpha^\circ{}_1 \mathbf{a}_1^\dagger}) |\{0\}_1\rangle = \exp\left(\frac{|\alpha_1^\circ|^2}{2}\right) |\{0\}_1\rangle$$

Hence, Eq. (11.48) yields

$$|\Psi_{\text{Full}}(t)\rangle = \exp\left\{\frac{|\alpha_1^\circ|^2}{2}\right\} \exp\{f_{11}(t)\alpha_1^\circ \mathbf{a}_1^\dagger\}|\{0\}_1\rangle \prod_{k\neq 1} \exp\left(f_{k1}(t)\alpha_1^\circ \mathbf{a}_k^\dagger\right)|\{0\}_k\rangle$$

or

$$|\Psi_{\text{Full}}(t)\rangle = (e^{|\alpha_1^\circ|^2/2}) \prod_k \exp\{\alpha_k(t)\mathbf{a}_k^\dagger\}|\{0\}_k\rangle \tag{11.51}$$

with

$$\alpha_k(t) = f_{k1}(t)\alpha_1^\circ \tag{11.52}$$

Again, just as for passing from Eqs. (11.49) to (11.50), that is,

$$|\{0\}_k\rangle = \exp\{-\alpha_k(t)\mathbf{a}_k\}|\{0\}_k\rangle$$

Eq. (11.51) may be written without modifications as

$$|\Psi_{\text{Full}}(t)\rangle = (e^{|\alpha_1^\circ|^2/2}) \prod_k \exp\{\alpha_k(t)\mathbf{a}_k^\dagger\} \exp\{-\alpha_k(t)\mathbf{a}_k\}|\{0\}_k\rangle \tag{11.53}$$

Now, keeping in mind that, according to Eq. (7.66),

$$\exp\left(\frac{|\alpha_k(t)|^2}{2}\right) \exp\{\alpha_k(t)\mathbf{a}_k^\dagger\} \exp\{-\alpha_k(t)\mathbf{a}_k\}|\{0\}_k\rangle = |\{\alpha(t)\}_k\rangle$$

where $|\{\alpha(t)\}_k\rangle$ is a time-dependent coherent state obeying the eigenvalue equation

$$\mathbf{a}_k|\{\alpha(t)\}_k\rangle = \alpha_k(t)|\{\alpha(t)\}_k\rangle \tag{11.54}$$

with

$$\langle\{\alpha(t)\}_k|\{\alpha(t)\}_k\rangle = 1 \tag{11.55}$$

then, Eq. (11.53) may be transformed into

$$|\Psi_{\text{Full}}(t)\rangle = (e^{|\alpha_1^\circ|^2/2}) \prod_k \exp\left(\frac{-|\alpha_k(t)|^2}{2}\right) |\{\alpha(t)\}_k\rangle$$

or

$$|\Psi_{\text{Full}}(t)\rangle = (e^{|\alpha_1^\circ|^2/2}) \prod_k |\{\tilde{\alpha}(t)\}_k\rangle \tag{11.56}$$

with

$$|\{\tilde{\alpha}(t)\}_k\rangle = \{e^{-|\alpha_k(t)|^2/2}\}|\{\alpha(t)\}_k\rangle \tag{11.57}$$

so that, due to Eq. (11.55),

$$\langle\{\tilde{\alpha}(t)\}_k|\{\tilde{\alpha}(t)\}_k\rangle = \langle\{\alpha(t)\}_k|\{\alpha(t)\}_k\rangle \, \{e^{-|\alpha_k(t)|^2}\} = \{e^{-|\alpha_k(t)|^2}\} \tag{11.58}$$

Now, owing to Eq. (11.56), the time-dependent average value of $\mathbf{a}_j^\dagger \mathbf{a}_j$ reads

$$\langle\Psi_{\text{Full}}(t)|\mathbf{a}_j^\dagger \mathbf{a}_j|\Psi_{\text{Full}}(t)\rangle = (e^{-|\alpha_1^\circ|^2}) \prod_l \langle\{\tilde{\alpha}(t)\}_l|\mathbf{a}_j^\dagger \mathbf{a}_j \prod_k |\{\tilde{\alpha}(t)\}_k\rangle$$

Next, via the properties of tensor products, this expression reduces to

$$\langle \Psi_{\text{Full}}(t)|\mathbf{a}_j^\dagger \mathbf{a}_j|\Psi_{\text{Full}}(t)\rangle = (e^{-|\alpha_1^\circ|^2})\langle\{\tilde{\alpha}(t)\}_j|\mathbf{a}_j^\dagger \mathbf{a}_j|\{\tilde{\alpha}(t)\}_j\rangle \prod_{k\neq j}\langle\{\tilde{\alpha}(t)\}_k|\{\tilde{\alpha}(t)\}_k\rangle \tag{11.59}$$

or, noting Eq. (11.57),

$$\begin{aligned}\langle \Psi_{\text{Full}}(t)|\mathbf{a}_j^\dagger \mathbf{a}_j|\Psi_{\text{Full}}(t)\rangle &= e^{-|\alpha_1^\circ|^2}\langle\{\alpha(t)\}_j|\mathbf{a}_j^\dagger \mathbf{a}_j|\{\alpha(t)\}_j\rangle \\ &\quad\times\{e^{-|\alpha_j(t)|^2}\}\prod_{k\neq j}\langle\{\alpha(t)\}_k|\{\alpha(t)\}_k\rangle\{e^{-|\alpha_k(t)|^2}\}\end{aligned}$$

In addition, remark that

$$\langle\{\alpha(t)\}_j|\mathbf{a}_j^\dagger \mathbf{a}_j|\{\alpha(t)\}_j\rangle = |\alpha_k(t)|^2\langle\{\alpha(t)\}_j|\{\alpha(t)\}_j\rangle$$

Then, with the help of Eq. (11.55), the quantum average (11.59) reduces to

$$\langle \Psi_{\text{Full}}(t)|\mathbf{a}_j^\dagger \mathbf{a}_j|\Psi_{\text{Full}}(t)\rangle = F(t)|\alpha_j(t)|^2 \tag{11.60}$$

with

$$F(t) = (e^{-|\alpha_1^\circ|^2})\prod_k\{e^{-|\alpha_k(t)|^2}\} \tag{11.61}$$

Now, because of Eqs. (11.40) and (11.52), the time-dependent arguments $|\alpha_k(t)|^2$ appearing in Eqs. (11.60) and (11.61) yields

$$|\alpha_k(t)|^2 = \alpha_1^{\circ 2}\left|\sum_l P_{kl}P_{k1}e^{-i\lambda_l t}\right|^2 \tag{11.62}$$

or

$$|\alpha_k(t)|^2 = \alpha_1^{\circ 2}\left(\sum_l P_{kl}P_{k1}e^{-i\lambda_l t}\right)\left(\sum_r P_{kr}P_{k1}e^{i\lambda_r t}\right) \tag{11.63}$$

11.4 APPLICATION TO A LINEAR CHAIN

Now, consider a linear chain of quantum oscillators of the same kind where two neighbors are mutually coupled in the same way. An equivalent classical description of such a system would be that of Fig. 11.1.

Then, the Hamiltonian (11.1) simplifies to

$$\mathbf{H}_{\text{Full}} = \sum_i \hbar\omega^\circ \mathbf{a}_i^\dagger \mathbf{a}_i + \hbar\omega(\mathbf{a}_i^\dagger(\mathbf{a}_{i+1} + \mathbf{a}_{i-1}) + \mathbf{a}_i(\mathbf{a}_{i-1}^\dagger + \mathbf{a}_{i+1}^\dagger)) \tag{11.64}$$

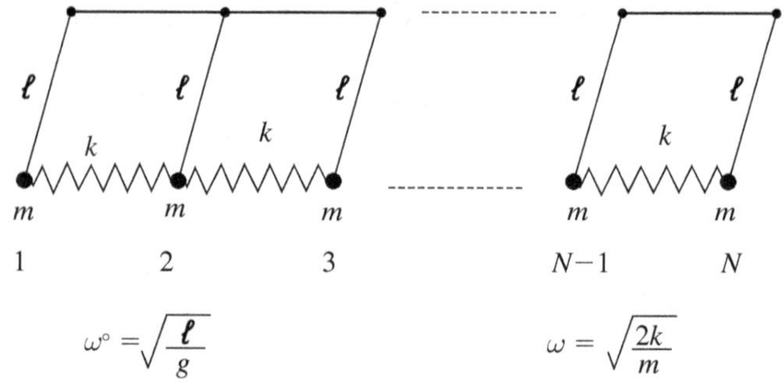

Figure 11.1 Classical model equivalent to the quantum one described by the Hamiltonian (11.64). A long chain of pendula of the same angular frequency ω° coupled by springs of angular frequency ω, where k is the force constant of the springs, l and m are, respectively, the lengths and the masses of the pendula, and g is the gravity acceleration constant.

In this special situation the matrix $\boxed{\Omega}$ given by Eq. (11.27) strongly simplifies. For instance, for five oscillators we have

$$\boxed{\Omega} = \begin{pmatrix} \omega^\circ & \omega & 0 & 0 & 0 \\ \omega & \omega^\circ & \omega & 0 & 0 \\ 0 & \omega & \omega^\circ & \omega & 0 \\ 0 & 0 & \omega & \omega^\circ & \omega \\ 0 & 0 & 0 & \omega & \omega^\circ \end{pmatrix} \tag{11.65}$$

Next, for matrices of dimension N having the same structure as that of (11.65), the eigenvalues and the corresponding eigenvectors may be given in closed form, the last ones constituting the Coulson formulas.[1] The eigenvalues λ_l are

$$\lambda_l = \omega^\circ + 2\omega\cos\left(\frac{l\pi}{N+1}\right) \tag{11.66}$$

The components P_{kl} of the eigenvectors are

$$P_{kl} = \sqrt{\frac{2}{N+1}}\sin\left(\frac{kl\pi}{N+1}\right) \tag{11.67}$$

Hence, owing to Eqs. (11.66) and (11.67), Eq. (11.62) transforms to

$$|\alpha_k(t)|^2 = \alpha_1^{\circ 2}\sum_l\sum_r \frac{2}{N+1}\sin\left(\frac{kl\pi}{N+1}\right)\sin\left(\frac{kr\pi}{N+1}\right)\sin^2\left(\frac{k\pi}{N+1}\right)f(t)$$

with

$$\begin{aligned} f(t) &= \cos\left\{\omega^\circ + 2\omega\cos\left(\frac{l\pi}{N+1}\right)\right\}\cos\left\{\omega^\circ + 2\omega\cos\left(\frac{r\pi}{N+1}\right)\right\} \\ &\quad + \sin\left\{\omega^\circ + 2\omega\cos\left(\frac{l\pi}{N+1}\right)\right\}\sin\left\{\omega^\circ + 2\omega\cos\left(\frac{r\pi}{N+1}\right)\right\} \end{aligned} \tag{11.68}$$

[1] C. A. Coulson, *Proc. Roy. Soc.*, **A169** (1939): 413; C. A. Coulson and H. C. Longuet-Higgins, *Proc. Roy. Soc.*, **A192** (1947): 16.

The mean value of the Hamiltonian averaged over the ket (11.56) reads

$$\langle \Psi_{\mathrm{Full}}(t)|\mathbf{H}_{\mathrm{Full}}|\Psi_{\mathrm{Full}}(t)\rangle = \sum_{k=1}^{N} \langle \mathbf{H}_k(t)\rangle + \langle \mathbf{H}_{k,k+1}(t)\rangle + \langle \mathbf{H}_{k,k-1}(t)\rangle \tag{11.69}$$

with, respectively,

$$\langle \mathbf{H}_k(t)\rangle = \hbar\omega^{\circ}\langle \Psi_{\mathrm{Full}}(t)|\mathbf{a}_k^{\dagger}\mathbf{a}_k|\Psi_{\mathrm{Full}}(t)\rangle \tag{11.70}$$

$$\langle \mathbf{H}_{k,k\pm1}(t)\rangle = \hbar\omega\langle \Psi_{\mathrm{Full}}(t)|(\mathbf{a}_k^{\dagger}\mathbf{a}_{k\pm1} + \mathbf{a}_k\mathbf{a}_{k\pm1}^{\dagger})|\Psi_{\mathrm{Full}}(t)\rangle \tag{11.71}$$

Due to Eq. (11.60), Eq. (11.61) yields

$$\langle \mathbf{H}_k(t)\rangle = \hbar\omega^{\circ}F(t)|\alpha_k(t)|^2 \tag{11.72}$$

or, due to Eq. (11.61) showing that $F(t)$ is the same for all oscillators,

$$\langle \mathbf{H}_k(t)\rangle = \hbar\omega^{\circ}|\alpha_k(t)|^2 \tag{11.73}$$

These results will be used later to show how such a system will evolve during time toward a stable situation when a coarse-grained analysis is performed. Such a stable situation will be understood to be a thermal equilibrium state.

11.5 CONCLUSION

We have found in this chapter that it is possible to find the dynamics of a very large set of identical harmonic oscillators coupled linearly in the ladder operators, and starting from an initial situation in which all the oscillators are in their Hamiltonian ground state, except one that is in a coherent state. It was shown that the system evolves in such a way that all the oscillators eventually become coherent, exchanging energy continuously. The interest of such a model is that it allows one in a subsequent chapter that this deterministic dynamics leads via a coarse-grained analysis and beyond a certain time to stationary situations that will appear to correspond to a thermal equilibrium state.

BIBLIOGRAPHY

P. Blaise, Ph. Durand, and O. Henri-Rousseau. *Physica A*, **209** (1994): 51.

CHAPTER 12

DENSITY OPERATORS FOR EQUILIBRIUM POPULATIONS OF OSCILLATORS

INTRODUCTION

In the last chapter, we studied the dynamics of a large set of quantum harmonic oscillators without using the density operator, which is, however, much the more suitable tool when dealing with a large population of particles. It is now time to incorporate the density operator formalism in our studies of large sets of oscillators. This will be unavoidable when we consider the thermal properties of a very large population of oscillators. The aim of the present chapter is to complete our previous studies by introducing concepts related to thermal equilibrium using the density operator. The chapter begins with the *Boltzmann theorem*, which states that the statistical entropy of a system involving a very large set of weakly coupled particles increases until statistical equilibrium is reached. It continues by applying the results of the previous chapter dealing with a very large set of weakly coupled harmonic oscillators to show, using a coarse-grained analysis, that the statistical entropy of the oscillator population obeys the Boltzmann theorem such that, when it has attained its maximum value, the coarse-grained energy analysis yields a Boltzmann distribution. Then, applying these results and the Boltzmann theorem, that is, the maximization of the statistical entropy at equilibrium, the chapter continues by obtaining the microcanonical and canonical density operators.

12.1 BOLTZMANN'S H-THEOREM

Now, we shall prove the *Boltzmann H-theorem*, which concerns the time evolution of the function $H(t)$ first considered by Boltzmann and is linked to the statistical entropy through

$$H(t) = -\frac{S(t)}{k_B} \tag{12.1}$$

Quantum Oscillators, First Edition. Olivier Henri-Rousseau and Paul Blaise.
© 2011 John Wiley & Sons, Inc. Published 2011 by John Wiley & Sons, Inc.

where k_B is Boltzmann's constant. Then, applied to a microcanonical system and according to Eq. (3.157), the Boltzmann $H(t)$ function reads

$$H(t) = \sum_{\mu} W_{\mu}(t) \ln W_{\mu}(t) \tag{12.2}$$

where $W_{\mu}(t)$ is the time-dependent probability for the microstate μ to be occupied. By differentiation of Eq. (12.2) with respect to the time, we get

$$\left(\frac{\partial H(t)}{\partial t}\right) = \sum_{\mu} \left(\frac{\partial W_{\mu}(t)}{\partial t}\right) \ln W_{\mu}(t) + \sum_{\mu} \left(\frac{\partial W_{\mu}(t)}{\partial t}\right) \tag{12.3}$$

Now, observe that

$$\sum_{\mu} \left(\frac{\partial W_{\mu}(t)}{\partial t}\right) = \frac{\partial}{\partial t} \left(\sum_{\mu} W_{\mu}(t)\right)$$

Now, for all times, the probabilities must remain normalized so that the sum of the probabilities must be equal to unity irrespective at the time t:

$$\sum_{\mu} W_{\mu}(t) = 1$$

Thus, the last right-hand-side term of Eq. (12.3) is zero so that it reduces to

$$\left(\frac{\partial H(t)}{\partial t}\right) = \sum_{\mu} \left(\frac{\partial W_{\mu}(t)}{\partial t}\right) \ln W_{\mu}(t) \tag{12.4}$$

Next, consider how the probability $W_{\mu}(t)$ changes with time. This variation is the result of a balance between gains and loss. The gains are given by all the possible jumps over the state $|\Psi_{\mu}\rangle$, eigenstate of $\mathbf{H}^{\circ}$ with energy E_{μ} at any time t, from all the other eigenstates $|\Psi_{\nu}\rangle$ of $\mathbf{H}^{\circ}$ with energy E_{ν}; thus, these gains are given by the sum of all the probabilities $W_{\nu}(t)$ that have states $|\Psi_{\nu}\rangle$ occupied, times the corresponding quantum probabilities $w_{\mu\nu}$ to jump from the initial states $|\Psi_{\nu}\rangle$ to the final one $|\Psi_{\mu}\rangle$, because of the small Hamiltonian $\mathbf{V}$ coupling $|\Psi_{\nu}\rangle$ to $|\Psi_{\mu}\rangle$. On the other hand, the loss is the sum of the transitions from the state $|\Psi_{\mu}\rangle$ to the other quantum states $|\Psi_{\nu}\rangle$ times the corresponding quantum probabilities $w_{\mu\nu}$ to jump from the initial state $|\Psi_{\mu}\rangle$ to the final ones $|\Psi_{\nu}\rangle$.

Thus, this gain–loss process leads one to write

$$\left(\frac{\partial W_{\mu}(t)}{\partial t}\right) = \sum_{\nu} W_{\nu}(t) w_{\mu\nu} - W_{\mu}(t) \sum_{\nu} w_{\mu\nu} \tag{12.5}$$

where the quantum transition probabilities $w_{\mu\nu}$ are given by Eq. (4.103), that is,

$$w_{\nu\mu} = \frac{2\pi}{\hbar} |\langle \Psi_{\nu} | \mathbf{V} | \Psi_{\mu} \rangle|^2 \delta(E_{\mu} - E_{\nu}) \tag{12.6}$$

and

$$w_{\mu\nu} = w_{\nu\mu} \qquad w_{\mu\nu} > 0 \tag{12.7}$$

Owing to Eq. (12.7), the gain–loss equation (12.5) simplifies to

$$\left(\frac{\partial W_\mu(t)}{\partial t}\right) = \sum_\nu w_{\mu\nu}(W_\nu(t) - W_\mu(t))$$

Thus, according to this result, Eq. (12.4) yields

$$\left(\frac{\partial H(t)}{\partial t}\right) = \sum_\mu \sum_\nu w_{\mu\nu}(W_\nu(t) - W_\mu(t)) \ln W_\mu(t) \tag{12.8}$$

or, after permutation of the μ and ν indices,

$$\left(\frac{\partial H(t)}{\partial t}\right) = \sum_\mu \sum_\nu w_{\mu\nu}(W_\mu(t) - W_\nu(t)) \ln W_\nu(t) \tag{12.9}$$

then, adding the half right-hand sides of the two equivalent Eqs. (12.8) and (12.9), one gets

$$\left(\frac{\partial H(t)}{\partial t}\right) = \frac{1}{2}\sum_\mu \sum_\nu w_{\mu\nu}(W_\nu(t) - W_\mu(t))(\ln W_\mu(t) - \ln W_\nu(t))$$

or

$$\left(\frac{\partial H(t)}{\partial t}\right) = \frac{1}{2}\sum_\mu \sum_\nu w_{\mu\nu}(W_\nu(t) - W_\mu(t)) \ln\left\{\frac{W_\mu(t)}{W_\nu(t)}\right\} \tag{12.10}$$

Now, observe that

$$\ln\left\{\frac{W_\mu(t)}{W_\nu(t)}\right\} < 0 \quad \text{if} \quad W_\nu(t) - W_\mu(t) > 0$$

$$\ln\left\{\frac{W_\mu(t)}{W_\nu(t)}\right\} > 0 \quad \text{if} \quad W_\nu(t) - W_\mu(t) < 0$$

Thus, the following inequality is verified irrespective of the difference $W_\nu(t) - W_\mu(t) < 0$,

$$[W_\nu(t) - W_\mu(t)] \ln\left\{\frac{W_\mu(t)}{W_\nu(t)}\right\} < 0 \quad \text{when} \quad W_\nu(t) \neq W_\mu(t)$$

Thus, since the transition probabilities cannot be negative, we have

$$\sum_\mu \sum_\nu w_{\mu\nu}(W_\nu(t) - W_\mu(t)) \ln\left\{\frac{W_\mu(t)}{W_\nu(t)}\right\} < 0 \quad \text{when} \quad W_\nu(t) \neq W_\mu(t) \tag{12.11}$$

Moreover, when all the probabilities $W_\mu(t)$ and $W_\nu(t)$ are equal irrespective of μ and ν, we have

$$\sum_\mu \sum_\nu w_{\mu\nu}(W_\nu(t) - W_\mu(t)) \ln\left\{\frac{W_\mu(t)}{W_\nu(t)}\right\} = 0 \quad \text{for all situations } W_\nu(t) = W_\mu(t) \tag{12.12}$$

Hence, owing to Eqs. (12.11) and (12.12), Eq. (12.10) becomes

$$\left(\frac{\partial H(t)}{\partial t}\right) < 0 \quad \text{when for some or all states } W_\nu(t) \neq W_\mu(t) \tag{12.13}$$

$$\left(\frac{\partial H(t)}{\partial t}\right) = 0 \quad \text{when for all states } W_\nu(t) = W_\mu(t) \tag{12.14}$$

Collecting Eqs. (12.14) and (12.15) leads to what is called the *Boltzmann H theorem*:

$$\boxed{\left(\frac{\partial H(t)}{\partial t}\right) \leqslant 0} \tag{12.15}$$

This theorem states that the function $H(t)$ either decreases or remains constant. Therefore, for a given situation where $H(t)$ remains constant, this implies that it has yet attained its minimum value H_{eq}, that is, the statistical equilibrium. Of course, one may assume that after a very long time, a physical system must always have reached statistical equilibrium

$$\left(\frac{\partial H(t)}{\partial t}\right) = 0 \quad \text{when} \quad t \to \infty$$

Note that, according to Eq. (12.14) corresponding to the equilibrium situation, it implies that all states labeled by μ or ν, must have the same probabilities to occur:

$$W_\nu(t) = W_\mu(t) = W_\mu^{\text{eq}} \quad \text{when} \quad t \to \infty \text{ for all } \mu \text{ and } \nu$$

Hence, because all the probabilities are the same at equilibrium and must be normalized and since Ω states exist that are accessible to the microcanonical system, all the equilibrium probabilities must be given by

$$W_\mu(t \to \infty) = W_\mu^{\text{eq}} = \frac{1}{\Omega} \tag{12.16}$$

We remark that, in view of the simple relation (12.1) between the H function and the statistical entropy, the Boltzmann H-theorem (12.15) implies the following inequality governing the time dependence of the statistical entropy:

$$\left(\frac{\partial S(t)}{\partial t}\right) \geqslant 0$$

where the symbol >0 holds for an irreversible evolution and $=0$ holds for an equilibrium situation. Hence, the statistical entropy either remains constant or increases irreversibly until it attains its equilibrium maximum value. Thus

$$dS(t) > 0 \quad \text{for irreversible process} \tag{12.17}$$

$$dS(t) = 0 \quad \text{at equilibrium} \tag{12.18}$$

The combined equations (12.17) and (12.18) lead to

$$dS(t) \geqslant 0$$

This last expression is just a special case of the *second law of thermodynamics* applied to a situation where there is no possibility of heat transfer dQ, the general formulation of which is

$$dS \geqslant \frac{dQ}{T}$$

Note that when the Boltzmann H-function has attained its minimum value corresponding to statistical equilibrium, that is, when the statistical entropy has attained its maximum equilibrium value, all the microcanonical equilibrium probabilities are the same and given by Eq. (12.16).

The statistical expression for the entropy (3.157) combined with the equilibrium probabilities (12.16) allows one to obtain a new interesting expression for the entropy at statistical equilibrium: Since in Eq. (3.156) the summation is performed over all the accessible states, this expression may be written explicitly as

$$S = -k_B \sum_{\mu=1}^{\Omega} W_\mu^{\text{eq}} \ln W_\mu^{\text{eq}}$$

or, due to Eq. (12.16),

$$S = -k_B \Omega \frac{1}{\Omega} \ln\left(\frac{1}{\Omega}\right)$$

hence

$$S = k_B \ln \Omega \tag{12.19}$$

Since the statistical entropy must increase in an irreversible way until it attains its maximum value, we have a tool for obtaining equilibrium density operators through their connection with statistical entropy by requiring the differential of the statistical entropy to be zero:

$$\beta = \frac{1}{k_B T}$$

12.2 EVOLUTION TOWARD EQUILIBRIUM OF A LARGE POPULATION OF WEAKLY COUPLED HARMONIC OSCILLATORS

12.2.1 Deterministic dynamics

In Chapter 11 the dynamics of a linear chain of harmonic oscillators was studied. The full Hamiltonian of this chain was

$$\mathbf{H}_{\text{Full}} = \sum_k \hbar\omega^\circ \mathbf{a}_k^\dagger \mathbf{a}_k + \hbar\omega \left(\mathbf{a}_k^\dagger(\mathbf{a}_{k+1} + \mathbf{a}_{k-1}) + \mathbf{a}_k(\mathbf{a}_{k-1}^\dagger + \mathbf{a}_{k+1}^\dagger)\right)$$

and the set of harmonic oscillators was assumed to start from an initial situation in which one of the oscillators (labeled 1) is in a coherent state while the others are in their ground state, that is,

$$|\Psi_{\text{Full}}(0)\rangle = |\{\alpha^\circ\}_1\rangle \prod_{j\neq 1} |\{0\}_j\rangle$$

with

$$\mathbf{a}_1 |\{\alpha^\circ\}_1\rangle = \alpha_1^\circ |\{\alpha^\circ\}_1\rangle$$

Observe that since $\hbar\omega << \hbar\omega^\circ$, the mean value of the Hamiltonian $\mathbf{H}_{\text{Full}}$ averaged over the initial state $|\Psi_{\text{Full}}(0)\rangle$ may be approximated by taking into account only its diagonal parts $\hbar\omega^\circ \mathbf{a}_k^\dagger \mathbf{a}_k$ leading us to write

$$\langle \Psi_{\text{Full}}(0)|\mathbf{H}_{\text{Full}}|\Psi_{\text{Full}}(0)\rangle \backsimeq \hbar\omega^\circ \langle\{\alpha^\circ\}_1|\mathbf{a}_1^\dagger \mathbf{a}_1|\{\alpha^\circ\}_1\rangle \prod_{j\neq 1} \langle\{0\}_j|\{0\}_j\rangle$$
$$+ \hbar\omega^\circ \langle\{\alpha^\circ\}_1|\{\alpha^\circ\}_1\rangle \sum_{k\neq 1} \langle\{0\}_k|\mathbf{a}_k^\dagger \mathbf{a}_k|\{0\}_k\rangle \prod_{j\neq 1,k} \langle\{0\}_j|\{0\}_j\rangle$$

which, after using the normalization properties of the coherent state and of the ground state, becomes

$$\langle \Psi_{\text{Full}}(0)|\mathbf{H}_{\text{Full}}|\Psi_{\text{Full}}(0)\rangle \backsimeq \hbar\omega^\circ \langle\{\alpha^\circ\}_1|\mathbf{a}_1^\dagger \mathbf{a}_1|\{\alpha^\circ\}_1\rangle = \hbar\omega^\circ \alpha_1^{\circ 2} \tag{12.20}$$

It has been shown in Chapter 11 that the energy α_1° initially on the coherent state $|\{\alpha^\circ\}_1\rangle$ with time passes on all the chain oscillators and also that the dimensionless average energies of each oscillator at time t are given by Eq. (11.73), that is,

$$\langle \mathbf{H}_k(t)\rangle \propto \hbar\omega^\circ |\alpha_k(t)|^2 \tag{12.21}$$

with, due to Eq. (11.68),

$$|\alpha_k(t)|^2 = \alpha_1^{\circ 2} \left(\frac{2}{N+1}\right)^2$$
$$\times \sum_l \sin\left(\frac{kl\pi}{N+1}\right) \sin\left(\frac{l\pi}{N+1}\right) \exp\left(-i(\omega^\circ + 2\omega)t\right) \tag{12.22}$$
$$\times \sum_r \sin\left(\frac{kr\pi}{N+1}\right) \sin\left(\frac{r\pi}{N+1}\right) \exp\left(i(\omega^\circ + 2\omega)t\right)$$

It may be of interest to look in a first step at the time evolution of the local energy of the oscillator initially excited. Figure 12.1 gives the time evolutions of the local energy computed by the aid of Eqs. (12.21) and (12.22) for four different chains of oscillators, where the first oscillator ($k = 1$) is excited at initial time in the first situation where the number N of oscillators is 2.

The calculations show, as required, the well-known energy exchange between two resonant oscillators. In the other situations, N is, respectively, equal to 10, 100, and 500. The numerical calculations show that the energy attains zero in the same way for all the set of oscillators according to a $\cos^2$ form and not to an exponential one. The difference between linear chains involving different numbers N of oscillators is because the time after which an energy returns to the initially excited oscillator is increasing with N after a time period T_θ, which depends on N, a small amount of the initial energy is returning to the first oscillator and then coming back and forth (see the situation for $N = 500$), and together spreading out progressively; T_θ appears from the calculations, when N is large, to be approximately given in the units used by

$$T_\theta \backsimeq N \tag{12.23}$$

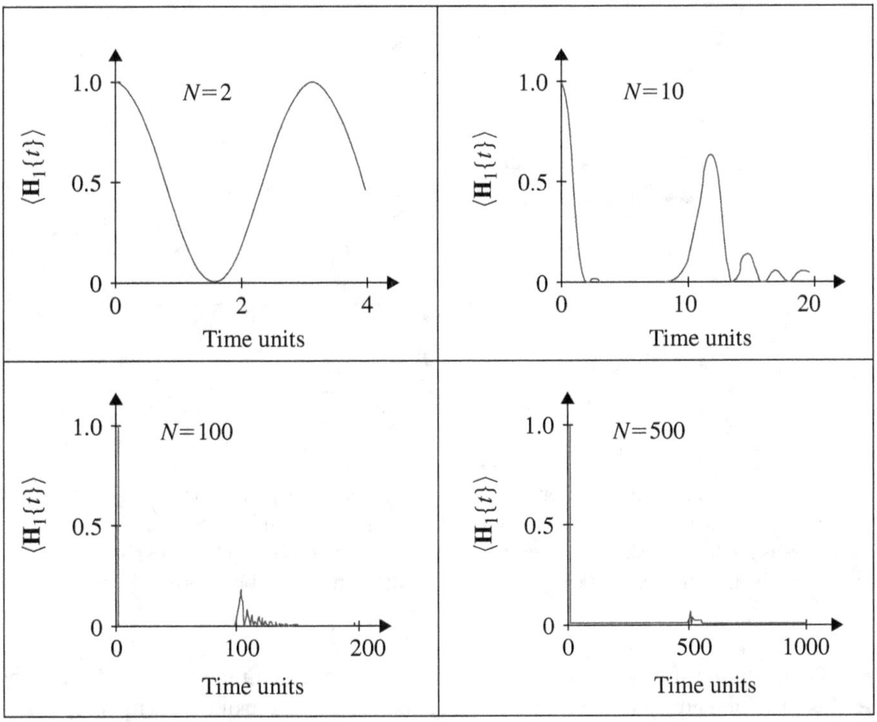

Figure 12.1 Time evolution of the local energy $\langle \mathbf{H}_1(t) \rangle$ of oscillator 1 of systems involving $N = 2$, 10, 100, and 500 oscillators computed by Eqs. (12.21) and (12.22). The time is expressed in units corresponding to the time required to attain the first zero value of the local energy.

12.2.2 Energy and entropy analysis

12.2.2.1 Energy distribution $n_i(E,t)$ of the local oscillators We consider, as time proceeds, the distribution of the energies of local oscillators, which can be considered as quasi-autonomous entities weakly coupled with their neighbors. The energy of each oscillator evolves continuously with time according to Eqs. (12.21) and (12.22). A fine-grained approach to these evolutions would be of little interest since the multiplicity of the time evolution details are not compatible with the impossibility of perfectly accurate observations. If we limit ourselves to make observations of only limited accuracy, which are in general insufficient to distinguish between neighboring energies, a coarse-grained approach, involving some lack of information concerning the evolution of the system, appears to be more suitable. Thus, although the energy distribution is continuous, we use a discrete analysis. Thus, use energy cells $i = 1, 2, 3, \ldots$, of a given width $\Delta\varepsilon_\gamma$, covering all the energies going from zero to the energy $\alpha_1^{\circ 2}\hbar\omega^\circ$ of the initial excited state, as pictured in Fig. 12.2.

More precisely, we take the width $\Delta\varepsilon_\gamma$ of these cells as a function of the initial excitation energy $\alpha_1^{\circ 2}\hbar\omega^\circ$ and of the number N of the degrees of freedom, that is,

$$\Delta\varepsilon_\gamma = \frac{1}{\gamma}\left(\frac{\alpha_k^2}{N}\right) \tag{12.24}$$

where $1/\gamma$ is the scale of the cell width $\Delta\varepsilon_\gamma$.

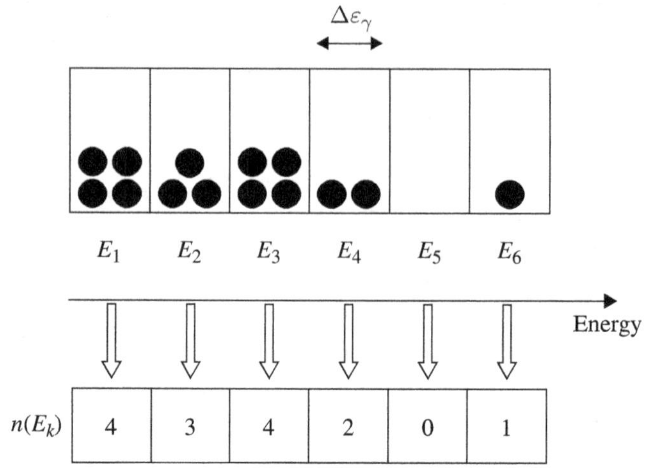

Figure 12.2 Pictorial representation of the coarse-grained analysis of the energy distribution of the oscillators inside energy cells of increasing energy E_i. The boxes indicate the energy cells, whereas the black disks represent the oscillators. The number $n_i(E_i)$ of oscillators having energy E_i is given in the bottom boxes. $\Delta\varepsilon_\gamma$ is the width of the energy cells given by Eq. (12.24).

Then, define the function $n_i(E, t)$, at time t, as the number of oscillators, the energies of which computed by Eqs. (12.21) and (12.22) lie inside the kth energy cell, as illustrated in Fig. 12.2.

12.2.2.2 Statistical entropy As observed above, as time proceeds, the time evolution of the mean energies of each oscillator shows some tendency to a spreading out of the energy over different oscillators. Thus, the information we have about the system is decreasing. Since the number $n_k(E, t)$ of oscillators present in the ith energy cell, and defined in Section 3.2, must depend on the width of the cell, the statistical entropy must depend also on this width, reflecting the knowledge we have concerning the oscillators set.

Start from the statistical definition of the entropy S:

$$S = -\sum_i P_i \ln P_i \tag{12.25}$$

where P_i is the probability of occurrence of the ith energy cell. Now, in the present situation, the probabilities P_i depend on the energy distribution $n_k(E, t)$ and, thus, are time dependent. They are given by

$$P_i(E, t) = \left(\frac{n_i(E, t)}{N}\right) \tag{12.26}$$

Then, Eq. (12.25) takes the form

$$S(t) = -\sum_i \left(\frac{n_i(E, t)}{N}\right) \ln\left(\frac{n_i(E, t)}{N}\right) \tag{12.27}$$

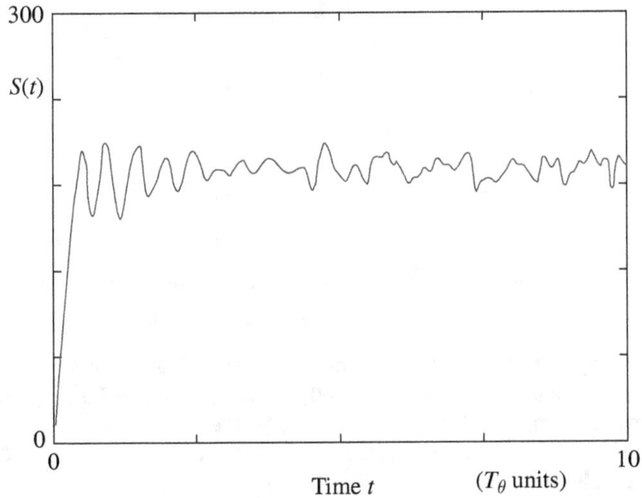

Figure 12.3 Time evolution of the entropy of a chain of $N = 100$ quantum harmonic oscillators. The time is in T_θ units, with T_θ given by Eq. (12.23). The initial excitation energy of the site $k = 1$ is $\alpha_1^2 = N$.

Now, consider the time behavior of the entropy of the set of oscillators. We show in Fig. 12.3 the results for a chain of $N = 100$ oscillators from computations performed using Eqs. (12.21) and (12.22) for local energies and using Eq. (12.27) for entropy.

We see that the entropy increases with time in a chaotic way and, after some time which is of the order of T_θ, fluctuates around a constant mean value. Thus, the time behavior, which then remains constant, is in agreement with the Boltzmann H-theorem.

Now, study the consequences of the stabilization of the entropy occurring after some transient time, that is, for large time t_∞.

12.2.3 Coarse-grained energy analysis

Figure 12.4 give the dependence of the $n_i(E, t)$ as a function of their energy for two different values of the scale factor γ and for an oscillator chain with $N = 1000$ oscillators. The site of excitation was the first oscillator of the chain, that is, $k = 1$, the energy excitation being $\alpha_k^2 = 1000$ and the time t_∞ being 1000 T_θ, that is, corresponding to a situation where in view of Fig. 12.3, the average of the entropy fluctuation has ceased to increase and has attained its maximum value.

Figure 12.4 exhibits an energy dependence of $n_i(E, t_\infty)$, namely a decreasing exponential of the form

$$n_i(E, t_\infty) \backsimeq \text{cst} \cdot e^{-\eta E} \tag{12.28}$$

where cst and η are constants.

This expression is confirmed by Fig. 12.5, which gives the energy dependence of the $n_i(E, t_\infty)$ for a population of $N = 1000$ oscillators, at different times t_∞, going from

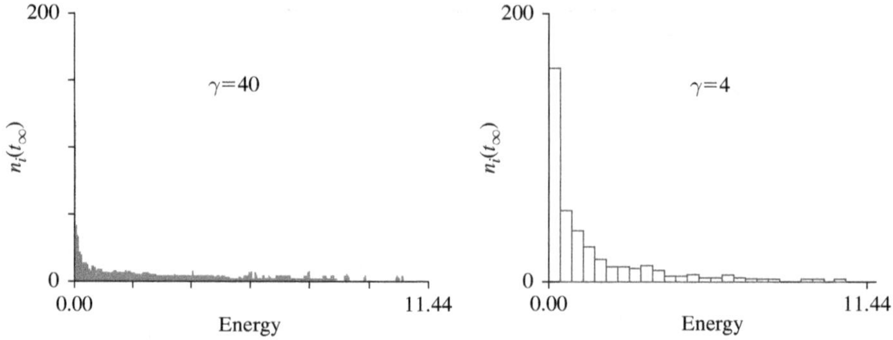

Figure 12.4 Energy distribution of a chain of $N = 1000$ oscillators for several values of the cell parameter γ. The analyzing time $t_\infty = 1000T_\theta$ with T_θ given by Eq. (12.23). The initial excitation energy of the site $k = 1$ is $\alpha_1^2 = N$. $n_i(E, t_\infty)$ is the number of oscillators having their energy calculated by Eqs. (12.21) and (12.22) within the energy cell i of width $\Delta\varepsilon_\gamma$ given by Eq. (12.24) according to Fig. 12.2.

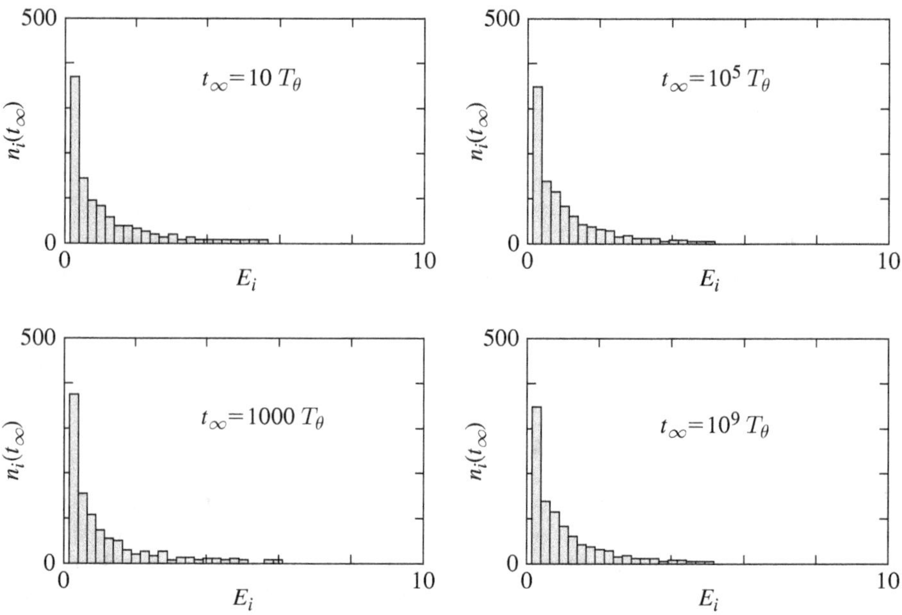

Figure 12.5 Energy distribution of $N = 1000$ coupled oscillators for $\gamma = 4$ and for time t_∞ going from $t_\infty = 10\ T_\theta$ to $t_\infty = 10^9\ T_\theta$.

$10\ T_\theta$, to $10^9\ T_\theta$, the excitation energy and the excitation sites remaining the same for all the calculations, since it shows that the coarse-grained exponential distribution of the energy is relatively stable with respect to time. Hence, the exponential distribution appears to be stable in form with respect to t_∞.

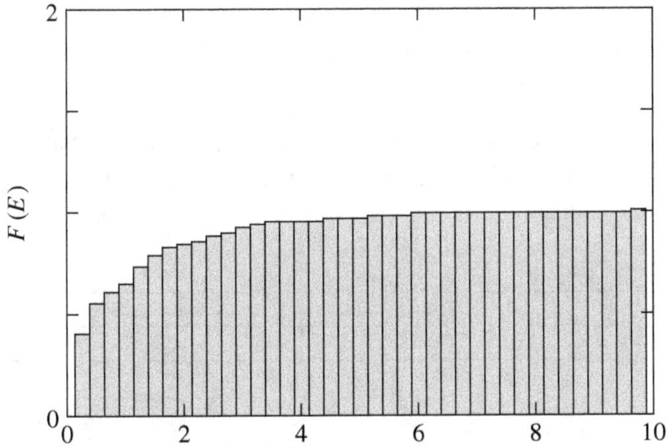

Figure 12.6 Staircase representation of the cumulative distribution functions of the probabilities (12.26).

The result (12.28) is very interesting since it is an illustration of the Boltzmann distribution (13.11) encountered in the previous section, that is,

$$W^{\text{eq}}(E_k) = \frac{e^{-\beta E_k}}{Z} \quad \text{with} \quad W^{\text{eq}}(E_k) = \frac{N_k^{\text{eq}}(E_k)}{N_{\text{Tot}}}$$

suggesting the following correspondence between the different terms of Eq. (12.28) and those of the latter equation:

$$n_k(E, t_\infty) \leftrightarrow N_k^{\text{eq}}(E_k) \qquad \eta \leftrightarrow \beta \qquad \text{cst} \leftrightarrow \frac{1}{Z}$$

12.2.4 Staircase representation of the B(t_∞) damping parameter

In order to examine the exponential form of the energy cell populations, it is convenient to use a *staircase representation* of the cumulative distribution functions of the probabilities (12.26) according to Fig. 12.6:

$$F(E) = \sum_{E_i=0}^{E} P_i(E_i, t_\infty) \tag{12.29}$$

Then, with a least-squares procedure, we may get the curve fitting of this staircase representation of the form

$$F(E) = C(t_\infty)(e^{-B(t_\infty)E} - 1) \tag{12.30}$$

which gives the best fitted $B(t_\infty)$ and $C(t_\infty)$ parameters. Now, the analytic function $f(E)$ we search for is, by definition,

$$f(E) = \frac{dF(E)}{dE} \tag{12.31}$$

Hence,

$$f(E) = A(t)e^{-B(t_\infty)E} \quad \text{with} \quad A(t_\infty) = C(t_\infty)B(t_\infty) \tag{12.32}$$

This staircase procedure allows one to extract the parameter $B(t_\infty)$ for the energy distribution of the oscillators. Of course, $B(t_\infty)$ depends on time because of the time dependence of the population $n_k(E, t_\infty)$ of the energy cells involved in the expression of the staircase cumulative distribution function. In the tabular data in (12.33), we show the values obtained for $B(t_\infty)$ obtained for a linear set of 7500 oscillators at time $t_\infty = 10^3\ T_\theta$, with $\gamma = 4$, when changing the site of the initial energy excitation of the oscillators chain:

k	1	100	200	300	400	500	600	700	800	900
$B(t_\infty)$	−0.76	−0.76	−0.75	−0.75	−0.76	−0.74	−0.76	−0.76	−0.75	−0.76

(12.33)

Inspection of this data shows that $B(t_\infty)$ is approximately the same for the different excitation sites, the dispersion around its average value being small.

A more detailed examination shows, however, that the coarse-grained exponential parameter B fluctuates around some average value. Calculations reproduced in Fig. 12.7 involving a given set of local oscillators undergoing the same initial excitation, and analyzed with the same energy cell width show (see Fig. 12.2) that the fitted parameter B is changing with time in a way that appears to be stochastic when discrete times are chosen for the numerical calculations.

The time average $\langle B\rangle$ of the fluctuating parameter $B(t_i)$ may be obtained from

$$\langle B\rangle = \sum_i \frac{B(t_i)}{N_{t_i}} \tag{12.34}$$

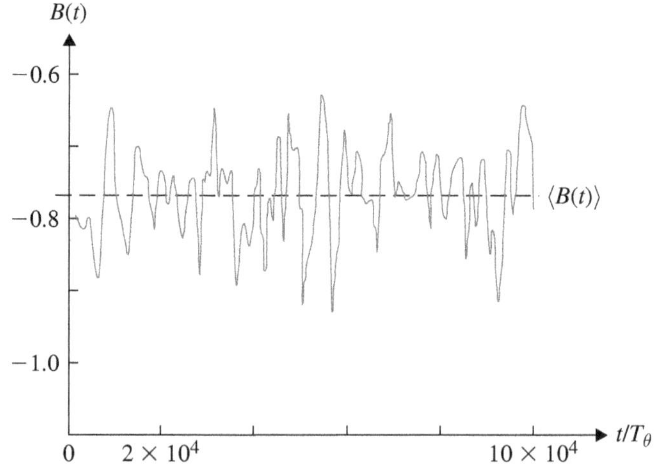

Figure 12.7 Time fluctuation of $B(t)$ around its mean value $\langle B(t)\rangle$ for a chain of $N = 100$ coupled quantum harmonic oscillators.

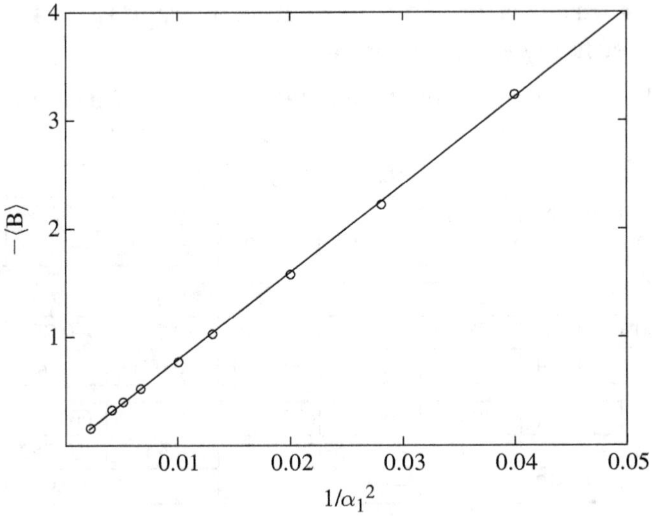

Figure 12.8 Linear regression $-\langle B\rangle$ as a function of $1/\alpha_1^{\circ 2}$ from the values of expression (12.33). The solid line is the regression curve corresponding to $-\langle B\rangle = 80.659 \times \frac{1}{\alpha_1^{\circ 2}} - 0.0179$ with a regression coefficient $r^2 = 0.999$.

where N_{t_k} is the number of samples t_k of the large time t_∞ at which the values $B(t_i)$ have been calculated using the staircase procedure involving Eqs. (12.29)–(12.32). The corresponding dispersion may be obtained using

$$\Delta B = \sqrt{\sum_i \frac{B(t_i)^2}{N_{t_i}} - \left(\sum_i \frac{B(t_i)}{N_{t_i}}\right)^2} \tag{12.35}$$

The calculation of $\langle B\rangle$ and ΔB is performed by selecting $N_{t_i} = 100$ time samples, uniformly distributed within a time interval equal to $10^2 T_\theta$, with T_θ given by Eq. (12.23). We emphasize that modifications in the selection of the time intervals or in the number of time samples do not affect sensitively the obtained statistical values.

A linear regression of the relative dispersion of $B(t_i)$ with respect to the inverse of $\alpha_1^{\circ 2}$, [i.e., due to Eq. (12.20), to the inverse of the energy amount in $\hbar\omega^\circ$ units, introduced at the initial time in the coherent state, $|\{\alpha^\circ\}_1\rangle]$ is reproduced in Fig. 12.8 using the numerical values given in (12.36):

$\alpha_i^{\circ 2}$	500	250	200	150	100	75	50	36	25	10
$\langle B\rangle$	−0.16	−0.32	−0.40	−0.52	−0.76	−1.02	−1.57	−2.22	−3.24	−8.13
$\frac{\Delta B}{\langle B\rangle}$	−0.10	−0.11	−0.11	−0.11	−0.10	−0.10	−0.10	−0.10	−0.10	−0.12

$\alpha_i^{\circ 2}$	3	1	0.3	0.1	3×10^{-2}	10^{-3}	10^{-4}			
$\langle B\rangle$	−27.4	−81.5	−270	−820	−2700	−80000	−8000000			
$\frac{\Delta B}{\langle B\rangle}$	−0.10	−0.11	−0.10	−0.10	−0.10	−0.12	−0.12			

(12.36)

12.2.5 Linear regressions of the relative average values of B and of entropy with respect to $1\sqrt{N}$

Now, consider the dependence of $\langle B \rangle$ and $\Delta B/\langle B \rangle$ on the number of oscillators N shown in (12.37).

N	50	64	75	100	125	250	375	500
$\langle B \rangle$	−0.743	−0.763	−0.757	−0.764	−0.773	−0.767	−0.764	−0.758
$\frac{\Delta B}{\langle B \rangle}$	−0.153	−0.136	−0.131	−0.100	−0.100	−0.060	−0.047	−0.043

N	675	750	825	1000	1100	1250	1300	1500
$\langle B \rangle$	−0.760	−0.761	−0.759	−0.760	−0.759	−0.758	−0.758	−0.757
$\frac{\Delta B}{\langle B \rangle}$	−0.044	−0.034	−0.035	−0.028	−0.028	−0.024	−0.025	−0.026

N	1600	1750	1800	2000	2750	3000	6000	
$\langle B \rangle$	−0.757	−0.758	−0.754	−0.755	−0.750	−0.7512	−0.752	
$\frac{\Delta B}{\langle B \rangle}$	−0.025	−0.022	−0.019	−0.021	−0.017	−0.016	−0.014	

(12.37)

From this data, Fig. 12.9 gives the dependence of the relative dispersion of B on $1/\sqrt{N}$.

Thus, by inspection of Fig. 12.9, the relative dispersion of $\langle B \rangle$ exhibits a linear dependence of the form

$$\frac{\Delta B}{\langle B \rangle} \propto \frac{1}{\sqrt{N}}$$

In connection with this relative dispersion, it may be of interest to consider in a similar way the relative dispersion of the statistical entropy of the linear chain. The

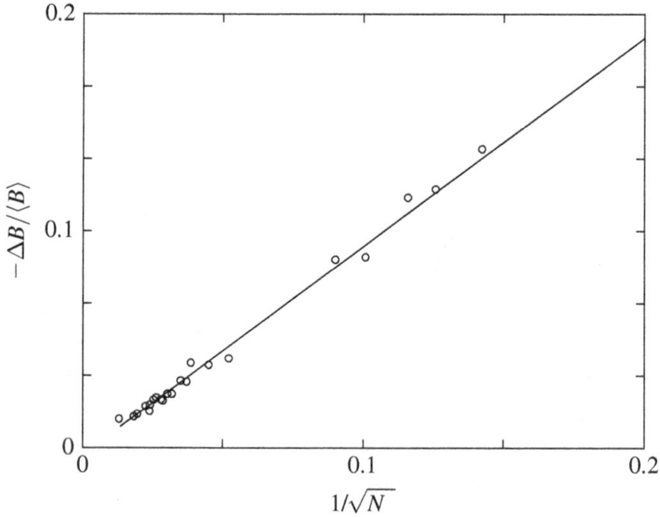

Figure 12.9 Linear regression of $\Delta B/\langle B \rangle$ of B with respect to $1/\sqrt{N}$ obtained according to the values of expression (12.37).

time-averaged entropy and the corresponding dispersion entropy may be, respectively, obtained from

$$\langle S\rangle = \sum_k \frac{S(t_k)}{N_{t_k}}$$

$$\Delta S = \sqrt{\sum_i \frac{S(t_i)^2}{N_{t_i}} - \left(\sum_i \frac{S(t_i)}{N_{t_i}}\right)^2}$$

where $S(t_k)$ may be calculated from Eq. (12.27). The tabular data in (12.38) reports the average values $\langle S\rangle$ and the relative dispersion $\Delta S/\langle S\rangle$ as a function of the number N of degrees freedom of:

N	50	75	100	150	200	250	300
$\langle S\rangle$	109	183	259	426	607	797	980
$\frac{\Delta S}{\langle S\rangle}$	0.08	0.06	0.06	0.04	0.04	0.04	0.04

N	400	500	600	700	900	1100	1200
$\langle S\rangle$	1400	1800	2200	2600	3600	4500	5000
$\frac{\Delta S}{\langle S\rangle}$	0.03	0.03	0.03	0.02	0.02	0.02	0.02

(12.38)

N	1350	1500	1800	2100	3100	3600
$\langle S\rangle$	5700	6400	7900	9400	14600	17300
$\frac{\Delta S}{\langle S\rangle}$	0.02	0.02	0.01	0.01	0.01	0.01

Figure 12.10 is gives the relative dispersion $\Delta S/\langle S\rangle$ of the entropy versus $1/\sqrt{N}$ from the values of (12.38).

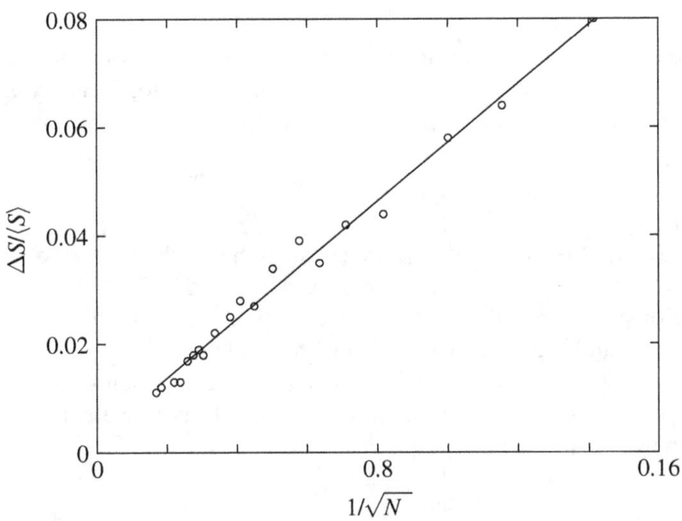

Figure 12.10 Relative dispersion $\Delta S/\langle S\rangle$ of the entropy S as a function of the number N of degrees of freedom. $\gamma = 4$, $k = 1$, $\alpha_i^{\circ 2} = N$, $t_\infty = 10^3 T_\theta$, $N_{t_k} = 10^2$. The full line corresponds to the linear regression $\Delta S/\langle S\rangle = 0.543(1/\sqrt{N}) + 0.003$ with a correlation coefficient $r^2 = 0.988$.

The linear regression leads also to a linear dependence of the form

$$\frac{\Delta S}{\langle S \rangle} \propto \frac{1}{\sqrt{N}} \tag{12.39}$$

12.3 MICROCANONICAL SYSTEMS

Consider a system formed by a set of N equivalent noninteracting quantum components enclosed in a constant volume V, which is adiabatically isolated from the medium, which does not exchange any energy, and whose number of components remains constant. Then, its total energy E_{Tot} is constant. Such a system where V, N, and E_{Tot} are constant is called a *microcanonical system*.

Denote by $\mathbf{H}^{\{k\}}$ the Hamiltonian of the kth quantum component. Its eigenvalue equation is

$$\{\mathbf{H}^{\{k\}}\}|\Phi_{n_\mu}^{\{k\}}\rangle = \{E_{n_\mu}^{\{k\}}\}|\Phi_{n_\mu}^{\{k\}}\rangle \tag{12.40}$$

Next, we may consider a ket of the whole system. This ket, which is called a *microstate*, is the tensor product over all the N equivalent quantum components of the kets appearing in Eq. (12.40), that is,

$$|\Psi_\mu^{\text{full}}\rangle = |\Phi_{n_\mu}^{\{1\}}\rangle|\Phi_{k_\mu}^{\{2\}}\rangle \cdots |\Phi_{m_\mu}^{\{N\}}\rangle$$

This may be written in the condensed form

$$|\Psi_\mu^{\text{full}}\rangle = \prod_{k=1}^{N} |\Phi_{n_\mu}^{\{k\}}\rangle \tag{12.41}$$

where μ indicates that the microstate is characterized by the set of N quantum indices characterizing the eigenstates of the N components. The total energy E_{Tot} of the system is the same whatever the microstate since it is assumed to be constant. Hence

$$E_{\text{Tot}} = \sum_k \{E_{n_\mu}^{\{k\}}\} \quad \text{which may be } \mu \tag{12.42}$$

Now, since by hypothesis, N and E_{Tot} are constant, hence the number of microstates that are possible is finite because the eigenvalues involved in Eq. (12.42) are discrete. The number of states accessible to the microcanonical system is called the *number of accessible states* and is generally represented by the letter Ω.

Now, suppose that the N equivalent quantum components are very weakly interacting through a very small Hamiltonian $\mathbf{H}_{\text{Int}}$. Hence, we can write the full Hamiltonian as

$$\mathbf{H}_{\text{Tot}} = \sum_k \{\mathbf{H}^{\{k\}}\} + \mathbf{H}_{\text{Int}}$$

Then, owing to this coupling Hamiltonian $\mathbf{H}_{\text{Int}}$, a transition probability exists per unit time $w_{\mu\nu}$ for passing from one microstate μ to another one ν.

In order to render the microcanonical system more intuitive, it is convenient to consider the components of the system as equivalent quantum harmonic oscillators, the Hamiltonians of angular frequency ω of which are

$$\{\mathbf{H}^{\{k\}}\} = \hbar\omega\left(\mathbf{a}_k^{\dagger}\mathbf{a}_k + \tfrac{1}{2}\right) \quad \text{with} \quad [\mathbf{a}_k, \mathbf{a}_k^{\dagger}] = 1$$

The eigenvalue equations of these individual Hamiltonians are

$$\mathbf{H}^{\{k\}}|\{n_k\}\rangle = \hbar\omega\left(n_k + \tfrac{1}{2}\right)|\{n_k\}\rangle \tag{12.43}$$

Now, the interaction Hamiltonian may be viewed as weakly exchanging the energy between two oscillators according to

$$\mathbf{H}_{\text{Int}} = \sum_k \sum_l \varepsilon_{kl}|\{n_k\}\rangle\langle(n_l \pm 1)| + \text{hc}$$

Then, the total energy (12.42) in the absence of coupling is

$$E_{\text{Tot}} = \hbar\omega \sum_k \left(n_k + \frac{1}{2}\right)$$

while the microstates (12.41) takes the special form

$$|\Psi_\mu^{\text{full}}\rangle = \prod_{k=1}^{N} |\{n_{k_\mu}\}\rangle$$

12.4 EQUILIBRIUM DENSITY OPERATORS FROM ENTROPY MAXIMIZATION

12.4.1 Microcanonical density operators

Consider the statistical entropy Eq. (3.149). For a microcanonical system it may be written in terms of the microcanonical density operator $\boldsymbol{\rho}_{\text{MC}}$, that is,

$$S = -k_B\,\mathbf{tr}\{\boldsymbol{\rho}_{\text{MC}} \ln \boldsymbol{\rho}_{\text{MC}}\} \tag{12.44}$$

Now, the density operator must be normalized so that, according to Eq. (3.140), it must satisfy

$$\mathbf{tr}\{\boldsymbol{\rho}_{\text{MC}}\} = 1 \tag{12.45}$$

Also, the Boltzmann H-theorem (12.15) requires that the statistical entropy is stationary and so must satisfy Eq. (12.18), that is,

$$dS = 0 \tag{12.46}$$

Hence, in order to get the density operator at equilibrium, one has to solve Eq. (12.46) with S given by Eq. (12.44), noting that the normalization condition (12.45) is satisfied. Differentiation of Eq. (12.44) leads to

$$dS = -k_B\,\mathbf{tr}\{(1 + \ln \boldsymbol{\rho}_{\text{MC}})\delta\boldsymbol{\rho}_{\text{MC}}\} \tag{12.47}$$

Again, in order to incorporate the normalization condition (12.45), which acts as a constraint on the differential (12.46), we use the *Lagrange multipliers method* given in Section 18.7, which imposes on Eq. (12.47) the following constraint equation:

$$-k_B\lambda\,\mathbf{tr}\{d\boldsymbol{\rho}_{\mathrm{MC}}\} = 0 \tag{12.48}$$

where λ is a Lagrange multiplier given at the end of the calculation. Thus, adding this constraint (12.48) to the differential (12.47), which must satisfy (12.46), one obtains for the maximization of the statistical entropy required by the stationary condition (12.46), the differential equation

$$-k_B\,\mathbf{tr}\{(1 + \ln \boldsymbol{\rho}_{\mathrm{MC}} + \lambda)\delta\boldsymbol{\rho}_{\mathrm{MC}}\} = 0$$

or, since k_B and $\delta\boldsymbol{\rho}_{\mathrm{MC}}$ differ from zero,

$$\mathbf{tr}\{(1 + \ln \boldsymbol{\rho}_{\mathrm{MC}} + \lambda)\} = 0$$

Now, since this equation must be satisfied regardless of the basis over which the trace is made, the condition simplifies to

$$1 + \ln \boldsymbol{\rho}_{\mathrm{MC}} + \lambda = 0$$

or

$$\ln \boldsymbol{\rho}_{\mathrm{MC}} = -(1 + \lambda)$$

and therefore

$$\boldsymbol{\rho}_{\mathrm{MC}} = e^{-(1+\lambda)} \tag{12.49}$$

Now, observe that the diagonal matrix element of the density operator calculated over any microstate $|\Psi_\mu\rangle$, is the probability to find, in equilibrium, the system in this microstate, that is,

$$\langle\Psi_\mu|\boldsymbol{\rho}_{\mathrm{MC}}|\Psi_\mu\rangle = W_\mu(t \to \infty) \tag{12.50}$$

However, comparing Eq. (12.49), the left-hand side of Eq. (12.50) reads

$$\langle\Psi_\mu|\boldsymbol{\rho}_{\mathrm{MC}}|\Psi_\mu\rangle = \langle\Psi_\mu|e^{-(1+\lambda)}|\Psi_\mu\rangle = e^{-(1+\lambda)} \tag{12.51}$$

whereas the right-hand side of Eq. (12.50) is given by Eq. (12.16), that is,

$$W_\mu(t \to \infty) = \frac{1}{\Omega} \tag{12.52}$$

where Ω is the number of accessible states of the microcanonical system, so that

$$\langle\Psi_\mu|\boldsymbol{\rho}_{\mathrm{MC}}|\Psi_\mu\rangle = \frac{1}{\Omega} \tag{12.53}$$

Moreover, due to Eqs. (12.50)–(12.52), it appears that

$$\lambda = \ln \Omega - 1$$

12.4.2 Canonical density operators

By definition, a canonical system is one in which the total energy is not known exactly but only as an average. Thus, in place of the E used for the microcanonical approach, one has to consider the average value of the total Hamiltonian performed over the canonical density operator $\boldsymbol{\rho}_B$, that is,

$$\langle \mathbf{H} \rangle = \mathbf{tr}\{\boldsymbol{\rho}_B \mathbf{H}\} \tag{12.54}$$

Hence, we have in the present situation, dealing with the canonical system, three equations involving the density operator: the first one defining the statistical entropy, the second one governing the normalization of the density operator and the last one allowing us to obtain the average value of the energy, that is,

$$-k_B\, \mathbf{tr}\{\boldsymbol{\rho}_B \ln \boldsymbol{\rho}_B\} = S \tag{12.55}$$

$$\mathbf{tr}\{\boldsymbol{\rho}_B\} = 1 \tag{12.56}$$

Observe that at equilibrium the density operator must satisfy $dS = 0$ so that in the absence of the two constraints (12.54) and (12.56), Eq. (12.55) would lead one to write

$$\mathbf{tr}\{(1 + \ln \boldsymbol{\rho}_B) d\boldsymbol{\rho}_B\} = 0 \tag{12.57}$$

However, owing to the two conditions (12.56) and (12.54), which have to be satisfied, one has, according to the Lagrange multiplier method, to add to Eq. (12.57) the following constraint equations:

$$\lambda\, \mathbf{tr}\{d\boldsymbol{\rho}_B\} = 0 \tag{12.58}$$

$$\beta\, \mathbf{tr}\{\mathbf{H} d\boldsymbol{\rho}_B\} = 0 \tag{12.59}$$

where λ and β are two Lagrange multipliers to be found at the end of the calculation. Hence, after incorporation of the constraints (12.58) and (12.59), the differential equation governing the density operator at equilibrium resulting from the condition $dS = 0$ yields

$$\mathbf{tr}\{(1 + \ln \boldsymbol{\rho}_B + \lambda + \beta \mathbf{H}) d\boldsymbol{\rho}_B\} = 0 \tag{12.60}$$

Because $d\boldsymbol{\rho}_B \neq 0$, and since Eq. (12.60) must be satisfied regardless the basis used to perform the trace, we have

$$1 + \ln \boldsymbol{\rho}_B + \lambda + \beta \mathbf{H} = 0$$

Thus, it appears that at equilibrium the density operator of the canonical system is

$$\boldsymbol{\rho}_B = e^{-(1+\lambda) - \beta \mathbf{H}}$$

That may be also written

$$\boldsymbol{\rho}_B = \frac{1}{Z}(e^{-\beta \mathbf{H}}) \quad \text{with} \quad \frac{1}{Z} = e^{-(1+\lambda)} \tag{12.61}$$

where Z is called the *partition function*. Equation (12.61) is the expression of *the canonical density operator*. Next, in order that Eq. (12.56) is verified, the following expression has to hold:

$$\frac{1}{Z}\mathbf{tr}\{e^{-\beta\mathbf{H}}\} = 1$$

so that the partition function reads

$$\boxed{Z = \mathbf{tr}\{(e^{-\beta\mathbf{H}})\}} \tag{12.62}$$

12.4.3 Boltzmann distribution law

Start from the canonical density operator (12.61), that is,

$$\boxed{\rho_B = \frac{1}{Z}(e^{-\beta\mathbf{H}})} \tag{12.63}$$

This operator, which depends only on the Hamiltonian, may be therefore expressed in the representation corresponding to the eigenvectors of this Hamiltonian. The eigenvalue equation of this operator is

$$\mathbf{H}|\Phi_i\rangle = E_i|\Phi_i\rangle$$

with, since the Hamiltonian is Hermitian,

$$\sum_i |\Phi_i\rangle\langle\Phi_i| = 1 \quad \text{and} \quad \langle\Phi_i|\Phi_k\rangle = \delta_{ik} \tag{12.64}$$

Postmultiply the right-hand side of Eq. (12.63) by the above closure relation:

$$\rho_B = \frac{1}{Z}\sum_i (e^{-\beta\mathbf{H}})|\Phi_i\rangle\langle\Phi_i|$$

Then, expand the exponential operator

$$\rho_B = \frac{1}{Z}\sum_i\sum_n\left(\frac{(-\beta)^n}{n!}\right)\mathbf{H}^n|\Phi_i\rangle\langle\Phi_i| \tag{12.65}$$

Again, observe that

$$\mathbf{H}^2|\Phi_i\rangle = \mathbf{H}E_i|\Phi_i\rangle = E_i\mathbf{H}|\Phi_i\rangle$$

and thus

$$\mathbf{H}^2|\Phi_i\rangle = E_iE_i|\Phi_i\rangle = E_i^2|\Phi_i\rangle$$

By recurrence we get

$$\mathbf{H}^n|\Phi_i\rangle = E_i^n|\Phi_i\rangle$$

Hence, the density operator (12.65) transforms to

$$\rho_B = \frac{1}{Z}\sum_i\sum_n\left(\frac{(-\beta)^n}{n!}\right)E_i^n|\Phi_i\rangle\langle\Phi_i|$$

Returning to the exponential, we have

$$\rho_B = \frac{1}{Z}\sum_i (e^{-\beta E_i})|\Phi_i\rangle\langle\Phi_i|$$

The matrix elements of this density operator performed over the eigenstates of the Hamiltonian are

$$\langle\Phi_i|\rho_B|\Phi_k\rangle = \frac{1}{Z}\sum_i \langle\Phi_i|(e^{-\beta E_i})|\Phi_i\rangle\langle\Phi_i|\Phi_k\rangle$$

or, using the orthonormality properties appearing in Eq. (12.64),

$$\langle\Phi_i|\rho_B|\Phi_k\rangle = \frac{1}{Z}\langle\Phi_i|(e^{-\beta E_i})|\Phi_i\rangle\delta_{k,i}$$

Thus, after simplification,

$$\langle\Phi_i|\rho_B|\Phi_i\rangle = \frac{1}{Z}(e^{-\beta E_i})$$

This last result may be also expressed as

$$\boxed{W_i = \frac{1}{Z}(e^{-\beta E_i})} \tag{12.66}$$

with

$$W_i = \langle\Phi_i|\rho_B|\Phi_i\rangle$$

Equation (12.66) is the *Boltzmann distribution*. Now, after imposing the sum of the probabilities to be unity, that is,

$$\sum_i W_i = 1$$

the partition function is

$$Z = \sum_j (e^{-\beta E_j}) \tag{12.67}$$

12.4.4 Thermal energy

In order to find the average value of the Hamiltonian of a system obeying a canonical distribution, start from Eq. (12.54), that is,

$$\langle\mathbf{H}\rangle = \mathbf{tr}\{\rho_B\mathbf{H}\}$$

with, in view of Eqs. (12.62) and 12.63),

$$\rho_B = \frac{1}{Z}(e^{-\beta\mathbf{H}}) \quad \text{with} \quad Z = \mathbf{tr}\{(e^{-\beta\mathbf{H}})\} \tag{12.68}$$

Hence, the canonical energy reads

$$\langle\mathbf{H}\rangle = \frac{1}{Z}\mathbf{tr}\{(e^{-\beta\mathbf{H}})\mathbf{H}\}$$

or

$$\langle \mathbf{H} \rangle = -\frac{1}{Z}\mathbf{tr}\left\{\left(\frac{\partial e^{-\beta \mathbf{H}}}{\partial \beta}\right)\right\}$$

Again, since the operations of trace and of partial derivation with respect to β commute, we have

$$\langle \mathbf{H} \rangle = -\frac{1}{Z}\left(\frac{\partial}{\partial \beta}\mathbf{tr}\{e^{-\beta \mathbf{H}}\}\right)$$

or, using the definition of Z appearing in (12.68),

$$\langle \mathbf{H} \rangle = -\frac{1}{Z}\left(\frac{\partial Z}{\partial \beta}\right)$$

so that

$$\boxed{\langle \mathbf{H} \rangle = -\left(\frac{\partial \ln Z}{\partial \beta}\right)} \tag{12.69}$$

Up to now, we do not know the expression of β.

12.4.5 Identification of β

One may consider a microcanonical system subdivided into two parts separated by a rigid diathermic wall allowing thermal energy transfers but forbidding passage of particles. The total number of accessible states Ω_{Tot} is the product of the number of accessible states Ω_1 and Ω_2 of each part, that is,

$$\Omega_{\text{Tot}} = \Omega_1 \Omega_2 \tag{12.70}$$

Furthermore, the total energy E_{Tot} of the whole system is the sum of the energies E_1 and E_2 of the two parts:

$$E_{\text{Tot}} = E_1 + E_2$$

Now, since the whole system is microcanonical, the total energy is a constant so that the energy exchanges between the two compartments through the diathermic wall obey

$$dE_1 = -dE_2 \tag{12.71}$$

Thus, the two variables are not independent. At equilibrium between the two compartments, the energies inside them must remain constant. This equilibrium condition requires that the total number of accessible states has attained its maximum value. When considering the energy-independent variable as E_1, this equilibrium condition leads us to write

$$\left(\frac{\partial \Omega_{\text{Tot}}}{\partial E_1}\right) = 0 \quad \text{at equilibrium}$$

Since the total number of accessible states is a monotonically increasing function of the energy, this equilibrium condition may be expressed in terms of the Neperian logarithm of Ω_{Tot}. Hence

$$\left(\frac{\partial \ln \Omega_{\text{Tot}}}{\partial E_1}\right) = 0 \quad \text{at equilibrium} \tag{12.72}$$

Then, due to Eq. (12.70), the partial derivative involved in this last equilibrium condition becomes

$$\left(\frac{\partial \ln \Omega_{\text{Tot}}}{\partial E_1}\right) = \left(\frac{\partial \ln \Omega_1}{\partial E_1}\right) + \left(\frac{\partial \ln \Omega_2}{\partial E_1}\right)$$

so that using Eqs. (12.71) and (12.72), one obtains

$$\left(\frac{\partial \ln \Omega_1}{\partial E_1}\right) = \left(\frac{\partial \ln \Omega_2}{\partial E_2}\right) \quad \text{at equilibrium}$$

Again, multiplying both members by Boltzmann's constant k_B, we have

$$k_B\left(\frac{\partial \ln \Omega_1}{\partial E_1}\right) = k_B\left(\frac{\partial \ln \Omega_2}{\partial E_2}\right) \quad \text{at equilibrium}$$

Moreover, observe that Eq. (12.19) allows us to write for the two compartments 1 and 2

$$S_i = k_B \ln \Omega_i \quad \text{with } i = 1, 2$$

where S_1 and S_2 are the entropies of the two compartments. These last equations allow one to express the above equilibrium condition by

$$\left(\frac{\partial S_1}{\partial E_1}\right) = \left(\frac{\partial S_2}{\partial E_2}\right) \quad \text{at equilibrium}$$

This equation governs the statistical equilibrium between the two compartments susceptible to exchange energy through the diathermic wall. This equilibrium is just a thermal equilibrium. Then, because the dimension of the energy is entropy times the temperature according to the second law of thermodynamics, that leads us to write

$$\left(\frac{\partial S_i}{\partial E_i}\right) = \frac{1}{T_i} \quad \text{with } i = 1, 2 \tag{12.73}$$

and thus, for this thermal equilibrium,

$$\frac{1}{T_1} = \frac{1}{T_2} \quad \text{that is, } T_1 = T_2$$

12.4.6 Alternative demonstration of the Boltzmann distribution

Now, we give another more physical demonstration of the Boltzmann distribution (12.66). For this purpose, one may consider, as above, a microcanonical system subdivided into two parts, separated by a rigid diathermic wall allowing thermal energy transfer but forbidding passage of particles. However, in this new approach, one of the two compartments, the left one, is very large with respect to the right one.

The total number Ω_{Tot} of accessible states is the product of Ω characterizing the large compartment and of Ω° characterizing the small one

$$\Omega_{\text{Tot}} = \Omega\Omega^\circ \tag{12.74}$$

Now, as above, the total energy, which remains constant regardless of the energy flux through the diathermic wall, is the sum of the energies of the two compartments, that is,

$$E_{\text{Tot}} = E + E^\circ \quad \text{with} \quad E >> E^\circ \tag{12.75}$$

where E is the energy of the large compartment and E° that of the small one. The number of accessible states Ω and Ω° are, respectively, functions of the energy of the two compartments that they characterize, that is,

$$\Omega(E) \qquad \text{and} \qquad \Omega^\circ(E^\circ)$$

Hence, Eq. (12.74) reads

$$\Omega_{\text{Tot}} = \Omega(E)\,\Omega^\circ(E^\circ)$$

Now, due to Eq. (12.75), the total number of accessible states may be written as a function of the independent variable E°:

$$\Omega_{\text{Tot}}(E^\circ) = \Omega(E_{\text{Tot}} - E^\circ)\Omega^\circ(E^\circ) \tag{12.76}$$

Next, distinguish between two situations labeled 1 and 2, corresponding to the case where there is either the energy E_1° or that E_2° in the small compartment. We shall now find the relative probability of these two energies. This relative probability must be equal to the ratio of the two total number of accessible states corresponding to these energy situations in the small compartment, that is,

$$\left\{\frac{W(E_1^\circ)}{W(E_2^\circ)}\right\} = \left(\frac{\Omega_{\text{Tot}}(E_1^\circ)}{\Omega_{\text{Tot}}(E_2^\circ)}\right) \tag{12.77}$$

Then, in view of Eq. (12.76) this ratio reads

$$\left\{\frac{W(E_1^\circ)}{W(E_2^\circ)}\right\} = \left(\frac{\Omega(E_{\text{Tot}} - E_1^\circ)\Omega^\circ(E_1^\circ)}{\Omega(E_{\text{Tot}} - E_2^\circ)\Omega^\circ(E_2^\circ)}\right) \tag{12.78}$$

For the small subsystem, its number of accessible states $\Omega^\circ(E_i^\circ)$ is just the degeneracy $g(E_i^\circ)$ of the energy E_i°, that is, the number of microstates of the small subsystem having the energy E_i°:

$$\Omega^\circ(E_i^\circ) = g(E_i^\circ)$$

Hence, the ratio (12.78) is

$$\left\{\frac{W(E_1^\circ)}{W(E_2^\circ)}\right\} = \left(\frac{\Omega(E_{\text{Tot}} - E_1^\circ)}{\Omega(E_{\text{Tot}} - E_2^\circ)}\right)\left(\frac{g(E_1^\circ)}{g(E_2^\circ)}\right) \tag{12.79}$$

Next, use the definition (12.19) of the entropy, that is,

$$S = k_B \ln \Omega$$

so that

$$\Omega(E_{\text{Tot}} - E_1^\circ) = e^{S(E_{\text{Tot}} - E_1^\circ)/k_B}$$

$$\Omega(E_{\text{Tot}} - E_2^\circ) = e^{S(E_{\text{Tot}} - E_2^\circ)/k_B}$$

hence, Eq. (12.79) transforms to

$$\left\{\frac{W(E_1^\circ)}{W(E_2^\circ)}\right\} = \left(\frac{e^{S(E_{\text{Tot}} - E_1^\circ)/k_B}}{e^{S(E_{\text{Tot}} - E_2^\circ)/k_B}}\right)\left(\frac{g(E_1^\circ)}{g(E_2^\circ)}\right) \tag{12.80}$$

Again, since

$$E >> E_1^\circ \qquad \text{and} \qquad E >> E_2^\circ$$

we see that the differences $(E_{\text{Tot}} - E_1^\circ)$ and $(E_{\text{Tot}} - E_2^\circ)$ are very small, so that the entropies appearing in Eq. (12.80) are very near their values when E_1° and E_2° are vanishing. That allows us to truncate up to first order the Taylor expansion of the entropies involved in Eq. (12.80), that is, to write

$$S(E_{\text{Tot}} - E_k^\circ) = S(E_{\text{Tot}}) - E_k^\circ\left(\frac{\partial S}{\partial E}\right)_{E=E_{\text{Tot}}} \qquad \text{with} \quad k = 1, 2 \tag{12.81}$$

Besides, owing to Eq. (12.73), it is possible to relate the partial derivative of the entropy with respect to the energy, to the absolute temperature T, via the thermodynamic relation

$$\left(\frac{\partial S}{\partial E}\right)_{E=E_{\text{Tot}}} = \frac{1}{T}$$

Then, Eq. (12.81) reads

$$S(E_{\text{Tot}} - E_k^\circ) = S(E_{\text{Tot}}) - \frac{E_k^\circ}{T} \quad \text{with} \quad k = 1, 2$$

Hence, the probability ratio (12.80) becomes

$$\left\{\frac{W(E_1^\circ)}{W(E_2^\circ)}\right\} = \left(\frac{e^{S(E_{\text{Tot}})/k_B}}{e^{S(E_{\text{Tot}})/k_B}}\left(\frac{e^{-E_1^\circ/k_BT}}{e^{-E_2^\circ/k_BT}}\right)\right)\left(\frac{g(E_1^\circ)}{g(E_2^\circ)}\right)$$

or, after simplification,

$$\left\{\frac{W(E_1^\circ)}{W(E_2^\circ)}\right\} = \left(\frac{e^{-E_1^\circ/k_BT}}{e^{-E_2^\circ/k_BT}}\right)\left(\frac{g(E_1^\circ)}{g(E_2^\circ)}\right) \tag{12.82}$$

Of course, when the degeneracies corresponding to the two situations are unity, this last equation reduces to

$$\left\{\frac{W(E_1^\circ)}{W(E_2^\circ)}\right\} = \left(\frac{e^{-E_1^\circ/k_BT}}{e^{-E_2^\circ/k_BT}}\right) \tag{12.83}$$

Moreover, since the probabilities are normalized, that is,

$$\sum_k W(E_k^\circ) = 1 \tag{12.84}$$

Eq. (12.82) implies that the probability for the system to have the energy E_i° is

$$W(E_i^\circ) = \frac{1}{Z}(e^{-E_i^\circ/k_B T})g(E_i^\circ) \tag{12.85}$$

with

$$Z = \sum_j (e^{-E_j^\circ/k_B T})g(E_j^\circ) \tag{12.86}$$

or, when the degeneracy is unity,

$$W(E_i^\circ) = \frac{1}{Z}(e^{-E_i^\circ/k_B T}) \tag{12.87}$$

with

$$Z = \sum_j (e^{-E_j^\circ/k_B T}) \tag{12.88}$$

At last, keeping in mind Eq. (12.66), that is,

$$W_i = \frac{1}{Z}(e^{-\beta E_i}) \tag{12.89}$$

and, by identification of Eqs. (12.85) and (12.89), it appears that the Lagrange parameter β is given by

$$\boxed{\beta = \frac{1}{k_B T}} \tag{12.90}$$

12.5 CONCLUSION

This chapter has focused attention on the theoretical fact that, at statistical equilibrium, the statistical entropy is maximum. This was approached via the Boltzmann H-theorem, proving that statistical entropy must increase until equilibrium, and numerically verified with the model of Chapter 11 dealing with a large set of weakly coupled harmonic oscillators, which showed that after the statistical entropy has attained its maximum, the energy distribution of the oscillators obeys the Boltzmann law. Finally, using the entropy maximization at statistical equilibrium, it was then possible to get the microcanonical density operator and the Boltzmann canonical density operator, allowing to get the thermal average energy as a function of the partition function whence it is possible to normalize the density operator. This latter canonical density operator will be extensively used in the following chapters in order to study the thermal properties of harmonic oscillators.

BIBLIOGRAPHY

P. Blaise, Ph. Durand, and O. Henri-Rousseau. *Physica A*, **209** (1994): 51.

B. Diu, C. Guthmann, D. Lederer, and B. Roulet. *Eléments de physique statistique*. Hermann: Paris, 1989.

Ch. Kittel and H. Kroemer. *Thermal Physics*, 2nd ed. W. H. Freeman: 1980.

H. Louisell. *Quantum Statistical Properties of Radiations*. Wiley: New York, 1973.

F. Reif. *Fundamentals of Statistical and Thermal Physics*. McGraw-Hill: New York, 1965.

F. Reif. *Berkeley Physics Course, Vol. 5, Statistical Physics*. McGraw-Hill: New York, 1967.

CHAPTER 13

THERMAL PROPERTIES OF HARMONIC OSCILLATORS

INTRODUCTION

Using the concepts encountered in the previous chapter, Chapter 13 is concerned with the thermal properties of oscillators and specially by thermal average energies, heat capacities, thermal fluctuations of energy, position, and momentum and thermal entropies. It ends by giving the detailed demonstration of the thermal average over Boltzmann density operators for harmonic oscillator, of very general functions of Boson operators, which admits as a special case the *Bloch's theorem* dealing with the thermal average of the *translation operator*.

13.1 BOLTZMANN DISTRIBUTION LAW INSIDE A LARGE POPULATION OF EQUIVALENT OSCILLATORS

Consider a set of N equivalent quantum harmonic oscillators with the same Hamiltonian

$$\mathbf{H}_k = \hbar\omega\left(\mathbf{a}_k^{\dagger}\mathbf{a}_k + \tfrac{1}{2}\right)$$

In the following we shall suppose that N is very large, its magnitude being, for instance, Avogadro's number. The eigenvalue equation of these Hamiltonians $\mathbf{H}_k$ is

$$\mathbf{H}_k|\{n\}_k\rangle = E_k^{\circ}|\{n\}_k\rangle$$

with, neglecting the same zero-point energies,

$$E_k^{\circ} = n_k\hbar\omega \tag{13.1}$$

Now, assume that this set of oscillators cannot exchange energy with the neigborhood, so that the total energy E_{Tot} of the set is constant and suppose that each oscillator may exchange energy with the other ones.

In any configuration, among a multitude, the total energy E_{Tot} of the set is

$$E_{\text{Tot}} = \sum_k E_k^{\circ}N_k \tag{13.2}$$

Quantum Oscillators, First Edition. Olivier Henri-Rousseau and Paul Blaise.
© 2011 John Wiley & Sons, Inc. Published 2011 by John Wiley & Sons, Inc.

where N_k is the number of oscillators having the same eigenvalue energy E_k° defined by Eq. (13.1). Of course, the total number N of oscillators is the sum over the numbers N_k , that is,

$$N_{\text{Tot}} = \sum_k N_k \tag{13.3}$$

Since the number of oscillators and the total energy are constant, one has, respectively,

$$dE_{\text{Tot}} = 0 \qquad \text{and} \qquad dN_{\text{Tot}} = 0$$

Thus, Eqs. (13.2) and (13.3) lead to

$$\sum_k E_k^\circ dN_k = 0 \qquad \text{and} \qquad \sum_k dN_k = 0 \tag{13.4}$$

The statistical weight of a configuration corresponding to a situation where there are N_1 oscillators having the energy E_1, N_2 oscillators having the energy E_2, and so on is given by the statistical distribution

$$W(N_1, N_2, \dots) = \frac{N_{\text{Tot}}!}{\prod_k N_k!} \tag{13.5}$$

where the N_k are constrained to verify simultaneously Eqs. (13.4).

Figure 13.1 gives for a set of $N_{\text{Tot}} = 21$ oscillators, the values of $W(N_1, N_2, \dots)$ calculated by Eq. (13.5), subjected to the constraints of Eqs. (13.4), when applied to eight possible distributions of the total energy $E_{\text{Tot}} = 21\hbar\omega$.

Inspection of Fig. 13.1 shows that some configurations are more probable than others. The most probable is that corresponding to the situation where there are less and less oscillators when the energy increases.

We shall now show that the most probable configuration is that corresponding to the situation where the number of oscillators having a given energy is exponentially decreasing with energy. Thus, we write Eq. (13.5) in logarithm form, that is,

$$\ln W(N_1, N_2, \dots) = \ln (N_{\text{Tot}}!) - \sum_k \ln (N_k!) \tag{13.6}$$

Now, in order to find the most probable configurations, the differential of Eq. (13.6) must be zero, that is,

$$d \ln W(N_1, N_2, \dots) = -\sum_k \left(\frac{\partial \ln (N_k!)}{\partial N_k} \right) dN_k = 0 \tag{13.7}$$

Next, in order to take into account the two constraints (13.4) on the $N_k{}^{\text{eq}}$, one must use the Lagrange multipliers method described in Section 18.7 leading one to write in place of Eq. (13.7) the following equation:

$$-\sum_k \left(\frac{\partial \ln (N_k{}^{\text{eq}}!)}{\partial N_k{}^{\text{eq}}} \right) dN_k{}^{\text{eq}} - \beta \sum_k E_k \, dN_k{}^{\text{eq}} + \alpha \sum_k dN_k{}^{\text{eq}} = 0$$

Since this last expression must hold for each k, we see that they are as many following equations as they are of k:

$$-\left\{ \left(\frac{\partial \ln (N_k{}^{\text{eq}}!)}{\partial N_k{}^{\text{eq}}} \right) + \beta E_k - \alpha \right\} dN_k{}^{\text{eq}} = 0 \tag{13.8}$$

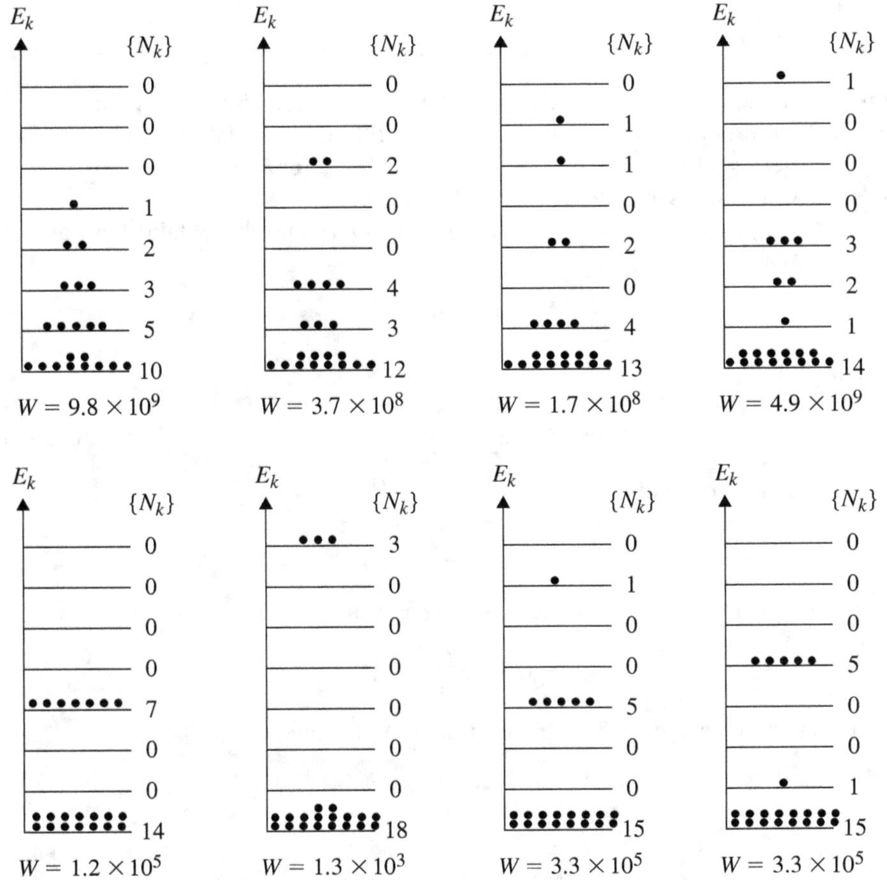

Figure 13.1 Values of $W(N_1, N_2, \ldots)$ calculated by Eqs. (13.5) and for $N_{\text{Tot}} = 21$, $E_{\text{Tot}} = 21\hbar\omega$, for eight different configurations verifying Eqs. (13.4). For each configuration, the eight lowest energy levels E_k of the quantum harmonic oscillators are reproduced, with for each of them, as many dark circles as they are (N_k) of oscillators having the corresponding energy E_k.

In order to calculate the partial derivative of Eq. (13.6) with respect to $N_k{}^{\text{eq}}$, it is convenient, if the numbers N and $N_k{}^{\text{eq}}$ are very large, to use the Stirling approximation

$$\ln(N_k{}^{\text{eq}}!) \backsimeq N_k{}^{\text{eq}} \ln(N_k{}^{\text{eq}}) - N_k{}^{\text{eq}}$$

Then, the partial derivative of Eq. (13.6) of interest reads

$$\left(\frac{\partial \ln(N_k{}^{\text{eq}}!)}{\partial N_k{}^{\text{eq}}}\right) \backsimeq \ln(N_k{}^{\text{eq}})$$

Hence, Eq. (13.8) is

$$(-\ln N_k{}^{\text{eq}} - \beta E_k + \alpha) dN_k{}^{\text{eq}} = 0$$

Moreover, since $dN_k{}^{\text{eq}} \neq 0$, it yields

$$-\ln N_k{}^{\text{eq}} - \beta E_k + \alpha = 0$$

so that

$$N_k{}^{\mathrm{eq}} = e^{\alpha} e^{-\beta E_k} \tag{13.9}$$

It is this distribution that is the closest to the one described by the configuration of Fig. 13.1 corresponding to the situation leading to $W = 9.8 \times 10^9$ and where $N_0 = 10$ for $E_0 = 0, N_1 = 5$ for $E_1 = 1, N_2 = 3$ for $E_2 = 2, N_3 = 2$ for $E_3 = 3, N_4 = 1$ for $E_4 = 4$, and $N_k = 0$ for the higher levels.

The expression for the Lagrange multiplier α may be obtained by aid of Eqs. (13.3) and (13.9) yielding

$$N_{\mathrm{Tot}} = e^{\alpha} \sum_k e^{-\beta E_k}$$

so that

$$e^{\alpha} = \frac{N_{\mathrm{Tot}}}{Z} \tag{13.10}$$

where Z is the partition function:

$$Z = \sum_k e^{-\beta E_k}$$

As a consequence, the Lagrange parameter α appears to be

$$\alpha = \ln\left(\frac{N_{\mathrm{Tot}}}{Z}\right)$$

Moreover, with the help of Eq. (13.10), Eq. (13.9) becomes

$$N_k{}^{\mathrm{eq}} = \frac{N_{\mathrm{Tot}}}{Z} e^{-\beta E_k}$$

or

$$N_k{}^{\mathrm{eq}} = N_{\mathrm{Tot}} W^{\mathrm{eq}}(E_k)$$

where $W^{\mathrm{eq}}(E_k)$ is the Boltzmann probability to find oscillators having the energy E_k, which is given by

$$\boxed{W^{\mathrm{eq}}(E_k) = \frac{e^{-\beta E_k}}{Z}} \tag{13.11}$$

Recall that the value of the Lagrange parameter β appearing in the exponential and decreasing with the energy levels E_k has been found above to be given by Eq. (12.90).

13.2 THERMAL PROPERTIES OF HARMONIC OSCILLATORS

13.2.1 Canonical density operators of harmonic oscillators

Consider the canonical density operator ρ_B of a quantum harmonic oscillator defined by Eq. (12.61), that is,

$$\rho_B = \frac{1}{Z}(e^{-\beta \mathbf{H}}) \tag{13.12}$$

where Z is the partition function given by Eq. (12.62), that is,

$$Z = \mathbf{tr}\{(e^{-\beta \mathbf{H}})\} \tag{13.13}$$

β is the thermal Lagrange parameter given by Eq. (12.90), that is,

$$\beta = \frac{1}{k_B T} \tag{13.14}$$

and $\mathbf{H}$ is the Hamiltonian of the harmonic oscillator given by Eq. (5.9), that is,

$$\mathbf{H} = \hbar\omega\left(\mathbf{a}^\dagger\mathbf{a} + \tfrac{1}{2}\right) \quad \text{with} \quad [\mathbf{a}, \mathbf{a}^\dagger] = 1 \tag{13.15}$$

Owing to Eqs. (13.12) and (13.15), the canonical density operator of the harmonic oscillator reads

$$\boldsymbol{\rho}_B = \frac{1}{Z}(e^{-\beta \mathbf{a}^\dagger\mathbf{a}\hbar\omega} e^{-\beta\hbar\omega/2}) \tag{13.16}$$

so that the partition function (13.13) yields

$$Z = (e^{-\beta\hbar\omega/2})\mathbf{tr}\{(e^{-\beta \mathbf{a}^\dagger\mathbf{a}\hbar\omega})\} \tag{13.17}$$

Now, to perform the trace involved in this last equation, it is convenient to use the basis of eigenstates of $\mathbf{a}^\dagger\mathbf{a}$, that is,

$$\mathbf{a}^\dagger\mathbf{a}|(n)\rangle = n|(n)\rangle \quad \text{with} \quad \langle(n)|(m)\rangle = \delta_{nm} \tag{13.18}$$

Hence, owing to Eq. (13.15), the partition function (13.13) takes the form

$$Z = (e^{-\beta\hbar\omega/2})\sum_n \langle(n)|(e^{-\beta \mathbf{a}^\dagger\mathbf{a}\hbar\omega})|(n)\rangle$$

Expanding the exponential operator gives

$$Z = (e^{-\beta\hbar\omega/2})\sum_n\sum_k \langle(n)|\left(\frac{(-\beta\hbar\omega)^k(\mathbf{a}^\dagger\mathbf{a})^k}{k!}\right)|(n)\rangle \tag{13.19}$$

Moreover, due to Eq. (13.18) one obtains by recurrence

$$(\mathbf{a}^\dagger\mathbf{a})^k|(n)\rangle = n^k|(n)\rangle$$

so that Eq. (13.19) transforms to

$$Z = (e^{-\beta\hbar\omega/2})\sum_n\sum_k \langle(n)|\left(\frac{(-\beta\hbar\omega)^k n^k}{k!}\right)|(n)\rangle$$

Hence, after coming back to the exponential

$$Z = (e^{-\beta\hbar\omega/2})\sum_n \langle(n)|(e^{-\beta n\hbar\omega})|(n)\rangle$$

and using the normality property of the kets

$$Z = (e^{-\beta\hbar\omega/2})\sum_n (e^{-\beta n\hbar\omega})$$

we have

$$Z = (e^{-\beta\hbar\omega/2})\sum_n y^n \quad \text{with} \quad y = (e^{-\beta\hbar\omega}) \tag{13.20}$$

Now, observe that at temperatures T, which are not very far from the room temperature, the following inequality is generally satisfied for harmonic oscillators describing molecular vibrations:

$$\hbar\omega > k_B T$$

so that, due to Eq. (13.14),

$$\beta\hbar\omega > 1 \qquad \text{and thus} \qquad e^{-\beta\hbar\omega} < 1$$

In this special situation, the series involved in Eq. (13.20) is convergent and given by

$$\sum_n y^n = \frac{1}{1-y} \qquad \text{with} \quad y < 1$$

Hence, the partition function (13.20) becomes

$$Z = \left(\frac{e^{-\beta\hbar\omega/2}}{1 - e^{-\beta\hbar\omega}}\right) = \left(\frac{e^{-\hbar\omega/2k_BT}}{1 - e^{-\hbar\omega/k_BT}}\right) \tag{13.21}$$

a result that may also be written

$$Z = \left(\frac{1}{e^{\beta\hbar\omega/2} - e^{-\beta\hbar\omega/2}}\right) = \frac{1}{2\sinh(\hbar\omega/2)}$$

Moreover, the canonical density operator (13.16) becomes after simplification

$$\boldsymbol{\rho}_B = (1 - e^{-\beta\hbar\omega})(e^{-\beta\mathbf{a}^\dagger\mathbf{a}\hbar\omega}) \tag{13.22}$$

a result that may be also written

$$\boxed{\boldsymbol{\rho}_B = (1 - e^{-\lambda})(e^{-\lambda\mathbf{a}^\dagger\mathbf{a}})} \tag{13.23}$$

and, comparing Eq. (13.14),

$$\boxed{\lambda = \frac{\hbar\omega}{k_BT} = \beta\hbar\omega} \tag{13.24}$$

13.2.2 Thermal energy

Now, consider the mean thermal average energy of a quantum harmonic oscillator that is the average of the Hamiltonian over the canonical density operator, that is,

$$\langle\mathbf{H}\rangle = \mathbf{tr}\{\boldsymbol{\rho}_B\mathbf{H}\} \tag{13.25}$$

which, due to Eqs. (13.12) and (13.23), reads either

$$\langle\mathbf{H}\rangle = \hbar\omega(1 - e^{-\lambda})\,\mathbf{tr}\left\{(e^{-\lambda\mathbf{a}^\dagger\mathbf{a}})\left(\mathbf{a}^\dagger\mathbf{a} + \tfrac{1}{2}\right)\right\} \tag{13.26}$$

or, due to Eq. (13.22),

$$\langle\mathbf{H}\rangle = \hbar\omega(1 - e^{-\beta\hbar\omega})\,\mathbf{tr}\{(e^{-\beta\mathbf{a}^\dagger\mathbf{a}\hbar\omega})\mathbf{a}^\dagger\mathbf{a}\} + \frac{\hbar\omega}{2} \tag{13.27}$$

However, observe it is unnecessary to separately calculate the partition function and the trace involved in Eq. (13.27), since it has been shown that the thermal average energy (13.25) of a system whatever its Hamiltonian may be, is given by Eq. (12.69), that is,

$$\langle \mathbf{H} \rangle = -\left(\frac{\partial \ln Z}{\partial \beta}\right) \tag{13.28}$$

so that it is possible to get the thermal average value of the energy (13.25) using Eq. (13.28). Hence, start from Eq. (13.21) giving $\ln Z$, that is,

$$\ln (Z) = -\beta\frac{\hbar\omega}{2} - \ln (1 - e^{-\beta\hbar\omega})$$

so that, by differentiation, one obtains

$$\left(\frac{\partial \ln Z}{\partial \beta}\right) = -\frac{\hbar\omega}{2} - \left(\frac{\hbar\omega\, e^{-\beta\hbar\omega}}{1 - e^{-\beta\hbar\omega}}\right)$$

or, after rearranging,

$$\left(\frac{\partial \ln Z}{\partial \beta}\right) = -\left(\frac{\hbar\omega}{e^{\beta\hbar\omega} - 1} + \frac{\hbar\omega}{2}\right)$$

Thus, comparing Eq. (13.14), the thermal average energy (13.28) becomes

$$\boxed{\langle \mathbf{H} \rangle = \left(\frac{\hbar\omega}{e^{\hbar\omega/k_BT} - 1} + \frac{\hbar\omega}{2}\right)} \tag{13.29}$$

which is the Planck expression of the average energy of a quantum oscillator belonging to a population of quantum harmonic oscillators in thermal equilibrium. Of course, the total average energy of a population of N oscillators is

$$\langle \mathbf{H}_{\mathrm{Tot}} \rangle = N\, \langle \mathbf{H} \rangle \tag{13.30}$$

Moreover, by comparison of Eqs. (13.27) and (13.29), it yields

$$(1 - e^{-\beta\hbar\omega})\mathbf{tr}\{(e^{-\beta\hbar\omega\mathbf{a}^\dagger\mathbf{a}})\mathbf{a}^\dagger\mathbf{a}\} = \frac{1}{e^{\beta\hbar\omega} - 1} \tag{13.31}$$

or

$$(1 - e^{-\lambda})\mathbf{tr}\{(e^{-\lambda\mathbf{a}^\dagger\mathbf{a}})\mathbf{a}^\dagger\mathbf{a}\ \} = \frac{1}{e^{\lambda} - 1} \tag{13.32}$$

13.2.3 Boltzmann distribution

Now, consider the diagonal matrix elements of the density operator as given by Eq. (12.85), that is,

$$P_n = \langle (n)|\boldsymbol{\rho}_B|(n)\rangle \tag{13.33}$$

Then, comparing Eq. (13.12), the right-hand-side matrix elements read

$$\langle (n)|\boldsymbol{\rho}_B|(n)\rangle = \frac{1}{Z}\langle (n)|(e^{-\beta\mathbf{H}})|(n)\rangle$$

or, due to Eq. (13.22),

$$\langle (n)|\boldsymbol{\rho}_B|(n)\rangle = (1 - e^{-\beta\hbar\omega})\langle (n)|(e^{-\beta\hbar\omega \mathbf{a}^\dagger \mathbf{a}})|(n)\rangle$$

Hence, after using the eigenvalue equation

$$\mathbf{a}^\dagger \mathbf{a}|(n)\rangle = n|(n)\rangle$$

the matrix elements become

$$\langle (n)|\boldsymbol{\rho}_B|(n)\rangle = (1 - e^{-\beta\hbar\omega})\langle (n)|e^{-n\beta\hbar\omega}|(n)\rangle$$

or

$$\langle (n)|\boldsymbol{\rho}_B|(n)\rangle = (1 - e^{-\beta\hbar\omega})(e^{-n\beta\hbar\omega})$$

Hence, Eq. (13.33) yields

$$\boxed{P_n = (1 - e^{-\beta\hbar\omega})(e^{-n\beta\hbar\omega})} \tag{13.34}$$

This last result, which is the Boltzmann distribution of the energy level of harmonic oscillators, that is, the probability for them to be occupied at any temperature, may be put in correspondence with the result (12.28) obtained in the coarse-grained analysis where an exponential decreasing with energy of the probability occupation appears.

13.2.4 Thermal average of the occupation number

Now, observe that, due to Eq. (13.31), and since the occupation number is defined by

$$\mathbf{n} \equiv \mathbf{a}^\dagger \mathbf{a} \tag{13.35}$$

it appears that its thermal average is

$$\langle \mathbf{n} \rangle = \frac{1}{e^{\beta\hbar\omega} - 1} \tag{13.36}$$

Next, comparing Eq. (13.36),

$$1 + \langle \mathbf{n} \rangle = 1 + \frac{1}{e^{\beta\hbar\omega} - 1} = \frac{e^{\beta\hbar\omega}}{e^{\beta\hbar\omega} - 1}$$

the ratio of $\langle \mathbf{n} \rangle$ and $1 + \langle \mathbf{n} \rangle$ yields

$$\frac{\langle \mathbf{n} \rangle}{1 + \langle \mathbf{n} \rangle} = e^{-\beta\hbar\omega} \tag{13.37}$$

Besides, from Eq. (13.37) it reads

$$(1 - e^{-\beta\hbar\omega}) = 1 - \frac{\langle \mathbf{n} \rangle}{1 + \langle \mathbf{n} \rangle} = \frac{1}{1 + \langle \mathbf{n} \rangle} \tag{13.38}$$

while the *nth* power of (13.37) takes the form

$$(e^{-n\beta\hbar\omega}) = \left(\frac{\langle \mathbf{n} \rangle}{1 + \langle \mathbf{n} \rangle}\right)^n \tag{13.39}$$

so that it results from Eqs. (13.39) and (13.38) that

$$(1 - e^{-\beta\hbar\omega})(e^{-n\beta\hbar\omega}) = \frac{1}{1 + \langle \mathbf{n} \rangle}\left(\frac{\langle \mathbf{n} \rangle}{1 + \langle \mathbf{n} \rangle}\right)^n$$

Hence, the Boltzmann distribution function (13.34) becomes

$$\boxed{P_n = \frac{\langle \mathbf{n} \rangle^n}{(1 + \langle \mathbf{n} \rangle)^{n+1}}} \tag{13.40}$$

a result that is widely used in the area of the theory of lasers.

13.2.5 Heat capacity

Now, consider the thermal capacity at constant volume C_v which is, by definition, the time derivative of the total average energy of a population of N oscillators:

$$C_v = \left(\frac{\partial \langle \mathbf{H}_{\text{Tot}}(T) \rangle}{\partial T} \right)_v \tag{13.41}$$

where $\mathbf{H}_{\text{Tot}}$ is given by Eq. (13.30) so that

$$C_v = N \left(\frac{\partial \langle \mathbf{H}(T) \rangle}{\partial T} \right)_v$$

Then, due to Eq. (13.29), Eq. (13.41) reads

$$C_v = N\hbar\omega \frac{\partial}{\partial T} \left(\frac{1}{e^{\hbar\omega/k_B T} - 1} \right)$$

and thus, on differentiation

$$C_v = N\hbar\omega \left(\frac{-1}{(e^{\hbar\omega/k_B T} - 1)^2} \right) e^{\hbar\omega/k_B T} \left(\frac{-\hbar\omega}{k_B T^2} \right)$$

or

$$C_v = Nk_B \left(\frac{\hbar\omega}{k_B T} \right)^2 \frac{e^{\hbar\omega/k_B T}}{(e^{\hbar\omega/k_B T} - 1)^2} \tag{13.42}$$

Figure 13.2 discusses the evolution with temperature of the thermal capacity C_v for a mole of oscillators of angular frequency $\omega = 1000\,\text{cm}^{-1}$.

13.2.6 Thermal fluctuations

13.2.6.1 Thermal energy fluctuation Now, the thermal fluctuation of the energy of N oscillators is

$$\Delta E_{\text{Tot}} = \sqrt{\langle \mathbf{H}_{\text{Tot}}{}^2 \rangle - \langle \mathbf{H}_{\text{Tot}} \rangle^2} \tag{13.43}$$

with

$$\langle \mathbf{H}_{\text{Tot}}{}^2 \rangle = N^2 \langle \mathbf{H}^2 \rangle$$

Thus Eq. (13.30), becomes

$$\Delta E_{\text{Tot}} = N\sqrt{\langle \mathbf{H}^2 \rangle - \langle \mathbf{H} \rangle^2}$$

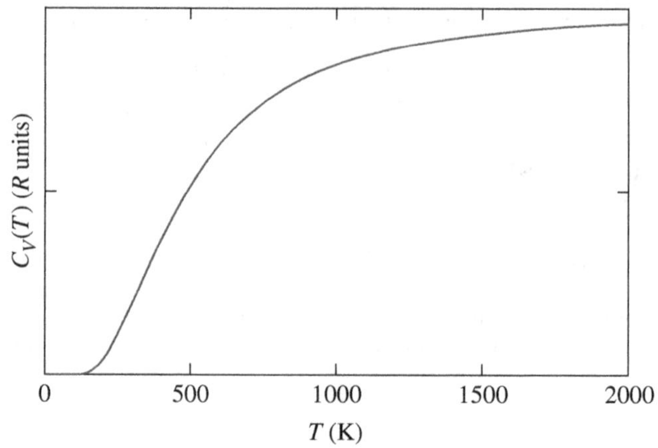

Figure 13.2 Thermal capacity C_v in R units for a mole of oscillators of angular frequency $\omega = 1000\,\text{cm}^{-1}$.

Recall that the thermal average of the Hamiltonian may be obtained by Eq. (12.69), that is,

$$\langle \mathbf{H} \rangle = -\left(\frac{1}{Z} \frac{\partial Z}{\partial \beta} \right) \tag{13.44}$$

Now, the thermal average of $\mathbf{H}^2$ may be found from

$$\langle \mathbf{H}^2 \rangle = \mathbf{tr}\{\rho_B \mathbf{H}^2\}$$

so that, due to Eq. (13.12), we have

$$\langle \mathbf{H}^2 \rangle = \frac{1}{Z} \mathbf{tr}\{(e^{-\beta \mathbf{H}}) \mathbf{H}^2\} \tag{13.45}$$

Next, observe that the product of operators appearing under the trace may be written

$$(e^{-\beta \mathbf{H}}) \mathbf{H}^2 = \left(\frac{\partial^2 e^{-\beta \mathbf{H}}}{\partial \beta^2} \right) \tag{13.46}$$

so that Eq. (13.45) reads

$$\langle \mathbf{H}^2 \rangle = \frac{1}{Z} \mathbf{tr} \left\{ \left(\frac{\partial^2}{\partial \beta^2} \right) \left(e^{-\beta \mathbf{H}} \right) \right\}$$

or, since the partial derivative commutes with the trace operation,

$$\langle \mathbf{H}^2 \rangle = \frac{1}{Z} \left(\frac{\partial^2}{\partial \beta^2} \right) \mathbf{tr}\{(e^{-\beta \mathbf{H}})\} \tag{13.47}$$

Again, since the partition function is given by Eqs. (13.13) and (13.14), that is,

$$Z = \mathbf{tr}\{(e^{-\beta \mathbf{H}})\} \quad \text{and} \quad \beta = \frac{1}{k_B T} \tag{13.48}$$

Eq. (13.47) reads

$$\langle \mathbf{H}^2 \rangle = \left\{ \frac{1}{Z} \left(\frac{\partial^2 Z}{\partial \beta^2} \right) \right\} \tag{13.49}$$

Now, observe that the following equation is satisfied:

$$\left(\frac{\partial}{\partial\beta}\left(\frac{1}{Z}\frac{\partial Z}{\partial\beta}\right)\right) = \left(\frac{\partial}{\partial\beta}\frac{1}{Z}\right)\left(\frac{\partial Z}{\partial\beta}\right) + \left\{\frac{1}{Z}\left(\frac{\partial^2 Z}{\partial\beta^2}\right)\right\}$$

which yields

$$\left(\frac{\partial}{\partial\beta}\left(\frac{1}{Z}\frac{\partial Z}{\partial\beta}\right)\right) = -\frac{1}{Z^2}\left(\frac{\partial Z}{\partial\beta}\right)\left(\frac{\partial Z}{\partial\beta}\right) + \left\{\frac{1}{Z}\left(\frac{\partial^2 Z}{\partial\beta^2}\right)\right\}$$

Hence, equating the last right-hand side of this last equation and the right-hand side of Eq. (13.49) leads to

$$\langle\mathbf{H}^2\rangle = \left(\frac{\partial}{\partial\beta}\left(\frac{1}{Z}\frac{\partial Z}{\partial\beta}\right)\right) + \frac{1}{Z^2}\left(\frac{\partial Z}{\partial\beta}\right)^2$$

or, because of Eq. (13.44), to

$$\langle\mathbf{H}^2\rangle = -\left(\frac{\partial\langle\mathbf{H}\rangle}{\partial\beta}\right) + \langle\mathbf{H}\rangle^2 \tag{13.50}$$

Hence, the thermal energy fluctuation (13.43) is

$$\Delta E_{\text{Tot}} = N\sqrt{-\left(\frac{\partial\langle\mathbf{H}\rangle}{\partial\beta}\right)}$$

or

$$\Delta E_{\text{Tot}} = N\sqrt{-\left(\frac{\partial\langle\mathbf{H}\rangle}{\partial T}\right)\left(\frac{\partial T}{\partial\beta}\right)}$$

and thus, due to the definition (13.41) of the heat capacity at constant volume C_v,

$$\Delta E_{\text{Tot}} = N\sqrt{-\frac{C_v}{N}\left(\frac{\partial T}{\partial\beta}\right)} \tag{13.51}$$

Again, owing to Eq. (13.14) leading to

$$T = \frac{1}{k_B\beta} \tag{13.52}$$

the partial derivative of the absolute temperature with respect to β reads

$$\left(\frac{\partial T}{\partial\beta}\right) = \frac{\partial}{\partial\beta}\left(\frac{1}{k_B\beta}\right) = -\left(\frac{1}{k_B\beta^2}\right)$$

and thus, thanks to (13.52),

$$\left(\frac{\partial T}{\partial\beta}\right) = -k_B T^2$$

Thus, owing to this result, and to the expression (13.42) for the heat capacity, Eq. (13.51) leads to

$$\Delta E_{\text{Tot}} = \sqrt{N}\sqrt{k_B\left(\frac{\hbar\omega}{k_BT}\right)^2\frac{e^{\hbar\omega/k_BT}}{(e^{\hbar\omega/k_BT}-1)^2}k_BT^2}$$

or, after simplification,

$$\Delta E_{\text{Tot}} = \sqrt{N}\hbar\omega \frac{e^{\hbar\omega/2k_BT}}{(e^{\hbar\omega/k_BT} - 1)}$$

Besides, keeping in mind that, due to Eq. (13.29), and when the zero-point energy is ignored, the thermal average (13.30) reduces to

$$\langle \mathbf{H}_{\text{Tot}} \rangle = N \frac{\hbar\omega}{(e^{\hbar\omega/k_BT} - 1)}$$

the relative energy fluctuation becomes

$$\frac{\Delta E_{\text{Tot}}}{\langle \mathbf{H}_{\text{Tot}} \rangle} = \frac{1}{\sqrt{N}} e^{\hbar\omega/2k_BT}$$

At high temperature, the argument of the exponential being very small, the relative fluctuation reduces to

$$\frac{\Delta E_{\text{Tot}}}{\langle \mathbf{H}_{\text{Tot}} \rangle} \rightarrow \frac{1}{\sqrt{N}}$$

It must be emphasized that the inverse dependence of the relative fluctuation with respect to the number N of oscillators is the same as that of (12.39) yet encountered in the previous section, dealing with a coarse-grained analysis of a large set of coupled harmonic oscillators.

13.2.6.2 Thermal number occupation fluctuation Starting from Eq. (13.50), that is,

$$\langle \mathbf{H}^2 \rangle = -\left(\frac{\partial \langle \mathbf{H} \rangle}{\partial \beta} \right) + \langle \mathbf{H} \rangle^2 \tag{13.53}$$

and passing to Boson operators using Eq. (5.9), reads

$$\left\langle \left(\mathbf{a}^\dagger \mathbf{a} + \frac{1}{2} \right)^2 \right\rangle = -\frac{1}{\hbar\omega} \left(\frac{\partial \langle (\mathbf{a}^\dagger \mathbf{a} + \frac{1}{2}) \rangle}{\partial \beta} \right) + \left\langle \left(\mathbf{a}^\dagger \mathbf{a} + \frac{1}{2} \right) \right\rangle^2$$

Now, when the zero-point energy is ignored, Eq. (13.53) remains true so that it is possible to write

$$\langle (\mathbf{a}^\dagger \mathbf{a})^2 \rangle = -\frac{1}{\hbar\omega} \left(\frac{\partial \langle \mathbf{a}^\dagger \mathbf{a} \rangle}{\partial \beta} \right) + \langle \mathbf{a}^\dagger \mathbf{a} \rangle^2$$

or, due to Eq. (13.35),

$$\langle \mathbf{n}^2 \rangle = -\frac{1}{\hbar\omega} \left(\frac{\partial \langle \mathbf{n} \rangle}{\partial \beta} \right) + \langle \mathbf{n} \rangle^2 \tag{13.54}$$

Hence, comparing Eq. (13.36), that is,

$$\langle \mathbf{n} \rangle = \left(\frac{1}{e^{\beta\hbar\omega} - 1} \right) \tag{13.55}$$

Eq. (13.54) reads

$$\langle \mathbf{n}^2 \rangle = -\frac{1}{\hbar\omega} \left(\frac{\partial}{\partial \beta} \left(\frac{1}{e^{\beta\hbar\omega} - 1} \right) \right) + \left(\frac{1}{e^{\beta\hbar\omega} - 1} \right)^2 \tag{13.56}$$

Next, evaluating the partial derivative of the first right-hand-side term leads to

$$\left(\frac{\partial}{\partial\beta}\left(\frac{1}{e^{\beta\hbar\omega}-1}\right)\right) = -\hbar\omega\left(\frac{e^{\beta\hbar\omega}}{(e^{\beta\hbar\omega}-1)^2}\right)$$

so that Eq. (13.56) simplifies to

$$\langle\mathbf{n}^2\rangle = \left(\frac{e^{\beta\hbar\omega}+1}{(e^{\beta\hbar\omega}-1)^2}\right)$$

or

$$\langle\mathbf{n}^2\rangle = \frac{e^{\beta\hbar\omega}-1+2}{(e^{\beta\hbar\omega}-1)^2}$$

and thus

$$\langle\mathbf{n}^2\rangle = \frac{1}{e^{\beta\hbar\omega}-1} + \frac{2}{(e^{\beta\hbar\omega}-1)^2}$$

Hence, comparing Eq. (13.55), we have

$$\langle\mathbf{n}^2\rangle = \langle\mathbf{n}\rangle + 2\langle\mathbf{n}\rangle^2 \tag{13.57}$$

Now, by definition of the $\Delta\mathbf{n}$ thermal fluctuation

$$\Delta\mathbf{n} = \sqrt{\langle\mathbf{n}^2\rangle - \langle\mathbf{n}\rangle^2}$$

and with Eq. (13.57) this fluctuation becomes

$$\boxed{\Delta\mathbf{n} = \sqrt{\langle\mathbf{n}\rangle^2 + \langle\mathbf{n}\rangle}} \tag{13.58}$$

a result that is widely used in the area of the theory of lasers. Equation (13.58) may be also written

$$\Delta\mathbf{n} = \langle\mathbf{n}\rangle\sqrt{1+\frac{1}{\langle\mathbf{n}\rangle}}$$

Then, when

$$\langle\mathbf{n}\rangle >> 1$$

the argument of the square root may be expanded up to first order in $1/\langle\mathbf{n}\rangle$ according to

$$\sqrt{1+\frac{1}{\langle\mathbf{n}\rangle}} \backsimeq 1 + \frac{1}{2\langle\mathbf{n}\rangle}$$

so that in this limit

$$\Delta\mathbf{n} = \langle\mathbf{n}\rangle + \tfrac{1}{2}$$

Hence, in this limit, the relative fluctuations read

$$\frac{\Delta\mathbf{n}}{\langle\mathbf{n}\rangle} \backsimeq 1 + \frac{1}{2\langle\mathbf{n}\rangle} \backsimeq 1$$

13.2.6.3 Thermal average of Q, Q², and the potential We now consider the thermal equilibrium value of the position operator $\mathbf{Q}$ and of its square $\mathbf{Q}^2$, which are given, respectively, by the following thermal average over the Boltzmann density operator $\boldsymbol{\rho}_B$:

$$\langle \mathbf{Q}(T) \rangle = \mathbf{tr}\{\boldsymbol{\rho}_B \mathbf{Q}\} \qquad \text{and} \qquad \langle \mathbf{Q}(T)^2 \rangle = \mathbf{tr}\{\boldsymbol{\rho}_B \mathbf{Q}^2\} \tag{13.59}$$

Recall that within the raising and lowering operators picture of oscillators, $\mathbf{Q}$ is given by Eq. (5.6), that is,

$$\mathbf{Q} = \sqrt{\frac{\hbar}{2m\omega}}(\mathbf{a}^\dagger + \mathbf{a}) \tag{13.60}$$

whereas the Boltzmann density operator is given by Eqs. (13.23) and (13.24):

$$\boldsymbol{\rho}_B = (1 - e^{-\lambda})(e^{-\lambda \mathbf{a}^\dagger \mathbf{a}}) \tag{13.61}$$

Hence, the thermal average defined by the first equation of (13.59) is, therefore,

$$\langle \mathbf{Q}(T) \rangle = (1 - e^{-\lambda})\sqrt{\frac{\hbar}{2m\omega}}\mathbf{tr}\{(e^{-\lambda \mathbf{a}^\dagger \mathbf{a}})(\mathbf{a}^\dagger + \mathbf{a})\}$$

Performing the trace over the eigenstates $|\{n\}\rangle$ of $\mathbf{a}^\dagger \mathbf{a}$ gives

$$\langle \mathbf{Q}(T) \rangle = (1 - e^{-\lambda})\sqrt{\frac{\hbar}{2m\omega}}\sum_n \langle \{n\}|(e^{-\lambda \mathbf{a}^\dagger \mathbf{a}})(\mathbf{a}^\dagger + \mathbf{a})|\{n\}\rangle \tag{13.62}$$

Moreover, since $\mathbf{a}^\dagger \mathbf{a}$ is Hermitian

$$\mathbf{a}^\dagger \mathbf{a}|\{n\}\rangle = n|\{n\}\rangle$$

with

$$\langle \{m\}|\{n\}\rangle = \delta_{mn} \tag{13.63}$$

the two following Hermitian conjugate eigenvalue equations involved in Eq. (13.62) are verified:

$$(e^{-\lambda \mathbf{a}^\dagger \mathbf{a}})|\{n\}\rangle = (e^{-\lambda n})|\{n\}\rangle \qquad \text{and} \qquad \langle \{n\}|(e^{-\lambda \mathbf{a}^\dagger \mathbf{a}}) = \langle \{n\}|(e^{-\lambda n}) \tag{13.64}$$

Hence, the right-hand side of (13.62) becomes

$$\langle \{n\}|(e^{-\lambda \mathbf{a}^\dagger \mathbf{a}})(\mathbf{a}^\dagger + \mathbf{a})|\{n\}\rangle = (e^{-\lambda n})\langle \{n\}|(\mathbf{a}^\dagger + \mathbf{a})|\{n\}\rangle \tag{13.65}$$

Again, owing to Eqs. (5.53) and of its Hermitian conjugate, that is,

$$\mathbf{a}|\{n\}\rangle = \sqrt{n}|\{n-1\}\rangle \qquad \text{and thus} \qquad \langle \{n\}|\mathbf{a}^\dagger = \sqrt{n}\langle \{n-1\}|$$

and due to the orthogonality (13.63), Eq. (13.65) becomes

$$\langle \{n\}|(\mathbf{a}^\dagger + \mathbf{a})|\{n\}\rangle = \sqrt{n}\langle \{n-1\}|\{n\}\rangle + \sqrt{n}\langle \{n\}|\{n-1\}\rangle = 0$$

so that Eq. (13.65) transforms to

$$\langle \{n\}|(e^{-\lambda \mathbf{a}^\dagger \mathbf{a}})(\mathbf{a}^\dagger + \mathbf{a})|\{n\}\rangle = 0 \tag{13.66}$$

Hence, the equilibrium thermal average value (13.62) is zero.

Next pass to the thermal average value of $\mathbf{Q}^2$, which, according to Eqs. (13.60) and (13.61), is

$$\langle \mathbf{Q}(T)^2 \rangle = (1 - e^{-\lambda})\frac{\hbar}{2m\omega}\mathbf{tr}\{(e^{-\lambda \mathbf{a}^\dagger \mathbf{a}})(\mathbf{a}^\dagger + \mathbf{a})^2\}$$

Expanding the square using $[\mathbf{a}, \mathbf{a}^\dagger] = 1$, gives

$$\langle \mathbf{Q}(T)^2 \rangle = (1 - e^{-\lambda})\frac{\hbar}{2m\omega}(\mathbf{tr}\{(e^{-\lambda \mathbf{a}^\dagger \mathbf{a}})(2\mathbf{a}^\dagger \mathbf{a} + 1)\} + \mathbf{tr}\{(e^{-\lambda \mathbf{a}^\dagger \mathbf{a}})((\mathbf{a}^\dagger)^2 + (\mathbf{a})^2)\}) \tag{13.67}$$

Now, observe that, owing to Eq. (13.32) the first right-hand-side term of Eq. (13.67) is

$$(1 - e^{-\lambda})\mathbf{tr}\{(e^{-\lambda \mathbf{a}^\dagger \mathbf{a}})(2\mathbf{a}^\dagger \mathbf{a} + 1)\} = (2\langle \mathbf{n} \rangle + 1) \tag{13.68}$$

with

$$\langle \mathbf{n} \rangle = \frac{1}{e^{\lambda} - 1} \tag{13.69}$$

Now, perform trace over the basis of the eigenstates of $\mathbf{a}^\dagger \mathbf{a}$ appearing on the last right-hand-side term of Eq. (13.67)

$$\mathbf{tr}\{(e^{-\lambda \mathbf{a}^\dagger \mathbf{a}})((\mathbf{a}^\dagger)^2 + (\mathbf{a})^2)\} = \sum_n \langle \{n\}|(e^{-\lambda \mathbf{a}^\dagger \mathbf{a}})((\mathbf{a}^\dagger)^2 + (\mathbf{a})^2)|\{n\}\rangle$$

which, due to the last equation of (13.64), this trace reads

$$\mathbf{tr}\{(e^{-\lambda \mathbf{a}^\dagger \mathbf{a}})((\mathbf{a}^\dagger)^2 + (\mathbf{a})^2)\} = \sum_n (e^{-\lambda n})\langle \{n\}|((\mathbf{a}^\dagger)^2 + (\mathbf{a})^2)|\{n\}\rangle \tag{13.70}$$

Then, using Eqs. (5.71) and its Hermitian conjugate leads to the following Hermitian conjugate linear transformations:

$$(\mathbf{a})^2|\{n\}\rangle = \sqrt{n(n-1)}|\{n-2\}\rangle \qquad \text{and} \qquad \langle \{n\}|(\mathbf{a}^\dagger)^2 = \sqrt{n(n-1)}\langle \{n-2\}|$$

and by orthogonality of the eigenstates of $\mathbf{a}^\dagger \mathbf{a}$, Eq. (13.70) gives

$$\mathbf{tr}\{(e^{-\lambda \mathbf{a}^\dagger \mathbf{a}})((\mathbf{a}^\dagger)^2 + (\mathbf{a})^2)\} = \sum_n 2((e^{-\lambda n})\sqrt{n(n-1)}\delta_{n,n-2})$$

and thus

$$\mathbf{tr}\{e^{-\lambda \mathbf{a}^\dagger \mathbf{a}}((\mathbf{a}^\dagger)^2 + (\mathbf{a})^2)\} = 0 \tag{13.71}$$

Hence, comparing Eqs. (13.68) and (13.71), Eq. (13.67) becomes simply

$$\boxed{\langle \mathbf{Q}(T)^2 \rangle = \frac{\hbar}{2m\omega}(2\langle \mathbf{n} \rangle + 1)} \tag{13.72}$$

or, using Eq. (13.69)

$$\langle \mathbf{Q}(T)^2 \rangle = \frac{\hbar}{2m\omega}\left(\frac{2}{e^{\lambda} - 1} + 1\right)$$

that is,

$$\langle \mathbf{Q}(T)^2 \rangle = \frac{\hbar}{2m\omega}\left(\frac{1 + e^{\lambda}}{e^{\lambda} - 1}\right)$$

Again, multiplying both numerator and denominator by the same quantity $\exp(-\lambda/2)$ to get

$$\langle \mathbf{Q}(T)^2 \rangle = \frac{\hbar}{2m\omega}\left(\frac{e^{\lambda/2}+e^{-\lambda/2}}{e^{\lambda/2}-e^{-\lambda/2}}\right) \tag{13.73}$$

with

$$\coth\left(\frac{\lambda}{2}\right) = \left(\frac{e^{\lambda/2}+e^{-\lambda/2}}{e^{\lambda/2}-e^{-\lambda/2}}\right) \tag{13.74}$$

$$\langle \mathbf{Q}(T)^2 \rangle = \frac{\hbar}{2m\omega}\coth\left(\frac{\lambda}{2}\right)$$

to obtain finally, by aid of Eq. (13.24), that is,

$$\lambda = \frac{\hbar\omega}{k_B T} \tag{13.75}$$

the following expression:

$$\langle \mathbf{Q}(T)^2 \rangle = \frac{\hbar}{2m\omega}\coth\left(\frac{\hbar\omega}{2k_B T}\right) \tag{13.76}$$

Hence, the thermal average $\langle \mathbf{V}(T) \rangle$ of the potential operator

$$\langle \mathbf{V}(T) \rangle = \tfrac{1}{2}m\omega^2 \langle \mathbf{Q}(T)^2 \rangle \tag{13.77}$$

becomes, comparing Eq. (13.76),

$$\langle \mathbf{V}(T) \rangle = \frac{\hbar\omega}{4}\coth\left(\frac{\hbar\omega}{2k_B T}\right) \tag{13.78}$$

Next, when the absolute temperature is such that $k_B T >> \hbar\omega$, so that, due to Eq. (13.75), $\lambda << 1$, the coth appearing in Eq. (13.73), yields after Taylor expansion up to first order

$$\left(\frac{e^{\lambda/2}+e^{-\lambda/2}}{e^{\lambda/2}-e^{-\lambda/2}}\right) \backsimeq \frac{(1+\lambda/2)+(1-\lambda/2)}{(1+\lambda/2)-(1-\lambda/2)} = \frac{2}{\lambda}$$

Then, the coth function reads with the help of Eq. (13.75):

$$\coth\left(\frac{\hbar\omega}{2k_B T}\right) \backsimeq \frac{k_B T}{2\hbar\omega} \quad \text{when} \quad k_B T > \hbar\omega$$

so that, for this high-temperature limit, Eqs. (13.76) and (13.77) simplify to

$$\langle \mathbf{Q}(T)^2 \rangle \backsimeq \frac{\hbar}{2m\omega}\left(\frac{2k_B T}{\hbar\omega}\right) = \frac{k_B T}{m\omega^2} \tag{13.79}$$

$$\langle \mathbf{V}(T) \rangle = \frac{k_B T}{2} \tag{13.80}$$

In the case of very low temperatures, corresponding to $\hbar\omega >> k_B T$, due to (13.75), when $\lambda >> 1$

$$\left(\frac{e^{\lambda/2}+e^{-\lambda/2}}{e^{\lambda/2}-e^{-\lambda/2}}\right) \backsimeq \left(\frac{e^{\lambda/2}}{e^{\lambda/2}}\right) \backsimeq 1$$

the coth function reduces to unity, that is,

$$\coth\left(\frac{\hbar\omega}{2k_BT}\right) \backsimeq 1 \quad \text{when} \quad \hbar\omega > k_BT$$

thus, in the very low temperature limit, Eqs. (13.76) and (13.77) reduce to

$$\langle \mathbf{Q}(0)^2 \rangle = \frac{\hbar}{2m\omega}$$

$$\langle \mathbf{V}(0) \rangle = \frac{\hbar\omega}{4} \tag{13.81}$$

In Eq. (13.81), one may recognize the mean value of the potential of the harmonic oscillator averaged over the ground state $|\{0\}\rangle$ of the harmonic oscillator Hamiltonian. Finally, the fluctuation of the position coordinate at any temperature T, which is defined by

$$\Delta\mathbf{Q}(T) = \sqrt{\langle \mathbf{Q}(T)^2 \rangle - \langle \mathbf{Q}(T) \rangle^2}$$

becomes, in view of Eqs. (13.62), (13.66), and (13.76),

$$\boxed{\Delta\mathbf{Q}(T) = \sqrt{\frac{\hbar}{2m\omega}}\sqrt{\coth\left(\frac{\hbar\omega}{2k_BT}\right)}} \tag{13.82}$$

13.2.6.4 Thermal average of P, P², and the kinetic operator In like manner as for $\mathbf{Q}(T)$ given by Eq. (13.62), one would obtain for the thermal average of the momentum

$$\langle \mathbf{P}(T) \rangle = 0 \tag{13.83}$$

and for the thermal average of the squared momentum, an expression similar to that (13.72) obtained for $\mathbf{Q}(T)^2$, that is, in the present situation

$$\langle \mathbf{P}(T)^2 \rangle = \frac{\hbar m\omega}{2}(2\langle \mathbf{n} \rangle + 1) \tag{13.84}$$

or, similarly to Eq. (13.76),

$$\langle \mathbf{P}(T)^2 \rangle = \frac{\hbar m\omega}{2}\coth\left(\frac{\hbar\omega}{2k_BT}\right) \tag{13.85}$$

Via this last expression, the thermal average value of the kinetic energy yields, respectively, for very high and very low temperatures is given by

$$\langle \mathbf{T}(T) \rangle = \frac{\hbar\omega}{4}\coth\left(\frac{\hbar\omega}{2k_BT}\right) \tag{13.86}$$

$$\langle \mathbf{T}(T) \rangle = \frac{k_BT}{2} \quad \text{when} \quad k_BT > \hbar\omega \tag{13.87}$$

$$\langle \mathbf{T}(0) \rangle = \frac{\hbar\omega}{4} \tag{13.88}$$

so that, comparing Eqs. (13.83) and (13.85), the fluctuation of $\mathbf{P}(T)$ is

$$\boxed{\Delta\mathbf{P}(T) = \sqrt{\frac{\hbar m\omega}{2}}\sqrt{\coth\left(\frac{\hbar\omega}{2k_BT}\right)}} \tag{13.89}$$

13.2.6.5 Verification of the virial theorem Now, for harmonic oscillators, the thermal average of the kinetic and potential energies obey the virial theorem (2.89), and since we deal with mean values averaged over linear combinations of harmonic oscillator Hamiltonian eigenstates (which are necessarily stationary states), it is not surprising to find that Eqs. (13.78) and (13.86) also obey this theorem (2.89) since

$$\langle\mathbf{T}(T)\rangle = \langle\mathbf{V}(T)\rangle = \frac{\hbar\omega}{4}\coth\left(\frac{\hbar\omega}{2k_BT}\right) \tag{13.90}$$

Furthermore, since the thermal averaged Hamiltonian is the sum of the thermal average kinetic and potential operators, it follows from Eq. (13.90) that the form of the virial theorem (2.89) holds also for any temperature

$$\langle\mathbf{H}(T)\rangle = 2\langle\mathbf{T}(T)\rangle = 2\langle\mathbf{V}(T)\rangle = \frac{\hbar\omega}{2}\coth\left(\frac{\hbar\omega}{2k_BT}\right)$$

whereas, comparing Eqs. (13.80) and (13.87), its high-temperature limit is

$$\langle\mathbf{T}(T)\rangle = \langle\mathbf{V}(T)\rangle = \frac{k_BT}{2} \tag{13.91}$$

and, due to Eqs. (13.81) and (13.88), its low temperature yields

$$\langle\mathbf{T}(0)\rangle = \langle\mathbf{V}(0)\rangle = \frac{\hbar\omega}{2} \tag{13.92}$$

We remark that Eq. (13.92) is in agreement with the results (5.99) and (5.100) found for the average values of the kinetic and potential operators when the harmonic oscillator is in the ground state $|\{0\}\rangle$ of its Hamiltonian.

13.2.6.6 Equipartition theorem The fact that at high temperatures the thermal average values of the potential and of the kinetic operators are equal and given by Eq. (13.91) is an illustration of the *equipartition theorem* of classical statistical mechanics, according to which the thermal energy is quadratic with respect to the independent variables and is $k_BT/2$ for each degree of freedom. Now, we prove this theorem in a general way.

Suppose that the energy E of the system is quadratic with respect to N classical different continuous independent variables q_k, that is,

$$E = \sum_{k=1}^{N} E_k \quad \text{with} \quad E_k = \lambda_k {q_k}^2 \tag{13.93}$$

For each energy term E_k, its thermal average value may be obtained by Eq. (12.69):

$$\langle E_k(T)\rangle = -\left(\frac{\partial \ln Z_k}{\partial\beta}\right) \tag{13.94}$$

where Z_k is the partition function, which for continuous variable may be got from Eq. (12.67) by passing from the discrete sum to the corresponding integral according to

$$Z_k = \int_{-\infty}^{+\infty} e^{-\beta\lambda_k q_k{}^2} dq_k$$

which yields after integration

$$Z_k = \frac{1}{2}\sqrt{\frac{\pi}{\beta\lambda_k}}$$

so that

$$\ln Z_k = \ln\left(\frac{1}{2}\sqrt{\frac{\pi}{\lambda_k}}\right) - \frac{1}{2}\ln\beta$$

Hence, the thermal average (13.94) becomes

$$\langle E_k(T)\rangle = \frac{1}{2\beta} = \frac{k_B T}{2}$$

that is the *equipartition theorem* of classical statistical mechanics. As a consequence of this result and due to Eq. (13.93), the thermal average of the total energy $\langle E(T)\rangle$ is the number N of different independent variables times $k_B T/2$:

$$\langle E(T)\rangle = N\frac{k_B T}{2} = \frac{RT}{2}$$

where R is the ideal gas constant

13.2.6.7 Thermal Heisenberg uncertainty relation Now, we consider the thermal fluctuations of the position and momentum operators. Owing to Eqs. (13.82) and (13.89), the product of the thermal average of the uncertainty relation reads

$$\Delta\mathbf{P}(T)\,\Delta\mathbf{Q}(T) = \frac{\hbar}{2}\coth\left(\frac{\hbar\omega}{2k_B T}\right)$$

or, in view of the expression of the coth function,

$$\Delta\mathbf{P}(T)\,\Delta\mathbf{Q}(T) = \frac{\hbar}{2}\left(\frac{e^{\hbar\omega/2k_B T} + e^{-\hbar\omega/2k_B T}}{e^{\hbar\omega/2k_B T} - e^{-\hbar\omega/2k_B T}}\right) \qquad (13.95)$$

When the absolute temperature approaches zero, the arguments of the decreasing exponential also narrow to zero. Thus, after simplification, one obtains the limit

$$\Delta\mathbf{P}(T)\,\Delta\mathbf{Q}(T) \to \frac{\hbar}{2} \quad \text{when} \quad T \to 0$$

As required by Eq. (5.96), this limit corresponds to the lowest Heisenberg uncertainty (5.97) obtained for the ground state of the harmonic oscillator.

Also, when the absolute temperature is very large, that is, when $k_B T > \hbar\omega$, Taylor expansions of the exponentials the arguments of which are very small may be limited to first order, that is,

$$e^{\pm\hbar\omega/2k_B T} = 1 \pm \hbar\omega/2k_B T$$

so that, for this high-temperature limit, Eq. (13.95) reduces to

$$\Delta\mathbf{P}(T)\,\Delta\mathbf{Q}(T) = \frac{\hbar}{2}\left(\frac{2}{\hbar\omega/2k_BT}\right)$$

or

$$\Delta\mathbf{P}(T)\,\Delta\mathbf{Q}(T) = 2\hbar\left(\frac{k_BT}{\hbar\omega}\right)$$

13.2.7 Coherent-state density operator at thermal equilibrium

13.2.7.1 Density operator from the Lagrange multipliers method We now determine the expression for the density operator of a coherent state at thermal equilibrium. Thus, it is convenient to work in the same way as when obtaining the canonical and microcanonical density operators (12.49) and (12.63) using the Lagrange multipliers method. Thus, consider a population of equivalent harmonic oscillators for which one knows the entropy and the average value of the Hamiltonian **H** of the position operator **Q** and of its conjugate momentum **P**. Then, the normalization condition of the density operator $\boldsymbol{\rho}_c$, the expression of the statistical entropy S in terms of $\boldsymbol{\rho}_c$, and the average values of **H** , **Q**, and **P** lead, respectively, to

$$\mathbf{tr}\{\boldsymbol{\rho}_c \ln \boldsymbol{\rho}_c\} = S \tag{13.96}$$

$$\mathbf{tr}\{\boldsymbol{\rho}_c\} = 1 \tag{13.97}$$

$$\mathbf{tr}\{\boldsymbol{\rho}_c\mathbf{H}\} = \langle\mathbf{H}\rangle \tag{13.98}$$

$$\mathbf{tr}\{\boldsymbol{\rho}_c\mathbf{Q}\} = \langle\mathbf{Q}\rangle \tag{13.99}$$

$$\mathbf{tr}\{\boldsymbol{\rho}_c\mathbf{P}\} = \langle\mathbf{P}\rangle \tag{13.100}$$

Just as for Eq. (12.47), the equation dealing with the maximization $dS = 0$ of the statistical entropy S is

$$\mathbf{tr}\{(1 + \ln \boldsymbol{\rho}_c)\delta\boldsymbol{\rho}_c\} = 0$$

Moreover, due to the constraints linked to Eqs. (13.96)–(13.100), leading to

$$\mathbf{tr}\{\delta\boldsymbol{\rho}_c\} = 0 \tag{13.101}$$

$$\mathbf{tr}\{\mathbf{H}\delta\boldsymbol{\rho}_c\} = 0 \tag{13.102}$$

$$\mathbf{tr}\{\mathbf{Q}\delta\boldsymbol{\rho}_c\} = 0 \tag{13.103}$$

$$\mathbf{tr}\{\mathbf{P}\delta\boldsymbol{\rho}_c\} = 0 \tag{13.104}$$

one has, according to the Lagrange multipliers method, to multiply each of them by Lagrange multipliers according to

$$\lambda_0\,\mathbf{tr}\{\delta\boldsymbol{\rho}_c\} = 0 \tag{13.105}$$

$$\beta\,\mathbf{tr}\{\mathbf{H}\delta\boldsymbol{\rho}_c\} = 0 \tag{13.106}$$

$$\lambda_1\,\mathbf{tr}\{\mathbf{Q}\delta\boldsymbol{\rho}_c\} = 0 \tag{13.107}$$

$$\lambda_2\,\mathbf{tr}\{\mathbf{P}\delta\boldsymbol{\rho}_c\} = 0 \tag{13.108}$$

where λ_0, β, λ_1, and λ_2 are, respectively, the Lagrange parameters associated to the constraints (13.101)–(13.104).

Next, collecting the constraints multiplied by the corresponding Lagrange multipliers, we have maximizing the statistical entropy

$$\mathbf{tr}\{(1 + \ln\boldsymbol{\rho}_c + \lambda_0 + \beta\mathbf{H} + \lambda_1\mathbf{Q} + \lambda_2\mathbf{P})\delta\boldsymbol{\rho}_c\} = 0$$

Hence, since this last equation must be satisfied irrespective of the basis on which the trace is performed, we have

$$1 + \ln\{\boldsymbol{\rho}_c\} + \lambda_0 + \beta\mathbf{H} + \lambda_1\mathbf{P} + \lambda_2\mathbf{Q} = 0$$

or, by integration,

$$\boldsymbol{\rho}_c = e^{-(1+\lambda_0)-\beta\mathbf{H}+\lambda_1\mathbf{Q}+\lambda_2\mathbf{P}}$$

or, since λ_0 is a scalar,

$$\boldsymbol{\rho}_c = e^{-(1+\lambda_0)}e^{-(\beta\mathbf{H}-\lambda_1\mathbf{Q}-\lambda_2\mathbf{P})} \tag{13.109}$$

where ω is the angular frequency of the oscillator and m its reduced mass. Again, express the position operator and its momentum conjugate and also the Hamiltonian in which the zero-point energy is ignored, in terms of the Boson operators according to

$$\mathbf{Q} = \sqrt{\frac{\hbar}{2m\omega}}(\mathbf{a}^\dagger + \mathbf{a}) \qquad \text{and} \qquad \mathbf{P} = i\sqrt{\frac{\hbar m\omega}{2}}(\mathbf{a}^\dagger - \mathbf{a})$$

$$\mathbf{H} = \hbar\omega\,\mathbf{a}^\dagger\mathbf{a}$$

so that the argument of the last exponential of the right-hand side of Eq. (13.109) is

$$\lambda_1\mathbf{PQ} + \lambda_2\mathbf{P} = i\lambda_2 m\omega\sqrt{\frac{\hbar}{2m\omega}}(\mathbf{a}^\dagger - \mathbf{a}) + \lambda_1\sqrt{\frac{\hbar}{2m\omega}}(\mathbf{a}^\dagger + \mathbf{a})$$

or

$$\lambda_1\mathbf{PQ} + \lambda_2\mathbf{P} = \sqrt{\frac{\hbar}{2m\omega}}\{\mathbf{a}^\dagger(\lambda_1 + i\lambda_2 m\omega) + \mathbf{a}(\lambda_1 - i\lambda_2 m\omega)\}$$

Now, let

$$\lambda = \hbar\omega\beta$$

$$\alpha = \frac{1}{\lambda}\sqrt{\frac{\hbar}{2m\omega}}(\lambda_1 + i\lambda_2 m\omega) \qquad \text{and} \qquad \alpha^* = \frac{1}{\lambda}\sqrt{\frac{\hbar}{2m\omega}}(\lambda_1 - i\lambda_2 m\omega)$$

so that

$$\beta\mathbf{H} + \lambda_1\mathbf{Q} + \lambda_2\mathbf{P} = -\lambda(\mathbf{a}^\dagger\mathbf{a} + \alpha^*\mathbf{a} + \alpha\mathbf{a}^\dagger)$$

13.2.7.2 Some properties In terms of these new scalar and operator variables, the density operator (13.109) takes the form

$$\boldsymbol{\rho}_c = e^{-(1+\lambda_0)} e^{-\lambda(\mathbf{a}^\dagger \mathbf{a} + \alpha \mathbf{a}^\dagger + \alpha^* \mathbf{a})}$$

Next, in order to normalize, as required, the density operator, assume that

$$e^{-(1+\lambda_0)} = (1 - e^{-\lambda}) e^{-\lambda|\alpha|^2} \tag{13.110}$$

Then, the density operator, which will appear later to be normalized, reads

$$\boldsymbol{\rho}_c = (1 - e^{-\lambda}) e^{-\lambda(\mathbf{a}^\dagger \mathbf{a} + \alpha \mathbf{a}^\dagger + \alpha^* \mathbf{a} + |\alpha|^2)}$$

or

$$\boxed{\boldsymbol{\rho}_c = (1 - e^{-\lambda}) e^{-\lambda(\mathbf{a}^\dagger + \alpha)(\mathbf{a} + \alpha^*)}} \tag{13.111}$$

Next, perform the following canonical transformation:

$$\mathbf{A}(\alpha) \boldsymbol{\rho}_c \mathbf{A}(\alpha)^{-1} = (1 - e^{-\lambda}) \mathbf{A}(\alpha) e^{-\lambda(\mathbf{a}^\dagger + \alpha)(\mathbf{a} + \alpha^*)} \mathbf{A}(\alpha)^{-1} \tag{13.112}$$

with

$$\mathbf{A}(\alpha) = (e^{\alpha^* \mathbf{a}^\dagger - \alpha \mathbf{a}})$$

Next, due to Eqs. (7.9) and (7.10), which read

$$\mathbf{A}(\alpha) \{\mathbf{f}(\mathbf{a}, \mathbf{a}^\dagger)\} \mathbf{A}(\alpha)^{-1} = \{\mathbf{f}(\mathbf{a} - \alpha^*, \mathbf{a}^\dagger - \alpha)\}$$

Eq. (13.112) yields

$$\mathbf{A}(\alpha) \boldsymbol{\rho}_c \mathbf{A}(\alpha)^{-1} = (1 - e^{-\lambda})(e^{-\lambda \mathbf{a}^\dagger \mathbf{a}})$$

a result that, according to Eq. (13.22), is the Boltzmann density operator, leading to

$$\mathbf{A}(\alpha) \boldsymbol{\rho}_c \mathbf{A}(\alpha)^{-1} = \boldsymbol{\rho}_B \tag{13.113}$$

Observe that, since the Boltzmann density operator is normalized, and since a canonical transformation does not modify the normalization, it appears that $\boldsymbol{\rho}_c$ has been, indeed, normalized by the assumption.

Now, the coherent-state density operator reduces at zero temperature to the pure coherent-state density operator built up from a coherent state. For this purpose, inverse Eq. (13.113), so that

$$\boldsymbol{\rho}_c = \mathbf{A}(\alpha)^{-1} \mathbf{A}(\alpha) \boldsymbol{\rho}_c \mathbf{A}(\alpha)^{-1} \mathbf{A}(\alpha) = \mathbf{A}(\alpha)^{-1} \boldsymbol{\rho}_B \mathbf{A}(\alpha)$$

or, due to Eq. (13.22),

$$\boldsymbol{\rho}_c = (1 - e^{-\lambda}) \mathbf{A}(\alpha)^{-1} (e^{-\lambda \mathbf{a}^\dagger \mathbf{a}}) \mathbf{A}(\alpha)$$

Now, insert between the Boltzmann density operator and the translation operator a closure relation over the eigenstates of $\mathbf{a}^\dagger \mathbf{a}$, that is,

$$\boldsymbol{\rho}_c = (1 - e^{-\lambda}) \mathbf{A}(\alpha)^{-1} (e^{-\lambda \mathbf{a}^\dagger \mathbf{a}}) \sum_n |\{n\}\rangle\langle\{n\}| \mathbf{A}(\alpha)$$

Then, with the eigenvalue equation of $\mathbf{a}^\dagger \mathbf{a}$, the coherent-state density operator reads

$$\boldsymbol{\rho}_c = (1 - e^{-\lambda}) \mathbf{A}(\alpha)^{-1} \sum_n (e^{-\lambda n}) |\{n\}\rangle\langle\{n\}| \mathbf{A}(\alpha)$$

Next, if the temperature vanishes, λ which is given by Eq. (13.24), that is,

$$\lambda = \frac{\hbar\omega}{k_B T}$$

becomes infinite, so that

$$e^{-\lambda} \to 0$$

$$e^{-\lambda n} = e^{-n\hbar\omega/k_B T} \to 0 \quad \text{if} \quad n \neq 0$$

$$e^{-\lambda n} = e^{-n\hbar\omega/k_B T} = 1 \quad \text{if} \quad n = 0$$

Hence, the sum over the n eigenstates of $\mathbf{a}^\dagger\mathbf{a}$ reduces to the ground state, so that the coherent density operator reduces to

$$\{\boldsymbol{\rho}_c(T = 0)\} = \mathbf{A}(\alpha)^{-1}|\{0\}\rangle\langle\{0\}|\mathbf{A}(\alpha)$$

or, comparing Eq. (6.92),

$$\{\boldsymbol{\rho}_c(T = 0)\} = |\{\tilde{\alpha}\}\rangle\langle\{\tilde{\alpha}\}|$$

with

$$\mathbf{a}|\{\tilde{\alpha}\}\rangle = -\alpha|\{\tilde{\alpha}\}\rangle$$

Hence, when the absolute temperature vanishes, the density operator $\{\boldsymbol{\rho}_c(T=0)\}$ reduces to a coherent state $|\{\tilde{\alpha}\}\rangle$ of eigenvalue $-\alpha$, and so is the reason for its name.

13.2.8 Entropy of oscillators at thermal equilibrium

To get now an expression for the classical entropy of a population of oscillators at thermal equilibrium, which is the purpose of the present subsection, one has first to find an expression for the differential of the partition function in terms of the differential changes in the statistical parameter β and of the thermal average differential work $\langle dW\rangle$.

Hence, we first calculate $\langle dW\rangle$ and start from the differential expression of the 1D mechanical work along the x abscissa, that is,

$$dW = -F(x)\,dx \quad \text{with} \quad F(x) = -\left(\frac{\partial E(x)}{\partial x}\right)dx$$

and thus, when the energy $E(x)$ is quantized and defined by the energy levels $E_n(x)$,

$$dW_n = \left(\frac{\partial E_n(x)}{\partial x}\right)dx$$

the thermal average of the differential work is the average over the Boltzmann distribution of the different dW_n, that is,

$$\langle dW\rangle = \frac{1}{Z_\mu}\sum_n e^{-\beta E_n(x)}\left(\frac{\partial E_n(x)}{\partial x}\right)dx \quad \text{with} \quad Z_\mu = \sum_n e^{-\beta E_n(x)} \quad \text{and} \quad \beta = \frac{1}{k_B T}$$

where Z_μ is the partition function of a single oscillator. This expression may be also written

$$\langle dW \rangle = -\frac{1}{Z_\mu \beta} \sum_n \left(\frac{\partial e^{-\beta E_n(x)}}{\partial x} \right) dx$$

Moreover, since the sum and the partial derivative commute,

$$\langle dW \rangle = -\frac{1}{Z_\mu \beta} \frac{\partial}{\partial x} \sum_n e^{-\beta E_n(x)} \, dx = -\frac{1}{Z_\mu \beta} \left(\frac{\partial Z_\mu}{\partial x} \right) dx = -\frac{1}{\beta} \left(\frac{\partial \ln Z_\mu}{\partial x} \right) dx \tag{13.114}$$

Next, the total differential of $\ln Z_\mu(x, \beta)$ viewed as a function of the independent variables x and β reads

$$d\{\ln Z_\mu(x, \beta)\} = \left(\frac{\partial \ln Z_\mu}{\partial x} \right) dx + \left(\frac{\partial \ln Z_\mu}{\partial \beta} \right) d\beta \tag{13.115}$$

The thermal average Hamiltonian $\langle \mathbf{H} \rangle$, that is, the thermal energy, is given by Eq. (12.69):

$$\langle \mathbf{H} \rangle = -\left(\frac{\partial \ln Z_\mu}{\partial \beta} \right) \tag{13.116}$$

Hence, due to Eqs. (13.114) and (13.116), the total differential (13.115) yields

$$d\{\ln Z_\mu(x, \beta)\} = -\beta \langle dW \rangle - \langle \mathbf{H} \rangle d\beta$$

or

$$d\{\ln Z_\mu(x, \beta)\} = -\beta \langle dW \rangle - d\{\langle \mathbf{H} \rangle \beta\} + \beta d\langle \mathbf{H} \rangle$$

and thus

$$d\{\ln Z_\mu + (\langle \mathbf{H} \rangle \beta)\} = \beta\{d\langle \mathbf{H} \rangle - \langle dW \rangle\} \tag{13.117}$$

Then, recognizing in the difference between $d\langle \mathbf{H} \rangle$ and $\langle dW \rangle$ the differential heat exchange dQ, and using for β, Eq. (13.14), Eq. (13.117) reads

$$d\left\{\ln Z_\mu + \frac{\langle \mathbf{H} \rangle}{k_B T}\right\} = \frac{dQ}{k_B T}$$

Now, multiplying both terms of this last equation by the Boltzmann constant k_B and recognizing on the left-hand side the thermodynamical expression of the differential entropy dS, this expression becomes

$$k_B d\left\{\ln Z_\mu + \frac{\langle \mathbf{H} \rangle}{k_B T}\right\} = \frac{dQ}{T} = dS$$

Hence, the canonical entropy takes the form

$$\boxed{S = k_B \ln Z_\mu + \frac{\langle \mathbf{H} \rangle}{T}} \tag{13.118}$$

Equation (13.118) holds for one particle at thermal equilibrium, Z_μ and $\langle \mathbf{H} \rangle$ being, respectively, the partition function and the thermal average of this single particle. For N particles and because the partition function is the sum over exponentials, the partition function Z must be the Nth power of Z_μ. However, since the particles are indistinguishable, according to Chapter 2 because of the Heisenberg uncertainty relations, this power must be divided by $N!$ in order to avoid redundancies due to indistinguishable situations. Therefore, for N particles, Eq. (13.118) becomes

$$S = k_B \frac{\ln (Z_\mu)^{nN^\circ}}{(nN^\circ)!} + \frac{nN^\circ \langle \mathbf{H} \rangle}{T} \quad \text{with} \quad N = nN^\circ \tag{13.119}$$

where N° is the Avogadro number and n the number of moles. Again, after using, respectively, for Z_μ and $\langle E \rangle$, Eqs. (13.21) and (13.29), the entropy (13.119) yields

$$\boxed{S = n\left\{\frac{R}{(nN^\circ)!} \ln\left\{\frac{(e^{\hbar\omega/2k_BT})}{1 - e^{\hbar\omega/k_BT}}\right\} + N^\circ \frac{\hbar\omega}{T}\left(\frac{1}{e^{\hbar\omega k_BT} - 1} + \frac{1}{2}\right)\right\}} \tag{13.120}$$

where R is the ideal gas constant.

13.2.9 Oscillator Helmholtz potential

In thermodynamics, the Helmholtz thermodynamic potential is defined by

$$F = U - TS$$

where U is the internal energy. Then, for a population of oscillators, one may assimilate U to the oscillator thermal energy, and thus it is possible to write

$$U = \langle \mathbf{H} \rangle$$

so that, using for the entropy Eq. (13.118) the thermodynamic potential reads after simplification

$$F = -k_B T \ln Z_\mu \tag{13.121}$$

where it must be remembered that the partition function Z_μ is related to the Boltzmann density operator via Eq. (13.13), that is,

$$Z_\mu = \mathbf{tr}\{e^{-\beta \mathbf{H}}\} \quad \text{with} \quad \beta = \frac{1}{k_B T}$$

Hence, it appears from Eq. (13.121) that

$$\boxed{e^{-\beta F} = Z_\mu = \mathbf{tr}\{e^{-\beta \mathbf{H}}\}} \tag{13.122}$$

For a population of harmonic oscillators in thermal equilibrium, Eq. (13.122) reads with the help of Eq. (13.21)

$$e^{-\beta F} = \frac{e^{-\lambda/2}}{1 - e^{-\lambda}} \quad \text{with} \quad \lambda = \frac{\hbar\omega}{k_B T}$$

so that the thermodynamic potential yields

$$F = \frac{1}{\beta} \ln(1 - e^{-\lambda}) + \frac{\lambda}{2\beta}$$

or

$$F = k_B T \ln(1 - e^{-\hbar\omega/k_B T}) + \frac{\hbar\omega}{2} \tag{13.123}$$

13.2.10 Anharmonic oscillators dilatation with temperature

The dilatation of a solid with temperature is a well-known physical observation. This thermal dilatation is a result of anharmonicity we desire to treat here, where the dilation with temperature will be obtained in a numerical way for 1D oscillator. Hence, consider the thermal average value of the Q coordinate of an anharmonic oscillator performed over the Boltzmann density operator.

First, the Hamiltonian of an anharmonic oscillator is given by

$$\mathbf{H} = \hbar\omega\left(\mathbf{a}^\dagger\mathbf{a} + \tfrac{1}{2}\right) + \lambda\hbar\omega(\mathbf{a}^\dagger + \mathbf{a})^3$$

Its eigenvalue equation is

$$\mathbf{H}|\Psi_k\rangle = E_k|\Psi_k\rangle \tag{13.124}$$

with

$$\langle\Psi_k|\Psi_l\rangle = \delta_{kl}$$

For a given value of λ, this equation may be numerically solved working in the basis where $\mathbf{a}^\dagger\mathbf{a}$ is diagonal. In this basis, the expansion of the eigenkets of $\mathbf{H}$ is given by

$$|\Psi_k\rangle = \sum_n C_{kn}|\{n\}\rangle \quad \text{with} \quad \mathbf{a}^\dagger\mathbf{a}\;|\{n\}\rangle = n|\{n\}\rangle \tag{13.125}$$

The thermal average of the $\mathbf{Q}$ coordinate is

$$\langle\mathbf{Q}(T)\rangle = \mathbf{tr}\{\rho_B\mathbf{Q}\}$$

where the Boltzmann density operator is given by Eq. (13.13)

$$\rho_B = \frac{1}{Z}(e^{-\beta\mathbf{H}}) \quad \text{with} \quad \beta = \frac{1}{k_BT} \qquad \text{and} \qquad Z = \mathbf{tr}\{e^{-\beta\mathbf{H}}\}$$

and where $\mathbf{Q}$ is given in terms of the Boson operators by Eq. (5.6), that is,

$$\mathbf{Q} = \sqrt{\frac{\hbar}{2m\omega}}(\mathbf{a}^\dagger + \mathbf{a})$$

Writing explicitly the thermal average of $\mathbf{Q}$ over the basis where the Hamiltonian $\mathbf{H}$ is diagonal gives

$$\langle\mathbf{Q}(T)\rangle = \frac{1}{Z}\sqrt{\frac{\hbar}{2m\omega}}\sum_k\langle\Psi_k|(e^{-\beta\mathbf{H}})(\mathbf{a}^\dagger + \mathbf{a})|\Psi_k\rangle$$

Then, according to Eq. (13.124), the action on the bra of the exponential operator gives

$$\langle\mathbf{Q}(T)\rangle = \frac{1}{Z}\sqrt{\frac{\hbar}{2m\omega}}\sum_k \exp\left(-\frac{E_k}{k_BT}\right)\langle\Psi_k|(\mathbf{a}^\dagger + \mathbf{a})|\Psi_k\rangle$$

Next, introduce after $(\mathbf{a}^\dagger + \mathbf{a})$ the closure relation built on the eigenstates of $\mathbf{a}^\dagger\mathbf{a}$ to get

$$\langle\mathbf{Q}(T)\rangle = \frac{1}{Z}\sqrt{\frac{\hbar}{2m\omega}}\sum_k\sum_n \exp\left(-\frac{E_k}{k_BT}\right)\langle\Psi_k|(\mathbf{a}^\dagger + \mathbf{a})|\{n\}\rangle\langle\{n\}|\,\Psi_k\rangle$$

or

$$\langle \mathbf{Q}(T) \rangle = \frac{1}{Z}\sqrt{\frac{\hbar}{2m\omega}} \sum_k \sum_n \exp\left(-\frac{E_k}{k_B T}\right) C_{nk} \langle \Psi_k | (\mathbf{a}^\dagger + \mathbf{a}) | \{n\} \rangle$$

with

$$C_{n,k} = \langle \{n\} | \Psi_k \rangle \tag{13.126}$$

Again, using the result of the action of **a** and $\mathbf{a}^\dagger$ on an eigenket of $\mathbf{a}^\dagger \mathbf{a}$ leads according to Eqs. (5.53) and (5.63) to

$$\langle \mathbf{Q}(T) \rangle = \frac{1}{Z}\sqrt{\frac{\hbar}{2m\omega}} \sum_k \sum_n \exp\left(-\frac{E_k}{k_B T}\right) \times C_{nk} \left(\sqrt{n+1} \langle \Psi_k | \{n+1\} \rangle + \sqrt{n} \langle \Psi_k | \{n-1\} \rangle \right)$$

or, using in turn Eq. (13.126), the orthogonality of the eigenkets of $\mathbf{a}^\dagger \mathbf{a}$ and the result of the action of **a** and $\mathbf{a}^\dagger$ on an eigenket of $\mathbf{a}^\dagger \mathbf{a}$ leads to

$$\langle \mathbf{Q}(T) \rangle = \frac{1}{Z}\sqrt{\frac{\hbar}{2m\omega}} \sum_k \sum_n \exp\left(-\frac{E_k}{k_B T}\right) C_{nk} \left\{ C_{k,n+1}\sqrt{n+1} + C_{k,n-1}\sqrt{n} \right\} \tag{13.127}$$

Equation (13.127) allows one to compute the variation with temperature of the average value of the elongation of the anharmonic oscillator from the knowledge of the **H** eigenvalues E_k and of the expansion coefficients C_{kn} of the corresponding eigenvectors.

Figure 13.3 gives the temperature evolution of $\langle \mathbf{Q}(T) \rangle$ calculated in this way by the aid of Eq. (13.127) from the E_k and C_{kn} computed with the help of Eqs. (9.50) and (9.51).

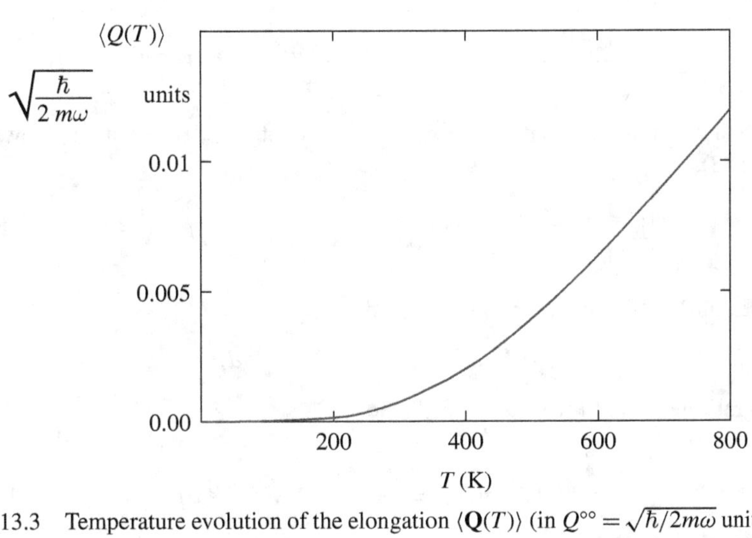

Figure 13.3 Temperature evolution of the elongation $\langle \mathbf{Q}(T) \rangle$ (in $Q^{\circ\circ} = \sqrt{\hbar/2m\omega}$ units) of an anharmonic oscillator. Anharmonic parameter $\beta = 0.017\hbar\omega$; number of basis states 75.

In three dimensions, the cube of Eq. (13.127) allows one to obtain the temperature dependence of the dilatation of a solid modelized by a 3D anharmonic oscillator. Observe that, according to Eqs. (13.62) and (13.66), which hold when the anharmonicity of the oscillator is missing, the average value of $\mathbf{Q}$ is zero for all quantum numbers so that the thermal average $\langle \mathbf{Q}(T) \rangle$ vanishes whatever the temperature.

13.3 HELMHOLTZ POTENTIAL FOR ANHARMONIC OSCILLATORS

Consider the Hamiltonian of an anharmonic oscillator of the form

$$\mathbf{H} = \mathbf{H}^\circ + \mathbf{V} \tag{13.128}$$

where $\mathbf{H}^\circ$ is the Hamiltonian of the harmonic oscillator and $\mathbf{V}$ the anharmonic Hamiltonian perturbation. Then, according to Eq. (13.12) the unnormalized Boltzmann density operator of the harmonic and anharmonic oscillators read, respectively,

$$\boldsymbol{\rho}^\circ \propto e^{-\beta \mathbf{H}^\circ} \quad \text{and} \quad \boldsymbol{\rho} \propto e^{-\beta \mathbf{H}} \tag{13.129}$$

Now, the partial differential of these density operators with respect to β read, respectively,

$$\left(\frac{\partial \boldsymbol{\rho}^\circ}{\partial \beta}\right) \propto -\mathbf{H}^\circ e^{-\beta \mathbf{H}^\circ} \quad \text{and} \quad \left(\frac{\partial \boldsymbol{\rho}}{\partial \beta}\right) \propto -\mathbf{H} e^{-\beta \mathbf{H}} \tag{13.130}$$

Next, in order to express $\boldsymbol{\rho}$ in terms of $\boldsymbol{\rho}^\circ$, first calculate

$$\left(\frac{\partial (e^{\beta \mathbf{H}^\circ} e^{-\beta \mathbf{H}})}{\partial \beta}\right) = \left(\frac{\partial e^{\beta \mathbf{H}^\circ}}{\partial \beta}\right) e^{-\beta \mathbf{H}} + e^{\beta \mathbf{H}^\circ} \left(\frac{\partial e^{-\beta \mathbf{H}}}{\partial \beta}\right) \tag{13.131}$$

which, due to (13.130), yields

$$\left(\frac{\partial (e^{\beta \mathbf{H}^\circ} e^{-\beta \mathbf{H}})}{\partial \beta}\right) = \mathbf{H}^\circ e^{\beta \mathbf{H}^\circ} e^{-\beta \mathbf{H}} - e^{\beta \mathbf{H}^\circ} \mathbf{H} e^{-\beta \mathbf{H}}$$

Or, since $\mathbf{H}^\circ$ commutes with the exponential constructed from it, and owing to Eq. (13.128),

$$\left(\frac{\partial (e^{\beta \mathbf{H}^\circ} e^{-\beta \mathbf{H}})}{\partial \beta}\right) = e^{\beta \mathbf{H}^\circ} (\mathbf{H}^\circ - \mathbf{H}) e^{-\beta \mathbf{H}} = -e^{\beta \mathbf{H}^\circ} \mathbf{V} e^{-\beta \mathbf{H}}$$

Due to Eq. (13.129), the latter equation leads to

$$d\{e^{\beta \mathbf{H}^\circ} \boldsymbol{\rho}(\beta)\} = -e^{\beta \mathbf{H}^\circ} \mathbf{V} e^{-\beta \mathbf{H}} \, d\beta$$

the integration of which from zero to β reads

$$\int_0^\beta d\{e^{\beta' \mathbf{H}^\circ} \boldsymbol{\rho}(\beta')\} = -\int_0^\beta e^{\beta' \mathbf{H}^\circ} \mathbf{V} e^{-\beta' \mathbf{H}} \, d\beta' \tag{13.132}$$

Now, observe that when $\beta = 0$, it appears from (13.129) that

$$e^{\beta \mathbf{H}^\circ} \boldsymbol{\rho}(\beta) = e^{\beta \mathbf{H}^\circ} e^{-\beta \mathbf{H}} = 1$$

so that the integration of (13.132) reads

$$e^{\beta \mathbf{H}^\circ} \boldsymbol{\rho}(\beta) - 1 = -\int_0^\beta e^{-(\beta-\beta')\mathbf{H}^\circ} \mathbf{V} e^{-\beta' \mathbf{H}} \, d\beta'$$

or, after premultiplying both members by $\exp\{-\beta \mathbf{H}^\circ\}$ and using the last expression of (13.129),

$$\boldsymbol{\rho}(\beta) = e^{-\beta \mathbf{H}^\circ} - \int_0^\beta e^{-(\beta-\beta')\mathbf{H}^\circ} \mathbf{V} \boldsymbol{\rho}(\beta) \, d\beta'$$

The first-order solution of this last integral is

$$\boldsymbol{\rho}(\beta) = e^{-\beta \mathbf{H}^\circ} - \int_0^\beta e^{-(\beta-\beta')\mathbf{H}^\circ} \mathbf{V} e^{-\beta' \mathbf{H}^\circ} \, d\beta'$$

whereas the second-order solution is

$$\boldsymbol{\rho}(\beta) = e^{-\beta \mathbf{H}^\circ} - \int_0^\beta e^{-(\beta-\beta')\mathbf{H}^\circ} \mathbf{V} e^{-\beta' \mathbf{H}^\circ} d\beta' + \int_0^\beta \int_0^{\beta'} e^{-(\beta-\beta')\mathbf{H}^\circ} \mathbf{V} e^{-(\beta'-\beta'')\mathbf{H}^\circ} \times \mathbf{V} e^{-\beta'' \mathbf{H}^\circ} d\beta' d\beta'' \tag{13.133}$$

The solution (13.133) is dealing with a density operator that is unnormalized. But that is of no importance if one is interested in the Helmholtz energy F, which is related, via Eq. (13.122), to the Boltzmann density operator through

$$e^{-\beta F} = \mathbf{tr}\{e^{-\beta \mathbf{H}}\} = \mathbf{tr}\{\boldsymbol{\rho}(\beta)\} \tag{13.134}$$

an expression that is true whatever the normalization of the density operator. Hence, one gets

$$e^{-\beta F} = \mathbf{tr}\{e^{-\beta \mathbf{H}^\circ}\} - \int_0^\beta \mathbf{tr}\{e^{-(\beta-\beta')\mathbf{H}^\circ} \mathbf{V} e^{-\beta' \mathbf{H}^\circ}\} d\beta' + \int_0^\beta \int_0^{\beta'} \mathbf{tr}\{e^{-(\beta-\beta')\mathbf{H}^\circ} \mathbf{V} e^{-(\beta'-\beta'')\mathbf{H}^\circ} \mathbf{V} e^{-\beta'' \mathbf{H}^\circ}\} d\beta' d\beta'' \tag{13.135}$$

Now, due to the invariance of the trace with respect to a circular permutation within it, it appears that

$$\mathbf{tr}\{e^{-(\beta-\beta')\mathbf{H}^\circ} \mathbf{V} e^{-\beta' \mathbf{H}^\circ}\} = \mathbf{tr}\{e^{-\beta' \mathbf{H}^\circ} e^{-(\beta-\beta')\mathbf{H}^\circ} \mathbf{V}\} = \mathbf{tr}\{e^{-\beta \mathbf{H}^\circ} \mathbf{V}\} \tag{13.136}$$

$$\begin{aligned}\mathbf{tr}\{e^{-(\beta-\beta')\mathbf{H}^\circ}\mathbf{V}e^{-(\beta'-\beta'')\mathbf{H}^\circ}\mathbf{V}e^{-\beta''\mathbf{H}^\circ}\} &= \mathbf{tr}\{e^{-\beta''\mathbf{H}^\circ}e^{-(\beta-\beta')\mathbf{H}^\circ}\mathbf{V}e^{-(\beta'-\beta'')\mathbf{H}^\circ}\mathbf{V}\} \\ &= \mathbf{tr}\{e^{-\beta\mathbf{H}^\circ}e^{-(\beta''-\beta')\mathbf{H}^\circ}\mathbf{V}e^{-(\beta'-\beta'')\mathbf{H}^\circ}\mathbf{V}\}\end{aligned} \tag{13.137}$$

Moreover, perform the first trace over the eigenstates $|(n)\rangle$ of $\mathbf{H}^\circ$ to get

$$\mathbf{tr}\{e^{-\beta\mathbf{H}^\circ}\mathbf{V}\} = \sum_n \langle (n)|e^{-\beta\mathbf{H}^\circ}\mathbf{V}|(n)\rangle = \sum_n e^{-n\beta\hbar\omega}\langle (n)|\mathbf{V}|(n)\rangle \tag{13.138}$$

whereas working in the same way for the trace (13.137) and after inserting a closure relation over the basis $\{|(n)\rangle\}$ after the first operator $\mathbf{V}$ yields

$$\mathbf{tr}\{e^{-\beta\mathbf{H}^\circ}e^{-(\beta''-\beta')\mathbf{H}^\circ}\mathbf{V}e^{-(\beta'-\beta'')\mathbf{H}^\circ}\mathbf{V}\}$$
$$= \sum_n\sum_m \langle (n)|e^{-\beta\mathbf{H}^\circ}e^{-(\beta''-\beta')\mathbf{H}^\circ}\mathbf{V}|(m)\rangle\langle (m)|e^{-(\beta'-\beta'')\mathbf{H}^\circ}\mathbf{V}|(n)\rangle$$

or

$$\mathbf{tr}\{e^{-\beta\mathbf{H}^\circ}e^{-(\beta''-\beta')\mathbf{H}^\circ}\mathbf{V}e^{-(\beta'-\beta'')\mathbf{H}^\circ}\mathbf{V}\}$$
$$= \sum_n\sum_m e^{-n\beta\hbar\omega}e^{-n(\beta''-\beta')\hbar\omega}\langle (n)|\mathbf{V}|(m)\rangle e^{-m(\beta'-\beta'')\hbar\omega}\langle (m)|\mathbf{V}|(n)\rangle \tag{13.139}$$

Due to Eqs. (13.136) and (13.137) and to Eqs. (13.138) and (13.139), Eq. (13.135) giving the Helmholtz free energy becomes

$$\begin{aligned}e^{-\beta F} &= \mathbf{tr}\{e^{-\beta\mathbf{H}^\circ}\} - \sum_n e^{-n\beta\hbar\omega}\langle (n)|\mathbf{V}|(n)\rangle\int_0^\beta d\beta' \\ &\quad + \sum_n\sum_m e^{-n\beta\hbar\omega}|\langle (m)|\mathbf{V}|(n)\rangle^2|\int_0^\beta\int_0^{\beta'} e^{(n-m)\hbar\omega(\beta'-\beta'')}d\beta' d\beta''\end{aligned} \tag{13.140}$$

Now, observe the latter integral may be written

$$\int_0^\beta\int_0^{\beta'} e^{(n-m)\hbar\omega(\beta'-\beta'')}d\beta' d\beta'' = \frac{1}{2}\int_0^\beta e^{(n-m)\hbar\omega\eta}d\eta\int_0^\beta d\beta''$$

leading to

$$\int_0^\beta\int_0^{\beta'} e^{(n-m)\hbar\omega(\beta'-\beta'')}d\beta' d\beta'' = \frac{\beta}{2\hbar\omega}\frac{e^{(n-m)\beta\hbar\omega}-1}{n-m}$$

As a consequence, Eq. (13.140) takes the form

$$\boxed{e^{-\beta F} = e^{-\beta F^\circ} - \beta\sum_n e^{-n\beta\hbar\omega}\langle (n)|\mathbf{V}|(n)\rangle + \frac{\beta}{2\hbar\omega}\sum_n\sum_m |\langle (m)|\mathbf{V}|(n)\rangle^2|\frac{e^{-m\beta\hbar\omega}-e^{-n\beta\hbar\omega}}{(n-m)}} \tag{13.141}$$

with, in a similar way as in Eq. (13.134),

$$e^{-\beta F^\circ} = \mathbf{tr}\{e^{-\beta \mathbf{H}^\circ}\}$$

From Eq. (13.141), it may be shown[1] that

$$\boxed{F \leqslant F^\circ + \langle \mathbf{V} \rangle_0 \quad \text{with} \quad \langle \mathbf{V} \rangle_0 = \frac{\mathbf{tr}\{\mathbf{V} e^{-\beta \mathbf{H}^\circ}\}}{\mathbf{tr}\{e^{-\beta \mathbf{H}^\circ}\}}}$$

a result that allows one to find physical average values by minimization procedure. Besides, Eq. (13.141) may be applied to an anharmonic oscillator in which $\mathbf{V}$ is given by

$$\mathbf{V} = \left(\frac{\hbar}{2m\omega}\right)^{3/2} (\mathbf{a}^\dagger + \mathbf{a})^3$$

with the help of Eqs. (9.41)–(9.48) and using Eq. (13.21) allowing one to write

$$e^{-\beta F^\circ} = \mathbf{tr}\{e^{-\beta \mathbf{H}^\circ}\} = Z = \frac{e^{-\lambda/2}}{1 - e^{-\lambda}} \quad \text{with} \quad \lambda = \frac{\hbar\omega}{k_B T}$$

13.4 THERMAL AVERAGE OF BOSON OPERATOR FUNCTIONS

Now, we shall obtain the general expression for the average of any function of Boson operators over the Boltzmann equilibrium density operator. We shall obtain a general expression that reduces to the Bloch theorem when the function of Boson operators is either the position operator or its conjugate momentum. If the demonstration is somewhat tedious, it has the merit of avoiding the mathematical complications required to obtain its simplified form, which is the Bloch theorem.

13.4.1 Calculation of thermal average

In this section we derive the expression of the thermal average of any function of Boson operators over the canonical density operator of an harmonic oscillator, that is,

$$\langle \mathbf{F}(\mathbf{a}^\dagger, \mathbf{a}) \rangle = \mathbf{tr}\{\{\mathbf{F}(\mathbf{a}^\dagger, \mathbf{a})\}\rho_B(\mathbf{a}^\dagger, \mathbf{a})\}$$

which, in view of Eqs. (13.23) and (13.24), reads

$$\langle \mathbf{F}(\mathbf{a}^\dagger, \mathbf{a}) \rangle = (1 - e^{-\lambda})\mathbf{tr}\{\{\mathbf{F}(\mathbf{a}^\dagger, \mathbf{a})\}(e^{-\lambda \mathbf{a}^\dagger \mathbf{a}})\} \tag{13.142}$$

$$\lambda = \frac{\hbar\omega}{k_B T} \tag{13.143}$$

[1] R. P. Feynman. *Statistical Mechanics: A Set of Lectures*, 2nd ed. Perseus Books: New York, 1998.

13.4.1.1 From the basic equation (13.142) to a more tractable one Tracing on the right-hand side, over the eigenstates $|(n)\rangle$ of $\mathbf{a}^\dagger\mathbf{a}$, transforms Eq. (13.142) to

$$\frac{\langle \mathbf{F}(\mathbf{a}^\dagger, \mathbf{a})\rangle}{(1-e^{-\lambda})} = \sum_n \langle (n)|\{\mathbf{F}(\mathbf{a}^\dagger, \mathbf{a})\}(e^{-\lambda \mathbf{a}^\dagger \mathbf{a}})|(n)\rangle \tag{13.144}$$

with

$$\mathbf{a}^\dagger \mathbf{a}|(n)\rangle = n|(n)\rangle \tag{13.145}$$

Now, observe that the only terms of $\mathbf{F}(\mathbf{a}^\dagger, \mathbf{a})$, which may contribute to the diagonal matrix elements involved on the right-hand side of this last equation, are those having the same power of $\mathbf{a}^\dagger$ and $\mathbf{a}$.

Accordingly, in the trace above, we are free to change $\mathbf{a}$ into $k\mathbf{a}$ and $\mathbf{a}^\dagger$ into $\mathbf{a}^\dagger/k$ where k is some real scalar (that will be defined later). Hence, since the product $\mathbf{a}^\dagger\mathbf{a}$ involved in the density operator is not affected by this change, Eq. (13.144) yields

$$\frac{\langle \mathbf{F}(\mathbf{a}^\dagger, \mathbf{a})\rangle}{(1-e^{-\lambda})} = \sum_n \langle (n)|\{\mathbf{F}(\mathbf{a}^\dagger/k, k\mathbf{a})\}(e^{-\lambda \mathbf{a}^\dagger \mathbf{a}})|(n)\rangle$$

Also we write this last equation in the following more complex form:

$$\frac{\langle \mathbf{F}(\mathbf{a}^\dagger, \mathbf{a})\rangle}{(1-e^{-\lambda})} = \sum_n \sum_m \langle (n)|\{\mathbf{F}(\mathbf{a}^\dagger/k, k\mathbf{a})\}(e^{-\lambda \mathbf{a}^\dagger \mathbf{a}})|(m)\rangle \delta_{nm} \tag{13.146}$$

Moreover, due to Eqs. (5.69) and (5.70),

$$|(m)\rangle = \left(\frac{(\mathbf{a}^\dagger)^m}{\sqrt{m!}}\right)|(0)\rangle \qquad \text{and} \qquad \langle (n)| = \langle (0)|\left(\frac{(\mathbf{a})^n}{\sqrt{n!}}\right) \tag{13.147}$$

Equation (13.146) transforms to

$$\frac{\langle \mathbf{F}(\mathbf{a}^\dagger, \mathbf{a})\rangle}{(1-e^{-\lambda})} = \sum_n \sum_m \langle (0)|\left(\frac{(\mathbf{a})^n}{\sqrt{n!}}\right)\{\mathbf{F}(\mathbf{a}^\dagger/k, k\mathbf{a})\}(e^{-\lambda \mathbf{a}^\dagger \mathbf{a}})\left(\frac{(\mathbf{a}^\dagger)^m}{\sqrt{m!}}\right)|(0)\rangle \delta_{nm} \tag{13.148}$$

Now, since the Kronecker symbol δ_{nm} appearing on the right-hand side of Eq (13.148) may be viewed as the scalar product of two eigenstates of any operator $\mathbf{b}^\dagger\mathbf{b}$, of Boson operators that commute with $\mathbf{a}^\dagger$ and $\mathbf{a}$, that is,

$$\delta_{nm} = \langle\{n\}|\{m\}\rangle \quad \text{with} \quad \mathbf{b}^\dagger\mathbf{b}|\{n\}\rangle = n|\{n\}\rangle \tag{13.149}$$

with

$$[\mathbf{a}, \mathbf{b}] = 0 \qquad [\mathbf{a}^\dagger, \mathbf{b}] = 0 \qquad [\mathbf{a}, \mathbf{b}^\dagger] = 0$$

Next, using for these Boson operators equations similar to those of (13.147),

$$|\{m\}\rangle = \left(\frac{(\mathbf{b}^\dagger)^m}{\sqrt{m!}}\right)|\{0\}\rangle \qquad \text{and} \qquad \langle\{n\}| = \langle\{0\}|\left(\frac{(\mathbf{b})^n}{\sqrt{n!}}\right)$$

the Kronecker symbol appearing in (13.149) becomes

$$\delta_{nm} = \langle\{0\}|\left(\frac{(\mathbf{b})^n}{\sqrt{n!}}\right)\left(\frac{(\mathbf{b}^\dagger)^m}{\sqrt{m!}}\right)|\{0\}\rangle \tag{13.150}$$

so that Eq. (13.148) reads

$$\frac{\langle \mathbf{F}(\mathbf{a}^\dagger, \mathbf{a})\rangle}{(1-e^{-\lambda})} = \sum_n \sum_m \langle\{0\}|\langle(0)| \left(\frac{(\mathbf{b})^n}{\sqrt{n!}}\right)\left(\frac{(\mathbf{a})^n}{\sqrt{n!}}\right) \times \{\mathbf{F}(\mathbf{a}^\dagger/k, k\mathbf{a})\}(e^{-\lambda \mathbf{a}^\dagger \mathbf{a}})\left(\frac{(\mathbf{a}^\dagger)^m}{\sqrt{m!}}\right)\left(\frac{(\mathbf{b}^\dagger)^m}{\sqrt{m!}}\right)|(0)\rangle|\{0\}\rangle$$

Now, one may replace $\mathbf{b}$ by $\mu\mathbf{b}$ and $\mathbf{b}^\dagger$ by $\mathbf{b}^\dagger/\mu$, where μ is a real scalar, without modifying the right-hand-side average value, so that Eq. (13.150) becomes

$$\frac{\langle \mathbf{F}(\mathbf{a}^\dagger, \mathbf{a})\rangle}{(1-e^{-\lambda})} = \sum_n \sum_m \langle\{0\}|\langle(0)| \left(\frac{(\mu\mathbf{b})^n}{\sqrt{n!}}\right)\left(\frac{(\mathbf{a})^n}{\sqrt{n!}}\right) \times \{\mathbf{F}(\mathbf{a}^\dagger/k, k\mathbf{a})\}(e^{-\lambda \mathbf{a}^\dagger \mathbf{a}})\left(\frac{(\mathbf{a}^\dagger)^m}{\sqrt{m!}}\right)\left(\frac{(\mathbf{b}^\dagger/\mu)^m}{\sqrt{m!}}\right)|(0)\rangle|\{0\}\rangle$$

or, rearranging and simplifying the notation for the ket or bra products,

$$\frac{\langle \mathbf{F}(\mathbf{a}^\dagger, \mathbf{a})\rangle}{(1-e^{-\lambda})} = \sum_n \sum_m \langle\{0\}(0)| \left(\frac{(\mu\mathbf{ba})^n}{n!}\right) \times \{\mathbf{F}(\mathbf{a}^\dagger/k, k\mathbf{a})\}(e^{-\lambda \mathbf{a}^\dagger \mathbf{a}})\left(\frac{(\mathbf{a}^\dagger\mathbf{b}^\dagger/\mu)^m}{m!}\right)|(0)\{0\}\rangle$$

Then, pass to exponentials

$$\frac{\langle \mathbf{F}(\mathbf{a}^\dagger, \mathbf{a})\rangle}{(1-e^{-\lambda})} = \langle\{0\}(0)|(e^{\mu\mathbf{ba}})\{\mathbf{F}(\mathbf{a}^\dagger/k, k\mathbf{a})\}(e^{-\lambda \mathbf{a}^\dagger \mathbf{a}})(e^{\mathbf{a}^\dagger\mathbf{b}^\dagger/\mu})|(0)\{0\}\rangle$$

Furthermore, introduce after the function of Boson operators the unity operator defined by

$$1 = (e^{-\mu\mathbf{ba}})(e^{\mu\mathbf{ba}})$$

leading to

$$\frac{\langle \mathbf{F}(\mathbf{a}^\dagger, \mathbf{a})\rangle}{(1-e^{-\lambda})} = \langle\{0\}(0)|(e^{\mu\mathbf{ba}})\{\mathbf{F}(\mathbf{a}^\dagger/k, k\mathbf{a})\}(e^{-\mu\mathbf{ba}})(e^{\mu\mathbf{ba}})(e^{-\lambda \mathbf{a}^\dagger \mathbf{a}})(e^{\mathbf{a}^\dagger\mathbf{b}^\dagger/\mu})|(0)\{0\}\rangle \tag{13.151}$$

Now, according to Eq. (7.5), it appears that

$$(e^{\mu\mathbf{ba}})\{\mathbf{F}(\mathbf{a}^\dagger/k, k\mathbf{a})\}(e^{-\mu\mathbf{ba}}) = \{\mathbf{F}((\mathbf{a}^\dagger + \mu\mathbf{b})/k, k\mathbf{a})\} \tag{13.152}$$

and, comparing Eq. (7.106), that is,

$$(e^{-\lambda \mathbf{a}^\dagger \mathbf{a}})(e^{y\mathbf{a}^\dagger})|(0)\rangle = (e^{y\mathbf{a}^\dagger(e^{-\lambda})})|(0)\rangle$$

with

$$y = \frac{\mathbf{b}^\dagger}{\mu}$$

it appears that, since $\mathbf{b}^\dagger$ is dimensionless and does not act on the ket $|(0)\rangle$ because acting on $|\{0\}\rangle$ one finds

$$(e^{-\lambda \mathbf{a}^\dagger \mathbf{a}})(e^{\mathbf{a}^\dagger \mathbf{b}^\dagger/\mu})|(0)\rangle = \{e^{\mathbf{a}^\dagger \mathbf{b}^\dagger e^{-\lambda}/\mu}\}|(0)\rangle \tag{13.153}$$

Hence, in view of Eqs. (13.152) and (13.153), the average value (13.151) simplifies to

$$\frac{\langle \mathbf{F}(\mathbf{a}^\dagger, \mathbf{a})\rangle}{(1-e^{-\lambda})} = \langle\{0\}(0)|\{\mathbf{F}((\mathbf{a}^\dagger + \mu\mathbf{b})/k, k\mathbf{a})\}(e^{\mu \mathbf{b}\mathbf{a}})\{e^{\xi \mathbf{a}^\dagger \mathbf{b}^\dagger}\}|(0)\{0\}\rangle \tag{13.154}$$

with

$$\xi = e^{-\lambda}/\mu \tag{13.155}$$

13.4.1.2 Action of the product of exponential operators involved in Eq. (13.154) on $|(0)\{0\}\rangle$ It is now required to find the action of the product of the two exponential operators involved on the right-hand side of Eq. (13.154) on the ground state $|(0)\{0\}\rangle$ of $\mathbf{a}^\dagger\mathbf{a}\,\mathbf{b}^\dagger\mathbf{b}$. Hence, one has to find a function of ladder operators $\mathbf{G}(\mu, \mathbf{a}^\dagger, \mathbf{b}^\dagger, \mathbf{a}, \mathbf{b})$ satisfying

$$(e^{\mathbf{G}(\mu,\mathbf{a}^\dagger,\mathbf{b}^\dagger,\mathbf{a},\mathbf{b})})|(0)\{0\}\rangle = (e^{\mu \mathbf{b}\mathbf{a}})\{e^{\xi \mathbf{a}^\dagger \mathbf{b}^\dagger}\}|(0)\{0\}\rangle \tag{13.156}$$

For this purpose, differentiate both members of Eq. (13.156) with respect to μ, yielding

$$\exp(\mathbf{G})\left(\frac{\partial \mathbf{G}}{\partial \mu}\right)|(0)\{0\}\rangle = \mathbf{b}\mathbf{a}(e^{\mu \mathbf{b}\mathbf{a}})\{e^{\xi \mathbf{a}^\dagger \mathbf{b}^\dagger}\}|(0)\{0\}\rangle$$

or

$$\exp(\mathbf{G})\left(\frac{\partial \mathbf{G}}{\partial \mu}\right)|(0)\{0\}\rangle = \mathbf{b}\mathbf{a}\ \exp\{\mathbf{G}\}|(0)\{0\}\rangle$$

Then, premultiplying both terms by $\exp(-\mathbf{G})$ we have

$$\left(\frac{\partial \mathbf{G}}{\partial \mu}\right)|(0)\{0\}\rangle = \exp\{-\mathbf{G}\}\mathbf{b}\mathbf{a}\ \exp\{\mathbf{G}\}|(0)\{0\}\rangle$$

Again, insert between $\mathbf{b}$ and $\mathbf{a}$ the unity operator built up from $\exp\{-\mathbf{G}\}$, that is,

$$\left(\frac{\partial \mathbf{G}}{\partial \mu}\right)|(0)\{0\}\rangle = \exp\{-\mathbf{G}\}\ \mathbf{b}\{\mathbf{G}\}\exp\{-\mathbf{G}\}\mathbf{a}\ \exp\{\mathbf{G}\}|(0)\{0\}\rangle \tag{13.157}$$

and apply Eq. (7.60), that is,

$$\mathbf{a}\mathbf{f}(\mathbf{a}, \mathbf{a}^\dagger) - \mathbf{f}(\mathbf{a}, \mathbf{a}^\dagger)\mathbf{a} = \left(\frac{\partial \mathbf{f}(\mathbf{a}, \mathbf{a}^\dagger)}{\partial \mathbf{a}^\dagger}\right)$$

to the function

$$\mathbf{f}(\mathbf{a}, \mathbf{a}^\dagger) = \exp\{\mathbf{G}\}$$

Hence

$$\mathbf{a}\exp\{\mathbf{G}\} - \exp\{\mathbf{G}\}\mathbf{a} = \left(\frac{\partial \exp\{\mathbf{G}\}}{\partial \mathbf{a}^\dagger}\right)$$

or

$$\mathbf{a}\exp(\mathbf{G}) = \exp(\mathbf{G})\mathbf{a} + \exp\{\mathbf{G}\}\left(\frac{\partial \mathbf{G}}{\partial \mathbf{a}^\dagger}\right)$$

Then, premultiplying both terms by $\exp(-\mathbf{G})$ we have after simplification

$$\exp\{-\mathbf{G}\}\mathbf{a}\exp\{\mathbf{G}\} = \mathbf{a} + \left(\frac{\partial \mathbf{G}}{\partial \mathbf{a}^\dagger}\right) \tag{13.158}$$

In a similar way one would obtain

$$\exp\{-\mathbf{G}\}\mathbf{b}\exp\{\mathbf{G}\} = \mathbf{b} + \left(\frac{\partial \mathbf{G}}{\partial \mathbf{b}^\dagger}\right) \tag{13.159}$$

As a consequence of Eqs. (13.158) and (13.159), Eq. (13.157) becomes

$$\left(\frac{\partial \mathbf{G}}{\partial \mu}\right)|(0)\{0\}\rangle = \left(\mathbf{b} + \left(\frac{\partial \mathbf{G}}{\partial \mathbf{b}^\dagger}\right)\right)\left(\mathbf{a} + \left(\frac{\partial \mathbf{G}}{\partial \mathbf{a}^\dagger}\right)\right)|(0)\{0\}\rangle$$

Then, performing the product involved on the right-hand side gives

$$\left(\frac{\partial \mathbf{G}}{\partial \mu}\right)|(0)\{0\}\rangle = \left(\mathbf{ba} + \left(\frac{\partial \mathbf{G}}{\partial \mathbf{b}^\dagger}\right)\left(\frac{\partial \mathbf{G}}{\partial \mathbf{a}^\dagger}\right) + \mathbf{b}\left(\frac{\partial \mathbf{G}}{\partial \mathbf{a}^\dagger}\right) + \left(\frac{\partial \mathbf{G}}{\partial \mathbf{b}^\dagger}\right)\mathbf{a}\right)|(0)\{0\}\rangle \tag{13.160}$$

Now, observe that since

$$\mathbf{b}|\{0\}\rangle = \mathbf{a}|(0)\rangle = 0 \tag{13.161}$$

we have

$$\mathbf{ba}|(0)\{0\}\rangle = \mathbf{b}|\{0\}\rangle\mathbf{a}|(0)\rangle = 0$$

so that Eq. (13.160) yields

$$\left(\frac{\partial \mathbf{G}}{\partial \mu}\right)|(0)\{0\}\rangle = \left(\left(\frac{\partial \mathbf{G}}{\partial \mathbf{b}^\dagger}\right)\left(\frac{\partial \mathbf{G}}{\partial \mathbf{a}^\dagger}\right) + \mathbf{b}\left(\frac{\partial \mathbf{G}}{\partial \mathbf{a}^\dagger}\right)\right)|(0)\{0\}\rangle \tag{13.162}$$

Again, using in turn Eq. (7.60),

$$\mathbf{bf}(\mathbf{b},\mathbf{b}^\dagger) - \mathbf{f}(\mathbf{b},\mathbf{b}^\dagger)\mathbf{b} = \left(\frac{\partial \mathbf{f}(\mathbf{b},\mathbf{b}^\dagger)}{\partial \mathbf{b}^\dagger}\right) \quad \text{with} \quad \mathbf{f}(\mathbf{b},\mathbf{b}^\dagger) = \left(\frac{\partial \mathbf{G}}{\partial \mathbf{a}^\dagger}\right)$$

so that

$$\mathbf{b}\left(\frac{\partial \mathbf{G}}{\partial \mathbf{a}^\dagger}\right) = \left(\frac{\partial \mathbf{G}}{\partial \mathbf{a}^\dagger}\right)\mathbf{b} + \left(\frac{\partial^2 \mathbf{G}}{\partial \mathbf{b}^\dagger \partial \mathbf{a}^\dagger}\right)$$

and since, due to Eq. (13.161),

$$\left(\frac{\partial \mathbf{G}}{\partial \mathbf{a}^\dagger}\right)\mathbf{b}|\{0\}\rangle = 0$$

Eq. (13.162) reads

$$\left(\frac{\partial \mathbf{G}}{\partial \mu}\right)|(0)\{0\}\rangle = \left(\left(\frac{\partial \mathbf{G}}{\partial \mathbf{a}^\dagger}\right)\left(\frac{\partial \mathbf{G}}{\partial \mathbf{b}^\dagger}\right) + \left(\frac{\partial^2 \mathbf{G}}{\partial \mathbf{b}^\dagger\, \partial \mathbf{a}^\dagger}\right)\right)|(0)\{0\}\rangle \tag{13.163}$$

or

$$\left(\frac{\partial \mathbf{G}(\mathbf{a}^\dagger, \mathbf{b}^\dagger, \mu)}{\partial \mu}\right) |(0)\{0\}\rangle = \left(\frac{\partial \mathbf{G}}{\partial \mathbf{a}^\dagger}\right)\left(\frac{\partial \mathbf{G}}{\partial \mathbf{b}^\dagger}\right)\left(\frac{\partial^2 \mathbf{G}(\mathbf{a}^\dagger, \mathbf{b}^\dagger, \mu)}{\partial \mathbf{b}^\dagger\, \partial \mathbf{a}^\dagger}\right) |(0)\{0\}\rangle \quad (13.164)$$

Next, in order to solve this partial differential equation involving only $\mathbf{a}^\dagger$, $\mathbf{b}^\dagger$, and μ, one may seek a solution at an expression of the following form:

$$\mathbf{G}(\mathbf{a}^\dagger, \mathbf{b}^\dagger, \mu) = A(\mu) + B(\mu)\mathbf{a}^\dagger \mathbf{b}^\dagger \quad (13.165)$$

where $\mathbf{a}$ and $\mathbf{b}$ disappear, whereas $A(\mu)$ and $B(\mu)$ are unknown scalar coefficients. Now, in order to get the expression of the function (13.165) for the special situation where $\mu = 0$, use the fact that in this special situation, Eq. (13.156) reduces to

$$(e^{\mathbf{G}(0,\mathbf{a}^\dagger,\mathbf{b}^\dagger)})|(0)\{0\}\rangle = \{e^{\xi \mathbf{a}^\dagger \mathbf{b}^\dagger}\}|(0)\{0\}\rangle \quad (13.166)$$

Thereby, since the arguments of the exponentials appearing on the right- and on the left-hand-side operators of this last equation must be the same, we have

$$\mathbf{G}(0, \mathbf{a}^\dagger, \mathbf{b}^\dagger) = \xi \mathbf{a}^\dagger \mathbf{b}^\dagger$$

Thus, the comparison of this last expression with Eq. (13.165) in which $\mu = 0$ leads, respectively, to

$$A(0) = 0 \qquad \text{and} \qquad B(0) = \xi \quad (13.167)$$

Furthermore, due to Eq. (13.165), it appears that the partial derivative of $\mathbf{G}$ with respect to μ reads

$$\left(\frac{\partial \mathbf{G}(\mu, \mathbf{a}^\dagger, \mathbf{b}^\dagger)}{\partial \mu}\right) = \left(\frac{\partial A(\mu)}{\partial \mu}\right) + \left(\frac{\partial B(\mu)}{\partial \mu}\right)\mathbf{a}^\dagger \mathbf{b}^\dagger \quad (13.168)$$

while, the crossed second-order partial derivative of $\mathbf{G}$ with respect to $\mathbf{a}^\dagger$ and $\mathbf{b}^\dagger$ yields

$$\left(\frac{\partial^2 \mathbf{G}(\mu, \mathbf{a}^\dagger, \mathbf{b}^\dagger)}{\partial \mathbf{b}^\dagger \partial \mathbf{a}^\dagger}\right) = B(\mu) + \{B(\mu)\}^2 \mathbf{a}^\dagger \mathbf{b}^\dagger \quad (13.169)$$

At last, due to Eq. (13.165)

$$\left(\frac{\partial \mathbf{G}(\mu, \mathbf{a}^\dagger, \mathbf{b}^\dagger)}{\partial \mathbf{a}^\dagger}\right) = B(\mu)\mathbf{b}^\dagger \qquad \text{and} \qquad \left(\frac{\partial \mathbf{G}(\mu, \mathbf{a}^\dagger, \mathbf{b}^\dagger)}{\partial \mathbf{b}^\dagger}\right) = B(\mu)\mathbf{a}^\dagger$$

so that, since $\mathbf{a}^\dagger$ and $\mathbf{b}^\dagger$ commute,

$$\left(\frac{\partial \mathbf{G}(\mu, \mathbf{a}^\dagger, \mathbf{b}^\dagger)}{\partial \mathbf{a}^\dagger}\right)\left(\frac{\partial \mathbf{G}(\mu, \mathbf{a}^\dagger, \mathbf{b}^\dagger)}{\partial \mathbf{b}^\dagger}\right) = \{B(\mu)\}^2 \mathbf{a}^\dagger \mathbf{b}^\dagger \quad (13.170)$$

Hence, due to Eqs. (13.168)–(13.170), Eq. (13.164) takes the form

$$\left(\left(\frac{\partial A(\mu)}{\partial \mu}\right) + \left(\frac{\partial B(\mu)}{\partial \mu}\right)\mathbf{a}^\dagger \mathbf{b}^\dagger\right) |(0)\{0\}\rangle = (\{B(\mu)\} + \{B(\mu)\}^2 \mathbf{a}^\dagger \mathbf{b}^\dagger)|(0)\{0\}\rangle$$

so that one obtains by identification

$$\left(\frac{\partial B(\mu)}{\partial \mu}\right) = \{B(\mu)\}^2 \quad (13.171)$$

$$\left(\frac{\partial A(\mu)}{\partial \mu}\right) = B(\mu) \tag{13.172}$$

By integration Eq. (13.171) yields

$$\mu = -\left(\frac{1}{B(\mu)} - \frac{1}{B(0)}\right)$$

or, in view of the boundary condition appearing in (13.167),

$$\mu = -\left(\frac{1}{B(\mu)} - \frac{1}{\xi}\right)$$

and thus, after rearranging,

$$B(\mu) = \frac{\xi}{1 - \xi\mu} \tag{13.173}$$

Next, insert this result into Eq. (13.172) to get

$$\left(\frac{\partial A(\mu)}{\partial \mu}\right) = \frac{\xi}{1 - \xi\mu}$$

which by integration yields

$$A(\mu) - A(0) = \xi \int_0^{\mu} \frac{d\mu'}{1 - \xi\mu'}$$

and thus, due to the first boundary condition of Eq. (13.167), and after calculation of the integral

$$A(\mu) = -\ln(1 - \xi\mu) \tag{13.174}$$

Hence, comparing Eq. (13.173), the function (13.165) becomes

$$\mathbf{G}(\mathbf{a}^\dagger, \mathbf{b}^\dagger, \mu) = -\ln(1 - \xi\mu) + \frac{\xi}{1 - \xi\mu}\mathbf{a}^\dagger\mathbf{b}^\dagger$$

so that Eq. (13.156) is

$$(e^{\mathbf{G}(\mu, \mathbf{a}^\dagger, \mathbf{b}^\dagger)})|(0)\{0\}\rangle = \exp\left\{\frac{\xi\mathbf{a}^\dagger\mathbf{b}^\dagger}{1 - \xi\mu} - \ln(1 - \xi\mu)\right\}|(0)\{0\}\rangle$$

or

$$(e^{\mathbf{G}(\mu, \mathbf{a}^\dagger, \mathbf{b}^\dagger)})|(0)\{0\}\rangle = \exp\left\{\frac{\xi\mathbf{a}^\dagger\mathbf{b}^\dagger}{1 - \xi\mu}\right\}\left(\frac{1}{(1 - \xi\mu)}\right)|(0)\{0\}\rangle$$

Now, observe that, due to Eq. (13.165), the left-hand side of this last equation is the same as that of (13.156), so that of the identification of the corresponding right-hand sides gives

$$(e^{\mu\mathbf{ba}})\{e^{\xi\mathbf{a}^\dagger\mathbf{b}^\dagger}\}|(0)\{0\}\rangle = \left(\frac{1}{(1 - \xi\mu)}\right)\exp\left\{\frac{\xi\mathbf{a}^\dagger\mathbf{b}^\dagger}{1 - \xi\mu}\right\}|(0)\{0\}\rangle \tag{13.175}$$

At last, coming back from ξ to λ by the aid of Eq. (13.155) leading to

$$\left(\frac{1}{(1 - \xi\mu)}\right) = \frac{1}{1 - e^{-\lambda}}$$

Eq. (13.175) transforms to

$$(e^{\mu\mathbf{ba}})\{e^{\mathbf{a}^\dagger\mathbf{b}^\dagger e^{-\lambda}/\mu}\}|(0)\{0\}\rangle = \frac{1}{(1-e^{-\lambda})}\{e^{\xi\mathbf{a}^\dagger\mathbf{b}^\dagger/(1-\xi\mu)}\}|(0)\{0\}\rangle \tag{13.176}$$

13.4.1.3 Final step for the thermal average value As a consequence of Eq. (13.176), the thermal average value (13.154) becomes

$$\langle\mathbf{F}(\mathbf{a}^\dagger,\mathbf{a})\rangle = \langle\{0\}(0)|\{\mathbf{F}((\mathbf{a}^\dagger+\mu\mathbf{b})/k, k\mathbf{a})\}(e^{\mathbf{a}^\dagger\mathbf{b}^\dagger(\xi/(1-\xi\mu))})|(0)\{0\}\rangle \tag{13.177}$$

Again, observe that, owing to Eq. (13.155), it yields

$$\frac{\xi}{1-\xi\mu} = \frac{1}{\mu}\left(\frac{e^{-\lambda}}{1-e^{-\lambda}}\right) \tag{13.178}$$

Hence, due to Eqs. (13.14) and (13.36) we have

$$\langle\mathbf{n}\rangle = \frac{1}{e^{\lambda}-1} = \frac{e^{-\lambda}}{1-e^{-\lambda}} \quad \text{with} \quad \lambda = \frac{\hbar\omega}{k_B T} \tag{13.179}$$

where $\langle\mathbf{n}\rangle$ is the thermal average of the occupation number, that is, of $\mathbf{a}^\dagger\mathbf{a}$ or of $\mathbf{b}^\dagger\mathbf{b}$, that is,

$$\langle\mathbf{n}\rangle = (1-e^{-\lambda})\mathbf{tr}\{(e^{-\lambda\mathbf{a}^\dagger\mathbf{a}})\mathbf{a}^\dagger\mathbf{a}\} = (1-e^{-\lambda})\mathbf{tr}\{(e^{-\lambda\mathbf{b}^\dagger\mathbf{b}})\mathbf{b}^\dagger\mathbf{b}\}$$

Equation (13.178) reads

$$\frac{\xi}{1-\xi\mu} = \frac{\langle\mathbf{n}\rangle}{\mu} \tag{13.180}$$

so that the thermal average (13.177) yields

$$\langle\mathbf{F}(\mathbf{a}^\dagger,\mathbf{a})\rangle = \langle\{0\}(0)\{|\mathbf{F}((\mathbf{a}^\dagger+\mu\mathbf{b})/k, k\mathbf{a})\}(e^{\langle\mathbf{n}\rangle\mathbf{a}^\dagger\mathbf{b}^\dagger/\mu})|(0)\{0\}\rangle \tag{13.181}$$

Now, observe that

$$\langle\{0\}(0)| = \langle\{0\}(0)|(e^{-\langle\mathbf{n}\rangle\mathbf{a}^\dagger\mathbf{b}^\dagger/\mu}) \tag{13.182}$$

that is because, after its expansion, the right-hand side reads

$$\langle\{0\}(0)|(e^{-\langle\mathbf{n}\rangle\mathbf{a}^\dagger\mathbf{b}^\dagger/\mu}) = \sum_m \langle\{0\}(0)|\left(\frac{\langle\mathbf{n}\rangle}{\mu}\right)^m \frac{(\mathbf{a}^\dagger)^m(\mathbf{b}^\dagger)^m(-1)^m}{m!}$$

or, after action of each operator within its own subspace,

$$\langle\{0\}(0)|(e^{-\langle\mathbf{n}\rangle\mathbf{a}^\dagger\mathbf{b}^\dagger/\mu}) = \sum_m (-1)^m\left(\frac{\langle\mathbf{n}\rangle}{\mu}\right)^m \frac{1}{m!}\langle\{0\}|(\mathbf{b}^\dagger)^m\langle(0)|(\mathbf{a})^m$$

then, using Eq. (5.73) leading to

$$\langle\{0\}|(\mathbf{b}^\dagger)^m = \langle(0)|(\mathbf{a}^\dagger)^m = \delta_{m,0}$$

it appears, Q.E.D.

$$\langle\{0\}(0)|(e^{-\langle\mathbf{n}\rangle\mathbf{a}^\dagger\mathbf{b}^\dagger/\mu}) = \langle\{0\}(0)|1$$

Therefore, Eq. (13.181) becomes

$$\langle \mathbf{F}(\mathbf{a}^\dagger, \mathbf{a}) \rangle = \langle \{0\}(0)|(e^{-\langle \mathbf{n} \rangle \mathbf{a}^\dagger \mathbf{b}^\dagger/\mu})\{\mathbf{F}((\mathbf{a}^\dagger + \mu \mathbf{b})/k, k\mathbf{a})\}(e^{\langle \mathbf{n} \rangle \mathbf{a}^\dagger \mathbf{b}^\dagger/\mu})|(0)\{0\}\rangle \tag{13.183}$$

Moreover, keeping in mind theorem (1.77) applying to some function $\mathbf{F}(\mathbf{B})$ of operator $\mathbf{B}$, that is,

$$e^{\xi \mathbf{A}}\, \mathbf{F}(\mathbf{B}) e^{-\xi \mathbf{A}} = \mathbf{F}(e^{\xi \mathbf{A}} \mathbf{B} e^{-\xi \mathbf{A}})$$

where ξ is a c-number and $\mathbf{A}$ is an operator that does not commute with $\mathbf{B}$, apply it to the canonical transformation appearing on the right-hand side of Eq. (13.183) by taking

$$\mathbf{A} = \mathbf{a}^\dagger \mathbf{b}^\dagger \qquad \xi = \frac{\langle \mathbf{n} \rangle}{\mu}$$

and

$$\mathbf{B} = \frac{\mathbf{a}^\dagger + \mu \mathbf{b}}{k} \qquad \text{or} \qquad \mathbf{B} = k\mathbf{a}$$

Then, this canonical transformation reads

$$\begin{aligned}&(e^{-\langle \mathbf{n} \rangle \mathbf{a}^\dagger \mathbf{b}^\dagger/\mu})\{\mathbf{F}((\mathbf{a}^\dagger + \mu \mathbf{b})/k, k\mathbf{a})\}(e^{\langle \mathbf{n} \rangle \mathbf{a}^\dagger \mathbf{b}^\dagger/\mu}) \\ &\quad = \mathbf{F}\left((e^{-\langle \mathbf{n} \rangle \mathbf{a}^\dagger \mathbf{b}^\dagger/\mu})\left(\frac{\mathbf{a}^\dagger + \mu \mathbf{b}}{k}\right)(e^{\langle \mathbf{n} \rangle \mathbf{a}^\dagger \mathbf{b}^\dagger/\mu}), k(e^{-\langle \mathbf{n} \rangle \mathbf{a}^\dagger \mathbf{b}^\dagger/\mu})\mathbf{a}(e^{\langle \mathbf{n} \rangle \mathbf{a}^\dagger \mathbf{b}^\dagger/\mu})\right)\end{aligned} \tag{13.184}$$

Besides, since $\mathbf{a}^\dagger$ commutes with $\mathbf{b}^\dagger$, it is clear that

$$(e^{-\langle \mathbf{n} \rangle \mathbf{a}^\dagger \mathbf{b}^\dagger/\mu})\, \mathbf{a}^\dagger (e^{\langle \mathbf{n} \rangle \mathbf{a}^\dagger \mathbf{b}^\dagger/\mu}) = \mathbf{a}^\dagger$$

Moreover, applying theorem (7.7), that is,

$$e^{-\xi \mathbf{a}^\dagger} \mathbf{F}(\mathbf{a}, \mathbf{a}^\dagger) e^{\xi \mathbf{a}^\dagger} = \mathbf{F}(\mathbf{a} + \xi, \mathbf{a}^\dagger)$$

with taking

$$\xi = \frac{\langle \mathbf{n} \rangle}{\mu}\mathbf{a}^\dagger \qquad \text{or} \qquad \xi = \frac{\langle \mathbf{n} \rangle}{\mu}\mathbf{b}^\dagger$$

and keeping in mind the following commutators

$$[\mathbf{a}, \mathbf{b}^\dagger] = [\mathbf{a}^\dagger, \mathbf{b}^\dagger] = [\mathbf{a}^\dagger, \mathbf{b}] = [\mathbf{a}, \mathbf{b}] = 0$$

one finds, respectively,

$$(e^{-\langle \mathbf{n} \rangle \mathbf{a}^\dagger \mathbf{b}^\dagger/\mu})\mathbf{b}(e^{\langle \mathbf{n} \rangle \mathbf{a}^\dagger \mathbf{b}^\dagger/\mu}) = \mathbf{b} + \frac{\langle \mathbf{n} \rangle}{\mu}\mathbf{a}^\dagger \tag{13.185}$$

$$(e^{-\langle \mathbf{n} \rangle \mathbf{a}^\dagger \mathbf{b}^\dagger/\mu})\mathbf{a}(e^{\langle \mathbf{n} \rangle \mathbf{a}^\dagger \mathbf{b}^\dagger/\mu}) = \mathbf{a} + \frac{\langle \mathbf{n} \rangle}{\mu}\mathbf{b}^\dagger \tag{13.186}$$

As a consequence of Eqs. (13.185) and (13.186), the canonical transformation (13.184) becomes

$$(e^{\langle\mathbf{n}\rangle\mathbf{ab}/\mu})\{\mathbf{F}((\mathbf{a}^\dagger + \mu\mathbf{b})/k, k\mathbf{a})\}(e^{\langle\mathbf{n}\rangle\mathbf{a}^\dagger\mathbf{b}^\dagger/\mu})$$
$$= \left\{\mathbf{F}\left(\left(\frac{(1+\langle\mathbf{n}\rangle)}{k}\mathbf{a}^\dagger + \frac{\mu}{k}\mathbf{b}\right), \left(k\mathbf{a}+k\frac{\langle\mathbf{n}\rangle}{\mu}\mathbf{b}^\dagger\right)\right)\right\} \tag{13.187}$$

It is now necessary to find the expressions for the unknown scalars μ and k involved in this equation. For this purpose, we may write Eq. (13.187) in terms of two Boson operators $\mathbf{c}$ and $\mathbf{c}^\dagger$, which are linear combinations of $\mathbf{a}$ and $\mathbf{a}^\dagger$ and $\mathbf{b}$ and $\mathbf{b}^\dagger$ according to

$$(e^{\langle\mathbf{n}\rangle\mathbf{ab}/\mu})\{\mathbf{F}((\mathbf{a}^\dagger + \mu\mathbf{b})/k, k\mathbf{a})\}(e^{\langle\mathbf{n}\rangle\mathbf{a}^\dagger\mathbf{b}^\dagger/\mu}) = \{\mathbf{F}(\mathbf{c}^\dagger, \mathbf{c})\} \tag{13.188}$$

with

$$\mathbf{c} = C_1\mathbf{a}+C_2\mathbf{b}^\dagger \qquad \text{and} \qquad \mathbf{c}^\dagger = C_1^*\mathbf{a}^\dagger + C_2^*\mathbf{b} \tag{13.189}$$

so that after identification with the right-hand side of Eq. (13.187), one obtains, respectively, for the coefficients C_1 and C_2 of Eq. (13.189)

$$C_1 = k \qquad \text{and} \qquad C_2 = k\frac{\langle\mathbf{n}\rangle}{\mu} \tag{13.190}$$

$$C_1^* = \frac{(1+\langle\mathbf{n}\rangle)}{k} \qquad \text{and} \qquad C_2^* = \frac{\mu}{k} \tag{13.191}$$

Then, since the scalars k, μ, and $\langle\mathbf{n}\rangle$ appearing in Eqs. (13.190) and (13.191) are real, we have

$$C_1 = C_1^* \qquad \text{and} \qquad C_2 = C_2^*$$

so that Eqs. (13.190) and (13.191) read

$$k = \frac{(1+\langle\mathbf{n}\rangle)}{k} \qquad \text{and} \qquad k\frac{\langle\mathbf{n}\rangle}{\mu} = \frac{\mu}{k}$$

leading to

$$k = \sqrt{1+\langle\mathbf{n}\rangle} \qquad \text{and} \qquad \mu = k\sqrt{\langle\mathbf{n}\rangle}$$

Furthermore, introducing these expressions for k and μ into Eq. (13.187), we have

$$(e^{\langle\mathbf{n}\rangle\mathbf{ab}/\mu})\{\mathbf{F}((\mathbf{a}^\dagger + \mu\mathbf{b})/k, k\mathbf{a})\}(e^{\langle\mathbf{n}\rangle\mathbf{a}^\dagger\mathbf{b}^\dagger/\mu})$$
$$= \mathbf{F}\left(\sqrt{1+\langle\mathbf{n}\rangle}\mathbf{a}^\dagger + \sqrt{\langle\mathbf{n}\rangle}\mathbf{b}, \sqrt{1+\langle\mathbf{n}\rangle}\mathbf{a} + \sqrt{\langle\mathbf{n}\rangle}\mathbf{b}^\dagger\right)$$

Hence, using this result allows one to transform the thermal average (13.183) into

$$\langle\mathbf{F}(\mathbf{a}^\dagger, \mathbf{a})\rangle = \langle\{0\}(0)|\{\mathbf{F}(\sqrt{1+\langle\mathbf{n}\rangle}\mathbf{a}^\dagger + \sqrt{\langle\mathbf{n}\rangle}\mathbf{b}, \sqrt{1+\langle\mathbf{n}\rangle}\mathbf{a} + \sqrt{\langle\mathbf{n}\rangle}\mathbf{b}^\dagger)\}|(0)\{0\}\rangle$$

which due to the definition of $\langle \mathbf{F}(\mathbf{a}^\dagger, \mathbf{a})\rangle$ given by Eq. (13.142) reads

$$(1 - e^{-\lambda})\mathbf{tr}\{(e^{-\lambda \mathbf{a}^\dagger \mathbf{a}})\{\mathbf{F}(\mathbf{a}^\dagger, \mathbf{a})\}\}$$
$$= \langle\{0\}(0)|\{\mathbf{F}(\sqrt{1+\langle\mathbf{n}\rangle}\mathbf{a}^\dagger + \sqrt{\langle\mathbf{n}\rangle}\mathbf{b}, \sqrt{1+\langle\mathbf{n}\rangle}\mathbf{a} + \sqrt{\langle\mathbf{n}\rangle}\mathbf{b}^\dagger)\}|(0)\{0\}\rangle \tag{13.192}$$

13.4.2 Thermal average of translation operators and Bloch theorem

Now, suppose that the operator function to be averaged and given by Eq. (13.192) is a translation operator, that is,

$$\langle \mathbf{A}(T)\rangle = (1 - e^{-\lambda})\mathbf{tr}\{(e^{-\lambda \mathbf{a}^\dagger \mathbf{a}})\{\mathbf{A}(\mathbf{a}^\dagger, \mathbf{a})\}\} \tag{13.193}$$

with

$$\mathbf{A}(\mathbf{a}^\dagger, \mathbf{a}) = \{e^{\alpha \mathbf{a}^\dagger - \alpha^* \mathbf{a}}\}$$

and thus

$$\langle \mathbf{A}(T)\rangle = (1 - e^{-\lambda})\mathbf{tr}\{(e^{-\lambda \mathbf{a}^\dagger \mathbf{a}})\{e^{\alpha \mathbf{a}^\dagger - \alpha^* \mathbf{a}}\}\} \tag{13.194}$$

Then, using Glauber's theorem (1.79), in order to factorize the right-hand-side exponential operators

$$\{e^{\alpha \mathbf{a}^\dagger - \alpha^* \mathbf{a}}\} = (e^{\alpha \mathbf{a}^\dagger})(e^{-\alpha^* \mathbf{a}})e^{-[\alpha \mathbf{a}^\dagger, -\alpha^* \mathbf{a}]/2} \tag{13.195}$$

with

$$e^{-[\alpha \mathbf{a}^\dagger, -\alpha^* \mathbf{a}]/2} = e^{|\alpha|^2[\mathbf{a}^\dagger, \mathbf{a}]/2} = (e^{-|\alpha|^2/2}) \tag{13.196}$$

using Eqs. (13.195) and (13.196), the thermal average (13.194) becomes

$$\langle \mathbf{A}(T)\rangle = (1 - e^{-\lambda})e^{-|\alpha|^2/2}\mathbf{tr}\{(e^{-\lambda \mathbf{a}^\dagger \mathbf{a}})(e^{\alpha \mathbf{a}^\dagger})(e^{-\alpha^* \mathbf{a}})\} \tag{13.197}$$

Now, apply theorem (13.192) to Eq. (13.197) in order to find the expression for its thermal average. Then, ignoring momentously the phase factor $e^{-|\alpha|^2/2}$, we have

$$(1 - e^{-\lambda})\mathbf{tr}\{(e^{-\lambda \mathbf{a}^\dagger \mathbf{a}})(e^{\alpha \mathbf{a}^\dagger})(e^{-\alpha^* \mathbf{a}})\}$$
$$= \langle(0)|\langle\{0\}|(e^{\alpha(\sqrt{\langle\mathbf{n}\rangle}\mathbf{a}^\dagger + \sqrt{1+\langle\mathbf{n}\rangle}\mathbf{b})})(e^{-\alpha^*(\sqrt{\langle\mathbf{n}\rangle}\mathbf{a} + \sqrt{1+\langle\mathbf{n}\rangle}\mathbf{b}^\dagger)})|\{0\}_2\rangle|\{0\}_1\rangle$$

Then, factorizing both exponentials, each involving commuting operators, gives

$$(1 - e^{-\lambda})\mathbf{tr}\{(e^{-\lambda \mathbf{a}^\dagger \mathbf{a}})(e^{\alpha \mathbf{a}^\dagger})(e^{-\alpha^* \mathbf{a}})\}$$
$$= \langle(0)|\langle\{0\}|(e^{\alpha\sqrt{\langle\mathbf{n}\rangle}\mathbf{a}^\dagger})(e^{\alpha\sqrt{1+\langle\mathbf{n}\rangle}\mathbf{b}})(e^{-\alpha^*\sqrt{\langle\mathbf{n}\rangle}\mathbf{a}})(e^{-\alpha^*\sqrt{1+\langle\mathbf{n}\rangle}\mathbf{b}^\dagger})|\{0\}\rangle|(0)\rangle$$

Again, working within the two different subspaces leads to

$$(1 - e^{-\lambda})\mathbf{tr}\{(e^{-\lambda \mathbf{a}^\dagger \mathbf{a}})(e^{\alpha(t)\mathbf{a}^\dagger})(e^{-\alpha^*(t)\mathbf{a}})\}$$
$$= \langle(0)|\{(e^{\alpha\sqrt{\langle\mathbf{n}\rangle}\mathbf{a}^\dagger})(e^{-\alpha^*\sqrt{\langle\mathbf{n}\rangle}\mathbf{a}})\}|(0)\rangle\langle\{0\}|\{(e^{\alpha\sqrt{1+\langle\mathbf{n}\rangle}\mathbf{b}})(e^{-\alpha^*\sqrt{1+\langle\mathbf{n}\rangle}\mathbf{b}^\dagger})\}|\{0\}\rangle \tag{13.198}$$

Next, expand the two exponentials of the last right-hand-side matrix element of this last equation to get

$$\langle\{0\}|\{(e^{\alpha\sqrt{1+\langle\mathbf{n}\rangle}\mathbf{b}})(e^{-\alpha^*\sqrt{1+\langle\mathbf{n}\rangle}\mathbf{b}^\dagger})\}|\{0\}\rangle$$

$$= \sum_k \sum_l (-1)^l \frac{\alpha^k \alpha^{*l}}{k!l!} (\sqrt{1+\langle\mathbf{n}\rangle})^{k+l} \langle(0)|\{(\mathbf{b})^k (\mathbf{b}^\dagger)^l\}|\{0\}\rangle \quad (13.199)$$

Then, comparing Eqs. (5.67) and (5.68), that is,

$$\langle\{0\}|(\mathbf{b})^k = \langle\{k\}|\sqrt{k!} \qquad \text{and} \qquad (\mathbf{b}^\dagger)^l\ |\{0\}\rangle = \sqrt{l!}|\{l\}\rangle$$

it appears that

$$\langle\{0\}|(\mathbf{b})^k (\mathbf{b}^\dagger)^l |(0)\rangle = \sqrt{k!}\sqrt{l!}\langle\{k\}|\{l\}\rangle = l!$$

so that the double sum of the matrix elements involved on the right-hand side of Eq. (13.199) reduces after simplifications to

$$\langle\{0\}|\{(e^{\alpha\sqrt{1+\langle\mathbf{n}\rangle}\mathbf{b}})(e^{-\alpha^*\sqrt{1+\langle\mathbf{n}\rangle}\mathbf{b}^\dagger})\}|\{0\}\rangle = \sum_l (-1)^l \frac{|\alpha|^{2l}}{l!} (\langle\mathbf{n}\rangle + 1)^l$$

Therefore, coming back to the exponentials, this last equation becomes

$$\langle\{0\}|\{(e^{\alpha\sqrt{1+\langle\mathbf{n}\rangle}\mathbf{b}})(e^{-\alpha^*\sqrt{1+\langle\mathbf{n}\rangle}\mathbf{b}^\dagger})\}|\{0\}\rangle = \exp\{-|\alpha|^2(\langle\mathbf{n}\rangle + 1)\} \quad (13.200)$$

Now, expand the exponential of the first matrix element of the right-hand side of Eq. (13.198), that is,

$$\langle(0)|\{(e^{\alpha\sqrt{\langle\mathbf{n}\rangle}\mathbf{a}^\dagger})(e^{-\alpha^*\sqrt{\langle\mathbf{n}\rangle}\mathbf{a}})\}|(0)\rangle$$

$$= \sum_k \sum_l (-1)^l \frac{\alpha^k \alpha^{*l}}{k!l!} (\sqrt{\langle\mathbf{n}\rangle})^{k+l} \langle(0)|\{(\mathbf{a}^\dagger)^k (\mathbf{a})^l\}|(0)\rangle \quad (13.201)$$

Then, owing to Eq. (5.55), we have

$$\langle(0)|(\mathbf{a}^\dagger)^k = 0 \quad \text{except if } k = 0$$

$$(\mathbf{a})^l|(0)\rangle = 0 \quad \text{except if } l = 0$$

This follows that Eq. (13.201) reduces to

$$\langle(0)|\{(e^{\alpha\sqrt{\langle\mathbf{n}\rangle}\mathbf{a}^\dagger})(e^{-\alpha^*\sqrt{\langle\mathbf{n}\rangle}\mathbf{a}})\}|(0)\rangle = 1 \quad (13.202)$$

As a consequence of Eqs. (13.200) and (13.202), the thermal average (13.198) becomes

$$(1 - e^{-\lambda})\mathbf{tr}\{(e^{-\lambda\mathbf{a}^\dagger\mathbf{a}})(e^{\alpha\mathbf{a}^\dagger})(e^{-\alpha^*\mathbf{a}})\} = \exp\{-|\alpha|^2(\langle\mathbf{n}\rangle + 1)\} \quad (13.203)$$

with $\langle\mathbf{n}\rangle$ given by Eq. (13.179). Again, using Glauber's theorem yields

$$(1 - e^{-\lambda})e^{-|\alpha|^2/2}\mathbf{tr}\{(e^{-\lambda\mathbf{a}^\dagger\mathbf{a}})(e^{\alpha\mathbf{a}^\dagger - \alpha^*\mathbf{a}})\}e^{|\alpha|^2} = \exp\{-|\alpha|^2(\langle\mathbf{n}\rangle + 1)\}$$

so that after simplification

$$\boxed{\langle\mathbf{A}(T)\rangle = (1 - e^{-\lambda})\mathbf{tr}\{(e^{-\lambda\mathbf{a}^\dagger\mathbf{a}})(e^{\alpha\mathbf{a}^\dagger - \alpha^*\mathbf{a}})\} = \exp\left\{-|\alpha|^2\left(\langle\mathbf{n}\rangle + \tfrac{1}{2}\right)\right\}} \quad (13.204)$$

13.4.2.1 Bloch theorem Of course, one would obtain in a similar way as for Eq. (13.203)

$$(1 - e^{-\lambda})\mathbf{tr}\{(e^{-\lambda \mathbf{a}^\dagger \mathbf{a}})(e^{\alpha \mathbf{a}^\dagger + \alpha^* \mathbf{a}})\} = \exp\left\{|\alpha|^2 \left(\langle \mathbf{n} \rangle + \tfrac{1}{2}\right)\right\} \tag{13.205}$$

Next, if we denote and keep in mind Eqs. (5.6) allowing one to pass from the Boson operators to the position operator $\mathbf{Q}$ according to

$$\mathbf{Q} = \alpha(\mathbf{a}^\dagger + \mathbf{a}) \quad \text{with} \quad \alpha = \sqrt{\frac{\hbar}{2m\omega}}$$

it appears that if $\alpha(t)$ is real, the left-hand side of Eq. (13.205) reads

$$(1 - e^{-\lambda})\mathbf{tr}\{(e^{-\lambda \mathbf{a}^\dagger \mathbf{a}})(e^{\alpha(\mathbf{a}^\dagger + \mathbf{a})})\} = \mathbf{tr}\{\rho_B e^{\mathbf{Q}}\} = \langle e^{\mathbf{Q}} \rangle \tag{13.206}$$

so that Eq. (13.205) yields

$$\langle e^{\mathbf{Q}} \rangle = \exp\left\{\frac{\hbar}{2m\omega}(\langle \mathbf{n} \rangle + \tfrac{1}{2})\right\} \tag{13.207}$$

Now, observe that the thermal average of $\mathbf{Q}(T)^2$ defined by

$$\langle \mathbf{Q}(T)^2 \rangle = \mathbf{tr}\{\rho_B \mathbf{Q}(T)^2\}$$

that is,

$$\langle \mathbf{Q}(T)^2 \rangle = \frac{\hbar}{2m\omega}(1 - e^{-\lambda})\mathbf{tr}\{(e^{-\lambda \mathbf{a}^\dagger \mathbf{a}})(\mathbf{a}^\dagger + \mathbf{a})^2\}$$

is given by Eq. (13.72), that is,

$$\boxed{\langle \mathbf{Q}(T)^2 \rangle = \frac{\hbar}{2m\omega}(2\langle \mathbf{n} \rangle + 1)} \tag{13.208}$$

As a consequence of Eqs. (13.207) and (13.208),

$$\boxed{\langle e^{\mathbf{Q}} \rangle = e^{\langle \mathbf{Q}^2 \rangle /2}}$$

This last result is the *Bloch theorem.* In a similar way, one would obtain for the momentum

$$\boxed{\langle e^{\mathbf{P}} \rangle = e^{\langle \mathbf{P}^2 \rangle /2}}$$

13.5 CONCLUSION

Using the canonical operator, it was possible in this chapter to find many thermal properties of quantum harmonic oscillators such as the fundamental Planck law, the thermal average of kinetic and potential energies, the heat capacities, the energy fluctuations, and the part of the *Sackur and Tetrode law* dealing with entropy. Finally, we gave some complex demonstrations of the thermal average energy of ladder operator functions, one consequence of which is the Bloch theorem. The most important results dealing with thermal average of simple operators characterizing harmonic oscillators are reported as follows:

Thermal average over Boltzmann density operators
Boltzmann density operators:
$\rho_B = \frac{1}{Z}(e^{-\beta \mathbf{H}}) \quad \text{with} \quad \beta = \frac{1}{k_B T}$
Partition function:
$Z = \left(\frac{e^{-\beta\hbar\omega/2}}{1 - e^{-\beta\hbar\omega}} \right)$
Average Hamiltonian:
$\langle \mathbf{H} \rangle = \left(\frac{\hbar\omega}{2} + \frac{\hbar\omega}{e^{\hbar\omega/k_B T} - 1} \right)$
Heat capacity:
$C_v = N k_B \left(\frac{\hbar\omega}{k_B T} \right)^2 \frac{e^{\hbar\omega/k_B T}}{(e^{\hbar\omega/k_B T} - 1)^2}$
Energy fluctuation:
$\Delta E_{\text{Tot}} = \sqrt{N}\hbar\omega \frac{e^{\hbar\omega/2k_B T}}{(e^{\hbar\omega/k_B T} - 1)}$
Average of $\mathbf{Q}^2$:
$\langle \mathbf{Q}(T)^2 \rangle = \frac{\hbar}{2m\omega} \coth\left(\frac{\hbar\omega}{2k_B T} \right)$
Entropy:
$S = n \left\{ \frac{R}{(nN^\circ)!} \ln\left\{ \frac{(e^{\hbar\omega/2k_B T})}{1 - e^{\hbar\omega/k_B T}} \right\} + N^\circ \frac{\hbar\omega}{T} \left(\frac{1}{e^{\hbar\omega/k_B T} - 1} + \frac{1}{2} \right) \right\}$

whereas we give hereafter some important theorems dealing with the thermal average of exponential operators involving the ladder operators:

Theorems dealing with thermal averages
Thermal average of operators over Boltzmann density operator:
$(1-e^{-\lambda})\mathbf{tr}\{(e^{-\lambda \mathbf{a}^\dagger \mathbf{a}})\{\mathbf{F}(\mathbf{a}^\dagger, \mathbf{a})\}\}$ $= \langle\{0\}(0)\|\{\mathbf{F}(\sqrt{1+\langle \mathbf{n}\rangle}\mathbf{a}^\dagger + \sqrt{\langle \mathbf{n}\rangle}\mathbf{b}, \sqrt{1+\langle \mathbf{n}\rangle}\mathbf{a} + \sqrt{\langle \mathbf{n}\rangle}\mathbf{b}^\dagger)\}\|(0)\{0\}\rangle$
Thermal average of the translation operator:
$(1-e^{-\lambda})\mathbf{tr}\{(e^{-\lambda \mathbf{a}^\dagger \mathbf{a}})(e^{\alpha \mathbf{a}^\dagger - \alpha^* \mathbf{a}})\} = \exp\{-\|\alpha\|^2(\langle \mathbf{n}\rangle + \frac{1}{2})\}$
Bloch's theorem:
$(1-e^{-\lambda})\mathbf{tr}\{(e^{-\lambda \mathbf{a}^\dagger \mathbf{a}})e^{\mathbf{Q}}\} = \exp\{(1-e^{-\lambda})\mathbf{tr}\{(e^{-\lambda \mathbf{a}^\dagger \mathbf{a}})\mathbf{Q}^2/2\}\}$ $(1-e^{-\lambda})\mathbf{tr}\{(e^{-\lambda \mathbf{a}^\dagger \mathbf{a}})e^{\mathbf{P}}\} = \exp\{(1-e^{-\lambda})\mathbf{tr}\{(e^{-\lambda \mathbf{a}^\dagger \mathbf{a}})\mathbf{P}^2/2\}\}$

BIBLIOGRAPHY

B. Diu, C. Guthmann, D. Lederer, and B. Roulet. *Physique statistique*. Hermann: Paris, 1988.
Ch. Kittel and H. Kroemer. *Thermal Physics*, 2nd ed. W. H. Freeman: New York, 1980.
H. Louisell. *Quantum Statistical Properties of Radiations*. Wiley: New York, 1973.
F. Reif. *Fundamentals of Statistical and Thermal Physics*. McGraw-Hill: New York, 1965.

PART V

QUANTUM NORMAL MODES OF VIBRATION

Part IV was essentially devoted to large sets of weakly coupled harmonic oscillators, allowing one to obtain many thermal properties. However, other kinds of large sets of oscillators exist involving couplings that allow one to separate them so as to get decoupled harmonic oscillators, that is, the normal modes of the oscillator system. The aim of Part V is to treat normal modes.

CHAPTER 14

QUANTUM ELECTROMAGNETIC MODES

INTRODUCTION

The first chapter of this part, Chapter 14, will treat the quantum modes of the electromagnetic field and the last one the normal modes of molecular systems of 1D solids. The purpose of the present chapter is to study the quantum electromagnetic modes. First, we shall see how to get the classical electromagnetic modes in reciprocal space. Second, we shall show how to pass from these classical modes to the corresponding quantum ones. Then, applying some results encountered in the previous chapters dealing with the properties of quantum harmonic oscillators, it will be possible to introduce the notion of light corpuscles of a given angular frequency, called *photons*, which are the excitation degrees of the normal modes. Besides, applying the thermal properties of quantum oscillators we have obtained previously, it will be also possible to get different important results dealing with the thermal properties of light such as, for instance, the *Planck black-body radiation law* or the *Stefan–Boltzmann law.*

14.1 MAXWELL EQUATIONS

14.1.1 Maxwell equations within the geometrical space

We start from the Maxwell equations governing the electric field $\mathbf{E}(\mathbf{r}, t)$ and the magnetic field $\mathbf{B}(\mathbf{r}, t)$, that is,

$$\nabla \cdot \mathbf{E}(\mathbf{r}, t) = \frac{1}{\varepsilon^{\circ}} \rho(\mathbf{r}, t) \tag{14.1}$$

$$\nabla \cdot \mathbf{B}(\mathbf{r}, t) = 0 \tag{14.2}$$

$$\nabla \times \mathbf{E}(\mathbf{r}, t) = -\left(\frac{\partial \mathbf{B}(\mathbf{r}, t)}{\partial t}\right) \tag{14.3}$$

Quantum Oscillators, First Edition. Olivier Henri-Rousseau and Paul Blaise.
© 2011 John Wiley & Sons, Inc. Published 2011 by John Wiley & Sons, Inc.

$$\nabla \times \mathbf{B}(\mathbf{r}, t) = \frac{1}{c^2}\left(\frac{\partial \mathbf{E}(\mathbf{r}, t)}{\partial t}\right) + \frac{1}{\varepsilon^\circ c^2}\mathbf{J}(\mathbf{r}, t) \tag{14.4}$$

where $\mathbf{r}$ is the position vector, t the time, ε° the vacuum permittivity, c the velocity of light, $\rho(\mathbf{r}, t)$ the charge density at position $\mathbf{r}$ and time t, and $\mathbf{J}(\mathbf{r}, t)$ the current density related to $\rho(\mathbf{r}, t)$ through the charge conservation law:

$$\frac{\partial \rho(\mathbf{r}, t)}{\partial t} + \nabla \cdot \mathbf{J}(\mathbf{r}, t) = 0 \tag{14.5}$$

In the absence of charge

$$\rho(\mathbf{r}, t) = \mathbf{J}(\mathbf{r}, t) = 0$$

so that the four Maxwell equations governing the electric and magnetic fields reduce to

$$\nabla \cdot \mathbf{E}(\mathbf{r}, t) = 0 \tag{14.6}$$

$$\nabla \cdot \mathbf{B}(\mathbf{r}, t) = 0 \tag{14.7}$$

$$\nabla \times \mathbf{E}(\mathbf{r}, t) = -\left(\frac{\partial \mathbf{B}(\mathbf{r}, t)}{\partial t}\right) \tag{14.8}$$

$$\nabla \times \mathbf{B}(\mathbf{r}, t) = \frac{1}{c^2}\left(\frac{\partial \mathbf{E}(\mathbf{r}, t)}{\partial t}\right) \tag{14.9}$$

Now, the scalar and vector potentials $V(\mathbf{r}, t)$and $\mathbf{A}(\mathbf{r}, t)$ may be defined from the electric and magnetic fields via

$$\mathbf{B}(\mathbf{r}, t) = \nabla \times \mathbf{A}(\mathbf{r}, t) \tag{14.10}$$

and

$$\mathbf{E}(\mathbf{r}, t) = -\left(\frac{\partial \mathbf{A}(\mathbf{r}, t)}{\partial t}\right) - \nabla V(\mathbf{r}, t) \tag{14.11}$$

In the Coulomb gauge $V(\mathbf{r}, t)$ and $\mathbf{A}(\mathbf{r}, t)$ are chosen in such a way as

$$\nabla \cdot \mathbf{A}(\mathbf{r}, t) = 0 \qquad \text{and} \qquad \nabla V(\mathbf{r}, t) = 0 \tag{14.12}$$

so that in this gauge Eq. (14.11) simplifies to

$$\mathbf{E}(\mathbf{r}, t) = -\left(\frac{\partial \mathbf{A}(\mathbf{r}, t)}{\partial t}\right) \tag{14.13}$$

14.1.2 Maxwell equations within reciprocal space

We made Fourier transforms allowing one to pass from geometric to reciprocal spaces:

$$\mathbf{B}(\mathbf{k}, t) = \left(\frac{1}{2\pi}\right)^{3/2} \int \mathbf{B}(\mathbf{r}, t) e^{-i\mathbf{k}\cdot\mathbf{r}} d^3\mathbf{r} \tag{14.14}$$

$$\mathbf{E}(\mathbf{k}, t) = \left(\frac{1}{2\pi}\right)^{3/2} \int \mathbf{E}(\mathbf{r}, t) e^{-i\mathbf{k}\cdot\mathbf{r}} d^3\mathbf{r} \tag{14.15}$$

$$\mathbf{A}(\mathbf{k}, t) = \left(\frac{1}{2\pi}\right)^{3/2} \int \mathbf{A}(\mathbf{r}, t) e^{-i\mathbf{k}\cdot\mathbf{r}} d^3\mathbf{r} \tag{14.16}$$

According to Eqs. (18.47) and (18.53) of Section 18.6, and Eqs. (14.14)–(14.16), Eqs. (14.6)–(14.9) read

$$i\mathbf{k} \cdot \mathbf{E}(\mathbf{k}, t) = 0 \tag{14.17}$$

$$i\mathbf{k} \cdot \mathbf{B}(\mathbf{k}, t) = 0 \tag{14.18}$$

$$i\mathbf{k} \times \mathbf{E}(\mathbf{k}, t) = -\left(\frac{\partial \mathbf{B}(\mathbf{k}, t)}{\partial t}\right) \tag{14.19}$$

$$i\mathbf{k} \times \mathbf{B}(\mathbf{k}, t) = \frac{1}{c^2}\left(\frac{\partial \mathbf{E}(\mathbf{k}, t)}{\partial t}\right) \tag{14.20}$$

and Eqs. (14.12) and (14.13) yield

$$i\mathbf{k} \cdot \mathbf{A}(\mathbf{k}, t) = 0 \tag{14.21}$$

$$\mathbf{E}(\mathbf{k}, t) = -\left(\frac{\partial \mathbf{A}(\mathbf{k}, t)}{\partial t}\right) \tag{14.22}$$

The passage from geometric space to reciprocal space allows one to transform the Maxwell equations (14.6)–(14.9) and Eqs. (14.12) and (14.13), which are partial differential equations, to the new ones (14.17)–(14.20), which form, for each point $\mathbf{k}$ of the reciprocal space, a infinite set of differential equations governing $\mathbf{E}(\mathbf{k}, t)$ and $\mathbf{B}(\mathbf{k}, t)$.

Now, according to the Helmholtz theorem of vectorial analysis, any vector $\mathbf{F}(\mathbf{k}, t)$ may be always decomposed according to

$$\mathbf{F}(\mathbf{k}, t) = \mathbf{F}_{//}(\mathbf{k}, t) + \mathbf{F}_{\perp}(\mathbf{k}, t)$$

with

$$i\mathbf{k} \times \mathbf{F}_{//}(\mathbf{k}, t) = 0$$

$$i\mathbf{k} \cdot \mathbf{F}_{\perp}(\mathbf{k}, t) = 0 \tag{14.23}$$

Thus, due to Eq. (14.17) and Eqs. (14.18)–(14.21), the fields $\mathbf{E}(\mathbf{k}, t), \mathbf{B}(\mathbf{k}, t)$, and $\mathbf{A}(\mathbf{k}, t)$ only involve components perpendicular to the wave vector $\mathbf{k}$, leading one to write

$$\mathbf{E}_{//}(\mathbf{k}, t) = 0 \quad \text{and} \quad \mathbf{E}(\mathbf{k}, t) = \mathbf{E}_{\perp}(\mathbf{k}, t) \tag{14.24}$$

$$\mathbf{B}_{//}(\mathbf{k}, t) = 0 \quad \text{and} \quad \mathbf{B}(\mathbf{k}, t) = \mathbf{B}_{\perp}(\mathbf{k}, t) \tag{14.25}$$

$$\mathbf{A}_{//}(\mathbf{k}, t) = 0 \quad \text{and} \quad \mathbf{A}(\mathbf{k}, t) = \mathbf{A}_{\perp}(\mathbf{k}, t) \tag{14.26}$$

14.1.3 Linear combinations of $\mathbf{E}_\perp(\mathbf{k}, t)$ and $\mathbf{B}_\perp(\mathbf{k}, t)$ acting as normal modes

Now, the direct product of Eq. (14.19) by the vector **k**, that is,

$$\mathbf{k}\left(\frac{\partial \mathbf{B}_\perp(\mathbf{k},t)}{\partial t}\right) = -i\mathbf{k} \times (\mathbf{k} \times \mathbf{E}_\perp(\mathbf{k},t))$$

reads, since **k** does not depend on time,

$$\left(\frac{\partial\{\mathbf{k}\times\mathbf{B}_\perp(\mathbf{k},t)\}}{\partial t}\right) = -i\mathbf{k} \times (\mathbf{k} \times \mathbf{E}_\perp(\mathbf{k},t)) \tag{14.27}$$

Then, applying to the right-hand side of Eq. (14.27) theorem (18.88) of Section 18.8, that is,

$$\mathbf{V} \times (\mathbf{W} \times \mathbf{U}) = (\mathbf{V} \cdot \mathbf{U})\mathbf{W} - (\mathbf{V} \cdot \mathbf{W})\mathbf{U} \tag{14.28}$$

where **V**, **W**, and **U** are vectors, yields

$$\mathbf{k} \times (\mathbf{k} \times \mathbf{E}_\perp(\mathbf{k},t)) = (\mathbf{k} \cdot \mathbf{E}_\perp(\mathbf{k},t))\mathbf{k} - (\mathbf{k} \cdot \mathbf{k})\mathbf{E}_\perp(\mathbf{k},t) \tag{14.29}$$

Thus, keeping in mind that, according to Eqs. (14.23) and (14.24) that

$$\mathbf{k} \cdot \mathbf{E}_\perp(\mathbf{k},t) = 0 \tag{14.30}$$

the latter equation and (14.29) and (14.30) allow one to transform Eq. (14.27) into

$$\left(\frac{\partial\{\mathbf{k}\times\mathbf{B}_\perp(\mathbf{k},t)\}}{\partial t}\right) = ik^2\mathbf{E}_\perp(\mathbf{k},t) \tag{14.31}$$

with

$$k^2 = \mathbf{k} \cdot \mathbf{k}$$

Moreover, introducing the unit vector $\hat{\kappa}$ through

$$\mathbf{k} = k\hat{\kappa} \tag{14.32}$$

and, after simplification by k, Eq. (14.31) transforms to

$$\left(\frac{\partial\{\hat{\kappa}\times\mathbf{B}_\perp(\mathbf{k},t)\}}{\partial t}\right) = ik\mathbf{E}_\perp(\mathbf{k},t) \tag{14.33}$$

On the other hand, owing to Eqs. (14.24), (14.25), and (14.32), the partial differential equation (14.20) becomes

$$\left(\frac{\partial \mathbf{E}_\perp(\mathbf{k},t)}{\partial t}\right) = ic^2k\{\hat{\kappa}\times\mathbf{B}_\perp(\mathbf{k},t)\} \tag{14.34}$$

Then, adding and subtracting Eqs. (14.33) and (14.34), we have

$$\left(\frac{\partial\{\mathbf{E}_\perp(\mathbf{k},t)+c\hat{\kappa}\times\mathbf{B}_\perp(\mathbf{k},t)\}}{\partial t}\right) = i\omega(k)\{\mathbf{E}_\perp(\mathbf{k},t)+c\hat{\kappa}\times\mathbf{B}_\perp(\mathbf{k},t)\} \tag{14.35}$$

$$\left(\frac{\partial\{\mathbf{E}_\perp(\mathbf{k},t)-c\hat{\kappa}\times\mathbf{B}_\perp(\mathbf{k},t)\}}{\partial t})\right) = -i\omega(k)\{\mathbf{E}_\perp(\mathbf{k},t)-c\hat{\kappa}\times\mathbf{B}_\perp(\mathbf{k},t)\} \tag{14.36}$$

where

$$\omega(k) = ck \tag{14.37}$$

Observe that the two equations (14.35) and (14.36), involving linear combinations of the electrical and magnetic fields in reciprocal space act as those governing decoupled normal modes.

14.1.4 Dimensionless normal modes

Now, introduce the two following dimensionless fields defined by:

$$\{\boldsymbol{\alpha}_\perp(k,t)\} = -iK(k)\left\{\frac{\mathbf{E}_\perp(\mathbf{k},t) - c\hat{\kappa}\times\mathbf{B}_\perp(\mathbf{k},t)}{2}\right\} \tag{14.38}$$

$$\{\boldsymbol{\beta}_\perp(k,t)\} = -iK(k)\left\{\frac{\mathbf{E}_\perp(\mathbf{k},t) + c\hat{\kappa}\times\mathbf{B}_\perp(\mathbf{k},t)}{2}\right\} \tag{14.39}$$

where $K(k)$ are real constants allowing dimensionless $\boldsymbol{\alpha}_\perp(\mathbf{k},t)$ and $\boldsymbol{\beta}_\perp(\mathbf{k},t)$ to be dimensionless. Then, because in the Euclidian space $\mathbf{E}_\perp(\mathbf{r},t)$ and $\mathbf{B}_\perp(\mathbf{r},t)$ are real, and due to Eq. (18.43) of Section 18.6, it follows from Eqs. (14.14) and (14.15) that

$$\mathbf{E}_\perp(\mathbf{k},t)^* = \mathbf{E}_\perp(-\mathbf{k},t) \qquad \text{and} \qquad \mathbf{B}_\perp(\mathbf{k},t)^* = \mathbf{B}_\perp(-\mathbf{k},t) \tag{14.40}$$

so that

$$\{\boldsymbol{\alpha}_\perp(\mathbf{k},t)\}^* = \{\boldsymbol{\alpha}_\perp(-\mathbf{k},t)\} \qquad \text{and} \qquad \{\boldsymbol{\beta}_\perp(\mathbf{k},t)\}^* = \{\boldsymbol{\beta}_\perp(-\mathbf{k},t)\}$$

These properties allow one to find the relation between $\boldsymbol{\alpha}_\perp(\mathbf{k},t)$ and $\boldsymbol{\beta}_\perp(\mathbf{k},t)$ defined by Eqs. (14.38) and (14.39) in the following way: Because $K(k)$ is real, the conjugate complex of $\boldsymbol{\alpha}_\perp(\mathbf{k},t)$ given by Eq. (14.38) reads

$$\{\boldsymbol{\alpha}_\perp(\mathbf{k},t)\}^* = iK(k)\left\{\frac{\mathbf{E}_\perp(\mathbf{k},t)^* - c\hat{\kappa}\times B_\perp(k,t)^*}{2}\right\}$$

which transforms, in view of Eq. (14.40), into

$$\{\boldsymbol{\alpha}_\perp(\mathbf{k},t)\}^* = iK(k)\left\{\frac{\mathbf{E}_\perp(-\mathbf{k},t) - c\hat{\kappa}\times\mathbf{B}_\perp(-\mathbf{k},t)}{2}\right\}$$

Hence, changing $\mathbf{k}$ into $-\mathbf{k}$, and thus, according to Eq. (14.32), $\hat{\kappa}$ into $-\hat{\kappa}$, yields

$$\{\boldsymbol{\alpha}_\perp(-\mathbf{k},t)\}^* = iK(k)\left\{\frac{\mathbf{E}_\perp(\mathbf{k},t) + c\hat{\kappa}\times\mathbf{B}_\perp(\mathbf{k},t)}{2}\right\}$$

so that, by comparison of this result with (14.39), we have

$$\{\boldsymbol{\beta}_\perp(\mathbf{k},t)\} = -\{\boldsymbol{\alpha}_\perp(-\mathbf{k},t)\}^* \tag{14.41}$$

Hence, comparing Eqs. (14.38) and (14.39), the partial differential equations (14.35) and (14.36) yield

$$\boxed{\left(\frac{\partial\boldsymbol{\alpha}_\perp(\mathbf{k},t)}{\partial t}\right) = -i\omega(k)\{\boldsymbol{\alpha}_\perp(\mathbf{k},t)\} \qquad \text{and} \qquad \left(\frac{\partial\boldsymbol{\alpha}_\perp(\mathbf{k},t)^*}{\partial t}\right) = i\omega(k)\{\boldsymbol{\alpha}_\perp(\mathbf{k},t)\}^*} \tag{14.42}$$

which, after integration, lead to

$$\{\boldsymbol{\alpha}_\perp(\mathbf{k},t)\} = \{\boldsymbol{\alpha}_\perp(\mathbf{k})\}(e^{-i\omega(k)t}) \qquad \text{and} \qquad \{\boldsymbol{\alpha}_\perp(\mathbf{k},t)\}^* = \{\boldsymbol{\alpha}_\perp(\mathbf{k})\}(e^{i\omega(k)t}) \tag{14.43}$$

Thus, from Eqs. (14.42) and (14.43), the dimensionless variables $\boldsymbol{\alpha}_\perp(\mathbf{k}, t)$ or $\boldsymbol{\alpha}_\perp(\mathbf{k}, t)^*$ characterized by different values of the wave vector k, are decoupled, so that they may be viewed as the normal modes of the field within the reciprocal space.

14.1.5 Fields in terms of dimensionless normal modes

Next, one obtains by inversion of Eqs. (14.38), (14.39), and (14.41)

$$\{E_\perp(\mathbf{k}, t)\} = i\left\{\frac{\{\boldsymbol{\alpha}_\perp(\mathbf{k}, t)\} - \{\boldsymbol{\alpha}_\perp(-\mathbf{k}, t)\}^*}{K(k)}\right\} \tag{14.44}$$

$$\{\mathbf{B}_\perp(\mathbf{k}, t)\} = i\left\{\frac{\hat{\kappa} \times \boldsymbol{\alpha}_\perp(\mathbf{k}, t) + \hat{\kappa} \times \boldsymbol{\alpha}_\perp(-\mathbf{k}, t)^*}{cK(k)}\right\} \tag{14.45}$$

or, in view of Eq. (14.43),

$$\{\mathbf{E}_\perp(\mathbf{k}, t)\} = i\left\{\frac{\boldsymbol{\alpha}_\perp(\mathbf{k})e^{-i\omega(k)t} - \boldsymbol{\alpha}_\perp(-\mathbf{k})^* e^{i\omega(k)t}}{K(k)}\right\} \tag{14.46}$$

$$\{\mathbf{B}_\perp(\mathbf{k}, t)\} = i\left\{\frac{\hat{\kappa} \times \boldsymbol{\alpha}_\perp(\mathbf{k})e^{-i\omega(k)t} + \hat{\kappa} \times \boldsymbol{\alpha}_\perp(-\mathbf{k})^* e^{i\omega(k)t}}{cK(k)}\right\} \tag{14.47}$$

On the other hand, to get the expression of the fields within the geometrical space, perform the inverse Fourier transforms of (14.14) and (14.15) to get

$$\mathbf{E}_\perp(\mathbf{r}, t) = \left(\frac{1}{2\pi}\right)^{3/2} \int \mathbf{E}_\perp(\mathbf{k}, t) e^{i\mathbf{k}.\mathbf{r}} d^3\mathbf{k}$$

$$\mathbf{B}_\perp(\mathbf{r}, t) = \left(\frac{1}{2\pi}\right)^{3/2} \int \mathbf{B}_\perp(\mathbf{k}, t) e^{i\mathbf{k}.\mathbf{r}} d^3\mathbf{k}$$

which, due to Eqs. (14.46) and (14.47), take, respectively, the forms

$$\mathbf{E}_\perp(\mathbf{r}, t) = i\int E_{\omega_k}(\boldsymbol{\alpha}_\perp(\mathbf{k})(e^{i\mathbf{k}.\mathbf{r}})e^{-i\omega(k)t} - \boldsymbol{\alpha}_\perp(-\mathbf{k})^*(e^{i\mathbf{k}.\mathbf{r}})e^{i\omega(k)t})d^3\mathbf{k} \tag{14.48}$$

$$\mathbf{B}_\perp(\mathbf{r}, t) = i\int B_{\omega_k}((\hat{\kappa} \times \boldsymbol{\alpha}_\perp(\mathbf{k}))(e^{i\mathbf{k}.\mathbf{r}})e^{-i\omega(k)t} + (\hat{\kappa} \times \boldsymbol{\alpha}_\perp(-\mathbf{k})^*)(e^{i\mathbf{k}.\mathbf{r}})e^{i\omega(k)t})d^3\mathbf{k} \tag{14.49}$$

with, according to Eq. (14.37),

$$E_{\omega_k} = \left(\frac{1}{2\pi}\right)^{3/2} \frac{1}{K(k)} \quad \text{and} \quad B_{\omega_k} = \left(\frac{1}{2\pi}\right)^{3/2} \frac{1}{cK(k)} \tag{14.50}$$

Again, let $\mathbf{k} \to -\mathbf{k}$ inside the last part of the integrals (14.48) and (14.49), leading therefore to

$$\boldsymbol{\alpha}_\perp(-\mathbf{k})^* \to \boldsymbol{\alpha}_\perp(\mathbf{k})^*$$

$$\kappa = \frac{\mathbf{k}}{k} \to -\kappa \qquad \text{and} \qquad e^{i\mathbf{k}.\mathbf{r}} \to e^{-i\mathbf{k}.\mathbf{r}}$$

(14.48) and (14.49) transform into

$$\boxed{\mathbf{E}_\perp(\mathbf{r},t) = i\int E_{\omega_k}\{\boldsymbol{\alpha}_\perp(\mathbf{k})e^{i\mathbf{k}.\mathbf{r}}e^{-i\omega(k)t} - \boldsymbol{\alpha}_\perp(\mathbf{k})^*\mathbf{e}^{-i\mathbf{k}.\mathbf{r}}e^{i\omega(k)t}\}d^3\mathbf{k}} \tag{14.51}$$

$$\boxed{\mathbf{B}_\perp(\mathbf{r},t) = i\int B_{\omega_k}\{(\hat{\kappa}\times\boldsymbol{\alpha}_\perp(\mathbf{k}))e^{i\mathbf{k}.\mathbf{r}}e^{-i\omega(k)t} - (\hat{\kappa}\times\boldsymbol{\alpha}_\perp(\mathbf{k})^*)e^{-i\mathbf{k}.\mathbf{r}}e^{i\omega(k)t}\}d^3\mathbf{k}} \tag{14.52}$$

Next, keeping in mind Eq. (14.22), that is,

$$\mathbf{E}_\perp(\mathbf{k},t) = -\left(\frac{\partial\mathbf{A}_\perp(\mathbf{k},t)}{\partial t}\right) \tag{14.53}$$

it appears that, due to Eq. (14.46), and in order to satisfy Eq. (14.53), $\mathbf{A}_\perp(\mathbf{k},t)$ must obey

$$\mathbf{A}_\perp(\mathbf{k},t) = \left\{\frac{\boldsymbol{\alpha}_\perp(\mathbf{k})e^{-i\omega(k)t} + \boldsymbol{\alpha}_\perp(-\mathbf{k})^*e^{i\omega(k)t}}{\omega(k)K(k)}\right\} \tag{14.54}$$

which reads, at an initial time,

$$\mathbf{A}_\perp(\mathbf{k},0) = \left\{\frac{\boldsymbol{\alpha}_\perp(\mathbf{k}) + \boldsymbol{\alpha}_\perp(-\mathbf{k})^*}{\omega(k)K(k)}\right\} \tag{14.55}$$

Furthermore, due to Eq. (14.54) the potential vector working within the geometrical space, that is, the inverse Fourier transform of $\mathbf{A}_\perp(\mathbf{k},t)$, yields

$$\mathbf{A}_\perp(\mathbf{r},t) = \left(\frac{1}{2\pi}\right)^{3/2}\int \mathbf{A}_\perp(\mathbf{k},t)e^{i\mathbf{k}.\mathbf{r}}d^3\mathbf{k}$$

and transforms, after changing as above $\mathbf{k}$ into $-\mathbf{k}$ inside the last integral, into

$$\boxed{\mathbf{A}_\perp(\mathbf{r},t) = \int A_{\omega_k}(\boldsymbol{\alpha}_\perp(\mathbf{k})e^{i\mathbf{k}.\mathbf{r}}e^{-i\omega(k)t} + \boldsymbol{\alpha}_\perp(\mathbf{k})^*e^{-i\mathbf{k}.\mathbf{r}}e^{i\omega(k)t})d^3\mathbf{k}} \tag{14.56}$$

with, in view of Eq. (14.37),

$$A_{\omega_k} = \left(\frac{1}{2\pi}\right)^{3/2}\frac{1}{\omega(k)K(k)} \tag{14.57}$$

14.2 ELECTROMAGNETIC FIELD HAMILTONIAN

Now, consider the classical Hamiltonian $H(t)$ of the electromagnetic field, that is, its energy, which is given in the absence of charge by

$$H(t) = \int\left\{\frac{\varepsilon^\circ\mathbf{E}_\perp(\mathbf{r},t)^2 + \mu^{\circ-1}\mathbf{B}_\perp(\mathbf{r},t)^2}{2}\right\}d^3\mathbf{r} \tag{14.58}$$

where ε° and μ° are, respectively, the electrical susceptibility and the magnetic permeability of the vacuum related to the velocity of light c through

$$\mu^\circ\varepsilon^\circ c^2 = 1 \tag{14.59}$$

Now, in the absence of charge, since an isolated electromagnetic field cannot exchange energy, H must remain constant. Hence, t may be omitted in Eq. (14.58), so that

$$H = \varepsilon^\circ \int \left\{ \frac{\mathbf{E}_\perp(\mathbf{r})^2 + c^2 \mathbf{B}_\perp(\mathbf{r})^2}{2} \right\} d^3\mathbf{r} \tag{14.60}$$

Moreover, since $\mathbf{E}_\perp(\mathbf{r})$ and $\mathbf{B}_\perp(\mathbf{r})$ are real, the Parseval–Plancherel identity (18.44) of Section 18.6, allows one to write

$$\int \mathbf{E}_\perp(\mathbf{r})^2 d^3\mathbf{r} = \int |\mathbf{E}_\perp(\mathbf{k})|^2 d^3\mathbf{k}$$

$$\int \mathbf{B}_\perp(\mathbf{r})^2 d^3\mathbf{r} = \int |\mathbf{B}_\perp(\mathbf{k})|^2 d^3\mathbf{k}$$

so that, energy (14.60) yields in reciprocal space

$$H = \varepsilon^\circ \int \left\{ \frac{|\mathbf{E}_\perp(\mathbf{k})|^2 + c^2 |\mathbf{B}_\perp(\mathbf{k})|^2}{2} \right\} d^3\mathbf{k} \tag{14.61}$$

Next, in view of Eq. (14.46), the squared absolute value of the electric field appearing in Eq. (14.61) reads

$$|\mathbf{E}_\perp(\mathbf{k})|^2 = \left\{ \frac{(\boldsymbol{\alpha}_\perp(\mathbf{k})^* - \boldsymbol{\alpha}_\perp(-\mathbf{k})) \cdot (\boldsymbol{\alpha}_\perp(\mathbf{k}) - \boldsymbol{\alpha}_\perp(-\mathbf{k})^*)}{K(k)^2} \right\}$$

or, performing the product without changing the order of the factors, for reasons that will become obvious when passing to quantum mechanics,

$$|\mathbf{E}_\perp(\mathbf{k})|^2 = \left\{ \frac{\boldsymbol{\alpha}_\perp(\mathbf{k})^* \cdot \boldsymbol{\alpha}_\perp(\mathbf{k}) + \boldsymbol{\alpha}_\perp(-\mathbf{k}) \cdot \boldsymbol{\alpha}_\perp(-\mathbf{k})^* - \boldsymbol{\alpha}_\perp(\mathbf{k})^* \cdot \boldsymbol{\alpha}_\perp(-\mathbf{k})^* - \boldsymbol{\alpha}_\perp(-\mathbf{k}) \cdot \boldsymbol{\alpha}_\perp(\mathbf{k})}{K(k)^2} \right\} \tag{14.62}$$

so that, after passing from the vectors $\boldsymbol{\alpha}_\perp(\pm\mathbf{k})$ to their corresponding scalar $\alpha(\pm\mathbf{k})$

$$|\mathbf{E}_\perp(\mathbf{k})|^2 = \left\{ \frac{\alpha_\perp(\mathbf{k})^* \alpha_\perp(\mathbf{k}) + \alpha(-\mathbf{k})\alpha(-\mathbf{k})^* - \alpha_\perp(\mathbf{k})^* \alpha(-\mathbf{k})^* - \alpha(-\mathbf{k})\alpha_\perp(\mathbf{k})}{K(k)^2} \right\}$$

Now, in view of Eq. (14.47), when ignoring time, the squared absolute value of the magnetic field appearing in Eq. (14.61), reads

$$|\mathbf{B}_\perp(\mathbf{k})|^2 = \frac{1}{c^2} \left\{ \frac{(\hat{\kappa} \times \boldsymbol{\alpha}_\perp(\mathbf{k})^* + \hat{\kappa} \times \boldsymbol{\alpha}_\perp(-\mathbf{k})) \cdot (\hat{\kappa} \times \boldsymbol{\alpha}_\perp(\mathbf{k}) + \hat{\kappa} \times \boldsymbol{\alpha}_\perp(-\mathbf{k})^*)}{K(k)^2} \right\}$$

yielding

$$c^2 |\mathbf{B}_\perp(\mathbf{k})|^2 = \left\{ \frac{(\hat{\kappa} \times \boldsymbol{\alpha}_\perp(\mathbf{k})^*) \cdot (\hat{\kappa} \times \boldsymbol{\alpha}_\perp(\mathbf{k})) + (\hat{\kappa} \times \boldsymbol{\alpha}_\perp(\mathbf{k})^*) \cdot (\hat{\kappa} \times \boldsymbol{\alpha}_\perp(-\mathbf{k})^*)}{K(k)^2} \right\}$$

$$+ \left\{ \frac{(\hat{\kappa} \times \boldsymbol{\alpha}_\perp(-\mathbf{k})) \cdot (\hat{\kappa} \times \boldsymbol{\alpha}_\perp(\mathbf{k})) + (\hat{\kappa} \times \boldsymbol{\alpha}_\perp(-\mathbf{k})) \cdot (\hat{\kappa} \times \boldsymbol{\alpha}_\perp(-\mathbf{k})^*)}{K(k)^2} \right\} \tag{14.63}$$

Now, observe that the cross product of the dimensionless unit vector $\hat{\kappa} = \mathbf{k}/k$ by the vector $\boldsymbol{\alpha}_\perp(\mathbf{k})$ is a new vector $\boldsymbol{\alpha}(\mathbf{k})$, the modulus of which $\alpha(\mathbf{k})$ remains $|\boldsymbol{\alpha}_\perp(\mathbf{k})|$, but which is orthogonal to the plane $\hat{\kappa}$, according to

$$\hat{\kappa} \times \boldsymbol{\alpha}_\perp(\mathbf{k}) = \frac{\mathbf{k} \times \boldsymbol{\alpha}_\perp(\mathbf{k})}{k} = \boldsymbol{\alpha}(\mathbf{k})$$

Hence, the first scalar product involved on the first right-hand side of Eq. (14.63), reads

$$(\hat{\kappa} \times \boldsymbol{\alpha}_\perp(\mathbf{k})^*) \cdot (\hat{\kappa} \times \boldsymbol{\alpha}_\perp(\mathbf{k})) = \boldsymbol{\alpha}(\mathbf{k})^* \cdot \boldsymbol{\alpha}(\mathbf{k})$$

or, since the modulus $\alpha(\mathbf{k})$ of $\boldsymbol{\alpha}_\perp(\mathbf{k})$ is the same as that of $\boldsymbol{\alpha}(\mathbf{k})$

$$(\hat{\kappa} \times \boldsymbol{\alpha}_\perp(\mathbf{k})^*) \cdot (\hat{\kappa} \times \boldsymbol{\alpha}_\perp(\mathbf{k})) = \alpha(\mathbf{k})^* \alpha(\mathbf{k}) \tag{14.64}$$

In like manner

$$(\hat{\kappa} \times \boldsymbol{\alpha}_\perp(\mathbf{k})^*) \cdot (\hat{\kappa} \times \boldsymbol{\alpha}_\perp(-\mathbf{k})^*) = \alpha(\mathbf{k})^* \alpha(-\mathbf{k})^*$$

$$(\hat{\kappa} \times \boldsymbol{\alpha}_\perp(-\mathbf{k})) \cdot (\hat{\kappa} \times \boldsymbol{\alpha}_\perp(\mathbf{k})) = \alpha(-\mathbf{k}) \alpha(\mathbf{k})$$

$$(\hat{\kappa} \times \boldsymbol{\alpha}_\perp(-\mathbf{k})) \cdot (\hat{\kappa} \times \boldsymbol{\alpha}_\perp(-\mathbf{k})^*) = \alpha(-\mathbf{k}) \alpha(-\mathbf{k})^* \tag{14.65}$$

Therefore, comparing Eqs. (14.64) to (14.65), Eq. (14.63) reads

$$c^2 |\mathbf{B}_\perp(\mathbf{k})|^2 = \left\{ \frac{\alpha(\mathbf{k})^* \alpha(\mathbf{k}) + \alpha(-\mathbf{k})\, \alpha(-\mathbf{k})^* + \alpha(\mathbf{k})^*\, \alpha(-\mathbf{k})^* + \alpha(-\mathbf{k})\, \alpha(\mathbf{k})}{K(k)^2} \right\} \tag{14.66}$$

As a consequence of Eqs. (14.62) and (14.66), the field energy (14.61) becomes after simplification

$$H = \varepsilon^\circ \int \left\{ \frac{\alpha(\mathbf{k})^* \alpha(\mathbf{k}) + \alpha(-\mathbf{k})\, \alpha(-\mathbf{k})^*}{K(k)^2} \right\} d^3\mathbf{k} \tag{14.67}$$

Now, it is convenient to write the classical Hamiltonian as an expression involving Planck's constant and the factor $\frac{1}{2}$, which will be of interest when passing to quantum mechanics. Thus, we write Eq. (14.67) as

$$H = \int \hbar\omega(k) \left\{ \frac{\alpha(\mathbf{k})^* \alpha(\mathbf{k}) + \alpha(-\mathbf{k})\, \alpha(-\mathbf{k})^*}{2} \right\} d^3\mathbf{k} \tag{14.68}$$

with, by identification of (14.67) and (14.68), $K(k)$ reads

$$K(k) = \sqrt{\frac{2\varepsilon^\circ}{\hbar\omega(k)}} \tag{14.69}$$

Finally, changing $-\mathbf{k}$ into $\mathbf{k}$, inside the last right-hand side of Eq. (14.68), that does not modify anything since this change concerns a scalar product, the classical Hamiltonian (14.68) takes the final form

$$\boxed{H = \int \hbar\omega(k) \left\{ \frac{\alpha(\mathbf{k})^* \alpha(\mathbf{k}) + \alpha(\mathbf{k})\, \alpha(\mathbf{k})^*}{2} \right\} d^3\mathbf{k}} \tag{14.70}$$

14.3 POLARIZED NORMAL MODES

Now, in view of Eq. (14.69), the constants defined by Eqs. (14.50) and (14.57) read

$$E_{\omega_k} = \left(\frac{1}{2\pi}\right)^{3/2} \sqrt{\frac{\hbar\omega(k)}{2\varepsilon^\circ}} \quad \text{and} \quad B_{\omega_k} = \left(\frac{1}{2\pi}\right)^{3/2} \sqrt{\frac{\hbar\omega(k)}{2\varepsilon^\circ c^2}} \tag{14.71}$$

$$A_{\omega_k} = \left(\frac{1}{2\pi}\right)^{3/2} \sqrt{\frac{\hbar}{2\varepsilon^\circ \omega(k)}} \tag{14.72}$$

Next, recall that according to Eqs. (14.24)–(14.26), the fields $\mathbf{E}_\perp(\mathbf{k}, t)$, $\mathbf{B}_\perp(\mathbf{k}, t)$, and $\mathbf{A}_\perp(\mathbf{k}, t)$ are transverse to the wave vector $\mathbf{k}$, and thus to the corresponding unit vector $\hat{\kappa} = \mathbf{k}/k$ [defined by Eq. (14.32)], so that each field, at any point $\mathbf{k}$ of the reciprocal space, may be considered as the sum of two perpendicular combinations both orthogonal to $\mathbf{k}$:

$$\mathbf{E}_\perp(\mathbf{k}, t) = \mathbf{E}_{\boldsymbol{\varepsilon}}(\mathbf{k}, t) + \mathbf{E}_{\boldsymbol{\varepsilon}'}(\mathbf{k}, t) \tag{14.73}$$

$$\mathbf{B}_\perp(\mathbf{k}, t) = \mathbf{B}_{\boldsymbol{\varepsilon}}(\mathbf{k}, t) + \mathbf{B}_{\boldsymbol{\varepsilon}'}(\mathbf{k}, t) \tag{14.74}$$

$$\mathbf{A}_\perp(\mathbf{k}, t) = \mathbf{A}_{\boldsymbol{\varepsilon}}(\mathbf{k}, t) + \mathbf{A}_{\boldsymbol{\varepsilon}'}(\mathbf{k}, t) \tag{14.75}$$

These two perpendicular combinations characterized by $\hat{\boldsymbol{\varepsilon}}_\mathbf{k}$ and $\hat{\boldsymbol{\varepsilon}}'_\mathbf{k}$ are the two polarized components of the different fields in the reciprocal space, the two polarization vectors $\hat{\boldsymbol{\varepsilon}}_\mathbf{k}$ and $\hat{\boldsymbol{\varepsilon}}'_\mathbf{k}$ giving the directions of the polarized components of the fields being perpendicular to the unit vector $\hat{\kappa}$ characterizing the vector $\mathbf{k}$ and thus satisfying

$$\hat{\boldsymbol{\varepsilon}}_\mathbf{k} \cdot \hat{\boldsymbol{\varepsilon}}'_\mathbf{k} = \hat{\boldsymbol{\varepsilon}}_\mathbf{k} \cdot \hat{\kappa} = \hat{\kappa} \cdot \hat{\boldsymbol{\varepsilon}}'_\mathbf{k} = 0$$

$$\hat{\boldsymbol{\varepsilon}}_\mathbf{k} \cdot \hat{\boldsymbol{\varepsilon}}_\mathbf{k} = \hat{\boldsymbol{\varepsilon}}'_\mathbf{k} \cdot \hat{\boldsymbol{\varepsilon}}'_\mathbf{k} = \hat{\kappa} \cdot \hat{\kappa} = 1$$

$$\hat{\kappa} = \frac{\mathbf{k}}{|\mathbf{k}|} = \frac{\mathbf{k}}{k}$$

Hence, the polarized modes $\mathbf{E}_{\boldsymbol{\varepsilon}}(\mathbf{k}, t)$, $\mathbf{A}_{\boldsymbol{\varepsilon}}(\mathbf{k}, t)$, and $\mathbf{B}_{\boldsymbol{\varepsilon}}(\mathbf{k}, t)$ are given by

$$\mathbf{E}_{\boldsymbol{\varepsilon}}(\mathbf{k}, t) = i\sqrt{\frac{\hbar\omega(k)}{2\varepsilon^\circ}}\hat{\boldsymbol{\varepsilon}}_\mathbf{k}(\alpha_{\varepsilon\perp}(\mathbf{k}, t) - \alpha_{\varepsilon\perp}(-\mathbf{k}, t)^*) \tag{14.76}$$

$$\mathbf{A}_{\boldsymbol{\varepsilon}}(\mathbf{k}, t) = \sqrt{\frac{\hbar}{2\varepsilon^\circ\omega(k)}}\, \hat{\boldsymbol{\varepsilon}}_\mathbf{k}(\alpha_{\varepsilon\perp}(\mathbf{k}, t) + \alpha_{\varepsilon\perp}(-\mathbf{k}, t)^*) \tag{14.77}$$

$$\mathbf{B}_{\boldsymbol{\varepsilon}}(\mathbf{k}, t) = i\sqrt{\frac{\hbar\omega(k)}{2\varepsilon^\circ c^2}}(\hat{\kappa} \times \hat{\boldsymbol{\varepsilon}}_\mathbf{k})(\alpha_{\boldsymbol{\varepsilon}\perp}(\mathbf{k}, t) + \alpha_{\boldsymbol{\varepsilon}\perp}(-\mathbf{k}, t)^*) \tag{14.78}$$

with

$$\alpha_{\boldsymbol{\varepsilon}\perp}(\mathbf{k}, t) = \hat{\boldsymbol{\varepsilon}}_\mathbf{k} \cdot \boldsymbol{\alpha}_\perp(\mathbf{k}) \tag{14.79}$$

while the polarized modes $\mathbf{E}_{\varepsilon'}(\mathbf{k}, t)$, $\mathbf{A}_{\varepsilon'}(\mathbf{k}, t)$, and $\mathbf{B}_{\varepsilon'}(\mathbf{k}, t)$, which are orthogonal to $\mathbf{E}_{\varepsilon}(\mathbf{k}, t)$, $\mathbf{A}_{\varepsilon}(\mathbf{k}, t)$, and $\mathbf{B}_{\varepsilon}(\mathbf{k}, t)$, appear to be given by similar expressions by changing in Eqs. (14.76)–(14.78) the unit vector $\hat{\boldsymbol{\varepsilon}}_{\mathbf{k}}$ into $\hat{\boldsymbol{\varepsilon}}'_{\mathbf{k}}$ and the corresponding index ε by that ε', that is,

$$\mathbf{E}_{\varepsilon'}(\mathbf{k}, t) = i\sqrt{\frac{\hbar\omega(k)}{2\varepsilon^{\circ}}}\hat{\boldsymbol{\varepsilon}}'_{\mathbf{k}}(\alpha_{\varepsilon'\perp}(\mathbf{k}, t) - \alpha_{\varepsilon'\perp}(-\mathbf{k}, t)^*)$$

$$\mathbf{A}_{\varepsilon'}(\mathbf{k}, t) = \sqrt{\frac{\hbar}{2\varepsilon^{\circ}\omega(k)}}\,\hat{\boldsymbol{\varepsilon}}'_{\mathbf{k}}(\alpha_{\varepsilon'\perp}(\mathbf{k}, t) + \alpha_{\varepsilon'\perp}(-\mathbf{k}, t)^*)$$

$$\mathbf{B}_{\varepsilon'}(\mathbf{k}, t) = i\sqrt{\frac{\hbar\omega(k)}{2\varepsilon^{\circ}c^2}}(\hat{\kappa} \times \hat{\boldsymbol{\varepsilon}}'_{\mathbf{k}})(\alpha_{\varepsilon'\perp}(k, t) + \alpha_{\varepsilon'\perp}(-k, t)^*)$$

with

$$\alpha_{\varepsilon'\perp}(\mathbf{k}, t) = \hat{\boldsymbol{\varepsilon}}'_{\mathbf{k}} \cdot \boldsymbol{\alpha}_{\perp}(\mathbf{k}) \tag{14.80}$$

Next, in order to get the electromagnetic field in Euclidian geometric space, take the Fourier transforms of Eqs. (14.73)–(14.75), after changing $\mathbf{k}$ into $-\mathbf{k}$ within the last integrals [as for the passage from Eqs. (14.48) and (14.49) to Eqs. (14.51) and (14.52)], then, after using Eqs. (14.71) and (14.72), the following description of the electromagnetic field with the geometric space is obtained:

$$\begin{aligned}\mathbf{E}_{\perp}(\mathbf{r}, t) &= i\int E_{\omega_k} d^3\mathbf{k} \\ &\times \hat{\boldsymbol{\varepsilon}}_{\mathbf{k}}(\alpha_{\varepsilon\perp}(\mathbf{k}, t)e^{i\mathbf{k}\cdot\mathbf{r}} - \alpha_{\varepsilon\perp}(\mathbf{k}, t)^* e^{-i\mathbf{k}\cdot\mathbf{r}}) \\ &+ \hat{\boldsymbol{\varepsilon}}'_{\mathbf{k}}(\alpha_{\varepsilon'\perp}(\mathbf{k}, t)e^{i\mathbf{k}\cdot\mathbf{r}} - \alpha_{\varepsilon'\perp}(\mathbf{k}, t)^* e^{-i\mathbf{k}\cdot\mathbf{r}})\}\end{aligned} \tag{14.81}$$

$$\begin{aligned}\mathbf{A}_{\perp}(\mathbf{r}, t) &= \int A_{\omega_k} d^3\mathbf{k} \\ &\times \{\hat{\boldsymbol{\varepsilon}}_{\mathbf{k}}(\alpha_{\varepsilon\perp}(\mathbf{k}, t)e^{i\mathbf{k}\cdot\mathbf{r}} + \alpha_{\varepsilon\perp}(\mathbf{k}, t)^* e^{-i\mathbf{k}\cdot\mathbf{r}}) \\ &+ \hat{\boldsymbol{\varepsilon}}'_{\mathbf{k}}(\alpha_{\varepsilon'\perp}(\mathbf{k}, t)e^{i\mathbf{k}\cdot\mathbf{r}} + \alpha_{\varepsilon'\perp}(\mathbf{k}, t)^* e^{-i\mathbf{k}\cdot\mathbf{r}})\}\end{aligned} \tag{14.82}$$

$$\begin{aligned}\mathbf{B}_{\perp}(\mathbf{r}, t) &= i\int B_{\omega_k} d^3\mathbf{k} \\ &\times \{(\hat{\kappa}\times\hat{\boldsymbol{\varepsilon}}_{\mathbf{k}})\{\alpha_{\varepsilon\perp}(\mathbf{k}, t)e^{i\mathbf{k}\cdot\mathbf{r}} - \alpha_{\varepsilon\perp}(\mathbf{k}, t)^* e^{-i\mathbf{k}\cdot\mathbf{r}}\} \\ &+ (\hat{\kappa}\times\hat{\boldsymbol{\varepsilon}}'_{\mathbf{k}})\{\alpha_{\varepsilon'\perp}(\mathbf{k}, t)e^{i\mathbf{k}\cdot\mathbf{r}} - \alpha_{\varepsilon'\perp}(\mathbf{k}, t)^* e^{-i\mathbf{k}\cdot\mathbf{r}}\}\}\end{aligned} \tag{14.83}$$

14.4 NORMAL MODES OF A CAVITY

14.4.1 Fields and corresponding Hamiltonians

Now, suppose that the electromagnetic field is enclosed in a cubic box of length L. Then, as seen above, not all possible values of the $\mathbf{k}$ wave vectors are permitted, only certain discrete values depending on the boundary conditions. Then, the continuous wave vectors $\mathbf{k}$ appearing in Eqs. (14.81)–(14.83), now transform into discrete wave vectors $\mathbf{k}_n$. Moreover, if the length L is so large that $L >> \lambda_{\text{Max}}$ where $\lambda_{\text{Max}} = 2\pi/|\mathbf{k}_{\text{Min}}|$ and where $|\mathbf{k}_{\text{Min}}|$ is the smallest modulus of the wave vector, it is possible to neglect the effects occurring near the walls of the container and thus to describe the electromagnetic field in terms of a set of discrete components so that the discrete wave vector must satisfy

$$\mathbf{k}_n = \left(\frac{2\pi}{L}\right)(n_x\hat{\mathbf{x}} + n_y\hat{\mathbf{y}} + n_z\hat{\mathbf{z}})$$

where $\hat{\mathbf{x}}$, $\hat{\mathbf{y}}$, and $\hat{\mathbf{z}}$ are the unit vectors along the Cartesian coordinates. In a similar way, the angular frequency $\omega(k)$ defined by Eq. (14.37) depending continuously on the modulus k of the wave vector $\mathbf{k}$, transforms to discrete angular frequency ω_n according to

$$\omega(k) = ck \rightarrow \omega_n = ck_n$$

with

$$k_n = |\mathbf{k}_n| = \left(\frac{2\pi}{L}\right)\sqrt{n_x^2 + n_y^2 + n_z^2}$$

Hence, the continuous variables $\alpha_{\boldsymbol{\varepsilon}\perp}(\mathbf{k}, t)$ and $\alpha_{\boldsymbol{\varepsilon}'\perp}(\mathbf{k}, t)$ defined by Eqs. (14.79) and (14.80) transform into discontinuous variables $\alpha_{\boldsymbol{\varepsilon}\perp}(t)$ and $\alpha_{\boldsymbol{\varepsilon}'\perp}(t)$:

$$\alpha_{\boldsymbol{\varepsilon}\perp}(\mathbf{k}, t) \rightarrow \alpha_{n\varepsilon\perp}(t) \qquad \text{and} \qquad \alpha_{\boldsymbol{\varepsilon}'\perp}(\mathbf{k}, t) \rightarrow \alpha_{n\varepsilon'\perp}(t)$$

Then, after such transformation, the classical Hamiltonian of the electromagnetic field (14.70) involving an integral must transform into the following one involving now a sum, according to

$$H = \sum_n \hbar\omega_n \left\{\frac{\alpha_n^*\alpha_n + \alpha_n\alpha_n^*}{2}\right\} \tag{14.84}$$

Now, observe that the change when passing to an infinite space to a finite one of volume $V = L^3$ must to be compatible with the transformation of Eqs. (14.70) into Eq. (14.84), it is required that the electromagnetic fields (14.81)–(14.83) must be each multiplied by the factor $(2\pi/L)^{3/2}$. Then, one obtains from Eqs. (14.81)–(14.83), respectively, the following expressions:

$$\begin{aligned}\mathbf{E}_\perp(\mathbf{r}, t) = i\sum_n \sqrt{\frac{\hbar\omega_n}{2\varepsilon^\circ V}} \\ \times \{\hat{\boldsymbol{\varepsilon}}_{\mathbf{k}_n}(\alpha_{n\varepsilon\perp}(t)e^{i\mathbf{k}_n\cdot\mathbf{r}} - \alpha_{n\varepsilon\perp}^*(t)e^{-i\mathbf{k}_n\cdot\mathbf{r}}) \\ + \hat{\boldsymbol{\varepsilon}}'_{\mathbf{k}_n}(\alpha_{n\varepsilon'\perp}(t)e^{i\mathbf{k}_n\cdot\mathbf{r}} - \alpha_{n\varepsilon'\perp}^*(t)e^{-i\mathbf{k}_n\cdot\mathbf{r}})\}\end{aligned} \tag{14.85}$$

$$\mathbf{A}_{\perp}(\mathbf{r},t) = \sum_n \sqrt{\frac{\hbar}{2\varepsilon^\circ \omega_n V}}$$
$$\times \{\hat{\boldsymbol{\varepsilon}}_{\mathbf{k}_n}(\alpha_{n\varepsilon\perp}(t)e^{i\mathbf{k}_n\cdot\mathbf{r}} + \alpha^*_{n\varepsilon\perp}(t)e^{-i\mathbf{k}_n\cdot\mathbf{r}})$$
$$+ \hat{\boldsymbol{\varepsilon}}'_{\mathbf{k}_n}(\alpha_{n\varepsilon'\perp}(t)e^{i\mathbf{k}_n\cdot\mathbf{r}} + \alpha^*_{n\varepsilon'\perp}(t)e^{-i\mathbf{k}_n\cdot\mathbf{r}})\} \tag{14.86}$$

$$\mathbf{B}_{\perp}(\mathbf{r},t) = i\sum_n \sqrt{\frac{\hbar\omega_n}{2\varepsilon^\circ V c^2}}$$
$$\times \{(\hat{\kappa} \times \hat{\boldsymbol{\varepsilon}}_{\mathbf{k}_n})(\alpha_{n\varepsilon\perp}(t)e^{i\mathbf{k}_n\cdot\mathbf{r}} - \alpha^*_{n\varepsilon\perp}(t)e^{-i\mathbf{k}_n\cdot\mathbf{r}}).$$
$$+ (\hat{\kappa} \times \hat{\boldsymbol{\varepsilon}}'_{\mathbf{k}_n})(\alpha_{n\varepsilon'\perp}(t)e^{i\mathbf{k}_n\cdot\mathbf{r}} - \alpha^*_{n\varepsilon'\perp}(t)e^{-i\mathbf{k}_n\cdot\mathbf{r}})\} \tag{14.87}$$

We emphasize that the discrete dimensionless polarized components $\alpha_{n\varepsilon\perp}(t)$ and $\alpha_{n\varepsilon'\perp}(t)$, just as the components $\alpha_{\varepsilon\perp}(\mathbf{k},t)$ and $\alpha_{\varepsilon'\perp}(\mathbf{k},t)$ are variables, the magnitudes of which may alter when passing from some wave vector to another one.

14.4.2 Modes density

The total number n of electrical modes inside the cavity is equal to the number of discrete wave vectors $\mathbf{k}_n$ times 2, because of the two polarization orientations of each wave vector. Its differential reads

$$dn = 2dn_x\, dn_y\, dn_z \tag{14.88}$$

where n_x, n_y, and n_z, which are momentarily considered as continuous variables, are related to the components of the wave vectors through

$$k_x = \left(\frac{2\pi n_x}{L}\right) \qquad k_y = \left(\frac{2\pi n_y}{L}\right) \qquad k_z = \left(\frac{2\pi n_z}{L}\right) \tag{14.89}$$

with

$$k_x = \mathbf{k}\cdot\hat{\mathbf{x}} \qquad k_y = \mathbf{k}\cdot\hat{\mathbf{y}} \qquad k_z = \mathbf{k}\cdot\hat{\mathbf{z}}$$

Then, owing to (14.89), Eq. (14.88) reads

$$dn = 2V\left(\frac{1}{2\pi}\right)^3 dk_x\, dk_y\, dk_z \quad \text{with} \quad V = L^3$$

Again, after passing to spherical coordinates, defined in Fig. 14.1, it becomes

$$dn = 2V\left(\frac{1}{2\pi}\right)^3 k^2\, dk \sin\theta\, d\theta\, d\phi \tag{14.90}$$

with

$$k = \sqrt{(k_x)^2 + (k_y)^2 + (k_z)^2}$$

Moreover, passing from the variable k to the corresponding angular frequency $\omega = kc$, Eq. (14.90) yields

$$dn = 2V\left(\frac{1}{2\pi c}\right)^3 \omega^2 d\omega\, d\Omega \tag{14.91}$$

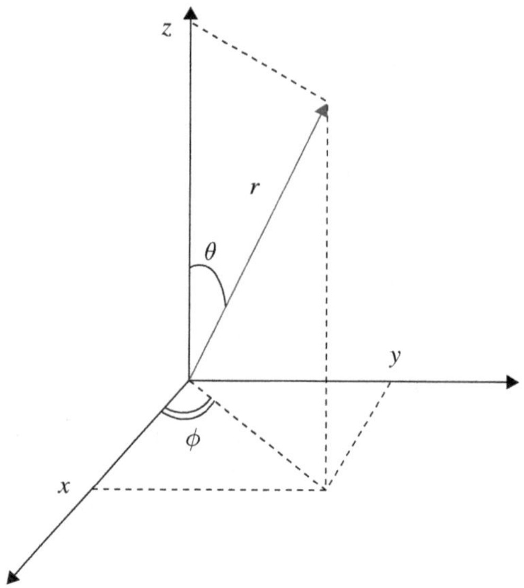

Figure 14.1 Polar spheric coordinates: $x = r\sin\theta\cos\phi$, $y = r\sin\theta\sin\phi$, and $z = r\cos\theta$; and $0 \le r < \infty$, $0 \le \theta \le \pi$, and $0 \le \phi \le 2\pi$. r is the radial coordinate, θ and ϕ are respectively the inclination and azimuth angles.

$$d\Omega = \sin\theta\, d\theta\, d\phi \tag{14.92}$$

the ω derivative of which being

$$\boxed{\left(\frac{dn}{d\Omega}\right) = Vg(\omega)\, d\omega \quad \text{with} \quad g(\omega) = \frac{2\omega^2}{(2\pi c)^3}} \tag{14.93}$$

We find the number of modes of the electromagnetic field, which lie in the range between ω and $d\omega$. This may be obtained by summing dn given by Eq. (14.91) over the Ω angle variables, allowing one to get the radial density of modes $d\rho(\omega)/d\omega$ within the spherical shell lying between ω and $\omega + d\omega$, that is,

$$\left(\frac{d\rho(\omega)}{d\omega}\right) = 2V\left(\frac{1}{2\pi c}\right)^3 \omega^2 \int d\Omega$$

or, due to (14.92)

$$\left(\frac{d\rho(\omega)}{d\omega}\right) = 2V\left(\frac{1}{2\pi c}\right)^3 \omega^2 \int_0^{\pi} \sin\theta\, d\theta \int_0^{2\pi} d\phi$$

Thus, after integration over the angular variables, it yields

$$\left(\frac{d\rho(\omega)}{d\omega}\right) = 8\pi V\left(\frac{1}{2\pi c}\right)^3 \omega^2 = V\left(\frac{1}{\pi^2 c^3}\right)\omega^2$$

or,

$$\left(\frac{d\rho(\omega)}{d\omega}\right) = Vf(\omega) \quad \text{with} \quad f(\omega) = \left(\frac{1}{\pi^2 c^3}\right)\omega^2 \tag{14.94}$$

14.5 QUANTIZATION OF THE ELECTROMAGNETIC FIELDS

14.5.1 Standard Lagrangian within the Coulomb gauge

14.5.1.1 The Lagrangian In order to quantize the electromagnetic field, it is convenient to refer to the standard Lagrangian of a system of charged particles interacting with the electromagnetic field. Thus, differing from the previous sections, it is convenient to take into account the charged particles by considering the *standard Lagrangian* $L(\mathbf{r}, t)$ of such a system within the Coulomb gauge, which is given by

$$L(\mathbf{r}, t) = \sum_{\alpha} \frac{1}{2} m_\alpha \dot{\mathbf{r}}_\alpha^2(t) - V_C(t) + \int L_C(\mathbf{r}, t)\, d^3\mathbf{r} \tag{14.95}$$

with, respectively,

$$L_C(\mathbf{r}, t) = \varepsilon^\circ \left\{ \frac{\dot{\mathbf{A}}(\mathbf{r}, t) - c^2(\nabla \times \mathbf{A}(\mathbf{r}, t))^2}{2} \right\} + \mathbf{J}(\mathbf{r}, \dot{\mathbf{r}}, t) \cdot \mathbf{A}(\mathbf{r}, t) \tag{14.96}$$

Consider the αth particle. In these equations m_α is the mass of the particle, $\dot{\mathbf{r}}_\alpha(t)$ the time derivative of the position coordinate $\mathbf{r}_\alpha(t)$:

$$\dot{\mathbf{r}}_\alpha(t) = \left(\frac{\partial \mathbf{r}_\alpha(t)}{\partial t} \right)$$

whereas $\dot{\mathbf{A}}(\mathbf{r}, t)$ is the time derivative of the vector potential at the $\mathbf{r}$ position, that is,

$$\dot{\mathbf{A}}(\mathbf{r}, t) = \left(\frac{\partial \mathbf{A}(\mathbf{r}, t)}{\partial t} \right)$$

and $\mathbf{J}(\mathbf{r}, t)$ is the current density defined by

$$\mathbf{J}(\mathbf{r}, \dot{\mathbf{r}}, t) = \sum_{\beta} q_\beta \dot{\mathbf{r}}_\beta(t) \delta(\mathbf{r} - \mathbf{r}_\beta(t)) \tag{14.97}$$

in which q_β is the electrical charge of the charged β particle, V_C is the Coulomb potential defined by

$$V_C(t) = \sum_{\alpha} \varepsilon_{\text{Coul}}{}^\alpha + \frac{1}{4\pi\varepsilon^\circ} \sum_{\alpha>\beta} \sum_{\beta} \frac{q_\alpha q_\beta}{|\mathbf{r}_\alpha(t) - \mathbf{r}_\beta(t)|}$$

with

$$\varepsilon_{\text{Coul}}{}^\alpha = \frac{q_\alpha^2}{2\varepsilon^\circ} \left(\frac{1}{2\pi} \right)^3 \int \frac{1}{k^2} d^3\mathbf{k}$$

The Lagrangian (14.95) may be considered as a very general postulate of the electromagnetic theory from which it is possible to deduce the Maxwell equations (14.1)–(14.4) and the Lorentz force law

$$m_\alpha \ddot{\mathbf{r}}_\alpha = q_\alpha \{\mathbf{E}(\mathbf{r}_\alpha(t), t) + \dot{\mathbf{r}}_\alpha(t) \times \mathbf{B}(\mathbf{r}_\alpha(t), t)\}$$

keeping in mind that all the other symbols have the same meaning as above.

Now, when passing from the geometric space to the reciprocal one, the standard Lagrangian may be shown[1] to transform into

$$\widetilde{L}(\mathbf{k},t)=\sum_\alpha \frac{1}{2}m_\alpha \dot{\mathbf{r}}_\alpha^2(t)-\int \frac{|\rho(\mathbf{k},t)|^2}{2\varepsilon^\circ k^2}d^3\mathbf{k}+\int L_C(\mathbf{k},t)\,d^3\mathbf{k} \tag{14.98}$$

with

$$\begin{aligned} L_C(\mathbf{k},t) &= \varepsilon^\circ\left\{\frac{(\dot{\mathbf{A}}(\mathbf{k},t)^*\cdot\dot{\mathbf{A}}(\mathbf{k},t))-c^2k^2(\mathbf{A}(\mathbf{k},t)^*\cdot\mathbf{A}(\mathbf{k},t))}{2}+cc\right\} \\ &\quad+\left\{\frac{\mathbf{J}(\mathbf{k},t)^*\cdot\mathbf{A}(\mathbf{k},t)+\mathbf{J}(\mathbf{k},t)\cdot\mathbf{A}(\mathbf{k},t)^*}{2}\right\} \end{aligned} \tag{14.99}$$

Here $\mathbf{A}(\mathbf{k},t)$ is the Fourier transform (14.16) of $\mathbf{A}(\mathbf{r},t)$, $\dot{\mathbf{A}}(k,t)$ its time derivative, and $\mathbf{J}(\mathbf{k},t)$ is the Fourier transform of $\mathbf{J}(\mathbf{r},t)$ defined by

$$\mathbf{J}(\mathbf{k},t)=\left(\frac{1}{2\pi}\right)^{3/2}\int \mathbf{J}(\mathbf{r},t)e^{-i\mathbf{k}\cdot\mathbf{r}}d^3\mathbf{r}$$

and $\rho(\mathbf{k},t)$ is the Fourier transform of $\rho(\mathbf{r},t)$:

$$\rho(\mathbf{k},t)=\left(\frac{1}{2\pi}\right)^{3/2}\int \rho(\mathbf{r},t)e^{-i\mathbf{k}\cdot\mathbf{r}}d^3\mathbf{r}$$

where $\rho(\mathbf{r},t)$ is the charge density:

$$\rho(\mathbf{r},t)=\sum_\beta q_\beta\delta(\mathbf{r}-\mathbf{r}_\beta(t))$$

14.5.1.2 Conjugate momentum of $\mathbf{r}_\alpha(t)$ and $A(k,t)$ In the Lagrange formalism, the conjugate momentum $\mathbf{p}_\alpha(t)$ of $\mathbf{r}_\alpha(t)$ is the partial derivative of the Lagrangian with respect to $\dot{\mathbf{r}}_\alpha(t)$:

$$\boxed{\mathbf{p}_\alpha(t)=\left(\frac{\partial L(\mathbf{r},t)}{\partial \dot{\mathbf{r}}_\alpha(t)}\right)}$$

Hence, in the present situation where the Lagrangian is given by Eqs. (14.95) and (14.96), the conjugate momentum becomes

$$\mathbf{p}_\alpha(t)=m_\alpha\dot{\mathbf{r}}_\alpha(t)+\int\left(\frac{\partial \mathbf{J}(\mathbf{r},t)}{\partial\dot{\mathbf{r}}_\alpha(t)}\right)d^3\mathbf{r}$$

or, due to Eq. (14.97), and after commuting the volume integral with the sum over β

$$\mathbf{p}_\alpha(t)=m_\alpha\dot{\mathbf{r}}_\alpha(t)+\sum_\beta\int\left(q_\beta\frac{\partial\dot{\mathbf{r}}_\beta(t)\delta(\mathbf{r}-\mathbf{r}_\beta(t))}{\partial\dot{\mathbf{r}}_\alpha(t)}\cdot\mathbf{A}(\mathbf{r},t)\right)d^3\mathbf{r}$$

so that

$$\mathbf{p}_\alpha(t)=m_\alpha\dot{\mathbf{r}}_\alpha(t)+q_\alpha\int\delta(\mathbf{r}-\mathbf{r}_\alpha(t))\cdot\mathbf{A}(\mathbf{r},t)d^3\mathbf{r}$$

[1]C. Cohen-Tannoudji, J. Dupont-Roc, and G. Grynberg. *Atom-Photon Interactions: Basic Processes and Applications*. Wiley Science Paperback Series: New York, 1998.

which simplifies to

$$\mathbf{p}_\alpha(t) = m_\alpha \dot{\mathbf{r}}_\alpha(t) + q_\alpha \mathbf{A}(\mathbf{r}_\alpha(t), t) \tag{14.100}$$

or, more simply,

$$\mathbf{p}(t) = m_\alpha \dot{\mathbf{r}}_\alpha(t) + q\mathbf{A}(\mathbf{r}(t), t)$$

On the other hand, in the reciprocal space, the conjugate momentum $\pi(\mathbf{k}, t)^*$ of $\mathbf{A}(\mathbf{k}, t)$ is, by definition, the partial derivative of the Lagrangian with respect to $\dot{\mathbf{A}}(\mathbf{k}, t)$, that is, the time derivative of $\mathbf{A}(\mathbf{k}, t)$, yielding

$$\pi(\mathbf{k}, t)^* = \left(\frac{\partial \widetilde{L}(\mathbf{k}, t)}{\partial \dot{\mathbf{A}}(\mathbf{k}, t)}\right)$$

where the Lagrangian $\widetilde{L}(\mathbf{k}, t)$ involved in the partial derivative is given by Eq. (14.98), so that due to this equation, it becomes

$$\pi(\mathbf{k}, t)^* = \left(\frac{\partial \int L_C(\mathbf{k}', t)\, d^3\mathbf{k}'}{\partial \dot{\mathbf{A}}(\mathbf{k}', t)}\right)$$

or, in view of Eq. (14.99),

$$\pi(\mathbf{k}, t)^* = \varepsilon^\circ \left(\frac{\partial \int \dot{\mathbf{A}}(\mathbf{k}', t)^* \cdot \dot{\mathbf{A}}(\mathbf{k}', t)/d^3\mathbf{k}'}{\partial \dot{\mathbf{A}}(\mathbf{k}, t)}\right)$$

so that

$$\boxed{\pi(\mathbf{k}, t)^* = \varepsilon^\circ \dot{\mathbf{A}}(\mathbf{k}, t)^*} \qquad \text{and thus} \qquad \boxed{\pi(\mathbf{k}, t) = \varepsilon^\circ \dot{\mathbf{A}}(\mathbf{k}, t)} \tag{14.101}$$

Now, recall that the vector potential $\mathbf{A}_\perp(\mathbf{k}, t)$ being perpendicular to $\mathbf{k}$, may be decomposed into two polarized vectors according to Eq. (14.75), that is,

$$\mathbf{A}_\perp(\mathbf{k}, t) = \mathbf{A}_{\boldsymbol{\varepsilon}}(\mathbf{k}, t) + \mathbf{A}_{\boldsymbol{\varepsilon}'}(\mathbf{k}, t)$$

which is also true for the time derivative of $\mathbf{A}_\perp(\mathbf{k}, t)$, that is,

$$\dot{\mathbf{A}}_\perp(\mathbf{k}, t) = \dot{\mathbf{A}}_{\boldsymbol{\varepsilon}}(\mathbf{k}, t) + \dot{\mathbf{A}}_{\boldsymbol{\varepsilon}'}(\mathbf{k}, t)$$

Hence, the conjugate momentum appearing in (14.101) may be decomposed in the same way according to

$$\pi_\perp(\mathbf{k}, t) = \pi_\varepsilon(\mathbf{k}, t) + \pi_{\varepsilon'}(\mathbf{k}, t)$$

14.5.2 Quantization in the Schrödinger picture

14.5.2.1 Field quantization in infinite space It is possible to find the quantum operators corresponding to the classical electromagnetic field, keeping in mind that within the Schrödinger picture, the operators do not depend on time. Quantizing the system of material particles, requires, for each α charged particle, to impose on the x, y, and z components $(\mathbf{r}_\alpha)_k$ of $\mathbf{r}_\alpha$, and on the x, y, and z components, and also on the corresponding components $(\mathbf{p}_{\alpha'})_j$ of $\mathbf{p}_{\alpha'}$ the condition that they are operators obeying the commutation rules:

$$\boxed{[\{(\mathbf{r}_\alpha)_k{}^{\mathrm{SP}}\}, \{(\mathbf{p}_{\alpha'})_j{}^{\mathrm{SP}}\}] = i\hbar\delta_{\alpha\alpha'}\, \delta_{jk}} \tag{14.102}$$

In a similar way, it is convenient to undertake the quantization of the electromagnetic field in the reciprocal space, by assuming that the potential vector $\mathbf{A}_{\boldsymbol{\varepsilon}}(\mathbf{k})$ and its conjugate momentum $\boldsymbol{\pi}_{\varepsilon'}(\mathbf{k}')^*$ become, respectively, operators $\mathbf{A}_{\boldsymbol{\varepsilon}}(\mathbf{k})$ and $\boldsymbol{\pi}_{\boldsymbol{\varepsilon}'}(\mathbf{k}')^\dagger$ obeying the commutation rule

$$\boxed{[\{\mathbf{A}_\varepsilon(\mathbf{k})^{\mathrm{SP}}\}, \{\boldsymbol{\pi}_{\varepsilon'}(\mathbf{k}')^{\mathrm{SP}}\}^\dagger] = i\hbar\, \delta_{\varepsilon\varepsilon'}\, \delta(\mathbf{k}-\mathbf{k}')} \tag{14.103}$$

Next, since, due to Eq. (14.22), the conjugate momentum $\boldsymbol{\pi}_{\boldsymbol{\varepsilon}}(\mathbf{k})$ of $\mathbf{A}_{\boldsymbol{\varepsilon}}(\mathbf{k})$ defined by Eq. (14.101) is related to the electric field $\mathbf{E}_{\boldsymbol{\varepsilon}}(\mathbf{k})$ through

$$\{\boldsymbol{\pi}_\varepsilon(\mathbf{k})^{\mathrm{SP}}\} = -\varepsilon^\circ\{\mathbf{E}_\varepsilon(\mathbf{k})^{\mathrm{SP}}\} \tag{14.104}$$

We have, after changing the classical field $\mathbf{E}_{\varepsilon'}(\mathbf{k}')^*$, the complex conjugate of $\mathbf{E}_{\varepsilon'}(\mathbf{k}')$ into the quantum operator $\mathbf{E}_{\varepsilon'}(\mathbf{k}')^{\mathrm{SP}\dagger}$ the Hermitian conjugate of $\mathbf{E}_{\varepsilon'}(\mathbf{k}')^{\mathrm{SP}}$, one obtains from Eq. (14.103) the following commutator between $\mathbf{A}_{\boldsymbol{\varepsilon}}(\mathbf{k})^{\mathrm{SP}}$ and its conjugate momentum $-\varepsilon^\circ\mathbf{E}_{\boldsymbol{\varepsilon}'}(\mathbf{k}')^{\mathrm{SP}\dagger}$:

$$[\{\mathbf{A}_{\boldsymbol{\varepsilon}}(\mathbf{k})^{\mathrm{SP}}\}, \{\mathbf{E}_{\varepsilon'}(\mathbf{k}')^{\mathrm{SP}}\}^\dagger] = -i\hbar\, \frac{1}{\varepsilon^\circ}\delta_{\varepsilon\varepsilon'}\, \delta(\mathbf{k}-\mathbf{k}') \tag{14.105}$$

Moreover, having obtained the quantum commutation rule dealing with the time-independent SP quantum operators describing the electromagnetic field in the reciprocal space, it is possible to obtain the corresponding SP operators in the geometric space, by performing the following time-independent transformations, analogous to the time-dependent ones (14.81)–(14.83), applied to the classical fields

$$\begin{aligned}\mathbf{E}_\perp(\mathbf{r})^{\mathrm{SP}} &= i\int E_{\omega_k} d^3\mathbf{k}\\ &\quad\times \{\hat{\boldsymbol{\varepsilon}}_{\mathbf{k}}(\mathbf{a}_\varepsilon(\mathbf{k})e^{i\mathbf{k}_n\cdot\mathbf{r}} - \mathbf{a}_\varepsilon(\mathbf{k})^\dagger e^{-i\mathbf{k}_n\cdot\mathbf{r}}) + \hat{\boldsymbol{\varepsilon}}'_{\mathbf{k}}(\mathbf{a}_{\varepsilon'}(\mathbf{k})e^{i\mathbf{k}_n\cdot\mathbf{r}} - \mathbf{a}_{\varepsilon'}(\mathbf{k})^\dagger e^{-i\mathbf{k}_n\cdot\mathbf{r}})\}\end{aligned} \tag{14.106}$$

$$\begin{aligned}\mathbf{A}_\perp(\mathbf{r})^{\mathrm{SP}} &= \int A_{\omega_k} d^3\mathbf{k}\\ &\quad\times \{\hat{\boldsymbol{\varepsilon}}_{\mathbf{k}}(\mathbf{a}_\varepsilon(\mathbf{k})e^{i\mathbf{k}_n\cdot\mathbf{r}} + \mathbf{a}_\varepsilon(\mathbf{k})^\dagger e^{-i\mathbf{k}_n\cdot\mathbf{r}}) + \hat{\boldsymbol{\varepsilon}}'_{\mathbf{k}}(\mathbf{a}_{\varepsilon'}(\mathbf{k})e^{i\mathbf{k}_n\cdot\mathbf{r}} + \mathbf{a}_{\varepsilon'}(\mathbf{k})^\dagger e^{-i\mathbf{k}_n\cdot\mathbf{r}})\}\end{aligned} \tag{14.107}$$

$$\begin{aligned}\mathbf{B}_\perp(\mathbf{r})^{\mathrm{SP}} &= i\int B_{\omega_k} d^3\mathbf{k}\\ &\quad\times \{(\hat{\kappa}\times\hat{\boldsymbol{\varepsilon}}_{\mathbf{k}})\{\mathbf{a}_\varepsilon(\mathbf{k})e^{i\mathbf{k}_n\cdot\mathbf{r}} - \mathbf{a}_\varepsilon(\mathbf{k})^\dagger e^{-i\mathbf{k}_n\cdot\mathbf{r}}\}\\ &\quad+(\hat{\kappa}\times\hat{\boldsymbol{\varepsilon}}'_{\mathbf{k}})\{\mathbf{a}_{\varepsilon'}(\mathbf{k})e^{i\mathbf{k}_n\cdot\mathbf{r}} - \mathbf{a}_{\varepsilon'}(\mathbf{k})^\dagger e^{-i\mathbf{k}_n\cdot\mathbf{r}}\}\}\end{aligned} \tag{14.108}$$

In these equations, the time-independent operators $\mathbf{a}_\varepsilon(\mathbf{k})$ and their Hermitian conjugates $\mathbf{a}_\varepsilon(\mathbf{k})^\dagger$ replace, respectively, the time-dependent normal modes $\alpha_{\varepsilon\perp}(\mathbf{k},t)$ and $\alpha_{\varepsilon\perp}(\mathbf{k},t)^*$ appearing in Eqs. (14.81)–(14.83), with, in order to satisfy Eq. (14.105), the following commutation rule between $\mathbf{a}_\varepsilon(\mathbf{k})$ and $\mathbf{a}_\varepsilon(\mathbf{k})^\dagger$:

$$[\mathbf{a}_\varepsilon(\mathbf{k}), \mathbf{a}_{\varepsilon'}(\mathbf{k}')^\dagger] = \delta_{\varepsilon\varepsilon'}\delta(\mathbf{k}-\mathbf{k}') \tag{14.109}$$

Moreover, due to Eq. (14.70) giving the energy of the electromagnetic field expressed in terms of $\boldsymbol{\alpha}_{\varepsilon\perp}(\mathbf{k})$ and of its complex conjugate $\boldsymbol{\alpha}_{\varepsilon\perp}(\mathbf{k})^*$, the corresponding Hamiltonian operator $\mathbf{H}^{SP}$ may be obtained from this last expression by replacing, respectively, in it $\boldsymbol{\alpha}_{\varepsilon\perp}(\mathbf{k})$, and $\boldsymbol{\alpha}_{\varepsilon\perp}(\mathbf{k})^*$ by $\mathbf{a}_\varepsilon(\mathbf{k})$ and $\mathbf{a}_\varepsilon(\mathbf{k})^\dagger$:

$$\mathbf{H}^{SP} = \int \hbar\omega(k) \left\{ \frac{\mathbf{a}_\varepsilon(\mathbf{k})^\dagger \mathbf{a}_\varepsilon(\mathbf{k}) + \mathbf{a}_\varepsilon(\mathbf{k})\mathbf{a}_\varepsilon(\mathbf{k})^\dagger}{2} \right\} d^3\mathbf{k}$$

Then, using the commutator (14.109) the Hamiltonian transforms to

$$\mathbf{H}^{SP} = \int \mathbf{H}(\mathbf{k})^{SP} d^3\mathbf{k} \quad \text{with} \quad \mathbf{H}(\mathbf{k})^{SP} = \hbar\omega(k)\left(\mathbf{a}_\varepsilon(\mathbf{k})^\dagger \mathbf{a}_\varepsilon(\mathbf{k}) + \frac{1}{2}\right) \tag{14.110}$$

14.5.2.2 Field quantization inside a cavity In a cavity of volume V, the operators describing the fields corresponding to the classical fields defined by Eqs. (14.85)–(14.87) may be obtained by proceeding in passing from Eqs. (14.85)–(14.87) to Eqs. (14.106)–(14.108):

$$\begin{aligned}\mathbf{E}_\perp(\mathbf{r})^{SP} &= i\sum_n \sqrt{\frac{\hbar\omega_n}{2\varepsilon^\circ V}} \\ &\times \{\hat{\boldsymbol{\varepsilon}}_{\mathbf{k}_n}(\mathbf{a}_{n\varepsilon}e^{i\mathbf{k}_n\cdot\mathbf{r}} - \mathbf{a}^\dagger_{n\varepsilon}e^{-i\mathbf{k}_n\cdot\mathbf{r}}) \\ &+ \hat{\boldsymbol{\varepsilon}}'_{\mathbf{k}_n}(\mathbf{a}_{n\varepsilon'}e^{i\mathbf{k}_n\cdot\mathbf{r}} - \mathbf{a}^\dagger_{n\varepsilon'}e^{-i\mathbf{k}_n\cdot\mathbf{r}})\}\end{aligned} \tag{14.111}$$

$$\begin{aligned}\mathbf{A}_\perp(\mathbf{r})^{SP} &= \sum_n \sqrt{\frac{\hbar}{2\varepsilon^\circ\omega_n V}} \\ &\times \{\hat{\boldsymbol{\varepsilon}}_{\mathbf{k}_n}(\mathbf{a}_{n\varepsilon}e^{i\mathbf{k}_n\cdot\mathbf{r}} + \mathbf{a}^\dagger_{n\varepsilon}e^{-i\mathbf{k}_n\cdot\mathbf{r}}) \\ &+ \hat{\boldsymbol{\varepsilon}}'_{\mathbf{k}_n}(\mathbf{a}_{n\varepsilon'}e^{i\mathbf{k}_n\cdot\mathbf{r}} + \mathbf{a}^\dagger_{n\varepsilon'}e^{-i\mathbf{k}_n\cdot\mathbf{r}})\}\end{aligned} \tag{14.112}$$

$$\begin{aligned}\mathbf{B}_\perp(\mathbf{r})^{SP} &= i\sum_n \sqrt{\frac{\hbar\omega_n}{2\varepsilon^\circ V c^2}} \\ &\times \{(\hat{\kappa} \times \hat{\boldsymbol{\varepsilon}}_{\mathbf{k}_n})\{\mathbf{a}_{n\varepsilon}e^{i\mathbf{k}_n\cdot\mathbf{r}} - \mathbf{a}^\dagger_{n\varepsilon}e^{-i\mathbf{k}_n\cdot\mathbf{r}}\} \\ &+ (\hat{\kappa} \times \hat{\boldsymbol{\varepsilon}}'_{\mathbf{k}_n})(\mathbf{a}_{n\varepsilon'}e^{i\mathbf{k}_n\cdot\mathbf{r}} - \mathbf{a}^\dagger_{n\varepsilon'}e^{-i\mathbf{k}_n\cdot\mathbf{r}})\}\end{aligned} \tag{14.113}$$

where the classical variables $\alpha_{n\varepsilon\perp}$ and their complex conjugates $\alpha^*_{n\varepsilon\perp}$ have been replaced, respectively, by the operators $\mathbf{a}_{n\varepsilon}$ and their Hermitian conjugates $\mathbf{a}^\dagger_{n\varepsilon}$ required to obey the commutation rules

$$[\mathbf{a}_{n\varepsilon}, \mathbf{a}^\dagger_{m\varepsilon'}] = \delta_{\varepsilon\varepsilon'}\delta_{nm} \tag{14.114}$$

The commutators of the Cartesian components $\mathbf{A}_\perp(\mathbf{r})_l{}^{SP}$ and $\mathbf{E}_\perp(\mathbf{r})_l{}^{SP}$ with $l = x, y, z$ of operators $\mathbf{A}_\perp(\mathbf{r})^{SP}$ and $\mathbf{E}_\perp(\mathbf{r})^{SP}$ may be proved to be given by

$$[\mathbf{A}_\perp(\mathbf{r})_l{}^{SP}, \mathbf{E}_\perp(\mathbf{r})_j{}^{SP}] = -i\hbar\varepsilon^{\circ -1}\, \delta^T_{lj}(\mathbf{r} - \mathbf{r}') \tag{14.115}$$

where the last right-hand-side term is the transverse Dirac delta function defined by

$$\delta^T_{lj}(\mathbf{r} - \mathbf{r}') = \left(\frac{1}{2\pi}\right)^3 \int (e^{i\mathbf{k}\cdot(\mathbf{r}-\mathbf{r}')}) \left(\delta_{lj} - \frac{k_l k_j}{k^2}\right) d^3\mathbf{k}$$

in which the k_i or k_j are the Cartesian components of the $\mathbf{k}$ vector. Observe that when $i=j$, the commutator (14.115) reduces to

$$[\mathbf{A}_{\perp}(\mathbf{r})_l{}^{\mathrm{SP}}, \mathbf{E}_{\perp}(\mathbf{r})_l{}^{\mathrm{SP}}] = -i\hbar\varepsilon^{\circ -1}\left(\frac{1}{2\pi}\right)^3 \int (e^{i\mathbf{k}\cdot(\mathbf{r}-\mathbf{r}')})\left(1-\frac{k_l^2}{k^2}\right) d^3\mathbf{k}$$

Moreover, the Hamiltonian operator $\mathbf{H}^{\mathrm{SP}}$ corresponding to the electromagnetic energy (14.84), takes the form

$$\mathbf{H}^{\mathrm{SP}} = \sum_n \mathbf{H}_n{}^{\mathrm{SP}} \quad \text{with} \quad \mathbf{H}_n{}^{\mathrm{SP}} = \hbar\omega_n\left(\mathbf{a}_{n\varepsilon}^{\dagger}\mathbf{a}_{n\varepsilon} + \frac{1}{2}\right) \tag{14.116}$$

14.5.2.3 Eigenvalue equation of the electromagnetic field Hamiltonians We observe that the continuous and discrete sums of Hamiltonians given, respectively, by Eq. (14.110) or (14.116) have the same structure as that (5.9) of the quantum harmonic oscillator, and involve the basic commutation rules (14.109) and (14.114), which have also the same structure as that (5.5) dealing with the usual quantum harmonic oscillator. As a consequence, all that has been found for the quantum harmonic oscillator holds also for the Hamiltonians involved in these equations, so that the following eigenvalue equations equivalent to (5.40) read in the present situation

$$\left(\{\mathbf{a}_\varepsilon(\mathbf{k})^{\dagger}\mathbf{a}_\varepsilon(\mathbf{k})\} + \tfrac{1}{2}\right)|\{l_\varepsilon(\mathbf{k})\}\rangle = \left(l_\varepsilon(\mathbf{k}) + \tfrac{1}{2}\right)|\{l_\varepsilon(\mathbf{k})\}\rangle \tag{14.117}$$

and

$$\left(\{\mathbf{a}_{n\varepsilon}^{\dagger}\mathbf{a}_{n\varepsilon}\} + \tfrac{1}{2}\right)|\{l_{n\varepsilon}\}\rangle = \left(l_{n\varepsilon} + \tfrac{1}{2}\right)|\{l_{n\varepsilon}\}\rangle \tag{14.118}$$

with for each $l_\varepsilon(\mathbf{k})$ or $l_{n\varepsilon}$

$$l_{n\varepsilon} = 0, 1, 2, \ldots \qquad \text{and} \qquad l_\varepsilon(\mathbf{k}) = 0, 1, 2, \ldots$$

and where the $|\{l_\varepsilon(\mathbf{k})\}\rangle$ and $|\{l_{n\varepsilon}\}\rangle$ are, respectively, the eigenvectors of $\{\mathbf{a}_\varepsilon(\mathbf{k})^{\dagger}\mathbf{a}_\varepsilon(\mathbf{k})\}$ and $\{\mathbf{a}_{n\varepsilon}^{\dagger}\mathbf{a}_{n\varepsilon}\}$ obeying the orthonormality properties

$$\langle\{l_{n\varepsilon}\}|\{j_{n\varepsilon}\}\rangle = \delta_{l_{n\varepsilon}\, j_{n\varepsilon}} \qquad \text{and} \qquad \langle\{l_\varepsilon(\mathbf{k})\}|\{j_\varepsilon(\mathbf{k})\}\rangle = \delta_{l_\varepsilon(\mathbf{k})\, j_\varepsilon(\mathbf{k})} \tag{14.119}$$

Hence, due to Eqs. (14.110) and (14.116), the eigenvalue equations (14.117) and (14.118) read

$$\mathbf{H}(\mathbf{k})|\{l_\varepsilon(\mathbf{k})\}\rangle = \hbar\omega(k)\left(l_\varepsilon(\mathbf{k}) + \tfrac{1}{2}\right)|\{l_\varepsilon(\mathbf{k})\}\rangle \tag{14.120}$$

$$\mathbf{H}_n|\{l_{n\varepsilon}\}\rangle = \hbar\omega_n\left(l_{n\varepsilon} + \tfrac{1}{2}\right)|\{l_{n\varepsilon}\}\rangle \tag{14.121}$$

The quantum numbers $l_\varepsilon(\mathbf{k})$ or $l_{n\varepsilon}$ are, respectively, the excitation degrees of the continuous mode characterized by the wave vector $\mathbf{k}$ and that of the nth discrete mode within the polarization ε. Hence, when the field is in one state $|\{l_\varepsilon(\mathbf{k})\}\rangle$ of the continuous situation, or in one $|\{l_{n\varepsilon}\}\rangle$ of the discrete case, the corresponding quantum numbers $l_\varepsilon(\mathbf{k})$ or $l_{n\varepsilon}$ may be viewed through Eqs. (14.120) and (14.121), as the number of energy packets $\hbar\omega(k)$ or $\hbar\omega_n$ inside the corresponding modes of the electromagnetic field. These energy packets may be considered as light corpuscles, which are called *photons*.

Now, owing to the correspondence between quantum electromagnetic modes and quantum harmonic oscillators, it is clear that Eqs. (5.53) and (5.63) may be applied to the ladder operators of the electromagnetic field to yield, respectively, for the continuous and discrete case

$$\mathbf{a}_\varepsilon(\mathbf{k})|\{l_\varepsilon(\mathbf{k})\}\rangle = \sqrt{l_\varepsilon(\mathbf{k})}|\{l_\varepsilon(\mathbf{k}) - 1\}\rangle$$

$$\mathbf{a}_\varepsilon(\mathbf{k})^\dagger|\{l_\varepsilon(\mathbf{k})\}\rangle = \sqrt{l_\varepsilon(\mathbf{k}) + 1}|\{l_\varepsilon(\mathbf{k}) + 1\}\rangle$$

$$\mathbf{a}_{n\varepsilon}|\{l_{n\varepsilon}\}\rangle = \sqrt{l_{n\varepsilon}}|\{l_{n\varepsilon} - 1\}\rangle \qquad (14.122)$$

$$\mathbf{a}_{n\varepsilon}^\dagger|\{l_{n\varepsilon}\}\rangle = \sqrt{l_{n\varepsilon} + 1}|\{l_{n\varepsilon} + 1\}\rangle \qquad (14.123)$$

14.5.3 Heisenberg picture fields

When passing to the Heisenberg picture, the SP time-independent ladder operators of the field $\mathbf{a}_{n\varepsilon}$ and $\mathbf{a}_{n\varepsilon}^\dagger$, or $\mathbf{a}_\varepsilon(\mathbf{k})$ and $\mathbf{a}_\varepsilon^\dagger(\mathbf{k})$, become time dependent, that is, $\mathbf{a}_{n\varepsilon}(t)$ and $\mathbf{a}_{n\varepsilon}^\dagger(t)$ or $\mathbf{a}_\varepsilon(\mathbf{k}, t)$ and $\mathbf{a}_\varepsilon^\dagger(\mathbf{k}, t)$. For the situation of electromagnetic fields enclosed in a box, the time dependence of $\mathbf{a}_{n\varepsilon}(t)$ is given by the Heisenberg equations (3.94) involving the Hamiltonian $\mathbf{H}_{n\varepsilon}$, which read

$$i\hbar\left(\frac{\partial \mathbf{a}_{n\varepsilon}(t)}{\partial t}\right) = [\mathbf{a}_{n\varepsilon}(t), \mathbf{H}_{n\varepsilon}]$$

or, due to Eq. (14.116), giving the expression of the total Hamiltonian $\mathbf{H}$ of the field, which is the same in the Heisenberg and Schrödinger pictures when an isolated electromagnetic field is considered

$$i\left(\frac{\partial \mathbf{a}_{n\varepsilon}(t)}{\partial t}\right) = \omega_n[\mathbf{a}_{n\varepsilon}(t), \mathbf{a}_{n\varepsilon}^\dagger(t)\mathbf{a}_{n\varepsilon}(t)]$$

Again, using the commutator (14.114), which reads

$$[\mathbf{a}_{n\varepsilon}(t), \mathbf{a}_{n\varepsilon}^\dagger(t)] = 1$$

this equation transforms to

$$\left(\frac{\partial \mathbf{a}_{n\varepsilon}(t)}{\partial t}\right) = -i\omega_n\mathbf{a}_{n\varepsilon}(t)$$

the solution of which is

$$\mathbf{a}_{n\varepsilon}(t) = \mathbf{a}_{n\varepsilon}(0)e^{-i\omega_n t}$$

On the other hand, for the free space, the Heisenberg equation governing the $\mathbf{a}_\varepsilon(\mathbf{k}, t)$ is given by

$$i\hbar\left(\frac{\partial \mathbf{a}_\varepsilon(\mathbf{k}, t)}{\partial t}\right) = [\mathbf{a}_\varepsilon(\mathbf{k}, t), \mathbf{H}(\mathbf{k})]$$

where the total Hamiltonian of the field is now given by Eq. (14.110), the solution of this equation being

$$\mathbf{a}_\varepsilon(\mathbf{k}, t) = \mathbf{a}_\varepsilon(\mathbf{k},0)e^{-i\omega(k)t}$$

On the other hand, in the free space, the HP operators corresponding to the SP fields (14.111)–(14.113), become

$$\{\mathbf{E}_\perp(\mathbf{r},t)^{\mathrm{HP}}\} = i\int E_{\omega_k} d^3\mathbf{k}$$
$$\times\{\hat{\boldsymbol{\varepsilon}}_{\mathbf{k}}(\mathbf{a}_\varepsilon(\mathbf{k})e^{i\mathbf{k}\cdot\mathbf{r}}e^{-i\omega(k)t} - \mathbf{a}_\varepsilon(\mathbf{k})^\dagger e^{-i\mathbf{k}\cdot\mathbf{r}}e^{i\omega(k)t})$$
$$+\hat{\boldsymbol{\varepsilon}}'_{\mathbf{k}}(\mathbf{a}_{\varepsilon'}(\mathbf{k})e^{i\mathbf{k}\cdot\mathbf{r}}e^{-i\omega(k)t} - \mathbf{a}_{\varepsilon'}(\mathbf{k}')^\dagger e^{-i\mathbf{k}_n\cdot\mathbf{r}}e^{i\omega(k)t})\} \quad (14.124)$$

$$\{\mathbf{A}_\perp(\mathbf{r},t)^{\mathrm{HP}}\} = \int A_{\omega_k} d^3\mathbf{k}$$
$$\times\{\hat{\boldsymbol{\varepsilon}}_{\mathbf{k}}(\mathbf{a}_\varepsilon(\mathbf{k})e^{i\mathbf{k}\cdot\mathbf{r}}e^{-i\omega(k)t} + \mathbf{a}_\varepsilon(\mathbf{k})^\dagger e^{-i\mathbf{k}\cdot\mathbf{r}}e^{i\omega(k)t})$$
$$+\hat{\boldsymbol{\varepsilon}}'_{\mathbf{k}}(\mathbf{a}_{\varepsilon'}(\mathbf{k})e^{i\mathbf{k}\cdot\mathbf{r}}e^{-i\omega(k)t} + \mathbf{a}_\varepsilon(\mathbf{k})^\dagger e^{-i\mathbf{k}\cdot\mathbf{r}}e^{i\omega(k)t})\}$$

$$\{\mathbf{B}_\perp(\mathbf{r},t)^{\mathrm{HP}}\} = i\int B_{\omega_k} d^3\mathbf{k}$$
$$\times\{(\hat{\kappa}\times\hat{\boldsymbol{\varepsilon}}_{\mathbf{k}})\{\mathbf{a}_\varepsilon(\mathbf{k})e^{i\mathbf{k}\cdot\mathbf{r}}e^{-i\omega(k)t} - \mathbf{a}_\varepsilon(\mathbf{k})^\dagger e^{-i\mathbf{k}\cdot\mathbf{r}}e^{i\omega(k)t}\}$$
$$+(\hat{\kappa}\times\hat{\boldsymbol{\varepsilon}}'_{\mathbf{k}})\{\mathbf{a}_{\varepsilon'}(\mathbf{k})e^{i\mathbf{k}\cdot\mathbf{r}}e^{-i\omega(k)t} - \mathbf{a}_{\varepsilon'}(\mathbf{k})^\dagger e^{-i\mathbf{k}\cdot\mathbf{r}}e^{i\omega(k)t}\}\}$$

whereas within a cavity of volume V the HP operators corresponding to the SP fields (14.106)–(14.108) take the form

$$\{\mathbf{E}_\perp(\mathbf{r},t)^{\mathrm{HP}}\} = i\sum_{n\varepsilon}\sqrt{\frac{\hbar\omega_n}{2\varepsilon^\circ V}}$$
$$\times\{\hat{\boldsymbol{\varepsilon}}_{\mathbf{k}_n}(\mathbf{a}_{n\varepsilon}e^{i\mathbf{k}_n\cdot\mathbf{r}}e^{-i\omega_n t} - \mathbf{a}^\dagger_{n\varepsilon}e^{-i\mathbf{k}_n\cdot\mathbf{r}}e^{i\omega_n t})$$
$$+\hat{\boldsymbol{\varepsilon}}'_{\mathbf{k}_n}(\mathbf{a}_{n\varepsilon'}e^{i\mathbf{k}_n\cdot\mathbf{r}}e^{-i\omega_n t} - \mathbf{a}^\dagger_{n\varepsilon'}e^{-i\mathbf{k}_n\cdot\mathbf{r}}e^{i\omega_n t})\} \quad (14.125)$$

$$\{\mathbf{A}_\perp(\mathbf{r},t)^{\mathrm{HP}}\} = \sum_{n\varepsilon}\sqrt{\frac{\hbar}{2\varepsilon^\circ\omega_n V}}$$
$$\times\{\hat{\boldsymbol{\varepsilon}}_{\mathbf{k}_n}(\mathbf{a}_{n\varepsilon}e^{i\mathbf{k}_n\cdot\mathbf{r}}e^{-i\omega_n t} + \mathbf{a}^\dagger_{n\varepsilon}e^{-i\mathbf{k}_n\cdot\mathbf{r}}e^{i\omega_n t})$$
$$+\hat{\boldsymbol{\varepsilon}}'_{\mathbf{k}_n}(\mathbf{a}_{n\varepsilon'}e^{i\mathbf{k}_n\cdot\mathbf{r}}e^{-i\omega_n t} + \mathbf{a}^\dagger_{n\varepsilon'}e^{-i\mathbf{k}_n\cdot\mathbf{r}}e^{i\omega_n t})\} \quad (14.126)$$

$$\{\mathbf{B}_\perp(\mathbf{r},t)^{\mathrm{HP}}\} = i\sum_{n\varepsilon}\sqrt{\frac{\hbar\omega_n}{2\varepsilon^\circ Vc^2}}$$
$$\times\{(\hat{\kappa}\times\hat{\boldsymbol{\varepsilon}}_{\mathbf{k}_n})\{\mathbf{a}_{n\varepsilon}e^{i\mathbf{k}_n\cdot\mathbf{r}}e^{-i\omega_n t} - \mathbf{a}^\dagger_{n\varepsilon}e^{-i\mathbf{k}_n\cdot\mathbf{r}}e^{i\omega_n t}\}$$
$$+(\hat{\kappa}\times\hat{\boldsymbol{\varepsilon}}'_{\mathbf{k}_n})\{\mathbf{a}_{n\varepsilon'}e^{i\mathbf{k}_n\cdot\mathbf{r}}e^{-i\omega_n t} - \mathbf{a}^\dagger_{n\varepsilon'}e^{-i\mathbf{k}_n\cdot\mathbf{r}}e^{i\omega_n t}\}\} \quad (14.127)$$

14.5.4 Average values of electromagnetic field operators

14.5.4.1 Analogies between $A_\perp(r,t)^{SP}$ and Q and between $E_\perp(r,t)^{SP}$ and P

Observe that the SP operators describing the electric and magnetic potential vector fields defined by Eqs. (14.111) and (14.112) may be written

$$\{\mathbf{E}_\perp(\mathbf{r})^{SP}\} = \sum_{n\varepsilon}\{\mathbf{E}_{n\varepsilon}(\mathbf{r})^{SP} + \mathbf{E}_{n\varepsilon'}(\mathbf{r})^{SP}\} \qquad \text{and}$$

$$\{\mathbf{A}_\perp(\mathbf{r})^{SP}\} = \sum_{n\varepsilon}\{\mathbf{A}_{n\varepsilon}(\mathbf{r})^{SP} + \mathbf{A}_{n\varepsilon'}(\mathbf{r})^{SP}\}$$

with, respectively,

$$\{\mathbf{A}_{n\varepsilon}(\mathbf{r})^{SP}\} = \hat{\boldsymbol{\varepsilon}}_{\mathbf{k}_n}\sqrt{\frac{\hbar}{2\varepsilon^\circ\omega_n V}}\{\mathbf{a}_{n\varepsilon}(e^{i\mathbf{k}_n\cdot\mathbf{r}}) + \mathbf{a}_{n\varepsilon}^\dagger(e^{-\mathbf{k}_n\cdot\mathbf{r}})\}$$

$$\{\mathbf{E}_{n\varepsilon}(\mathbf{r})^{SP}\} = i\hat{\boldsymbol{\varepsilon}}_{\mathbf{k}_n}\sqrt{\frac{\hbar\omega_n}{2\varepsilon^\circ V}}\{\mathbf{a}_{n\varepsilon}(e^{i\mathbf{k}_n\cdot\mathbf{r}}) - \mathbf{a}_{n\varepsilon}^\dagger(e^{-i\mathbf{k}_n\cdot\mathbf{r}})\} \tag{14.128}$$

and similar expressions for $\mathbf{A}_{n\varepsilon'}(\mathbf{r})^{SP}$ and $\mathbf{E}_{n\varepsilon'}(\mathbf{r})^{SP}$ by changing $\hat{\boldsymbol{\varepsilon}}$ into $\hat{\boldsymbol{\varepsilon}}'$.

Besides, keeping in mind the expressions of the operator $\mathbf{Q}$ and of its conjugate momentum $\mathbf{P}$ given, respectively, by Eqs. (5.6) and (5.7), that is,

$$\mathbf{Q} = \sqrt{\frac{\hbar}{2M\omega}}(\mathbf{a}^\dagger + \mathbf{a}) \qquad \text{and} \qquad \mathbf{P} = i\sqrt{\frac{M\hbar\omega}{2}}(\mathbf{a}^\dagger - \mathbf{a})$$

where $\mathbf{a}^\dagger$ and $\mathbf{a}$ are, respectively, the lowering and raising operators of the oscillator, whereas M is its reduced mass and ω its angular frequency, it is of interest to remark the analogy between the vector potential operator $\mathbf{A}_{n\varepsilon}(\mathbf{r})^{SP}$ of the electromagnetic field and the coordinate operator $\mathbf{Q}$ of the quantum harmonic oscillator and between the electric field operator $\mathbf{E}_{n\varepsilon}(\mathbf{r})^{SP}$ and the momentum operator $\mathbf{P}$ conjugate of $\mathbf{Q}$.

14.5.4.2 Mean values performed over Hamiltonian eigenstates

Now, write the average value of the electric field operator $\mathbf{E}_{n\varepsilon}(\mathbf{r})^{SP}$ on the eigenstates $|\{l_{n\varepsilon}\}\rangle$ defined by the eigenvalue equation (14.121), that is,

$$\langle\{l_{n\varepsilon}\}|\{\mathbf{E}_{n\varepsilon}(\mathbf{r})^{SP}\}|\{l_{n\varepsilon}\}\rangle = i\hat{\boldsymbol{\varepsilon}}_{\mathbf{k}_n}\sqrt{\frac{\hbar\omega_n}{2\varepsilon^\circ V}}\langle\{l_{n\varepsilon}\}|(\mathbf{a}_{n\varepsilon}(e^{i\mathbf{k}_n\cdot\mathbf{r}}) - \mathbf{a}_{n\varepsilon}^\dagger(e^{-i\mathbf{k}_n\cdot\mathbf{r}}))|\{l_{n\varepsilon}\}\rangle$$

Then, due to Eqs. (14.122) and (14.123), it reads

$$\begin{aligned}\langle\{l_{n\varepsilon}\}|\{\mathbf{E}_{n\varepsilon}(\mathbf{r})^{SP}\}|\{l_{n\varepsilon}\}\rangle &= i\hat{\boldsymbol{\varepsilon}}_{\mathbf{k}_n}\sqrt{\frac{\hbar\omega_n}{2\varepsilon^\circ V}}\{(e^{i\mathbf{k}_n\cdot\mathbf{r}})\sqrt{l_{n\varepsilon}}\langle\{l_{n\varepsilon}\}|\{l_{n\varepsilon}-1\}\rangle \\ &\quad - (e^{-i\mathbf{k}_n\cdot\mathbf{r}})\sqrt{l_{n\varepsilon}+1}\langle\{l_{n\varepsilon}\}|\{l_{n\varepsilon}+1\}\rangle\}\end{aligned}$$

or, due to the orthogonality relations (14.119)

$$\langle\{l_{n\varepsilon}\}|\{\mathbf{E}_{n\varepsilon}(\mathbf{r})^{SP}\}|\{l_{n\varepsilon}\}\rangle = 0 \tag{14.129}$$

In a similar way, one would obtain

$$\langle\{l_{n\varepsilon}\}|\{\mathbf{A}_{n\varepsilon}(\mathbf{r})^{SP}\}|\{l_{n\varepsilon}\}\rangle = 0 \tag{14.130}$$

Results (14.129) and (14.130) are for the electromagnetic fields the equivalent of those (5.85) and (5.93) dealing with harmonic oscillator, that is,

$$\langle\{n\}|\mathbf{Q}|\{n\}\rangle = 0 \qquad \text{and} \qquad \langle\{n\}|\mathbf{P}|\{n\}\rangle = 0$$

14.5.4.3 Mean values performed over coherent states The average values of the electromagnetic fields over coherent states $|\{\alpha_{n\varepsilon}\}\rangle$ obeying the eigenvalue equation

$$\mathbf{a}_{n\varepsilon}|\{\alpha_{n\varepsilon}\}\rangle = \alpha_{n\varepsilon}|\{\alpha_{n\varepsilon}\}\rangle \quad \text{with} \quad \langle\{\alpha_{n\varepsilon}\}|\{\alpha_{n\varepsilon}\}\rangle = 1$$

read

$$\langle\{\alpha_{n\varepsilon}\}|\{\mathbf{E}_{n\varepsilon}(\mathbf{r})^{\mathrm{SP}}\}|\{\alpha_{n\varepsilon}\}\rangle = i\hat{\boldsymbol{\varepsilon}}_{\mathbf{k}_n}\sqrt{\frac{\hbar\omega_n}{2\varepsilon^\circ V}}\{\alpha_{n\varepsilon}(e^{i\mathbf{k}_n\cdot\mathbf{r}}) - \alpha^*_{n\varepsilon}(e^{-i\mathbf{k}_n\cdot\mathbf{r}})\} \tag{14.131}$$

where the coherent states are described by the following expansions, which are analogous to that of (6.16),

$$|\{\alpha_{n\varepsilon}\}\rangle = e^{-|\alpha_{n\varepsilon}|^2/2}\sum_{l_{n\varepsilon}}\frac{(\alpha_{n\varepsilon})^{l_{n\varepsilon}}}{\sqrt{l_{n\varepsilon}!}}|\{l_{n\varepsilon}\}\rangle \tag{14.132}$$

In a similar way, one would obtain for the magnetic potential vector averaged over coherent states

$$\langle\{\alpha_{n\varepsilon}\}|\{\mathbf{A}_{n\varepsilon}(\mathbf{r})^{\mathrm{SP}}\}|\{\alpha_{n\varepsilon}\}\rangle = \hat{\boldsymbol{\varepsilon}}_{\mathbf{k}_n}\sqrt{\frac{\hbar}{2\varepsilon^\circ\omega_n V}}\{\alpha_{n\varepsilon}(e^{i\mathbf{k}_n\cdot\mathbf{r}}) + \alpha^*_{n\varepsilon}(e^{-i\mathbf{k}_n\cdot\mathbf{r}})\} \tag{14.133}$$

Of course, when passing to the Heisenberg picture, the Schrödinger picture time-independent operators given by Eqs. (14.131) and (14.133) become time dependent, so that, owing to Eqs. (14.125) and (14.126), they take, respectively, the forms

$$\begin{aligned}\langle\{\alpha_{n\varepsilon}\}|\{\mathbf{A}_{n\varepsilon}(\mathbf{r},t)^{\mathrm{HP}}\}|\{\alpha_{n\varepsilon}\}\rangle &= \hat{\boldsymbol{\varepsilon}}_{\mathbf{k}_n}\sqrt{\frac{\hbar}{2\varepsilon^\circ\omega_n V}}\{\alpha_{n\varepsilon}(e^{i\mathbf{k}_n\cdot\mathbf{r}})(e^{-i\omega_n t}) \\ &\quad + \alpha^*_{n\varepsilon}(e^{-i\mathbf{k}_n\cdot\mathbf{r}})(e^{i\omega_n t})\}\end{aligned} \tag{14.134}$$

and

$$\boxed{\langle\{\alpha_{n\varepsilon}\}|\{\mathbf{E}_{n\varepsilon}(\mathbf{r},t)^{\mathrm{HP}}\}|\{\alpha_{n\varepsilon}\}\rangle = i\hat{\boldsymbol{\varepsilon}}_{\mathbf{k}_n}\sqrt{\frac{\hbar\omega_n}{2\varepsilon^\circ V}}\{\alpha_{n\varepsilon}(e^{i\mathbf{k}_n\cdot\mathbf{r}})(e^{-i\omega_n t}) - \alpha^*_{n\varepsilon}(e^{-i\mathbf{k}_n\cdot\mathbf{r}})(e^{i\omega_n t})\}} \tag{14.135}$$

Observe that the Heisenberg equations (14.134) and (14.135), and those corresponding to the other polarization $\hat{\boldsymbol{\varepsilon}}'_{\mathbf{k}_n}$, have the same structure as the corresponding components (14.85) and (14.86) defined by Eqs. (14.125) and (14.126), appearing in classical electromagnetic theory. Moreover, circular polarized light may be introduced with the help of a 2D coherent state, by aid of an equation similar to (6.61).

It must be emphasized that, except for the fact that the commutators

$$[\{\mathbf{A}_{n\varepsilon}(\mathbf{r})^{\mathrm{SP}}\}, \{\mathbf{E}_{n\varepsilon}(\mathbf{r})^{\mathrm{SP}}\}^\dagger] \qquad \text{and} \qquad [\{\mathbf{A}_{\varepsilon}(\mathbf{k})^{\mathrm{SP}}\}, \{\mathbf{E}_{\varepsilon'}(\mathbf{k}')^{\mathrm{SP}}\}^\dagger]$$

are more complicated than those between $\mathbf{Q}$ and $\mathbf{P}$, a large part of the relations that have been found for quantum harmonic oscillators hold also for electromagnetic fields. The only differences lie in the presence of the phase factors $e^{-i\mathbf{k}_n\cdot\mathbf{r}}$ and $e^{i\mathbf{k}_n\cdot\mathbf{r}}$ and also in the changes occurring when passing from $\mathbf{Q}$ to $\mathbf{A}_{n\varepsilon}(\mathbf{r})^{\mathrm{SP}}$ and from $\mathbf{P}$ to $\mathbf{E}_{n\varepsilon}(\mathbf{r})^{\mathrm{SP}}$, which are

$$\sqrt{\frac{\hbar}{2M\omega}} \rightarrow \hat{\boldsymbol{\varepsilon}}_{\mathbf{k}_n}\sqrt{\frac{\hbar}{2\varepsilon^\circ\omega_n V}} \qquad \text{and} \qquad \sqrt{\frac{M\hbar\omega}{2}} \rightarrow -\hat{\boldsymbol{\varepsilon}}_{\mathbf{k}_n}\sqrt{\frac{\hbar\omega_n}{2\varepsilon^\circ V}}$$

Owing to the deep analogies between the electric field and the operators $\mathbf{P}$ and also between the vector potential and the operator $\mathbf{Q}$, it is possible to apply many results found above for operators $\mathbf{Q}$ and $\mathbf{P}$, of quantum harmonic oscillators to the electric field and the vector potential operators. For instance, applying Eq. (6.50), it would appear that, in the Heisenberg picture, the average value of the squared electric field reads

$$\begin{aligned}&\langle\{\alpha_{n\varepsilon}\}|\{\mathbf{E}_{n\varepsilon}(\mathbf{r},t)^{\mathrm{HP}}\}^2|\{\alpha_{n\varepsilon}\}\rangle\\ &\quad = -(\hat{\boldsymbol{\varepsilon}}_{\mathbf{k}_n})^2\frac{\hbar\omega_n}{2\varepsilon^\circ V}\{(\alpha_{n\varepsilon}e^{-i\mathbf{k}_n\cdot\mathbf{r}}e^{i\omega_n t} - \alpha^*_{n\varepsilon}e^{i\mathbf{k}_n\cdot\mathbf{r}}e^{-i\omega_n t})^2 - 1\}\end{aligned}$$

so that the relative dispersion $\Delta\mathbf{E}_{n\varepsilon}(t)/\langle\mathbf{E}_{n\varepsilon}(t)\rangle$ of the HP electric field when it is in a coherent state reads

$$\frac{\Delta\mathbf{E}_{n\varepsilon}(t)}{\langle\mathbf{E}_{n\varepsilon}(t)\rangle} = \frac{\sqrt{\langle\{\alpha_{n\varepsilon}\}|(\mathbf{E}_{n\varepsilon}(\mathbf{r},t)^{\mathrm{HP}})^2|\{\alpha_{n\varepsilon}\}\rangle - \langle\{\alpha_{n\varepsilon}\}|\mathbf{E}_{n\varepsilon}(\mathbf{r},t)^{\mathrm{HP}}|\{\alpha_{n\varepsilon}\}\rangle^2}}{\langle\{\alpha_{n\varepsilon}\}|\mathbf{E}_{n\varepsilon}(\mathbf{r},t)^{\mathrm{HP}}|\{\alpha_{n\varepsilon}\}\rangle} \tag{14.136}$$

Figure 14.2 gives the time dependence of the HP electric field averaged over different coherent states $|\{\alpha_{n\varepsilon}\}\rangle$ of increasing eigenvalues $\alpha_{n\varepsilon}$ and also the corresponding relative fluctuations (14.136) indicated by the thickness of the time dependence field function. As expected, the relative fluctuation is lowered when $\alpha_{n\varepsilon}$ is increasing, so that for $\alpha_{n\varepsilon} = 20$, it still vanishes. This example illustrates how the electric field operator averaged over a coherent state approaches the classical electric field when the coherent state parameter becomes very large.

It must be also observed that the *ondulatory nature of light* is described by the *quantum linear operators* describing the electromagnetic field, whereas the corresponding *corpuscular nature of light* is under the dependence of the *kets* (which are related to waves through wave mechanics) over which these operators are averaged. This is summarized in the following tabular expression (14.137):

Physical Behaviour	Quantum Entities	Examples
Wave	Hermitian operators	$\mathbf{E}(\mathbf{r},t)^{\mathrm{HP}}$, $\mathbf{B}(\mathbf{r},t)^{\mathrm{HP}}$, $\mathbf{A}(\mathbf{r},t)^{\mathrm{HP}}$
Corpuscle	Kets	$\|\{\alpha_{n\varepsilon}\}\rangle$, $\|\{\alpha_\varepsilon(\mathbf{k})\}\rangle$, $\|\{l_{n\varepsilon}\}\rangle$, $\|\{l_\varepsilon(\mathbf{k})\}\rangle$
Corpuscle	Wavefunctions	$\langle\{\mathbf{r}\}\|\{\alpha_\varepsilon(k)\}\rangle$, $\langle\{\mathbf{r}\}\|\{l_{n\varepsilon}\}\rangle$

(14.137)

More precisely, the number of electromagnetic particles within a given mode, that is, the number of photons of the corresponding angular frequency, may be obtained directly from the quantum number appearing in the eigenvalue equation (14.121) if the mode is an eigenstate of the Hamiltonian corresponding to this mode, or by a number proportional to the transition probability:

$$\{P_{n\varepsilon}^{\{l\alpha\}}\} = |\langle\{l_{n\varepsilon}\}|\{\alpha_{n\varepsilon}\}\rangle|^2$$

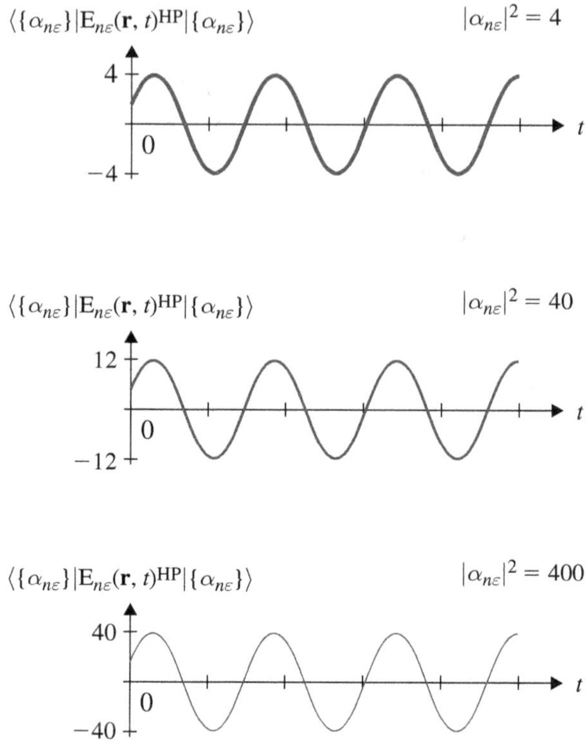

Figure 14.2 HP electric field averaged over different coherent states of increasing eigenvalue $\alpha_{n\varepsilon}$ and their corresponding relative dispersion pictured by the thickness of the time dependence field function.

or

$$\{P_{n\varepsilon}^{\{l\alpha\}}\} = e^{-|\alpha_{n\varepsilon}|^2} \sum_{l_{n\varepsilon}} \frac{|\alpha_{n\varepsilon}|^{2l_{n\varepsilon}}}{l_{n\varepsilon}!}$$

if the mode is in the coherent state (14.132).

Now, of course, it is possible to average the SP or HP electric field operator over squeezed states such as those (8.57) met in Section 8.2. All the results obtained in this section for the mean values of $\mathbf{Q}$, $\mathbf{Q}^2$, and $\Delta\mathbf{Q}$ averaged over the squeezed states and given, respectively, by Eqs. (8.79), (8.85), and (8.86), can be easily transposed to those of the electric field, its square, and its fluctuation.

14.5.5 Electromagnetic field spectrum

All the results obtained in the previous sections of this chapter hold irrespective of the angular frequency ω or of the corresponding frequency $\nu = \omega/2\pi$ of the electromagnetic field. As it may be observed by inspection of Fig. 14.3, they apply to γ rays involved in radioactivity to X and ultraviolet (UV) rays, to visible light,

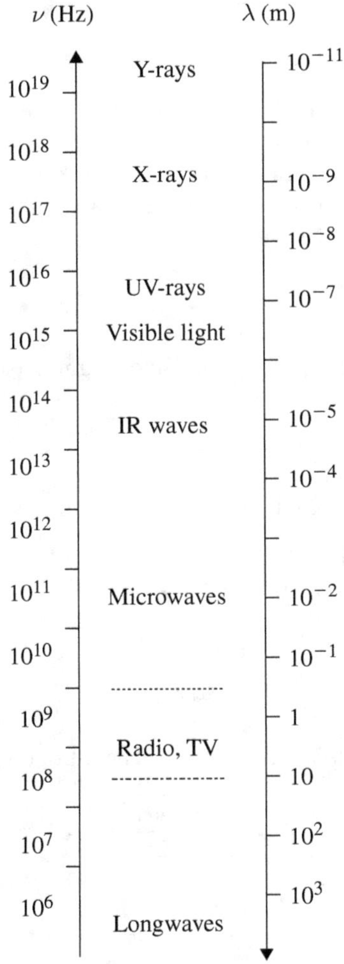

Figure 14.3 Electromagnetic field spectrum.

to infrared, and to microwaves and beneath to radar and radio waves. Clearly, by inspection of Fig. 14.3 the frequency ν may vary over a very wide range, since being susceptible to be greater than 10^{19} Hz, for γ rays and around 10^6 Hz for radio long waves, whereas the corresponding wavelength $\lambda = c/\nu$ (where c is the velocity of light around 3.10^8 m s^{-1}), may be smaller than 10^{-11} m for γ rays and around 10^3 m for long radio waves.

14.5.6 Long wavelength approximation for electric field

We start from Eq. (14.135) in order to obtain the mean value of the polarized electromagnetic field along $\hat{\boldsymbol{\varepsilon}}_{\mathbf{k}}$ averaged over a coherent state $|\{\alpha_\varepsilon\}\rangle$:

$$\langle\{\alpha_{n\varepsilon}\}|\mathbf{E}_{n\varepsilon}(\mathbf{r},t)^{\mathrm{HP}}|\{\alpha_{n\varepsilon}\}\rangle = i\hat{\boldsymbol{\varepsilon}}_{\mathbf{k}_n} \int \sqrt{\frac{\hbar\omega_n}{2\varepsilon^\circ}}\{\alpha_{n\varepsilon} e^{i\mathbf{k}_n\cdot\mathbf{r}} e^{-i\omega_n t} - \alpha_{n\varepsilon}{}^* e^{-i\mathbf{k}_n\cdot\mathbf{r}} e^{i\omega_n t}\} d^3\mathbf{k} \tag{14.138}$$

where this coherent state is defined by

$$\mathbf{a}_{n\varepsilon}|\{\alpha_\varepsilon\}\rangle = \alpha_{n\varepsilon}|\{\alpha_\varepsilon\}\rangle \quad \text{with} \quad \langle\{\alpha_{n\varepsilon}\}|\{\alpha_{n\varepsilon}\}\rangle = 1$$

Now, from inspection of Fig. 14.3, the wavelengths of the electromagnetic radiations used in molecular spectroscopy go from 3×10^{-4} for infrared to 3×10^{-7} meters for ultraviolet. Hence, since the modulus k_n of the vector $\mathbf{k}$ involved in the scalar products $\mathbf{k} \cdot \mathbf{r}$ appearing in the arguments of the exponentials encountered in Eqs. (14.134) and (14.135), is the inverse of the wavelength

$$k_n = \frac{2\pi}{\lambda_n}$$

it follows that for this range of wavelengths the magnitude of the wave vector $\mathbf{k}_n$ lies in the interval

$$2 \times 10^4 \le k_n \le 2 \times 10^7 \text{ rad}\cdot\text{m}^{-1}$$

Hence, taking $|\mathbf{r}|$ as 10 atomic radii, that is, a few angstroms, for example, 2×10^{-9} m, we have

$$10^{-5} \le \mathbf{k}_n \cdot \mathbf{r} \le 10^{-2}$$

so that

$$e^{\pm i\mathbf{k}_n\cdot\mathbf{r}} \backsimeq 1$$

which is the *long wavelength approximation*. Then, Eq. (14.138) simplifies to

$$\langle\{\alpha_{n\varepsilon}\}|\{\mathbf{E}_{n\varepsilon}(t)^{\text{HP}}\}|\{\alpha_\varepsilon\}\rangle = i\hat{\boldsymbol{\varepsilon}}_{\mathbf{k}_n}\sqrt{\frac{\hbar\omega_n}{2\varepsilon^\circ}}\{\alpha_{n\varepsilon}(e^{-i\omega_n t}) - \alpha_{n\varepsilon}{}^*(e^{i\omega_n t})\}$$

or, taking α_ε as real,

$$\langle\{\alpha_{n\varepsilon}\}|\{\mathbf{E}_{n\varepsilon}(t)^{\text{HP}}\}|\{\alpha_{n\varepsilon}\}\rangle = i\{\mathbf{E}(\omega_n)\}(e^{-i\omega_n t} - e^{i\omega_n t}) = 2\{\mathbf{E}(\omega_n)\}\sin\omega_n t$$

with

$$\mathbf{E}(\omega_n) = \hat{\varepsilon}_{\mathbf{k}_n}\sqrt{\frac{\hbar\omega_n}{2\varepsilon^\circ}}\alpha_\varepsilon \qquad (14.139)$$

a result that, after introducing a phase $-\pi/2$, reads

$$\langle\{\alpha_{n\varepsilon}\}|\mathbf{E}_{n\varepsilon}(t)^{\text{HP}}|\{\alpha_{n\varepsilon}\}\rangle = \mathbf{E}(\omega_n)(e^{i\omega_n t} + e^{-i\omega_n t}) \qquad (14.140)$$

As it appears, the mean value of the electric field averaged over the coherent state given by Eq. (14.140) appears to be an infinite sum of time-dependent electric fields depending continuously on ω and given by

$$\boxed{\mathbf{E}(\omega_n, t) = \mathbf{E}(\omega_n)(e^{i\omega_n t} + e^{-i\omega^n t})} \qquad (14.141)$$

Of course, the long wavelength approximation holds for microwaves, the wavelengths of which are greater than those of infrared radiations.

14.6 SOME THERMAL PROPERTIES OF THE QUANTUM FIELDS

14.6.1 Black-body radiation law

Kirchhoff in 1859 asked how does the intensity of the electromagnetic radiation emitted by a black body (a perfect absorber, also known as a cavity radiator) depend on the frequency of the radiation (i.e., the color of the light) and the temperature of the body. The answer was given by Planck who described the experimentally observed black-body spectrum well.

We consider the electromagnetic radiation in thermal equilibrium inside an enclosure of volume V whose walls are maintained at absolute temperature T. In this situation, photons corresponding to excitation degrees of the different modes of the electromagnetic radiation, which are continuously absorbed and reemitted by the walls. Thus, due to this mechanism, the radiation inside the container depends on the temperature of the walls. Of course, it is not necessary to investigate the details of the mechanism that brings about thermal equilibrium since general arguments of statistical mechanics suffice by the aid of a coarse-grained analysis to describe the thermal equilibrium situation.

If we regard the radiation as a collection of photons, the total number of them inside the enclosure is not fixed but depends on the temperature of the walls. The different modes of the field are specified by equations such as (14.125)–(14.127). Moreover, the radiation field existing in thermal equilibrium inside the enclosure is completely described by the thermal averages of the number occupation of each mode of the field, or, equivalently, by the corresponding thermal energy averages. Hence, the density of energy $U(\omega, T)$ of the electromagnetic field in the range between ω and $\omega + d\omega$, may be obtained from the expression of the thermal average energies $\langle \mathbf{H}(\omega, T)\rangle$ of the electromagnetic modes of angular frequency ω, by multiplying them by the density of modes $g(\omega)$ obtained above, that is,

$$U(\omega, T) = g(\omega)\langle \mathbf{H}(\omega, T)\rangle \tag{14.142}$$

For each mode of the field, and due to similarity between the Hamiltonian (14.116) and that of the usual quantum harmonic oscillator (5.9), it is clear that the thermal average energy is given by Eq. (13.29), that is,

$$\langle \mathbf{H}(\omega, T)\rangle = \frac{\hbar\omega}{e^{\hbar\omega/k_B T} - 1} + \frac{\hbar\omega}{2} \tag{14.143}$$

Now, we already saw that the density $g(\omega)$ of modes of the electromagnetic field between ω and $(\omega + d\omega)$ per unit volume is given by (14.93), that is,

$$g(\omega) = \frac{2\omega^2}{(2\pi c)^3} \tag{14.144}$$

Hence, discarding the zero-point energy in Eq. (14.143), the energy density (14.142) of the electromagnetic field becomes

$$U(\omega, T) = \frac{2\hbar}{(2\pi c)^3}\left(\frac{\omega^3}{e^{\hbar\omega/k_B T} - 1}\right) \tag{14.145}$$

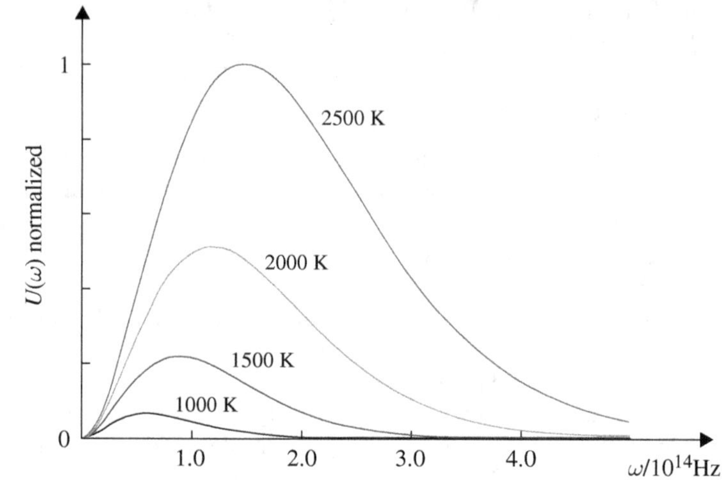

Figure 14.4 Energy density $U(\omega)$ within a cavity for different temperatures. The $U(\omega)$ are normalized with respect to the maximum of the curve at 2500 K.

that is *the Planck black-body radiation law*, discovered by Max Planck, governing at equilibrium temperature, the energy density of the electromagnetic field enclosed in a cavity at temperature T. This energy density is reproduced in Fig. 14.4 for different temperatures.

Planck's law (14.145) is one of the most fundamental equations in physics and is experimentally well verified.

The total electromagnetic energy within the cavity may be obtained by integrating the energy density by unit volume over ω and then multiplying it by the total volume V°, that is,

$$U_{\mathrm{Tot}}(T) = V^\circ \int_0^\infty U(\omega, T)\, d\omega$$

so that comparing, Eq. (14.145)

$$U_{\mathrm{Tot}}(T) = V^\circ \frac{2\hbar}{(2\pi c)^3} \int_0^\infty \left(\frac{\omega^3}{e^{\hbar\omega/k_B T} - 1} \right) d\omega \tag{14.146}$$

Then, changing the variable

$$x = \frac{\hbar\omega}{k_B T}$$

we have

$$\omega^3 = \left(\frac{k_B T}{\hbar} \right)^3 x^3 \qquad \text{and} \qquad d\omega = \left(\frac{k_B T}{\hbar} \right) dx$$

the total energy (14.146) takes the form

$$U_{\mathrm{Tot}}(T) = V^{\circ}\frac{2}{(2\pi c)^3}\left(\frac{k_B{}^4}{\hbar^3}\right)T^4\int_0^{\infty}\left(\frac{x^3}{e^x-1}\right)dx \tag{14.147}$$

Moreover, since the integral involves a dimensionless variable, it must lead to a dimensionless number that has to be finite because of the presence of $\exp x$ in the denominator of the integrand. Hence, it appears that the total energy is of the form

$$U_{\mathrm{Tot}}(T) \propto T^4$$

which shows that the total energy of the electromagnetic field inside the cavity is proportional to the fourth power of the absolute temperature. This is the *Stefan-Boltzmann law*. To go further in the calculation of $U_{\mathrm{Tot}}(T)$, we require the integral involved in Eq. (14.147), which has the value

$$\int_0^{\infty}\left(\frac{x^3}{e^x-1}\right)dx = \frac{\pi^4}{15}$$

so that the Stefan–Boltzmann law reads more precisely

$$U_{\mathrm{Tot}}(T) = \sigma T^4$$

where σ is the Stefan–Boltzmann constant given by

$$\sigma = \frac{\pi}{60}\frac{k_B{}^4}{(c\hbar)^3}V^{\circ}$$

We emphasize that all the above results dealing with the black-body radiation hold for the whole electromagnetic spectrum, in particular for the spectrum of the cosmic microwave background[2] appearing in Fig. 14.5.

14.6.2 Einstein coefficients

The Planck radiation law allows one to find the ratio of the Einstein absorption and emission coefficients. To get this, consider two energy levels of energy E_1 and E_2 with $E_1 < E_2$, subjected to an electromagnetic field at thermal equilibrium, obeying therefore the black-body radiation law, with this field being able to induce changes in the time-dependent population $N_1(t)$ and $N_2(t)$ (Fig. 14.6).

The time dependence of the response to E_2 is given by the kinetic equation

$$\left(\frac{dN_2(t)}{dt}\right) = -A_{21}(\omega)N_2(t) - B_{21}(\omega)U(\omega)N_2(t) \tag{14.148}$$

Here $U(\omega)$ is the energy density of the electromagnetic field at angular frequency ω given by Eq. (14.145) and corresponding to the resonant situation

$$\omega = \frac{E_2 - E_1}{\hbar} \tag{14.149}$$

[2]From J. C. Mather, et al., *Astrophys. J.*, **354** (1990): L37–L49.

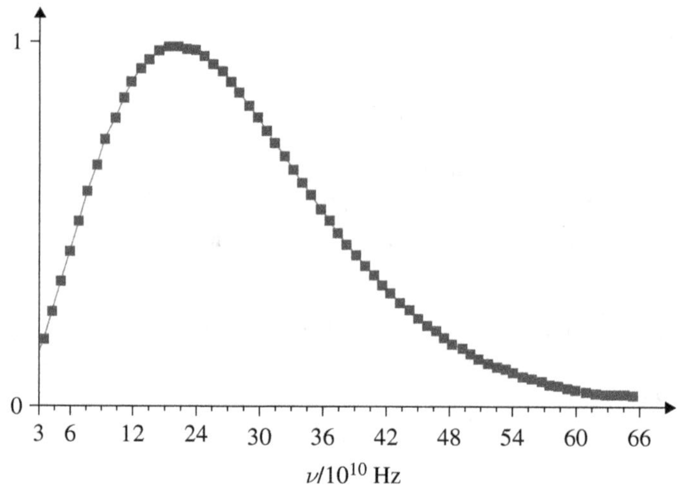

Figure 14.5 Spectrum of the cosmic microwave background (squares) superposed on a 2.735 K black-body emission (full line). The intensities are normalized to the maximum of the curve.

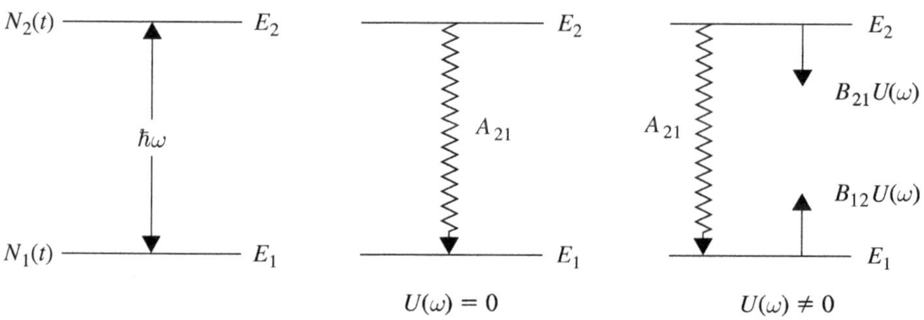

Figure 14.6 Einstein coefficients for two energy levels.

while $A_{21}(\omega)$ is the spontaneous emission coefficient of Einstein and $B_{21}(\omega)$ the corresponding induced emission coefficient at the angular frequency ω.

Now, the depopulation of the ground state E_1 is

$$\left(\frac{dN_1(t)}{dt}\right) = -B_{12}(\omega)U(\omega)N_1(t) \tag{14.150}$$

and $B_{12}(\omega)$ the induced Einstein absorption coefficient obeying

$$B_{21}(\omega) = B_{12}(\omega) \tag{14.151}$$

After some time has occurred, which is large with respect to the characteristic times of the system, that is, $t \to \infty$, a steady state must obtain so that

$$\left(\frac{dN_2(\infty)}{dt}\right) = \left(\frac{dN_1(\infty)}{dt}\right) = 0 \tag{14.152}$$

Beyond this time and owing to Eqs. (14.148) and (14.150), the steady-state condition (14.152) leads to

$$A_{21}(\omega)N_2(\infty) + B_{21}(\omega)U(\omega)N_2(\infty) = B_{21}(\omega)U(\omega, T)N_1(\infty)$$

and thus, after rearranging, to

$$\left\{\frac{N_2(\infty)}{N_1(\infty)}\right\} = \frac{B_{21}(\omega)U(\omega, T)}{A_{21}(\omega) + B_{21}(\omega)U(\omega, T)} \tag{14.153}$$

However, under the steady conditions, this ratio must obey the equilibrium Boltzmann distributions ratio defined by Eqs. (12.83) so that

$$\left\{\frac{N_2(\infty)}{N_1(\infty)}\right\} = \frac{e^{-E_2/k_BT}}{e^{-E_1/k_BT}} \tag{14.154}$$

or, owing to Eq. (14.149),

$$\left\{\frac{N_2(\infty)}{N_1(\infty)}\right\} = e^{-\hbar\omega/k_BT}$$

Hence, by identification of Eqs. (14.153) and (14.154) one obtains

$$\frac{B_{21}(\omega)U(\omega, T)}{A_{21}(\omega) + B_{21}(\omega)U(\omega, T)} = e^{-\hbar\omega/k_BT} \tag{14.155}$$

or

$$B_{21}(\omega)U(\omega, T) = e^{-\hbar\omega/k_BT}(A_{21}(\omega) + B_{21}(\omega)U(\omega, T))$$

and

$$B_{21}(\omega)U(\omega, T)(1 - e^{-\hbar\omega/k_BT}) = e^{-\hbar\omega/k_BT}A_{21}(\omega)$$

so that, the ratio of the two Einstein coefficients reads

$$\left(\frac{A_{21}(\omega)}{B_{21}(\omega)}\right) = U(\omega, T)\frac{(1 - e^{-\hbar\omega/k_BT})}{e^{-\hbar\omega/k_BT}}$$

and the ratio of the induced emission coefficients $B_{21}(\omega)$ times the energy density $U(\omega)$ with the spontaneous emission coefficient $A_{21}(\omega)$ yields

$$\left(\frac{B_{21}(\omega)U(\omega, T)}{A_{21}(\omega)}\right) = \frac{1}{(e^{\lambda_B} - 1)} \quad \text{with } \lambda_B = \frac{\hbar\omega}{k_BT} \tag{14.156}$$

or, due to Eq. (13.36),

$$\left(\frac{B_{21}(\omega)U(\omega, T)}{A_{21}(\omega)}\right) = \langle \mathbf{n}(\lambda_B)\rangle$$

where $\langle \mathbf{n}(\lambda_B)\rangle$ is the thermal average of the occupation number at the absolute temperature T and at the angular frequency ω given by

$$\langle \mathbf{n}(\lambda_B)\rangle = \frac{1}{(e^{\lambda_B} - 1)} \tag{14.157}$$

Now, the energy density of the electromagnetic field is given by Eq. (14.145), that is,

$$U(\omega, T) = \frac{2\hbar}{(2\pi c)^3}\left(\frac{\omega^3}{e^{\hbar\omega/k_BT} - 1}\right)$$

so that the Einstein coefficients ratio becomes

$$\left(\frac{A_{21}(\omega)}{B_{21}(\omega)}\right) = \frac{2\hbar}{(2\pi c)^3}\omega^3 \frac{(1 - e^{-\hbar\omega/k_B T})}{e^{-\hbar\omega/k_B T}(e^{\hbar\omega/k_B T} - 1)}$$

or, after simplification,

$$\boxed{\left(\frac{A_{21}(\omega)}{B_{21}(\omega)}\right) = \frac{2\hbar}{(2\pi c)^3}\omega^3 = 2\hbar\left(\frac{\nu}{c}\right)^3} \tag{14.158}$$

Hence, the ratio of the spontaneous and induced Einstein coefficients increases with the third power of the frequency.

14.7 CONCLUSION

In this chapter the classical normal modes of the electromagnetic field were obtained by transforming the Maxwell equation from the geometric to reciprocal space. Then, showing that in reciprocal space the conjugate momentum of the vector potential is deeply related to the electric field, it was possible to quantize the electromagnetic field by assuming a commutation rule between the operators corresponding to the vector potential and the electric field of the same kind as that assumed for the position coordinate and its conjugate momentum. It was then possible to find for each electromagnetic mode of the reciprocal space that a Hamiltonian exists that has the same structure in terms of self-conjugate Hermitian ladder operators as that of the usual quantum harmonic oscillator, thereby allowing one to apply all the results obtained for that oscillators to the quantum electromagnetic modes, particularly all those dealing with the Hamiltonian eigenvalue equation and with coherent and squeezed states. It was also shown that the degree of excitation of the Hamiltonian eigenstates of normal modes of a given frequency may be viewed as the number of light corpuscles, that is, the number of photons having this frequency. Moreover, it is apparent that this corpuscular property is related to the electromagnetic field kets, and thus, keeping in mind the link between quantum mechanics and wave mechanics, to quantum wavefunctions describing the field. At the opposite, it became clear that the wave behavior of the electromagnetic field is the reflection of the Hermitian operators describing these fields. Moreover, applying to the electromagnetic normal modes the thermal properties of oscillators, it was possible to find the Planck black-body radiation law and the Stefan–Boltzmann law, and to get the relation between the spontaneous and induced Einstein emission coefficients.

BIBLIOGRAPHY

C. Cohen-Tannoudji, J. Dupont-Roc, and G. Grinberg. *Photons and Atoms*. Wiley: New York, 1997.

R. Loudon. *The Quantum Theory of Light*. Oxford University Press: New York, 1983.

H. Louisell. *Quantum Statistical Properties of Radiations*. Wiley: New York, 1973.

CHAPTER 15

QUANTUM MODES IN MOLECULES AND SOLIDS

INTRODUCTION

Chapters of Parts II, III, and IV studied the properties of a single harmonic oscillator (Parts II and III) or a large population of such oscillators (Part IV). But, if one wishes to apply these properties to molecules or solids, it is first necessary to extract from these complex systems their *normal vibrational modes*, where for each of them all the atoms of such extended systems may be classically viewed as oscillating back and forth at the same angular frequency and at the same phase. Hence, as in the last chapter, which dealt with the normal modes of electromagnetic fields, the aim of the present chapter is to describe a method for determining the normal modes of a molecule and those of solids, the last approach leading after quantization of the normal modes to the concept of *phonons*, that is, to the quantum vibrational energy of a normal mode considered as a quasi-particle in a way that evokes the photons of the electromagnetic field modes.

15.1 MOLECULAR NORMAL MODES

15.1.1 Obtainment of the normal modes

Consider a set of N harmonic oscillators of the same reduced masses m that are linearly coupled through the potential

$$V(t) = V^{\circ}(t) + V_{\mathrm{Int}}(t)$$

with, respectively,

$$V^{\circ}(t) = \frac{1}{2}\sum_{i} k_{ii}x_i^2(t)$$

$$V_{\mathrm{Int}}(t) = \frac{1}{2}\sum_{i}\sum_{j\neq i} k_{ij}(x_i(t) - x_j(t))^2$$

Quantum Oscillators, First Edition. Olivier Henri-Rousseau and Paul Blaise.
© 2011 John Wiley & Sons, Inc. Published 2011 by John Wiley & Sons, Inc.

where $x_i(t)$ is the time-dependent elongation of the ith oscillator. The force acting on the ith oscillator is

$$m\left(\frac{d^2x_i(t)}{dt^2}\right) = -\left(\frac{\partial V(t)}{\partial x_i}\right) \tag{15.1}$$

Besides

$$-\left(\frac{\partial V(t)}{\partial x_i}\right) = -k_{ii}x_i(t) - \sum_{j\neq i} k_{ij}(x_i(t) - x_j(t))$$

or

$$-\left(\frac{\partial V(t)}{\partial x_i}\right) = -(k_{ii} + \sum_{j\neq i} k_{ij})x_i(t) + \sum_{j\neq i} k_{ij}x_j(t)$$

and thus

$$-\left(\frac{\partial V(t)}{\partial x_i}\right) = -\sum_j K_{ij}x_j(t) \tag{15.2}$$

with

$$K_{ii} = \sum_j k_{ij} \qquad \text{and} \qquad K_{ij} = -k_{ij}$$

Hence, owing to (15.2), the dynamics equations (15.1) yield

$$m\left(\frac{d^2x_i(t)}{dt^2}\right) = -\sum_j K_{ij}x_j(t) \tag{15.3}$$

which may be written in a matrix form according to

$$\boxed{\mathbf{M}}\{\ddot{\mathbf{X}}(t)\} + \boxed{\mathbf{K}}\{\mathbf{X}(t)\} = \{\mathbf{0}\} \tag{15.4}$$

where $\{\mathbf{0}\}$ is the zero column vector, $\{\mathbf{X}(t)\}$ and $\{\ddot{\mathbf{X}}(t)\}$ are column vectors formed, respectively, by the set of positions $x_i(t)$ and accelerations $\ddot{x}_i(t)$:

$$\{\ddot{\mathbf{X}}(t)\} = \begin{pmatrix} \ddot{x}_1(t) \\ \ddot{x}_2(t) \\ \vdots \\ \ddot{x}_N(t) \end{pmatrix} \qquad \text{and} \qquad \{\mathbf{X}(t)\} = \begin{pmatrix} x_1(t) \\ x_2(t) \\ \vdots \\ x_N(t) \end{pmatrix} \tag{15.5}$$

whereas $\boxed{\mathbf{K}}$ is the matrix of the force constants K_{ij}

$$\boxed{\mathbf{K}} = \begin{pmatrix} K_{11} & K_{12} & \dots & K_{1N} \\ K_{21} & K_{22} & \dots & \dots \\ \dots & \dots & \dots & \dots \\ K_{N1} & \dots & \dots & K_{NN} \end{pmatrix}$$

and $\boxed{\mathbf{M}}$ is the diagonal matrix

$$\boxed{\mathbf{M}} = m\boxed{\mathbf{1}}$$

where $\boxed{\mathbf{1}}$ is the unity matrix.

Premultiply Eq. (15.4) by the inverse of the masses matrix

$$\boxed{\mathbf{M}}^{-1}\boxed{\mathbf{M}}\{\ddot{\mathbf{X}}(t)\} + \boxed{\mathbf{M}}^{-1}\boxed{\mathbf{K}}\{\mathbf{X}(t)\} = \{\mathbf{0}\}$$

After simplification that gives

$$\{\ddot{\mathbf{X}}(t)\} + \boxed{\mathbf{D}}\{\mathbf{X}(t)\} = \{\mathbf{0}\} \tag{15.6}$$

where

$$\boxed{\mathbf{D}} = \boxed{\mathbf{M}}^{-1}\boxed{\mathbf{K}} \tag{15.7}$$

Next, introduce the diagonalization transformation of the matrix defined by Eq. (15.7) through

$$\boxed{\boldsymbol{\lambda}} = \boxed{\mathbf{O}}^{-1}\boxed{\mathbf{D}}\boxed{\mathbf{O}} \tag{15.8}$$

where $\boxed{\mathbf{O}}$ is the eigenvector matrix, whereas $\boxed{\boldsymbol{\lambda}}$ is the eigenvalue matrix that is diagonal. Introduce within Eq. (15.6) between the matrix and the vector the diagonal unity matrix defined by

$$\boxed{\mathbf{1}} = \boxed{\mathbf{O}}\boxed{\mathbf{O}}^{-1}$$

that is,

$$\{\ddot{\mathbf{X}}(t)\} + \boxed{\mathbf{D}}\boxed{\mathbf{O}}\boxed{\mathbf{O}}^{-1}\{\mathbf{X}(t)\} = \{\mathbf{0}\}$$

Again, premultiply this last equation by the eigenvector matrix

$$\boxed{\mathbf{O}}^{-1}\{\ddot{\mathbf{X}}(t)\} + \boxed{\mathbf{O}}^{-1}\boxed{\mathbf{D}}\boxed{\mathbf{O}}\boxed{\mathbf{O}}^{-1}\{\mathbf{X}(t)\} = \{\mathbf{0}\}$$

Then, in view of Eq. (15.8), this expression simplifies to

$$\{\ddot{\mathbf{Y}}(t)\} + \boxed{\boldsymbol{\lambda}}\{\mathbf{Y}(t)\} = \{\mathbf{0}\} \tag{15.9}$$

with, respectively,

$$\{\ddot{\mathbf{Y}}(t)\} = \boxed{\mathbf{O}}^{-1}\{\ddot{\mathbf{X}}(t)\}$$

$$\{\mathbf{Y}(t)\} = \boxed{\mathbf{O}}^{-1}\{\mathbf{X}(t)\} \tag{15.10}$$

and where the λ_l are the eigenvalues of matrix (15.7). The linear transformation (15.10) reads

$$y_l(t) = \sum_k a_{lk}\, x_k(t)$$

where the a_{lk} are the components of the transformation matrix, whereas the $y_l(t)$ are the components of the column vector $\{\mathbf{Y}(t)\}$. Equation (15.9) where the transformation matrix is diagonal, summarizes N-decoupled differential equations of the form

$$\ddot{y}_l(t) + \lambda_l\, y_l(t) = 0 \tag{15.11}$$

that may be written

$$\left(\frac{d^2 y_l(t)}{dt^2}\right) = -\Omega_l^2 y_l(t) \quad \text{with} \quad \Omega_l^2 = \lambda_l \tag{15.12}$$

The solutions of the decoupled differential equations are

$$y_l(t) = y_l(0)\sin(\Omega_l t + \Phi_l)$$

where Φ_l are phases. The $y_l(t)$ are the *normal modes of vibration* of the oscillator system in which all the parts of the system oscillate at the same angular frequency Ω_{ll} with the same phase Φ_l. Next, after multiplying both terms of Eq. (15.9) by the mass matrix involved in Eq. (15.4)

$$\boxed{\mathbf{M}}\{\ddot{\mathbf{Y}}(t)\} + \boxed{\mathbf{M}}\boxed{\boldsymbol{\lambda}}\{\mathbf{Y}(t)\} = \{\mathbf{0}\}$$

one obtains N decoupled equations of the form

$$m\,\ddot{y}_l(t) + m\Omega_l^2 y_l(t) = 0 \tag{15.13}$$

Now, within the normal modes description, the full classical Hamiltonian $\tilde{H}$, which is by definition

$$\tilde{H} = \tilde{T} + \tilde{V}$$

may be written as the sum of decoupled Hamiltonians:

$$\tilde{H} = \sum_l \tilde{H}_l \quad \text{with} \quad \tilde{H}_l = \left(\frac{1}{2m}\,p_l(t)^2 + \frac{1}{2}m\Omega_l^2 y_l(t)^2\right) \tag{15.14}$$

with

$$p_l(t) = m\dot{y}_l(t)$$

15.1.2 Quantization of the normal modes

In the Schrödinger picture, the operators do not change with time. Hence, in order to pass to quantum mechanics, we have to perform the change:

$$y_l(t) \to \mathbf{Q}_l \qquad \text{and} \qquad p_l(t) \to \mathbf{P}_l$$

where $\mathbf{Q}_l$ and $\mathbf{P}_l$ are the time-independent operators corresponding, respectively, to the normal classical variables $y_l(t)$ and $p_l(t)$, which obey the commutation rule

$$[\mathbf{Q}_l\,,\mathbf{P}_k] = i\hbar\delta_{lk}$$

Then, the classical Hamiltonian (15.14) transforms to a Hamiltonian operator $\mathbf{H}$ given by

$$\mathbf{H} = \sum_l \mathbf{H}_l$$

with

$$\mathbf{H}_l = \frac{\mathbf{P}_l^2}{2m} + \frac{1}{2}m\Omega_l^2\mathbf{Q}_l^2$$

The Hamiltonians $\mathbf{H}_l$ are those of quantum harmonic oscillators so that all that has been found above for quantum harmonic oscillators may apply to the $\mathbf{H}_l$, in a way similar, for instance, to that used in passing from 1D to 3D harmonic oscillators. Hence, we write

$$\mathbf{H}_l = \hbar\Omega_l \left(\mathbf{a}_l^\dagger \mathbf{a}_l + \tfrac{1}{2}\right)$$

where

$$[\mathbf{a}_k, \mathbf{a}_l^\dagger] = \delta_{kl}$$

$$\mathbf{Q}_l = \sqrt{\frac{\hbar}{2M\Omega_l}}(\mathbf{a}_l^\dagger + \mathbf{a}_l)$$

$$\mathbf{P}_l = i\sqrt{\frac{\hbar M\Omega_l}{2}}(\mathbf{a}_l^\dagger - \mathbf{a}_l)$$

Of course, all the results obtained for quantum oscillators and for their thermal properties hold for the normal modes of molecules.

15.1.3 Application to a system of two coupled oscillators

Applied to a situation where there are, for instance, only two oscillators, the column vectors (15.5) corresponding to the elongations and to their respective accelerations are given by

$$\{\mathbf{X}(t)\} = \begin{pmatrix} x_1(t) \\ x_2(t) \end{pmatrix} \quad \text{and} \quad \{\ddot{\mathbf{X}}(t)\} = \begin{pmatrix} \ddot{x}_1(t) \\ \ddot{x}_2(t) \end{pmatrix} \tag{15.15}$$

$$\boxed{\mathbf{M}} = \begin{pmatrix} m & 0 \\ 0 & m \end{pmatrix} \quad \text{and} \quad \boxed{\mathbf{K}} = \begin{pmatrix} k_{11} + k_{12} & -k_{12} \\ -k_{21} & k_{22} + k_{21} \end{pmatrix} \tag{15.16}$$

Now, let us look at the matrix $\boxed{\mathbf{D}}$ given by Eq. (15.7), that is,

$$\boxed{\mathbf{D}} = \boxed{\mathbf{M}}^{-1}\boxed{\mathbf{K}} \tag{15.17}$$

Observe that since the matrix $\boxed{\mathbf{M}}$ is diagonal, its inverse is also diagonal and given by

$$\boxed{\mathbf{M}}^{-1} = \begin{pmatrix} 1/m & 0 \\ 0 & 1/m \end{pmatrix}$$

Thus, owing to Eq. (15.16), Eq. (15.7) takes the form

$$\boxed{\mathbf{D}} = \begin{pmatrix} 1/m & 0 \\ 0 & 1/m \end{pmatrix} \begin{pmatrix} k_{11} + k_{12} & -k_{12} \\ -k_{21} & k_{22} + k_{21} \end{pmatrix}$$

Performing the matrix product gives

$$\boxed{\mathbf{D}} = \begin{pmatrix} \omega_{11}^2 & -\omega_{12}^2 \\ -\omega_{21}^2 & \omega_{22}^2 \end{pmatrix} \tag{15.18}$$

with

$$\omega_{11} = \sqrt{\frac{k_{11}+k_{12}}{m}} \quad \text{and} \quad \omega_{22} = \sqrt{\frac{k_{22}+k_{21}}{m}}$$

$$\omega_{12} = \sqrt{\frac{k_{12}}{m}} \quad \text{and} \quad \omega_{21} = \sqrt{\frac{k_{21}}{m}}$$

Then, the diagonalization transformation (15.8), that is,

$$\boxed{\mathbf{O}}^{-1}\boxed{\mathbf{D}}\,\boxed{\mathbf{O}} - \boxed{\boldsymbol{\lambda}} = 0$$

Now, when passing to the components D_{ij} of the matrix, it reads

$$\sum_j D_{ij}C_j^{\pm} - \lambda^{\pm}C_i^{\pm} = 0 \tag{15.19}$$

where $\lambda^{\pm}$ are the two unknown eigenvalues of the $\boxed{\boldsymbol{\lambda}}$ diagonal matrix to be found, whereas the $C_j^{\pm}$ are the unknown components for the matrix

$$\boxed{\mathbf{O}} = \begin{pmatrix} C_1^+ & C_1^- \\ C_2^+ & C_2^- \end{pmatrix}$$

Since the $\lambda^{\pm}$ and $C_j^{\pm}$ are unknown, Eq. (15.19) corresponds to the following set of simultaneous equations:

$$(D_{11} - \lambda)C_1 + D_{12}C_2 = 0 \tag{15.20}$$

$$D_{21}C_1 + (D_{22} - \lambda)C_2 = 0 \tag{15.21}$$

Since the C_i are different from zero, these two last equations are satisfied if the following determinant is zero:

$$\begin{vmatrix} (D_{11} - \lambda) & D_{12} \\ D_{21} & (D_{22} - \lambda) \end{vmatrix} = 0$$

Expansion of the determinant following the usual rule leads to the second-order equation in λ:

$$\lambda^2 - (D_{11} + D_{22})\lambda + (D_{11}D_{22} - D_{12}D_{21}) = 0$$

The two solutions for λ are

$$\lambda^{\pm} = \tfrac{1}{2}[(D_{11} + D_{22}) \pm \sqrt{(D_{11} + D_{22})^2 - 4(D_{11}D_{22} - D_{12}D_{21})}] \tag{15.22}$$

with, in view of Eq. (15.18),

$$D_{ij} = \omega_{ij}^2 \quad \text{with} \quad i,j = 1,2$$

When the two values of $\lambda^{\pm}$ have been obtained by the aid of Eqs. (15.22) and (15.21) in terms of ω_{ij}, it is possible with the help of Eq. (15.20) to find the expression of the components of the orthogonal matrix, that is,

$$C_1^- = \frac{D_{12}\, C_2^-}{(\lambda^- - D_{11})} \qquad \text{and} \qquad C_1^+ = \frac{D_{12}\, C_2^+}{(\lambda^+ - D_{11})}$$

Observe that the orthogonal transformation

$$\boxed{\mathbf{O}}\boxed{\mathbf{D}}\boxed{\mathbf{O}}^{-1} = \boxed{\boldsymbol{\lambda}} \tag{15.23}$$

takes on, in the present situation, the following form:

$$\begin{pmatrix} C_1^+ & C_1^- \\ C_2^+ & C_2^- \end{pmatrix}\begin{pmatrix} \omega_{11}^2 & \omega_{12}^2 \\ \omega_{21}^2 & \omega_{22}^2 \end{pmatrix}\begin{pmatrix} C_1^+ & C_2^+ \\ C_1^- & C_2^- \end{pmatrix} = \begin{pmatrix} \lambda_1^+ & 0 \\ 0 & \lambda_1^- \end{pmatrix}$$

15.1.4 Identification of symmetric molecules normal modes

When a molecule presents different symmetry elements, it may be of interest to classify its normal modes according to the different *irreducible representations* of the *symmetry point group* to which it belongs. That is particularly important in molecular vibrational spectroscopy. Section 18.9 gives some information on the symmetry point groups and on the irreducible representations giving in a compact form how the symmetry operations act.

Equation (18.134) in Section 18.9 allows one to analyze the reducible representation Γ of any molecule belonging to a given point group in terms of the irreducible representation of that point group. This may be seen by studying how the atomic coordinates transform under the different symmetry operations of the point group. To illustrate that, such a procedure is now applied to the H_2O molecule, which admits two symmetry planes σ_v and σ_v', one belonging to the plane of the molecule and the other to the plane orthogonal to the first one and separating the molecule into two symmetrical parts, and also a rotational axis of symmetry C_2 passing through the intersection of the two planes, as shown in Fig. 15.1.

Then, it is shown that the reducible representation of H_2O is given by (18.122) in Section 18.9, that is,

$\mathbf{C}_{2v}$	E	C_2	σ_v	σ_v'
Γ°	9	−1	1	3

where the numbers 9, −1, 1, and 3 are the characters for the four symmetry classes corresponding to the E, C_2, σ_v, and σ_v' symmetry elements. Because of these symmetry elements and of the identity symmetry element E the H_2O molecule belongs to the C_{2v} point group, the *character table* of which is given by tabular data in (18.110) in Section 18.9, i.e.

$\mathbf{C}_{2v}$	E	C_2	σ_v	σ_v'	Rot and Trans
Γ_{A_1}	1	1	1	1	T_z
Γ_{A_2}	1	1	−1	−1	R_z
Γ_{B_1}	1	−1	1	−1	R_y, T_x
Γ_{B_2}	1	−1	−1	1	R_x, T_y

(15.24)

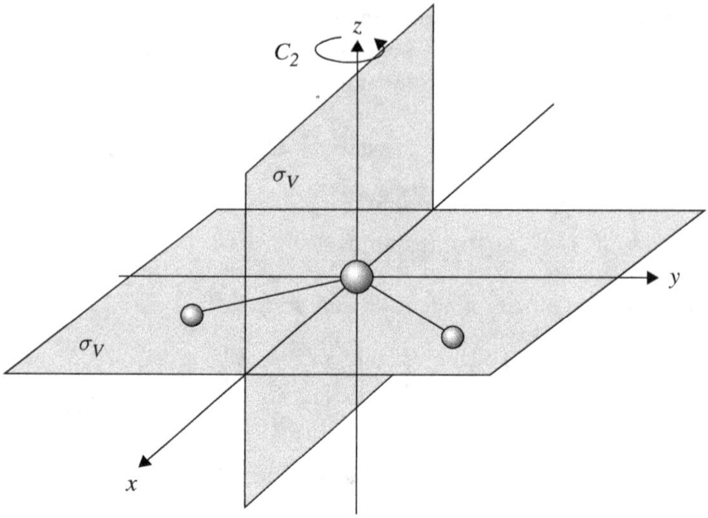

Figure 15.1 Symmetry elements for a C_{2v} molecule.

where the numbers in (15.24) are the characters $\{\chi^{\Gamma_k}(R_r)\}$ of the different irreducible representations Γ_k, that is, Γ_{A_1}, Γ_{A_2}, Γ_{B_1}, and Γ_{B_2} for the different symmetry classes R_r, that is, in the present situation E, C_2, σ_v, and σ'_v. The presence in (15.24), on the lines corresponding, respectively, to Γ_{A_1}, Γ_{B_1}, and Γ_{B_2}, of the notations T_z, T_x, and T_y corresponding to the translations along the three Cartesian coordinates, means that these translations transform, according to these irreducible representations, the explanation being the same for R_z, R_y, and R_x corresponding to the rotations around the z, y, and x axis. Then, by application of Eq. (18.127) of Section 18.9, the reducible representation of the H_2O molecule appears to be

$$\Gamma^\circ = a_{A_1}\Gamma_{A_1} \oplus a_{A_2}\Gamma_{A_2} \oplus a_{B_1}\Gamma_{B_1} \oplus a_{B_2}\Gamma_{B_2} \tag{15.25}$$

where a_{A_1}, a_{A_2}, a_{B_1}, and a_{B_2} are numbers that indicate how often the corresponding irreducible representations Γ_k, that is, Γ_{A_1}, Γ_{A_2}, Γ_{B_1}, and Γ_{B_2} appear. Moreover, applying Eq. (18.134) of Section 18.9, that is,

$$\boxed{a_{\Gamma_k} = \frac{1}{g}\sum_r \{\chi^{\Gamma_k}(R_r)\}\{\chi^{\Gamma^\circ}(R_r)\}}$$

the components of the reducible representation (15.25) may be obtained using

$$\begin{aligned}
a_{A_1} &= \tfrac{1}{4}\{(9 \times 1) \oplus (-1 \times 1) + (1 \times 1) \oplus (3 \times 1)\} = 3 \\
a_{A_2} &= \tfrac{1}{4}\{(9 \times 1) \oplus (-1 \times 1) \oplus (1 \times -1) \oplus (3 \times -1)\} = 1 \\
a_{B_1} &= \tfrac{1}{4}\{(9 \times 1) \oplus (-1 \times -1) \oplus (1 \times 1) \oplus (3 \times -1)\} = 2 \\
a_{B_2} &= \tfrac{1}{4}\{(9 \times 1) \oplus (-1 \times -1) \oplus (1 \times -1) \oplus (3 \times 1)\} = 3
\end{aligned}$$

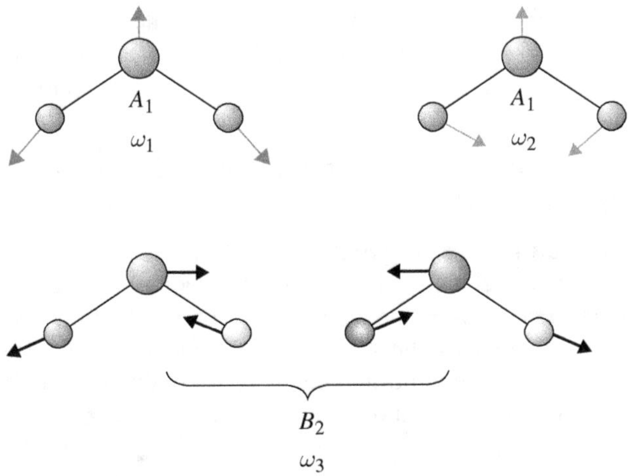

Figure 15.2 Three normal modes of a C_{2V} molecule.

Hence, the reducible representation (15.25) becomes

$$\Gamma = 3\Gamma_{A_1} \oplus \Gamma_{A_2} \oplus 2\Gamma_{B_1} \oplus 3\Gamma_{B_2} \tag{15.26}$$

From inspection of the C_{2v} table of characters, it appears that the representations of the rotations R_x, R_y, and R_z and translations T_x, T_y, and T_z are, respectively, given by

$$\Gamma_{\text{Rot}} = \Gamma_{A_2} \oplus \Gamma_{B_1} \oplus \Gamma_{B_2} \tag{15.27}$$

$$\Gamma_{\text{Tr}} = \Gamma_{A_1} \oplus \Gamma_{B_1} \oplus \Gamma_{B_2} \tag{15.28}$$

Then, the Γ_{Vib} normal modes representation is the difference between the Γ reducible representation (15.26) and those Γ_{Rot} and Γ_{Tr} given, respectively, by Eqs. (15.27) and (15.28) so that

$$\Gamma_{\text{Vib}} = 2\Gamma_{A_1} \oplus \Gamma_{B_2}$$

Thus, it appears that one of the three normal modes of H_2O belongs to the irreducible representation Γ_{B_2} is symmetric with respect to the C_2 and σ_v symmetry operations and antisymmetric with respect to σ_v' operation, whereas the two other vibrational modes are fully symmetric since they belong to the irreducible representation Γ_{A_1}. This is shown in Fig 15.2.

15.2 PHONONS AND NORMAL MODES IN SOLIDS

Having determined the normal modes of molecules, we must determine those of solids. This is the aim of the present section. As for molecules, we shall begin by seeking the classical normal modes of the solid and then continue by quantizing them. However, the procedure to get the normal modes of solids will appear to completely differ from that we have used for molecules. The method used for solids will proceed from Fourier transforms, allowing one to pass from a *geometric space* description to a new one in the *reciprocal space*, so that the quantization rules will be introduced for the normal mode coordinate and momentum components belonging to the reciprocal space.

15.2.1 Determination of the classical normal modes of a long chain of oscillators

15.2.1.1 Basic equations in geometric space We shall limit ourselves to a 1D approach to the solid normal modes first considered from a classical viewpoint. Consider an infinite linear chain of harmonic oscillators of angular frequency ω_0 of mass m, coupled to each neighbor via the same force constant $m\omega^2$, the distance between two successive oscillators at equilibrium being L. Then, the force acting on the nth oscillator obeys the following equation:

$$m\left(\frac{d^2q_n(t)}{dt^2}\right) = -m\omega_0^2 q_n(t) - m\omega^2\{(q_n(t) - q_{n+1}(t)) + (q_n(t) - q_{n-1}(t))\} \quad (15.29)$$

where $q_n(t)$ is the time-dependent displacement of the nth oscillator with respect to its equilibrium position. The solutions of this equation are

$$q_n(t) = \{e^{iknL}e^{-i\Omega(k)t} + e^{-iknL}e^{i\Omega(k)t}\} \quad (15.30)$$

where k is a continuous variable having the dimensions of inverse length, whereas $\Omega(k)$ is given by

$$\Omega(k) = \sqrt{\omega_0^2 + \omega^2(2 - e^{ikL} - e^{-ikL})} \quad (15.31)$$

That may be easily verified as follows. First, start from the second time derivative of $q_n(t)$ assumed to obey Eq. (15.30), which, due to

$$\frac{\partial e^{\pm i\Omega(k)t}}{\partial t} = -\omega^2 e^{\pm i\Omega(k)t}$$

reads

$$\left(\frac{d^2q_n(t)}{dt^2}\right) = -\Omega(k)^2\{e^{iknL}e^{-i\Omega(k)t} + e^{-iknL}e^{i\Omega(k)t}\}$$

Then, using Eq. (15.31) yields

$$\left(\frac{d^2q_n(t)}{dt^2}\right) = -\{\omega_0^2 + \omega^2(2 - e^{ikL} - e^{-ikL})\}\{e^{iknL}e^{-i\Omega(k)t} + e^{-iknL}e^{i\Omega(k)t}\} \quad (15.32)$$

and, owing to the fact that

$$e^{\pm ikL}e^{iknL} = e^{ik(n\pm 1)L} \qquad \text{and} \qquad e^{\pm ikL}e^{-iknL} = e^{ik(n\mp 1)L}$$

it appears, with the help of Eq. (15.30), that

$$(2 - e^{ikL} - e^{-ikL})\{e^{iknL}e^{-i\Omega(k)t} + e^{-iknL}e^{i\Omega(k)t}\} = 2q_n(t) - (q_{n+1}(t) + q_{n-1}(t))$$

so that, introducing this result into Eq. (15.32) and after simplification and multiplication by m, Eq. (15.29) is obtained.

Next, passing in Eq. (15.31) from the imaginary exponentials to the corresponding trigonometric functions leads to

$$\Omega(k) = \sqrt{\omega_0^2 + 4\omega^2 \sin^2\left(\frac{kL}{2}\right)} \tag{15.33}$$

so that

$$e^{2in^\circ\pi} = \cos(2n^\circ\pi) + i\sin(2n^\circ\pi) = 1 \quad \text{when} \quad n^\circ = \pm1, \pm2\ldots$$

Therefore, if

$$k' = k + \frac{2n^\circ\pi}{L}$$

then

$$e^{ik'L} = e^{ikL}e^{2in^\circ\pi} = e^{ikL}$$

Hence, it appears that Eq. (15.31) reads

$$\Omega(k) = \Omega\left(k + \frac{2n^\circ\pi}{L}\right)$$

so that all the information for $\Omega(k)$ is confined within the following k interval:

$$-\frac{\pi}{L} \leq k \leq \frac{\pi}{L} \tag{15.34}$$

which is called the *first Brillouin zone*.

Next, from the angular frequency $\Omega(k)$, one may get the phase velocity $v_\phi(k)$ and the group velocity $v_G(k)$ defined, respectively, by

$$v_\phi(k) = \frac{\Omega(k)}{k} \qquad \text{and} \qquad v_G(k) = \frac{d\Omega(k)}{dk}$$

Now, observe that it is possible to write the following infinite sum involving the $q_n(t)$ governed by Eq. (15.29) via $q_{n\pm1}(t)$ times e^{-iknL}:

$$\sum_{n=-\infty}^{+\infty} q_{n\pm1}(t)e^{-iknL} = e^{\pm ikL} \sum_{n=-\infty}^{+\infty} q_{n\pm1}(t)e^{-ik(n\pm1)L}$$

Then, changing in the right-hand-side sum the $n \pm 1$ terms into new ones n does not modify anything since the sum is infinite so that

$$\sum_{n=-\infty}^{+\infty} q_{n\pm1}(t)e^{-iknL} = e^{\pm ikL} \sum_{n=-\infty}^{+\infty} q_n(t)e^{-iknL} \tag{15.35}$$

15.2.1.2 *Normal modes within the reciprocal space*

Now, introduce the following discrete Fourier expansions:

$$\xi(k,t) = \sum_{n=-\infty}^{+\infty} q_n(t)e^{-iknL} \tag{15.36}$$

$$\zeta(k,t) = \sum_{n=-\infty}^{+\infty} p_n(t) e^{-iknL} \tag{15.37}$$

where k is a continuous variable having the dimension of the inverse length, that is, a 1D wave vector, whereas the $p_n(t)$ are the momentum coordinates corresponding to the position coordinates $q_n(t)$, that is,

$$p_n(t) = m\left(\frac{dq_n(t)}{dt}\right) \tag{15.38}$$

Owing to Eq. (15.38), Eq. (15.37) yields

$$\zeta(k,t) = m\sum_{n=-\infty}^{+\infty}\left(\frac{dq_n(t)}{dt}\right)e^{-iknL}$$

or, due to Eq. (15.36),

$$\zeta(k,t) = m\left(\frac{d\xi(k,t)}{dt}\right) \tag{15.39}$$

Note that, owing to Eqs. (15.36) and (15.37),

$$\xi(-k,t) = \xi(k,t)^* \qquad \text{and} \qquad \zeta(-k,t) = \zeta(k,t)^*$$

Next, since k is continuous whereas n is discrete, the inverse transformations of Eqs. (15.36) and (15.37) are the following integral Fourier transforms working within the first Brillouin zone (15.34), that is,

$$q_n(t) = \frac{L}{2\pi}\int_{-\pi/L}^{\pi/L} \xi(k,t)e^{iknL}\,dk \tag{15.40}$$

$$p_n(t) = \frac{L}{2\pi}\int_{-\pi/L}^{\pi/L} \zeta(k,t)e^{iknL}\,dk \tag{15.41}$$

where, according to Eq. (15.34), k runs from $-\pi/L$ to $+\pi/L$. Besides, the second time derivative of Eq. (15.36) reads

$$\frac{d^2\xi(k,t)}{dt^2} = \sum_{n=-\infty}^{+\infty}\left(\frac{d^2q_n(t)}{dt^2}\right)e^{-iknL}$$

or, in view of Eq. (15.29),

$$\left(\frac{d^2\xi(k,t)}{dt^2}\right) = -\sum_{n=-\infty}^{+\infty}\{\omega_0^2 + \omega^2\{(q_n(t) - q_{n+1}(t)) + (q_n(t) - q_{n-1}(t))\}\}e^{-iknL}$$

and thus

$$\left(\frac{d^2\xi(k,t)}{dt^2}\right) = -\omega_0^2 - \omega^2 \left\{ 2 \sum_{n=-\infty}^{+\infty} q_n(t)e^{-iknL} - \sum_{n=-\infty}^{+\infty} q_{n+1}(t)e^{-iknL} - \sum_{n=-\infty}^{+\infty} q_{n-1}(t)e^{-iknL} \right\} \tag{15.42}$$

Next, keeping in mind Eq. (15.35), that is,

$$\sum_{n=-\infty}^{+\infty} q_{n\pm1}(t)e^{-iknL} = e^{\pm ikL} \sum_{n=-\infty}^{+\infty} q_n(t)e^{-iknL} \tag{15.43}$$

It is possible to transform Eq. (15.42) into

$$\left(\frac{d^2\xi(k,t)}{dt^2}\right) = -\omega_0^2 - \omega^2 \left\{ 2 \sum_{n=-\infty}^{+\infty} q_n(t)e^{-iknL} - e^{+ikL} \sum_{n=-\infty}^{+\infty} q_n(t)e^{-iknL} - e^{-ikL} \sum_{n=-\infty}^{+\infty} q_n(t)e^{-iknL} \right\}$$

so that, owing to Eq. (15.36), it simplifies to

$$\left(\frac{d^2\xi(k,t)}{dt^2}\right) = -\{\omega_0^2 + \omega^2(2 - e^{ikL} - e^{-ikL})\}\xi(k,t)$$

or, in view of Eq. (15.31),

$$\boxed{\left(\frac{d^2\xi(k,t)}{dt^2}\right) = -\Omega(k)^2\xi(k,t)} \tag{15.44}$$

Clearly, irrespective of the value of the continuous 1D wave vector k, the second-order time derivative of $\xi(k,t)$ depends on $\xi(k,t)$ for the same value of k in a form that is that of an harmonic oscillator, so that the $\xi(k,t)$ act as *normal modes*.

Now, pass to the conjugate variables of these normal modes. The second time derivative of Eq. (15.39) reads

$$\left(\frac{(d^2\zeta(k,t)}{dt^2}\right) = m\frac{d^3\xi(k,t)}{dt^3}$$

or

$$\left(\frac{d^2\zeta(k,t)}{dt^2}\right) = m\frac{d}{dt}\left(\frac{d^2\xi(k,t)}{dt^2}\right)$$

and thus, in view of Eq. (15.44),

$$\left(\frac{d^2\zeta(k,t)}{dt^2}\right) = -m\Omega(k)^2\frac{d}{dt}(\xi(k,t))$$

so that, owing to Eq. (15.39),

$$\boxed{\left(\frac{d^2\zeta(k,t)}{dt^2}\right) = -\Omega(k)^2\zeta(k,t)} \tag{15.45}$$

a result that has the same form as that of (15.44)

15.2.2 Quantization of the long chain of oscillators

15.2.2.1 Mode quantization Now, within the Schrödinger picture of quantum mechanics, one has to consider $\xi(k,t)$ and $\zeta(k,t)$ as operators $\xi(k)$ and $\zeta(k)$, which do not depend on time, because of this chosen picture, and which obey the commutation rule

$$[\xi(k), \zeta(k)] = i\hbar \tag{15.46}$$

Then, introduce the two following dimensionless Hermitian self-conjugate operators, analogous to (5.6) and (5.7) used for the quantum oscillators, that is,

$$\boldsymbol{\xi}(k) = \sqrt{\frac{\hbar}{2m\Omega(k)}}(\mathbf{a}^\dagger(k) + \mathbf{a}(k)) \tag{15.47}$$

$$\boldsymbol{\zeta}(k) = i\sqrt{\frac{\hbar m\Omega(k)}{2}}(\mathbf{a}^\dagger(k) - \mathbf{a}(k)) \tag{15.48}$$

with, as a result of Eq. (15.46), the commutator

$$[\mathbf{a}(k), \mathbf{a}^\dagger(k)] = 1 \tag{15.49}$$

Just as $\xi(k,t)$ and $\zeta(k,t)$ have been transformed into time-independent operators $\xi(k)$ and $\zeta(k)$, the coordinates and momenta defined by Eqs. (15.40) and (15.41) become time-independent operators $\mathbf{q}_n$ and $\mathbf{p}_n$:

$$\mathbf{q}_n = \frac{L}{2\pi}\int_{-\pi/L}^{\pi/L} \xi(k)e^{iknL}dk \qquad \text{and} \qquad \mathbf{p}_n = \frac{L}{2\pi}\int_{-\pi/L}^{\pi/L} \zeta(k)e^{iknL}dk$$

which, by analogy with Eqs. (15.36) and (15.37), take the form

$$\boldsymbol{\xi}(k) = \sum_{n=-\infty}^{+\infty} \mathbf{q}_n e^{-iknL} \qquad \text{and} \qquad \boldsymbol{\zeta}(k) = \sum_{n=-\infty}^{+\infty} \mathbf{p}_n e^{-iknL} \tag{15.50}$$

15.2.2.2 Hamiltonian obtainment The full Hamiltonian operator of the linear set of coupled oscillators related to the classical dynamic equation (15.29) reads

$$\mathbf{H}_{\text{Tot}} = \sum_{n=-\infty}^{+\infty} \mathbf{H}_n + \mathbf{H}_{\text{Int}} \tag{15.51}$$

with, respectively,

$$\mathbf{H}_n = \left\{\frac{\mathbf{p}_n^2}{2m} + \frac{1}{2}m\omega_0^2\mathbf{q}_n^2\right\} \tag{15.52}$$

$$\mathbf{H}_{\text{Int}} = \sum_{n=-\infty}^{+\infty}\left\{\frac{1}{2}m\omega^2(\mathbf{q}_n - \mathbf{q}_{n+1})^2\right\} \tag{15.53}$$

Even if the operators $\mathbf{p}_n$ and $\mathbf{q}_n$ are real, it is convenient, for reasons that will appear later, to write the full Hamiltonian (15.51) using Eq. (15.52) and (15.53) as

$$\mathbf{H}_{\text{Tot}} = \frac{1}{2m}\sum_{n=-\infty}^{+\infty} |\mathbf{p}_n|^2 + \frac{m\omega_0^2}{2}\sum_{n=-\infty}^{+\infty} |\mathbf{q}_n|^2 + \frac{m\omega^2}{2}\sum_{n=-\infty}^{+\infty} |\mathbf{q}_n - \mathbf{q}_{n+1}|^2 \qquad (15.54)$$

Moreover, owing to Eq. (15.43), which holds not only for the scalars $q_n(t)$ but also for the operators $\mathbf{q}_n$, it reads

$$\sum_{n=-\infty}^{+\infty} (\mathbf{q}_n - \mathbf{q}_{n+1})e^{-iknL} = \sum_{n=-\infty}^{+\infty} \mathbf{q}_n\, e^{-iknL} - \sum_{n=-\infty}^{+\infty} \mathbf{q}_{n+1}\, e^{-iknL}$$

a result that transforms, according to Eq. (15.43), into

$$\sum_{n=-\infty}^{+\infty} (\mathbf{q}_n - \mathbf{q}_{n+1})e^{-iknL} = \sum_{n=-\infty}^{+\infty} \mathbf{q}_n\, e^{-iknL} - e^{ikL}\sum_{n=-\infty}^{+\infty} \mathbf{q}_n\, e^{-iknL}$$

or

$$\sum_{n=-\infty}^{+\infty} (\mathbf{q}_n - \mathbf{q}_{n+1})e^{-iknL} = (1 - e^{ikL})\sum_{n=-\infty}^{+\infty} \mathbf{q}_n\, e^{-iknL} \qquad (15.55)$$

Then, apply the Bessel–Parseval relation (18.34) of Section 18.6 for a periodic function $f(k)$, where the C_n are the expansion coefficients within the interval $-L/2, L/2$, that is,

$$\sum_{n=-\infty}^{+\infty} |C_n|^2 = \frac{L}{2\pi}\int_{-\pi/L}^{\pi/L} |f(k)|^2\, dk$$

leading to the following functions appearing in (15.56):

Eqs.	$f(k)$	$\sum_{n=-\infty}^{+\infty} C_n e^{-iknL}$
(15.36)	$\xi(k)$	$\sum_{n=-\infty}^{+\infty} \mathbf{q}_n e^{-iknL}$
(15.37)	$\zeta(k)$	$\sum_{n=-\infty}^{+\infty} \mathbf{p}_n e^{-iknL}$
(15.55)	$(1 - e^{ikL})\xi(k)$	$\sum_{n=-\infty}^{+\infty} (\mathbf{q}_n - \mathbf{q}_{n+1})e^{-iknL}$

(15.56)

Then, one obtains, respectively, for the three sums involved in Eq. (15.54), the following relations:

$$\sum_{n=-\infty}^{+\infty} |\mathbf{q}_n|^2 = \frac{L}{2\pi}\int_{-\pi/L}^{\pi/L} |\xi(k)|^2\, dk$$

$$\sum_{n=-\infty}^{+\infty} |\mathbf{p}_n|^2 = \frac{L}{2\pi} \int_{-\pi/L}^{\pi/L} |\boldsymbol{\zeta}(k)|^2 \, dk$$

$$\sum_{n=-\infty}^{+\infty} |\mathbf{q}_n - \mathbf{q}_{n+1}|^2 = \frac{L}{2\pi} \int_{-\pi/L}^{\pi/L} |(1 - e^{ikL})\boldsymbol{\xi}(k)|^2 \, dk$$

As a consequence of these three equations, the Hamiltonian (15.54) reads

$$\mathbf{H}_{\mathrm{Tot}} = \frac{L}{2\pi} \int_{-\pi/L}^{\pi/L} \left\{ \frac{1}{2m} |\boldsymbol{\zeta}(k)|^2 + \frac{m\omega_0^2}{2} |\boldsymbol{\xi}(k)|^2 + \frac{m\omega^2}{2} |(1 - e^{ikL})\boldsymbol{\xi}(k)|^2 \right\} dk \quad (15.57)$$

Since the last squared modulus involved on the right-hand side of Eq. (15.57) is

$$|(1 - e^{ikL})\boldsymbol{\xi}(k)|^2 = 2|\boldsymbol{\xi}(k)|^2(1 - \cos kL)$$

or, using the usual trigonometric relations,

$$|(1 - e^{ikL})\boldsymbol{\xi}(k)|^2 = 4|\boldsymbol{\xi}(k)|^2 \sin^2\left(\frac{kL}{2}\right)$$

the Hamiltonian (15.57) becomes,

$$\mathbf{H}_{\mathrm{Tot}} = \frac{L}{2\pi} \int_{-\pi/L}^{\pi/L} \left\{ \frac{m}{2} \left(\omega_0^2 + 4\omega^2 \sin^2\left(\frac{kL}{2}\right) \right) |\boldsymbol{\xi}(k)|^2 + \frac{1}{2m} |\boldsymbol{\zeta}(k)|^2 \right\} dk$$

or, due to Eq. (15.33),

$$\mathbf{H}_{\mathrm{Tot}} = \frac{L}{2\pi} \int_{-\pi/L}^{\pi/L} \left\{ \frac{m}{2} \Omega(k)^2 |\boldsymbol{\xi}(k)|^2 + \frac{1}{2m} |\boldsymbol{\zeta}(k)|^2 \right\} dk$$

this latter expression for the total Hamiltonian may also be written as an integral over the Hamiltonian functions of k varying continuously, that is,

$$\boxed{\mathbf{H}_{\mathrm{Tot}} = \frac{L}{2\pi} \int_{-\pi/L}^{\pi/L} \{\mathbf{H}(k)\} \, dk} \quad (15.58)$$

with

$$\{\mathbf{H}(k)\} = \frac{m}{2} \Omega(k)^2 |\boldsymbol{\xi}(k)|^2 + \frac{1}{2m} |\boldsymbol{\zeta}(k)|^2$$

Finally, the Hamiltonians $\mathbf{H}(k)$ may be transformed by the aid of Eqs. (15.47) and (15.48) into

$$\boxed{\{\mathbf{H}(k)\} = \hbar\Omega(k) \left(\mathbf{a}(k)^\dagger \mathbf{a}(k) + \tfrac{1}{2} \right)} \quad (15.59)$$

Thus, these Hamiltonians have the same structure as (5.9) of a single quantum harmonic oscillator so that their eigenvalue equations must be of the same kind as that of the Hamiltonian (5.9), that is, as that of Eq. (5.42). Hence

$$\{\mathbf{H}(k)\}|\{n(k)\}\rangle = \left(n(k) + \tfrac{1}{2}\right)\hbar\Omega(k)|\{n(k)\}\rangle$$

where $|\{n(k)\}\rangle$ are the eigenkets of the Hamiltonians $\mathbf{H}(k)$, whereas $n(k) = 0, 1, 2, \ldots$ are the number of vibrational quanta within the normal mode k, which may be viewed as the excitation degrees of the modes of angular frequency $\Omega(k)$. Besides, since each Hamiltonian (15.59) is Hermitian as this (5.9), the eigenkets $|\{n(k)\}\rangle$ form, of course, for each value of the angular frequency, an orthonormalized basis defined by

$$\langle\{m(k)\}|\{n(k)\}\rangle = \delta_{m(k),n(k)} \qquad \text{and} \qquad \sum_{n(k)} |\{n(k)\}\rangle\langle\{n(k)\}| = 1$$

Moreover, as in Eq. (5.12), working for single quantum harmonic oscillator, one may introduce, for each normal mode of the solid, an occupation number operator defined by

$$\mathbf{N}(k) = \mathbf{a}(k)^{\dagger}\mathbf{a}(k)$$

the eigenvalue equation of which is

$$\mathbf{N}(k)|\{n(k)\}\rangle = n(k)|\{n(k)\}\rangle \quad \text{with} \quad n(k) = 0, 1, 2, \ldots$$

Since the $n(k)$ may be viewed as the number of vibrational quanta corresponding to the normal mode k, these vibrational quanta are called *phonons* in solid-state physics.

Moreover, in the Heisenberg picture, each lowering operator of the different normal modes obeys the Heisenberg equation

$$i\hbar\left(\frac{d\mathbf{a}(k,t)^{\text{HP}}}{dt}\right) = [\mathbf{a}(k,t)^{\text{HP}}, \mathbf{H}(k)]$$

Thereby, using Eqs. (15.49) and (15.59), and proceeding in a similar way as for passing from Eqs. (5.150) to (5.151), one would obtain

$$\mathbf{a}(k,t)^{\text{HP}} = \mathbf{a}(k,0)^{\text{HP}}e^{-i\Omega(k)t}$$

Finally, as in the usual quantum harmonic oscillator, one may obtain for each normal mode at any temperature T, the following thermal average (13.32):

$$\langle\mathbf{n}(k)\rangle = \langle\mathbf{a}(k)^{\dagger}\mathbf{a}(k)\rangle = (1 - e^{-\lambda(k)})\mathbf{tr}\{e^{-\lambda(k)\mathbf{a}(k)^{\dagger}\mathbf{a}(k)}\mathbf{a}(k)^{\dagger}\mathbf{a}(k)\}$$

leading to the result

$$\langle\mathbf{n}(k)\rangle = \frac{1}{e^{\lambda(k)} - 1}$$

with

$$\lambda(k) = \frac{\hbar\Omega(k)}{k_B T}$$

which gives the mean number of phonons of k wave vector at temperature T, which is analogous to that (13.36). In a similar way, one would obtain for each normal mode of the solid, the thermal average energy having an expression of the same form as that of (13.29), that is,

$$\langle \mathbf{H}(k) \rangle = \left(\frac{\hbar\Omega(k)}{e^{\hbar\omega/k_B T} - 1} + \frac{\hbar\Omega(k)}{2} \right)$$

and for the heat capacity, an expression of the same form as that of (13.42), that is,

$$C_v(k, T) = N k_B \left(\frac{\hbar\Omega(k)}{k_B T} \right)^2 \frac{e^{\hbar\Omega(k)/k_B T}}{(e^{\hbar\Omega(k)/k_B T} - 1)^2}$$

15.3 EINSTEIN AND DEBYE MODELS OF HEAT CAPACITY

15.3.1 Einstein model

Before the introduction of quantum ideas, it was not possible to understand why the molar specific heat of solids should fall at low temperature, below the classical equipartition value $3R$, with R the ideal gas constant. In 1907, Einstein clarified this mystery using Planck's hypothesis concerning the quantization of energy oscillators. In his model, Einstein used the rough assumption that all the oscillators of solids as having the same characteristic angular frequency $\omega°$. Then, the heat capacity of the solid is equal to the total number of freedom degrees of vibration of the solid, times the heat capacity of each oscillator. If there are $N = n\mathcal{N}$ atoms (where n is the number of moles and $\mathcal{N}$ the Avogadro number) in the solid, the total number of degrees of freedom is $3N - 6 \backsimeq 3N$ since N is very large. Now, the heat capacity of N oscillators is given by Eq. (13.42). Hence, in the Einstein model, the *heat capacity of the solid* reads

$$C_v(T) = 3N k_B \left(\frac{\hbar\omega°}{k_B T} \right)^2 \frac{e^{\hbar\omega°/k_B T}}{(e^{\hbar\omega°/k_B T} - 1)^2} \tag{15.60}$$

and the *molar heat capacity of the solid* reads

$$\bar{C}_v(T) = \frac{C_v(T)}{n} = 3R \left(\frac{T_E}{T} \right)^2 \frac{e^{T_E/T}}{(e^{T_E/T} - 1)^2} \tag{15.61}$$

where T_E is the *Einstein temperature* defined by

$$T_E = \frac{\hbar\omega°}{k_B}$$

If Eq. (15.61) of the Einstein model reproduces the general sigmoid form of the experimental evolution with the absolute temperature of the heat capacity, however, the experimental specific heat approaches zero slower than that predicted by this equation since it obeys an empirical law of the form

$$\bar{C}_v(T)_{\text{Exp}} \backsimeq T^3$$

The reason for this discrepancy is the crude assumption that all atoms of the solid vibrate with the same characteristic angular frequency. It is clear that there are always some modes of oscillation corresponding to a sufficiently large group of atoms moving collectively with so small an angular frequency that these modes may contribute more appreciably to the specific heat than that predicted from the Einstein assumption, thus preventing the heat capacity $\overline{C}_v(T)$ from decreasing quite as rapidly.

15.3.2 Debye model

To improve the Einstein model, which relies on the approximation that all the oscillators of the solid may be viewed as having the same angular frequency, Debye supposed that the angular frequencies of the vibrational modes vary, the number $\sigma(\omega)$ of these modes lying between ω and $\omega + d\omega$, being assumed to be those of the normal modes inside a closed cavity of volume V. This number $\sigma(\omega)$ is given by an expression that may be obtained just as that used to pass from Eqs. (14.88) to (14.94), in counting the number of normal modes of the electromagnetic field in a cavity, that is,

$$\sigma(\omega) = 3V\frac{\omega^2}{2\pi^2 c_s^3} \tag{15.62}$$

where c_s is the effective sonic velocity. Moreover, Debye assumed a cut-off value ω_D in such a way as the $3N$ degrees of freedom of vibration result from

$$3N = \int_0^{\omega_D} \sigma(\omega)\, d\omega \tag{15.63}$$

This approximation may be compared to experimental results obtained for a metal from X-ray scattering measurements at 300 K (see Fig. 15.3).

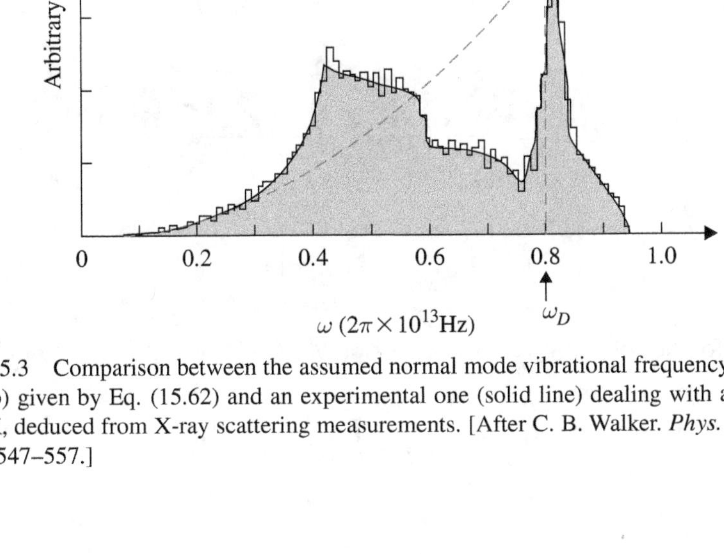

Figure 15.3 Comparison between the assumed normal mode vibrational frequency distribution $\sigma(\omega)$ given by Eq. (15.62) and an experimental one (solid line) dealing with aluminum at 300 K, deduced from X-ray scattering measurements. [After C. B. Walker. *Phys. Rev.*, **103** (1956): 547–557.]

Then, using Eq. (15.62), Eq. (15.63) reads

$$3N = \frac{3V}{2\pi^2 c_s^3}\int_0^{\omega_D} \omega^2 d\omega = V\frac{\omega_D^3}{2\pi^2 c_s^3}$$

From this result, one may express the cut-off angular frequency ω_D and the volume V using

$$\omega_D = c_s\left(6\pi^2\frac{N}{V}\right)^{1/3} \tag{15.64}$$

and

$$V = 6\pi^2 N\left(\frac{c_s}{\omega_D}\right)^3$$

Then, using the Debye cut-off approximation, the heat capacity of the solid reads

$$C_v(T) = \int_0^{\omega_D} \sigma(\omega)C_v(\omega, T)\, d\omega$$

Hence, using Eq. (15.60) for a single oscillator, we have with $N = 1$ and Eq. (15.62)

$$C_v(T) = \frac{3Vk_B}{2\pi^2 c_s^3}\int_0^{\omega_D} \omega^2\left(\frac{\hbar\omega}{k_B T}\right)^2 \frac{e^{\hbar\omega/k_B T}}{(e^{\hbar\omega/k_B T}-1)^2} d\omega \tag{15.65}$$

Again, using the notation

$$x = \frac{\hbar\omega}{k_B T} \qquad \text{and thus} \qquad \omega = \frac{x k_B T}{\hbar}$$

Eq. (15.65) becomes

$$C_v(T) = \frac{3Vk_B}{2\pi^2}\left(\frac{k_B T}{c_s\hbar}\right)^3 \int_0^{x_D} \frac{e^x}{(e^x-1)^2} x^4\, dx$$

where

$$x_D = \frac{\hbar\omega_D}{k_B T_D} \tag{15.66}$$

or, using for the volume V the last expression of Eq. (15.64),

$$C_v(T) = nR\frac{3^2}{(x_D)^3}\int_0^{x_D} \frac{e^x}{(e^x-1)^2} x^4\, dx \tag{15.67}$$

In the high-temperature limit $T \gg T_D$ and $x = \hbar\omega/k_B T \ll 1$. Then, it is possible to expand the exponential, but only up to second order, in order to avoid indeterminate behavior. Then, using the second-order expansions

$$e^x \backsimeq \left(1 + x + \frac{x^2}{2}\right) \qquad \text{and} \qquad e^{2x} \backsimeq \left(1 + 2x + \frac{4x^2}{2}\right)$$

we have

$$(e^x - 1)^2 \backsimeq \left(1 + 2x + \frac{4x^2}{2}\right) - 2\left(1 + x + \frac{x^2}{2}\right) + 1 = x^2$$

so that, since $x << 1$,

$$\frac{e^x}{(e^x - 1)^2} \backsimeq \frac{1}{x^2} + \frac{2}{x} + 2 \backsimeq \frac{1}{x^2}$$

Then, the heat capacity (15.67) reduces in this high-temperature limit to

$$C_v(T) = 3nR\frac{3}{(x_D)^3}\int_0^{x_D} x^2\,dx = 3nR \quad \text{for} \quad T >> T_D$$

This result is the *Dulong and Petit* law governing the heat capacity of solids at high temperature, a limiting result that can also be obtained by the Einstein model. However, the more interesting in Eq. (15.67) is its limiting case of very low temperature, corresponding to $x = \hbar\omega/k_B T >> 1$. In this low-temperature region, the upper limit x_D of the integral appearing in Eq. (15.67) can be replaced by infinity even if x_D is maintained in the constant appearing in front of the integral sign:

$$C_v(T) = nR\frac{3^2}{(x_D)^3}\int_0^{\infty} \frac{e^x}{(e^x - 1)^2}x^4\,dx \tag{15.68}$$

The dimensionless integral is then a constant that does not depend on the temperature and which may be found to be

$$\int_0^{\infty} \frac{e^x}{(e^x - 1)^2}x^4 dx = \frac{4}{15}\pi^4$$

Hence, in the low-temperature limit, and due to Eq. (15.66), Eq. (15.68) yields

$$C_v(T) = nR\frac{4\pi^4}{15}\left(\frac{T}{T_D}\right)^3 \tag{15.69}$$

where T_D is the Debye temperature given by

$$T_D = \frac{\hbar\omega_D}{k_B}$$

Observe that the low-temperature limit (15.69) of the Debye heat capacity reproduces satisfactorily the experimental T^3 dependence, as shown in Fig. 15.4.

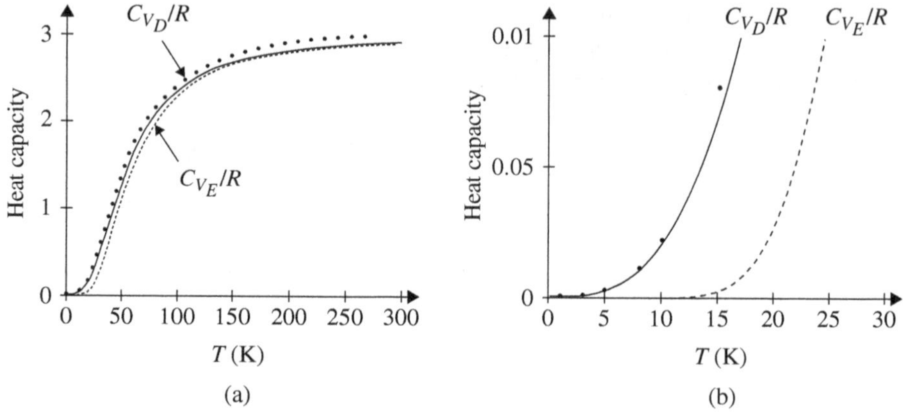

Figure 15.4 Temperature dependence of experimental (*Handbook of Physics and Chemistry*, 72 ed.) heat capacities (dots) of silver as compared to the Einstein (C_{V_E}) and the Debye (C_{V_D}) models as a function of the absolute temperature T. $T_E = 181$ K, $T_D = 225$ K.

15.4 CONCLUSION

In this chapter, we showed how to get the normal modes of a molecule and of 1D solids, by assuming that molecular or solid oscillators involve coupling linear in the elongations of the coupled oscillators. The classical coupled molecular oscillators were decoupled, leading to normal modes having the same properties as harmonic oscillators, which behave on quantization as the harmonic oscillators studied in Parts II, III, and IV. Next, when the molecules involve the symmetry elements of a symmetry group, it was shown using point-group theory, how to determine to what irreducible representation of the symmetry group belong the different molecular normal modes. Then, considering 1D solids, it was shown how, on passing from the geometric to reciprocal space that it is possible to get the solid normal modes acting as usual harmonic oscillators and thus allowing us to apply to them all the results met for single harmonic oscillators and thus, particularly, to find some solid thermal properties such as, for instance, their heat capacities, either in the context of the Einstein model or that of Debye.

BIBLIOGRAPHY

C. Cohen-Tannoudji, B. Diu, and F. Laloe. *Quantum Mechanics*. Wiley-Interscience: Hoboken, NJ, 2006.

F. Reif. *Fundamentals of Statistical and Thermal Physics*. McGraw-Hill: New York, 1965.

E. Wilson, J. Decius, and P. Cross. *Molecular Vibrations*. McGraw-Hill: New York, 1955.

PART VI

DAMPED HARMONIC OSCILLATORS

CHAPTER 16

DAMPED OSCILLATORS

INTRODUCTION

In Chapter 5 the energy levels of an isolated quantum harmonic oscillator was found, and in Chapter 13 the properties of a population of quantum harmonic oscillators at thermal equilibrium were derived, a thermal situation that must occur whatever the initial condition. Furthermore, in Chapter 11 we studied a linear chain of quantum harmonic oscillators linearly coupled in the rotating-wave approximation, the initial situation being where one of the oscillators is in a coherent state, whereas the other ones are in the ground state of their Hamiltonian. We found that the system, if it evolves in a deterministic way, leads, however, after a sufficient time to a continuous distribution of the energy between the different oscillators. Then, it was possible using a coarse-grained analysis of the energy distribution between the different oscillators to show in Chapter 13 that the mean statistical entropy of the system increases until it has attained a stable maximum value and that, when this maximum has been attained, the mean distribution of the energy levels of the different oscillators then obeys the thermal equilibrium Boltzmann law.

However, the coarse-grained analysis of the irreversible mean evolution of a quantum oscillator toward the thermal equilibrium distribution has not yet been studied. The purpose of the present chapter is to treat this question. This chapter begins with an exposition of the quantum model generally used to treat the irreversible behavior of an oscillator embedded in a thermal bath. Then, second-order time-dependent perturbation theory (i.e., second order with respect to the coupling between the damped oscillator and the bath) is used to calculate the master equation governing the time derivative of the reduced density operator of a driven damped quantum harmonic oscillator. To go beyond the previous perturbative approach, a short subsequent section is devoted, without demonstration, to the results of the Louisell and Walker models, which gives in closed form an expression for the time evolution of the reduced density operator of the driven damped quantum harmonic oscillator, which may be viewed as a result of the integration of a master equation, which would have been obtained up to infinite order of perturbation instead of second order as in the above master equation. In the next section, we transform the master equation to its corresponding antinormal expression (see Chapter 7), which has the form of a Fokker–Planck equation and which may then transform, using the inverse of the antinormal order operator, into a second-order partial differential equation having a structure analogous to the Fokker–Planck equation of Brownian oscillators met in the

Quantum Oscillators, First Edition. Olivier Henri-Rousseau and Paul Blaise.
© 2011 John Wiley & Sons, Inc. Published 2011 by John Wiley & Sons, Inc.

area of statistical irreversible classical mechanics. In a subsequent section, the quantum Langevin equation governing the irreversible evolution of the average values of Boson operators is derived. Finally, using this Langevin equation, we to get what may be considered as the IP time evolution operator governing the dynamics of a driven damped harmonic oscillator.

16.1 QUANTUM MODEL FOR DAMPED HARMONIC OSCILLATORS

16.1.1 Hamiltonian

The full Hamiltonian of the driven damped quantum harmonic oscillator may be written

$$\mathbf{H}_{\mathrm{Tot}} = \mathbf{H}^{\circ} + \mathbf{H}_{\mathrm{Dr}} + \mathbf{V} + \mathbf{H}_{\theta} \tag{16.1}$$

Here, $\mathbf{H}^{\circ}$ is the harmonic part of the Hamiltonian of the oscillator of interest, whereas $\mathbf{H}_{\mathrm{Dr}}$ is the driven part of the Hamiltonian of this oscillator. $\mathbf{H}_{\theta}$ is the Hamiltonian of the thermal bath and $\mathbf{V}$ the coupling Hamiltonian between the thermal bath and the oscillator, which will be damped by this bath.

The harmonic part of the Hamiltonian of the oscillator of interest is, of course,

$$\mathbf{H}^{\circ} = \left(\frac{\mathbf{P}^2}{2M} + \frac{1}{2} M \omega^{\circ 2} \mathbf{Q}^2 \right) \tag{16.2}$$

where M is the reduced mass of the oscillator, ω° is the corresponding angular frequency, $\mathbf{Q}$ is the coordinate operator, and $\mathbf{P}$ its conjugate momentum obeying

$$[\mathbf{Q}, \mathbf{P}] = i\hbar$$

The part of the Hamiltonian driving the oscillator is

$$\mathbf{H}_{\mathrm{Dr}} = k^{\circ} \mathbf{Q} \tag{16.3}$$

where k° is a constant.

Now, the thermal bath may be simulated by an infinite set of quantum harmonic oscillators of reduced masses m_l and of angular frequencies ω_l, which are varying in a quasi-continuous way. Thus, the Hamiltonian of the bath may be written

$$\mathbf{H}_{\theta} = \left(\sum_l \frac{\mathbf{p}_l^2}{2m_l} + \frac{1}{2} m_l \omega_l^2 \mathbf{q}_l^2 \right) \tag{16.4}$$

where $\mathbf{q}_l$ is the position operator of the lth oscillator, whereas $\mathbf{p}_l$ is the conjugate momentum obeying

$$[\mathbf{q}_k, \mathbf{p}_l] = i\hbar \delta_{kl}$$

The Hamiltonian coupling the driven oscillator to the thermal bath may be assumed to be given by

$$\mathbf{V} = \sum_l k_l \mathbf{Q} \mathbf{q}_l \tag{16.5}$$

where the k_l are the coupling constants between the driven oscillator and the lth oscillator of the bath. Of course, the operators characterizing the driven oscillator and the bath oscillators commute, that is,

$$[\mathbf{Q}, \mathbf{p}_l] = [\mathbf{P}, \mathbf{q}_l] = 0$$

In the following, one will pass from the discrete expression of the thermal bath (16.4) to the following continuous one:

$$\mathbf{H}_\theta = \int g(\omega) \left\{ \frac{\mathbf{p}(\omega)^2}{2m(\omega)} + \frac{1}{2} m(\omega)\omega^2 \mathbf{q}(\omega)^2 \right\} d\omega$$

$$\mathbf{V} = \int g(\omega) k(\omega) \mathbf{Q} \mathbf{q}(\omega) \, d\omega$$

In these continuous expressions, $g(\omega)$ is the *density of modes* of angular frequency ω and reduced mass $m(\omega)$, whereas $k(\omega)$ are the corresponding frequency-dependent coupling constants. Finally, $\mathbf{q}(\omega)$ and $\mathbf{p}(\omega)$ are the frequency-dependent position and conjugate momentum obeying

$$[\mathbf{q}(\omega), \mathbf{p}(\omega')] = i\hbar\delta(\omega - \omega')$$

Now, we pass from the position and momentum operators of the different oscillators to the corresponding Boson operators by means of the usual transformations (5.6) and (5.7):

$$\mathbf{Q} = \sqrt{\frac{\hbar}{2M\omega^\circ}} (\mathbf{a}^\dagger + \mathbf{a}) \qquad \mathbf{P} = i\sqrt{\frac{\hbar M\omega^\circ}{2}} (\mathbf{a}^\dagger - \mathbf{a}) \tag{16.6}$$

$$\mathbf{q}_l = \sqrt{\frac{\hbar}{2m_l\omega_l}} (\mathbf{b}_l^\dagger + \mathbf{b}_l) \qquad \mathbf{p}_l = i\sqrt{\frac{\hbar m_l\omega_l}{2}} (\mathbf{b}_l^\dagger - \mathbf{b}_l) \tag{16.7}$$

in which $\mathbf{a}$, $\mathbf{a}^\dagger$, $\mathbf{b}_l^\dagger$, and $\mathbf{b}_l$ are the dimensionless Boson operators obeying the commutation rules of the same kind as that of (5.5)

$$[\mathbf{a}, \mathbf{a}^\dagger] = 1 \qquad [\mathbf{b}_k, \mathbf{b}_l^\dagger] = \delta_{kl}$$

$$[\mathbf{a}_l^\dagger, \mathbf{b}] = [\mathbf{a}, \mathbf{b}_l^\dagger] = [\mathbf{a}^\dagger, \mathbf{b}_l^\dagger] = [\mathbf{a}, \mathbf{b}_l] = 0 \tag{16.8}$$

In the Boson operator picture and after neglecting the zero-point energy, which is irrelevant the harmonic part of the driven oscillator defined by Eq. (16.2) takes the form (5.9), that is,

$$\mathbf{H} = \hbar\omega^\circ \mathbf{a}^\dagger \mathbf{a} \tag{16.9}$$

whereas the driven part (16.3) of the Hamiltonian (16.1) becomes

$$\mathbf{H}_{\mathrm{Dr}} = \alpha^\circ \hbar\omega^\circ (\mathbf{a}^\dagger + \mathbf{a}) \tag{16.10}$$

with

$$\alpha^\circ = k^\circ \sqrt{\frac{\hbar}{2M\omega^\circ}}$$

Moreover, due to (16.7) and (16.8), the thermal bath Hamiltonian (16.4) yields, after neglecting the zero-point energies of the different oscillators,

$$\mathbf{H}_\theta = \sum_k \hbar\omega_k \mathbf{b}_k^\dagger \mathbf{b}_k \tag{16.11}$$

Thus, owing to Eqs. (16.6) and (16.7), the Hamiltonian (16.5) coupling the driven oscillator to the thermal bath yields

$$\mathbf{V} = \sqrt{\frac{\hbar}{2M\omega^\circ}}(\mathbf{a}^\dagger + \mathbf{a})\sum_k k_k \sqrt{\frac{\hbar}{2m_k\omega_k}}(\mathbf{b}_k^\dagger + \mathbf{b}_k)$$

or

$$\mathbf{V} = \hbar \sum_k K_k(\mathbf{a}^\dagger\mathbf{b}_k^\dagger + \mathbf{a}\mathbf{b}_k + \mathbf{a}^\dagger\mathbf{b}_k + \mathbf{a}\mathbf{b}_k^\dagger) \tag{16.12}$$

where K_k are dimensionless coupling constants given by

$$K_k = \frac{k_k}{2}\sqrt{\frac{1}{Mm_k\omega_k\omega^\circ}}$$

In the *rotating-wave approximation*, it is usual to neglect the double creations and annihilations induced by the terms $\mathbf{a}^\dagger\mathbf{b}_k^\dagger$ and $\mathbf{a}\mathbf{b}_k$ and to single out those interactions in which exist simultaneously an excitation of one of the interacting oscillators and a deexcitation of the last one. Hence, we write in place of Eq. (16.12)

$$\mathbf{V} = \sum_k \hbar K_k\{(\mathbf{a}^\dagger\mathbf{b}_k) + (\mathbf{a}\mathbf{b}_k^\dagger)\}$$

which may be generalized to

$$\mathbf{V} = \sum_k \{\hbar K_k(\mathbf{a}^\dagger\mathbf{b}_k) + \hbar K_k^*(\mathbf{a}\mathbf{b}_k^\dagger)\} \tag{16.13}$$

Now, consider the various density operators of the system. Since the thermal bath involves a very large number of oscillators, its density operator $\boldsymbol{\rho}_\theta(t)$ may be assumed to be unperturbed by the single driven oscillator to which it is coupled, so that it may be assumed to be constant, leading one to write

$$\boldsymbol{\rho}_\theta(t) = \boldsymbol{\rho}_\theta(t_0)$$

where t_0 is an initial time. This thermal density operator will be viewed as the product of the density operators $\boldsymbol{\rho}_j$ of the different oscillators forming the bath, each being in thermal equilibrium and thus described by a Boltzmann density operator

$$\boldsymbol{\rho}_\theta = \prod_j \boldsymbol{\rho}_j \tag{16.14}$$

Moreover, the density operators of the thermal bath oscillators may be assumed to be given at all times by canonical density operators of the form (13.23)

$$\boldsymbol{\rho}_j = (1 - e^{-\lambda_j})(e^{-\lambda_j \mathbf{b}_j^\dagger \mathbf{b}_j}) \tag{16.15}$$

with, according to Eq. (13.24),

$$\lambda_j = \frac{\hbar\omega_j}{k_B T} \qquad \varepsilon_j = (1 - e^{-\lambda_j}) \tag{16.16}$$

where T is the absolute temperature. Now, in the Schrödinger picture, the full density operator of the system at an initial time $t = t_0$ may be considered as the density operator $\boldsymbol{\rho}(t_0)$ of the driven oscillator at this time, multiplied by that of the bath $\rho_\theta(t_0)$ at this same time, that is,

$$\boldsymbol{\rho}_{\text{Tot}}(t_0)^{\text{SP}} = \boldsymbol{\rho}(t_0)^{\text{SP}}\,\boldsymbol{\rho}_\theta \tag{16.17}$$

where k_B is Boltzmann's constant. Next, in the Schrödinger picture, and according to Eq. (3.170), the dynamics of the full density operator is governed by the Liouville equation

$$i\hbar\left(\frac{\partial \boldsymbol{\rho}_{\text{Tot}}(t)^{\text{SP}}}{\partial t}\right) = [\mathbf{H}_{\text{Tot}}, \rho_{\text{Tot}}(t)^{\text{SP}}] \tag{16.18}$$

subject to the boundary condition (16.17). Notice that once the expression of $\boldsymbol{\rho}_{\text{Tot}}(t)^{\text{SP}}$ has been obtained, it is possible to get the time dependence of the density operator of the driven damped harmonic oscillator by performing the partial trace of the time-dependent full density operator over the thermal bath:

$$\boldsymbol{\rho}(t)^{\text{SP}} = \mathbf{tr}_\theta\{\boldsymbol{\rho}_{\text{Tot}}(t)^{\text{SP}}\} \tag{16.19}$$

Due to Eq. (16.1), the Schrödinger–Liouville equation (16.18) reads

$$i\hbar\left(\frac{\partial \boldsymbol{\rho}_{\text{Tot}}(t)^{\text{SP}}}{\partial t}\right) = [\mathbf{H}^\circ, \boldsymbol{\rho}_{\text{Tot}}(t)^{\text{SP}}] + [\,\mathbf{H}_{\text{Dr}}, \boldsymbol{\rho}_{\text{Tot}}(t)^{\text{SP}}]$$
$$+ [\mathbf{V}, \boldsymbol{\rho}_{\text{Tot}}(t)^{\text{SP}}] + [\mathbf{H}_\theta, \boldsymbol{\rho}_{\text{Tot}}(t)^{\text{SP}}]$$

16.1.2 Interaction picture

We make the following partition of the Hamiltonian (16.1):

$$\mathbf{H}_{\text{Tot}} = \mathbf{H}^\circ_{\text{Tot}} + \mathbf{V} + \mathbf{H}_{\text{Dr}} \quad \text{with} \quad \mathbf{H}^\circ_{\text{Tot}} = \mathbf{H} + \mathbf{H}_\theta \tag{16.20}$$

Within this partition, the operators $\mathbf{V}$ and $\mathbf{H}_{\text{Dr}}$ become, respectively, in the interaction picture with respect to $\mathbf{H}^\circ_{\text{Tot}}$

$$\widetilde{\mathbf{V}}(t)^{\text{IP}} = \mathbf{U}^\circ_{\text{Tot}}(t)^{-1}\mathbf{V}\mathbf{U}^\circ_{\text{Tot}}(t) \tag{16.21}$$

$$\widetilde{\mathbf{H}}_{\text{Dr}}(t) = \mathbf{U}^\circ_{\text{Tot}}(t)^{-1}\,\mathbf{H}_{\text{Dr}}\mathbf{U}^\circ_{\text{Tot}}(t) \tag{16.22}$$

with

$$\mathbf{U}^\circ_{\text{Tot}}(t) = \exp\left(\frac{-i\mathbf{H}^\circ_{\text{Tot}}t}{\hbar}\right) \tag{16.23}$$

Next, in view of Eqs. (16.9), (16.11), and (16.20), the time evolution operator Eq. (16.23) transforms to

$$\mathbf{U}^\circ_{\text{Tot}}(t) = \exp\left\{-i\left(\mathbf{a}^\dagger\mathbf{a}\,\omega^\circ t + \sum_j \mathbf{b}_j^\dagger\mathbf{b}_j\,\omega_j t\right)\right\} \tag{16.24}$$

Then, since according to Eq. (16.8) the Boson operators of the thermal bath commute with those of the driven harmonic oscillator, this time evolution operator (16.23) factorizes into

$$\mathbf{U}^{\circ}_{\text{Tot}}(t) = e^{-i\mathbf{a}^{\dagger}\mathbf{a}\,\omega^{\circ}t} \prod_j e^{-i\mathbf{b}_j^{\dagger}\mathbf{b}_j\,\omega_j t} \tag{16.25}$$

or

$$\mathbf{U}^{\circ}_{\text{Tot}}(t) = \mathbf{U}^{\circ}(t)\mathbf{U}^{\circ}_{\theta}(t) \tag{16.26}$$

with, respectively,

$$\mathbf{U}^{\circ}(t) = e^{-i\mathbf{a}^{\dagger}\mathbf{a}\omega^{\circ}t} \tag{16.27}$$

$$\mathbf{U}^{\circ}_{\theta}(t) = \prod_j \mathbf{U}^{\circ}_j(t) \tag{16.28}$$

and

$$\mathbf{U}^{\circ}_j(t) = e^{-i\mathbf{b}_j^{\dagger}\mathbf{b}_j\,\omega_j t} \tag{16.29}$$

Moreover, owing to Eqs. (16.13), (16.26), and (16.28), the IP coupling Hamiltonian (16.21) reads

$$\widetilde{\mathbf{V}}(t)^{\text{IP}} = \hbar\mathbf{U}^{\circ}(t)^{-1}\left\{\prod_j \mathbf{U}^{\circ}_j(t)^{-1}\left\{\sum_k (\mathbf{a}^{\dagger}\mathbf{b}_k K_k + \mathbf{a}\mathbf{b}_k^{\dagger}K_k^*)\right\}\mathbf{U}^{\circ}_j(t)\right\}\mathbf{U}^{\circ}(t)$$

or

$$\widetilde{\mathbf{V}}(t)^{\text{IP}} = \hbar\mathbf{U}^{\circ}(t)^{-1}\mathbf{a}^{\dagger}\mathbf{U}^{\circ}(t)\prod_j\left\{\mathbf{U}^{\circ}_j(t)^{-1}\left\{\sum_k K_k\mathbf{b}_k\right\}\mathbf{U}^{\circ}_j(t)\right\} + \text{hc}$$

or, writing explicitly the evolution operators dealing with the thermal bath using Eqs. (16.27) and (16.29),

$$\widetilde{\mathbf{V}}(t)^{\text{IP}} = (e^{i\mathbf{a}^{\dagger}\mathbf{a}\omega^{\circ}t})\mathbf{a}^{\dagger}(e^{-i\mathbf{a}^{\dagger}\mathbf{a}\omega^{\circ}t})\prod_j\left\{e^{i\mathbf{b}_j^{\dagger}\mathbf{b}_j\,\omega_j t}\left\{\sum_k \hbar K_k\mathbf{b}_k\right\}e^{-i\mathbf{b}_j^{\dagger}\mathbf{b}_j\,\omega_j t}\right\} + \text{hc}$$

so that, following the action of each operator within their respective subspaces, we have

$$\begin{aligned}\widetilde{\mathbf{V}}(t)^{\text{IP}} &= (e^{i\mathbf{a}^{\dagger}\mathbf{a}\omega^{\circ}t})\mathbf{a}^{\dagger}(e^{-i\mathbf{a}^{\dagger}\mathbf{a}\omega^{\circ}t})\left\{\sum_j \hbar K_j(e^{i\mathbf{b}_j^{\dagger}\mathbf{b}_j\,\omega_j t})\mathbf{b}_j(e^{-i\mathbf{b}_j^{\dagger}\mathbf{b}_j\,\omega_j t})\right\}\\ &\quad\times\prod_{k\neq j}(e^{i\mathbf{b}_k^{\dagger}\mathbf{b}_k\,\omega_k t})(e^{-i\mathbf{b}_k^{\dagger}\mathbf{b}_k\,\omega_k t}) + \text{hc}\end{aligned}$$

which reduces to

$$\widetilde{\mathbf{V}}(t)^{\text{IP}} = e^{i\mathbf{a}^{\dagger}\mathbf{a}\omega^{\circ}t}\mathbf{a}^{\dagger}e^{-i\mathbf{a}^{\dagger}\mathbf{a}\omega^{\circ}t}\left\{\sum_j \hbar K_j e^{i\mathbf{b}_j^{\dagger}\mathbf{b}_j\,\omega_j t}\mathbf{b}_j e^{-i\mathbf{b}_j^{\dagger}\mathbf{b}_j\,\omega_j t}\right\} + \text{hc} \tag{16.30}$$

since

$$\prod_{j \neq k} e^{i\mathbf{b}_k^\dagger \mathbf{b}_k\, \omega_k t} e^{-i\mathbf{b}_k^\dagger \mathbf{b}_k \omega_k t} = 1$$

Now, applying theorems (7.21) and (7.22) to the Boson operators of the oscillator of interest and to those of the thermal bath, that is,

$$(e^{i\mathbf{a}^\dagger \mathbf{a}\omega^\circ t})\mathbf{a}^\dagger(e^{-i\mathbf{a}^\dagger \mathbf{a}\omega^\circ t}) = \mathbf{a}^\dagger(e^{i\omega^\circ t}) \tag{16.31}$$

$$(e^{i\mathbf{b}_j^\dagger \mathbf{b}_j\, \omega_j t})\mathbf{b}_j(e^{-i\mathbf{b}_j^\dagger \mathbf{b}_j\, \omega_j t}) = \mathbf{b}_j(e^{-i\omega_j t})$$

it appears that the IP coupling between the driven oscillator and the bath takes the form

$$\widetilde{\mathbf{V}}(t)^{\mathrm{IP}} = \sum_j \{\hbar K_j \mathbf{a}^\dagger e^{i\omega^\circ t}\mathbf{b}_j e^{-i\omega_j t} + \hbar K_j^* \mathbf{a} e^{-i\omega^\circ t}\mathbf{b}_j^\dagger e^{i\omega_j t}\} \tag{16.32}$$

Now due to Eq. (16.26), the IP expression (16.22) of the driving Hamiltonian (16.10), which depends only on $\mathbf{a}^\dagger$ and $\mathbf{a}$, reads

$$\widetilde{\mathbf{H}}_{\mathrm{Dr}}(t)^{\mathrm{IP}} = \alpha^\circ \hbar\omega^\circ \mathbf{U}^\circ_{\mathrm{Tot}}(t)^{-1}(\mathbf{a}+\mathbf{a}^\dagger)\mathbf{U}^\circ_{\mathrm{Tot}}(t)\prod_j \mathbf{U}^\circ_j(t)^{-1}\mathbf{U}^\circ_j(t)$$

or, since

$$\mathbf{U}^\circ_j(t)^{-1}\mathbf{U}^\circ_j(t) = 1$$

we have on simplification

$$\widetilde{\mathbf{H}}_{\mathrm{Dr}}(t)^{\mathrm{IP}} = \alpha^\circ \hbar\omega^\circ \mathbf{U}^\circ(t)^{-1}(\mathbf{a}+\mathbf{a}^\dagger)\mathbf{U}^\circ(t)$$

a result that also reads

$$\widetilde{\mathbf{H}}_{\mathrm{Dr}}(t) = \mathbf{U}^\circ(t)^{-1}\mathbf{H}_{\mathrm{Dr}}\mathbf{U}^\circ(t) \tag{16.33}$$

the inverse canonical transformation being

$$\widetilde{\mathbf{H}}_{\mathrm{Dr}} = \mathbf{U}^\circ(t)\,\mathbf{H}_{\mathrm{Dr}}(t)\mathbf{U}^\circ(t)^{-1} \tag{16.34}$$

Furthermore, according to Eq. (16.10), because the different operators act within their own vector subspace, the driven part of the Hamiltonian (16.22) reads, after simplification,

$$\widetilde{\mathbf{H}}_{\mathrm{Dr}}(t)^{\mathrm{IP}} = \alpha^\circ \hbar\omega^\circ \mathbf{U}^\circ(t)^{-1}(\mathbf{a}+\mathbf{a}^\dagger)\mathbf{U}^\circ(t) \tag{16.35}$$

so that, due to Eq. (16.27),

$$\widetilde{\mathbf{H}}_{\mathrm{Dr}}(t)^{\mathrm{IP}} = \alpha^\circ \hbar\omega^\circ (e^{i\mathbf{a}^\dagger \mathbf{a}\omega^\circ t}\mathbf{a}e^{-i\mathbf{a}^\dagger \mathbf{a}\omega^\circ t} + e^{i\mathbf{a}^\dagger \mathbf{a}\omega^\circ t}\mathbf{a}^\dagger e^{-i\mathbf{a}^\dagger \mathbf{a}\omega^\circ t})$$

which, using (16.31), yields

$$\widetilde{\mathbf{H}}_{\mathrm{Dr}}(t)^{\mathrm{IP}} = \alpha^\circ \hbar\omega^\circ (\mathbf{a}e^{-i\omega^\circ t} + \mathbf{a}^\dagger e^{i\omega^\circ t}) \tag{16.36}$$

In this same picture, the IP density operator $\widetilde{\rho}(t)^{\mathrm{IP}}$ of the oscillator of interest reads

$$\widetilde{\rho}(t)^{\mathrm{IP}} = \mathbf{U}^\circ_{\mathrm{Tot}}(t)^{-1}\rho(t)^{SP}\,\mathbf{U}^\circ_{\mathrm{Tot}}(t) \tag{16.37}$$

where $\rho(t)^{\text{SP}}$ is the Schrödinger picture density operator (16.19). Now, again, due to Eq. (16.26), Eq. (16.37) becomes

$$\widetilde{\rho}(t)^{\text{IP}} = \mathbf{U}^\circ(t)^{-1}\mathbf{U}^\circ_\theta(t)^{-1}\rho(t)^{\text{SP}}\mathbf{U}^\circ(t)\mathbf{U}^\circ_\theta(t)$$

so that, since the IP density operator $\widetilde{\rho}(t)^{\text{IP}}$ is that of the oscillator of interest and not that of the thermal bath, it transforms to

$$\widetilde{\rho}(t)^{\text{IP}} = \mathbf{U}^\circ(t)^{-1}\rho(t)^{\text{SP}}\mathbf{U}^\circ(t)\mathbf{U}^\circ_\theta(t)^{-1}\mathbf{U}^\circ_\theta(t) = \mathbf{U}^\circ(t)^{-1}\rho(t)^{\text{SP}}\mathbf{U}^\circ(t)$$

16.1.3 IP Liouville equation

In the chosen interaction picture, the Liouville equation governing the time dependence of the full density operator is, owing to Eq. (3.193),

$$i\hbar\left(\frac{\partial\tilde{\boldsymbol{\rho}}_{\text{Tot}}(t)^{\text{IP}}}{\partial t}\right) = [(\widetilde{\mathbf{H}}_{\text{Dr}}(t)^{\text{IP}} + \widetilde{\mathbf{V}}(t)^{\text{IP}}), \tilde{\boldsymbol{\rho}}_{\text{Tot}}(t)^{\text{IP}}]$$

Now, performing the trace over the thermal bath on this last expression,

$$i\hbar\,\mathbf{tr}_\theta\left\{\left(\frac{\partial\tilde{\boldsymbol{\rho}}_{\text{Tot}}(t)^{\text{IP}}}{\partial t}\right)\right\} = \mathbf{tr}_\theta\{[(\widetilde{\mathbf{H}}_{\text{Dr}}(t)^{\text{IP}} + \widetilde{\mathbf{V}}(t)^{\text{IP}}), \tilde{\boldsymbol{\rho}}_{\text{Tot}}(t)^{\text{IP}}]\}$$

yields

$$i\hbar\left(\frac{\partial\boldsymbol{\rho}_{\text{Dr}}(t)^{\text{IP}}}{\partial t}\right) = \mathbf{tr}_\theta\{[\widetilde{\mathbf{H}}_{\text{Dr}}(t)^{\text{IP}}, \tilde{\boldsymbol{\rho}}_{\text{Tot}}(t)^{\text{IP}}]\} + \mathbf{tr}_\theta\{[\widetilde{\mathbf{V}}(t)^{\text{IP}}, \tilde{\boldsymbol{\rho}}_{\text{Tot}}(t)^{\text{IP}}]\} \tag{16.38}$$

with

$$\boldsymbol{\rho}_{\text{Dr}}(t)^{\text{IP}} = \mathbf{tr}_\theta\{\tilde{\boldsymbol{\rho}}_{\text{Tot}}(t)^{\text{IP}}\} \tag{16.39}$$

Next, since the IP Hamiltonian $\widetilde{\mathbf{H}}_{\text{Dr}}(t)^{\text{IP}}$ does not involve the thermal bath, the first right-hand-side term of Eq. (16.38) reads

$$\mathbf{tr}_\theta\{[\widetilde{\mathbf{H}}_{\text{Dr}}(t)^{\text{IP}}, \boldsymbol{\rho}_{\text{Tot}}(t)^{\text{IP}}]\} = [\widetilde{\mathbf{H}}_{\text{Dr}}(t)^{\text{IP}}, \mathbf{tr}_\theta\{\tilde{\boldsymbol{\rho}}_{\text{Tot}}(t)^{\text{IP}}\}]$$

And thus, due to Eq. (16.39),

$$\mathbf{tr}_\theta\{[\widetilde{\mathbf{H}}_{\text{Dr}}(t)^{\text{IP}}, \boldsymbol{\rho}_{\text{Tot}}(t)^{\text{IP}}]\} = [\widetilde{\mathbf{H}}_{\text{Dr}}(t)^{\text{IP}}, \boldsymbol{\rho}(t)^{\text{IP}}]$$

Hence, Eq. (16.38) reads

$$i\hbar\left(\frac{\partial\boldsymbol{\rho}_{\text{Dr}}(t)^{\text{IP}}}{\partial t}\right) = [\widetilde{\mathbf{H}}_{\text{Dr}}(t)^{\text{IP}}, \boldsymbol{\rho}(t)^{\text{IP}}] + \mathbf{tr}_\theta\{[\widetilde{\mathbf{V}}(t)^{\text{IP}}, \boldsymbol{\rho}_{\text{Tot}}(t)^{\text{IP}}]\}$$

A result that may also be written as

$$\boxed{i\hbar\left(\frac{\partial\boldsymbol{\rho}_{\text{Dr}}(t)^{\text{IP}}}{\partial t}\right) = [\widetilde{\mathbf{H}}_{\text{Dr}}(t)^{\text{IP}}, \boldsymbol{\rho}_{\text{Dr}}(t)^{\text{IP}}] + i\hbar\left(\frac{\partial\widetilde{\boldsymbol{\rho}}(t)^{\text{IP}}}{\partial t}\right)} \tag{16.40}$$

with

$$\boxed{i\hbar\left(\frac{\partial\widetilde{\boldsymbol{\rho}}(t)^{\text{IP}}}{\partial t}\right) = \mathbf{tr}_\theta\{[\widetilde{\mathbf{V}}(t)^{\text{IP}}, \boldsymbol{\rho}_{\text{Tot}}(t)^{\text{IP}}]\}} \tag{16.41}$$

16.2 SECOND-ORDER SOLUTION OF EQ. (16.41)

Equation (16.41) is the IP Liouville equation governing the dynamics of the undriven oscillator of interest interacting with the thermal bath, the integration of which, up to second order in $\widetilde{\mathbf{V}}(t)^{\mathrm{IP}}$, is the aim of the present section. The formal integration of Eq. (16.41) leads to the integral equation

$$\widetilde{\rho}(t)^{\mathrm{IP}} = \widetilde{\rho}(t_0)^{\mathrm{IP}} + \left(\frac{1}{i\hbar}\right)\int_{t_0}^{t} dt' \mathbf{tr}_\theta\{[\widetilde{\mathbf{V}}(t'-t_0)^{\mathrm{IP}}, \widetilde{\boldsymbol{\rho}}_{\mathrm{Tot}}(t')^{\mathrm{IP}}]\} \tag{16.42}$$

Up to second order in $\widetilde{\mathbf{V}}(t)^{\mathrm{IP}}$, the integral equation (16.42) reads

$$\widetilde{\rho}(t)^{\mathrm{IP}} - \widetilde{\rho}(t_0)^{\mathrm{IP}} = I_1 + I_2 \tag{16.43}$$

with, respectively,

$$I_1 = \left(\frac{1}{i\hbar}\right)\int_{t_0}^{t} dt' \mathbf{tr}_\theta\{[\widetilde{\mathbf{V}}(t'-t_0)^{\mathrm{IP}}, \widetilde{\rho}(t_0)^{\mathrm{IP}}]\} \tag{16.44}$$

$$I_2 = \left(\frac{1}{i\hbar}\right)^2\int_{t_0}^{t} dt' \int_{t_0}^{t'} dt'' \mathbf{tr}_\theta\{[\widetilde{\mathbf{V}}(t'-t_0)^{\mathrm{IP}}, [\widetilde{\mathbf{V}}(t''-t_0)^{\mathrm{IP}}, \widetilde{\rho}(t_0)^{\mathrm{IP}}]]\} \tag{16.45}$$

and where, according to Eq. (3.195),

$$\mathbf{V}(\tau - t_0)^{\mathrm{IP}} = e^{i\mathbf{H}^\circ(\tau-t_0)/\hbar}\mathbf{V}e^{-i\mathbf{H}^\circ(\tau-t_0)/\hbar} \quad \text{with} \quad \tau = t' \text{ or } t''$$

16.2.1 Making explicit Eq. (16.43)

To go further, we prove that the integral (16.44) involved in Eq. (16.43) is zero and for this purpose start from the commutator involved in Eq. (16.44):

$$\begin{aligned}&\mathbf{tr}_\theta\{[\widetilde{\mathbf{V}}(t'-t_0)^{\mathrm{IP}}, \widetilde{\boldsymbol{\rho}}_{\mathrm{Tot}}(t_0)^{\mathrm{IP}}]\}\\ &= \mathbf{tr}_\theta\{\widetilde{\mathbf{V}}(t'-t_0)^{\mathrm{IP}}\widetilde{\boldsymbol{\rho}}_{\mathrm{Tot}}(t_0)^{\mathrm{IP}}\} - \mathbf{tr}_\theta\{\widetilde{\boldsymbol{\rho}}_{\mathrm{Tot}}(t_0)^{\mathrm{IP}}\widetilde{\mathbf{V}}(t'-t_0)^{\mathrm{IP}}\}\end{aligned} \tag{16.46}$$

Owing to Eq. (16.32) the IP coupling Hamiltonian appearing in Eq. (16.46) is

$$\widetilde{\mathbf{V}}(t'-t_0)^{\mathrm{IP}} = \mathbf{a}^\dagger e^{i\omega^\circ(t'-t^\circ)}\sum_l \mathbf{b}_l\hbar K_l e^{-i\omega_l(t'-t^\circ)} + \mathrm{hc}$$

or, on simplification by taking $t = t' - t_0$,

$$\widetilde{\mathbf{V}}(t)^{\mathrm{IP}} = \mathbf{a}^\dagger e^{i\omega^\circ t}\sum_l \mathbf{b}_l\hbar K_l e^{-i\omega_l t} + \mathrm{hc} \tag{16.47}$$

Due to Eq. (16.17), the two right-hand sides of Eq. (16.46) yield, respectively,

$$\begin{aligned}&\mathbf{tr}_\theta\{\widetilde{\mathbf{V}}(t)^{\mathrm{IP}}\widetilde{\boldsymbol{\rho}}_{\mathrm{Tot}}(t_0)^{\mathrm{IP}}\}\\ &= \mathbf{tr}_\theta\left\{\mathbf{a}^\dagger e^{i\omega^\circ t}\sum_l \mathbf{b}_l\hbar K_l e^{-i\omega_l t}\widetilde{\rho}(t_0)^{\mathrm{IP}}\widetilde{\boldsymbol{\rho}}_\theta\right\} + \mathbf{tr}_\theta\left\{\mathbf{a}e^{-i\omega^\circ t}\sum_l \mathbf{b}_l^\dagger\hbar K_l^* e^{+i\omega_l t}\widetilde{\rho}(t_0)^{\mathrm{IP}}\widetilde{\boldsymbol{\rho}}_\theta\right\}\end{aligned}$$

$$\mathbf{tr}_\theta\{\widetilde{\rho}_{\text{Tot}}(t_0)^{\text{IP}}\widetilde{\mathbf{V}}(t)^{\text{IP}}\}$$
$$= \mathbf{tr}_\theta\left\{\widetilde{\rho}(t_0)^{\text{IP}}\widetilde{\rho}_\theta\mathbf{a}^\dagger e^{i\omega^\circ t}\sum_l \mathbf{b}_l\hbar K_l e^{-i\omega_l t}\right\} + \mathbf{tr}_\theta\left\{\widetilde{\rho}(t_0)^{\text{IP}}\widetilde{\rho}_\theta\mathbf{a}e^{-i\omega^\circ t}\sum_l \mathbf{b}_l^\dagger\hbar K_l^* e^{+i\omega_l t}\right\}$$

Moreover, since the trace operation over the thermal bath affects neither the lowering and raising operators $\mathbf{a}$ and $\mathbf{a}^\dagger$ of the oscillator of interest nor its IP density operator $\widetilde{\rho}(t)^{\text{IP}}$, these two last equations become, respectively,

$$\mathbf{tr}_\theta\{\widetilde{\mathbf{V}}(t)^{\text{IP}}\widetilde{\rho}_{\text{Tot}}(t_0)^{\text{IP}}\}$$
$$= \mathbf{tr}_\theta\left\{\sum_l \mathbf{b}_l\widetilde{\rho}_\theta\hbar K_l e^{-i\omega_l t}\right\}\mathbf{a}^\dagger\widetilde{\rho}(t_0)^{\text{IP}}e^{i\omega^\circ t} + \mathbf{tr}_\theta\left\{\sum_l \mathbf{b}_l^\dagger\widetilde{\rho}_\theta\hbar K_l^* e^{+i\omega_l t}\right\}\mathbf{a}\widetilde{\rho}(t_0)^{\text{IP}}e^{-i\omega^\circ t} \tag{16.48}$$

$$\mathbf{tr}_\theta\{\widetilde{\rho}_{\text{Tot}}(t_0)^{\text{IP}}\widetilde{\mathbf{V}}(t)^{\text{IP}}\}$$
$$= \widetilde{\rho}(t_0)^{\text{IP}}\mathbf{a}^\dagger e^{i\omega^\circ t}\mathbf{tr}_\theta\left\{\widetilde{\rho}_\theta\sum_l \mathbf{b}_l\hbar K_l e^{-i\omega_l t}\right\} + \widetilde{\rho}(t_0)^{\text{IP}}\mathbf{a}e^{-i\omega^\circ t}\mathbf{tr}_\theta\left\{\widetilde{\rho}_\theta\sum_l \mathbf{b}_l^\dagger\hbar K_l^* e^{+i\omega_l t}\right\} \tag{16.49}$$

Next, owing to Eq. (16.15), it appears that

$$\rho_\theta\sum_l \mathbf{b}_l = \prod_j \varepsilon_j(e^{-\lambda_j\mathbf{b}_j^\dagger\mathbf{b}_j})\sum_l \mathbf{b}_l$$

so that since each Boson operator of the thermal bath works within its specific state space, this last expression transforms to

$$\rho_\theta\sum_l \mathbf{b}_l = \sum_l \varepsilon_l(e^{-\lambda_l\mathbf{b}_l^\dagger\mathbf{b}_l})\mathbf{b}_l\prod_{j\neq l}\varepsilon_j(e^{-\lambda_j\mathbf{b}_j^\dagger\mathbf{b}_j})$$

Moreover, tracing over the thermal bath for this last term may be realized in the basis $\{|(n_l)\rangle\}$ defined by the eigenvalue equations dealing with the thermal bath, that is,

$$\mathbf{b}_l^\dagger\mathbf{b}_l|(n_l)\rangle = n_l|(n_l)\rangle \tag{16.50}$$

with

$$\langle(n_l)|(m_l)\rangle = \delta_{n_l m_l} \qquad \text{and} \qquad \sum|(n_l)\rangle\langle(n_l)| = 1 \tag{16.51}$$

leading us to write this partial trace according to

$$\mathbf{tr}_\theta\left\{\rho_\theta\sum_l \mathbf{b}_l\right\} = \sum_l\sum_{n_l}\varepsilon_l\langle(n_l)|(e^{-\lambda_l\mathbf{b}_l^\dagger\mathbf{b}_l})\mathbf{b}_l\,|(n_l)\rangle\prod_{j\neq l}\varepsilon_j\sum_{n_j}\langle(n_j)|(e^{-\lambda_j\mathbf{b}_j^\dagger\mathbf{b}_j})|(n_j)\rangle$$

Again, observe that since the Boltzmann density operators are normalized through the normalization constants ε_j, one has for each oscillator j

$$\varepsilon_j\sum_{n_j}\langle(n_j)|(e^{-\lambda_j\mathbf{b}_j^\dagger\mathbf{b}_j})|(n_j)\rangle = 1 \tag{16.52}$$

so that the above equation reduces to

$$\mathbf{tr}_\theta\left\{\rho_\theta \sum_l \mathbf{b}_l\right\} = \sum_l \varepsilon_l \sum_{n_l} \langle (n_l)|(e^{-\lambda_l \mathbf{b}_l^\dagger \mathbf{b}_l})\mathbf{b}_l|(n_l)\rangle \tag{16.53}$$

Again, owing to the action of $\mathbf{b}_l^\dagger \mathbf{b}_l$ and $\mathbf{b}_l$ on the eigenkets of $\mathbf{b}_l^\dagger \mathbf{b}_l$, which obey equations similar to that (5.53), that is,

$$\mathbf{b}_l|(n_l)\rangle = \sqrt{n_l}|(n_l - 1)\rangle \tag{16.54}$$

the following result is verified:

$$\langle (n_l)|(e^{-\lambda_l \mathbf{b}_l^\dagger \mathbf{b}_l})\mathbf{b}_l|(n_l)\rangle = \sqrt{n_l}\langle (n_l)|(e^{-\lambda_l \mathbf{b}_l^\dagger \mathbf{b}_l})|(n_l - 1)\rangle$$

so that, using the eigenvalue equation (16.50), it yields

$$\langle (n_l)|(e^{-\lambda_l \mathbf{b}_l^\dagger \mathbf{b}_l})\mathbf{b}_l|(n_l)\rangle = \sqrt{n_l}(e^{-\lambda_l(n_l-1)})\langle (n_l)|(n_l - 1)\rangle$$

or, owing to the orthogonality of the kets appearing in (16.51)

$$\langle (n_l)|(e^{-\lambda_l \mathbf{b}_l^\dagger \mathbf{b}_l})\mathbf{b}_l|(n_l)\rangle = \sqrt{n_l}(e^{-\lambda_l(n_l-1)})\delta_{n_l,n_l-1} = 0 \tag{16.55}$$

then Eq. (16.53) reduces to

$$\mathbf{tr}_\theta\left\{\rho_\theta \sum_l \mathbf{b}_l\right\} = 0 \tag{16.56}$$

Of course, one would obtain in like manner

$$\mathbf{tr}_\theta\left\{\sum_l \mathbf{b}_l^\dagger \rho_\theta\right\} = 0 \tag{16.57}$$

Thus, as a consequence of Eqs. (16.56) and (16.57), Eqs. (16.48) and (16.49) read

$$\mathbf{tr}_\theta\{\widetilde{\mathbf{V}}(t)^{\mathrm{IP}}\widetilde{\boldsymbol{\rho}}_{\mathrm{Tot}}(t_0)^{\mathrm{IP}}\} = \mathbf{tr}_\theta\{\widetilde{\boldsymbol{\rho}}_{\mathrm{Tot}}(t_0)^{\mathrm{IP}}\widetilde{\mathbf{V}}(t)^{\mathrm{IP}}\} = 0$$

so that Eq. (16.44) yields

$$I_1 = 0$$

The last result implies that Eq. (16.43) reduces to

$$\widetilde{\boldsymbol{\rho}}(t)^{\mathrm{IP}} - \widetilde{\boldsymbol{\rho}}(t_0)^{\mathrm{IP}} = I_2$$

$$= \left(\frac{1}{i\hbar}\right)^2 \int_{t_0}^{t} dt' \int_{t_0}^{t'} dt''\, \mathbf{tr}_\theta\{[\widetilde{\mathbf{V}}(t' - t_0)^{\mathrm{IP}}, [\widetilde{\mathbf{V}}(t'' - t_0)^{\mathrm{IP}}, \widetilde{\boldsymbol{\rho}}_{\mathrm{Tot}}(t_0)^{\mathrm{IP}}]]\}$$

which, writing explicitly the double commutator appearing in this last equation, takes the form

$$\widetilde{\rho}(t)^{\mathrm{IP}} - \widetilde{\rho}(t_0)^{\mathrm{IP}}$$

$$= +\left(\frac{1}{i\hbar}\right)^2 \int_{t_0}^{t} dt' \int_{t_0}^{t'} dt''\, \mathbf{tr}_\theta\{\widetilde{\mathbf{V}}(t'-t_0)^{\mathrm{IP}}\widetilde{\mathbf{V}}(t''-t_0)^{\mathrm{IP}}\widetilde{\boldsymbol{\rho}}_{\mathrm{Tot}}(t_0)^{\mathrm{IP}}\}$$

$$-\left(\frac{1}{i\hbar}\right)^2 \int_{t_0}^{t} dt' \int_{t_0}^{t'} dt''\, \mathbf{tr}_\theta\{\widetilde{\mathbf{V}}(t''-t_0)^{\mathrm{IP}}\widetilde{\boldsymbol{\rho}}_{\mathrm{Tot}}(t_0)^{\mathrm{IP}}\widetilde{\mathbf{V}}(t'-t_0)^{\mathrm{IP}}\}$$

$$-\left(\frac{1}{i\hbar}\right)^2 \int_{t_0}^{t} dt' \int_{t_0}^{t'} dt''\, \mathbf{tr}_\theta\{\widetilde{\mathbf{V}}(t'-t_0)^{\mathrm{IP}}\widetilde{\boldsymbol{\rho}}_{\mathrm{Tot}}(t_0)^{\mathrm{IP}}\widetilde{\mathbf{V}}(t''-t_0)^{\mathrm{IP}}\}$$

$$+\left(\frac{1}{i\hbar}\right)^2 \int_{t_0}^{t} dt' \int_{t_0}^{t'} dt''\, \mathbf{tr}_\theta\{\widetilde{\boldsymbol{\rho}}_{\mathrm{Tot}}(t_0)^{\mathrm{IP}}\widetilde{\mathbf{V}}(t''-t_0)^{\mathrm{IP}}\widetilde{\mathbf{V}}(t'-t_0)^{\mathrm{IP}}\} \quad (16.58)$$

Note that in order to calculate Eq. (16.58) it is not possible to use the invariance of the trace with respect to a circular permutation in order to get some traces in terms of others because $\mathbf{tr}_\theta$ is a partial trace over the thermal bath, because this invariance holds only if the trace operation is performed over a basis belonging to the complete space involving both the bath and the oscillator embedded in it.

16.2.2 Calculation of the first average values involved in Eq. (16.58)

Now, one has to find the result of the traces involved in Eq. (16.58). For this purpose, begin with the first one of them, which, in view of Eq. (16.47), reads

$$\left(\frac{1}{i\hbar}\right)^2 \mathbf{tr}_\theta\{\widetilde{\mathbf{V}}(t'-t_0)^{\mathrm{IP}}\widetilde{\mathbf{V}}(t''-t_0)^{\mathrm{IP}}\widetilde{\boldsymbol{\rho}}_{\mathrm{Tot}}(t_0)^{\mathrm{IP}}\}$$

$$= +\mathbf{tr}_\theta\left\{\left(\mathbf{a}^\dagger e^{i\omega^\circ t'}\sum_l \mathbf{b}_l K_l e^{-i\omega_l t'}\right)\left(\mathbf{a} e^{-i\omega^\circ t''}\sum_k \mathbf{b}_k^\dagger K_k^* e^{i\omega_k t''}\right)\widetilde{\rho}(t_0)^{\mathrm{IP}}\boldsymbol{\rho}_\theta\right\}$$

$$+\mathbf{tr}_\theta\left\{\left(\mathbf{a} e^{-i\omega^\circ t'}\sum_l \mathbf{b}_l^\dagger K_l^* e^{i\omega_l t'}\right)\left(\mathbf{a}^\dagger e^{i\omega^\circ t''}\sum_k \mathbf{b}_k K_k e^{-i\omega_k t''}\right)\widetilde{\rho}(t_0)^{\mathrm{IP}}\boldsymbol{\rho}_\theta\right\}$$

$$+\mathbf{tr}_\theta\left\{\left(\mathbf{a}^\dagger e^{i\omega^\circ t'}\sum_l \mathbf{b}_l K_l e^{-i\omega_l t'}\right)\left(\mathbf{a}^\dagger e^{+i\omega^\circ t''}\sum_k \mathbf{b}_k K_k e^{-i\omega_k t''}\right)\widetilde{\rho}(t_0)^{\mathrm{IP}}\boldsymbol{\rho}_\theta\right\}$$

$$+\mathbf{tr}_\theta\left\{\left(\mathbf{a} e^{-i\omega^\circ t'}\sum_l \mathbf{b}_l^\dagger K_l^* e^{i\omega_l t'}\right)\left(\mathbf{a} e^{-i\omega^\circ t''}\sum_k \mathbf{b}_k^\dagger K_k^* e^{i\omega_k t''}\right)\widetilde{\rho}(t_0)^{\mathrm{IP}}\boldsymbol{\rho}_\theta\right\}$$

or

$$\left(\frac{1}{i\hbar}\right)^2 \mathbf{tr}_\theta\{\widetilde{\mathbf{V}}(t'-t_0)^{\mathrm{IP}}\widetilde{\mathbf{V}}(t''-t_0)^{\mathrm{IP}}\widetilde{\boldsymbol{\rho}}_{\mathrm{Tot}}(t_0)^{\mathrm{IP}}\}$$

$$= +\mathbf{a}^\dagger\mathbf{a}e^{i\omega^\circ(t'-t'')}\widetilde{\boldsymbol{\rho}}(t_0)^{\mathrm{IP}}\,\mathbf{tr}_\theta\left\{\sum_l\sum_k K_l K_k^* \mathbf{b}_l\mathbf{b}_k^\dagger e^{-i\omega_l t'}e^{+i\omega_k t''}\rho_\theta\right\}$$
$$+\mathbf{a}\mathbf{a}^\dagger e^{-i\omega^\circ(t'-t'')}\widetilde{\boldsymbol{\rho}}(t_0)^{\mathrm{IP}}\,\mathbf{tr}_\theta\left\{\sum_l\sum_k K_l^* K_k \mathbf{b}_l^\dagger\mathbf{b}_k e^{+i\omega_l t'}e^{-i\omega_k t''}\rho_\theta\right\}$$
$$+\mathbf{a}^\dagger\mathbf{a}^\dagger e^{i\omega^\circ(t'+t'')}\widetilde{\boldsymbol{\rho}}(t_0)^{\mathrm{IP}}\,\mathbf{tr}_\theta\left\{\sum_l\sum_k K_l K_k \mathbf{b}_l\mathbf{b}_k e^{-i\omega_l t'}e^{-i\omega_k t''}\rho_\theta\right\}$$
$$+\mathbf{a}\mathbf{a}e^{-i\omega^\circ(t'+t'')}\widetilde{\boldsymbol{\rho}}(t_0)^{\mathrm{IP}}\,\mathbf{tr}_\theta\left\{\sum_l\sum_k K_l^* K_k^* \mathbf{b}_l^\dagger\mathbf{b}_k^\dagger e^{+i\omega_l t'}e^{+i\omega_k t''}\rho_\theta\right\} \quad (16.59)$$

Next, notice that the trace over the thermal bath is the product of the traces dealing with the different oscillators of the bath, that is,

$$\mathbf{tr}_\theta\{\cdots\} = \prod_l \mathbf{tr}_l\{\cdots\}$$

Thus, after separation of the double sums into two parts, one where $k \neq l$ and the other where $k = l$, Eq. (16.59) reads

$$\left(\frac{1}{i\hbar}\right)^2 \mathbf{tr}_\theta\{\widetilde{\mathbf{V}}(t'-t_0)\widetilde{\mathbf{V}}(t''-t_0)\widetilde{\boldsymbol{\rho}}_{\mathrm{Tot}}(t_0)\} = \mathbf{A}(t'-t^\circ)+\mathbf{B}(t'-t^\circ) \quad (16.60)$$

with, respectively,

$$\mathbf{A}(t'-t^\circ)$$
$$= +\mathbf{a}^\dagger\mathbf{a}\widetilde{\boldsymbol{\rho}}(t_0)^{\mathrm{IP}}e^{i\omega^\circ(t'-t'')}\left\{\sum_l\sum_{k\neq l} K_l K_k^*\,\mathbf{tr}_l\{\rho_l\mathbf{b}_l\}\mathbf{tr}_k\{\rho_k\mathbf{b}_k^\dagger\}e^{-i\omega_l t'}e^{+i\omega_k t''}\right\}\prod_{j\neq l,k}\varepsilon_j\mathbf{tr}_j\{\rho_j\}$$
$$+\mathbf{a}\mathbf{a}^\dagger\widetilde{\boldsymbol{\rho}}(t_0)^{\mathrm{IP}}e^{-i\omega^\circ(t'-t'')}\left\{\sum_l\sum_{k\neq l} K_l^* K_k\,\mathbf{tr}_l\{\rho_l\mathbf{b}_l^\dagger\}\mathbf{tr}_k\{\rho_k\mathbf{b}_k\}e^{+i\omega_l t'}e^{-i\omega_k t''}\right\}\prod_{j\neq l,k}\varepsilon_j\mathbf{tr}_j\{\rho_j\}$$
$$+\mathbf{a}^\dagger\mathbf{a}^\dagger\widetilde{\boldsymbol{\rho}}(t_0)^{\mathrm{IP}}e^{i\omega^\circ(t'+t'')}\left\{\sum_l\sum_{k\neq l} K_l K_k\,\mathbf{tr}_l\{\rho_l\mathbf{b}_l\}\mathbf{tr}_k\{\rho_k\mathbf{b}_k\}e^{-i\omega_l t'}e^{-i\omega_k t''}\right\}\prod_{j\neq l,k}\varepsilon_j\mathbf{tr}_j\{\rho_j\}$$
$$+\mathbf{a}\mathbf{a}\widetilde{\boldsymbol{\rho}}(t_0)^{\mathrm{IP}}e^{-i\omega^\circ(t'+t'')}\left\{\sum_l\sum_{k\neq l} K_l^* K_k^*\,\mathbf{tr}_l\{\rho_l\mathbf{b}_l^\dagger\}\mathbf{tr}_k\{\rho_k\mathbf{b}_k^\dagger\}e^{i\omega_l t'}e^{+i\omega_k t''}\right\}\prod_{j\neq l,k}\varepsilon_j\mathbf{tr}_j\{\rho_j\}$$
$$(16.61)$$

and

$$\begin{aligned}\mathbf{B}(t'-t^\circ) = &+\mathbf{a}^\dagger\mathbf{a}\widetilde{\rho}(t_0)^{\mathrm{IP}}e^{i\omega^\circ(t'-t'')}\left\{\sum_l |K_l|^2\mathbf{tr}_l\{\boldsymbol{\rho}_l\mathbf{b}_l\mathbf{b}_l^\dagger\}e^{-i\omega_l(t'-t'')}\right\}\prod_{j\neq l}\mathbf{tr}_j\{\boldsymbol{\rho}_j\}\\ &+\mathbf{a}\mathbf{a}^\dagger\widetilde{\rho}(t_0)^{\mathrm{IP}}e^{-i\omega^\circ(t'-t'')}\left\{\sum_l |K_l|^2\mathbf{tr}_l\{\boldsymbol{\rho}_l\mathbf{b}_l^\dagger\mathbf{b}_l\}e^{i\omega_l(t'-t'')}\right\}\prod_{j\neq l}\mathbf{tr}_j\{\boldsymbol{\rho}_j\}\\ &+\mathbf{a}^\dagger\mathbf{a}^\dagger\widetilde{\rho}(t_0)^{\mathrm{IP}}e^{i\omega^\circ(t'+t'')}\left\{\sum_l |K_l|^2\mathbf{tr}_l\{\boldsymbol{\rho}_l\mathbf{b}_l\mathbf{b}_l\}e^{-i\omega_l(t'+t'')}\right\}\prod_{j\neq l}\mathbf{tr}_j\{\boldsymbol{\rho}_j\}\\ &+\mathbf{a}\mathbf{a}\widetilde{\rho}(t_0)^{\mathrm{IP}}e^{-i\omega^\circ(t'+t'')}\left\{\sum_l |K_l|^2\mathbf{tr}_l\{\boldsymbol{\rho}_l\mathbf{b}_l^\dagger\mathbf{b}_l^\dagger\}e^{+i\omega_l(t'+t'')}\right\}\prod_{j\neq l}\mathbf{tr}_j\{\boldsymbol{\rho}_j\}\end{aligned} \tag{16.62}$$

where

$$\boldsymbol{\rho}_l = \varepsilon_l(e^{-\lambda_l\mathbf{b}_l^\dagger\mathbf{b}_l}) \tag{16.63}$$

Moreover, observe that the thermal averages involved in Eqs. (16.61) and (16.62) are given by the following equations:

$$\mathbf{tr}_l\{\boldsymbol{\rho}_l\} = \varepsilon_l\ \mathbf{tr}_l\{(e^{-\lambda_l\mathbf{b}_l^\dagger\mathbf{b}_l})\} = 1 \tag{16.64}$$

Now, owing to Eqs. (16.55), (16.56), and (16.63), the trace over $\boldsymbol{\rho}_l\mathbf{b}_l$ is zero whatever l may be, that is,

$$\mathbf{tr}_l\{\boldsymbol{\rho}_l\mathbf{b}_l\} = \mathbf{tr}_l\{\boldsymbol{\rho}_l\mathbf{b}_l^\dagger\} = 0 \tag{16.65}$$

Furthermore, after writing explicitly the trace over the eigenkets of $\mathbf{b}_l^\dagger\mathbf{b}_l$, the thermal averages of $\mathbf{b}_l\mathbf{b}_l$ read

$$\mathbf{tr}_l\{\boldsymbol{\rho}_l\mathbf{b}_l\mathbf{b}_l\} = \varepsilon_l\sum_{n_l}\langle(n_l)|(e^{-\lambda_l\mathbf{b}_l^\dagger\mathbf{b}_l})\mathbf{b}_l\mathbf{b}_l|(n_l)\rangle$$

or, using twice Eq. (16.54),

$$\mathbf{tr}_l\{\boldsymbol{\rho}_l\mathbf{b}_l\mathbf{b}_l\} = \varepsilon_l\sum_{n_l}\sqrt{n_l(n_l-1)}\langle(n_l)|(e^{-\lambda_l\mathbf{b}_l^\dagger\mathbf{b}_l})|(n_l-2)\rangle$$

and thus, due to Eq. (16.50),

$$\mathbf{tr}_l\{\boldsymbol{\rho}_l\mathbf{b}_l\mathbf{b}_l\} = \varepsilon_l\sum_{n_l}\sqrt{n_l(n_l-1)}(e^{-\lambda_l(n_l-2)})\langle(n_l)|(n_l-2)\rangle$$

hence, due to Eqs. (1.71) and (1.73), owing to the orthogonality properties (16.51) of the eigenkets of $\mathbf{b}_l^\dagger \mathbf{b}_l$,

$$\mathbf{tr}_l\{\boldsymbol{\rho}_l \mathbf{b}_l \mathbf{b}_l\} = 0 \tag{16.66}$$

In like manner, using similar reasoning, we have

$$\mathbf{tr}_l\{\boldsymbol{\rho}_l \mathbf{b}_l^\dagger \mathbf{b}_l^\dagger\} = 0 \tag{16.67}$$

Now, in view of Eq. (16.63), the thermal average of the number occupation $\mathbf{b}_l^\dagger \mathbf{b}_l$ reads

$$\mathbf{tr}_l\{\boldsymbol{\rho}_l \mathbf{b}_l^\dagger \mathbf{b}_l\} = \varepsilon_l \mathbf{tr}_l\{(e^{-\lambda_l \mathbf{b}_l^\dagger \mathbf{b}_l})\mathbf{b}_l^\dagger \mathbf{b}_l\}$$

or, due to Eq. (13.32),

$$\mathbf{tr}_l\{\boldsymbol{\rho}_l \mathbf{b}_l^\dagger \mathbf{b}_l\} = \langle \mathbf{n}_l \rangle \tag{16.68}$$

where

$$\langle \mathbf{n}_l \rangle = \frac{1}{e^{\hbar\omega_l/kT} - 1} \tag{16.69}$$

Furthermore, the last thermal average of interest is

$$\mathbf{tr}_l\{\boldsymbol{\rho}_l \mathbf{b}_l \mathbf{b}_l^\dagger\} = \varepsilon_l \, \mathbf{tr}_l\{(e^{-\lambda_l \mathbf{b}_l^\dagger \mathbf{b}_l})\mathbf{b}_l \mathbf{b}_l^\dagger\}$$

or, using the commutation rule of Boson operators $[\mathbf{b}, \mathbf{b}^\dagger] = 1$,

$$\mathbf{tr}_l\{\boldsymbol{\rho}_l \mathbf{b}_l \mathbf{b}_l^\dagger\} = \varepsilon_l \, \mathbf{tr}_l\{(e^{-\lambda_l \mathbf{b}_l^\dagger \mathbf{b}_l})(\mathbf{b}_l^\dagger \mathbf{b}_l + 1)\}$$

so that, due to Eq. (16.68),

$$\mathbf{tr}_l\{\boldsymbol{\rho}_l \mathbf{b}_l \mathbf{b}_l^\dagger\} = \langle \mathbf{n}_l + 1 \rangle \equiv \langle \mathbf{n}_l \rangle + 1 \tag{16.70}$$

Hence, as a consequence of Eqs. (16.65), (16.66), (16.67), and (16.70), Eq. (16.61) appears to be zero, that is,

$$\mathbf{A}(t' - t^\circ) = 0$$

Therefore, owing to this result and according to Eqs. (16.64), (16.66), (16.67), (16.68), and (16.70), Eq. (16.60) takes on the simplified form

$$\begin{aligned}
&\left(\frac{1}{i\hbar}\right)^2 \mathbf{tr}_\theta\{\widetilde{\mathbf{V}}(t' - t_0)^{\mathrm{IP}} \widetilde{\mathbf{V}}(t'' - t_0)^{\mathrm{IP}} \widetilde{\boldsymbol{\rho}}_{\mathrm{Tot}}(t_0)^{\mathrm{IP}}\} \\
&= \mathbf{a}^\dagger \mathbf{a} \widetilde{\boldsymbol{\rho}}(t_0)^{\mathrm{IP}} \left\{ \sum_l |K_l|^2 \langle \mathbf{n}_l + 1 \rangle \, e^{i(\omega^\circ - \omega_l)(t' - t'')} \right\} \\
&\quad + \mathbf{a}\mathbf{a}^\dagger \widetilde{\boldsymbol{\rho}}(t_0)^{\mathrm{IP}} \left\{ \sum_l |K_l|^2 \langle \mathbf{n}_l \rangle e^{-i(\omega^\circ - \omega_l)(t' - t'')} \right\}
\end{aligned} \tag{16.71}$$

16.2.3 Calculation of the other average values involved in Eq. (16.58)

Now, we have to get the other average values involved in Eq. (16.58). For instance, in view of Eq. (16.47), the third one reads

$$\left(\frac{1}{i\hbar}\right)^2 \mathbf{tr}_\theta\{\widetilde{\mathbf{V}}(t'-t_0)^{\mathrm{IP}}\widetilde{\boldsymbol{\rho}}_{\mathrm{Tot}}(t_0)^{\mathrm{IP}}\widetilde{\mathbf{V}}(t''-t_0)^{\mathrm{IP}}\}$$

$$= +\mathbf{tr}_\theta\left\{\left(\mathbf{a}^\dagger e^{i\omega^\circ t'}\sum_l \mathbf{b}_l K_l e^{-i\omega_l t'}\right)\widetilde{\rho}(t_0)^{\mathrm{IP}}\boldsymbol{\rho}_\theta\left(\mathbf{a}e^{-i\omega^\circ t''}\sum_k \mathbf{b}_k^\dagger K_k^* e^{i\omega_k t''}\right)\right\}$$

$$+\mathbf{tr}_\theta\left\{\left(\mathbf{a}e^{-i\omega^\circ t'}\sum_l \mathbf{b}_l^\dagger K_l^* e^{+i\omega_l t'}\right)\widetilde{\rho}(t_0)^{\mathrm{IP}}\boldsymbol{\rho}_\theta\left(\mathbf{a}^\dagger e^{+i\omega^\circ t''}\sum_k \mathbf{b}_k K_k e^{-i\omega_k t''}\right)\right\}$$

$$+\mathbf{tr}_\theta\left\{\left(\mathbf{a}^\dagger e^{i\omega^\circ t'}\sum_l \mathbf{b}_l K_l e^{-i\omega_l t'}\right)\widetilde{\rho}(t_0)^{\mathrm{IP}}\boldsymbol{\rho}_\theta\left(\mathbf{a}^\dagger e^{+i\omega^\circ t''}\sum_k \mathbf{b}_k K_k e^{-i\omega_k t''}\right)\right\}$$

$$+\mathbf{tr}_\theta\left\{\left(\mathbf{a}e^{-i\omega^\circ t'}\sum_l \mathbf{b}_l^\dagger K_l^* e^{+i\omega_l t'}\right)\widetilde{\rho}(t_0)^{\mathrm{IP}}\boldsymbol{\rho}_\theta\left(\mathbf{a}e^{-i\omega^\circ t''}\sum_k \mathbf{b}_k^\dagger K_k^* e^{i\omega_k t''}\right)\right\}$$

After rearrangement, it transforms to

$$\left(\frac{1}{i\hbar}\right)^2 \mathbf{tr}_\theta\{\widetilde{\mathbf{V}}(t'-t_0)^{\mathrm{IP}}\widetilde{\boldsymbol{\rho}}_{\mathrm{Tot}}(t_0)^{\mathrm{IP}}\widetilde{\mathbf{V}}(t''-t_0)^{\mathrm{IP}}\}$$

$$= +\mathbf{tr}_\theta\left\{\mathbf{a}^\dagger\widetilde{\rho}(t_0)^{\mathrm{IP}}\mathbf{a}e^{-i\omega^\circ(t'-t'')}\sum_l\sum_k K_l K_k^*\boldsymbol{\rho}_\theta\mathbf{b}_l\mathbf{b}_k^\dagger e^{+i\omega_k t'}e^{-i\omega_l t''}\right\}$$

$$+\mathbf{tr}_\theta\left\{\mathbf{a}\widetilde{\rho}(t_0)^{\mathrm{IP}}\mathbf{a}^\dagger e^{i\omega^\circ(t'-t'')}\sum_l\sum_k K_l^* K_k\boldsymbol{\rho}_\theta\mathbf{b}_l^\dagger\mathbf{b}_k e^{-i\omega_k t'}e^{+i\omega_l t''}\right\}$$

$$+\mathbf{tr}_\theta\left\{\mathbf{a}^\dagger\widetilde{\rho}(t_0)^{\mathrm{IP}}\mathbf{a}^\dagger e^{+i\omega^\circ(t''+t')}\sum_l\sum_k K_l K_k\boldsymbol{\rho}_\theta\mathbf{b}_l\mathbf{b}_k e^{-i\omega_k t'}e^{-i\omega_l t''}\right\}$$

$$+\mathbf{tr}_\theta\left\{\mathbf{a}\widetilde{\rho}(t_0)^{\mathrm{IP}}\mathbf{a}e^{-i\omega^\circ(t''+t')}\sum_l\sum_k K_l^* K_k^*\boldsymbol{\rho}_\theta\mathbf{b}_l^\dagger\mathbf{b}_k^\dagger e^{+i\omega_k t'}e^{+i\omega_l t''}\right\}$$

Then, in like manner as passing from Eqs. (16.59) to (16.71), we have

$$\left(\frac{1}{i\hbar}\right)^2 \mathbf{tr}_\theta\{\widetilde{\mathbf{V}}(t'-t_0)^{\mathrm{IP}}\widetilde{\rho}_{\mathrm{Tot}}(t_0)^{\mathrm{IP}}\widetilde{\mathbf{V}}(t''-t_0)^{\mathrm{IP}}\}$$

$$= \mathbf{a}^\dagger\widetilde{\rho}(t_0)^{\mathrm{IP}}\mathbf{a}\left(\sum_l |K_l|^2\langle\mathbf{n}_l+1\rangle e^{+i(\omega^\circ-i\omega_l)(t'-t'')}\right)$$

$$+\mathbf{a}\widetilde{\rho}(t_0)^{\mathrm{IP}}\mathbf{a}^\dagger\left(\sum_l |K_l|^2\langle\mathbf{n}_l\rangle e^{-i(\omega^\circ-i\omega_l)(t'-t'')}\right) \tag{16.72}$$

In the same way, one would find for the other two average values of Eq. (16.58)

$$\left(\frac{1}{i\hbar}\right)^2 \mathbf{tr}_\theta\{\widetilde{\mathbf{V}}(t''-t_0)^{\mathrm{IP}}\widetilde{\rho}_{\mathrm{Tot}}(t_0)^{\mathrm{IP}}\widetilde{\mathbf{V}}(t'-t_0)^{\mathrm{IP}}\}$$

$$= \mathbf{a}^\dagger\widetilde{\rho}(t_0)^{\mathrm{IP}}\mathbf{a}\left(\sum_l |K_l|^2\langle\mathbf{n}_l+1\rangle e^{i(\omega^\circ-i\omega_l)(t'-t'')}\right)$$

$$+\mathbf{a}\widetilde{\rho}(t_0)^{\mathrm{IP}}\mathbf{a}^\dagger\left(\sum_l |K_l|^2\langle\mathbf{n}_l\rangle e^{-i(\omega^\circ-i\omega_l)(t'-t'')}\right) \tag{16.73}$$

and

$$\left(\frac{1}{i\hbar}\right)^2 \mathbf{tr}_\theta\{\widetilde{\rho}_{\mathrm{Tot}}(t_0)^{\mathrm{IP}}\widetilde{\mathbf{V}}(t''-t_0)^{\mathrm{IP}}\widetilde{\mathbf{V}}(t'-t_0)^{\mathrm{IP}}\}$$

$$=\widetilde{\rho}(t_0)^{\mathrm{IP}}\mathbf{a}^\dagger\mathbf{a}\left(\sum_l |K_l|^2\langle\mathbf{n}_l+1\rangle\, e^{i(\omega^\circ-i\omega_l)(t'-t'')}\right)$$

$$+\widetilde{\rho}(t_0)^{\mathrm{IP}}\mathbf{a}\mathbf{a}^\dagger\left(\sum_l |K_l|^2\langle\mathbf{n}_l\rangle e^{-i(\omega^\circ-i\omega_l)(t'-t'')}\right) \tag{16.74}$$

16.2.4 Time derivative of the IP density operator

16.2.4.1 Basic equation for the variation of the IP density operator Next, owing to Eqs. (16.71)–(16.74), Eq. (16.58) becomes

$$\widetilde{\rho}(t_0+\Delta t)^{\mathrm{IP}} - \widetilde{\rho}(t_0)^{\mathrm{IP}}$$

$$= -\mathbf{a}^\dagger\mathbf{a}\widetilde{\rho}(t_0)^{\mathrm{IP}}\sum_l |K_l|^2\langle\mathbf{n}_l+1\rangle \int_{t_0}^{t_0+\Delta t} dt' \int_{t_0}^{t'} dt''\; e^{+i(\omega^\circ-\omega_l)(t'-t'')}$$

$$-\mathbf{a}\mathbf{a}^\dagger\widetilde{\rho}(t_0)^{\mathrm{IP}}\sum_l |K_l|^2\langle\mathbf{n}_l\rangle \int_{t_0}^{t_0+\Delta t} dt' \int_{t_0}^{t'} dt''\; e^{-i(\omega^\circ-\omega_l)(t'-t'')}$$

$$+\mathbf{a}^\dagger\widetilde{\rho}(t_0)^{\mathrm{IP}}\mathbf{a}\sum_l |K_l|^2\langle\mathbf{n}_l+1\rangle \int_{t_0}^{t_0+\Delta t} dt' \int_{t_0}^{t'} dt''\; e^{+i(\omega^\circ-\omega_l)(t'-t'')}$$

$$+\mathbf{a}\widetilde{\rho}(t_0)^{\mathrm{IP}}\mathbf{a}^\dagger\sum_l |K_l|^2\langle\mathbf{n}_l\rangle \int_{t_0}^{t_0+\Delta t} dt' \int_{t_0}^{t'} dt''\; e^{-i(\omega^\circ-\omega_l)(t'-t'')}$$

$$+\mathbf{a}^\dagger\widetilde{\rho}(t_0)^{\mathrm{IP}}\mathbf{a}\sum_l |K_l|^2\langle\mathbf{n}_l+1\rangle \int_{t_0}^{t_0+\Delta t} dt' \int_{t_0}^{t'} dt''\; e^{+i(\omega^\circ-\omega_l)(t'-t'')}$$

$$+\mathbf{a}\widetilde{\rho}(t_0)^{\mathrm{IP}}\mathbf{a}^\dagger \sum_l |K_l|^2 \langle \mathbf{n}_l \rangle \int_{t_0}^{t_0+\Delta t} dt' \int_{t_0}^{t'} dt'' \; e^{-i(\omega^\circ-\omega_l)(t'-t'')}$$

$$-\widetilde{\rho}(t_0)^{\mathrm{IP}}\mathbf{a}^\dagger\mathbf{a} \sum_l |K_l|^2 \langle \mathbf{n}_l + 1 \rangle \int_{t_0}^{t_0+\Delta t} dt' \int_{t_0}^{t'} dt'' \; e^{+i(\omega^\circ-\omega_l)(t'-t'')}$$

$$-\widetilde{\rho}(t_0)^{\mathrm{IP}}\mathbf{a}\mathbf{a}^\dagger \sum_l |K_l|^2 \langle \mathbf{n}_l \rangle \int_{t_0}^{t_0+\Delta t} dt' \int_{t_0}^{t'} dt'' \; e^{-i(\omega^\circ-\omega_l)(t'-t'')} \tag{16.75}$$

where

$$\Delta t = t - t_0$$

In Eq. (16.75), the mean occupation number $\langle \mathbf{n}_l \rangle$ and the coupling terms K_l implicitly concern the angular frequencies ω_l so that it is convenient to rewrite this equation as follows:

$$\begin{aligned}
&\widetilde{\rho}(t_0 + \Delta t)^{\mathrm{IP}} - \widetilde{\rho}(t_0)^{\mathrm{IP}} \\
&\quad = -\mathbf{a}^\dagger\mathbf{a}\widetilde{\rho}(t_0)^{\mathrm{IP}} \int_{t_0}^{t_0+\Delta t} dt' \int_{t_0}^{t'} dt'' \left\{ \sum_l |K_l|^2 \langle \mathbf{n}_l + 1 \rangle e^{+i(\omega^\circ-\omega_l)(t'-t'')} \right\} \\
&\qquad -\mathbf{a}^\dagger\mathbf{a}\widetilde{\rho}(t_0)^{\mathrm{IP}} \int_{t_0}^{t_0+\Delta t} dt' \int_{t_0}^{t'} dt'' \left\{ \sum_l |K_l|^2 \langle \mathbf{n}_l + 1 \rangle e^{+i(\omega^\circ-\omega_l)(t'-t'')} \right\} \\
&\qquad +\mathbf{a}^\dagger\widetilde{\rho}(t_0)^{\mathrm{IP}}\mathbf{a} \int_{t_0}^{t_0+\Delta t} dt' \int_{t_0}^{t'} dt'' \left\{ \sum_l |K_l|^2 \langle \mathbf{n}_l + 1 \rangle e^{+i(\omega^\circ-\omega_l)(t'-t'')} \right\} \\
&\qquad +\mathbf{a}\widetilde{\rho}(t_0)^{\mathrm{IP}}\mathbf{a}^\dagger \int_{t_0}^{t_0+\Delta t} dt' \int_{t_0}^{t'} dt'' \left\{ \sum_l |K_l|^2 \langle \mathbf{n}_l \rangle e^{-i(\omega^\circ-\omega_l)(t'-t'')} \right\} \\
&\qquad +\mathbf{a}^\dagger\widetilde{\rho}(t_0)^{\mathrm{IP}}\mathbf{a} \int_{t_0}^{t_0+\Delta t} dt' \int_{t_0}^{t'} dt'' \left\{ \sum_l |K_l|^2 \langle \mathbf{n}_l + 1 \rangle e^{+i(\omega^\circ-\omega_l)(t'-t'')} \right\} \\
&\qquad +\mathbf{a}\widetilde{\rho}(t_0)^{\mathrm{IP}}\mathbf{a}^\dagger \int_{t_0}^{t_0+\Delta t} dt' \int_{t_0}^{t'} dt'' \left\{ \sum_l |K_l|^2 \langle \mathbf{n}_l \rangle e^{-i(\omega^\circ-\omega_l)(t'-t'')} \right\} \\
&\qquad -\widetilde{\rho}(t_0)^{\mathrm{IP}}\mathbf{a}^\dagger\mathbf{a} \int_{t_0}^{t_0+\Delta t} dt' \int_{t_0}^{t'} dt'' \left\{ \sum_l |K_l|^2 \langle \mathbf{n}_l + 1 \rangle e^{+i(\omega^\circ-\omega_l)(t'-t'')} \right\} \\
&\qquad -\widetilde{\rho}(t_0)^{\mathrm{IP}}\mathbf{a}\mathbf{a}^\dagger \int_{t_0}^{t_0+\Delta t} dt' \int_{t_0}^{t'} dt'' \left\{ \sum_l |K_l|^2 \langle \mathbf{n}_l \rangle e^{-i(\omega^\circ-\omega_l)(t'-t'')} \right\}
\end{aligned} \tag{16.76}$$

16.2.4.2 Markov approximation Observe that in Eq. (16.2.4.1) integrals appear of the form

$$\mathbf{I} = \int_{t_0}^{t_0+\Delta t} dt' \int_{t_0}^{t'} dt'' \sum_l A_l e^{+i(\omega^\circ - \omega_l)(t'-t'')} \tag{16.77}$$

where A_l is given either by

$$A_l = |K_l|^2 \langle \mathbf{n}_l \rangle \tag{16.78}$$

or by

$$A_l = |K_l|^2 \langle \mathbf{n}_l + 1 \rangle \tag{16.79}$$

Next, make in the integral (16.77) the following changes of variable

$$\tau \equiv t' - t'' \qquad \text{and} \qquad \xi \equiv t - t'$$

leading to

$$t' = \Delta t + t_0 - \xi \qquad \text{and} \qquad dt'dt'' = d\xi\, d\tau$$

Then, the double integrals (16.77) becomes

$$\mathbf{I} = \int_{t_0}^{t_0+\Delta t} d\xi \int_{t_0}^{t_0+\Delta t-\xi} d\tau \left\{ \sum_l A_l e^{+i(\omega^\circ - \omega_l)\tau} \right\} \tag{16.80}$$

Notice that the infinite sum appearing in this last equation involves imaginary exponentials, the time-independent arguments of which are quasi-continuously varying, so that this sum must vanish when the time τ becomes greater than the *correlation time* τ_c:

$$\sum_l A_l e^{+i(\omega^\circ - \omega_l)\tau} \simeq 0 \quad \text{if} \quad \tau > \tau_c \tag{16.81}$$

Next, examine in details Eq. (16.80) at the upper limit $(t_0 + \Delta t - \xi)$ of the integral over the τ variable. Owing to the approximation (16.81), the contribution of the integrand to the integration over τ is negligible for $\tau > \tau_c$, so that this integration limit may be extended from $(t_0 + \Delta t - \xi)$ to infinity without any sensible changes (see Fig.16.1). Such an approximation, which implies some lack of memory, is known, in the statistical physics of irreversible processes, as the *Markov approximation*. Hence, we may write the following approximate equation:

$$\int_{t_0}^{t_0+\Delta t-\xi} d\tau \left\{ \sum_l A_l e^{+i(\omega^\circ - \omega_l)\tau} \right\} \approx \int_{t_0}^{\infty} d\tau \left\{ \sum_l A_l e^{+i(\omega^\circ - \omega_l)\tau} \right\}$$

So, the integral (16.80) may be approximated by

$$\mathbf{I} \approx \int_{t_0}^{t_0+\Delta t} d\xi \int_{t_0}^{\infty} d\tau \left\{ \sum_l A_l e^{+i(\omega^\circ - \omega_l)\tau} \right\}$$

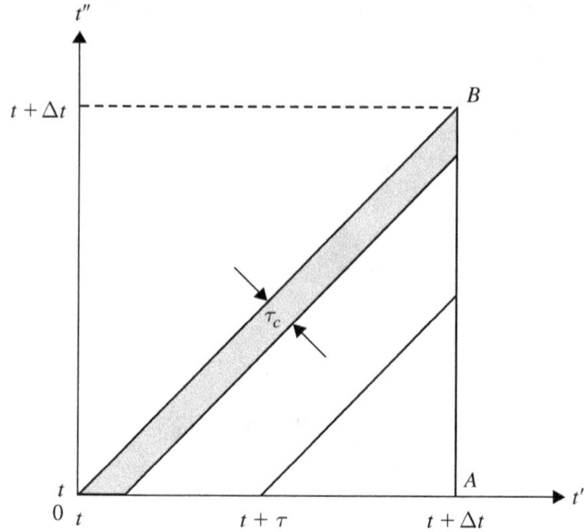

Figure 16.1 Integration area over t' and t''.

which, after integration over ξ of these integrals, yields

$$\mathbf{I} \approx \Delta t \int_{t_0}^{\infty} \left\{ \sum_l A_l e^{+i(\omega^\circ - \omega_l)\tau} \right\} d\tau \tag{16.82}$$

16.2.4.3 Time variation of the IP density operator Now, consider the variation of the IP density operator over the time interval Δt^{IP}:

$$\frac{\Delta\widetilde{\rho}(t)^{\mathrm{IP}}}{\Delta t} = \frac{\widetilde{\rho}(t_0 + \Delta t)^{\mathrm{IP}} - \widetilde{\rho}(t_0)^{\mathrm{IP}}}{\Delta t}$$

Due to Eqs. (16.78) and (16.79) and with the help of Eq. (16.82) yielding approximately the value of the integrals (16.77), Eq. (16.2.4.1) reads

$$\begin{aligned}
\frac{\Delta\widetilde{\rho}(t)^{\mathrm{IP}}}{\Delta t} = & -\mathbf{a}^\dagger \mathbf{a}\widetilde{\rho}(t_0)^{\mathrm{IP}} \sum_l \left\{ |K_l|^2 \langle \mathbf{n}_l + 1 \rangle \int_0^\infty e^{i(\omega^\circ - \omega_l)\tau} d\tau \right\} \\
& - \mathbf{a}\mathbf{a}^\dagger \widetilde{\rho}(t_0)^{\mathrm{IP}} \sum_l \left\{ |K_l|^2 \langle \mathbf{n}_l \rangle \int_0^\infty e^{-i(\omega^\circ - \omega_l)\tau} d\tau \right\} \\
& + \mathbf{a}^\dagger \widetilde{\rho}(t_0)^{\mathrm{IP}} \mathbf{a} \sum_l \left\{ |K_l|^2 \langle \mathbf{n}_l + 1 \rangle \int_0^\infty e^{-i(\omega^\circ - \omega_l)\tau} d\tau \right\} \\
& + \mathbf{a}\widetilde{\rho}(t_0)^{\mathrm{IP}} \mathbf{a}^\dagger \sum_l \left\{ |K_l|^2 \langle \mathbf{n}_l \rangle \int_0^\infty e^{i(\omega^\circ - \omega_l)\tau} d\tau \right\}
\end{aligned}$$

$$+\mathbf{a}^{\dagger}\widetilde{\rho}(t_0)^{\mathrm{IP}}\mathbf{a}\sum_l\left\{|K_l|^2\langle\mathbf{n}_l+1\rangle\int_0^{\infty}e^{i(\omega^{\circ}-\omega_l)\tau}d\tau\right\}$$

$$+\mathbf{a}\widetilde{\rho}(t_0)^{\mathrm{IP}}\mathbf{a}^{\dagger}\sum_l\left\{|K_l|^2\langle\mathbf{n}_l\rangle\int_0^{\infty}e^{-i(\omega^{\circ}-\omega_l)\tau}d\tau\right\}$$

$$-\widetilde{\rho}(t_0)^{\mathrm{IP}}\mathbf{a}^{\dagger}\mathbf{a}\sum_l\left\{|K_l|^2\langle\mathbf{n}_l+1\rangle\int_0^{\infty}e^{-i(\omega^{\circ}-\omega_l)\tau}d\tau\right\}$$

$$-\widetilde{\rho}(t_0)^{\mathrm{IP}}\mathbf{a}\mathbf{a}^{\dagger}\sum_l\left\{|K_l|^2\langle\mathbf{n}_l\rangle\int_0^{\infty}e^{i(\omega^{\circ}-\omega_l)\tau}d\tau\right\} \tag{16.83}$$

16.2.5 IP master equation for the density operator

16.2.5.1 Continuous approximation for the thermal bath It is now convenient to make the approximation of considering the set of oscillators of the thermal bath as continuous and thus to pass in Eq. (16.83) from the sums over the thermal bath oscillator to integrals over the continuous angular frequency variable concerning these oscillators, according to

$$\sum_l |K_l|^2\langle\mathbf{n}_l\rangle \to \int_{-\infty}^{+\infty} g(\omega)|K(\omega)|^2\langle\mathbf{n}(\omega)\rangle d\omega \tag{16.84}$$

Here, $g(\omega)$ is the *mode density* of the thermal bath, $K(\omega)$ is the coupling between oscillators of angular frequency ω, whereas $\langle\mathbf{n}(\omega)\rangle$ is the mean number occupation of the oscillator of angular frequency ω which, due to Eq. (16.69), is

$$\langle\mathbf{n}_l(\omega_l)\rangle = \frac{1}{e^{\hbar\omega_l/kT}-1} \tag{16.85}$$

Owing to this approximation, Eq. (16.83) becomes

$$\begin{aligned}\frac{\Delta\widetilde{\rho}(t)^{\mathrm{IP}}}{\Delta t} = &-\mathbf{a}^{\dagger}\mathbf{a}\widetilde{\rho}(t_0)\Omega_1 - \mathbf{a}\mathbf{a}^{\dagger}\widetilde{\rho}(t_0)\Omega_0^*\\ &+\mathbf{a}^{\dagger}\widetilde{\rho}(t_0)\mathbf{a}\ \Omega_1^* + \mathbf{a}\widetilde{\rho}(t_0)\mathbf{a}^{\dagger}\ \Omega_0\\ &+\mathbf{a}^{\dagger}\widetilde{\rho}(t_0)\mathbf{a}\ \Omega_1 + \mathbf{a}\widetilde{\rho}(t_0)\mathbf{a}^{\dagger}\Omega_0^*\\ &-\widetilde{\rho}(t_0)\mathbf{a}^{\dagger}\mathbf{a}\ \Omega_1^* - \widetilde{\rho}(t_0)\mathbf{a}\mathbf{a}^{\dagger}\Omega_0\end{aligned} \tag{16.86}$$

where

$$\Omega_0 \equiv \int_{-\infty}^{+\infty} g(\omega)|K(\omega)|^2\langle\mathbf{n}(\omega)\rangle\int_0^{\infty} d\tau\ e^{i(\omega^{\circ}-\omega)\tau}d\omega \tag{16.87}$$

$$\Omega_1 \equiv \int_{-\infty}^{+\infty} g(\omega)|K(\omega)|^2\langle\mathbf{n}(\omega)+1\rangle\int_0^{\infty} d\tau\ e^{i(\omega^{\circ}-\omega)\tau}d\omega \tag{16.88}$$

16.2.5.2 Calculation of Ω_0 and Ω_1 One has now to calculate the double integrals (16.87) and (16.88). For this purpose, observe that these integrals are of the form (18.63)

$$\Omega = \int_{-\infty}^{+\infty} f(\omega) \left(\int_0^{\infty} e^{-i(\omega-\omega^\circ)\tau} d\tau \right) d\omega$$

with

$$f(\omega) = g(\omega)|K(\omega)|^2 \langle \mathbf{n}(\omega) \rangle \qquad \text{or} \qquad g(\omega)|K(\omega)|^2 \langle \mathbf{n}(\omega) + 1 \rangle$$

Then, keeping in mind that, as shown in Section 18.6, Eq. (18.63) leads to Eq. (18.71), that is,

$$\Omega = -i \left\{ \int_{-\infty}^{+\infty} f(\omega) \mathcal{P} \left(\frac{1}{\omega - \omega^\circ} \right) d\omega \right\} d\omega - \int_{-\infty}^{+\infty} f(\omega) \pi \delta(\omega - \omega^\circ) \, d\omega$$

where $\mathcal{P}$ denotes the Cauchy principal part, it appears that Eq. (16.87) reads

$$\Omega_0 = \int_{-\infty}^{+\infty} g(\omega)|K(\omega)|^2 \langle \mathbf{n}(\omega) \rangle \delta(\omega - \omega^\circ) \, d\omega - i \left\{ \int_{-\infty}^{+\infty} g(\omega)|K(\omega)|^2 \mathcal{P} \left(\frac{1}{\omega - \omega^\circ} \right) d\omega \right\}$$

or

$$\Omega_0 = \langle \mathbf{n}(\omega^\circ) \rangle \left(\frac{\gamma}{2} + i \, \Delta\omega \right) \tag{16.89}$$

where $\Delta\omega$ is an angular frequency shift and γ a damping parameter given, respectively, by

$$\Delta\omega \equiv - \left\{ \int_{-\infty}^{+\infty} g(\omega)|K(\omega)|^2 \mathcal{P} \left(\frac{1}{\omega - \omega^\circ} \right) d\omega \right\} \tag{16.90}$$

$$\gamma \equiv 2\pi g(\omega^\circ)|K(\omega^\circ)|^2 \tag{16.91}$$

In like manner, Eq. (16.88) is

$$\Omega_1 = \langle \mathbf{n}(\omega^\circ) + 1 \rangle \left(\frac{\gamma}{2} + i \, \Delta\omega \right) \tag{16.92}$$

16.2.5.3 New expression for time variation of IP density operator Owing to Eq. (16.89) and (16.92), Eq. (16.86) transforms to

$$\frac{\Delta\widetilde{\rho}(t)^{\mathrm{IP}}}{\Delta t} = -\mathbf{a}^{\dagger}\mathbf{a}\widetilde{\rho}(t_0)^{\mathrm{IP}}\left(\frac{\gamma}{2}+i\,\Delta\omega\right)\langle\mathbf{n}+1\rangle \tag{16.93}$$

$$\begin{aligned}
&-\mathbf{a}\mathbf{a}^{\dagger}\widetilde{\rho}(t_0)^{\mathrm{IP}}\left(\frac{\gamma}{2}-i\,\Delta\omega\right)\langle\mathbf{n}\rangle\\
&+\mathbf{a}^{\dagger}\widetilde{\rho}(t_0)^{\mathrm{IP}}\mathbf{a}\left(\frac{\gamma}{2}-i\,\Delta\omega\right)\langle\mathbf{n}+1\rangle\\
&+\mathbf{a}\widetilde{\rho}(t_0)^{\mathrm{IP}}\mathbf{a}^{\dagger}\left(\frac{\gamma}{2}+i\,\Delta\omega\right)\langle\mathbf{n}\rangle\\
&+\mathbf{a}^{\dagger}\widetilde{\rho}(t_0)^{\mathrm{IP}}\mathbf{a}\left(\frac{\gamma}{2}+i\,\Delta\omega\right)\langle\mathbf{n}+1\rangle\\
&+\mathbf{a}\widetilde{\rho}(t_0)^{\mathrm{IP}}\mathbf{a}^{\dagger}\left(\frac{\gamma}{2}-i\,\Delta\omega\right)\langle\mathbf{n}\rangle\\
&-\widetilde{\rho}(t_0)^{\mathrm{IP}}\mathbf{a}^{\dagger}\mathbf{a}\left(\frac{\gamma}{2}-i\,\Delta\omega\right)\langle\mathbf{n}+1\rangle\\
&-\widetilde{\rho}(t_0)^{\mathrm{IP}}\mathbf{a}\mathbf{a}^{\dagger}\left(\frac{\gamma}{2}+i\,\Delta\omega\right)\langle\mathbf{n}\rangle
\end{aligned} \tag{16.94}$$

with

$$\langle\mathbf{n}\rangle \equiv \langle\mathbf{n}(\omega^{\circ})\rangle \tag{16.95}$$

16.2.5.4 IP master equation of undriven damped density operator Now, in order to pass from the infinitesimal change in a time interval Δt of the IP time density operator given by Eq. (16.94) to a partial time derivative, take

$$\Delta t = (t' - t_0) \to 0$$

leading to

$$\left(\frac{\Delta\widetilde{\rho}(t)^{\mathrm{IP}}}{\Delta t}\right) \to \left(\frac{\partial\widetilde{\rho}(t)^{\mathrm{IP}}}{\partial t}\right) \tag{16.96}$$

then, according to the transformation Eq. (16.96), Eq. (16.94) yields

$$\begin{aligned}
\left(\frac{\partial\widetilde{\rho}(t)^{\mathrm{IP}}}{\partial t}\right) &= -\mathbf{a}^{\dagger}\mathbf{a}\widetilde{\rho}(t)^{\mathrm{IP}}\left(\frac{\gamma}{2}+i\,\Delta\omega\right)\langle\mathbf{n}+1\rangle\\
&-\mathbf{a}\mathbf{a}^{\dagger}\widetilde{\rho}(t)^{\mathrm{IP}}\left(\frac{\gamma}{2}-i\,\Delta\omega\right)\langle\mathbf{n}\rangle\\
&+\mathbf{a}^{\dagger}\widetilde{\rho}(t)^{\mathrm{IP}}\mathbf{a}\left(\frac{\gamma}{2}-i\,\Delta\omega\right)\langle\mathbf{n}+1\rangle\\
&+\mathbf{a}\widetilde{\rho}(t)^{\mathrm{IP}}\mathbf{a}^{\dagger}\left(\frac{\gamma}{2}+i\,\Delta\omega\right)\langle\mathbf{n}\rangle\\
&+\mathbf{a}^{\dagger}\widetilde{\rho}(t)^{\mathrm{IP}}\mathbf{a}\left(\frac{\gamma}{2}+i\,\Delta\omega\right)\langle\mathbf{n}+1\rangle\\
&+\mathbf{a}\widetilde{\rho}(t)^{\mathrm{IP}}\mathbf{a}^{\dagger}\left(\frac{\gamma}{2}-i\,\Delta\omega\right)\langle\mathbf{n}\rangle
\end{aligned}$$

$$
\begin{aligned}
&- \widetilde{\rho}(t)^{\mathrm{IP}}\mathbf{a}^{\dagger}\mathbf{a}\left(\frac{\gamma}{2} - i\,\Delta\omega\right)\langle\mathbf{n}+1\rangle \\
&- \widetilde{\rho}(t)^{\mathrm{IP}}\mathbf{a}\mathbf{a}^{\dagger}\left(\frac{\gamma}{2} + i\,\Delta\omega\right)\langle\mathbf{n}\rangle
\end{aligned}
$$

which, after rearranging, becomes

$$
\begin{aligned}
\left(\frac{\partial\widetilde{\rho}(t)^{\mathrm{IP}}}{\partial t}\right) &= +i[\widetilde{\rho}(t)^{\mathrm{IP}}, \mathbf{a}^{\dagger}\mathbf{a}]\Delta\omega \\
&\quad - \frac{\gamma}{2}(\mathbf{a}^{\dagger}\mathbf{a}\widetilde{\rho}(t)^{\mathrm{IP}} + \widetilde{\rho}(t)^{\mathrm{IP}}\mathbf{a}^{\dagger}\mathbf{a} - 2\,\mathbf{a}\widetilde{\rho}(t)^{\mathrm{IP}}\mathbf{a}^{\dagger}) \\
&\quad - \langle\mathbf{n}\rangle\gamma(\mathbf{a}^{\dagger}\mathbf{a}\widetilde{\rho}(t)^{\mathrm{IP}} + \widetilde{\rho}(t)^{\mathrm{IP}}\mathbf{a}\mathbf{a}^{\dagger} - \mathbf{a}^{\dagger}\,\widetilde{\rho}(t)^{\mathrm{IP}}\mathbf{a} - \mathbf{a}\widetilde{\rho}(t)^{\mathrm{IP}}\mathbf{a}^{\dagger}) \qquad (16.97)
\end{aligned}
$$

This last equation represents the coarse-grained time evolution of the reduced IP density operator of the driven damped harmonic oscillator, which is named the IP master equation of this oscillator.

16.2.5.5 IP master equation of driven damped density operator Now, in order to pass from the IP Liouville equation (16.97) dealing with the undriven damped harmonic oscillator to the corresponding one dealing with the driven damped harmonic oscillator, use Eq. (16.40)

$$
\boxed{i\hbar\left(\frac{\partial\widetilde{\rho}_{\mathrm{Dr}}(t)^{\mathrm{IP}}}{\partial t}\right) = [\widetilde{\mathbf{H}}_{\mathrm{Dr}}(t)^{\mathrm{IP}}, \rho_{\mathrm{Dr}}(t)^{\mathrm{IP}}] + i\hbar\left(\frac{\partial\widetilde{\rho}(t)^{\mathrm{IP}}}{\partial t}\right)} \qquad (16.98)
$$

Next, in the independent variations approximation, it may be assumed that the instantaneous action of the damping on the oscillator of interest is the same whether the oscillator is driven or undriven, so that one may write in this spirit

$$
\begin{aligned}
\left(\frac{\partial\widetilde{\rho}(t)^{\mathrm{IP}}}{\partial t}\right) &= +i[\widetilde{\rho}_{\mathrm{Dr}}(t)^{\mathrm{IP}}, \mathbf{a}^{\dagger}\mathbf{a}]\Delta\omega \\
&\quad - \frac{\gamma}{2}(\mathbf{a}^{\dagger}\mathbf{a}\widetilde{\rho}_{\mathrm{Dr}}(t)^{\mathrm{IP}} + \widetilde{\rho}_{\mathrm{Dr}}(t)^{\mathrm{IP}}\mathbf{a}^{\dagger}\mathbf{a} - 2\,\mathbf{a}\widetilde{\rho}_{\mathrm{Dr}}(t)^{\mathrm{IP}}\mathbf{a}^{\dagger}) \\
&\quad - \langle\mathbf{n}\rangle\gamma(\mathbf{a}^{\dagger}\mathbf{a}\widetilde{\rho}_{\mathrm{Dr}}(t)^{\mathrm{IP}} + \widetilde{\rho}_{\mathrm{Dr}}(t)^{\mathrm{IP}}\mathbf{a}\mathbf{a}^{\dagger} - \mathbf{a}^{\dagger}\,\widetilde{\rho}_{\mathrm{Dr}}(t)^{\mathrm{IP}}\mathbf{a} - \mathbf{a}\widetilde{\rho}_{\mathrm{Dr}}(t)^{\mathrm{IP}}\mathbf{a}^{\dagger})
\end{aligned}
\qquad (16.99)
$$

Hence, using Eq. (16.36) and due to Eq. (16.99), Eq. (16.98) becomes

$$
\begin{aligned}
\left(\frac{\partial\widetilde{\rho}_{\mathrm{Dr}}(t)^{\mathrm{IP}}}{\partial t}\right) &= \frac{1}{i\hbar}[\widetilde{\mathbf{H}}_{\mathrm{Dr}}(t)^{\mathrm{IP}}, \rho_{\mathrm{Dr}}(t)^{\mathrm{IP}}] \\
&\quad + i[\widetilde{\rho}_{\mathrm{Dr}}(t)^{\mathrm{IP}}, \mathbf{a}^{\dagger}\mathbf{a}]\Delta\omega \\
&\quad - \frac{\gamma}{2}(\mathbf{a}^{\dagger}\mathbf{a}\widetilde{\rho}_{\mathrm{Dr}}(t)^{\mathrm{IP}} + \widetilde{\rho}_{\mathrm{Dr}}(t)^{\mathrm{IP}}\mathbf{a}^{\dagger}\mathbf{a} - 2\,\mathbf{a}\widetilde{\rho}_{\mathrm{Dr}}(t)^{\mathrm{IP}}\mathbf{a}^{\dagger}) \\
&\quad - \langle\mathbf{n}\rangle\gamma(\mathbf{a}^{\dagger}\mathbf{a}\widetilde{\rho}_{\mathrm{Dr}}(t)^{\mathrm{IP}} + \widetilde{\rho}_{\mathrm{Dr}}(t)^{\mathrm{IP}}\mathbf{a}\mathbf{a}^{\dagger} - \mathbf{a}^{\dagger}\,\widetilde{\rho}_{\mathrm{Dr}}(t)^{\mathrm{IP}}\mathbf{a} - \mathbf{a}\widetilde{\rho}_{\mathrm{Dr}}(t)^{\mathrm{IP}}\mathbf{a}^{\dagger})
\end{aligned}
\qquad (16.100)
$$

Equation (16.100) is the IP Liouville equation of the driven damped quantum harmonic oscillator.

16.2.6 Schrödinger picture master equation

Now, to return to the Schrödinger picture, keep in mind that, due to Eq. (3.184) allowing one to pass from the IP density operator $\widetilde{\rho}(t)^{\mathrm{IP}}$ to the corresponding SP one $\rho(t)^{\mathrm{SP}}$, that is

$$\rho(t)^{\mathrm{SP}} = \mathbf{U}^{\circ}(t)\widetilde{\rho}(t)^{\mathrm{IP}}\mathbf{U}^{\circ}(t)^{-1} \tag{16.101}$$

where

$$\mathbf{U}^{\circ}(t) = e^{-i\mathbf{H}^{\circ}t/\hbar} \tag{16.102}$$

the time derivative of the Schrödinger picture density operator is given by Eq. (3.190), that is,

$$\left(\frac{\partial \rho(t)^{\mathrm{SP}}}{\partial t}\right) = \frac{1}{i\hbar}[\mathbf{H}^{\circ}, \rho(t)^{\mathrm{SP}}] + \mathbf{U}^{\circ}(t)\left(\frac{\partial \widetilde{\rho}(t)^{\mathrm{IP}}}{\partial t}\right)\mathbf{U}^{\circ}(t)^{-1} \tag{16.103}$$

In the present situation, where the Hamiltonian $\mathbf{H}^{\circ}$ is given by Eq. (16.9), that is,

$$\mathbf{H}^{\circ} = \mathbf{a}^{\dagger}\mathbf{a}\hbar\omega^{\circ} \tag{16.104}$$

the time evolution operator appearing in Eqs. (16.101) and (16.103) is

$$\mathbf{U}^{\circ}(t) = e^{-i\mathbf{a}^{\dagger}\mathbf{a}\,\omega^{\circ}t} \tag{16.105}$$

Hence, returning to the Schrödinger picture, from Eq. (16.100), by using an equation of the form (16.103) we have

$$\begin{aligned}\left(\frac{\partial \widetilde{\rho}_{\mathrm{Dr}}(t)^{\mathrm{SP}}}{\partial t}\right) = &\left(\frac{1}{i\hbar}\right)[\mathbf{H}^{\circ}, \widetilde{\rho}_{\mathrm{Dr}}(t)^{\mathrm{SP}}] + \left(\frac{1}{i\hbar}\right)\mathbf{U}^{\circ}(t)[\widetilde{\mathbf{H}}_{\mathrm{Dr}}(t)^{\mathrm{IP}}, \widetilde{\rho}_{\mathrm{Dr}}(t)^{\mathrm{IP}}]\mathbf{U}^{\circ}(t)^{-1}\\ &- i\,\Delta\omega\mathbf{U}^{\circ}(t)[\mathbf{a}^{\dagger}\mathbf{a}, \widetilde{\rho}_{\mathrm{Dr}}(t)^{\mathrm{IP}}]\mathbf{U}^{\circ}(t)^{-1}\\ &- \frac{\gamma}{2}\mathbf{U}^{\circ}(t)(\mathbf{a}^{\dagger}\mathbf{a}\widetilde{\rho}_{\mathrm{Dr}}(t)^{\mathrm{IP}} + \widetilde{\rho}_{\mathrm{Dr}}(t)^{\mathrm{IP}}\mathbf{a}^{\dagger}\mathbf{a} - 2\,\mathbf{a}\widetilde{\rho}_{\mathrm{Dr}}(t)^{\mathrm{IP}}\mathbf{a}^{\dagger})\mathbf{U}^{\circ}(t)^{-1}\\ &- \langle\mathbf{n}\rangle\gamma\mathbf{U}^{\circ}(t)(\mathbf{a}^{\dagger}\mathbf{a}\widetilde{\rho}_{\mathrm{Dr}}(t)^{\mathrm{IP}} + \widetilde{\rho}_{\mathrm{Dr}}(t)^{\mathrm{IP}}\mathbf{a}\mathbf{a}^{\dagger}\\ &- \mathbf{a}^{\dagger}\,\widetilde{\rho}_{\mathrm{Dr}}(t)^{\mathrm{IP}}\mathbf{a} - \mathbf{a}\widetilde{\rho}_{\mathrm{Dr}}(t)^{\mathrm{IP}}\mathbf{a}^{\dagger})\mathbf{U}^{\circ}(t)^{-1}\end{aligned} \tag{16.106}$$

Of course, owing to the expression (16.104) of $\mathbf{H}^{\circ}$, it appears that

$$[\mathbf{H}^{\circ}, \rho_{\mathrm{Dr}}(t)^{\mathrm{SP}}] = [\mathbf{a}^{\dagger}\mathbf{a}, \rho_{\mathrm{Dr}}(t)^{\mathrm{SP}}]\hbar\omega^{\circ} \tag{16.107}$$

Next, inserting the unity operator built up from the evolution operator (16.102), we have

$$\begin{aligned}&\mathbf{U}^{\circ}(t)\widetilde{\mathbf{H}}_{\mathrm{Dr}}(t)^{\mathrm{IP}}\widetilde{\rho}_{\mathrm{Dr}}(t)^{\mathrm{IP}}\mathbf{U}^{\circ}(t)^{-1}\\ &\quad= \mathbf{U}^{\circ}(t)[\widetilde{\mathbf{H}}_{\mathrm{Dr}}(t)^{\mathrm{IP}}\mathbf{U}^{\circ}(t)^{-1}\mathbf{U}^{\circ}(t)\widetilde{\rho}_{\mathrm{Dr}}(t)^{\mathrm{IP}}]\mathbf{U}^{\circ}(t)^{-1}\end{aligned}$$

so that, using Eqs. (16.34) and (16.101), due to Eq. (16.104), we have

$$\mathbf{U}^{\circ}(t)\widetilde{\mathbf{H}}_{\mathrm{Dr}}(t)^{\mathrm{IP}}\widetilde{\rho}_{\mathrm{Dr}}(t)^{\mathrm{IP}}\mathbf{U}^{\circ}(t)^{-1} = \mathbf{H}_{\mathrm{Dr}}^{\mathrm{SP}}\widetilde{\rho}_{\mathrm{Dr}}(t)^{\mathrm{SP}}$$

In like manner

$$\mathbf{U}^{\circ}(t)\widetilde{\rho}_{\mathrm{Dr}}(t)^{\mathrm{IP}}\widetilde{\mathbf{H}}_{\mathrm{Dr}}(t)^{\mathrm{IP}}\mathbf{U}^{\circ}(t)^{-1} = \widetilde{\rho}_{\mathrm{Dr}}(t)^{\mathrm{SP}}\mathbf{H}_{\mathrm{Dr}}^{\mathrm{SP}}$$

so that the second right-hand-side commutator of Eq. (16.106) reads

$$\mathbf{U}^\circ(t)[\widetilde{\mathbf{H}}_{\mathrm{Dr}}(t)^{\mathrm{IP}}, \widetilde{\boldsymbol{\rho}}_{\mathrm{Dr}}(t)^{\mathrm{IP}}]\mathbf{U}^\circ(t)^{-1} = [\mathbf{H}_{\mathrm{Dr}}, \widetilde{\boldsymbol{\rho}}_{\mathrm{Dr}}(t)^{\mathrm{SP}}] \tag{16.108}$$

Hence, as a consequence of Eqs. (16.107) and (16.108), Eq. (16.106) becomes

$$\begin{aligned}\left(\frac{\partial \boldsymbol{\rho}_{\mathrm{Dr}}(t)^{\mathrm{SP}}}{\partial t}\right) &= -i\omega^\circ[\mathbf{a}^\dagger\mathbf{a}, \widetilde{\boldsymbol{\rho}}_{\mathrm{Dr}}(t)^{\mathrm{SP}}] - \frac{i}{\hbar}[\mathbf{H}_{\mathrm{Dr}}, \widetilde{\boldsymbol{\rho}}_{\mathrm{Dr}}(t)^{\mathrm{SP}}] \\ &\quad - i\,\Delta\omega\mathbf{U}^\circ(t)[\mathbf{a}^\dagger\mathbf{a}, \widetilde{\boldsymbol{\rho}}_{\mathrm{Dr}}(t)^{\mathrm{IP}}]\mathbf{U}^\circ(t)^{-1} \\ &\quad - \frac{\gamma}{2}\mathbf{U}^\circ(t)(\mathbf{a}^\dagger\mathbf{a}\widetilde{\boldsymbol{\rho}}_{\mathrm{Dr}}(t)^{\mathrm{IP}} + \widetilde{\boldsymbol{\rho}}_{\mathrm{Dr}}(t)^{\mathrm{IP}}\mathbf{a}^\dagger\mathbf{a} - 2\,\mathbf{a}\widetilde{\boldsymbol{\rho}}_{\mathrm{Dr}}(t)^{\mathrm{IP}}\mathbf{a}^\dagger)\mathbf{U}^\circ(t)^{\mathrm{IP}-1} \\ &\quad - \langle\mathbf{n}\rangle\gamma\mathbf{U}^\circ(t)(\mathbf{a}^\dagger\mathbf{a}\widetilde{\boldsymbol{\rho}}_{\mathrm{Dr}}(t)^{\mathrm{IP}} + \widetilde{\boldsymbol{\rho}}_{\mathrm{Dr}}(t)^{\mathrm{IP}}\mathbf{a}\mathbf{a}^\dagger - \mathbf{a}^\dagger\,\widetilde{\boldsymbol{\rho}}_{\mathrm{Dr}}(t)^{\mathrm{IP}}\mathbf{a} \\ &\quad - \mathbf{a}\widetilde{\boldsymbol{\rho}}_{\mathrm{Dr}}(t)^{\mathrm{IP}}\mathbf{a}^\dagger)\mathbf{U}^\circ(t)^{-1}\end{aligned} \tag{16.109}$$

Now, one has to get the result of the canonical transformations involved on the right-hand side of Eq. (16.109).

First, consider the canonical transformation over $\mathbf{a}^\dagger\mathbf{a}$ that, according to Eqs. (16.101) and (16.105), is

$$\mathbf{U}^\circ(t)\widetilde{\boldsymbol{\rho}}_{\mathrm{Dr}}(t)^{\mathrm{IP}}\mathbf{a}^\dagger\mathbf{a}\mathbf{U}^\circ(t)^{-1} = e^{-i\mathbf{a}^\dagger\mathbf{a}\omega^\circ t}\widetilde{\boldsymbol{\rho}}_{\mathrm{Dr}}(t)^{\mathrm{IP}}\mathbf{a}^\dagger\mathbf{a}e^{i\mathbf{a}^\dagger\mathbf{a}\omega^\circ t}$$

Now, the operator commutes with the exponential operator, hence this expression simplifies to

$$\mathbf{U}^\circ(t)\widetilde{\boldsymbol{\rho}}_{\mathrm{Dr}}(t)^{\mathrm{IP}}\mathbf{a}^\dagger\mathbf{a}\mathbf{U}^\circ(t)^{-1} = e^{-i\mathbf{a}^\dagger\mathbf{a}\omega^\circ t}\widetilde{\boldsymbol{\rho}}_{\mathrm{Dr}}(t)^{\mathrm{IP}}e^{i\mathbf{a}^\dagger\mathbf{a}\omega^\circ t}\mathbf{a}^\dagger\mathbf{a}$$

Then, Eqs. (16.101) and (16.105) allow one to transform this equation into

$$\mathbf{U}^\circ(t)\widetilde{\boldsymbol{\rho}}_{\mathrm{Dr}}(t)^{\mathrm{IP}}\mathbf{a}^\dagger\mathbf{a}\mathbf{U}^\circ(t)^{-1} = \widetilde{\boldsymbol{\rho}}_{\mathrm{Dr}}(t)^{\mathrm{SP}}\mathbf{a}^\dagger\mathbf{a} \tag{16.110}$$

In like manner, we have the following results for the other canonical transformations of interest:

$$\mathbf{U}^\circ(t)\mathbf{a}^\dagger\mathbf{a}\widetilde{\boldsymbol{\rho}}_{\mathrm{Dr}}(t)^{\mathrm{IP}}\mathbf{U}^\circ(t)^{-1} = \mathbf{a}^\dagger\mathbf{a}\,\widetilde{\boldsymbol{\rho}}_{\mathrm{Dr}}(t)^{\mathrm{SP}} \tag{16.111}$$

$$\mathbf{U}^\circ(t)\mathbf{a}\mathbf{a}^\dagger\widetilde{\boldsymbol{\rho}}_{\mathrm{Dr}}(t)^{\mathrm{IP}}\mathbf{U}^\circ(t)^{-1} = \mathbf{a}\mathbf{a}^\dagger\,\widetilde{\boldsymbol{\rho}}_{\mathrm{Dr}}(t)^{\mathrm{SP}} \tag{16.112}$$

$$\mathbf{U}^\circ(t)\widetilde{\boldsymbol{\rho}}_{\mathrm{Dr}}(t)^{\mathrm{IP}}\mathbf{a}\mathbf{a}^\dagger\mathbf{U}^\circ(t)^{-1} = \widetilde{\boldsymbol{\rho}}_{\mathrm{Dr}}(t)^{\mathrm{SP}}\mathbf{a}\mathbf{a}^\dagger \tag{16.113}$$

where $[\mathbf{a}, \mathbf{a}^\dagger] = 1$ has been used for the two last results.

Hence, collecting Eqs. (16.110)–(16.113) and using Eq. (16.10), the master equation (16.109) becomes after simplification

$$\begin{aligned}\left(\frac{\partial \boldsymbol{\rho}_{\mathrm{Dr}}(t)^{\mathrm{SP}}}{\partial t}\right) &= -i\alpha^\circ\omega^\circ\{[\mathbf{a}, \boldsymbol{\rho}_{\mathrm{Dr}}(t)^{\mathrm{SP}}] + [\mathbf{a}^\dagger, \boldsymbol{\rho}_{\mathrm{Dr}}(t)^{\mathrm{SP}}]\} \\ &\quad - i\omega^\circ[\mathbf{a}^\dagger\mathbf{a}, \boldsymbol{\rho}_{\mathrm{Dr}}(t)^{\mathrm{SP}}] - i\,\Delta\omega\,[\mathbf{a}^\dagger\mathbf{a}, \boldsymbol{\rho}_{\mathrm{Dr}}(t)^{\mathrm{SP}}] \\ &\quad - \frac{\gamma}{2}(\mathbf{a}^\dagger\mathbf{a}\boldsymbol{\rho}_{\mathrm{Dr}}(t)^{\mathrm{SP}} + \boldsymbol{\rho}_{\mathrm{Dr}}(t)^{\mathrm{SP}}\mathbf{a}^\dagger\mathbf{a} - 2\mathbf{a}\boldsymbol{\rho}_{\mathrm{Dr}}(t)^{\mathrm{SP}}\mathbf{a}^\dagger) \\ &\quad - \langle\mathbf{n}\rangle\gamma(\mathbf{a}^\dagger\mathbf{a}\boldsymbol{\rho}_{\mathrm{Dr}}(t)^{\mathrm{SP}} + \boldsymbol{\rho}_{\mathrm{Dr}}(t)^{\mathrm{SP}}\mathbf{a}\mathbf{a}^\dagger - \mathbf{a}^\dagger\,\boldsymbol{\rho}_{\mathrm{Dr}}(t)^{\mathrm{SP}}\mathbf{a} - \mathbf{a}\boldsymbol{\rho}_{\mathrm{Dr}}(t)^{\mathrm{SP}}\mathbf{a}^\dagger)\end{aligned} \tag{16.114}$$

Recall here that γ is the damping parameter induced by the irreversible influence of the thermal bath, whereas $\Delta\omega$ is an angular frequency shift induced by the bath, and where $\langle\mathbf{n}\rangle$ is the thermal average of the occupation number, which, owing to Eq. (16.69), is

$$\langle\mathbf{n}\rangle = \frac{1}{e^{\hbar\omega^\circ/kT} - 1}$$

16.2.7 Matrix representation of master equation (16.114) in basis of harmonic Hamiltonian

Now, consider the matrix representation of the master equation (16.114) in the basis $\{|(n)\rangle\}$ of the eigenkets of $\mathbf{a}^\dagger\mathbf{a}$, obeying

$$\mathbf{a}^\dagger\mathbf{a}|(n)\rangle = n|(n)\rangle \quad \text{with} \quad \langle(m)|(n)\rangle = \delta_{mn}$$

In this basis, a matrix element of Eq. (16.114) becomes

$$\begin{aligned}
\frac{\partial}{\partial t}\langle(m)|\boldsymbol{\rho}_{\mathrm{Dr}}(t)^{\mathrm{SP}}|(n)\rangle &= -i\alpha^\circ\langle(m)|\mathbf{a}\boldsymbol{\rho}_{\mathrm{Dr}}(t)^{\mathrm{SP}}|(n)\rangle - \langle(m)|\boldsymbol{\rho}_{\mathrm{Dr}}(t)^{\mathrm{SP}}\mathbf{a}|(n)\rangle \\
&\quad - i\alpha^\circ\{\langle(m)|\mathbf{a}^\dagger\boldsymbol{\rho}_{\mathrm{Dr}}(t)^{\mathrm{SP}}|(n)\rangle - \langle(m)|\boldsymbol{\rho}_{\mathrm{Dr}}(t)^{\mathrm{SP}}\mathbf{a}^\dagger|(n)\rangle\} \\
&\quad - i\{\langle(m)|\mathbf{a}^\dagger\mathbf{a}\boldsymbol{\rho}_{\mathrm{Dr}}(t)^{\mathrm{SP}}|(n)\rangle - \langle(m)|\boldsymbol{\rho}_{\mathrm{Dr}}(t)^{\mathrm{SP}}\mathbf{a}^\dagger\mathbf{a}|(n)\rangle\} \\
&\quad - \frac{\gamma}{2}\{\langle(m)|\mathbf{a}^\dagger\mathbf{a}\boldsymbol{\rho}_{\mathrm{Dr}}(t)^{\mathrm{SP}}|(n)\rangle + \langle(m)|\boldsymbol{\rho}_{\mathrm{Dr}}(t)^{\mathrm{SP}}\mathbf{a}^\dagger\mathbf{a}|(n)\rangle\} \\
&\quad + \gamma\{\langle(m)|\mathbf{a}\boldsymbol{\rho}_{\mathrm{Dr}}(t)^{\mathrm{SP}}\mathbf{a}^\dagger|(n)\rangle\} \\
&\quad - \langle\mathbf{n}\rangle\gamma\{\langle(m)|\mathbf{a}^\dagger\mathbf{a}\boldsymbol{\rho}_{\mathrm{Dr}}(t)^{\mathrm{SP}}|(n)\rangle + \langle(m)|\boldsymbol{\rho}_{\mathrm{Dr}}(t)^{\mathrm{SP}}\mathbf{a}^\dagger\mathbf{a}|(n)\rangle\} \\
&\quad - \langle\mathbf{n}\rangle\gamma\{\langle(m)|\mathbf{a}^\dagger\boldsymbol{\rho}_{\mathrm{Dr}}(t)^{\mathrm{SP}}\mathbf{a}|(n)\rangle + \langle(m)|\mathbf{a}\boldsymbol{\rho}_{\mathrm{Dr}}(t)^{\mathrm{SP}}\mathbf{a}^\dagger|(n)\rangle\}
\end{aligned} \tag{16.115}$$

Recall that the diagonal elements corresponding to $m = n$, are called *populations*, whereas the off-diagonal ones are called *coherences*.

To get expressions for the right-hand-side matrix elements appearing in Eq. (16.115) it is suitable to use Eqs. (5.53) and (5.63) giving the actions of $\mathbf{a}^\dagger$ and $\mathbf{a}$ on $|(n)\rangle$ and their Hermitian conjugates, that is,

$$\mathbf{a}^\dagger|(n)\rangle = \sqrt{n+1}|(n+1)\rangle \qquad \text{and} \qquad \mathbf{a}|(n)\rangle = \sqrt{n}|(n-1)\rangle$$

$$\langle(n)|\mathbf{a} = \langle(n+1)|\sqrt{n+1} \qquad \text{and} \qquad \langle(n)|\mathbf{a}^\dagger = \langle(n-1)|\sqrt{n}$$

Then, in view of these expressions, Eq. (16.115) becomes, after omitting the SP notation for the matrix elements,

$$\begin{aligned}
\left(\frac{\partial\boldsymbol{\rho}_{m,n}(t)}{\partial t}\right) &= -i(m-n)\{\boldsymbol{\rho}_{m,n}(t)\} \\
&\quad - i\alpha^\circ\{\sqrt{m+1}\{\boldsymbol{\rho}_{m+1,n}(t)\} - \sqrt{n}\{\boldsymbol{\rho}_{m,n-1}(t)\} \\
&\quad + \sqrt{m}\{\boldsymbol{\rho}_{m-1,n}(t)\} - \sqrt{n+1}\{\boldsymbol{\rho}_{m,n+1}(t)\}\}
\end{aligned}$$

$$
\begin{aligned}
&-\frac{\gamma}{2}\{(m+n)\{\rho_{m,n}(t)\} - 2\sqrt{(m+1)(n+1)}\{\rho_{m+1,n+1}(t)\}\} \\
&+\langle\mathbf{n}\rangle\gamma\{\sqrt{(m+1)(n+1)}\{\rho_{m+1,n+1}(t)\} \\
&-(m+n)\{\rho_{mn}(t)\} + \sqrt{mn}\{\rho_{m-1,n-1}(t)\}\}
\end{aligned}
\tag{16.116}
$$

with

$$\{\rho_{mn}(t)\} = \langle(m)|\rho_{\mathrm{Dr}}(t)^{\mathrm{SP}}|(n)\rangle$$

Equation (16.116) may be solved if one knows the initial condition for the various values of these matrix elements $\rho_{m,n}(0)^{\mathrm{SP}}$ at initial time $t = 0$, that is, the expression of the density operator $\rho_{\mathrm{Dr}}(0)^{\mathrm{SP}}$ of the driven damped harmonic oscillator at this initial time. This may be numerically performed, for instance, by the aid of the Runge–Kutta method. However, that may be avoided since, as we will see, an analytical expression of the reduced time evolution operator of the driven damped harmonic oscillator exists. This may be viewed as the integrated form of a generalization of the master equation (16.114) resulting from an infinite order expansion in the coupling $\widetilde{\mathbf{V}}(t)^{\mathrm{IP}}$ of Eq. (16.42).

16.3 FOKKER–PLANCK EQUATION CORRESPONDING TO (16.114)

However, before seeking such generalization of the master equation, it may be of interest to show how this master equation (16.114) may be transformed into a scalar partial of the same type as the Fokker–Planck equations encountered in the area of classical statistical mechanics treating irreversible processes dealing with Brownian oscillators. With this in mind, it is convenient to convert the SP master equation (16.114) into the antinormal order and thus, for this purpose, to first consider the action of $\mathbf{aa}^\dagger$ on the density operator in the following way:

$$\mathbf{aa}^\dagger\rho(t) = \mathbf{a}(\mathbf{a}^\dagger\rho(t) - \rho(t)\mathbf{a}^\dagger + \rho(t)\mathbf{a}^\dagger) \tag{16.117}$$

which reads

$$\mathbf{aa}^\dagger\rho(t) = \mathbf{a}([\mathbf{a}^\dagger, \rho(t)] + \rho(t)\mathbf{a}^\dagger)$$

Again, applying Eq. (7.59), that is,

$$[\mathbf{a}^\dagger, \{\mathbf{f}(\mathbf{a}, \mathbf{a}^\dagger)\}] = -\left(\frac{\partial \mathbf{f}(\mathbf{a}, \mathbf{a}^\dagger)}{\partial \mathbf{a}}\right)$$

to the function

$$\rho(t) = \rho(\mathbf{a}, \mathbf{a}^\dagger, t)$$

yields

$$[\mathbf{a}^\dagger, \rho(t)] = -\left(\frac{\partial \rho(t)}{\partial \mathbf{a}}\right) \tag{16.118}$$

so that Eq. (16.117) transforms to the antinormal order form

$$\boxed{\mathbf{a}\mathbf{a}^{\dagger}\rho(t) = -\mathbf{a}\left(\frac{\partial\rho^{a}(t)}{\partial\mathbf{a}}\right) + \mathbf{a}\rho^{a}(t)\mathbf{a}^{\dagger}} \tag{16.119}$$

Next, pass to the action of $\mathbf{a}^{\dagger}\mathbf{a}$ on $\rho(t)$, which may be written using the commutation rule $[\mathbf{a},\mathbf{a}^{\dagger}] = 1$,

$$\mathbf{a}^{\dagger}\mathbf{a}\rho(t) = (\mathbf{a}\mathbf{a}^{\dagger} - 1)\rho(t) = \mathbf{a}\mathbf{a}^{\dagger}\rho(t) - \rho(t)$$

Then, using Eq. (16.119), this expression takes the antinormal order form

$$\boxed{\mathbf{a}^{\dagger}\mathbf{a}\rho(t) = -\mathbf{a}\left(\frac{\partial\rho^{a}(t)}{\partial\mathbf{a}}\right) + \mathbf{a}\rho^{a}(t)\mathbf{a}^{\dagger} - \rho^{a}(t)} \tag{16.120}$$

Now, the commutation rule $[\mathbf{a},\mathbf{a}^{\dagger}] = 1$ of the Boson operators allows one to write

$$\rho(t)\mathbf{a}^{\dagger}\mathbf{a} = \rho(t)(\mathbf{a}\mathbf{a}^{\dagger} - 1) = \rho(t)\mathbf{a}\mathbf{a}^{\dagger} - \rho(t)$$

which reads after adding and substracting the same term $\rho(t)$

$$\rho(t)\mathbf{a}^{\dagger}\mathbf{a} = (\rho(t)\mathbf{a} - \mathbf{a}\rho(t) + \mathbf{a}\rho(t))\mathbf{a}^{\dagger} - \rho(t)$$

or

$$\rho(t)\mathbf{a}^{\dagger}\mathbf{a} = ([\rho(t),\mathbf{a}]\mathbf{a}^{\dagger} + \mathbf{a}\rho(t))\mathbf{a}^{\dagger} - \rho(t) \tag{16.121}$$

Then observing that Eq. (7.60) allows one to write

$$[\rho(t),\mathbf{a}] = -\left(\frac{\partial\rho(t)}{\partial\mathbf{a}^{\dagger}}\right) \tag{16.122}$$

the left-hand side of Eq. (16.121) takes the antinormal form

$$\boxed{\rho(t)\mathbf{a}^{\dagger}\mathbf{a} = -\left(\frac{\partial\rho^{a}(t)}{\partial\mathbf{a}^{\dagger}}\right)\mathbf{a}^{\dagger} + \mathbf{a}\rho^{a}(t)\mathbf{a}^{\dagger} - \rho^{a}(t)} \tag{16.123}$$

Moreover, using the following commutation rule

$$\rho(t)\mathbf{a}^{\dagger}\mathbf{a} = \rho(t)\mathbf{a}\mathbf{a}^{\dagger} - \rho(t)$$

on the left-hand side of Eq. (16.123), this expression leads to the antinormal form

$$\boxed{\rho(t)\mathbf{a}\mathbf{a}^{\dagger} = -\left(\frac{\partial\rho^{a}(t)}{\partial\mathbf{a}^{\dagger}}\right)\mathbf{a}^{\dagger} + \mathbf{a}\rho^{a}(t)\mathbf{a}^{\dagger}} \tag{16.124}$$

Next, to find the antinormal expression of $\mathbf{a}^{\dagger}\rho(t)\mathbf{a}$, write it by adding and subtracting the same term $\mathbf{a}\rho(t)$ according to

$$\mathbf{a}^{\dagger}\rho(t)\mathbf{a} = \mathbf{a}^{\dagger}(\rho(t)\mathbf{a} - \mathbf{a}\rho(t) + \mathbf{a}\rho(t))$$

so that

$$\mathbf{a}^{\dagger}\rho(t)\mathbf{a} = \mathbf{a}^{\dagger}([\rho(t),\mathbf{a}] + \mathbf{a}\rho(t))$$

and thus, owing to Eq. (16.122),

$$\mathbf{a}^\dagger \rho(t)\mathbf{a} = -\mathbf{a}^\dagger \left(\frac{\partial \rho(t)}{\partial \mathbf{a}^\dagger}\right) + \mathbf{a}^\dagger \mathbf{a}\rho(t) \tag{16.125}$$

At last, to come further in the quest of the antinormal form of $\mathbf{a}^\dagger \rho(t)\mathbf{a}$ denote

$$\mathbf{f}(t) \equiv \left(\frac{\partial \rho(t)}{\partial \mathbf{a}^\dagger}\right) \tag{16.126}$$

so that the first right-hand side of Eq. (16.125) reads

$$\mathbf{a}^\dagger \left(\frac{\partial \rho(t)}{\partial \mathbf{a}^\dagger}\right) = \mathbf{a}^\dagger \mathbf{f}(t)$$

which after adding and substracting the same term $\mathbf{f}(t)\mathbf{a}^\dagger$ gives

$$\mathbf{a}^\dagger \left(\frac{\partial \rho(t)}{\partial \mathbf{a}^\dagger}\right) = \mathbf{a}^\dagger \mathbf{f}(t) + \mathbf{f}(t)\mathbf{a}^\dagger - \mathbf{f}(t)\mathbf{a}^\dagger$$

or

$$\mathbf{a}^\dagger \left(\frac{\partial \rho(t)}{\partial \mathbf{a}^\dagger}\right) = [\mathbf{a}^\dagger, \mathbf{f}(t)] + \mathbf{f}(t)\mathbf{a}^\dagger$$

Then, applying Eq. (16.118) with $\mathbf{f}(t)$ in place of $\rho(t)$ gives

$$\mathbf{a}^\dagger \left(\frac{\partial \rho(t)}{\partial \mathbf{a}^\dagger}\right) = -\left(\frac{\partial \mathbf{f}(t)}{\partial \mathbf{a}}\right) + \mathbf{f}(t)\mathbf{a}^\dagger$$

or, after returning from $\mathbf{f}(t)$ to $\rho(t)$ by the aid of (16.126) we have

$$\mathbf{a}^\dagger \left(\frac{\partial \rho(t)}{\partial \mathbf{a}^\dagger}\right) = -\left(\frac{\partial^2 \rho(t)}{\partial \mathbf{a}\partial \mathbf{a}^\dagger}\right) + \left(\frac{\partial \rho(t)}{\partial \mathbf{a}^\dagger}\right)\mathbf{a}^\dagger$$

which transforms using of Eq. (16.125) into

$$\mathbf{a}^\dagger \rho(t)\mathbf{a} = \left(\frac{\partial^2 \rho(t)}{\partial \mathbf{a}\partial \mathbf{a}^\dagger}\right) - \left(\frac{\partial \rho(t)}{\partial \mathbf{a}^\dagger}\right)\mathbf{a}^\dagger + \mathbf{a}^\dagger \mathbf{a}\, \rho(t)$$

so that, due to Eq. (16.120) allowing to transform the last right-hand side, we have the final result for the antinormal form of $\mathbf{a}^\dagger \rho(t)\mathbf{a}$:

$$\boxed{\mathbf{a}^\dagger \rho(t)\mathbf{a} = \left(\frac{\partial^2 \rho^a(t)}{\partial \mathbf{a}\partial \mathbf{a}^\dagger}\right) - \left(\frac{\partial \rho^a(t)}{\partial \mathbf{a}^\dagger}\right)\mathbf{a}^\dagger - \mathbf{a}\left(\frac{\partial \rho^a(t)}{\partial \mathbf{a}}\right) + \mathbf{a}\rho^a(t)\mathbf{a}^\dagger - \rho^a(t)} \tag{16.127}$$

Hence, collecting Eqs. (16.120) and (16.123) and because $\mathbf{a}\rho(t)\mathbf{a}^\dagger$ is yet antinormal, the right-hand-side term involving $\gamma/2$ in the master equation (16.114) yields after simplification

$$2\mathbf{a}\rho(t)\mathbf{a}^\dagger - \mathbf{a}^\dagger \mathbf{a}\rho(t) - \rho(t)\mathbf{a}^\dagger \mathbf{a} = \left(\frac{\partial \rho^a(t)}{\partial \mathbf{a}^\dagger}\right)\mathbf{a}^\dagger + \mathbf{a}\left(\frac{\partial \rho^a(t)}{\partial \mathbf{a}}\right) + 2\rho^a(t) \tag{16.128}$$

or

$$\boxed{2\mathbf{a}\rho(t)\mathbf{a}^\dagger - \mathbf{a}^\dagger \mathbf{a}\rho(t) - \rho(t)\mathbf{a}^\dagger \mathbf{a} = \left(\frac{\partial(\rho^a(t)\mathbf{a}^\dagger)}{\partial \mathbf{a}^\dagger}\right) + \left(\frac{\mathbf{a}\partial(\rho^a(t))}{\partial \mathbf{a}}\right)} \tag{16.129}$$

Thus, with Eqs. (16.120), (16.124), and (16.127), one obtains after simplification of the last right-hand side of the master equation (16.114) involving $\langle \mathbf{n} \rangle \gamma$, the following antinormal expression:

$$\boxed{\mathbf{a}^{\dagger}\mathbf{a}\rho(t)+\rho(t)\mathbf{a}\mathbf{a}^{\dagger} - \mathbf{a}^{\dagger}\rho(t)\mathbf{a} - \mathbf{a}\rho(t)\mathbf{a}^{\dagger} = \frac{\partial^2 \rho^a(t)}{\partial \mathbf{a} \partial \mathbf{a}^{\dagger}}} \tag{16.130}$$

Now, the commutators multiplying $\alpha^{\circ}\omega^{\circ}$ appearing on the right-hand side of the master equation (16.114) may also be transformed into an antinormal form involving partial derivatives, by the aid of Eqs. (16.118) and (16.122), that is,

$$\boxed{[\mathbf{a}, \rho(t)] + [\mathbf{a}^{\dagger}, \rho(t)] = \left(\frac{\partial \rho^a(t)}{\partial \mathbf{a}^{\dagger}}\right) - \left(\frac{\partial \rho^a(t)}{\partial \mathbf{a}}\right)} \tag{16.131}$$

Finally, the commutator multiplying the term $i(\omega^{\circ} + \Delta\omega)$ appearing on the right-hand side of the master equation (16.114) may be written after adding and subtracting the same term $\mathbf{a}^{\dagger}\rho(t)\mathbf{a}$ as

$$[\mathbf{a}^{\dagger}\mathbf{a}, \rho(t)] = \mathbf{a}^{\dagger}\mathbf{a}\rho(t) - \rho(t)\mathbf{a}^{\dagger}\mathbf{a} + \mathbf{a}^{\dagger}\rho(t)\mathbf{a} - \mathbf{a}^{\dagger}\rho(t)\mathbf{a}$$

or

$$[\mathbf{a}^{\dagger}\mathbf{a}, \rho(t)] = \mathbf{a}^{\dagger}[\mathbf{a}, \rho(t)] + [\mathbf{a}^{\dagger}, \rho(t)]\mathbf{a}$$

so that, in view of Eqs. (16.118) and (16.122), it transforms to the antinormal form

$$\boxed{[\mathbf{a}^{\dagger}\mathbf{a}, \rho(t)] = \mathbf{a}^{\dagger}\left(\frac{\partial \rho^a(t)}{\partial \mathbf{a}^{\dagger}}\right) - \left(\frac{\partial \rho^a(t)}{\partial \mathbf{a}}\right)\mathbf{a}} \tag{16.132}$$

Thus, collecting Eqs. (16.129)–(16.132), the master equation (16.114) may be put into the following antinormal form:

$$\begin{aligned}
\left(\frac{\partial \rho^a(t)}{\partial t}\right) &= -i\alpha^{\circ}\omega^{\circ}\left\{\left(\frac{\partial \rho^a(t)}{\partial \mathbf{a}^{\dagger}}\right) - \left(\frac{\partial \rho^a(t)}{\partial \mathbf{a}}\right)\right\} \\
&\quad - i(\omega^{\circ} + \Delta\omega)\left\{\left(\frac{\partial \rho^a(t)}{\partial \mathbf{a}^{\dagger}}\right)\mathbf{a}^{\dagger} - \mathbf{a}\left(\frac{\partial \rho^a(t)}{\partial \mathbf{a}}\right)\right\} \\
&\quad + \frac{\gamma}{2}\left\{\left(\frac{\partial \rho^a(t)\mathbf{a}^{\dagger}}{\partial \mathbf{a}^{\dagger}}\right) + \left(\frac{\partial \rho^a(t)\mathbf{a}}{\partial \mathbf{a}}\right)\right\} + \langle \mathbf{n} \rangle \gamma \left(\frac{\partial^2 \rho^a(t)}{\partial \mathbf{a}\, \partial \mathbf{a}^{\dagger}}\right)
\end{aligned} \tag{16.133}$$

Now, it is possible to pass to the scalar representation of this equation using Eq. (7.41), which reads in the present situation

$$\hat{\mathbf{A}}^{-1}\{\rho^a(\mathbf{a}, \mathbf{a}^{\dagger}, t)\} = \{\rho^a(\alpha, \alpha^*, t)\}$$

so that the above equation (16.133) transforms to the following second-order partial differential equation:

$$\left(\frac{\partial\rho^a(t)}{\partial t}\right) = -i\alpha^\circ\omega^\circ\left\{\left(\frac{\partial\rho^a(t)}{\partial\alpha^*}\right) - \left(\frac{\partial\rho^a(t)}{\partial\alpha}\right)\right\}$$
$$- i(\omega^\circ + \Delta\omega)\left\{\alpha^*\left(\frac{\partial\rho^a(t)}{\partial\alpha^*}\right) - \alpha\left(\frac{\partial\rho^a(t)}{\partial\alpha}\right)\right\}$$
$$+\frac{\gamma}{2}\left\{\left(\frac{\partial\rho^a(t)\alpha^*}{\partial\alpha^*}\right) + \left(\frac{\partial\rho^a(t)\alpha}{\partial\alpha}\right)\right\} + \langle\mathbf{n}\rangle\gamma\left(\frac{\partial^2\rho^a(t)}{\partial\alpha\,\partial\alpha^*}\right) \quad (16.134)$$

This last equation is the *Fokker–Planck equation* corresponding to the master equation (16.114) governing the dynamics of the driven damped quantum harmonic oscillator to second order in the expansion of the coupling of the oscillator with the thermal bath.

16.4 NONPERTURBATIVE RESULTS FOR DENSITY OPERATOR

Recall that the master equation (16.114) from which results the Fokker–Planck equation (16.134) is a partial derivative equation of time, which takes into account, via Eq. (16.42), the irreversible influence of the thermal bath through a second-order expansion of the coupling between the oscillator and the thermal bath. However, a known closed expression for the density operator, at any time t, of driven damped harmonic oscillators, exists, which may be viewed as the result of the integration of a master equation of the same kind as (16.114) but that takes into account the coupling with the thermal bath, to infinite order in place of second order. The demonstration of this closed expression, due to Louisell and Walker[1] involves a very complicated treatment that is beyond the level of this book. Hence, in the present book, we shall only give the results of this treatment, leaving for the end of this chapter to show that it is possible to get also their result with the help of another treatment requiring knowledge of the IP time evolution of driven damped quantum harmonic oscillators.

16.4.1 Model

The Hamiltonian for the quantum harmonic oscillator weakly coupled linearly to a bath of oscillators is the same as above, that is,

$$\mathbf{H} = (\mathbf{a}^\dagger\mathbf{a} + \alpha^\circ(\mathbf{a}^\dagger + \mathbf{a}))\hbar\Omega + \sum_j(\mathbf{a}^\dagger\mathbf{b}_j\hbar\kappa_j + \mathbf{a}\mathbf{b}_j^\dagger\hbar\kappa_j^*)$$
$$+\sum_j \mathbf{b}_j^\dagger\mathbf{b}_j\hbar\omega_j \quad \text{with} \quad k = 0, 1 \qquad (16.135)$$

Just as for the master equation above, the density operator of the thermal bath is considered as the product of the Boltzmann density operators of the bath oscillators,

[1] W. Louisell and L. Walker. *Phys. Rev.*, **137** (1965): 204.

that is,

$$\boldsymbol{\rho}_\theta = \prod_k (1 - e^{-\lambda_k})(e^{-\lambda_k \mathbf{b}_k^\dagger \mathbf{b}_k}) \quad \text{with} \quad \mathbf{tr}_\theta\{\boldsymbol{\rho}_\theta\} = 1 \tag{16.136}$$

where

$$\lambda_k = \frac{\hbar\omega_k}{k_B T} \tag{16.137}$$

Louisell and Walker considered that, at an initial time, an equilibrium density operator, which is that of a coherent state at temperature T, is thus given by (13.111), so that

$$\boldsymbol{\rho}(0) = (1 - e^{-\lambda})e^{-\lambda(\mathbf{a}^\dagger - \alpha_c^*)(\mathbf{a} - \alpha_c)} \tag{16.138}$$

with

$$\lambda = \frac{\hbar\omega}{k_B T} \tag{16.139}$$

The reader should be aware that the dimensionless scalar parameter α_c characterizing the coherent density operator (16.138) has to be clearly distinguished from the dimensionless parameter α° reflecting the strength of the driving term in the Hamiltonian (16.135). Furthermore, the full density operator at initial time is taken as the product of the density operator (16.138) times that (16.136) of the thermal bath (16.138), that is,

$$\boldsymbol{\rho}_{\text{Tot}}(0) = (1 - e^{-\lambda})\{e^{-\lambda(\mathbf{a}^\dagger - \alpha_c^*)(\mathbf{a} - \alpha_c)}\} \prod_k (1 - e^{-\lambda_k})(e^{-\lambda_k \mathbf{b}_k^\dagger \mathbf{b}_k}) \tag{16.140}$$

The Liouville equation to be solved was

$$i\hbar\left(\frac{\partial \boldsymbol{\rho}_{\text{Tot}}(t)}{\partial t}\right) = [\mathbf{H}, \boldsymbol{\rho}_{\text{Tot}}(t)]$$

subject to the boundary condition (16.140), while the density operator of the damped oscillator was obtained from $\boldsymbol{\rho}_{\text{Tot}}(t)$ by performing a partial trace over the thermal bath eigenstates, according to

$$\boldsymbol{\rho}(t) = \mathbf{tr}_\theta\{\boldsymbol{\rho}_{\text{Tot}}(t)\}$$

16.4.2 Damped density operator at time *t*

For this model, and using a very long and complicated procedure involving the Markov approximation as for the master equation, Louisell and Walker have found that

$$\boldsymbol{\rho}(t) \cong \varepsilon\hat{\mathbf{N}}\{\exp\{-\varepsilon(\alpha^* - \phi(t))(\alpha - \phi^*(t))\}\} \tag{16.141}$$

where $\hat{\mathbf{N}}$ is the normal ordering operator, and α and α^* scalar complexes to be distinguished from α° and α_c, whereas $\phi(t)$ is given by

$$\phi(t) \cong \alpha_c\left\{\exp\left[-i(\omega^\circ + \Delta\omega)t - \frac{\gamma t}{2}\right]\right\} - i\int_0^t \alpha^\circ \exp\left[-i(\omega^\circ + \Delta\omega)t' - \frac{\gamma t'}{2}\right]dt' \tag{16.142}$$

Then, using the normal ordering operation and after integration of Eq. (16.142), they have obtained an expression for the density operator that is very similar to that of (16.138) of the initial situation, that is,

$$\rho(t) = \varepsilon \exp\{-\lambda(\mathbf{a}^\dagger - \phi(t))(\mathbf{a} - \phi^*(t))\} \tag{16.143}$$

with

$$\phi(t) = \alpha_c \left\{ \exp\left[-i(\omega + \Delta\omega)t - \frac{\gamma t}{2} \right] \right\} + \beta\{e^{-(t/2)} e^{+i(\omega^\circ + \Delta\omega)t} - 1\} \tag{16.144}$$

and where

$$\beta = \frac{\alpha^\circ(2\omega^{\circ 2} + i\gamma\omega^\circ)}{2(\omega^{\circ 2} + \gamma^2/4)} \tag{16.145}$$

Note in the passage from Eq. (16.141) to Eq. (16.143), the change of ε into λ, and that the expressions of the angular frequency shift $\Delta\omega$ and of the relaxation parameter γ are here the same as those in (16.90) and (16.91) encountered in the calculation of the master equation (16.114).

Moreover, the matrix elements of the time-dependent density operator (16.143) in the representation where $\mathbf{a}^\dagger\mathbf{a}$ is diagonal are

$$\begin{aligned} &\langle\{n\}|\rho(t)|\{m\}\rangle \\ &= \frac{(y-1)^n}{(y)^{n+1}} \left(\frac{\phi^*(t)}{y}\right)^{m-n} \left(\frac{n!}{m!}\right)^{1/2} \exp\left(-\frac{|\phi(t)|^2}{y}\right) \left\{L_n^{m-n}\left(-\frac{|\phi(t)|^2}{y(y-1)}\right)\right\} \\ &\text{if} \quad m \geq n, \end{aligned}$$

where $\{L_m^{n-m}(x)\}$ is the generalized Laguerre polynomial function of the variable x, with a similar relation when $n \geq m$ by permuting everywhere n and m.

16.4.3 Dynamics of averaged damped elongation

Now, as an application of the expression (16.143) of the density operator $\rho(t)$, we study how the average value of the position operator $\mathbf{Q}$ evolves with time when the oscillator is driven and damped.

16.4.3.1 Damped translation operator For this purpose, consider the special situation of a driven damped harmonic oscillator, starting at initial time from an undriven situation corresponding therefore to the situation $\alpha_c = 0$ in Eq. (16.138), that is, to

$$\rho_0(0) = (1 - e^{-\lambda})(e^{-\lambda\mathbf{a}^\dagger\mathbf{a}}) \tag{16.146}$$

Then, the density operator of the driven damped oscillator will be given by Eq. (16.143), that is,

$$\rho_0(t) = (1 - e^{-\lambda}) \exp\{-\lambda(\mathbf{a}^\dagger - \phi_0(t))(\mathbf{a} - \phi_0^*(t))\} \tag{16.147}$$

which involves a time-dependent argument (16.144) reducing to

$$\phi_0(t) = \beta\{e^{-\gamma t/2} e^{+i(\omega^\circ + \Delta\omega)t} - 1\} \tag{16.148}$$

Observe that in this special situation the density operator (16.147) has the same structure as that (13.11) obtained at any time using the Lagrange multiplier method, so that it appears to be the density operator of a coherent state at some temperature.

Now, observe that it is possible to consider the density operator (16.143), which here reads (16.147), as the result of the following canonical transformation:

$$\boldsymbol{\rho}_0(t) = \mathbf{A}(\phi_0(t))\boldsymbol{\rho}_0(0)\mathbf{A}(\phi_0(t))^{-1} \tag{16.149}$$

with

$$\mathbf{A}(\phi_0(t)) = (e^{\phi_0^*(t)\mathbf{a}^\dagger - \phi_0(t)\mathbf{a}}) \tag{16.150}$$

Now, owing to Eq. (7.9)

$$\mathbf{A}(\phi_0(t))\{\mathbf{f}(\mathbf{a}^\dagger, \mathbf{a})\}\mathbf{A}(\phi_0(t))^{-1}$$
$$= (e^{\phi_0^*(t)\mathbf{a}^\dagger - \phi_0(t)\mathbf{a}})\{\mathbf{f}(\mathbf{a}^\dagger, \mathbf{a})\}(e^{-\phi_0^*(t)\mathbf{a}^\dagger + \phi_0(t)\mathbf{a}}) = \{\mathbf{f}(\mathbf{a}^\dagger - \phi_0(t), \mathbf{a} - \phi_0^*(t))\} \tag{16.151}$$

so that, as required,

$$\mathbf{A}(\phi_0(t))(e^{-\lambda\mathbf{a}^\dagger\mathbf{a}})\mathbf{A}(\phi_0(t))^{-1} = e^{-\lambda(\mathbf{a}^\dagger - \phi_0(t))(\mathbf{a} - \phi_0^*(t))}$$

Hence, owing to Eq. (16.148), the damped translation operator (16.150) allowing to pass from the initial Boltzmann density operator (16.146) to the damped density operator at time t (16.147) is

$$\mathbf{A}(\phi_0(t)) = \exp\{\beta^*\{e^{-\gamma t/2}e^{i\omega^\circ t} - 1\}\mathbf{a}^\dagger - \beta\{e^{-\gamma t/2}e^{-i\omega^\circ t} - 1\}\mathbf{a}\} \tag{16.152}$$

16.4.3.2 Damped average elongation Besides, knowledge of the time-dependent density operator $\boldsymbol{\rho}(t)$ allows us to get the time dependence of the average value of the position operator $\mathbf{Q}$ according to

$$\langle \mathbf{Q}(t) \rangle = \mathbf{tr}\{\boldsymbol{\rho}(t)\mathbf{Q}\}$$

Owing to Eq. (16.149)) and due to Eq. (5.6) giving $\mathbf{Q}$ in terms of Boson operators, this equation yields

$$\langle \mathbf{Q}(t) \rangle = \sqrt{\frac{\hbar}{2M\omega^\circ}}\mathbf{tr}\{\mathbf{A}(\phi_o(t))\boldsymbol{\rho}_o(0)\mathbf{A}(\phi_o(t))^{-1}(\mathbf{a}^\dagger + \mathbf{a})\}$$

or, in view of Eq. (16.146),

$$\langle \mathbf{Q}(t) \rangle = \varepsilon\sqrt{\frac{\hbar}{2M\omega^\circ}}\mathbf{tr}\{\mathbf{A}(\phi_o(t))(e^{-\lambda\mathbf{a}^\dagger\mathbf{a}})\mathbf{A}(\phi_o(t))^{-1}(\mathbf{a}^\dagger + \mathbf{a})\}$$

Again, according to the invariance of the trace with respect to a circular permutation, we have

$$\langle \mathbf{Q}(t) \rangle = \varepsilon\sqrt{\frac{\hbar}{2M\omega^\circ}}\mathbf{tr}\{(e^{-\lambda\mathbf{a}^\dagger\mathbf{a}})\mathbf{A}(\phi_0(t))^{-1}(\mathbf{a}^\dagger + \mathbf{a})\mathbf{A}(\phi_0(t))\}$$

Then, theorem (7.9) allows us to transform this expression into

$$\langle \mathbf{Q}(t) \rangle = \varepsilon\sqrt{\frac{\hbar}{2M\omega^\circ}}\mathbf{tr}\{(e^{-\lambda\mathbf{a}^\dagger\mathbf{a}})(\mathbf{a}^\dagger + \phi_0(t) + \mathbf{a} + \phi_0^*(t))\}$$

Then, performing the trace over the eigenstates of $\mathbf{a}^\dagger\mathbf{a}$ gives

$$\langle \mathbf{Q}(t)\rangle = \varepsilon\sqrt{\frac{\hbar}{2M\omega^\circ}}\sum_n \langle\{n\}|(e^{-\lambda\mathbf{a}^\dagger\mathbf{a}})(\mathbf{a}^\dagger + \phi_0(t) + \mathbf{a} + \phi_0^*(t))|\{n\}\rangle \tag{16.153}$$

Moreover, keeping in mind Eq. (13.66), that is,

$$\langle\{n\}|(e^{-\lambda\mathbf{a}^\dagger\mathbf{a}})\{\mathbf{a}^\dagger + \mathbf{a}\}|\{n\}\rangle = 0$$

and using the orthonormality properties of the basis used for the trace, Eq. (16.153) becomes

$$\langle \mathbf{Q}(t)\rangle = \sqrt{\frac{\hbar}{2M\omega^\circ}}\varepsilon\sum_n \langle\{n\}|(e^{-\lambda\mathbf{a}^\dagger\mathbf{a}})|\{n\}\rangle(\phi_0(t) + \phi_0^*(t)) \tag{16.154}$$

Next observe that the trace of the Boltzmann density operator, which appears in this last equation, is just unity:

$$\varepsilon\,\mathbf{tr}\{e^{-\lambda\mathbf{a}^\dagger\mathbf{a}}\} = 1 = \varepsilon\sum_n \langle\{n\}|(e^{-\lambda\mathbf{a}^\dagger\mathbf{a}})|\{n\}\rangle$$

so that Eq. (16.154) simplifies to

$$\langle \mathbf{Q}(t)\rangle = \sqrt{\frac{\hbar}{2M\omega^\circ}}(\phi_0(t) + \phi_0^*(t))$$

Besides, in view of Eqs. (16.145) and (16.148), and after incorporating the shift $\Delta\omega$ into ω°, it becomes

$$\begin{aligned}\langle \mathbf{Q}(t)\rangle &= \alpha^\circ\sqrt{\frac{\hbar}{2M\omega^\circ}}\frac{1}{2(\omega^{\circ 2} + \gamma^2/4)} \\ &\times \{(2\omega^{\circ 2} + i\gamma\omega^\circ)\{e^{-\gamma t/2}e^{+i\omega^\circ t} - 1\} + (2\omega^{\circ 2} - i\gamma\omega^\circ)\{e^{-\gamma t/2}e^{-i\omega^\circ t} - 1\}\}\end{aligned}$$

so that

$$\begin{aligned}\langle \mathbf{Q}(t)\rangle &= \alpha^\circ\sqrt{\frac{\hbar}{2M\omega^\circ}}\left\{\left(\frac{2\omega^{\circ 2}}{\omega^{\circ 2} + \gamma^2/4}\right)(e^{-\gamma t/2}\cos\omega^\circ t - 1)\right. \\ &\left. - \left(\frac{\gamma\omega^\circ}{\omega^{\circ 2} + \gamma^2/4}\right)e^{-\gamma t/2}\sin\omega^\circ t\right\}\end{aligned} \tag{16.155}$$

We give in Fig. 16.2, the time evolution of the average position for the driven damped quantum harmonic oscillator. Note that in the very underdamped situation where $\omega^\circ \gg \gamma$, Eq. (16.155) reduces to

$$\langle \mathbf{Q}(t)\rangle = 2\alpha^\circ\sqrt{\frac{\hbar}{2M\omega^\circ}}(e^{-(\gamma t/2)}\cos\omega t - 1)$$

Furthermore, if at an initial time, we start from the density operator (16.146), the average value of the elongation reads

$$\langle \mathbf{Q}(t)\rangle = (1 - e^{-\lambda})\sqrt{\frac{\hbar}{2M\omega^\circ}}\mathbf{tr}\{(e^{-\lambda\mathbf{a}^\dagger\mathbf{a}})(\mathbf{a}^\dagger + \mathbf{a})\}$$

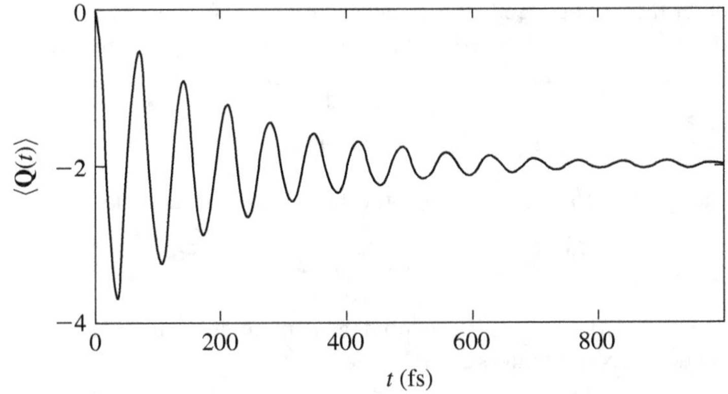

Figure 16.2 Time evolution of the average position for the driven damped quantum harmonic oscillator.

which is zero

$$\langle \mathbf{Q}(t=0) \rangle = 0$$

while at infinite time, according to Eq. (16.155), we have

$$\langle \mathbf{Q}(t=\infty) \rangle = -2\alpha^{\circ} \left(\frac{\omega^{\circ 2}}{\omega^{\circ 2} + \gamma^2/4} \right) \sqrt{\frac{\hbar}{2M\omega^{\circ}}}$$

16.5 LANGEVIN EQUATIONS FOR LADDER OPERATORS

16.5.1 Toward Mori's equation

Consider a harmonic oscillator with Hamiltonian $\mathbf{H}$ embedded in a thermal bath of Hamiltonian $\mathbb{H}_\theta$ to which it is coupled through the interaction Hamiltonian $\mathbf{V}$. The full Hamiltonian is

$$\mathbb{H} = \mathbf{H}^{\circ} + \mathbf{V} + \mathbb{H}_\theta$$

with $\mathbf{H}^{\circ}, \mathbf{V}$, and $\mathbb{H}_\theta$ given, respectively, by Eqs. (16.9), (16.11), and (16.13), that is,

$$\mathbf{H}^{\circ} = \mathbf{a}^{\dagger}\mathbf{a}\,\hbar\Omega \qquad \mathbb{H}_\theta = \sum_j \hbar\omega_j\, \mathbf{b}_j^{\dagger}\mathbf{b}_j \tag{16.156}$$

$$\mathbf{V} = \sum_j \{\hbar\kappa_j \mathbf{a}^{\dagger}\mathbf{b}_j + \hbar\kappa_j^{*}\mathbf{a}\mathbf{b}_j^{\dagger}\} \tag{16.157}$$

with

$$[\mathbf{a}, \mathbf{a}^{\dagger}] = 1 \qquad \text{and} \qquad [\mathbf{b}_j, \mathbf{b}_j^{\dagger}] = \delta_{ij} \tag{16.158}$$

The full Hamiltonian is, therefore,

$$\mathbb{H} = \hbar\Omega\mathbf{a}^{\dagger}\mathbf{a} + \sum_j \{\hbar\kappa_j\mathbf{a}^{\dagger}\mathbf{b}_j + \hbar\kappa_j^{*}\mathbf{a}\mathbf{b}_j^{\dagger}\} + \sum_j \hbar\omega_j\mathbf{b}_j^{\dagger}\mathbf{b}_j \quad \text{with} \quad k = 0, 1 \tag{16.159}$$

Now, consider the time-dependent Heisenberg picture of the Boson operator $\mathbf{a}(t)$ of the oscillator of interest embedded in the thermal bath. It is given by the Heisenberg equation (3.94), which reads in the present situation

$$i\hbar\left(\frac{\partial \mathbf{a}(t)}{\partial t}\right) = [\mathbf{a}, \mathbb{H}] \tag{16.160}$$

According to Eq. (16.159), the commutator involved in this equation is given by

$$[\mathbf{a}, \mathbb{H}] = \hbar\Omega[\mathbf{a}, \mathbf{a}\mathbf{a}^{\dagger}] + \sum_j \hbar\kappa_j[\mathbf{a}, \mathbf{a}^{\dagger}]\mathbf{b}_j$$

Next, due to Eq. (5.15), and keeping in mind the commutation rule of Boson operators, the right-hand-side commutators read

$$[\mathbf{a}, \mathbf{a}^{\dagger}\mathbf{a}] = \mathbf{a} \qquad [\mathbf{a}, \mathbf{a}^{\dagger}] = 1$$

so that Eq. (16.160) yields

$$\left(\frac{\partial \mathbf{a}(t)}{\partial t}\right) = -i\Omega\mathbf{a}(t) - i\sum_j \kappa_j \mathbf{b}_j(t) \tag{16.161}$$

Again, in order to obtain the right-hand-side unknowns $\mathbf{b}_j(t)$ appearing in Eq. (16.161), one has to solve the Heisenberg equation (3.94), which here takes on the form

$$i\hbar\left(\frac{\partial \mathbf{b}_j(t)}{\partial t}\right) = [\mathbf{b}_j(t), \mathbb{H}] \tag{16.162}$$

Now, owing to Eq. (16.159), the commutator involved in Eq. (16.162) is

$$[\mathbf{b}_j(t), \mathbb{H}] = \mathbf{a}\sum_l \{\hbar\kappa_l^*[\mathbf{b}_l(t), \mathbf{b}_l^{\dagger}] + \hbar\omega_l[\mathbf{b}_l(t), \mathbf{b}_l^{\dagger}\mathbf{b}_l]\}$$

while, due to (16.158), the right-hand-side commutators read

$$[\mathbf{b}_j(t), \mathbf{b}_j^{\dagger}] = 1 \qquad \text{and} \qquad \sum_j [\mathbf{b}_j(t), \mathbf{b}_j^{\dagger}\mathbf{b}_j] = \mathbf{b}_j(t)$$

Hence, Eq. (16.162) transforms to

$$\left(\frac{\partial \mathbf{b}_j(t)}{\partial t}\right) = -i(\omega_j\mathbf{b}_j(t) + \kappa_j^*\mathbf{a}(t)) \tag{16.163}$$

The solution of Eq. (16.163) without the second member $\kappa_j^*\mathbf{a}(t)$, that is,

$$\left(\frac{\partial \mathbf{b}_j(t)}{\partial t}\right) = -i(\omega_j\mathbf{b}_j(t))$$

is simply

$$\mathbf{b}^{\circ}{}_j(t) = \mathbf{b}^{\circ}{}_j(0)e^{-i\omega_j t}$$

Hence, the solution of the complete equation (16.163) reads

$$\mathbf{b}_j(t) = \mathbf{b}_j(0)e^{-i\omega_j t} - i\kappa_j^* e^{-i\omega_j t}\int_0^t \mathbf{a}(t')e^{i\omega_j t'}dt' \tag{16.164}$$

Furthermore, using this result, Eq. (16.161) takes the form

$$\left(\frac{\partial \mathbf{a}(t)}{\partial t}\right) = -i\Omega\mathbf{a}(t) - i\sum_j \mathbf{b}_j(0)\kappa_j e^{-i\omega_j t} - \sum_j |\kappa_j|^2 \int_0^t \mathbf{a}(t')e^{-i\omega_j(t-t')}dt'$$

Now, make the change of variable

$$\tilde{\mathbf{a}}(t) = \mathbf{a}(t)e^{i\Omega t} \tag{16.165}$$

Then, denoting $\mathbf{b}_j(0) = \mathbf{b}_j$, the above equation transforms to an equation that has the form of the Mori equation of statistical mechanics of irreversible processes:

$$\left(\frac{\partial \tilde{\mathbf{a}}(t)}{\partial t}\right) = -i\sum_j \mathbf{b}_j\kappa_j e^{-i(\omega_j-\Omega)t} - \sum_j |\kappa_j|^2 \int_0^t \tilde{\mathbf{a}}(t')e^{-i(\omega_j-\Omega)(t-t')}dt' \tag{16.166}$$

16.5.2 Toward the Langevin equation through the Markov approximation

Again, perform on the right-hand-side term of the Mori's equation (16.166) the Markov approximation according to which there is a loss of memory dealing with $\tilde{\mathbf{a}}(t')$ allowing to take in place of it $\tilde{\mathbf{a}}(t)$ and thereby to move it outside the integral. Then, after taking $\tau = t - t'$,

$$\int_0^t \tilde{\mathbf{a}}(t+\tau)e^{-i(\omega_j-\Omega)\tau}d\tau \simeq \tilde{\mathbf{a}}(t)\int_0^\infty e^{-i(\omega_j-\Omega)\tau}d\tau$$

Then, within the Markov approximation, the Mori equation (16.166) simplifies to

$$\left(\frac{\partial \tilde{\mathbf{a}}(t)}{\partial t}\right) = -i\sum_j \mathbf{b}_j\kappa_j e^{-i(\omega_j-\Omega)t} - \tilde{\mathbf{a}}(t)\sum_j |\kappa_j|^2 \int_0^\infty e^{-i(\omega_j-\Omega)\tau}d\tau$$

Now, assuming for the last sum a continuous variation of the thermal bath oscillators in the same way as in the above study dealing with the density operator of the driven damped harmonic oscillator [see Eq. (16.84)], that is,

$$\sum_j |\kappa_j|^2 f(\omega_j) \rightarrow \int_{-\infty}^{\infty} g(\omega)|\kappa(\omega)|^2 f(\omega)\,d\omega$$

where $f(\omega_j)$ is a function of ω_j and $g(\omega)$ is the density of modes, this last expression reads

$$\left(\frac{\partial \tilde{\mathbf{a}}(t)}{\partial t}\right) = -i\sum_j \mathbf{b}_j\kappa_j e^{-i(\omega_j-\Omega)t} - \tilde{\mathbf{a}}(t)\int_{-\infty}^{\infty} g(\omega)|\kappa(\omega)|^2 \left(\int_0^\infty e^{-i(\omega-\Omega)\tau}d\tau\right)d\omega \tag{16.167}$$

The last right-hand-side integral has the same form as that (18.63) of Section 18.6, which transforms to Eq. (18.71):

$$\int_{-\infty}^{\infty} f(\omega)\left(\int_{0}^{\infty} e^{-i(\omega-\Omega)\tau} d\tau\right) d\omega = -i\left\{\int_{-\infty}^{+\infty} f(\omega)\mathcal{P}\left(\frac{1}{\omega-\Omega}\right) d\omega\right\} d\omega$$

$$- \int_{-\infty}^{+\infty} f(\omega)\pi\delta(\omega-\Omega)\, d\omega$$

where $\mathcal{P}$ denotes the Cauchy principal part

$$\left(\frac{\partial\, \tilde{\mathbf{a}}(t)}{\partial t}\right) = -\,\tilde{\mathbf{a}}(t)\int_{-\infty}^{\infty} g(\omega)|\kappa(\omega)|^2 \left\{\pi\delta(\omega_j-\Omega) + iP\left(\frac{1}{\omega_j-\Omega}\right)\right\} d\omega$$

$$-\,i\sum_j \kappa_j \mathbf{b}_j e^{-i(\omega_j-\Omega)t}$$

This last equation simplifies to

$$\left(\frac{\partial\tilde{\mathbf{a}}(t)}{\partial t}\right) = -\tilde{\mathbf{a}}(t)\left(\frac{\gamma}{2} + i(\Omega+\Delta\Omega)\right) - i\sum_j \mathbf{b}_j\kappa_j e^{-i(\omega_j-\Omega)t}$$

with, respectively,

$$\gamma = 2\pi g(\omega)|\kappa(\omega)|^2 \qquad \text{and} \qquad \Delta\Omega = \int_{-\infty}^{\infty}\left(\frac{g(\omega)|\kappa(\omega)|^2}{(\omega-\Omega)}\right) d\omega \tag{16.168}$$

Now, one may incorporate into Ω the *Lamb shift* $\Delta\Omega$ via

$$\tilde{\Omega} = \Omega + \Delta\Omega \tag{16.169}$$

so that the above differential equation takes the simplified form

$$\left(\frac{\partial\tilde{\mathbf{a}}(t)}{\partial t}\right) = -\tilde{\mathbf{a}}(t)\left(\frac{\gamma}{2} + i\tilde{\Omega}\right) - i\sum_j \mathbf{b}_j\kappa_j e^{-i(\omega_j-\tilde{\Omega})t} \tag{16.170}$$

Then, returning to the initial IP Boson operator by the aid of Eq. (16.165), Eq. (16.170) becomes

$$\left(\frac{\partial\mathbf{a}(t)}{\partial t}\right) + \mathbf{a}(t)\left(\frac{\gamma}{2} + i\,\tilde{\Omega}\right) = -i\sum_j \mathbf{b}_j\kappa_j e^{-i\omega_j t} \tag{16.171}$$

This linear first-order differential equation, which is inhomogeneous since involving a second member, may be integrated in the following way by observing that the corresponding homogeneous equation (without second member) reads

$$\left(\frac{\partial\mathbf{a}^{\circ}(t)}{\partial t}\right) + \mathbf{a}^{\circ}(t)\left(\frac{\gamma}{2} + i\,\tilde{\Omega}\right) = 0 \tag{16.172}$$

which yields by integration

$$\mathbf{a}^{\circ}(t) = \mathbf{a}(0)e^{-(\gamma/2)t}e^{-i\tilde{\Omega}t}$$

Then, it is suitable to search for the solution of the complete inhomogeneous differential equation (16.171) an expression of the same form as that in (16.172) but in which the time-dependent function $\mathbf{u}(t)$ takes the place of $\mathbf{a}(0)$

$$\mathbf{a}(t) = \mathbf{u}(t)e^{-(\gamma/2)t}e^{-i\tilde{\Omega}t} \tag{16.173}$$

To find $\mathbf{u}(t)$ write the partial time derivative of Eq. (16.173)

$$\left(\frac{\partial \mathbf{a}(t)}{\partial t}\right) = \left(\frac{\partial \mathbf{u}(t)}{\partial t}\right) e^{-(\gamma/2)t}e^{-i\tilde{\Omega}t} - \mathbf{u}(t)\left(\frac{\gamma}{2} + i\tilde{\Omega}\right) e^{-(\gamma/2)t}e^{-i\tilde{\Omega}t}$$

so that it reads

$$\begin{aligned}\left(\frac{\partial \mathbf{a}(t)}{\partial t}\right) + \mathbf{a}(t)\left(\frac{\gamma}{2} + i\,\tilde{\Omega}\right) &= \left(\frac{\partial \mathbf{u}(t)}{\partial t}\right) e^{(-\gamma/2)t}e^{-i\tilde{\Omega}t} \\ &\quad - \mathbf{u}(t)\left(\frac{\gamma}{2} + i\tilde{\Omega}\right) e^{(-\gamma/2)t}e^{-i\tilde{\Omega}t} + \mathbf{u}(t)e^{(-\gamma/2)t}e^{-i\tilde{\Omega}t}\end{aligned}$$

and thus, after simplification,

$$\left(\frac{\partial \mathbf{a}(t)}{\partial t}\right) + \mathbf{a}(t)\left(\frac{\gamma}{2} + i\,\tilde{\Omega}\right) = \left(\frac{\partial \mathbf{u}(t)}{\partial t}\right) e^{(-\gamma/2)t}e^{-i\tilde{\Omega}t}$$

so that, by identification with Eq. (16.171), one obtains

$$\left(\frac{\partial \mathbf{u}(t)}{\partial t}\right) e^{-(\gamma/2)t}e^{-i\tilde{\Omega}t} = -i\sum_j \mathbf{b}_j\kappa_j e^{-i\omega_j t}$$

or

$$\left(\frac{\partial \mathbf{u}(t)}{\partial t}\right) = -i\sum_j \mathbf{b}_j\kappa_j e^{-i\omega_j t}e^{(\gamma/2)t}e^{i\tilde{\Omega}t}$$

The integration of this last equation reads

$$\mathbf{u}(t) = -i\sum_j \mathbf{b}_j\kappa_j e^{-(\gamma/2+i\tilde{\Omega})t}\int_o^t e^{(\gamma/2+i\tilde{\Omega})t'}e^{-i\omega_j t'}dt'$$

so that Eq. (16.173) becomes

$$\mathbf{a}(t) = \mathbf{a}(0)e^{(-\gamma/2)t}e^{-i\tilde{\Omega}t} - i\sum_j \mathbf{b}_j\kappa_j e^{-(\gamma/2+i\tilde{\Omega})t}\int_o^t e^{(\gamma/2+i\tilde{\Omega})t'}e^{-i\omega_j t'}dt' \tag{16.174}$$

or, after using Eq. (16.169),

$$\mathbf{a}(t) = \mathbf{a}(0)\Phi(t) + \sum_j \mathbf{b}_j\varphi_j(t) \tag{16.175}$$

with, respectively,

$$\Phi(t) = e^{-(\gamma/2)t}e^{-i(\Omega+\Delta\Omega)t} \tag{16.176}$$

$$\varphi_j(t) = -i\kappa_j\int_o^t e^{-(\gamma/2+i\tilde{\Omega})(t-t')}e^{-i\omega_j t'}dt' \tag{16.177}$$

Now, consider the average value of (16.175) over the Boltzmann density operator $\boldsymbol{\rho}_\theta$ of the thermal bath

$$\overline{\mathbf{a}(t)} = \mathbf{tr}_\theta\{\boldsymbol{\rho}_\theta \mathbf{a}(t)\} \tag{16.178}$$

with, as above, for Eqs. (16.14)–(16.16)

$$\boldsymbol{\rho}_\theta = \prod_k (1 - e^{-\lambda_k})(e^{-\lambda_k \mathbf{b}_k^\dagger \mathbf{b}_k}) \quad \text{with} \quad \lambda_k = \frac{\hbar\omega_k}{k_B T} \tag{16.179}$$

Owing to (16.179), the time-dependent average value (16.178) takes the form

$$\overline{\mathbf{a}(t)} = \mathbf{tr}_\theta \left\{ \prod_k (1 - e^{-\lambda_k})(e^{-\lambda_k \mathbf{b}_k^\dagger \mathbf{b}_k}) \left(\mathbf{a}(0)\Phi(t) + \sum_j \varphi_j(t)\mathbf{b}_j \right) \right\} \tag{16.180}$$

Again, after writing explicitly the trace over the thermal bath using Eq. (16.50) and because each kind of Boson operator acts in its specific space, Eq. (16.180) becomes

$$\overline{\mathbf{a}(t)} = \prod_k (1 - e^{-\lambda_k}) \sum_{n_k} \langle (n_k)|(e^{-\lambda_k \mathbf{b}_k^\dagger \mathbf{b}_k}) \left(\mathbf{a}(0)\Phi(t) + \sum_j \varphi_j(t)\mathbf{b}_j \right) |(n_k)\rangle$$

Moreover, using in turn this specific action of the Boson operators inside their own space, it transforms to

$$\begin{aligned} \overline{\mathbf{a}(t)} &= \prod_k (1 - e^{-\lambda_k}) \sum_{n_k} \langle (n_k)|(e^{-\lambda_k \mathbf{b}_k^\dagger \mathbf{b}_k})|(n_k)\rangle \mathbf{a}(0)\Phi(t) \\ &\quad + \sum_j \varphi_j(t) \sum_{n_j} (1 - e^{-\lambda_j}) \langle (n_j)| e^{-\lambda_j \mathbf{b}_j^\dagger \mathbf{b}_j} \mathbf{b}_j |(n_j)\rangle \\ &\quad \times \prod_{k \neq j} (1 - e^{-\lambda_k}) \sum_{n_k} \langle (n_k)|(e^{-\lambda_k \mathbf{b}_k^\dagger \mathbf{b}_k})|(n_k)\rangle \end{aligned} \tag{16.181}$$

Next, due to Eq. (16.52), that is,

$$(1 - e^{-\lambda_j}) \sum_{n_j} \langle (n_j)|(e^{-\lambda_j \mathbf{b}_j^\dagger \mathbf{b}_j})|(n_j)\rangle = 1$$

and, owing to Eq. (16.65), that is

$$(1 - e^{-\lambda_j}) \sum_{n_j} \langle (n_j)|\{(e^{-\lambda_j \mathbf{b}_j^\dagger \mathbf{b}_j})\mathbf{b}_j\}|(n_j)\rangle = 0$$

Eq. (16.181) simplifies to

$$\overline{\mathbf{a}(t)} = \mathbf{a}(0)\Phi(t) \tag{16.182}$$

or, in view of Eq. (16.176),

$$\boxed{\overline{\mathbf{a}(t)} = \mathbf{a}(0)e^{(-\gamma/2)t}e^{-i(\Omega+\Delta\Omega)t}} \tag{16.183}$$

the Hermitian conjugate of which is

$$\boxed{\overline{\mathbf{a}^\dagger(t)} = \mathbf{a}^\dagger(0)e^{(-\gamma/2)t}e^{i(\Omega+\Delta\Omega)t}} \tag{16.184}$$

Equations (16.183) and (16.184), which give the time dependence of the mean values of the Boson operators of the damped oscillator averaged over the thermal bath, are the quantum equivalents of the *Langevin equations* of classical statistical mechanics dealing with the irreversible behavior of harmonic oscillators. Keep in mind that the partial averaging over the thermal bath, which leads to operators, is denoted by an upper line, allowing one to distinguish it from the complete averaging denoted $\langle\ \rangle$, which leads to scalars. Moreover, it is impossible to admit of equations similar to Eqs. (16.183) and (16.184) but that deal with ladder operators in place of the average values, that is,

$$\mathbf{a}(t) = \mathbf{a}(0)e^{(-\gamma/2)t}e^{-i(\Omega+\Delta\Omega)t} \qquad \text{and} \qquad \mathbf{a}^{\dagger}(t) = \mathbf{a}^{\dagger}(0)e^{(-\gamma/2)t}e^{i(\Omega+\Delta\Omega)t} \tag{16.185}$$

This impossibility follows because if the equations (16.185) were true, they would imply the following commutator:

$$[\mathbf{a}(t), \mathbf{a}^{\dagger}(t)] = [\mathbf{a}(0), \mathbf{a}^{\dagger}(0)]e^{-\gamma t}$$

which is false since, at infinite time, it would lead erroneously to a zero value and therefore to commutativity of $\mathbf{a}$ and $\mathbf{a}^{\dagger}$ and thus to commutativity of the position and momentum operators from which $\mathbf{a}$ and $\mathbf{a}^{\dagger}$ are constructed using Eqs. (5.3) and (5.4).

16.6 EVOLUTION OPERATORS OF DRIVEN DAMPED OSCILLATORS

To end the present chapter devoted to the irreversible behavior of quantum oscillators, it may be of interest to find the IP time evolution operator of a driven damped quantum harmonic oscillator using Eqs. (16.183) and (16.184). This interaction picture operator plays a role in the quantum theory of the IR line shapes of weak H-bonded species. Besides, it will be seen that this time evolution operator allows one to get the expression (16.146) of the density operator of the driven damped harmonic oscillator obtained by Louisell and Walker, which has been given without providing their very complicated proof.

16.6.1 Time derivative equation to be solved within normal ordering formalism

For this purpose, consider as above the Hamiltonian $\mathbb{H}$ of a driven harmonic oscillator of Hamiltonian $\mathbf{H}$ embedded within a thermal bath defined by the Hamiltonian $\mathbb{H}_\theta$ to which it is coupled through the interacting Hamiltonian $\mathbf{V}$. It is given by

$$\mathbb{H} = \mathbf{H} + \mathbf{V} + \mathbb{H}_\theta$$

with

$$\mathbf{H} = \left(\frac{\mathbf{P}^2}{2M} + \frac{1}{2}M\Omega^2\mathbf{Q}^2\right) + b\mathbf{Q} \tag{16.186}$$

and where $\mathbf{V}$ and $\mathbb{H}_\theta$ are, respectively, given by Eqs. (16.4) and (16.5). In the area of the theory of the IR line shape of H-bonded species, the following expression has to be found:

$$\hat{\mathbf{P}}\{\exp[-ib\int_0^t \overline{\mathbf{Q}(t')}dt']\} \tag{16.187}$$

Here, $\hat{\mathbf{P}}$ is the Dyson time-ordering operator, $\mathbf{Q}(t)$ is the time-dependent coordinate in the Heisenberg picture governed by Eq. (3.94), that is,

$$\left(\frac{\partial \mathbf{Q}(t)}{\partial t}\right) = \frac{1}{i\hbar}[\mathbf{Q}(t), \mathbb{H}] \tag{16.188}$$

and $\overline{\mathbf{Q}(t)}$ is the Langevin averaged value of $\mathbf{Q}(t)$ when damped by the medium, which may be formally given by

$$\overline{\mathbf{Q}(t)} = \mathbf{tr}_\theta\{\boldsymbol{\rho}_\theta \mathbf{Q}(t)\}$$

In this last expression, $\boldsymbol{\rho}_\theta$ is, as above, the Boltzmann density operator of the thermal bath defined by (16.179). Of course, the time-dependent coordinate $\mathbf{Q}(t)$ is given by integration of the basic commutator equation (16.188),

Note that owing to Eqs. (3.86) and (3.87), Eq. (16.187) is the formal solution

$$\mathbf{U}(t) = \hat{\mathbf{P}}\left\{\exp\left[-ib\int_0^t \overline{\mathbf{Q}(t')}dt'\right]\right\} \tag{16.189}$$

of the dynamical equation

$$i\left(\frac{\partial \mathbf{U}(t)}{\partial t}\right) = b\overline{\mathbf{Q}(t)}\,\mathbf{U}(t) \tag{16.190}$$

with

$$\mathbf{U}(0) = 1 \tag{16.191}$$

In order to find the explicit expression of the evolution operator given by the formal Eq. (16.189), we pass to lowering and raising operators $\mathbf{a}$ and $\mathbf{a}^\dagger$ by the aid of Eqs. (5.6) and (5.7). Then, the coefficient b of Hamiltonian (16.186) transforms to the dimensionless coefficient α°:

$$\alpha^\circ = \frac{b}{\Omega}\sqrt{\frac{\hbar}{2M\Omega}}$$

Then, within the Boson operator representation, Eq. (16.189) becomes

$$\mathbf{U}(t, \mathbf{a}^\dagger, \mathbf{a}) = \hat{\mathbf{P}}\left\{\exp\left[-i\alpha^\circ\Omega\int_0^t \overline{\mathbf{a}^\dagger(t') + \mathbf{a}(t')}dt'\right]\right\} \tag{16.192}$$

in which $\mathbf{a}(t)$ may be obtained using the Heisenberg equation (3.94) reading, in the present situation,

$$\left(\frac{\partial \mathbf{a}(t)}{\partial t}\right) = \frac{1}{i\hbar}[\mathbf{a}(t), \mathbb{H}] \tag{16.193}$$

whereas $\mathbf{a}^\dagger(t)$ is given by the corresponding Hermitian conjugate. Next, the Langevin averaged values of the ladder operators have to be performed over the Boltzmann density operator $\boldsymbol{\rho}_\theta$ of the thermal bath using the partial trace

$$\overline{\mathbf{a}^\dagger(t) + \mathbf{a}(t)} = \mathbf{tr}_\theta\{\boldsymbol{\rho}_\theta(\mathbf{a}^\dagger(t) + \mathbf{a}(t))\} \tag{16.194}$$

Equation (16.192) is the formal solution of the Schrödinger equation:

$$i\left(\frac{\partial \mathbf{U}(t,\mathbf{a}^\dagger,\mathbf{a})}{\partial t}\right) = \alpha°\Omega\{\overline{\mathbf{a}^\dagger(t)} + \overline{\mathbf{a}(t)}\}\mathbf{U}(t,\mathbf{a}^\dagger,\mathbf{a}) \tag{16.195}$$

involving, of course, the same boundary condition as (16.191).

Now, observe that the average values involved on the right-hand side of Eq. (16.195) and obeying the Heisenberg equation (16.193) are given by the Langevin equations (16.183) and (16.184), which, after incorporating the Lamb shift into the angular frequency, reads

$$\overline{\mathbf{a}(t)} = \mathbf{a}(0)\, e^{(-\gamma/2)t}e^{-i(\Omega t} \qquad \text{and} \qquad \overline{\mathbf{a}^\dagger(t)} = \mathbf{a}^\dagger(0)e^{(-\gamma/2)t}e^{i(\Omega t} \tag{16.196}$$

As a consequence, the Schrödinger equation (16.195) becomes

$$i\left(\frac{\partial \mathbf{U}(t,\mathbf{a}^\dagger,\mathbf{a})}{\partial t}\right) = \alpha°\Omega(\mathbf{a}^\dagger\Phi(t)^* + \mathbf{a}\Phi(t))\mathbf{U}(t,\mathbf{a}^\dagger,\mathbf{a}) \tag{16.197}$$

where

$$\Phi(t) = e^{(-\gamma/2)t}e^{-i\Omega t} \tag{16.198}$$

and with the boundary condition (16.191).

16.6.2 Solution of Eq. (16.197)

To solve this differential equation, it is suitable to use the normal ordering technique, comparing Eqs. (7.20)–(7.122), according to which one has, respectively,

$$\widehat{\mathbf{N}}^{-1}\left\{\left(\frac{\partial \mathbf{U}(t,\mathbf{a}^\dagger,\mathbf{a})}{\partial t}\right)\right\} = \frac{\partial U^{\{n\}}(t,\alpha^*,\alpha)}{\partial t}$$

$$\widehat{\mathbf{N}}^{-1}\{\mathbf{a}^\dagger\mathbf{U}(t,\mathbf{a}^\dagger,\mathbf{a})\} = \alpha^*\{U^{\{n\}}(t,\alpha^*,\alpha)\}$$

$$\widehat{\mathbf{N}}^{-1}\{\mathbf{a}\,\mathbf{U}(t,\mathbf{a}^\dagger,\mathbf{a})\} = \left(\alpha + \frac{\partial}{\partial\alpha^*}\right)\{U^{\{n\}}(t,\alpha^*,\alpha)\} \tag{16.199}$$

where $\widehat{\mathbf{N}}$ is the normal ordering operator and $\widehat{\mathbf{N}}^{-1}$ its inverse. Then, Eq. (16.197) involving noncommutating operators transforms to the following partial differential equation:

$$\left(\frac{\partial U^{\{n\}}(t,\alpha^*,\alpha)}{\partial t}\right) = -i\alpha°\Omega\left\{\left(\alpha + \frac{\partial}{\partial\alpha^*}\right)\Phi(t) + \alpha^*\Phi(t)^*\right\}\{U^{\{n\}}(t,\alpha^*,\alpha)\} \tag{16.200}$$

with, due to Eq. (16.191), the boundary condition

$$\{U^{\{n\}}(0,\alpha^*,\alpha)\} = 1 \tag{16.201}$$

Now, make the change of variable

$$\{U^{\{n\}}(t,\alpha^*,\alpha)\} = (e^{G(t)}) \tag{16.202}$$

with, due to Eq. (16.201), the boundary condition

$$G(0) = 0 \tag{16.203}$$

and assume for $G(t)$ an expression of the form

$$G(t) = A_0(t) + A_1(t)\alpha + A_2(t)\alpha^* \tag{16.204}$$

where $A_0(t)$, $A_1(t)$, and $A_2(t)$ are unknown time-dependent coefficients that, owing to Eqs. (16.202) and (16.203), must satisfy the following boundary conditions:

$$A_0(0) = A_1(0) = A_2(0) = 0 \tag{16.205}$$

so that Eq. (16.202) becomes

$$\{U^{\{n\}}(t, \alpha^*, \alpha)\} = (e^{A_0(t)+A_1(t)\alpha+A_2(t)\alpha^*}) \tag{16.206}$$

Then, the partial time derivative of expression (16.202) reads

$$\left(\frac{\partial U^{\{n\}}(t, \alpha^*, \alpha)}{\partial t}\right) = \left(\frac{\partial G(t)}{\partial t}\right)(e^{G(t)}) = \left(\frac{\partial G(t)}{\partial t}\right)\{U^{\{n\}}(t, \alpha^*, \alpha)\}$$

while, due to Eq. (16.202), that of Eq. (16.206) yields

$$\left(\frac{\partial U^{\{n\}}(t, \alpha^*, \alpha)}{\partial t}\right) = \left\{\left(\frac{\partial A_0(t)}{\partial t}\right) + \alpha\left(\frac{\partial A_1(t)}{\partial t}\right) + \alpha^*\left(\frac{\partial A_2(t)}{\partial t}\right)\right\}\{U^{\{n\}}(t, \alpha^*, \alpha)\} \tag{16.207}$$

Then, by identification of Eqs. (16.200) and (16.207), one obtains, respectively,

$$\left(\frac{\partial A_0(t)}{\partial t}\right) = -i\alpha^\circ\Omega\Phi(t)A_2(t) \tag{16.208}$$

$$\left(\frac{\partial A_1(t)}{\partial t}\right) = -i\alpha^\circ\Omega\Phi(t) \qquad \left(\frac{\partial A_2(t)}{\partial t}\right) = -i\alpha^\circ\Omega\Phi(t)^* \tag{16.209}$$

so that the time-dependent coefficients involved in Eq. (16.204) appear to be interrelated via the following equations through

$$A_1(t) = -A_2^*(t) \equiv A(t) \tag{16.210}$$

16.6.3 Time evolution operator

As a consequence, owing to Eqs. (16.204) and (16.210), Eq. (16.202) takes the form

$$\{U^{\{n\}}(t, \alpha^*, \alpha)\} = e^{A_0(t)}(e^{A(t)\alpha - A^*(t)\alpha^*}) \tag{16.211}$$

In order to obtain the expressions of $A_0(t)$ and $A(t)$ one has to solve the two coupled equations (16.208) and (16.209), with $\Phi(t)$ given by Eq. (16.198), that is,

$$\left(\frac{\partial A_0(t)}{\partial t}\right) = i\alpha^\circ\Omega e^{(-\gamma/2)t}e^{-i\Omega t}A(t)^* \tag{16.212}$$

$$\left(\frac{\partial A(t)}{\partial t}\right) = -i\alpha^\circ\Omega e^{-\gamma t/2}e^{-i\Omega t} \tag{16.213}$$

subject to the boundary conditions (16.205), that is,

$$A(0) = A_0(0) = 0$$

After integration of Eq. (16.213), and owing to its specific boundary conditions, $A(t)$ is given by

$$A(t) = \beta^\circ(e^{-\gamma t/2}e^{-i\Omega t} - 1) \tag{16.214}$$

with

$$\beta^\circ = \alpha^\circ\Omega\left(\frac{\Omega + i\gamma/2}{\Omega^2 + (\gamma/2)^2}\right) \tag{16.215}$$

Again, inserting this result into Eq. (16.212) gives

$$A_0(t) = i(\alpha^\circ\Omega)^2\left(\frac{\Omega + i\gamma/2}{\Omega^2 + (\gamma/2)^2}\right)\int_0^t (e^{-\gamma t'/2}e^{-i\Omega t'})(e^{-\gamma t'/2}e^{i\Omega t'} - 1)dt'$$

or, with the ad hoc boundary condition $A(0) = 0$

$$A_0(t) = i(\alpha^\circ\Omega)^2\left(\frac{\Omega - i(\gamma/2)}{\Omega^2 + (\gamma/2)^2}\right)\left(\int_0^t e^{-\gamma t'}dt' - \int_0^t e^{-i(\Omega - i\gamma/2)t'}dt'\right) \tag{16.216}$$

and thus, after integration, we have

$$A_0(t) = |\beta^\circ|^2\left(-\frac{i}{\gamma}(e^{-\gamma t} - 1)\Omega - ie^{(-\gamma/2)t}\sin\Omega t - \frac{1}{2}e^{-\gamma t} + e^{-\gamma t/2}\cos\Omega t - \frac{1}{2}\right) \tag{16.217}$$

with

$$|\beta^\circ|^2 = \left(\frac{(\alpha^\circ\Omega)^2}{\Omega^2 + (\gamma/2)^2}\right) \tag{16.218}$$

Next, observe that, by action of the normal ordering operator $\widehat{\mathbf{N}}$, on Eq. (16.211), one obtains through Eq. (7.44), the time evolution operator

$$\widehat{\mathbf{N}}\{U^{\{n\}}(t, \alpha^*, \alpha)\} = \{\mathbf{U}(t, \mathbf{a}^\dagger, \mathbf{a})\}$$

with, due to Eq. (16.211),

$$\{\mathbf{U}(t, \mathbf{a}^\dagger, \mathbf{a})\} = (e^{A_0(t)})(e^{-A(t)^*\mathbf{a}^\dagger})(e^{A(t)\mathbf{a}}) \tag{16.219}$$

where $A_0(t)$ and $A(t)$ are, respectively, given by Eqs. (16.217) and (16.214). Then, using the Glauber–Weyl theorem (1.78), we can transform the product of exponential operators appearing in Eq. (16.219) into

$$(e^{-A(t)^*\mathbf{a}^\dagger})(e^{A(t)\mathbf{a}}) = \{e^{A(t)\mathbf{a} - A(t)^*\mathbf{a}^\dagger}\}[e^{-|A(t)|^2[\mathbf{a}^\dagger, \mathbf{a}]/2}]$$

so that Eq. (16.219) transforms to

$$\{\mathbf{U}(t, \mathbf{a}^\dagger, \mathbf{a})\} = e^{\xi(t)}\{e^{A(t)\mathbf{a} - A(t)^*\mathbf{a}^\dagger}\} = e^{\xi(t)}\{e^{-A(t)^*\mathbf{a}^\dagger + A(t)\mathbf{a}}\} \tag{16.220}$$

with

$$e^{\xi(t)} = (e^{A_0(t)})[e^{-|A(t)|^2[\mathbf{a}^\dagger, \mathbf{a}]/2}] = (e^{A_0(t) + |A(t)|^2/2}) \tag{16.221}$$

or, due to Eqs. (16.214) and (16.216),

$$\xi(t) = i|\beta^\circ|^2 \left(\frac{1}{\gamma}(e^{-\gamma t} - 1)\Omega + (e^{-\gamma t/2}) \sin \Omega t \right) \tag{16.222}$$

Hence, due to Eq. (16.189), it appears that Eq. (16.220) may be written in the form

$$\boxed{\begin{aligned}&\{\mathbf{U}(t, \mathbf{a}^\dagger, \mathbf{a})\}\\ &\quad = \hat{\mathbf{P}}\left\{\exp\left[-ib\int_0^t \langle \mathbf{Q}(t')\rangle\, dt'\right]\right\} = e^{\xi(t)}\{e^{A(t)\mathbf{a} - A(t)^*\mathbf{a}^\dagger}\} = e^{\xi(t)}\{e^{-A(t)^*\mathbf{a}^\dagger + A(t)\mathbf{a}}\}\end{aligned}} \tag{16.223}$$

where the last equality has been written mindful of the fact that in the exponential the position of $\mathbf{a}^\dagger$ and $\mathbf{a}$ is irrelevant. Observe that, due to Eq. (16.214),

$$A(t) = \beta^\circ(e^{-\gamma t/2}e^{-i\Omega t} - 1)$$

the evolution operator reads

$$\{\mathbf{U}(t, \mathbf{a}^\dagger, \mathbf{a})\} = e^{\xi(t)} \exp\{\beta^\circ(e^{-\gamma t/2}e^{-i\Omega t} - 1)\mathbf{a} - \beta^\circ(e^{-\gamma t/2}e^{i\Omega t} - 1)\mathbf{a}^\dagger\} \tag{16.224}$$

or, in view of Eqs. (16.148) and (16.150),

$$\{\mathbf{U}(t, \mathbf{a}^\dagger, \mathbf{a})\} = e^{\xi(t)}\{\mathbf{A}(-\phi_0(t))\} = e^{\xi(t)}\{\mathbf{A}(\phi_0(t))\}^{-1} \tag{16.225}$$

where $\{\mathbf{A}(\phi_0(t))\}$ is the damped translation operator (16.147) extracted from the Louisell and Walker density operator (16.150) through the canonical transformation (16.149). Of course, $\mathbf{U}(t, \mathbf{a}^\dagger, \mathbf{a})$ and $\mathbf{A}(\phi_0(t))$ being unitary, the inverse of $\mathbf{U}(t, \mathbf{a}^\dagger, \mathbf{a})$ is

$$\{\mathbf{U}(t, \mathbf{a}^\dagger, \mathbf{a})\}^{-1} = e^{-\xi(t)}\{\mathbf{A}(\phi_0(t))\} \tag{16.226}$$

16.6.4 Calculation of density operator

Now, make the following canonical transformation over the Boltzmann density operator by aid of the time evolution operator expressed by Eq. (16.220) leading to the density operator $\boldsymbol{\rho}(t)$ according to

$$\boldsymbol{\rho}(t) = (1 - e^{-\lambda})\{\mathbf{U}(t, \mathbf{a}^\dagger, \mathbf{a})\}(e^{-\lambda \mathbf{a}^\dagger \mathbf{a}})\{\mathbf{U}(t, \mathbf{a}^\dagger, \mathbf{a})\}^{-1} \tag{16.227}$$

where λ is the T temperature function

$$\lambda = \frac{\hbar\Omega}{k_B T}$$

Equation (16.227) transforms with the help of Eqs. (16.225) and (16.226) into

$$\boldsymbol{\rho}(t) = (1 - e^{-\lambda})e^{-\xi(t)}\{\mathbf{A}(\phi_0(t))\}^{-1}(e^{-\lambda \mathbf{a}^\dagger \mathbf{a}})\{\mathbf{A}(\phi_0(t))\}e^{\xi(t)}$$

or, on simplification,

$$\boldsymbol{\rho}(t) = (1 - e^{-\lambda})\{\mathbf{A}(\phi_0(t))\}^{-1}(e^{-\lambda \mathbf{a}^\dagger \mathbf{a}})\{\mathbf{A}(\phi_0(t))\}$$

Hence, owing to Eq. (16.147), this operator becomes

$$\boldsymbol{\rho}(t) = (1 - e^{-\lambda})(e^{-\lambda(\mathbf{a}^\dagger + \phi_0(t))(\mathbf{a} + \phi_0^*(t))})$$

which has the same structure as the density operator (16.151) of driven damped harmonic oscillators as obtained by Louisell and Walker. Observe that due to Eq. (16.223)

$$\boxed{\{\mathbf{U}(t,\mathbf{a}^\dagger,\mathbf{a})\} = \hat{\mathbf{P}}\left\{\exp\left[-ib\int_0^t \overline{\mathbf{Q}(t')}dt'\right]\right\}}$$

the expression (16.227) for this driven damped density operator, reads in terms of **P** and **Q**,

$$\rho(t) = (1-e^{-\lambda})\hat{\mathbf{P}}\left\{\exp\left[-ib\int_0^t \overline{\mathbf{Q}(t')}dt'\right]\right\}\left(\exp\left[-\left(\frac{\mathbf{P}^2}{2M}+\frac{1}{2}M\Omega^2\mathbf{Q}^2\right)\Big/k_BT\right]\right)$$
$$\times\,\hat{\mathbf{P}}\left\{\exp\left[ib\int_0^t \overline{\mathbf{Q}(t')}dt'\right]\right\}$$

where $\langle\mathbf{Q}(t)\rangle$ is the average Langevin value of the damped coordinate $\mathbf{Q}(t)$, that is,

$$\overline{\mathbf{Q}(t)} = \sqrt{\frac{\hbar}{2M\Omega}}\{\mathbf{a}^\dagger(0)e^{i\Omega t} + \mathbf{a}(0)e^{-i\Omega t}\}e^{-\gamma t/2}$$

16.7 CONCLUSION

This chapter, devoted to the irreversible behavior of damped oscillators, took into account the influence of the neighborhood by considering it as a thermal bath composed of a very large set of harmonic oscillators linearly coupled to an oscillator embedded in this bath. First, consider second order in the coupling with the bath. An iterative process for the integral equation governing the dynamics of the density operator was set up, yielding a master equation that is an operator equation giving the time derivative of the density operator of the damped oscillator in terms of a complicated combination of products of the ladder operators of this oscillator. We thus derived from this master equation, the corresponding Fokker–Planck equation, which is a time derivative of the scalar corresponding to the antinormal form of the damped oscillator density operator. Such an equation involving only scalars can be integrated (which was not done) in a similar way as the Franck–Condon equation encountered in the classical statistical mechanics of irreversible processes. Moreover, the Langevin equations governing the average values of the damped ladder operators have been found. Using these Langevin equations, we obtained the IP time evolution operator of a damped harmonic oscillator. Finally, using this time evolution operator, we could find for the time-dependent density operator of damped harmonic oscillators, a closed expression equivalent to that obtained by Louisell and Walker, and which may be viewed as the solution of the master equation. These important results are summarised as follows:

Equations govering the dynamics of driven damped harmonic oscillators
Master equation:
$\left(\frac{\partial \boldsymbol{\rho}_{\mathrm{Dr}}(t)^{\mathrm{SP}}}{\partial t}\right) = -i\alpha^{\circ}\omega^{\circ}\{[\mathbf{a}, \boldsymbol{\rho}_{\mathrm{Dr}}(t)^{\mathrm{SP}}] + [\mathbf{a}^{\dagger}, \boldsymbol{\rho}_{\mathrm{Dr}}(t)^{\mathrm{SP}}]\}$ $- i\omega^{\circ}[\mathbf{a}^{\dagger}\mathbf{a}, \boldsymbol{\rho}_{\mathrm{Dr}}(t)^{\mathrm{SP}}] - i\,\Delta\omega[\mathbf{a}^{\dagger}\mathbf{a}, \boldsymbol{\rho}_{\mathrm{Dr}}(t)^{\mathrm{SP}}]$ $-\frac{\gamma}{2}(\mathbf{a}^{\dagger}\mathbf{a}\boldsymbol{\rho}_{\mathrm{Dr}}(t)^{\mathrm{SP}} + \boldsymbol{\rho}_{\mathrm{Dr}}(t)^{\mathrm{SP}}\mathbf{a}^{\dagger}\mathbf{a} - 2\mathbf{a}\boldsymbol{\rho}_{\mathrm{Dr}}(t)^{\mathrm{SP}}\mathbf{a}^{\dagger})$ $- \langle\mathbf{n}\rangle\gamma(\mathbf{a}^{\dagger}\mathbf{a}\boldsymbol{\rho}_{\mathrm{Dr}}(t)^{\mathrm{SP}} + \boldsymbol{\rho}_{\mathrm{Dr}}(t)^{\mathrm{SP}}\mathbf{a}\mathbf{a}^{\dagger} - \mathbf{a}^{\dagger}\boldsymbol{\rho}_{\mathrm{Dr}}(t)^{\mathrm{SP}}\mathbf{a} - \mathbf{a}\boldsymbol{\rho}_{\mathrm{Dr}}(t)^{\mathrm{SP}}\mathbf{a}^{\dagger})$
Langevin equations:
$\overline{\mathbf{a}(t)} = \mathbf{a}(0)e^{(-\gamma/2)t}e^{-i(\Omega+\Delta\Omega)t}$ and $\overline{\mathbf{a}^{\dagger}(t)} = \mathbf{a}^{\dagger}(0)e^{(-\gamma/2)t}e^{i(\Omega+\Delta\Omega)t}$
Fokker–Planck equation:
$\left(\frac{\partial \rho^{a}(t)}{\partial t}\right) = -i\alpha^{\circ}\omega^{\circ}\left\{\left(\frac{\partial \rho^{a}(t)}{\partial \alpha^{*}}\right) - \left(\frac{\partial \rho^{a}(t)}{\partial \alpha}\right)\right\}$ $- i(\omega^{\circ} + \Delta\omega)\left\{\alpha^{*}\left(\frac{\partial \rho^{a}(t)}{\partial \alpha^{*}}\right) - \alpha\left(\frac{\partial \rho^{a}(t)}{\partial \alpha}\right)\right\}$ $+\frac{\gamma}{2}\left\{\left(\frac{\partial \rho^{a}(t)\alpha^{*}}{\partial \alpha^{*}}\right) + \left(\frac{\partial \rho^{a}(t)\alpha}{\partial \alpha}\right)\right\} + \langle\mathbf{n}\rangle\gamma\left(\frac{\partial^{2} \rho^{a}(t)}{\partial\alpha\,\partial\alpha^{*}}\right)$
Density operator:
$\rho(t) = (1 - e^{-\lambda})(e^{-\lambda(\mathbf{a}^{\dagger}+\phi_0(t))(\mathbf{a}+\phi_0^{*}(t))})$ with $\lambda = \frac{\hbar\omega^{\circ}}{k_B T}$ and $\phi_0(t) = \beta\{e^{-\gamma t/2}e^{+i(\omega^{\circ}+\Delta\omega)t} - 1\}$

BIBLIOGRAPHY

C. Cohen-Tannoudji, J. Dupont-Roc, and G. Grinberg. *Atom Photon Interactions. Basic Processes and Applications.* Wiley: New York, 1992.

H. Louisell. *Quantum Statistical Properties of Radiations.* Wiley: New York, 1973.

W. Louisell and L. Walker. *Phys. Rev.*, **137** (1965): 204.

PART VII

VIBRATIONAL SPECTROSCOPY

CHAPTER 17

APPLICATIONS TO OSCILLATOR SPECTROSCOPY

INTRODUCTION

All the properties of quantum oscillators obtained in Part III (Chapter 10), Part IV (Chapter 13), and Part VI (Chapter 16) are used in this last part (Chapter 17) to find some important results in vibrational spectroscopy, such as the IR selection rule for quantum harmonic oscillators, Fermi resonances, the tunneling effect through double-well potentials, and to study, in the linear response theory, the line shape of some physically realistic situations involving anharmonically coupled damped quantum harmonic oscillators met in H bonding such as Fermi resonances or Davydov coupling.

17.1 IR SELECTION RULES FOR MOLECULAR OSCILLATORS

17.1.1 Induced absorption and emission

In the last section of Chapter 4, we studied the Fermi golden rule, giving the probability of transition between two energy levels due to an energetic perturbation. When a molecule has a dipole moment, this dipole operator may interact with an electromagnetic field to induce transitions between two of its energy levels, characterizing it in the absence of a field. In spectroscopy, the electromagnetic field obeys the Maxwell equations. It was found in Chapter 15 that when the electric field is described by a coherent state, its quantum behavior is very nearly that of a classical one where the different modes of the field oscillate at angular frequency ω. Then, the perturbation generated by the interaction between the dipole of the molecule and the electric field is time dependent. Hence, in order to find the transition probability between the energy levels induced by the electromagnetic field, we have to reconsider the analysis of the Fermi golden rule by introducing in it the time dependence of the perturbation. That is the object of the present section.

Consider a molecular system having a dipole moment operator $\widehat{\mu}$ and described by a Hamiltonian $\mathbf{H}^{\circ}$ that interacts with an electromagnetic field $\mathbf{E}(\omega, t)$ of angular frequency ω through

$$\mathbf{V}(t) = \widehat{\mu}\mathbf{E}(\omega, t) \quad \text{with} \quad \mathbf{E}(\omega, t) = \mathbf{E}(\omega)(e^{-i\omega t} + e^{i\omega t}) \tag{17.1}$$

Quantum Oscillators, First Edition. Olivier Henri-Rousseau and Paul Blaise.
© 2011 John Wiley & Sons, Inc. Published 2011 by John Wiley & Sons, Inc.

where $\mathbf{E}(\omega)$ would be given by Eqs. (14.139) and (14.141). The full Hamiltonian $\mathbf{H}$ is the sum of the Hamiltonian $\mathbf{H}^{\circ}$ of the molecular system and of theHamiltonian $\mathbf{V}(t)$ coupling the molecular system to the electromagnetic field $\mathbf{E}(\omega, t)$ via the dipole moment operator $\widehat{\boldsymbol{\mu}}$:

$$\mathbf{H} = \mathbf{H}^{\circ} + \mathbf{V}(t)$$

Now, write the eigenvalue equation of $\mathbf{H}^{\circ}$:

$$\mathbf{H}^{\circ}|\Psi_k\rangle = E_k|\Psi_k\rangle \tag{17.2}$$

with

$$\langle\Psi_k|\Psi_l\rangle = \delta_{kl} \tag{17.3}$$

and seek the transition probability at time t for the system described by $\mathbf{H}$ to pass from any eigenstate of $\mathbf{H}^{\circ}$ to another because of the presence of $\mathbf{V}(t)$, that is,

$$|C(l,t|k,0)|^2 = |\langle\Psi_k(0)|\Psi_l(t)\rangle|^2$$

In the interaction picture with respect to $\mathbf{H}^{\circ}$ and according to Eq. (4.93), the transition probability yields

$$|C(l,t|k,0)|^2 = \left(\frac{1}{\hbar}\right)^2 \left|\int_0^t \langle\Psi_k(0)|(e^{i\mathbf{H}^{\circ}t'/\hbar})\,\mathbf{V}(t)\,(e^{-i\mathbf{H}^{\circ}t'/\hbar})|\Psi_l(0)\rangle dt'\right|^2$$

so that due to Eqs. (17.1) and (17.2)

$$|C(l,t|k,0)|^2 = \left(\frac{\mathbf{E}(\omega)}{\hbar}\right)^2 \left|\int_0^t \langle\Psi_k(0)|(e^{iE_kt'/\hbar})\,\widehat{\boldsymbol{\mu}}(e^{-i\omega t'} + e^{i\omega t'})\,(e^{-iE_lt'/\hbar})|\Psi_l(0)\rangle dt'\right|^2$$

and, after rearranging

$$|C(l,t|k,0)|^2 = \left(\frac{\mathbf{E}(\omega)}{\hbar}\right)^2 |\langle\Psi_k|\boldsymbol{\mu}|\Psi_l\rangle|^2 \left|\int_0^t (e^{i(\omega_{kl}-\omega)t'} + e^{i(\omega_{kl}+\omega)t'})\,dt'\right|^2 \tag{17.4}$$

with

$$\omega_{kl} = \frac{(E_k - E_l)}{\hbar}$$

Next, the integration of the time-dependent term of Eq. (17.4) reads

$$\int_0^t (e^{i(\omega_{kl}-\omega)t'} + e^{i(\omega_{kl}+\omega)t'})\,dt' = \frac{1}{i}\left(\frac{e^{i(\omega_{kl}-\omega)t'} - 1}{\omega_{kl} - \omega} + \frac{e^{i(\omega_{kl}+\omega)t'} - 1}{\omega_{kl} + \omega}\right) \tag{17.5}$$

Moreover, observe that if $\omega \simeq \omega_{kl}$, the first term of the right-hand side of Eq. (17.5) becomes very large with respect to the second one, so that the latter may be neglected, that is,

$$\int_0^t (e^{i(\omega_{kl}-\omega)t'} + e^{i(\omega_{kl}+\omega)t'})\,dt' \simeq \frac{1}{i}\left(\frac{e^{i(\omega_{kl}-\omega)t'} - 1}{\omega_{kl} - \omega}\right)$$

Hence, due to this approximation, Eq. (17.4) reads

$$|C(l,t|k,0)|^2 = \left(\frac{\mathbf{E}(\omega)}{\hbar}\right)^2 |\langle\Psi_k|\widehat{\boldsymbol{\mu}}|\Psi_l\rangle|^2 \left|\frac{e^{i(\omega_{kl}-\omega)t'}-1}{\omega_{kl}-\omega}\right|^2 \tag{17.6}$$

Furthermore, passing to the sine function leads for the time-dependent term of (17.6)

$$\left|\frac{e^{i(\omega_{kl}-\omega)t'}-1}{\omega_{kl}-\omega}\right|^2 = \frac{2}{(\omega_{kl}-\omega)^2}\{1-\cos((\omega_{kl}-\omega)t)\}$$

so that Eq. (17.6) transforms to

$$|C(l,t|k,0)|^2 = 4(\mathbf{E}(\omega))^2|\langle\Psi_k|\widehat{\boldsymbol{\mu}}|\Psi_l\rangle|^2\left(\frac{\sin^2((\omega_{kl}-\omega)t/2)}{\hbar^2(\omega_{kl}-\omega)^2}\right)$$

Then, by a procedure similar to that used in passing from Eq. (4.97) to Eq. (4.102), we have

$$|C(l,t|k,0)|^2 = \frac{2\pi}{\hbar}(\mathbf{E}(\omega))^2|\langle\Psi_k|\widehat{\boldsymbol{\mu}}|\Psi_l\rangle|^2\{\delta(E_k-E_l-\hbar\omega)\}t \tag{17.7}$$

Hence, the transition probability by unit time defined by

$$W(l,t|k,0) = \left(\frac{\partial|C(l,t|k,0)|^2}{\partial t}\right)$$

is

$$\boxed{W(l,t|k,0) = \frac{2\pi}{\hbar}(\mathbf{E}(\omega))^2|\langle\Psi_k|\widehat{\boldsymbol{\mu}}|\Psi_l\rangle|^2\delta(E_k-E_l-\hbar\omega)} \tag{17.8}$$

This result shows that the transition from the eigenstate $|\Psi_l\rangle$ of the molecular system Hamiltonian $\mathbf{H}^\circ$, to another one $|\Psi_k\rangle$, due to the presence of an electrical field, may occur only if the two following conditions are simultaneously verified:

$$\omega = \frac{|E_k - E_l|}{\hbar}$$

$$\langle\Psi_k|\widehat{\boldsymbol{\mu}}|\Psi_l\rangle \neq 0 \tag{17.9}$$

that is, if the angular frequency ω of the radiation satisfies the Bohr condition dealing with the energy levels $|\Psi_k\rangle$ and $|\Psi_k\rangle$, and if the corresponding off-diagonal matrix elements of the dipole moment operator are different from zero. That may be summarized in

$$\boxed{W(l,t|k,0) \neq 0 \quad \text{and} \quad \langle\Psi_k|\widehat{\boldsymbol{\mu}}|\Psi_l\rangle \neq 0}$$

17.1.2 Selection rules for harmonic oscillators

The dipole moment operator $\widehat{\boldsymbol{\mu}}$, which depends on the coordinate $\mathbf{Q}$, may be expanded, up to first order with respect to the equilibrium position to give

$$\widehat{\boldsymbol{\mu}} = \widehat{\boldsymbol{\mu}}(\mathbf{Q}) = \mu(0) + \left(\frac{\partial\widehat{\boldsymbol{\mu}}}{\partial\mathbf{Q}}\right)_{\mathbf{Q}=0}\mathbf{Q}$$

where $\mu(0)$ is the dipole moment operator at $\mathbf{Q}=0$, which is a scalar from the viewpoint of the quantum oscillator theory. Then, due to the orthogonality relation (17.3) between the kth and the lth eigenstates leads one to write

$$\langle \Psi_k | \mu(0) | \Psi_l \rangle = \mu(0) \langle \Psi_k | \Psi_l \rangle = 0$$

the off-diagonal matrix elements of the dipole moment operator appearing in Eq. (17.8) and corresponding necessarily to $k \neq l$, according to (17.9), read

$$\langle \Psi_k | \widehat{\boldsymbol{\mu}} | \Psi_l \rangle = \left(\frac{\partial \widehat{\boldsymbol{\mu}}}{\partial \mathbf{Q}} \right)_{\mathbf{Q}=0} \langle \Psi_k | \mathbf{Q} | \Psi_l \rangle \tag{17.10}$$

Next, when dealing with the vibrations of molecules, the partial derivative in front of the right-hand side of Eq. (17.10), which depends on the electronic part of the molecular system, has to be averaged over the electronic states $|\Phi_{\text{El}}\rangle$, and thus may be viewed as the scalar. Then, Eq. (17.10) reads

$$\langle \Psi_k | \widehat{\boldsymbol{\mu}} | \Psi_l \rangle = \langle \Phi_{\text{El}} | \left(\frac{\partial \widehat{\boldsymbol{\mu}}}{\partial \mathbf{Q}} \right)_{\mathbf{Q}=0} | \Phi_{\text{El}} \rangle \langle \Psi_k | \mathbf{Q} | \Psi_l \rangle \tag{17.11}$$

so that owing to Eq. (17.11), Eq. (17.8) yields

$$W(l, t | k, 0) = \frac{2\pi}{\hbar} \left| \langle \Phi_{\text{El}} | \left(\frac{\partial \widehat{\boldsymbol{\mu}}}{\partial \mathbf{Q}} \right)_{\mathbf{Q}=0} | \Phi_{\text{El}} \rangle \right|^2 (\mathbf{E}(\omega))^2 | \langle \Psi_k | \mathbf{Q} | \Psi_l \rangle |^2 \delta(E_k - E_l - \hbar\omega) \tag{17.12}$$

or

$$\boxed{W(l, t | k, 0) = \frac{2\pi}{\hbar} (\mu^\circ)^2 (\mathbf{E}(\omega))^2 | \langle \Psi_k | \mathbf{Q} | \Psi_l \rangle |^2 \delta(E_k - E_l - \hbar\omega)} \tag{17.13}$$

with

$$\mu^\circ \Leftrightarrow \langle \Phi_{\text{El}} | \left(\frac{\partial \widehat{\boldsymbol{\mu}}}{\partial \mathbf{Q}} \right)_{\mathbf{Q}=0} | \Phi_{\text{El}} \rangle$$

Apply now the result (17.13) to the special case of harmonic oscillators, the Hamiltonian of which is

$$\mathbf{H} = \hbar\omega^\circ \left(\mathbf{a}^\dagger \mathbf{a} + \tfrac{1}{2} \right)$$

Then, $|\Psi_k\rangle$ and $|\Psi_l\rangle$ are the eigenkets of this Hamiltonian, that is,

$$|\Psi_k\rangle = |\{k\}\rangle$$

while the energy levels E_k and E_l are the corresponding eigenvalues verifying

$$\mathbf{H} |\{k\}\rangle = E_k |\{k\}\rangle$$

with

$$E_k = \hbar\omega^\circ \left(k + \tfrac{1}{2} \right) \tag{17.14}$$

and

$$\langle \{k\} | \{l\} \rangle = \delta_{kl} \tag{17.15}$$

where ω° is the angular frequency of the harmonic oscillator. Then, the off-diagonal matrix elements involved in Eq. (17.13) read

$$\langle \Psi_k | \mathbf{Q} | \Psi_l \rangle = \langle \{k\} | \mathbf{Q} | \{l\} \rangle \tag{17.16}$$

Moreover, passing to Boson operators using Eq. (5.6), the off-diagonal matrix elements yield

$$\langle \Psi_k | \mathbf{Q} | \Psi_l \rangle = \sqrt{\frac{\hbar}{2M\omega^\circ}} \langle \{k\} | (\mathbf{a}^\dagger + \mathbf{a}) | \{l\} \rangle \tag{17.17}$$

Now, due to Eqs. (5.53) and (5.63), that is,

$$\mathbf{a}^\dagger | \{k\} \rangle = \sqrt{k+1} | \{k+1\} \rangle \qquad \text{and} \qquad \mathbf{a} | \{k\} \rangle = \sqrt{k} | \{k-1\} \rangle \tag{17.18}$$

the right-hand side of Eq. (17.17) becomes

$$\langle \Psi_k | (\mathbf{a}^\dagger + \mathbf{a}) | \Psi_l \rangle = \langle \{k\} | (\sqrt{l+1} | \{l+1\} \rangle + \sqrt{l} | \{l-1\} \rangle)$$

or, in view of the orthonormality properties of the eigenkets of the harmonic Hamiltonian,

$$\langle \Psi_k | (\mathbf{a}^\dagger + \mathbf{a}) | \Psi_l \rangle = (\sqrt{l+1}\delta_{k,l+1} + \sqrt{l}\delta_{k,l-1}) \tag{17.19}$$

so that, after using Eqs. (17.16) and (17.19), Eq. (17.13) reads

$$W(l,t|k,0) \propto \left| \langle \Phi_{\mathrm{El}} | \left(\frac{\partial \widehat{\mu}}{\partial \mathbf{Q}} \right)_{\mathbf{Q}=0} | \Phi_{\mathrm{El}} \rangle \right|^2 \{\delta_{k,l\pm 1}\} \{\delta((k-l)\omega^\circ - \omega)\}$$

Here it appears that a transition induced by the electromagnetic field occurs if the three following conditions are verified:

1. If the energy $\hbar\omega$ of the photon is equal to the difference in energy between E_k and E_l, that is, satisfies the *Bohr condition*

$$\omega = (k-l)\omega^\circ \tag{17.20}$$

2. If the change in the quantum numbers characterizing the energy levels of the harmonic oscillators obey

$$k = l \pm 1 \tag{17.21}$$

3. If the electronic matrix elements over the electronic state $|\Phi_{\mathrm{El}}\rangle$ of the partial derivative of the dipole moment of the diatomic molecule with respect to $\mathbf{Q}$ are different from zero

$$\langle \Phi_{\mathrm{El}} | \left(\frac{\partial \widehat{\mu}}{\partial \mathbf{Q}} \right)_{\mathbf{Q}=0} | \Phi_{\mathrm{El}} \rangle \neq 0 \tag{17.22}$$

Note that homonuclear diatomic molecules cannot have, because of symmetry, any dipole moment so that the partial derivative with respect to $\mathbf{Q}$ involved in Eq. (17.22) must be zero, leading therefore to the possibility of this kind of molecule absorbing or emitting IR radiations in the framework of the mechanism studied here.

Of course, the combined conditions (17.20) and (17.21) lead to the resonance condition

$$\omega = \omega^{\circ} \tag{17.23}$$

Recall that the angular vibrational frequency of diatomic molecules is roughly given by Eq. (9.31), which is in the range of IR electromagnetic radiations, so that Eq. (17.23) implies that IR radiations may be absorbed due to changes in the vibrational energy levels. Now, since Eq. (17.22) must be verified, the dipole moment operator $\widehat{\mu}$ cannot disappear, which is impossible for homonuclear diatomic molecules, which, being symmetric, cannot have a dipole moment operator. Hence, only heteronuclear diatomic molecules may change in their vibrational energy levels due to interaction with IR radiations.

Owing to the selection rule (17.23), it is clear that at zero temperature since the Boltzmann population of the harmonic oscillator energy levels is zero except for the ground state, the only absorption transition is that corresponding to a jump from the ground state $|\{0\}\rangle$ on the first excited state $|\{1\}\rangle$. Of course, the corresponding emission results from the jump from the excited state $|\{1\}\rangle$ to the ground state $|\{0\}\rangle$. Figure 17.1 illustrates this with the concomitant transitions of the electromagnetic mode compatible with the energy conservation known as the Bohr condition. On the other hand, at any finite temperature the different energy levels have some probability to be occupied, which is given by the Boltzmann distribution. Hence, different transitions corresponding to the following changes will take place:

$$|\{0\}\rangle \to |\{1\}\rangle \qquad |\{1\}\rangle \to |\{2\}\rangle \qquad |\{2\}\rangle \to |\{3\}\rangle \qquad \text{and so on}$$

However, since the angular frequency of these different transitions is the same in the harmonic approximation, it follows that at a temperature, the spectrum must form a single line of angular frequency ω.

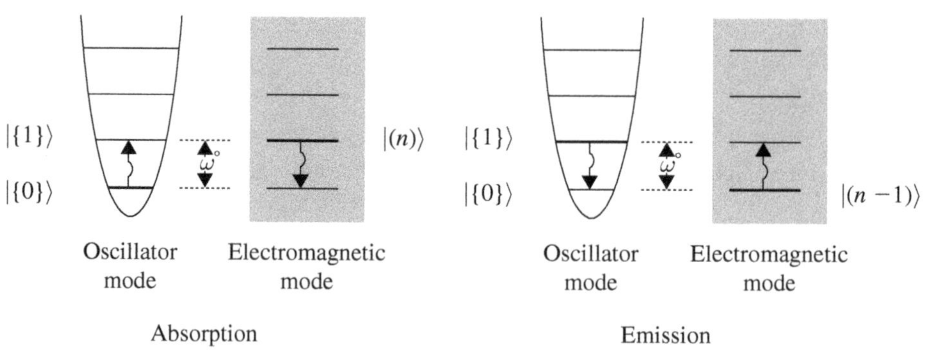

Figure 17.1 Absorption or emission by a quantum harmonic oscillator mode resulting from a resonant coupling with an electromagnetic mode of the same angular frequency ω°.

17.1.3 "Forbidden" transitions of weak intensity and anharmonicity

The IR transition selection rules [(17.20)–(17.23)], are very simple because they are dealing with harmonic oscillators. However, realistic molecular oscillators are generally anharmonic, their potentials being Morse-like. Hence, we shall examine what changes in the selection rules will be induced by the anharmonicity of the Morse-like potentials.

When the eigenvalue equation of the Hamiltonian **H** of Morse oscillators has been solved, it has been found that, in the basis of eigenkets of the harmonic Hamiltonian, the expansion of the lowest eigenstates of the Hamiltonian are given by (9.88), according to which the three lowest eigenstates of the Morse Hamiltonian may be crudely given by

$$|\Psi_0\rangle \simeq \sqrt{1-\varepsilon^2}|\{0\}\rangle + \varepsilon|\{1\}\rangle \tag{17.24}$$

$$|\Psi_{\mathrm{I}}\rangle = \xi|\{0\}\rangle + \sqrt{1-\xi^2-\eta^2}|\{1\}\rangle + \eta|\{2\}\rangle \tag{17.25}$$

$$|\Psi_{\mathrm{II}}\rangle = \zeta|\{1\}\rangle + \sqrt{1-\zeta^2}|\{2\}\rangle \tag{17.26}$$

with

$$\varepsilon << 1 \qquad \zeta << 1 \qquad \eta << 1 \qquad \xi << 1 \tag{17.27}$$

Furthermore, it has been found that the corresponding eigenvalues of the Morse Hamiltonian are weakly lowered with respect to those of the harmonic Hamiltonian, their difference being approximately given by

$$E_{\mathrm{I}} - E_0 \simeq \hbar\omega^\circ \qquad \text{and} \qquad E_{\mathrm{II}} - E_{\mathrm{I}} \simeq 2\hbar\omega^\circ \tag{17.28}$$

Equations (17.24)–(17.27) imply that the ground state $|\Psi_0\rangle$ of **H** works essentially as the ground state $|\{0\}\rangle$ of the harmonic Hamiltonian $\mathbf{H}^\circ$, whereas the first excited state $|\Psi_{\mathrm{I}}\rangle$ is essentially the first excited state $|\{1\}\rangle$ of $\mathbf{H}^\circ$, and that the second excited state $|\Psi_{\mathrm{II}}\rangle$ of **H** is centered on the second excited state $|\{2\}\rangle$ of $\mathbf{H}^\circ$.

Now, the off-diagonal matrix elements of the dipole moment operator between the ground state $|\Psi_0\rangle$ and the excited states $|\Psi_{\mathrm{I}}\rangle$ and $|\Psi_{\mathrm{II}}\rangle$ are, respectively,

$$\langle\Psi_{\mathrm{I}}|\mathbf{Q}|\Psi_0\rangle \qquad \langle\Psi_{\mathrm{II}}|\mathbf{Q}|\Psi_0\rangle \qquad \langle\Psi_{\mathrm{II}}|\mathbf{Q}|\Psi_{\mathrm{I}}\rangle \tag{17.29}$$

Passing to Boson operators by the aid of Eq. (5.6), the first matrix elements read

$$\langle\Psi_{\mathrm{I}}|\mathbf{Q}|\Psi_0\rangle = \sqrt{\frac{\hbar}{2m\omega^\circ}}\langle\Psi_{\mathrm{I}}|(\mathbf{a}^\dagger + \mathbf{a})|\Psi_0\rangle$$

or, using Eqs. (17.24) and (17.25),

$$\langle\Psi_{\mathrm{I}}|\mathbf{Q}|\Psi_0\rangle = \sqrt{\frac{\hbar}{2m\omega^\circ}}(\xi\langle\{0\}| + \sqrt{1-\xi^2-\eta^2}\langle\{1\}| + \eta\langle\{2\}|)$$
$$\times (\mathbf{a}^\dagger + \mathbf{a})(\sqrt{1-\varepsilon^2}|\{0\}\rangle + \varepsilon|\{1\}\rangle)$$

which, owing to the smallness of the coefficients ε, ξ, and η, the above equation may be simplified into

$$\langle\Psi_{\rm I}|\mathbf{Q}|\Psi_0\rangle = \sqrt{\frac{\hbar}{2m\omega^\circ}}\sqrt{(1-\xi^2-\eta^2)(1-\varepsilon^2)}\langle\{1\}|(\mathbf{a}^\dagger+\mathbf{a})|\{0\}\rangle$$

or, after using Eq. (17.18),

$$\langle\Psi_{\rm I}|\mathbf{Q}|\Psi_0\rangle = \sqrt{\frac{\hbar}{2m\omega^\circ}}\sqrt{(1-\xi^2-\eta^2)(1-\varepsilon^2)} \neq 0 \tag{17.30}$$

and, thus, due to the inequalities appearing in (17.27):

$$\boxed{\langle\Psi_{\rm I}|\mathbf{Q}|\Psi_0\rangle \backsimeq \sqrt{\frac{\hbar}{2m\omega^\circ}}} \tag{17.31}$$

On the other hand, because of Eqs. (17.24) and (17.26), the second matrix element appearing in (17.29) reads

$$\langle\Psi_{\rm II}|\mathbf{Q}|\Psi_0\rangle = \sqrt{\frac{\hbar}{2m\omega^\circ}}(\zeta\langle\{1\}| + \sqrt{1-\zeta^2}\langle\{2\}|)(\mathbf{a}^\dagger+\mathbf{a})(\sqrt{1-\varepsilon^2}|\{0\}\rangle + \varepsilon|\{1\}\rangle)$$

which, due to Eq. (17.18), gives

$$\langle\Psi_{\rm II}|\mathbf{Q}|\Psi_0\rangle = \sqrt{\frac{\hbar}{2m\omega^\circ}}\{\zeta\langle\{1\}| + (\sqrt{1-\zeta^2}\langle\{2\}|)(\sqrt{1-\varepsilon^2}|\{1\}\rangle + \varepsilon|\{0\}\rangle)\}$$

or after using the orthonormality properties of the basis $\{|\{n\}\rangle\}$ gives

$$\boxed{\langle\Psi_{\rm II}|\mathbf{Q}|\Psi_0\rangle = \sqrt{\frac{\hbar}{2m\omega^\circ}}\zeta\sqrt{1-\varepsilon^2} \neq 0} \tag{17.32}$$

Finally, the third matrix element appearing in (17.29) reads

$$\langle\Psi_{\rm II}|\mathbf{Q}|\Psi_{\rm I}\rangle = \sqrt{\frac{\hbar}{2m\omega^\circ}}(\zeta\langle\{1\}| + \sqrt{1-\zeta^2}\langle\{2\}|)(\mathbf{a}^\dagger+\mathbf{a})(\xi|\{0\}\rangle + \sqrt{1-\xi^2-\eta^2}|\{1\}\rangle + \eta|\{2\}\rangle)$$

or, after using Eq. (17.18), yields

$$\langle\Psi_{\rm II}|\mathbf{Q}|\Psi_{\rm I}\rangle = \sqrt{\frac{\hbar}{2m\omega^\circ}}(\sqrt{1-\zeta^2}\sqrt{1-\xi^2-\eta^2}\sqrt{2} + \zeta(\eta\sqrt{2}+\xi))$$

which, owing to the inequalities (17.27), transforms into

$$\boxed{\langle\Psi_{\rm II}|\mathbf{Q}|\Psi_{\rm I}\rangle \backsimeq \sqrt{2}\sqrt{\frac{\hbar}{2m\omega^\circ}}} \tag{17.33}$$

Thus, as a consequence of Eqs. (17.31) and (17.32), it appears, as shown by Fig. 17.2, that there are two transition probabilities of the kind of (17.13), which are starting from the ground state $|\Psi_0\rangle$, that is,

$$W(\Psi_{\rm I}, t|\Psi_0, 0) = \frac{2\pi}{\hbar}(\mu^\circ)^2(\mathbf{E}(\omega))^2|\langle\Psi_{\rm I}|\mathbf{Q}|\Psi_0\rangle|^2\delta(E_{\rm I}-E_0-\hbar\omega) \tag{17.34}$$

$$W(\Psi_{\rm II}, t|\Psi_0, 0) = \frac{2\pi}{\hbar}(\mu^\circ)^2(\mathbf{E}(\omega))^2|\langle\Psi_{\rm II}|\mathbf{Q}|\Psi_0\rangle|^2\delta(E_{\rm II}-E_0-\hbar\omega) \tag{17.35}$$

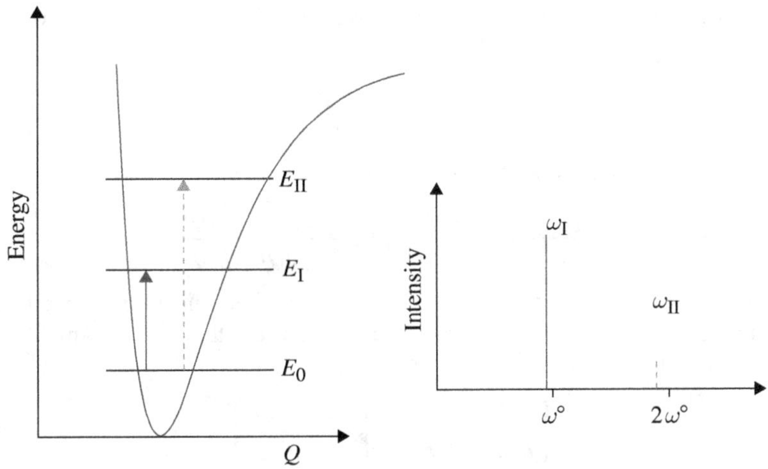

Figure 17.2 IR transitions in a Morse oscillator.

They correspond to the absorption of photons of angular frequencies ω_I and ω_{II}, which, according to Eq. (17.28), are given by

$$\omega_I = \frac{E_I - E_0}{\hbar} \backsimeq \omega^\circ \qquad \text{and} \qquad \omega_{II} = \frac{E_{II} - E_0}{\hbar} \backsimeq 2\omega^\circ$$

Thus, the spectrum resulting from Eqs. (17.34) and (17.35), which is shown on the right-hand side of Fig. 17.2, must be formed by two Dirac deltalike peaks corresponding to these two angular frequencies, the ratio of their intensities $I(\omega_I)$ and $I(\omega_{II})$ being given with the help of Eqs. (17.31) and (17.32) by

$$\frac{I(\omega_{II})}{I(\omega_I)} = \frac{W(\Psi_{II}, t|\Psi_0, 0)}{W(\Psi_I, t|\Psi_0, 0)} = \zeta^2(1 - \varepsilon) \backsimeq \zeta^2$$

where inequalities (17.27) involving the smallness of ξ and η have been used for the last result.

Hence, owing to the second inequality appearing in (17.27), that is, to the smallness of ζ, it appears that

$$I(\omega_{II}) << I(\omega_I)$$

Now, what we have obtained is also true at zero temperature where, according to the Boltzmann distribution, only the ground state is occupied. However, when the temperature is raised, all the probabilities P_n take finite values so that other transitions will occur. When the temperature is relatively low with respect to the gap $(E_I - E_0)/k_B$, according to the Boltzmann distribution, all the energy levels are practically unoccupied except the ground and the first excited states corresponding to the energies E_0 and E_I. As a consequence of a probability for the first excited energy level E_I to be occupied, the interaction of the molecule with the electromagnetic field will induce not only the transition from the ground state E_0 to the first excited energy level E_I but also from this first excited state E_I to the second excited energy level E_{II}, the angular frequencies corresponding to these last transitions being, respectively, given by ω_{I} and by

$$\omega_{\text{I,II}} = \frac{E_{\text{II}} - E_{\text{I}}}{\hbar}$$

The ratio of the corresponding transition intensities are

$$\left\{\frac{I(\omega_{\text{I,II}})}{I(\omega_{\text{I}})}\right\} = \left(\frac{P(E_{\text{I}})}{P(E_0)}\right)\left(\frac{W(\Psi_{\text{II}},t|\Psi_{\text{I}},0)}{W(\Psi_{\text{I}},t|\Psi_0,0)}\right) \tag{17.36}$$

where the $P(E_{\text{I}})$ and $P(E_0)$ are the Boltzmann thermal equilibrium probabilities for the oscillator to be either in the energy level E_{I} or in E_0, the ratio of which is given by Eq. (12.83), that is,

$$\left\{\frac{P(E_I)}{P(E_0)}\right\} = \exp\left\{-\frac{E_I - E_0}{k_B T}\right\} \tag{17.37}$$

while $W(\Psi_{\text{I}},t|\Psi_0,0)$ is given by Eq. (17.34) and $W(\Psi_{\text{II}},t|\Psi_{\text{I}},0)$ by an equation similar to those of (17.34) and (17.35) which reads

$$W(\Psi_{\text{II}},t|\Psi_{\text{I}},0) = \frac{2\pi}{\hbar}(\mu^\circ)^2(\mathbf{E}(\omega))^2|\langle\Psi_{\text{II}}|\mathbf{Q}|\Psi_{\text{I}}\rangle|^2\delta(E_{\text{II}} - E_{\text{I}} - \hbar\omega) \tag{17.38}$$

with, in view of Eq. (17.33),

$$|\langle\Psi_{\text{II}}|\mathbf{Q}|\Psi_{\text{I}}\rangle|^2 \backsimeq \frac{\hbar}{m\omega^\circ} \tag{17.39}$$

Hence, owing to Eqs. (17.31), (17.34), (17.37), (17.38), and (17.39), the ratio (17.36) yields

$$\left\{\frac{I(\omega_{\text{I,II}})}{I(\omega_{\text{I}})}\right\} \backsimeq 2\exp\left\{-\frac{E_{\text{I}} - E_0}{k_B T}\right\}$$

a result that shows that the intensity of the transition corresponding to $\omega_{\text{I,II}}$, which does not occur at zero temperature, is raised with temperature, leading therefore to being called a *hot band* (Fig. 17.3).

17.1.4 Isotope effect

Consider some heteronuclear diatomic molecules of the kind X – H, where $^1_1$H is an hydrogen atom and X some atom susceptible to be bonded to it, and also to its corresponding deuterated isotopic form X – D, where D is the deuterium $^2_1$H, an isotope of the hydrogen. The force constants $k_{\text{X-H}}$ and $k_{\text{X-D}}$ of the X – H and X – D isotopomers, which are of the form (9.8), respectively, given by

$$k_{\text{X-H}} = \left(\frac{\partial^2 U_{\text{X-H}}}{\partial Q^2}\right) \qquad \text{and} \qquad k_{\text{X-D}} = \left(\frac{\partial^2 U_{\text{X-D}}}{\partial Q^2}\right)$$

where $U_{\text{X-H}}$ and $U_{\text{X-D}}$ are the Morse potentials of these molecules. Quantum considerations concerning these molecules using the concept of adiabatic separation lead us to conclude that the potential in which the atoms of these molecules

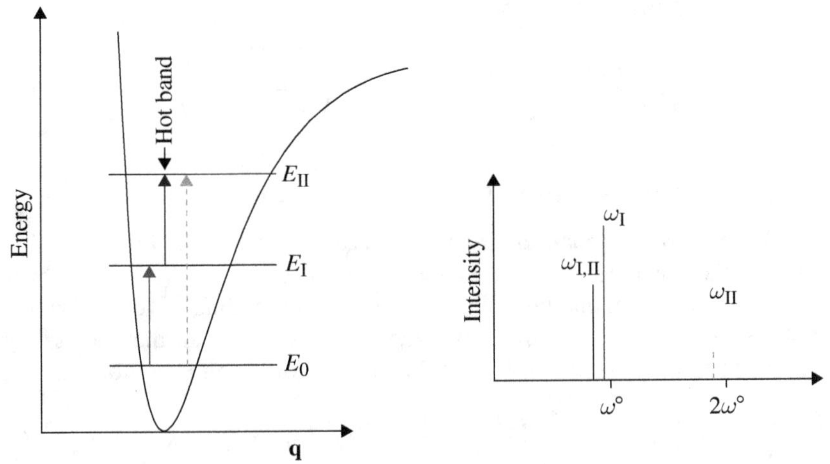

Figure 17.3 Appearance of a hot band in the IR spectrum of a Morse oscillator. (See color insert.)

oscillate must be the same for both X – H and X – D isotopomers so that the force constants of the two isotopomers are equal

$$k_{X-H} = k_{X-D} \tag{17.40}$$

Now, the only difference between these two isotopomers is their reduced masses, which, according to Eq. (5.2), read, respectively, for the X – H and X – D isotopomers

$$M_{X-H} = \left(\frac{m_X m_H}{m_X + m_H}\right) \quad \text{and} \quad M_{X-D} = \left(\frac{m_X m_D}{m_X + m_D}\right)$$

Moreover, since the masses of the hydrogen and of the deuterium are related through

$$m_D = 2m_H$$

the reduced mass of the X – D isotopomer takes the form

$$M_{X-D} = \left(\frac{2m_X m_H}{m_X + 2m_H}\right)$$

so that the ratio of the reduced masses of the X – H and X – D isotopomers becomes

$$\left(\frac{M_{X-H}}{M_{X-D}}\right) = \frac{1}{2}\left(\frac{m_X + 2m_H}{m_X + m_H}\right) \tag{17.41}$$

Next, due to Eq. (9.28), the angular frequencies of the two isotopomers are given by

$$\omega_{X-H} = \sqrt{\frac{k_{X-H}}{M_{X-H}}} \quad \text{and} \quad \omega_{X-D} = \sqrt{\frac{k_{X-D}}{M_{X-D}}}$$

so that their ratio appears to be

$$\left(\frac{\omega_{X-H}}{\omega_{X-D}}\right) = \sqrt{\frac{k_{X-H}}{M_{X-H}}}\sqrt{\frac{M_{X-D}}{k_{X-D}}}$$

or, due to Eqs. (17.40) and (17.41),

$$\left(\frac{\omega_{X-H}}{\omega_{X-D}}\right) = \sqrt{\frac{M_{X-D}}{M_{X-H}}} = \sqrt{2}\sqrt{\left(\frac{m_X + m_H}{m_X + 2m_H}\right)} \tag{17.42}$$

17.1.5 Line shape splitting by Fermi resonances

In the study of Fermi resonances, it has been seen that when the first excited state of some molecular normal mode is energetically close to the second excited state of another normal mode, some mixing occurs leading to two new vibrational states, which are delocalized over the two modes. Recall that, according to Eqs. (10.1)–(10.3), this system was described by the following Hamiltonian:

$$\mathbf{H} = \left(\frac{\mathbf{p}^2}{2m} + \frac{1}{2}m\omega^{\circ 2}\mathbf{q}^2\right) + \left(\frac{\mathbf{P}^2}{2M} + \frac{1}{2}M\,\omega_\delta^2\mathbf{Q}^2\right) + \lambda\mathbf{q}\mathbf{Q}^2 \quad \text{with} \quad \omega^\circ \simeq 2\Omega$$

where λ is a constant reflecting the strength of the anharmonic coupling. Then, passing to the corresponding Boson operators and in the context of *the exchange approximation*, this Hamiltonian reads

$$\mathbf{H} = \hbar\omega^\circ(\mathbf{a}^\dagger\mathbf{a} + \tfrac{1}{2}) + \hbar\omega_\delta(\mathbf{b}^\dagger\mathbf{b} + \tfrac{1}{2}) + \xi\hbar\omega_\delta(\mathbf{a}^\dagger(\mathbf{b})^2 + \mathbf{a}(\mathbf{b}^\dagger)^2) \tag{17.43}$$

with

$$\xi = \lambda\frac{\hbar}{2m\omega^\circ}\sqrt{\frac{\hbar}{2M\omega_\delta}}$$

Next, since $\omega^\circ \simeq 2\omega_\delta$, the first excited state $|\{1\}\rangle$ of $\mathbf{a}^\dagger\mathbf{a}$ is near the second excited state $|(2)\rangle$ of $\mathbf{b}^\dagger\mathbf{b}$, the operator proportional to ξ induces a mixing between the tensor product state built up from the first excited state $|\{1\}\rangle$ times the ground state $|(0)\rangle$ of $\mathbf{b}^\dagger\mathbf{b}$, and the tensorial state built up from the ground state $|\{0\}\rangle$ of $\mathbf{a}^\dagger\mathbf{a}$ times the second excited state $|(2)\rangle$ of $\mathbf{b}^\dagger\mathbf{b}$. Thus, in these two interacting energetic states approach, the matrix representation of the Hamiltonian is given by Eq. (10.12), that is,

$\boxed{\mathbf{H}}$ = (17.44)

	$\lvert\{1\}\rangle\lvert[0]\rangle$	$\lvert\{0\}\rangle\lvert[2]\rangle$
$\langle\{1\}\lvert\langle[0]\rvert$	$\hbar\omega^\circ + \left(\frac{1}{2}\hbar\omega^\circ + \frac{1}{2}\hbar\omega_\delta\right)$	$\sqrt{2}\xi\hbar\omega_\delta$
$\langle\{0\}\lvert\langle[2]\rvert$	$\sqrt{2}\xi\hbar\omega_\delta$	$2\hbar\omega_\delta + \left(\frac{1}{2}\hbar\omega^\circ + \frac{1}{2}\hbar\omega_\delta\right)$

The two eigenvalues of the matrix (17.44) are given by Eq. (10.15), that is,

$$\{E_\pm\} = \frac{1}{2}\left[\hbar(2\omega^\circ + 3\omega_\delta) \pm \hbar(\omega^\circ - 2\omega_\delta)\sqrt{1 + \frac{4(\sqrt{2}\xi\hbar\omega_\delta)^2}{(\hbar\omega^\circ - 2\hbar\omega_\delta)^2}}\right] \tag{17.45}$$

and the corresponding eigenkets are given by

$$|\Psi^\pm\rangle = \{C_1^\mp\}|(2)\rangle|\{0\}\rangle + \{C_2^\mp\}|(0)\rangle|\{1\}\rangle \tag{17.46}$$

where the expansion coefficients $C_k^\pm$ with $k = 1, 2$, which are real, are given by Eqs. (10.18) and (10.19). Furthermore, the tensorial ground state of the whole system of the two oscillators is simply

$$|\Psi_0\rangle = |(0)\rangle|\{0\}\rangle \tag{17.47}$$

its corresponding energy being the sum of their zero-point energies, that is,

$$E_0 = \tfrac{1}{2}\hbar\omega^\circ + \tfrac{1}{2}\hbar\omega_\delta \tag{17.48}$$

Then, as a consequence of Eqs. (17.45) and (17.48), the energy gaps between the two excited states and the ground state yield

$$(E_\pm^\circ - E_0) = \frac{1}{2}\left\{\hbar(\omega^\circ + 2\omega_\delta) \pm \hbar(\omega^\circ - 2\omega_\delta)\sqrt{1 + \frac{4(\sqrt{2}\xi\hbar\omega_\delta)^2}{(\hbar\omega^\circ - 2\hbar\omega_\delta)^2}}\right\} \tag{17.49}$$

Also, the corresponding transition moments of the high-frequency mode $\mathbf{q}$ involving the ground state $|\Psi_0\rangle$ and the two excited states $|\Psi^\pm\rangle$ read, respectively,

$$\langle\Psi^\pm|\mathbf{q}|\Psi_0\rangle = \sqrt{\frac{\hbar}{2m\omega^\circ}}\langle\Psi^\pm|(\mathbf{a}^\dagger + \mathbf{a})|\Psi_0\rangle$$

which, using Eq. (17.47), becomes

$$\langle\Psi^\pm|\mathbf{q}|\Psi_0\rangle = \sqrt{\frac{\hbar}{2m\omega^\circ}}\langle\Psi^\pm|(\mathbf{a}^\dagger + \mathbf{a})|(0)\rangle|\{0\}\rangle \tag{17.50}$$

Again, owing to the fact that the ladder operators act only on the kets corresponding to the high-frequency mode, and thus on $|\{0\}\rangle$ and not on $|(0)\rangle$, (17.18) leads to

$$(\mathbf{a}^\dagger + \mathbf{a})|(0)\rangle|\{0\}\rangle = |(0)\rangle|\{1\}\rangle$$

Hence, due to this result, and using Eq. (17.46), Eq. (17.50) yields

$$\langle\Psi^\pm|\mathbf{q}|\Psi_0\rangle = \sqrt{\frac{\hbar}{2m\omega^\circ}}(\{C_1^\mp\}\langle\{0\}|\langle(2)| + \{C_2^\mp\}\langle\{1\}|\langle(0)|)|(0)\rangle|\{1\}\rangle$$

or, using the orthonormality properties,

$$\langle\Psi^\pm|\mathbf{q}|\Psi_0\rangle = \sqrt{\frac{\hbar}{2m\omega^\circ}}\{C_2^\mp\} \tag{17.51}$$

Hence, owing to Eq. (17.51), it appears that two transition probabilities of the kind (17.13) exist

$$W(\Psi^\pm, t|\Psi_0, 0) = \frac{2\pi}{\hbar}(\mu^\circ)^2(\mathbf{E}(\omega))^2\left(\frac{\hbar}{2m\omega^\circ}(C_2^\mp)^2\right)\delta(E^\pm - E_0 - \hbar\omega) \tag{17.52}$$

so that the IR transition dealing with the high-frequency mode $\mathbf{q}$ involves a doublet, the angular frequencies of which are

$$\omega^+ = \frac{E^+ - E_0}{\hbar} \quad \text{and} \quad \omega^- = \frac{E^- - E_0}{\hbar}$$

in place of the singlet at the angular frequency ω°, the splitting between ω^+ and ω^- being, therefore, according to Eq. (17.45), given by

$$\omega^+ - \omega^- = 2(\omega^\circ - 2\omega_\delta)\sqrt{1 + \frac{4(\sqrt{2}\xi\hbar\omega_\delta)^2}{(\hbar\omega^\circ - 2\hbar\omega_\delta)^2}} \tag{17.53}$$

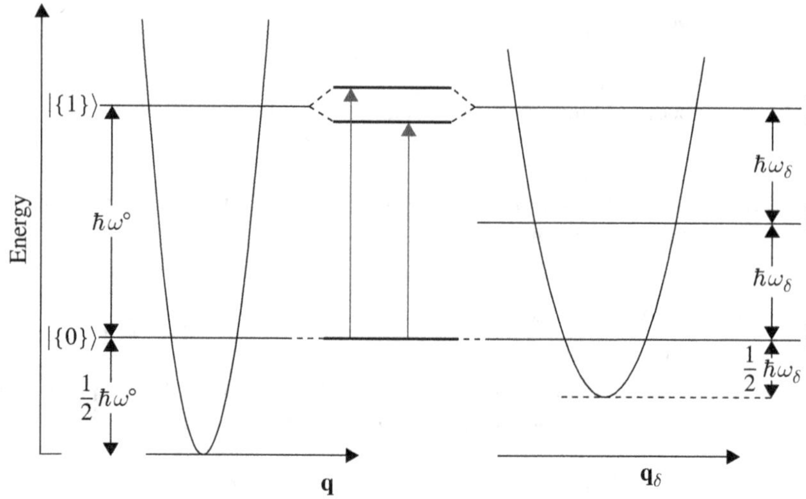

Figure 17.4 IR transition splitting by Fermi resonance.

Moreover, due to Eq. (17.52) the ratio of the intensities corresponding to ω^+ and ω^- reads

$$\frac{I(\omega^+)}{I(\omega^-)} = \frac{W(\Psi^+, t|\Psi_0, 0)}{W(\Psi^-, t|\Psi_0, 0)} = \frac{|C_2^+|^2}{|C_2^-|^2} \tag{17.54}$$

Furthermore, observe that the four coefficients $C_i^{\pm}$ are the matrix elements of the orthogonal matrix, which diagonalizes the 2 × 2 Hamiltonian matrix (17.44). Thus, since the inverse of an orthogonal matrix is equal to its transpose, we have

$$|C_2^-| = |C_1^+|$$

so that, due to Eq. (10.18) and (10.20), the ratio (17.54) reads

$$\frac{I(\omega^+)}{I(\omega^-)} = \frac{|C_2^+|^2}{|C_1^+|^2}$$

Figure 17.4 gives a schematic illustration of the IR doublet resulting from Fermi resonance. Fig. 17.5 illustrates the IR doublets of Fermi resonance for three situations: one at resonance ($2\omega_\delta = \omega^\circ = 3000\,\text{cm}^{-1}$) and two symmetric ones, out of resonance ($2\omega_\delta = \omega^\circ \pm 200\,\text{cm}^{-1} = 2800\,\text{cm}^{-1}$) for a coupling $\sqrt{2}\xi\omega_\delta = 120\,\text{cm}^{-1}$.

17.1.6 Line shapes splitting by tunneling effect

Now, consider the situation of molecular inversion such as that involved in the ammonia molecule, which was studied in Section 9.5. Recall that in this situation, the potential in which the inversion of the ammonia molecule takes place is a symmetric double-well one, through the barrier of which a tunneling is possible that affects more sensitively the two degenerate vibrational levels belonging to each minimum, which are nearest to the barrier energetic top. More specially, that leads to a splitting of the

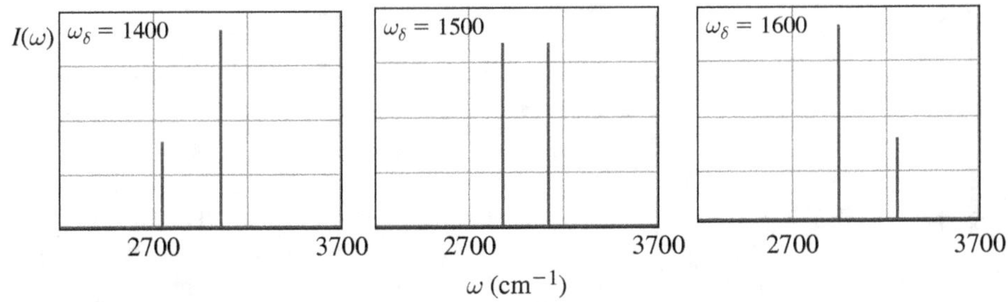

Figure 17.5 IR doublets of Fermi resonance for three situations: one at resonance ($2\omega_\delta = \omega^\circ = 3000\,\text{cm}^{-1}$) and two symmetric ones, out of resonance ($2\omega_\delta = \omega^\circ \pm 200\,\text{cm}^{-1} = 2800\,\text{cm}^{-1}$) for a coupling $\sqrt{2}\xi\omega_\delta = 120\,\text{cm}^{-1}$.

localized degenerate levels, which is greater for the first degenerate excited states than for the degenerate ground states. Let $|\Psi_0^{\pm}\rangle$ and $|\Psi_1^{\pm}\rangle$ be these four delocalized states of energies $E_0^{\pm}$ and $E_1^{\pm}$, resulting, respectively, from the splitting ofthe degenerate ground state E_0 and from the first excited state E_1. Then, the probability for the system to jump from $|\Psi_0^{\pm}\rangle$ to $|\Psi_1^{\pm}\rangle$ at time t, due to the influence of the electromagnetic field starting at an initial time, is governed by Eq. (17.13), which reads

$$W(\Psi_1^{\pm}, t|\Psi_0^{\pm}, 0) \propto |\langle\Psi_1^{\pm}|\mathbf{Q}|\Psi_0^{\pm}\rangle|^2\delta(E_1^{\pm} - E_0^{\pm} - \hbar\omega)$$

Then, insert on the right-hand-side matrix elements a closure relation over the eigenkets of $\mathbf{Q}$ to give

$$\langle\Psi_1^{\pm}|\mathbf{Q}|\Psi_0^{\pm}\rangle = \langle\Psi_1^{\pm}|\mathbf{Q}\int_{-\infty}^{+\infty}|\{Q\}\rangle\langle\{Q\}||\Psi_0^{\pm}\rangle\, dQ$$

or, due to Eq. (3.43),

$$\langle\Psi_1^{\pm}|\mathbf{Q}|\Psi_0^{\pm}\rangle = \int_{-\infty}^{+\infty}\Psi_1^{\pm}(Q)^* Q\Psi_0^{\pm}(Q)\, dQ \tag{17.55}$$

where $\Psi_1^{\pm}(Q)$ and $\Psi_0^{\pm}(Q)$ are the wavefunctions corresponding, respectively, to the kets $|\Psi_1^{\pm}\rangle$ and $|\Psi_0^{\pm}\rangle$. Now, observe that $\Psi_1^{+}(Q)$ and $\Psi_0^{+}(Q)$ are symmetric with respect to the origin, whereas $\Psi_1^{-}(Q)$ and $\Psi_0^{-}(Q)$ are anti-symmetric in the same way as Q. Hence, since to be different from zero the integral of Eq. (17.55) requires a symmetric integrand, it follows that the matrix elements that involve antisymmetric integrands must be zero:

$$\langle\Psi_1^{+}|\mathbf{Q}|\Psi_0^{+}\rangle = \langle\Psi_1^{-}|\mathbf{Q}|\Psi_0^{-}\rangle = 0$$

Hence, the transitions between wavefunctions of same symmetry through the antisymmetric coordinate operator $\mathbf{Q}$ will be forbidden so that the only two transitions that can occur will be

$$|\Psi_0^{+}\rangle \to |\Psi_0^{-}\rangle \qquad \text{and} \qquad |\Psi_0^{-}\rangle \to |\Psi_0^{+}\rangle$$

This is shown in Fig. 17.6.

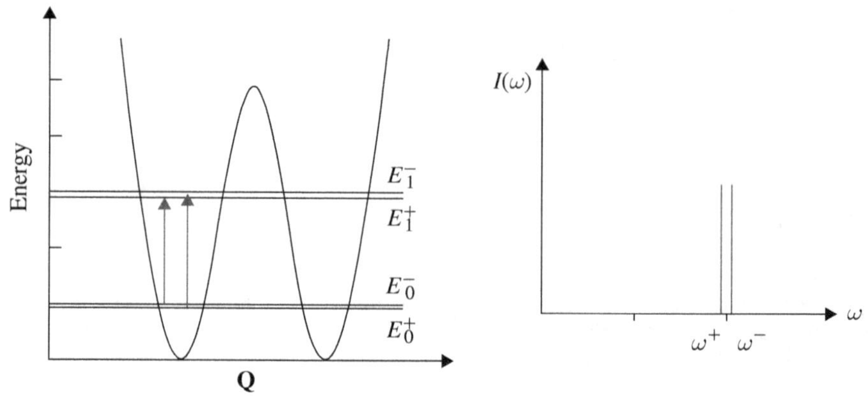

Figure 17.6 Tunnel effect splitting.

17.2 IR SPECTRA WITHIN THE LINEAR RESPONSE THEORY

In the previous section, the transition between vibrational energy levels induced by an electromagnetic electric field was studied for molecular systems where the irreversible influence of the medium leading to finite lifetimes for the energy levels was ignored. That allowed one to assume infinite lifetimes for the excited states leading to Dirac delta peak transitions between the energy levels. However, in order to be more physical, when the medium has to be taken into account, it becomes impossible to ignore the fact that the medium induces finite lifetimes for the excited states, which induce in turn some broadening of the transitions between energy levels. Then, to treat this broadening leading to realistic line shapes, it is necessary to use a time-dependent formalism susceptible to take into account the irreversible influence of the medium. When the electromagnetic field is weak, this formalism is the linear response theory, which is the aim of the present Section.

17.2.1 Spectral density

Consider a group of molecular oscillators of dipole moment $\widehat{\mu}$ that are lit by a monochromatic isotropic electromagnetic radiation. The interaction Hamiltonian $\mathbf{V}(t)$ of each molecule with the monochromatic electric field of angular frequency ω and of strength $\mathbf{E}(\omega)$ is

$$\mathbf{V}(t) = \widehat{\mu}\mathbf{E}(\omega)(e^{-i\omega t} + e^{i\omega t}) \tag{17.56}$$

In the following, it will be assumed that the strength $\mathbf{E}^\circ$ of the electric field is weak. Let $\mathbf{H}^\circ$ be the Hamiltonian of each molecular oscillator, the eigenvalue of which is

$$\mathbf{H}^\circ|\Psi_k\rangle = E_k|\Psi_k\rangle \quad \text{with} \quad \langle\Psi_f|\Psi_k\rangle = \delta_{fk} \tag{17.57}$$

and

$$\sum |\Psi_k\rangle\langle\Psi_k| = 1 \tag{17.58}$$

Of course, the full Hamiltonian **H** of the system formed by a single molecule interacting with the electromagnetic field is the sum of those given by Eqs. (17.56) and (17.57):

$$\mathbf{H} = \mathbf{H}^{\circ} + \mathbf{V}(t)$$

Now, according to Eq. (17.7), resulting from the first-order time-dependent perturbation method, the probability for a given molecular oscillator being at initial time in the ket $|\Psi_k\rangle$ to pass at time t to another ket $|\Psi_l\rangle$, is given by

$$|C(l,t|k,0)|^2 = \frac{2\pi}{\hbar}(\mathbf{E}(\omega))^2 |\langle\Psi_k|\widehat{\boldsymbol{\mu}}|\Psi_l\rangle|^2 \{\delta(E_k - E_l - \hbar\omega)\}t \tag{17.59}$$

Besides, the whole rate W of the total energy transfer between the molecular oscillator and the electromagnetic field is the sum of the time derivative of all these transition probabilities multiplied by the corresponding transferred energies $\hbar\omega_{kl}$ times the probabilities ρ_{kk} for the initial states $|\Psi_k\rangle$ to be occupied, that is,

$$W = \sum_k \sum_l \rho_{kk} \left(\frac{\partial |C(l,t|k,0)|^2}{\partial t}\right) \hbar\omega_{kl} \tag{17.60}$$

where

$$\omega_{kl} = \frac{(E_k - E_l)}{\hbar} \tag{17.61}$$

The probabilities ρ_{kk} are the average values of the Boltzmann density operator $\boldsymbol{\rho}_B$ given by Eq. (13.12) performed over the states $|\Psi_k\rangle$ defined by Eq. (17.57), that is,

$$\rho_{kk} = \langle\Psi_k|\boldsymbol{\rho}_B|\Psi_k\rangle = \frac{1}{Z}\langle\Psi_k|e^{-\beta\mathbf{H}^{\circ}}|\Psi_k\rangle = \frac{1}{Z}e^{-\beta E_k} \tag{17.62}$$

where Z is the partition function and β the usual thermal Lagrange parameter equal to the inverse of $k_B T$. Then, in view of Eqs. (17.59) and (17.62), the rate (17.60) becomes

$$W = W(\omega) = \frac{2\pi}{\hbar}(\mathbf{E}(\omega))^2 \sum_k \sum_l \langle\Psi_k|\boldsymbol{\rho}_B|\Psi_k\rangle |\langle\Psi_k|\boldsymbol{\mu}|\Psi_l\rangle|^2 \{\delta(\omega_{kl} - \omega)\}\hbar\omega_{kl} \tag{17.63}$$

On the other hand, since the spectral density $\mathbf{I}(\omega)$ is the intensity, at angular frequency ω, of the energy transfer between the field and the molecular oscillator, it follows that it is simply related to W through

$$W(\omega) = \mathbf{I}(\omega)\hbar\omega$$

so that, due to Eq. (17.63), the spectral density reads

$$\mathbf{I}(\omega) = \frac{2\pi}{\hbar}(\mathbf{E}(\omega))^2 \sum_k \sum_l \langle\Psi_k|\boldsymbol{\rho}_B|\Psi_k\rangle |\langle\Psi_k|\widehat{\boldsymbol{\mu}}|\Psi_l\rangle|^2 \{\delta(\omega_{kl} - \omega)\}$$

Again, passing from the Dirac delta function to its corresponding integral representation, by the aid of Eq. (18.62) of Section 18.6, and writing explicitly the squared modulus of the matrix element of the dipole moment operator,

$$\mathbf{I}(\omega) = \frac{2\pi}{\hbar}(\mathbf{E}(\omega))^2 \sum_k \sum_l \langle\Psi_k|\boldsymbol{\rho}_B|\Psi_k\rangle \langle\Psi_k|\widehat{\boldsymbol{\mu}}|\Psi_l\rangle \langle\Psi_l|\boldsymbol{\mu}|\Psi_k\rangle \frac{1}{2\pi}\int_{-\infty}^{\infty} e^{-i(\omega_{kl}-\omega)t}\,dt$$

or, due to Eq. (17.61),

$$\mathbf{I}(\omega) = \frac{2\pi}{\hbar}(\mathbf{E}(\omega))^2 \frac{1}{2\pi} \times \int_{-\infty}^{\infty} \sum_k \sum_l \langle \Psi_k | \boldsymbol{\rho} | \Psi_k \rangle \langle \Psi_k | \widehat{\boldsymbol{\mu}} | \Psi_l \rangle \langle \Psi_l | (e^{iE_l t/\hbar}) \widehat{\boldsymbol{\mu}} (e^{-iE_k t/\hbar}) | \Psi_k \rangle e^{-i\omega t} dt$$

and thus

$$\mathbf{I}(\omega) = \frac{2\pi}{\hbar}(\mathbf{E}(\omega))^2 \frac{1}{2\pi} \times \int_{-\infty}^{\infty} \sum_k \sum_l \langle \Psi_k | \boldsymbol{\rho} | \Psi_k \rangle \langle \Psi_k | \widehat{\boldsymbol{\mu}} | \Psi_l \rangle \langle \Psi_l | (e^{i\mathbf{H}^\circ t/\hbar}) \widehat{\boldsymbol{\mu}} (e^{-i\mathbf{H}^\circ t/\hbar}) | \Psi_k \rangle e^{-i\omega t} dt$$

This result reads either

$$\mathbf{I}(\omega) = \frac{2\pi}{\hbar}(\mathbf{E}(\omega))^2 \frac{1}{2\pi} \int_{-\infty}^{\infty} \sum_k \sum_l \langle \Psi_k | \boldsymbol{\rho} | \Psi_k \rangle \langle \Psi_k | \widehat{\boldsymbol{\mu}} | \Psi_l \rangle \langle \Psi_l(t) | \widehat{\boldsymbol{\mu}} | \Psi_k(t) \rangle e^{-i\omega t} dt \tag{17.64}$$

or

$$\mathbf{I}(\omega) = \frac{2\pi}{\hbar}(\mathbf{E}(\omega))^2 \frac{1}{2\pi} \times \int_{-\infty}^{\infty} \sum_k \langle \Psi_k | \boldsymbol{\rho}_B | \Psi_k \rangle \langle \Psi_k | \widehat{\boldsymbol{\mu}} \left\{ \sum_l | \Psi_l \rangle \langle \Psi_l | \right\} (e^{i\mathbf{H}^\circ t/\hbar}) \widehat{\boldsymbol{\mu}} (e^{-i\mathbf{H}^\circ t/\hbar}) | \Psi_k \rangle e^{-i\omega t} dt$$

or, after using the closure relation (17.58) and expressing the Boltzmann density operator $\boldsymbol{\rho}_B$ via (17.62), we have

$$\mathbf{I}(\omega) = \frac{2\pi}{\hbar}(\mathbf{E}(\omega))^2 \frac{1}{2\pi} \times \int_{-\infty}^{\infty} \sum_k \langle \Psi_k | \frac{1}{Z}(e^{-\beta \mathbf{H}^\circ}) | \Psi_k \rangle \langle \Psi_k | \widehat{\boldsymbol{\mu}} (e^{i\mathbf{H}^\circ t/\hbar}) \widehat{\boldsymbol{\mu}} (e^{-i\mathbf{H}^\circ t/\hbar}) | \Psi_k \rangle e^{-i\omega t} dt$$

Then, using the eigenvalues of the density operator, one obtains

$$\mathbf{I}(\omega) = \frac{2\pi}{\hbar}(\mathbf{E}(\omega))^2 \frac{1}{2\pi} \times \int_{-\infty}^{\infty} \sum_k \langle \Psi_k | \frac{1}{Z}(e^{-\beta E_k}) | \Psi_k \rangle \langle \Psi_k | \widehat{\boldsymbol{\mu}} (e^{i\mathbf{H}^\circ t/\hbar}) \widehat{\boldsymbol{\mu}} (e^{-i\mathbf{H}^\circ t/\hbar}) | \Psi_k \rangle e^{-i\omega t} dt$$

or, after commuting the scalar exponential with the bra and using the normalization condition appearing in (17.57) and returning to the Boltzmann density operator, we have

$$\mathbf{I}(\omega) = \frac{2\pi}{\hbar}(\mathbf{E}(\omega))^2 \frac{1}{2\pi} \int_{-\infty}^{\infty} \sum_k \langle \Psi_k | \boldsymbol{\rho}_B \widehat{\boldsymbol{\mu}} (e^{i\mathbf{H}^\circ t/\hbar}) \widehat{\boldsymbol{\mu}} (e^{-i\mathbf{H}^\circ t/\hbar}) | \Psi_k \rangle e^{-i\omega t} dt \tag{17.65}$$

or

$$\mathbf{I}(\omega) = \frac{2\pi}{\hbar}(\mathbf{E}(\omega))^2 \frac{1}{2\pi} \int_{-\infty}^{\infty} \sum_k \langle \Psi_k | \rho_B \hat{\boldsymbol{\mu}}(0) \hat{\boldsymbol{\mu}}(t) | \Psi_k \rangle e^{-i\omega t} dt \tag{17.66}$$

with $\hat{\boldsymbol{\mu}} \equiv \hat{\boldsymbol{\mu}}(0)$, and where $\widehat{\boldsymbol{\mu}}(t)$ is the Heisenberg picture dipole moment operator at time t

$$\widehat{\boldsymbol{\mu}}(t) = (e^{i\mathbf{H}^\circ t/\hbar}) \boldsymbol{\mu} (e^{-i\mathbf{H}^\circ t/\hbar}) \tag{17.67}$$

Finally, expressing the sum over k as a trace operation, Eq. (17.66) yields

$$\mathbf{I}(\omega) = \frac{2\pi}{\hbar}(\mathbf{E}(\omega))^2 \frac{1}{2\pi} \int_{-\infty}^{\infty} \mathbf{tr}\{\rho_B \hat{\boldsymbol{\mu}}(0) \hat{\boldsymbol{\mu}}(t)\} e^{-i\omega t} dt \tag{17.68}$$

Thus, this spectral density (SD) (17.68) may be viewed as proportional to the following Fourier transform:

$$\boxed{\mathbf{I}(\omega) \propto \int_{-\infty}^{\infty} \mathbf{G}(t) e^{-i\omega t} dt} \tag{17.69}$$

where $\mathbf{G}(t)$ is the autocorrelation function (ACF) of the dipole moment operator defined by

$$\boxed{\mathbf{G}(t) = \mathbf{tr}\{\rho_B \widehat{\boldsymbol{\mu}}(0) \widehat{\boldsymbol{\mu}}(t)\}} \tag{17.70}$$

Note that, according to the results of Section 18.6 for the Fourier transform of the complex ACF, the following equalities hold:

$$\mathbf{I}(\omega) = \int_{-\infty}^{\infty} \mathbf{G}(t) e^{-i\omega t} dt = 2\mathrm{Re} \int_{0}^{\infty} \mathbf{G}(t) e^{-i\omega t} dt = 2\mathrm{Re} \int_{0}^{\infty} \mathbf{G}(t)^* e^{i\omega t} dt \tag{17.71}$$

Equations (17.69) and (17.70) are the two most important results of the so-called *linear response theory*.

17.2.2 Non-Hermitian IR transition operators for oscillators

Next, apply the results of the previous section to the spectroscopy of oscillators. Then, on passing to Boson operators, the dipole moment operator at initial time (17.11) yields

$$\widehat{\boldsymbol{\mu}}(0) = \mu^\circ \sqrt{\frac{\hbar}{2M\omega^\circ}} (\mathbf{a}^\dagger + \mathbf{a})$$

or

$$\widehat{\boldsymbol{\mu}}(0) = \boldsymbol{\mu}(0) + \boldsymbol{\mu}(0)^\dagger \tag{17.72}$$

with, respectively,

$$\boldsymbol{\mu}(0) = \mu^{\circ}\sqrt{\frac{\hbar}{2M\omega^{\circ}}}\mathbf{a}^{\dagger} \qquad \text{and} \qquad \boldsymbol{\mu}(0)^{\dagger} = \mu^{\circ}\sqrt{\frac{\hbar}{2M\omega^{\circ}}}\mathbf{a}$$

Again, multiplying the right-hand side by the closure relation on the eigenkets of $\mathbf{a}^{\dagger}\mathbf{a}$ yields

$$\widehat{\boldsymbol{\mu}}(0) = \mu^{\circ}\sqrt{\frac{\hbar}{2M\omega^{\circ}}}\sum_{k=0}^{\infty}(\mathbf{a}^{\dagger} + \mathbf{a})|\{k\}\rangle\langle\{k\}| \tag{17.73}$$

Furthermore, using Eqs. (5.53) and (5.65), that is,

$$\mathbf{a}^{\dagger}|\{k\}\rangle = \sqrt{k+1}|\{k+1\}\rangle \qquad \text{and} \qquad \langle\{k\}|\mathbf{a} = \sqrt{k+1}\langle\{k+1\}|$$

and, acting with $\mathbf{a}^{\dagger}$ on the kets and $\mathbf{a}$ on the bras, Eq. (17.73) takes the form

$$\widehat{\boldsymbol{\mu}}(0) = \mu^{\circ}\sqrt{\frac{\hbar}{2M\omega^{\circ}}}\left\{\sum_{k=0}^{\infty}\sqrt{k+1}|\{k+1\}\rangle\langle\{k\}| + \sum_{k=0}^{\infty}\sqrt{k}|\{k-1\}\rangle\langle\{k\}|\right\} \tag{17.74}$$

Next, in order to remove from the right-hand side of Eq. (17.74) the ket $|\{-1\}\rangle$, which is without meaning, re-sum the last right-hand-side term to get

$$\widehat{\boldsymbol{\mu}}(0) = \mu^{\circ}\sqrt{\frac{\hbar}{2M\omega^{\circ}}}\sum_{k}\sqrt{k+1}\{|\{k+1\}\rangle\langle\{k\}| + |\{k\}\rangle\langle\{k+1\}|\}$$

Now, if one restricts oneself to the ground and the first excited states of the fast mode, this equation reduces to

$$\widehat{\boldsymbol{\mu}}(0) = \mu^{\circ}\sqrt{\frac{\hbar}{2M\omega^{\circ}}}\{|\{1\}\rangle\langle\{0\}| + |\{0\}\rangle\langle\{1\}|\} \tag{17.75}$$

Then, in this special situation, the two self-conjugate non-Hermitian operators appearing in Eq. (17.75) become

$$\boldsymbol{\mu}(0) = \mu^{\circ}\sqrt{\frac{\hbar}{2M\omega^{\circ}}}|\{1\}\rangle\langle\{0\}| \qquad \text{and} \qquad \boldsymbol{\mu}(0)^{\dagger} = \mu^{\circ}\sqrt{\frac{\hbar}{2M\omega^{\circ}}}|\{0\}\rangle\langle\{1\}| \tag{17.76}$$

The first one, $\boldsymbol{\mu}(0)$, induces the transition from the ground state to the first excited state, whereas its Hermitian conjugate, $\boldsymbol{\mu}(0)^{\dagger}$, induces the inverse transition from the first excited state to the ground state.

Moreover, due to Eqs. (17.70) and (17.76), the ACFs of the dipole moment operator corresponding, respectively, to the absorption and the emission processes yield

$$\boxed{\mathbf{G}_{\mathrm{Abs}}(t) = \mathbf{tr}\{\boldsymbol{\rho}_B\boldsymbol{\mu}(0)^{\dagger}\boldsymbol{\mu}(t)\}} \tag{17.77}$$

$$\mathbf{G}_{\mathrm{Emis}}(t) = \mathbf{tr}\{\boldsymbol{\rho}_B\boldsymbol{\mu}(0)\boldsymbol{\mu}(t)^{\dagger}\} \tag{17.78}$$

with

$$\boldsymbol{\rho}_B = \frac{1}{Z}(e^{-\beta\mathbf{H}^{\circ}}) \quad \text{with} \quad \beta = \frac{1}{k_B T} \tag{17.79}$$

Hence, since the Boltzmann density operator is Hermitian, it appears that the absorption and emission autocorrelation functions are interrelated through

$$\mathbf{G}_{\text{Abs}}(t) = \mathbf{G}_{\text{Emis}}(t)^* \tag{17.80}$$

17.3 IR SPECTRA OF WEAK H-BONDED SPECIES

17.3.1 General approach of spectral density

Using linear response theory, it is now possible to study the IR absorption line shape of weak H-bonded species in the absence of damping in the context of the strong anharmonic coupling theory. This study will begin by the special situation where the damping is missing, although the linear response theory could then be avoided, in order to give a simple introduction to the situations involving a relaxation where it is not yet possible to avoid linear response theory. Recall that within the strong anharmonic coupling theory, anharmonic coupling exists between the high-frequency mode and the H-bond bridge. In this case, owing to Eqs. (17.67), (17.76), and (17.79), and writing for the Hamiltonian of the H-bonded species $\mathbf{H}_{\text{Tot}}$ in place of $\mathbf{H}^\circ$, the absorption ACF (17.77) of the dipole moment operator of the high-frequency mode appearing in these equations reads

$$\mathbf{G}(t) \propto \mathbf{tr}\{(e^{-\mathbf{H}_{\text{Tot}}/k_BT})\boldsymbol{\mu}(0)^\dagger(e^{i\mathbf{H}_{\text{Tot}}t/\hbar})\boldsymbol{\mu}(0)(e^{-i\mathbf{H}_{\text{Tot}}t/\hbar})\}$$

$$\mathbf{G}(t) \propto \mathbf{tr}\{(e^{-\mathbf{H}_{\text{Tot}}/k_BT})|\{0\}\rangle\langle\{1\}|(e^{i\mathbf{H}_{\text{Tot}}t/\hbar})|\{1\}\rangle\langle\{0\}|(e^{-i\mathbf{H}_{\text{Tot}}t/\hbar})\} \tag{17.81}$$

Here, $\mathbf{H}_{\text{Tot}}$ is the full Hamiltonian (10.23) of the two interacting oscillators, which, due to Eqs. (10.24) and (10.27), reads

$$\mathbf{H}_{\text{Tot}} = \left(\frac{\mathbf{P}^2}{2M} + \frac{1}{2}M\Omega^2\mathbf{Q}^2\right) + \left(\frac{\mathbf{p}^2}{2m} + \frac{1}{2}m\omega^{\circ 2}\mathbf{q}^2\right) + m\omega^\circ b\mathbf{q}^2\mathbf{Q} + \frac{1}{2}mb^2\,\mathbf{q}^2\mathbf{Q}^2 \tag{17.82}$$

Here, $\mathbf{Q}$ and $\mathbf{P}$ are the position and momentum operators of the H-bond bridge and M and Ω the corresponding reduced mass and angular frequency, whereas $\mathbf{q}$, $\mathbf{p}$ are the position and momentum operators of the high-frequency mode and m and ω° the corresponding reduced mass and angular frequency. Furthermore, the $|\{0\}\rangle$ and $|\{1\}\rangle$ are the ground and first excited states of the high-frequency mode within the harmonic approximation, that is,

$$\left(\frac{\mathbf{p}^2}{2m} + \frac{1}{2}m\omega^{\circ 2}\mathbf{q}^2\right)|\{k\}\rangle = \left(k + \frac{1}{2}\right)\hbar\omega^\circ|\{k\}\rangle \tag{17.83}$$

with

$$\sum_k |\{k\}\rangle\langle\{k\}| = 1 \tag{17.84}$$

Next, introduce the eigenvalue equation of the harmonic Hamiltonian of the H-bond bridge

$$\left(\frac{\mathbf{P}^2}{2M} + \frac{1}{2}M\Omega^2\mathbf{Q}^2\right)|(m)\rangle = \left(m + \frac{1}{2}\right)\hbar\Omega|(m)\rangle \tag{17.85}$$

with

$$\sum_m |(m)\rangle\langle(m)| = 1 \tag{17.86}$$

From the eigenstates of the eigenvalue equations (17.83) and (17.85), it is possible to construct the following tensor product basis:

$$\{|\{k\}(m)\rangle\} \equiv \{|\{k\}\rangle|(m)\rangle\}$$

leading, with the help of Eqs. (17.84) and (17.86), to the following closure relation:

$$\sum_k \sum_m |\{k\}(m)\rangle\langle\{k\}(m)| = 1$$

Now, write the eigenvalue equation of the total Hamiltonian (17.82):

$$\mathbf{H}_{\text{Tot}}|\Theta_\mu\rangle = \hbar\omega_\mu|\Theta_\mu\rangle \quad \text{with} \quad \langle\Theta_\mu|\Theta_\nu\rangle = \delta_{\mu\upsilon} \qquad \text{and} \qquad \sum_\mu |\Theta_\mu\rangle\langle\Theta_\mu| = 1 \tag{17.87}$$

with

$$|\Theta_\mu\rangle = \sum_k \sum_m \{C_{\mu m}^{\{k\}}\}|\{k\}(m)\rangle \tag{17.88}$$

where the $C_{\mu m}^{\{k\}}$ are the expansion coefficients. Then, inserting twice in Eq. (17.81) the closure relation involved in (17.87) and after a circular permutation in the trace, this ACF becomes

$$\mathbf{G}(t) \propto$$

$$\mathbf{tr}\left\{|\{0\}\rangle\langle\{1\}|(e^{i\mathbf{H}_{\text{Tot}}t/\hbar})\sum_\mu|\Theta_\mu\rangle\langle\Theta_\mu|\{1\}\rangle\langle\{0\}|\sum_\nu|\Theta_\nu\rangle\langle\Theta_\nu|(e^{-i\mathbf{H}_{\text{Tot}}t/\hbar})(e^{-\mathbf{H}_{\text{Tot}}/k_BT})\right\}$$

Next, in view of the eigenvalue equation (17.87) and due to Eqs. (1.71) and (1.73), it reads

$$(e^{i\mathbf{H}_{\text{Tot}}t/\hbar})|\Theta_\mu\rangle = (e^{i\omega_\mu t})|\Theta_\mu\rangle \qquad \text{and} \qquad \langle\Theta_\nu|(e^{-i\mathbf{H}_{\text{Tot}}t/\hbar}) = \langle\Theta_\nu|(e^{-i\omega_\nu t})$$

so that the ACF becomes, after commuting $\exp\{-i\omega_\nu t\}$ with $\langle\Theta_\nu|$,

$$\mathbf{G}(t) \propto$$

$$\mathbf{tr}\left\{|\{0\}\rangle\langle\{1\}|\sum_\mu(e^{i\omega_\mu t})|\Theta_\mu\rangle\langle\Theta_\mu|\{1\}\rangle\langle\{0\}|\sum_\nu|\Theta_\nu\rangle(e^{-i\omega_\nu t})\langle\Theta_\nu|(e^{-\mathbf{H}_{Tot}/k_BT})\right\}$$

Again, using, with the help of Eqs. (1.71) and (1.73),

$$\langle\Theta_\nu|(e^{-\mathbf{H}_{\text{Tot}}/k_BT}) = \langle\Theta_\nu|(e^{-\hbar\omega_\nu/k_BT})$$

one obtains, after inserting between $|\{0\}\rangle$ and $\langle\{1\}|$ the closure relation (17.86)

$$\mathbf{G}(t) \propto \mathbf{tr}\left\{\langle\{1\}|\sum_{\mu}(e^{i\omega_\mu t})|\Theta_\mu\rangle\langle\Theta_\mu|\{1\}\rangle\sum_{m}|(m)\rangle\langle(m)|\langle\{0\}| \right.$$
$$\left.\times\sum_{\nu}|\Theta_\nu\rangle\langle\Theta_\nu|(e^{-i\omega_\nu t})(e^{-\hbar\omega_\nu/k_BT})|\{0\}\rangle\right\}$$

Furthermore, insert a closure relation between the eigenstates appearing in the eigenvalue equation (17.85), and then make explicit the trace with the help of the basis involved in this eigenvalue equation to get

$$\mathbf{G}(t) \propto \sum_{\nu}\sum_{n}\sum_{\mu}\sum_{m}$$
$$\times(e^{-\hbar\omega_\nu^\circ/k_BT})(e^{i\omega_\mu^\circ t}e^{-i\omega_\nu t})\langle(n)\{1\}|\Theta_\mu\rangle\langle\Theta_\mu|\{1\}(m)\rangle\langle(m)\{0\}|\Theta_\nu\rangle\langle\Theta_\nu|\{0\}(n)\rangle$$

Lastly, a finite lifetime of the excited state $|\{1\}\rangle$ may be incorporated by multiplying this ACF by a damping factor of the form $\exp\{-\gamma^\circ t\}$. Then, after Fourier transform, the following spectral density is obtained:

$$\mathbf{I}(\omega) \propto \sum_{\nu}\sum_{n}\sum_{\mu}\sum_{m}(e^{-\beta\hbar\omega_\nu})$$
$$\times 2\,\mathrm{Re}\int_{o}^{\infty} dt\, e^{-i\omega t}e^{-\gamma^\circ t}(e^{-i(\omega_\nu-\omega_\mu)t})[C_{\mu n}^{\{1\}}][C_{m\mu}^{\{1\}}][C_{\nu m}^{\{0\}}][C_{n\nu}^{\{0\}}] \qquad (17.89)$$

17.3.2 Spectral density within adiabatic approximation

Now, we show that the spectral density (17.89) may be reproduced satisfactorily for weak H bonds, working in the adiabatic approximation allowing one to separate the motions of the slow and high-frequency modes. That is because, for such situations, the difference in the angular frequencies of the slow and high-frequency mode is large enough, the first one being around 3000 cm^{-1}, and the last one lying around 100 cm^{-1}. Thereby, due to this adiabatic approximation of weak H-bonded system, the full Hamiltonian $\mathbf{H}_{\mathrm{Tot}}$ may be satisfactorily written, according to Eq. (10.88), as a sum of effective driven harmonic Hamiltonians $\{\mathbf{H}^{\{k\}}\}$ governing the H-bond bridge, which are depending on the excitation degree $\{k\}$ of the eigenket $|\{k\}\rangle$ of the high-frequency mode Hamiltonian:

$$\mathbf{H}_{\mathrm{Tot}} = \sum_{k}\{\mathbf{H}^{\{k\}}\}|\{k\}\rangle\langle\{k\}| \qquad (17.90)$$

where the effective Hamiltonians corresponding to the ground state $\{k\}=0$ and to the first excited state $\{k\}=1$, may be obtained from Eq. (10.89), which reads, when neglecting the zero-point energy of the fast mode

$$\{\mathbf{H}^{\{0\}}\} = \left(\frac{\mathbf{P}^2}{2M}+\frac{1}{2}M\Omega^2\mathbf{Q}^2\right) \qquad (17.91)$$

$$\{\mathbf{H}^{\{1\}}\} = \left(\frac{\mathbf{P}^2}{2M} + \frac{1}{2}M\Omega^2\mathbf{Q}^2\right) + b\mathbf{Q} + \hbar\omega^\circ - \alpha^{\circ 2}\hbar\Omega \tag{17.92}$$

Now, the absorption coefficient ACF (17.77) yields, from Eqs. (17.67) and (17.79),

$$\mathbf{G}^\circ(t) \propto \frac{1}{Z}\mathbf{tr}\{(e^{-\beta\mathbf{H}_{\text{Tot}}})|\{0\}\rangle\langle\{1\}|[e^{i\mathbf{H}_{\text{Tot}}t/\hbar}]|\{1\}\rangle\langle\{0\}|[e^{-i\mathbf{H}_{\text{Tot}}t/\hbar}]\} \tag{17.93}$$

where Z is the partition function. Then, in the adiabatic approximation, Eq. (17.93) takes the form

$$\mathbf{G}^\circ(t) \propto \frac{1}{Z}\mathbf{tr}\{(e^{-\beta\mathbf{H}^{\{0\}}})|\{0\}\rangle\langle\{1\}|[e^{i\mathbf{H}^{\{1\}}t/\hbar}]|\{1\}\rangle\langle\{0\}|[e^{-i\mathbf{H}^{\{0\}}t/\hbar}]\} \tag{17.94}$$

Now, since the effective Hamiltonians $\mathbf{H}^{\{k\}}$ given by Eqs. (17.91) and (17.92) are not operating on the kets $|\{k\}\rangle$, they commute with them, so that after performing a circular permutation within the trace, Eq. (17.94) becomes

$$\mathbf{G}^\circ(t) \propto \frac{1}{Z}\mathbf{tr}\{(e^{-\beta\mathbf{H}^{\{0\}}})\langle\{0\}|\{0\}\rangle\langle\{1\}|\{1\}\rangle[e^{i\mathbf{H}^{\{1\}}t/\hbar}][e^{-i\mathbf{H}^{\{0\}}t/\hbar}]\}$$

a result that, after using the orthonormality properties, reduces to

$$\mathbf{G}^\circ(t) \propto \frac{1}{Z}\mathbf{tr}\{(e^{-\beta\mathbf{H}^{\{0\}}})[e^{i\mathbf{H}^{\{1\}}t/\hbar}][e^{-i\mathbf{H}^{\{0\}}t/\hbar}]\} \tag{17.95}$$

or, due to Eqs. (17.91) and (17.92), becomes

$$\mathbf{G}^\circ(t) \propto \frac{1}{Z}e^{i\omega^\circ t}(e^{-i\alpha^{\circ 2}\Omega t})\mathbf{tr}\{(e^{-\beta\mathbf{H}^{\{0\}}})[\mathbf{U}^{\{1\}}(t)]^{-1}[\mathbf{U}^{\{0\}}(t)]\} \tag{17.96}$$

with

$$\mathbf{U}^{\{1\}}(t) = \left(\exp\left[-i\left(\frac{\mathbf{P}^2}{2M} + \frac{1}{2}M\Omega^2\mathbf{Q}^2 + \hbar b\mathbf{Q}\right)t/\hbar\right]\right) \tag{17.97}$$

and

$$\mathbf{U}^{\{0\}}(t) = \left(\exp\left[-i\left(\frac{\mathbf{P}^2}{2M} + \frac{1}{2}M\Omega^2\mathbf{Q}^2\right)t/\hbar\right]\right) \tag{17.98}$$

respectively. Now, make the following partition of the Hamiltonian (17.92)

$$\{\mathbf{H}^{\{1\}}\} = \{\mathbf{H}^{\{0\}}\} + \mathbf{V}$$

with

$$\mathbf{V} = b\mathbf{Q} + \hbar\omega^\circ - \alpha^{\circ 2}\hbar\Omega$$

Then, within the interaction picture, and according to Eq. (3.122) with $t_0 = 0$ for the initial time, the time evolution operator (17.97) may be written

$$\mathbf{U}^{\{1\}}(t) = [\mathbf{U}^{\{0\}}(t)][\mathbf{U}^{\circ\{1\}}(t)^{\text{IP}}] \tag{17.99}$$

with

$$\mathbf{U}^{\circ\{1\}}(t)^{\text{IP}} = \widehat{\mathbf{P}}\left\{\exp\left[-ib\int_0^t \mathbf{Q}(t')^{\text{IP}}dt'\right]\right\} \tag{17.100}$$

where $\widehat{\mathbf{P}}$ is the Dyson time-ordering operator and $\mathbf{Q}(t)^{\mathrm{IP}}$ the IP position operator, which due to Eq. (3.108) and according to Eqs. (17.98), reads

$$\mathbf{Q}(t)^{\mathrm{IP}} = \left\{\exp\left[i\left(\frac{\mathbf{P}^2}{2M}+\frac{1}{2}M\Omega^2\mathbf{Q}^2\right)t/\hbar\right]\right\}\mathbf{Q}(0)\left\{\exp\left[-i\left(\frac{\mathbf{P}^2}{2M}+\frac{1}{2}M\Omega^2\mathbf{Q}^2\right)t/\hbar\right]\right\} \tag{17.101}$$

Next, in view of Eqs. (17.98) and (17.100), the time evolution operator (17.99) reads

$$\mathbf{U}^{\{1\}}(t) = \left(\exp\left[-i\left(\frac{\mathbf{P}^2}{2M}+\frac{1}{2}M\Omega^2\mathbf{Q}^2\right)t/\hbar\right]\right)\widehat{\mathbf{P}}\left\{\exp\left[-ib\int_0^t \mathbf{Q}(t')^{\mathrm{IP}}dt'\right]\right\} \tag{17.102}$$

its inverse being

$$[\mathbf{U}^{\{1\}}(t)]^{-1} = \widehat{\mathbf{P}}\left\{\exp\left[ib\int_0^t \mathbf{Q}(t')^{\mathrm{IP}}dt'\right]\right\}\left(\exp\left[i\left(\frac{\mathbf{P}^2}{2M}+\frac{1}{2}M\Omega^2\mathbf{Q}^2\right)t/\hbar\right]\right)$$

Thus, after simplification, the product of evolution operators involved on the right-hand side of Eq. (17.96), reduces to

$$[\mathbf{U}^{\{1\}}(t)]^{-1}[\mathbf{U}^{\{0\}}(t)] = \widehat{\mathbf{P}}\left\{\exp\left[ib\int_0^t \mathbf{Q}(t')^{\mathrm{IP}}dt'\right]\right\} \tag{17.103}$$

so that the ACF (17.96) yields

$$\mathbf{G}^\circ(t) \propto e^{i\omega^\circ t}(e^{-i\alpha^{\circ 2}\Omega t})\frac{1}{Z}\mathbf{tr}\left\{(e^{-\beta\mathbf{H}^{\{0\}}})\widehat{\mathbf{P}}\left\{\exp\left[ib\int_0^t \mathbf{Q}(t')^{\mathrm{IP}}dt'\right]\right\}\right\} \tag{17.104}$$

Again, pass to the Boson operator representation where the IP coordinate (17.101) takes the form

$$\mathbf{Q}(t)^{\mathrm{IP}} = \sqrt{\frac{\hbar}{2M\Omega}}(e^{i\mathbf{a}^\dagger\mathbf{a}\Omega t})(\mathbf{a}^\dagger+\mathbf{a})(e^{-i\mathbf{a}^\dagger\mathbf{a}\Omega t})$$

or, after performing the canonical transformation comparing, Eqs. (7.21) and (7.22),

$$\mathbf{Q}(t)^{\mathrm{IP}} = \sqrt{\frac{\hbar}{2M\Omega}}(\mathbf{a}^\dagger e^{i\Omega t}+\mathbf{a}e^{-i\Omega t}) \tag{17.105}$$

Moreover, within the Boson operator representation, recall that the Boltzmann density operator is given by Eqs. (13.12) and (13.23), that is,

$$\frac{1}{Z}(e^{-\beta\mathbf{H}^{\{0\}}}) = (1-e^{-\lambda})(e^{-\lambda\mathbf{a}^\dagger\mathbf{a}}) \quad \text{with} \quad \lambda = \frac{\hbar\Omega}{k_BT} \tag{17.106}$$

As a consequence of Eqs. (17.105) and (17.106), the ACF (17.104) now reads

$$\mathbf{G}^\circ(t) \propto e^{i\omega^\circ t}(e^{-i\alpha^{\circ 2}\Omega t})(1-e^{-\lambda})\mathbf{tr}\left\{(e^{-\lambda\mathbf{a}^\dagger\mathbf{a}})\widehat{\mathbf{P}}\left\{\exp\left[i\alpha^\circ\int_0^t(\mathbf{a}^\dagger e^{i\Omega t'}+\mathbf{a}e^{-i\Omega t'})dt'\right]\right\}\right\} \tag{17.107}$$

with

$$\alpha^\circ = \frac{b}{\Omega}\sqrt{\frac{\hbar}{2M\Omega}} \tag{17.108}$$

Observe that in Eq. (17.107), the term on which the Dyson time-ordering operator acts is the IP time evolution operator of a driven harmonic oscillator, which has been proven to be given by Eqs. (7.142), that is,

$$\widehat{\mathbf{P}}\left\{\exp\left[i\alpha^\circ\int_0^t(\mathbf{a}^\dagger e^{i\Omega t'} + \mathbf{a}e^{-i\Omega t'})dt'\right]\right\} = (e^{-i\alpha^{\circ 2}\Omega t})(e^{i\alpha^{\circ 2}\sin\Omega t})(e^{\Phi^\circ(t)^*\mathbf{a}^\dagger - \Phi^\circ(t)\mathbf{a}}) \tag{17.109}$$

with $\Phi^\circ(t)$ given by Eq. (7.132), that is,

$$\Phi^\circ(t) = \alpha^\circ(e^{-i\Omega t} - 1) \tag{17.110}$$

Thus, owing to Eq. (17.109), the ACF (17.107) takes the form

$$\mathbf{G}^\circ(t) \propto e^{i\omega^\circ t}(e^{-i\alpha^{\circ 2}\Omega t})(e^{-i\alpha^{\circ 2}\Omega t})(e^{i\alpha^{\circ 2}\sin\Omega t})(1-e^{-\lambda})\mathbf{tr}\{(e^{-\lambda\mathbf{a}^\dagger\mathbf{a}})[e^{\Phi^\circ(t)^*\mathbf{a}^\dagger - \Phi^\circ(t)\mathbf{a}}]\} \tag{17.111}$$

Now, one may recognize that the thermal average appearing in Eq. (17.111) is a time-dependent translation operator of the form (13.193). Hence, keeping in mind that the thermal average of a translation operator is given by Eq. (13.204)

$$(1-e^{-\lambda})e^{-|\alpha|^2/2}\mathbf{tr}\{(e^{-\lambda\mathbf{a}^\dagger\mathbf{a}})(e^{\alpha\mathbf{a}^\dagger})(e^{-\alpha^*\mathbf{a}})\} = \exp\left\{-|\alpha|^2\left(\langle\mathbf{n}\rangle + \tfrac{1}{2}\right)\right\}$$

where $\langle\mathbf{n}\rangle$ is the thermal average of the number occupation given by Eq. (13.36), that is,

$$\langle\mathbf{n}\rangle = \frac{1}{e^\lambda - 1} = \frac{1}{e^{\hbar\Omega/k_BT} - 1} \tag{17.112}$$

it appears, taking $\Phi^\circ(t)$ in place of α, with therefore

$$|\Phi^\circ(t)|^2 = \alpha^{\circ 2}(e^{-i\Omega t} - 1)(e^{i\Omega t} - 1) = 2\alpha^{\circ 2}(1 - \cos\Omega t)$$

that the trace of interest appearing in the ACF (17.111) yields

$$(1-e^{-\lambda})\mathbf{tr}\{(e^{-\lambda\mathbf{a}^\dagger\mathbf{a}})[e^{\Phi^\circ(t)^*\mathbf{a}^\dagger - \Phi^\circ(t)\mathbf{a}}]\} = e^{2\alpha^{\circ 2}(\langle\mathbf{n}\rangle + 1/2)(\cos\Omega t - 1)}$$

As a consequence, the ACF (17.111) takes the final form

$$\mathbf{G}^\circ(t) \propto e^{i\omega^\circ t}(e^{-i2\alpha^{\circ 2}\Omega t})(e^{i\alpha^{\circ 2}\sin\Omega t})[e^{2\alpha^{\circ 2}(\langle\mathbf{n}\rangle + 1/2)(\cos\Omega t - 1)}] \tag{17.113}$$

Then, multiply this ACF by an exponential decay $\exp\{-\gamma^\circ t\}$ in which γ° is a relaxation parameter, in order to take into account the finite lifetime of the first excited state of the high-frequency mode.

$$\mathbf{G}(t) \propto e^{i\omega^\circ t}(e^{-i2\alpha^{\circ 2}\Omega t})(e^{i\alpha^{\circ 2}\sin\Omega t})[e^{2\alpha^{\circ 2}(\langle\mathbf{n}\rangle + 1/2)(\cos\Omega t - 1)}](e^{-\gamma^\circ t}) \tag{17.114}$$

Finally, make the Fourier transform to get the spectral density $\mathbf{I}(\omega)$, that is,

$$\mathbf{I}(\omega) \propto 2Re\int_0^\infty e^{i\omega^\circ t}(e^{-i2\alpha^{\circ 2}\Omega t})(e^{i\alpha^{\circ 2}\sin\Omega t})[e^{2\alpha^{\circ 2}(\langle\mathbf{n}\rangle + 1/2)(\cos\Omega t - 1)}](e^{-\gamma^\circ t})e^{-i\omega t}dt \tag{17.115}$$

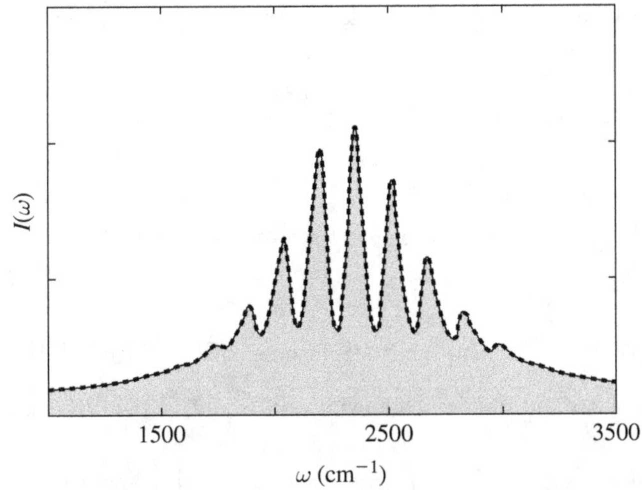

Figure 17.7 Comparison of the adiabatic (17.89) SD with the reference nonadiabatic (17.115) one: $\alpha^\circ = 1.00$, $T = 300$ K, $\omega^\circ = 3000\,\text{cm}^{-1}$, $\Omega = 150\,\text{cm}^{-1}$, $\gamma^\circ = 0.20\,\Omega$.

17.3.3 Accuracy of the SD (17.115) as compared to the nonadiabatic one (17.89)

It may be of interest to verify if the analytical SD (17.115), which has been obtained in the adiabatic approximation, may reproduce satisfactorily (17.89), which is working beyond this approximation.

Figure 17.7 gives a comparison between the approximate adiabatic SD (17.115) and the reference nonadiabatic one (17.89) for some basic parameters that may be expected for weak H bonds. Its examination shows that the adiabatic SD (dots) fits very accurately the reference nonadiabatic one (shown shaded).

17.3.4 Spectral density as Franck–Condon progression

Consider the translation operator times the phase factor $\exp\{i\alpha^{\circ 2}\sin\Omega t\}$, appearing in Eq. (17.111), that is,

$$(e^{i\alpha^{\circ 2}\sin\Omega t})[e^{\Phi^\circ(t)^*\mathbf{a}^\dagger - \Phi^\circ(t)\mathbf{a}}] \tag{17.116}$$

where $\Phi^\circ(t)$ is given by Eq. (17.110), that is,

$$\Phi^\circ(t) = \alpha^\circ(e^{-i\Omega t} - 1) \tag{17.117}$$

Next, write

$$\xi \equiv \Phi^{\circ\circ}(t) = \alpha^\circ e^{-i\Omega t} \qquad \text{and} \qquad \zeta \equiv -\alpha^\circ \tag{17.118}$$

so that

$$\Phi^\circ(t) = \xi + \zeta \tag{17.119}$$

Besides, observe that, owing to these definitions, it is possible to write although ζ is real

$$-\frac{1}{2}(\xi\zeta^* - \xi^*\zeta) = \alpha^{\circ 2}\left(\frac{e^{-i\Omega t} - e^{i\Omega t}}{2}\right) = i\alpha^{\circ 2}\sin\Omega t \tag{17.120}$$

Then, the expression (17.116)[1] becomes

$$(e^{i\alpha^{\circ 2}\sin\Omega t})[e^{\Phi^\circ(t)^*\mathbf{a}^\dagger - \Phi^\circ(t)\mathbf{a}}] = (e^{-(1/2)(\xi\zeta^* - \xi^*\zeta)})(e^{(\xi+\zeta)\mathbf{a}^\dagger - (\xi^*+\zeta^*)\mathbf{a}}) \tag{17.121}$$

Next, use Eq. (6.97), that is,

$$(e^{-(1/2)(\xi\zeta^* - \xi^*\zeta)})(e^{(\xi+\zeta)\mathbf{a}^\dagger - (\xi^*+\zeta^*)\mathbf{a}}) = (e^{\zeta\mathbf{a}^\dagger - \zeta^*\mathbf{a}})(e^{\xi\mathbf{a}^\dagger - \xi^*\mathbf{a}})$$

As a consequence of this relation and owing to Eqs. (17.117)–(17.120), Eq. (17.121) becomes

$$(e^{i\alpha^{\circ 2}\sin\Omega t})[e^{\Phi^\circ(t)^*\mathbf{a}^\dagger - \Phi^\circ(t)\mathbf{a}}] = [e^{-\alpha^\circ\mathbf{a}^\dagger + \alpha^\circ\mathbf{a}}][e^{\Phi^{\circ\circ}(t)^*\mathbf{a}^\dagger - \Phi^{\circ\circ}(t)\mathbf{a}}] \tag{17.122}$$

Thus, the ACF given by (17.111) in the previous representation becomes in this new equivalent one

$$\mathbf{G}^\circ(t) \propto e^{i\omega^\circ t}(e^{-i2\alpha^{\circ 2}\Omega t})(1 - e^{-\lambda})\mathbf{tr}\{(e^{-\lambda\mathbf{a}^\dagger\mathbf{a}})[e^{-\alpha^\circ\mathbf{a}^\dagger + \alpha^\circ\mathbf{a}}][e^{\Phi^{\circ\circ}(t)^*\mathbf{a}^\dagger - \Phi^{\circ\circ}(t)\mathbf{a}}]\} \tag{17.123}$$

Now, observe that, in view of Eqs. (17.117) and (17.118) and according to Eq. (7.31) giving the expression of $\Phi^{\circ\circ}(t)$, the last operator appearing on the right-hand side of Eq. (17.122) may be written as the result of an Heisenberg transformation on the translation operator, which is the inverse of the first operator appearing on the right-hand side of Eq. (17.122). That leads one to write

$$[e^{\Phi^{\circ\circ}(t)^*\mathbf{a}^\dagger - \Phi^{\circ\circ}(t)\mathbf{a}}] = (e^{i\mathbf{a}^\dagger\mathbf{a}\Omega t})[e^{\alpha^\circ\mathbf{a}^\dagger - \alpha^\circ\mathbf{a}}](e^{-i\mathbf{a}^\dagger\mathbf{a}\Omega t})$$

so that the ACF (17.123) becomes

$$\mathbf{G}^\circ(t) \propto e^{i\omega^\circ t}(e^{-i2\alpha^{\circ 2}\Omega t})(1 - e^{-\lambda})\mathbf{tr}\{(e^{-\lambda\mathbf{a}^\dagger\mathbf{a}})[\mathbf{A}(\alpha^\circ)]^{-1}[(e^{i\mathbf{a}^\dagger\mathbf{a}\Omega t})[\mathbf{A}(\alpha^\circ)](e^{-i\mathbf{a}^\dagger\mathbf{a}\Omega t})]\} \tag{17.124}$$

with

$$\mathbf{A}(\alpha^\circ) = (e^{\alpha^\circ\mathbf{a}^\dagger - \alpha^\circ\mathbf{a}}) \tag{17.125}$$

Next, the trace involved in Eq. (17.124) may be expressed using the basis of the eigenstates of $\mathbf{a}^\dagger\mathbf{a}$, that is,

$$\mathbf{a}^\dagger\mathbf{a}|(n)\rangle = n|(n)\rangle \quad \text{with} \quad \langle(m)|(n)\rangle = \delta_{mn} \tag{17.126}$$

$$\sum_n |(n)\rangle\langle(n)| = \hat{\mathbf{1}} \tag{17.127}$$

[1] Y. Maréchal and A. Witkowski. *J. Chem. Phys.*, **48** (1968): 3637.

Then, proceeding in this way and inserting the closure relation (17.127) between the two translation operators, the ACF (17.124) yields

$$\mathbf{G}^\circ(t) \propto e^{i\omega^\circ t}(e^{-i2\alpha^{\circ 2}\Omega t})(1-e^{-\lambda})\sum_n\sum_m \times \langle(n)|(e^{-\lambda \mathbf{a}^\dagger \mathbf{a}})\{[\mathbf{A}(\alpha^\circ)]^{-1}|(m)\rangle\langle(m)|e^{i\mathbf{a}^\dagger \mathbf{a}\Omega t}[\mathbf{A}(\alpha^\circ)]e^{-i\mathbf{a}^\dagger \mathbf{a}\Omega t}\}|(n)\rangle \tag{17.128}$$

Again, since, comparing Eq. (17.126), the kets $|(n)\rangle$ are the eigenstates of $\mathbf{a}^\dagger\mathbf{a}$ and Eq. (17.128), we have

$$\mathbf{G}^\circ(t) \propto e^{i\omega^\circ t}(e^{-i2\alpha^{\circ 2}\Omega t})(1-e^{-\lambda})\sum_n\sum_m (e^{-\lambda n}) \times \langle(n)|[\mathbf{A}(\alpha^\circ)]^{-1}|(m)\rangle\langle(m)|e^{im\Omega t}[\mathbf{A}(\alpha^\circ)]e^{-in\Omega t}|(n)\rangle \tag{17.129}$$

or

$$\mathbf{G}^\circ(t) \propto e^{i\omega^\circ t}(e^{-i2\alpha^{\circ 2}\Omega t})(1-e^{-\lambda})\sum_n\sum_m (e^{-\lambda n})[e^{i(m-n)\Omega t}]|\mathbf{A}_{mn}(\alpha^\circ)|^2 \tag{17.130}$$

with, owing to Eq. (17.125),

$$\mathbf{A}_{mn}(\alpha^\circ) = \langle(m)|(e^{\alpha^\circ \mathbf{a}^\dagger - \alpha^\circ \mathbf{a}})|(n)\rangle \tag{17.131}$$

which are the Franck–Condon factors given by Eq. (6.133).

Finally, owing to Eq. (17.69) the line shape may be obtained by Fourier transform of the ACF (17.130), that is,

$$\mathbf{I}^\circ(\omega) \propto (1-e^{-\lambda})\sum_n\sum_m (e^{-\lambda n})\int_{-\infty}^{+\infty} e^{i\omega^\circ t}(e^{-i2\alpha^{\circ 2}\Omega t})[e^{i(m-n)\Omega t}]|\mathbf{A}_{mn}(\alpha^\circ)|^2 e^{-i\omega t}\,dt$$

$$\mathbf{I}^\circ(\omega) \propto (1-e^{-\lambda})\sum_n\sum_m (e^{-n\hbar\Omega/k_BT})|\mathbf{A}_{mn}(\alpha^\circ)|^2\delta[\omega-[\omega^\circ-(n-m)-2\alpha^{\circ 2}]\Omega] \tag{17.132}$$

This line shape is pictured in Fig. 17.8 for absolute zero temperature.

Now, if a damping of the first excited state for the fast mode is incorporated as above for obtaining Eq. (17.115), then, because Eq. (17.130) has been deduced from Eq. (17.113), the Fourier transform of ACF (17.130) involving the damping, is given by

$$\mathbf{I}(\omega) \propto \varepsilon\sum_n\sum_m (e^{-n\hbar\Omega/k_BT})|\mathbf{A}_{mn}(\alpha^\circ)|^2\frac{\gamma^\circ}{(\omega-(\omega^\circ-(n-m)\Omega-2\alpha^{\circ 2}\Omega))^2+\gamma^{\circ 2}} \tag{17.133}$$

Clearly, the spectral density involving the damping of the high-frequency mode is a double sum over m and n of Lorentzians centered on $\omega = (\omega^\circ-(n-m)\Omega-2\alpha^{\circ 2}\Omega)$ and having the same half-width γ°, but different intensities, given by $e^{-\lambda n}|\mathbf{A}_{mn}(\alpha^\circ)|^2$.

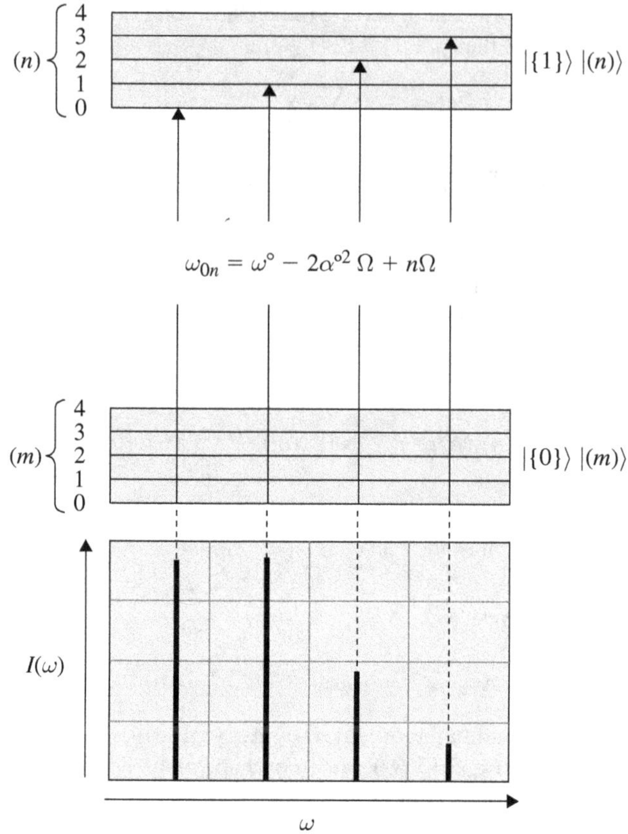

Figure 17.8 Spectral analysis at $T = 0$ K in the absence of indirect damping $\omega^\circ = 3000\,\text{cm}^{-1}$, $\Omega = 100\,\text{cm}^{-1}$, $\alpha^\circ = 1$, $\gamma^\circ = 0.025\,\Omega$, $\gamma = 0$.

17.4 SD OF DAMPED WEAK H-BONDED SPECIES

When the influence of the surroundings is ignored, starting from its intermediate expression (17.94), the ACF of the high-frequency dipole moment operator of a weak H-bond bridge is given by (17.95)

$$\mathbf{G}^\circ(t) \propto \mathbf{tr}\{\rho_B[e^{i\mathbf{H}^{\{1\}}t/\hbar}][e^{-i\mathbf{H}^{\{0\}}t/\hbar}]\} \tag{17.134}$$

with

$$\rho_B = \frac{1}{Z}(e^{-\beta\mathbf{H}^{\{0\}}})$$

an expression that has been transformed into Eq. (17.104), that is,

$$\mathbf{G}^\circ(t) \propto e^{i\omega^\circ t}e^{-i\alpha^{\circ 2}\Omega t}\frac{1}{Z}\mathbf{tr}\left\{(e^{-\beta\mathbf{H}^{\{0\}}})\hat{\mathbf{P}}\left\{\exp\left[ib\int_0^t \mathbf{Q}^\circ(t')dt'\right]\right\}\right\} \tag{17.135}$$

Now the question remains to know how the ACF (17.135) transforms when the irreversible influence of the medium on the H-bond bridge has to be taken into account. This leads one to write formally in place of Eq. (17.134) the following equation in which the medium is taken into account:

$$\mathbf{G}(t) \propto \mathbf{tr}\{\rho_B \mathbf{tr}_\theta\{[e^{i\mathbb{H}^{\{1\}}t/\hbar}][e^{-i\mathbb{H}^{\{0\}}t/\hbar}]\}\} \tag{17.136}$$

with

$$\mathbb{H}^{\{k\}} = \mathbf{H}^{\{k\}} + \mathbf{H}_\theta + \mathbf{V}$$

where $\mathbf{H}_\theta$ and $\mathbf{V}$ have the same meaning as in Eqs. (16.156) and (16.157), whereas $\mathbf{tr}_\theta$ is the trace to be performed over the thermal bath.

One possibility is to replace the IP time-dependent operator $\mathbf{Q}^\circ(t)^{\mathrm{IP}}$ governed by the deterministic dynamics Eq. (17.105) of the H-bond bridge coordinate in the absence of medium, by the time-dependent mean value $\langle \mathbf{Q}(t)\rangle$ corresponding to the time-dependent average of the position operator of a driven damped quantum harmonic oscillator. Such an approach is equivalent to assume that just after excitation of the high-frequency mode, the H-bond bridge coordinate operator, which becomes driven, is also damped by the surroundings. This leads one to write in place of the ACF (17.135)) the following expression:

$$\mathbf{G}(t) \propto e^{i\omega^\circ t} e^{-i\alpha^{\circ 2}\Omega t} \frac{1}{Z} \mathbf{tr}\left\{(e^{-\beta \mathbf{H}^{\{0\}}})\hat{\mathbf{P}}\left\{\exp\left[-ib\int_0^t \langle \mathbf{Q}(t')\rangle dt'\right]\right\}\right\} \tag{17.137}$$

which may be viewed as the result of a formal expression analogous to that of (17.94), however, with the coupling to the thermal bath incorporated into the Hamiltonian. Next, observe that the time-dependent expression over which the Dyson time-ordering operator $\hat{\mathbf{P}}$ is acting is given by Eq. (16.223), that is,

$$\hat{\mathbf{P}}\left\{\exp\left[-ib\int_0^t \langle \mathbf{Q}(t')\rangle dt'\right]\right\} = e^{\xi(t)}\{e^{A(t)\mathbf{a}-A(t)^*\mathbf{a}^\dagger}\} \tag{17.138}$$

where $A(t)$ is given by Eqs. (16.214)

$$A(t) = \beta^\circ[e^{-\gamma t/2}e^{-i\Omega t} - 1] \tag{17.139}$$

and $\xi(t)$ by Eq. (16.222), that is,

$$\xi(t) = i|\beta^\circ|^2\left(\frac{1}{\gamma}(e^{-\gamma t} - 1)\Omega + e^{-\gamma t/2}\sin \Omega t\right) \tag{17.140}$$

with, according to Eq. (16.215),

$$\beta^\circ = \alpha^\circ \Omega \frac{\Omega + i\gamma/2}{\Omega^2 + (\gamma/2)^2} \tag{17.141}$$

In these equations, γ is the damping parameter of the slow mode coordinate, which is acting through the irreversible quantum dynamics of the driven damped quantum harmonic oscillator, whereas α° is given by Eq. (17.108).

Then, due to Eq. (17.138) and writing explicitly the Boltzmann density operator (17.106), the ACF (17.137) transforms into

$$\mathbf{G}(t) \propto e^{i\omega^\circ t}(e^{-i\alpha^{\circ 2}\Omega t})e^{\xi(t)}(1 - e^{-\lambda})\mathbf{tr}\{(e^{\lambda \mathbf{a}^\dagger \mathbf{a}})e^{A(t)\mathbf{a}-A(t)^*\mathbf{a}^\dagger}\} \tag{17.142}$$

Next, using Eq. (13.203) in order to calculate the thermal average of the time-dependent translation operator, the ACF (17.142) reads

$$\mathbf{G}(t) \propto e^{i\omega^\circ t}(e^{-i\alpha^{\circ 2}\Omega t})e^{\xi(t)}\{e^{-|A(t)|^2(2\langle\mathbf{n}\rangle+1/2)}\} \tag{17.143}$$

with, as in Eq. (17.112),

$$\langle\mathbf{n}\rangle = \frac{1}{e^{\lambda}-1} \quad \text{with} \quad \lambda = \frac{\hbar\Omega}{k_B T} \tag{17.144}$$

At last, using Eqs. (17.139) and (17.140), the ACF (17.143) takes the final form that, making a small approximation, simplifies to

$$\begin{aligned}\mathbf{G}(t) \propto\ & (e^{i|\beta^\circ|^2(e^{-\gamma t}-1)\Omega/\gamma})e^{i\omega^\circ t}(e^{-i\alpha^{\circ 2}\Omega t})(e^{i|\beta^\circ|^2 e^{-\gamma t/2}\sin\Omega t}) \\ & \times (e^{\beta^{\circ 2}(\langle\mathbf{n}\rangle+1/2)(2e^{-\gamma t/2}\cos\Omega t - e^{-\gamma t}-1)})\end{aligned} \tag{17.145}$$

Note that when γ becomes very small, a first-order expansion of $\exp\{-\gamma t\}$ yields

$$\frac{1}{\gamma}(e^{-\gamma t}-1) \backsimeq \frac{1}{\gamma}(1-\gamma t-1) = -t$$

so that, due to Eq. (17.141), β° narrowing α°

$$(e^{i|\beta^\circ|^2(e^{-\gamma t}-1)\Omega/\gamma}) \backsimeq (e^{-i\alpha^{\circ 2}\Omega t})$$

Hence, for the underdamped situation, the ACF (17.145) reduces to

$$\boxed{\mathbf{G}(t) \propto e^{i\omega^\circ t}(e^{-i2\alpha^{\circ 2}\Omega t})(e^{i\beta^{\circ 2}e^{-\gamma t/2}\sin\Omega t})(e^{\beta^{\circ 2}(\langle\mathbf{n}\rangle+1/2)(2e^{-\gamma t/2}\cos\Omega t-e^{-\gamma t}-1)})} \tag{17.146}$$

17.5 APPROXIMATION FOR QUANTUM DAMPING

Owing to the convenient nature of the quantum theory of indirect damping and to the difficulty in generalizing it to complex situations, it is interesting to find a suitable approximation taking into account simply the quantum damping of oscillators. This is the aim of the present section.

17.5.1 New expression for ACF (17.146)

Start from the ACF (17.146) in which, for simplicity, we take $\beta^\circ = \alpha^\circ$, and then, pass from the $\sin\Omega t$ and $\cos\Omega t$ functions to the corresponding complex exponentials to get

$$\begin{aligned}\mathbf{G}(t) \propto\ & e^{i\omega^\circ t}(e^{-i2\alpha^{\circ 2}\Omega t})\exp\left\{\alpha^{\circ 2}\left(\frac{e^{-\gamma t/2}e^{i\Omega t} - e^{-\gamma t/2}e^{-i\Omega t}}{2}\right)\right\} \\ & \times \exp\left\{2\alpha^{\circ 2}\left\langle\mathbf{n}+\frac{1}{2}\right\rangle\left(\frac{e^{-\gamma t/2}e^{i\Omega t} + e^{-\gamma t/2}e^{-i\Omega t}}{2}\right) - \alpha^{\circ 2}\left\langle\mathbf{n}+\frac{1}{2}\right\rangle(e^{-\gamma t}-1)\right\}\end{aligned}$$

with the abridged notation

$$\left\langle\mathbf{n}+\tfrac{1}{2}\right\rangle \equiv \langle\mathbf{n}\rangle + \tfrac{1}{2}$$

Then, factorization of the last exponential leads after some reorganization to

$$\begin{aligned}\mathbf{G}(t) &\propto e^{i\omega^\circ t}(e^{-i2\alpha^{\circ 2}\Omega t})\exp\left\{\left(2\alpha^{\circ 2}\left\langle\mathbf{n}+\frac{1}{2}\right\rangle+\alpha^{\circ 2}\right)\left(\frac{e^{-\gamma t/2}e^{i\Omega t}}{2}\right)\right\}\\ &\times\exp\left\{\left(2\alpha^{\circ 2}\left\langle\mathbf{n}+\frac{1}{2}\right\rangle-\alpha^{\circ 2}\right)\left(\frac{e^{-\gamma t/2}e^{-i\Omega t}}{2}\right)\right\}\\ &\times\exp\left\{-\alpha^{\circ 2}\left\langle\mathbf{n}+\frac{1}{2}\right\rangle(e^{-\gamma t}-1)\right\}\end{aligned}$$

or

$$\mathbf{G}(t)\propto e^{i\omega^\circ t}(e^{-i2\alpha^{\circ 2}\Omega t})(e^{-\alpha^{\circ 2}\langle\mathbf{n}+1/2\rangle(e^{-\gamma t}-1)})\{(e^{\alpha^{\circ 2}\langle\mathbf{n}+1\rangle e^{-\gamma t/2}e^{i\Omega t}})(e^{\alpha^{\circ 2}\langle\mathbf{n}\rangle e^{-\gamma t/2}e^{-i\Omega t}})\}$$

Next, separate and expand each exponential:

$$\begin{aligned}\mathbf{G}(t) &\propto e^{i\omega^\circ t}(e^{-i2\alpha^{\circ 2}\Omega t})(e^{-\alpha^{\circ 2}\langle\mathbf{n}+1/2\rangle(e^{-\gamma t}-1)})\\ &\times\sum_k\sum_l\left(\frac{(\langle\mathbf{n}\rangle+1)^l}{l!}\right)\left(\frac{\langle\mathbf{n}\rangle^k}{k!}\right)(\alpha^\circ)^{2(l+k)}(e^{-\gamma(k+l)t/2})(e^{-i(k-l)\Omega t})\end{aligned} \tag{17.147}$$

Observe that, when the damping is missing, that is, when $\gamma=0$, Eq. (17.147) reduces to

$$\begin{aligned}\{\mathbf{G}(t,\gamma=0)\} &\propto e^{i\omega^\circ t}(e^{-i2\alpha^{\circ 2}\Omega t})(e^{-\alpha^{\circ 2}\langle\mathbf{n}+1/2\rangle(e^{-\gamma t}-1)})\\ &\times\sum_k\sum_l\left(\frac{(\langle\mathbf{n}\rangle+1)^l}{l!}\right)\left(\frac{\langle\mathbf{n}\rangle^k}{k!}\right)(\alpha^\circ)^{2(l+k)}(e^{i(l-k)\Omega t})\end{aligned} \tag{17.148}$$

Of course, this last expression must be equivalent to that (17.146) in which γ is taken to be zero, that is,

$$\{\mathbf{G}(t,\gamma=0)\}\propto e^{i\omega^\circ t}(e^{-i2\alpha^{\circ 2}\Omega t})(e^{i\alpha^{\circ 2}\sin\Omega t})[e^{2\alpha^{\circ 2}(\langle\mathbf{n}+1/2\rangle)(\cos(\Omega t)-1)}] \tag{17.149}$$

Besides, recall that this last expression is similar to Eq. (17.113), which has been shown previously to be also equivalent to the expression (17.130), that is,

$$\{\mathbf{G}(t,\gamma=0)\}\propto e^{i\omega^\circ t}(e^{-i2\alpha^{\circ 2}\Omega t})\frac{1}{Z}\sum_k\sum_l e^{-k\lambda}|A_{kl}(\alpha^\circ)|^2(e^{i(l-k)\Omega t}) \tag{17.150}$$

where

$$\lambda=\frac{\hbar\Omega}{k_BT}$$

while $A_{kl}(\alpha)$ are the Franck–Condon factors, that is, the matrix elements of the translation operator

$$\mathbf{A}(\alpha^\circ)=e^{\alpha^\circ\mathbf{a}^\dagger-\alpha^\circ\mathbf{a}} \tag{17.151}$$

in the basis where $\mathbf{a}^\dagger\mathbf{a}$ is diagonal, which are given by Eq. (17.131). Moreover, since, in the absence of damping, the three expressions (17.148), (17.149), and

(17.150) are equivalent, one gets by identification of Eqs. (17.148) and (17.150) and after simplification

$$(1-e^{-\lambda})\sum_k\sum_l e^{-k\lambda}|A_{kl}(\alpha^\circ)|^2$$
$$=(e^{-2\alpha^{\circ 2}\langle\mathbf{n}+1/2\rangle})\sum_k\sum_l\left(\frac{(\langle\mathbf{n}\rangle+1)^l}{l!}\right)\left(\frac{\langle\mathbf{n}\rangle^k}{k!}\right)(\alpha^{\circ 2(l+k)}) \quad (17.152)$$

Besides, since this result shows that the coefficients are the same for each component of angular frequency $(l-k)\Omega$, this last expression yields simply

$$(1-e^{-\lambda})e^{-k\lambda}|A_{kl}(\alpha^\circ)|^2=(e^{-2\alpha^{\circ 2}\langle\mathbf{n}+1/2\rangle})\left(\frac{(\langle\mathbf{n}\rangle+1)^l}{l!}\right)\left(\frac{\langle\mathbf{n}\rangle^k}{k!}\right)(\alpha^{\circ 2(l+k)}) \quad (17.153)$$

It may be observed that, at zero temperature,

$$(1-e^{-\lambda})e^{-k\lambda}\to\delta_{k,0} \qquad \text{and} \qquad \langle\mathbf{n}\rangle\to 0$$

so that, as required by Eq. (6.133) giving the Franck–Condon factors, Eq. (17.153) reduces to

$$|A_{0l}(\alpha^\circ)|^2=e^{-\alpha^{\circ 2}}\frac{\alpha^{\circ 2l}}{l!} \quad (17.154)$$

17.5.2 Approximation for Eq. (17.147)

Now, return to the situation where the quantum indirect damping is taken into account. By comparison of Eq. (17.147) dealing to the damped situation with that (17.148) corresponding to the undamped situation, it appears that incorporation of the damping in the model introduces an imaginary part in the energy levels, which is linearly increasing with the degree of excitation of the energy. Thus, and owing to the equivalence between Eqs. (17.148) and (17.150), it appears reasonable to suppose that the damping ought to induce the same consequence in the ACF (17.147), leading one therefore to write, in place of Eq. (17.147):

$$\mathbf{G}^{\text{eff}}(t)\propto(1-e^{-\lambda})e^{i\omega^\circ t}(e^{-i2\alpha^{\circ 2}\Omega t})\sum_k\sum_l e^{-k\lambda}|A_{kl}(\alpha^\circ)|^2(e^{-(k+l)\gamma t/2})(e^{-i(k-l)\Omega t}) \quad (17.155)$$

Besides, this assumption, the validity of which has to be verified a posteriori, implies a near equivalence between the exact Eq. (17.147) and its approximation Eq. (17.155), that is,

$$(1-e^{-\lambda})\sum_k\sum_l e^{-k\lambda}|A_{kl}(\alpha^\circ)|^2(e^{-(k+l)\gamma t/2})(e^{-i(k-l)\Omega t})$$
$$\backsimeq(e^{-\alpha^{\circ 2}\langle\mathbf{n}+1/2\rangle e^{-\gamma t}+1})\sum_k\sum_l\left(\frac{(\langle\mathbf{n}\rangle+1)^l}{l!}\right)\left(\frac{\langle\mathbf{n}\rangle^k}{k!}\right)\alpha^{\circ 2(l+k)}$$
$$\times(e^{-(k+l)\gamma t/2})(e^{-i(k-l)\Omega t}) \quad (17.156)$$

Next, if the above approximation (17.156) is satisfied, each term of the double sums has to satisfy

$$(1-e^{-\lambda})e^{-k\lambda}|A_{kl}(\alpha^{\circ})|^2$$
$$\backsimeq (e^{-\alpha^{\circ 2}\langle \mathbf{n}+1/2\rangle e^{-\gamma t}+1})\left(\frac{(\langle \mathbf{n}\rangle+1)^l}{l!}\right)\left(\frac{\langle \mathbf{n}\rangle^k}{k!}\right)\alpha^{\circ 2(l+k)}(e^{-(k+l)\gamma t/2})(e^{-i(k-l)\Omega t}) \tag{17.157}$$

Moreover, since the left-hand sides of Eqs. (17.157) and (17.153) are the same, the identification of the two corresponding right-hand-sides yield after simplification

$$\{e^{-\alpha^{\circ 2}\langle n+1/2\rangle e^{-\gamma t}}\} \backsimeq \{e^{-\alpha^{\circ 2}\langle n+1/2\rangle}\}$$

a result that is reasonable as the product γt is small.

17.5.3 ACF in terms of complex energy levels

Note that the approximate damped ACF (17.155) may be expressed in terms of complex energy levels by the aid of the following equation:

$$\mathbf{G}^{\text{eff}}(t) \propto e^{i\omega^{\circ}t}(e^{-i2\alpha^{\circ 2}\Omega t})(1-e^{-\lambda})\sum_n\sum_m(e^{-n\lambda})[e^{iE_n^{\{0\}*}t/\hbar}]|A_{mn}(\alpha^{\circ})|^2[e^{-iE_n^{\{0\}}t/\hbar}] \tag{17.158}$$

with, respectively,

$$E_n^{\{0\}} = n\hbar\Omega\left(1-i\frac{\gamma}{2\Omega}\right) \tag{17.159}$$

$$E_m^{\{1\}} = m\hbar\Omega\left(1-i\frac{\gamma}{2\Omega}\right)+\hbar\omega^{\circ}-2\alpha^{\circ 2}\hbar\Omega \tag{17.160}$$

Hence, from Eqs. (17.159) and (17.160), the imaginary part of the slow mode sublevels and, therefore, their "thicknesses," are linearly increasing with the excitation degree. Note also that, when $\gamma \to 0$, the ACF (17.158) reduces as required, to that of (17.130), that is,

$$\mathbf{G}^{\text{eff}}(t) \propto (1-e^{-\lambda})e^{i\omega^{\circ}t}(e^{-i2\alpha^{\circ 2}\Omega t})\sum_n\sum_m(e^{-\lambda n})|A_{mn}(\alpha^{\circ})|^2(e^{i(m-n)\Omega t})$$

17.5.4 SD in terms of Lorentzians

The approximate ACF (17.158) takes into account only the damping of the H-bond bridge. In order to introduce the damping of the fast mode, multiply, as above, for passing from Eq. (17.113) to Eq. (17.114), the ACF (17.158) by the exponential decay $e^{-\gamma^{\circ}t}$:

$$\mathbf{G}^{\text{eff}}(t) \propto \sum_n\sum_m|A_{mn}(\alpha^{\circ})|^2(e^{-n\lambda})(e^{i(\omega^{\circ}+m\Omega-2\alpha^{\circ 2}\Omega/2)t})(e^{-in\Omega t})$$
$$\times\left(\exp\left[-(m+n)\frac{\gamma}{2}t\right]\right)e^{-\gamma^{\circ}t} \tag{17.161}$$

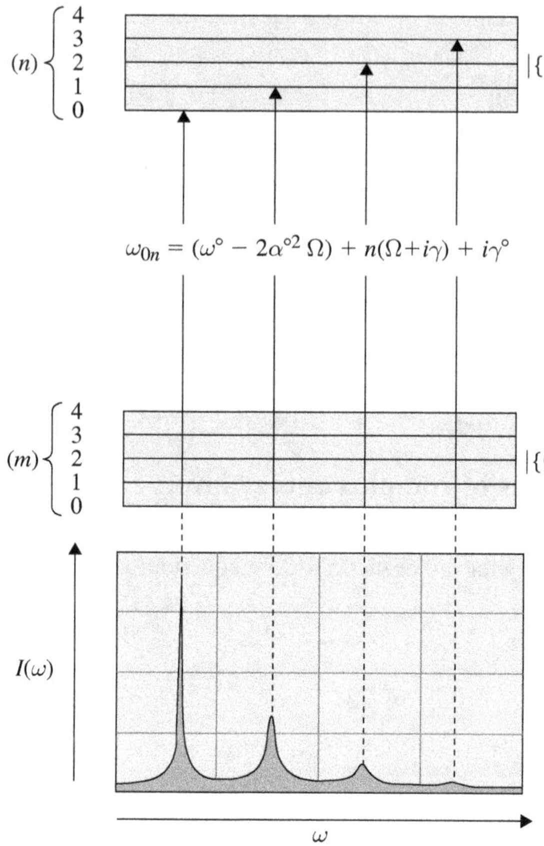

Figure 17.9 Spectral analysis at $T = 0\,\text{K}$ in the presence of damping. $\omega^\circ = 3000\,\text{cm}^{-1}$, $\Omega = 100\,\text{cm}^{-1}$, $\alpha^\circ = 1$, $\gamma^\circ = 0.025\Omega$, $\gamma = 0.10\Omega$.

so that the corresponding approached SD, which is the Fourier transform of (17.161), reads

$$\mathbf{I}^{\text{eff}}(\omega) \propto \sum_n \sum_m |A_{mn}(\alpha^\circ)|^2 (e^{-n\hbar\Omega/k_BT}) \times \frac{((m+n)(\gamma/2)+\gamma^\circ)}{\omega - (\omega^\circ + (m-n)\Omega - 2\alpha^{\circ 2}\Omega))^2 + ((m+n)(\gamma/2)+\gamma^\circ)^2} \tag{17.162}$$

Examination of Eq. (17.162) shows that the spectral density is the sum of different components, each of them being a superposition of Lorentzians involving different intensities and half-widths. This is shown in Fig. 17.9 for the special situation of zero temperature.

Now, by inspection of Eqs. (17.159) and (17.160), the H-bond bridge damping induces complex energies for the $|\{0\}\rangle|(m)\rangle$ and $|\{1\}\rangle|(n)\rangle$ H-bond bridge levels, their imaginary parts increasing linearly with excitation degree (m) or (n). This leads to linear broadening in (m) and (n) the $|\{0\}\rangle|(m)\rangle \rightarrow |\{1\}\rangle|(n)\rangle$ transitions, so that the half-widths of these transitions increase with (m) and (n), in a $(m+n)\gamma/2$ fashion. Of course, the $|\{0\}\rangle|(0)\rangle \rightarrow |\{1\}\rangle|(0)\rangle$ transition cannot be affected by the H-bond bridge damping because, according to Eqs. (17.159) and (17.160), it is given by a Dirac delta peak in the absence of the high-frequency mode damping.

17.5.5 Effective non-Hermitian Hamiltonians

From the complex energy levels (17.159) and (17.160), it is possible to pass to non-Hermitian effective Hamiltonians because these complex energy levels characterizing the initial and final states may be considered as the eigenvalues of the following effective non-Hermitian Hamiltonians:

$$\{\widetilde{\mathbf{H}}^{\{k\}}\}|(n)\rangle = \{E_n^{\{k\}}\}|(n)\rangle \tag{17.163}$$

that is,

$$\{\widetilde{\mathbf{H}}^{\{0\}}\}|(n)\rangle = n\left(1 - i\frac{\gamma}{2\Omega}\right)\hbar\Omega|(n)\rangle \tag{17.164}$$

$$\{\widetilde{\mathbf{H}}^{\{1\}}\}|(n)\rangle = \left(n\left(1 - i\frac{\gamma}{2\Omega}\right)\hbar\Omega + \hbar\omega^\circ - 2\alpha^{\circ 2}\hbar\Omega\right)|(n)\rangle \tag{17.165}$$

Hence, owing to the form of the eigenvalue equations (17.164) and (17.165), it may be inferred that the effective non-Hermitian Hamiltonians involved in them are respectively given by

$$\{\widetilde{\mathbf{H}}^{\{0\}}\} = \mathbf{a}^\dagger\mathbf{a}\left(1 - i\frac{\gamma}{2\Omega}\right)\hbar\Omega \tag{17.166}$$

$$\{\widetilde{\mathbf{H}}^{\{1\}}\} = \mathbf{a}^\dagger\mathbf{a}\left(1 - i\frac{\gamma}{2\Omega}\right)\hbar\Omega + \hbar\omega^\circ - 2\alpha^{\circ 2}\hbar\Omega \tag{17.167}$$

Note that, since the zero-point energy of the H-bond bridge harmonic Hamiltonian disappears in the eigenvalue equations (17.164) and (17.165), the corresponding factor $\frac{1}{2}$ disappears also in the effective Hamiltonians (17.166) and (17.167), which is unimportant since the line shape theory deals with transitions between eigenstates of these effective Hamiltonians.

17.6 DAMPED FERMI RESONANCES

17.6.1 First approach using non-Hermitian Hamiltonians

In Fermi resonances without damping, the Hamiltonian describing the interaction between the two excited states $|(0)\{1\}\rangle$ and $|(2)\{0\}\rangle$ [where $|\{k\}\rangle$ is the kth eigenstate of the harmonic oscillator Hamiltonian of angular frequency ω°, while $|(l)\rangle$ is the lth eigenstate of the harmonic oscillator Hamiltonian of angular frequency Ω, with $\omega^\circ \simeq 2\Omega$, to which the fast oscillator is anharmonically coupled] is given by Eq. (17.44). Thereby, after neglecting the zero-point energy, which does not play any role in the research of the line shapes, this Hamiltonian reads

$$\boxed{\mathbf{H}^\circ_{\text{Tot}}} \Longleftrightarrow \begin{pmatrix} \hbar\omega^\circ & \hbar\Lambda/2 \\ \hbar\Lambda/2 & 2\hbar\Omega \end{pmatrix} \tag{17.168}$$

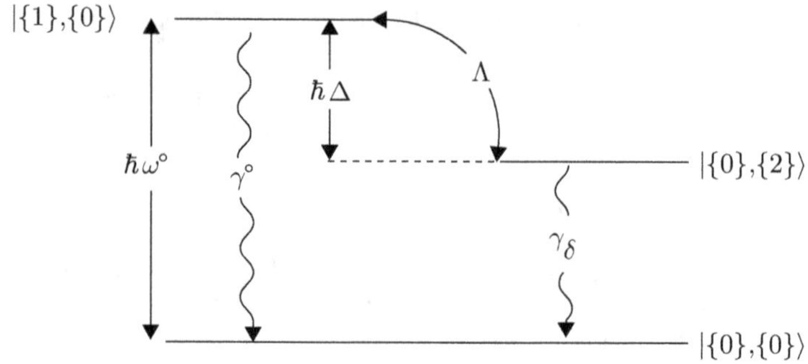

Figure 17.10 Damped Fermi resonance.

where $\hbar\Lambda/2$ is the anharmonic coupling between the two modes, which is given in terms of that ξ appearing in Eq. (17.44), by

$$\Lambda = 2\sqrt{2}\xi\Omega$$

Now, keeping in mind that according to Eqs. (17.159) and (17.160), the energy levels of a damped harmonic oscillator of angular frequency ω (for which the zero-point energy is ignored) are

$$E_n = n\hbar\omega - in\hbar\gamma \tag{17.169}$$

where γ is the damping parameter of the oscillator. Thus, in order to introduce in Eq. (17.168) the damping of the energy levels $\hbar\omega^\circ$ and $2\hbar\Omega$ corresponding, respectively, to the first excited state of the fast mode and to the second excited state of the low-frequency mode, it is convenient according to Eq. (17.169) to make the following change in the matrix (17.168):

$$\hbar\omega^\circ \to \hbar\omega^\circ - i\hbar\gamma^\circ \qquad \text{and} \qquad 2\hbar\Omega \to 2\hbar\Omega - i2\hbar\gamma^\circ$$

thus leading to

$$\boxed{\mathbf{H}_{\text{Tot}}} = \begin{pmatrix} \hbar\omega^\circ - i\hbar\gamma^\circ & \hbar\Lambda/2 \\ \hbar\Lambda/2 & \hbar\omega^\circ + \hbar\Delta - i\hbar\gamma_\delta \end{pmatrix} \quad \text{with} \quad \hbar\Delta = 2\hbar\Omega - \hbar\omega^\circ \tag{17.170}$$

Fig. 17.10 illustrates the different parameters involved in the Fermi resonance treatment. Now, write the eigenvalue equation of the Hamiltonian (17.170)

$$\mathbf{H}_{\text{Tot}}|\Psi^{\pm}\rangle = \{E^{\pm}\}|\Psi^{\pm}\rangle \tag{17.171}$$

where the $E^{\pm}$ are the complex eigenvalues, whereas the $|\Psi^{\pm}\rangle$ are the corresponding eigenkets given by

$$|\Psi^{\pm}\rangle = \{C_I^{\pm}\}|\{1\}(0)\rangle + \{C_{II}^{\pm}\}|\{0\}(2)\rangle \tag{17.172}$$

with

$$\{C_K^{\pm}\} = \langle\{1\}(0)|\Psi^{\pm}\rangle \quad \text{with} \quad K = 1, 2$$

Note that, the Hamiltonian (17.170) being non-Hermitian, its eigenvectors cannot be orthogonal, however, when the damping γ° becomes very weak with respect to the angular frequency ω°, the Hamiltonian eigenvectors remain approximately orthogonal, so that it is possible to write

$$\langle\Psi^+|\Psi^-\rangle \simeq 0 \tag{17.173}$$

$$|\Psi^+\rangle\langle\Psi^+| + |\Psi^-\rangle\langle\Psi^-| \simeq 1$$

Now, in order to study the role of Fermi resonances on IR line shapes when the damping is taken into account through (17.170), it becomes necessary to work within a time-dependent framework, and it is convenient to use the linear response theory starting from the ACF (17.77)

$$\mathbf{G}(t) = \mathbf{tr}\{\rho_B \boldsymbol{\mu}(0)^\dagger \boldsymbol{\mu}(t)\} \tag{17.174}$$

where the dipole moment operator of the fast mode is

$$\boldsymbol{\mu}(0) \propto |(0)\rangle|\{1\}\rangle\langle(0)|\langle\{0\}|$$

Next, by inversion of Eq. (17.172), the kets and bras appearing in this transition moment read

$$|(0)\{1\}\rangle = \{C_{\text{I}}^+\}|\Psi^+\rangle + \{C_{\text{II}}^-\}|\Psi^-\rangle \qquad \text{and} \qquad \langle(1)\{0\}| = \{C_{\text{I}}^+\}^*\langle\Psi^+| + \{C_{\text{II}}^-\}^*\langle\Psi^-|$$

Thus, the dipole moment operators of interest appearing in Eq. (17.174) read, respectively,

$$\boldsymbol{\mu}(t) \propto (e^{i\mathbf{H}_{\text{Tot}}t/\hbar})\{\{C_{\text{I}}^+\}|\Psi^+\rangle + \{C_{\text{II}}^-\}|\Psi^-\rangle\}\langle\{0\}(0)|(e^{-i\mathbf{H}_{\text{Tot}}t/\hbar})$$

$$\boldsymbol{\mu}(0)^\dagger \propto |(0)\{0\}\rangle\{\{C_{\text{I}}^+\}\langle\Psi^+| + \{C_{\text{II}}^-\}\langle\Psi^-|\} \tag{17.175}$$

Hence, after writing explicitly the Boltzmann density operator and the time dependence of the dipole moment operator via the Heisenberg canonical transformation (17.67), the ACF (17.174) yields

$$\begin{aligned}\mathbf{G}(t) \propto \mathbf{tr}\{&(e^{-\beta\mathbf{H}_{\text{Tot}}})|(0)\{0\}\rangle\{\{C_{\text{I}}^+\}^*\langle\Psi^+| + \{C_{\text{II}}^-\}^*\langle\Psi^-|\} \\ &\times (e^{+i\mathbf{H}_{\text{Tot}}t/\hbar})(\{C_{\text{I}}^+\}|\Psi^+\rangle + \{C_{\text{II}}^-\}|\Psi^-\rangle)\langle\{0\}(0)|(e^{-i\mathbf{H}_{\text{Tot}}t/\hbar})\}\end{aligned} \tag{17.176}$$

where $\mathbf{H}_{\text{Tot}}$ is the full Hamiltonian of the slow and fast anharmonically coupled oscillators. Now, since the zero-point energy of the ground state $|(0)\rangle|\{0\}\rangle$ may be ignored when dealing with IR line shapes, one may write

$$\mathbf{H}_{\text{Tot}}|(0)\{0\}\rangle = 0|(0)\{0\}\rangle$$

Moreover, using the usual expansion of the exponential for the application for the Boltzmann density operator on the ground state, the following expression is obtained:

$$(e^{-\beta\mathbf{H}_{\text{Tot}}})|(0)\{0\}\rangle = \left(1 - \beta\mathbf{H}_{\text{Tot}} + \frac{(\beta\mathbf{H}_{\text{Tot}})^2}{2!} + \cdots\right)|(0)\{0\}\rangle$$

so that, on neglecting the zero-point energy, we have

$$(e^{-\beta\mathbf{H}_{\text{Tot}}})|(0)\{0\}\rangle = |(0)\{0\}\rangle$$

In like manner, one obtains

$$\langle\{0\}(0)|(e^{-\beta \mathbf{H}_{\mathrm{Tot}}}) = \langle\{0\}(0)|$$

Thus, the ACF (17.176) reads

$$\begin{aligned}\mathbf{G}(t) \propto\ & \mathbf{tr}\{|(0)\{0\}\rangle\{\{C_{\mathrm{I}}^{+}\}^{*}\langle\Psi^{+}| + \{C_{\mathrm{II}}^{-}\}^{*}\langle\Psi^{-}|\}\\ & \times (e^{i\mathbf{H}_{\mathrm{Tot}}t/\hbar})(\{C_{\mathrm{I}}^{+}\}|\Psi^{+}\rangle + \{C_{\mathrm{II}}^{-}\}|\Psi^{-}\rangle)\langle\{0\}(0)|\}\end{aligned}$$

Again, using the invariance of the trace with respect to a circular permutation, and the normalization property of the ground state, that is,

$$\langle(0)\{0\}|(0)\{0\}\rangle = 1$$

one gets

$$\mathbf{G}(t) \propto \mathbf{tr}\{\{\{C_{\mathrm{I}}^{+}\}^{*}\langle\Psi^{+}| + \{C_{\mathrm{II}}^{-}\}^{*}\langle\Psi^{-}|\}(e^{i\mathbf{H}_{\mathrm{Tot}}t/\hbar})(\{C_{\mathrm{I}}^{+}\}|\Psi^{+}\rangle + \{C_{\mathrm{II}}^{-}\}|\Psi^{-}\rangle)\}$$

Now, the trace to be executed concerns only scalars, so that it disappears, and after reorganization, the ACF becomes

$$\begin{aligned}\mathbf{G}(t) \propto\ & |C_{\mathrm{I}}^{+}|^{2}\langle\Psi^{+}|(e^{i\mathbf{H}_{\mathrm{Tot}}t/\hbar})|\Psi^{+}\rangle + |C_{\mathrm{II}}^{-}|^{2}\langle\Psi^{-}|(e^{i\mathbf{H}_{\mathrm{Tot}}t/\hbar})|\Psi^{-}\rangle\\ & + \{C_{\mathrm{I}}^{+}\}^{*}\{C_{\mathrm{II}}^{-}\}\langle\Psi^{+}|(e^{i\mathbf{H}_{\mathrm{Tot}}t/\hbar})|\Psi^{-}\rangle + \{C_{\mathrm{I}}^{+}\}\{C_{\mathrm{II}}^{-}\}^{*}\langle\Psi^{-}|(e^{i\mathbf{H}_{\mathrm{Tot}}t/\hbar})|\Psi^{+}\rangle\end{aligned}$$

Then, due to the eigenvalue equation (17.171), we have

$$(e^{i\mathbf{H}_{\mathrm{Tot}}t/\hbar})|\Psi^{\pm}\rangle = (e^{iE^{\pm}t/\hbar})|\Psi^{\pm}\rangle$$

so that the ACF yields

$$\begin{aligned}\mathbf{G}(t) \propto\ & |C_{\mathrm{I}}^{+}|^{2}\langle\Psi^{+}|(e^{iE^{+}t/\hbar})|\Psi^{+}\rangle + |C_{\mathrm{II}}^{-}|^{2}\langle\Psi^{-}|(e^{iE^{-}t/\hbar})|\Psi^{-}\rangle\\ & + \{C_{\mathrm{I}}^{+}\}^{*}\{C_{\mathrm{II}}^{-}\}\langle\Psi^{+}|(e^{iE^{-}t/\hbar})|\Psi^{-}\rangle + \{C_{\mathrm{I}}^{+}\}\{C_{\mathrm{II}}^{-}\}^{*}\langle\Psi^{-}|(e^{iE^{+}t/\hbar})|\Psi^{+}\rangle\end{aligned}$$

or

$$\begin{aligned}\mathbf{G}(t) \propto\ & |C_{\mathrm{I}}^{+}|^{2}(e^{iE^{+}t/\hbar})\langle\Psi^{+}|\Psi^{+}\rangle + |C_{\mathrm{II}}^{-}|^{2}(e^{iE^{-}t/\hbar})\langle\Psi^{-}|\Psi^{-}\rangle\\ & + \{C_{\mathrm{I}}^{+}\}^{*}\{C_{\mathrm{II}}^{-}\}(e^{iE^{-}t/\hbar})\langle\Psi^{+}|\Psi^{-}\rangle + \{C_{\mathrm{I}}^{+}\}\{C_{\mathrm{II}}^{-}\}^{*}(e^{iE^{+}t/\hbar})\langle\Psi^{-}|\Psi^{+}\rangle\end{aligned}$$

and, thus, due to the pseudoorthogonality properties (17.174), and after assuming that the $|\Psi^{\pm}\rangle$ have been normalized,

$$\mathbf{G}(t) \propto |C_{\mathrm{I}}^{+}|^{2}(e^{+iE^{+}t/\hbar}) + |C_{\mathrm{II}}^{-}|^{2}(e^{+iE^{-}t/\hbar}) \tag{17.177}$$

Hence, we have the corresponding spectral density:

$$\mathbf{I}(\omega) \propto |C_{\mathrm{I}}^{+}|^{2}\int_{-\infty}^{+\infty}(e^{+iE^{+}t/\hbar})e^{-i\omega t}\,dt + |C_{\mathrm{II}}^{-}|^{2}\int_{-\infty}^{+\infty}(e^{+iE^{-}t/\hbar})e^{-i\omega t}\,dt \tag{17.178}$$

Now, if the damping is missing, γ° and γ_{δ} disappear in the Hamiltonian (17.170), which reduces to Eq. (17.168), the eigenvalues of which become $E^{\circ\pm}$, whereas the corresponding eigenvectors read

$$|\Psi^{\circ\pm}\rangle = \{C_{\mathrm{I}}^{\circ\pm}\}|(0)\{1\}\rangle + \{C_{\mathrm{II}}^{\circ\pm}\}|(2)\{0\}\rangle$$

then, the SD (17.178) reduces to

$$\mathbf{I}^{\circ}(\omega) \propto |C_{\mathrm{I}}^{\circ+}|^2 \int_{-\infty}^{+\infty} (e^{i\omega^{\circ+}t})e^{-i\omega t}dt + |C_{\mathrm{II}}^{\circ-}|^2 \int_{-\infty}^{+\infty} (e^{+i\omega^{\circ-}t})e^{-i\omega t}dt \qquad (17.179)$$

where

$$\omega^{\circ\pm} = \left(\frac{E^{\circ\pm}}{\hbar}\right)$$

and thus, due to Eq. (18.62) of Section 18.6,

$$\mathbf{I}^{\circ}(\omega) \propto |C_{\mathrm{I}}^{\circ+}|^2\{\delta(\omega - \omega^{\circ+})\} + |C_{\mathrm{II}}^{\circ-}|^2\delta\{(\omega - \omega^{\circ-})\} \qquad (17.180)$$

Different line shapes are calculated by Eq. (17.178) for different relaxation parameters; however, for the same values of the angular frequencies ω° and Ω and of the coupling parameter Λ are reported in Fig. 17.11 and compared to the corresponding two delta peaks obtained in the absence of relaxation by the aid of Eq. (17.180). By inspection of this figure, the relative magnitude of the two damping parameters strongly affects the line shapes, since it is susceptible to modify not only the overall form of the profiles but also the angular frequencies of the maxima of the two components.

17.6.2 Second approach using an analytical expression for the ACF

There is another approach to damped Fermi resonances involving, just as that leading to the SD (17.178) via non-Hermitian Hamiltonians, two damped nondegenerate interacting energy levels. This approach, which is more fundamental than that leading to (17.178), uses a formalism that is closely related to that of the quantum theory of damped oscillators studied in a previous chapter. It yields an analytical expression for the probability amplitude $\mathbf{F}(t)$ of any two nondegenerate damped interacting energy levels $|(0)\{1\}\rangle$ and $|(2)\{0\}\rangle$ lying above the ground state $|(0)\{0\}\rangle$ to remain in the excited state $|(0)\{1\}\rangle$. The meaning of this probability amplitude is close to that of the ACF (17.178), so that it may be used in place of the ACF (17.177) (which is a consequence of Eq. (17.178) for the calculation of the spectral density through a Fourier transform). Within this elaborate approach, this probability amplitude $\mathbf{F}(t)$ may be found in terms of the basic parameters $\Delta = (\omega^{\circ} - 2\Omega)$, Λ, γ_{δ}, and γ_0 given by[2]

$$\mathbf{F}(t) = \langle\{1\}|\langle(0)|(e^{+i\mathbf{H}_{\mathrm{Tot}}t/\hbar})|(0)\rangle|\{0\}\rangle = \frac{1}{2(y+ix)}e^{-\xi t/2}e^{i(\omega^{\circ}+\Delta)t}$$
$$\times \left\{[(\Delta + y) - i(\zeta - x)]e^{-xt/2}(e^{-i(\Delta+y)t/2}) - [(\Delta - y) - i(\zeta + x)]e^{xt/2}(e^{i(\Delta-y)t/2})\right\} \qquad (17.181)$$

with, respectively,

$$x^2 = \left(\frac{\delta - S}{2}\right) \qquad \text{and} \qquad y^2 = \left(\frac{\delta + S}{2}\right)$$

[2]M. Giry, B. Boulil, and O. Henri-Rousseau. *C. R. Acad. Sc. Paris*, **316** (1993): 455.

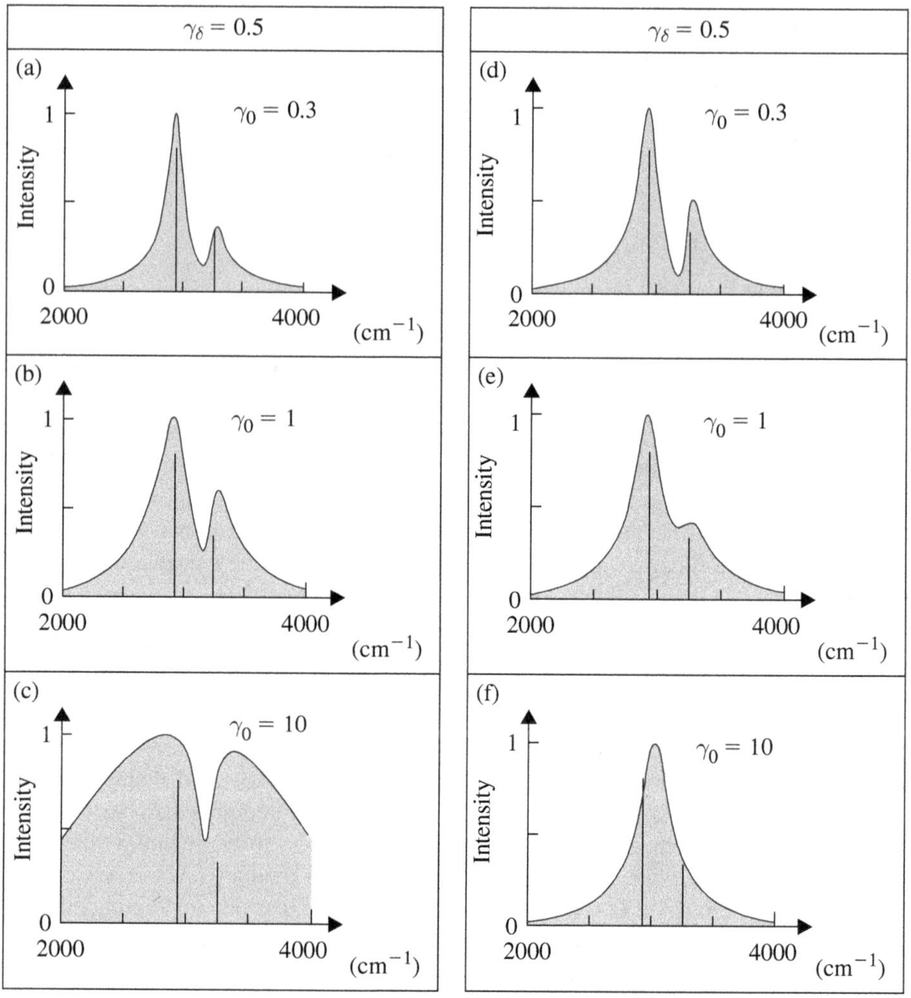

Figure 17.11 Influence of damping on line shapes involving Fermi resonance. Comparison between profiles calculated with the help of Eq. (17.179) to the corresponding Dirac delta peaks obtained from Eq. (17.180). $\omega^\circ = 3000\,\text{cm}^{-1}$, $\Lambda = 150\,\text{cm}^{-1}$, $2\omega_\delta = 3150\,\text{cm}^{-1}$.

$$\delta = \sqrt{S^2 + \Lambda^2\zeta^2} \qquad \text{and} \qquad S = \Delta^2 + \Lambda^2 - \zeta^2$$

$$\Delta = \omega^\circ - 2\Omega$$

$$\xi = \gamma_\delta + \gamma_0 \qquad \text{and} \qquad \zeta = \gamma_\delta - \gamma_0$$

The corresponding spectral densities given by the Fourier transform of Eq. (17.181) calculated with the help of the same parameters as those used in Fig. 17.11 are reported in Fig. 17.12 with the same Dirac delta peaks as those of Fig. 17.11. Comparison of the figures corroborates the above less rigorous SD approach.

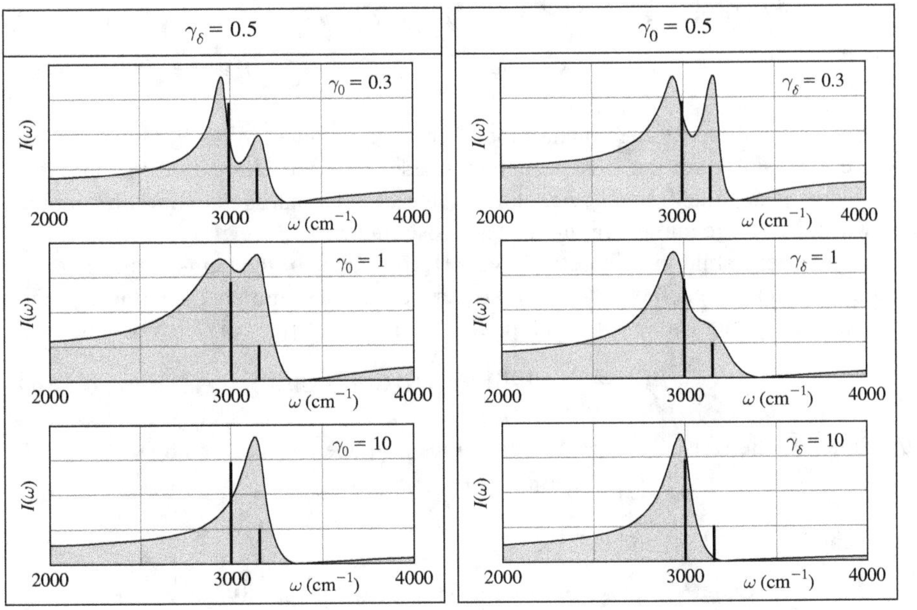

Figure 17.12 Influence of damping on line shapes involving Fermi resonance, calculated by Fourier transform of Eq. (17.181). $\omega^{\circ} = 3000\,\text{cm}^{-1}$, $\Lambda = 150\,\text{cm}^{-1}$, $2\omega_{\delta} = 3150\,\text{cm}^{-1}$.

17.7 H-BONDED IR LINE SHAPES INVOLVING FERMI RESONANCE

Now, consider the ν_{X-H} IR line shapes of weak H-bonded species in the strong anharmonic coupling theory for which a Fermi resonance occurs between the fast mode of angular frequency ω° in its ground state $|\{0\}\rangle$ and some bending mode of angular frequency ω_{δ} in its second excited state $|[2]\rangle$, and the quasi-resonant situation where the fast mode is in its first excited state $|\{1\}\rangle$ and the bending one in its ground state $|[0]\rangle$. Then, as seen in Section 10.5, it is possible, in the adiabatic approximation, to write the Hamiltonian of the system as the sum

$$\{\mathbf{H}_{\text{Fermi}}\} = \{\mathbf{H}_{\text{Fermi}}^{\{0\}}\} + \{\mathbf{H}_{\text{Fermi}}^{\{1\}}\} \tag{17.182}$$

Here the right-hand-side terms are the different effective Hamiltonians governing the H-bond bridge oscillator, the first one corresponding to the situation where both the high-frequency and the bending modes are in their ground states, and the last one to that where an energy exchange occurs between the first excitation of the fast mode and the double excitation of the bending one. More precisely, when neglecting the zero-point energy of the fast and bending modes, these Hamiltonians are

$$\{\mathbf{H}_{\text{Fermi}}^{\{0\}}\} = \left(\mathbf{a}^{\dagger}\mathbf{a} + \tfrac{1}{2}\right)\hbar\Omega \tag{17.183}$$

$$\{\mathbf{H}_{\text{Fermi}}^{\{1\}}\} = \begin{pmatrix} \left\{\left(\mathbf{a}^{\dagger}\mathbf{a} + \frac{1}{2}\right) + \alpha^{\circ}(\mathbf{a}^{\dagger} + \mathbf{a}) - \alpha^{\circ}\right\} \hbar\Omega + \hbar\omega^{\circ} & \sqrt{2}\xi\hbar\omega_{\delta} \\ \sqrt{2}\xi\hbar\omega_{\delta} & \left(\mathbf{a}^{\dagger}\mathbf{a} + \frac{1}{2}\right) \hbar\Omega + 2\hbar\omega_{\delta} \end{pmatrix} \tag{17.184}$$

where ξ is the dimensionless anharmonic coupling involved in the Fermi resonance, whereas $\mathbf{a}^{\dagger}$ and $\mathbf{a}$ are the Boson operators describing the H-bond bridge harmonic oscillator of angular frequency Ω, and α° the dimensionless anharmonic coupling between the bridge and the high-frequency oscillator.

In order to find the theoretical $\nu_{\text{X-H}}$ IR line shape of the high-frequency mode within the linear response theory, considers its non-Hermitian transition operator coupling the ground state $|\{0\}\rangle$ and the first excited state $|\{1\}\rangle$ of this mode, that is,

$$\boldsymbol{\mu}(0) = \sum_{n} |[0]\rangle|(n)\rangle|\{1\}\rangle\langle\{0\}|\langle(n)|\langle[0]| \tag{17.185}$$

At time t, in the Heisenberg picture, this transition operator is given by

$$\boldsymbol{\mu}(t) = (e^{i\{\mathbf{H}_{\text{Fermi}}\}t/\hbar})\boldsymbol{\mu}(0)(e^{-i\{\mathbf{H}_{\text{Fermi}}\}t/\hbar})$$

or, using Eqs. (17.184) and (17.185),

$$\boldsymbol{\mu}(t) = \sum_{n} (e^{i\{\mathbf{H}_{\text{Fermi}}^{\{1\}}\}t/\hbar})|[0](n)\{1\}\rangle\langle\{0\}(n)[0]|(e^{-i\{\mathbf{H}_{\text{Fermi}}^{\{0\}}\}t/\hbar})$$

with the shorthand notation

$$|[0](n)\{k\}\rangle = |[0]\rangle|(n)\rangle|\{k\}\rangle \quad \text{with} \quad k = 0, 1$$

Then, according to Eq. (17.77), the ACF $\mathbf{G}(t)$ for the $\nu_{\text{X-H}}$ absorption IR line shape yields

$$\mathbf{G}(t) \propto \sum_{m}\sum_{n} \mathbf{tr}\big\{\boldsymbol{\rho}_{\text{Tot}}|[0](m)\{1\}\rangle\langle\{0\}(m)[0]|(e^{i\{\mathbf{H}_{\text{Fermi}}^{\{1\}}\}t/\hbar}) \times |[0](n)\{1\}\rangle\langle\{0\}(n)[0]|(e^{-i\{\mathbf{H}_{\text{Fermi}}^{\{0\}}\}t/\hbar})\big\} \tag{17.186}$$

where $\boldsymbol{\rho}_{\text{Tot}}$ is the Boltzmann density operator involving the Hamiltonian (17.182), so that

$$\boldsymbol{\rho}_{\text{Tot}}|[0](n)\{1\}\rangle = \frac{1}{Z}(e^{-\{\mathbf{H}_{\text{Fermi}}^{\{0\}}\}/k_BT})|[0](n)\{1\}\rangle$$

where Z is the corresponding partition function. After simplification, using the fact that the effective Hamiltonians commute with the kets $|[l]\rangle$ and $|\{k\}\rangle$ and the corresponding bras, the ACF (17.186) simplifies to

$$\mathbf{G}(t) \propto \frac{1}{Z}\sum_{m}\sum_{n} \mathbf{tr}\big\{(e^{-\{\mathbf{H}_{\text{Fermi}}^{\{0\}}\}/k_BT})|(m)\rangle\langle(m)|(e^{i\{\mathbf{H}_{\text{Fermi}}^{\{1\}}\}t/\hbar})|(n)\rangle\langle(n)|(e^{-i\{\mathbf{H}_{\text{Fermi}}^{\{0\}}\}t/\hbar})\big\}$$

or, after suppressing the two closure relations in the basis $\{|(n)\rangle\}$ and ignoring the partition function Z, since one looks for $\mathbf{G}(t)$ with proportionality,

$$\mathbf{G}(t) \propto \mathbf{tr}\left\{(e^{-\{\mathbf{H}_{\text{Fermi}}^{\{0\}}\}/k_BT})\exp\left\{i\frac{\{\mathbf{H}_{\text{Fermi}}^{\{1\}}\}t}{\hbar}\right\}\exp\left\{-i\frac{\{\mathbf{H}_{\text{Fermi}}^{\{0\}}\}t}{\hbar}\right\}\right\}$$

Moreover, after using Eq. (17.183), performing the trace over the same basis, we have

$$\mathbf{G}(t) \propto \sum_n \langle (n)|(e^{-(\mathbf{a}^\dagger\mathbf{a}+1/2)\hbar\Omega/k_BT}) \exp\left\{ i\frac{\{\mathbf{H}^{\{1\}}_{\text{Fermi}}\}t}{\hbar} \right\} \exp\left\{ -i\frac{\{\mathbf{H}^{\{0\}}_{\text{Fermi}}\}t}{\hbar} \right\} |(n)\rangle \tag{17.187}$$

Again, using the fact that the ground-state effective Hamiltonians are diagonal with respect to $\mathbf{a}^\dagger\mathbf{a}$, of which the $|(n)\rangle$ are eigenkets, and after ignoring the zero-point energies of the oscillators in the Boltzmann distributions, Eq. (17.187) transforms to

$$\mathbf{G}(t) \propto \sum_n (e^{-(n+1/2)\lambda})\langle (n)| \exp\left\{ i\frac{\{\mathbf{H}^{\{1\}}_{\text{Fermi}}\}t}{\hbar} \right\} |(n)\rangle(e^{-i(n+1/2)\Omega t})$$

with

$$\lambda = \frac{\hbar\Omega}{k_BT} \tag{17.188}$$

Next, write the eigenvalue equation of the effective Hamiltonian (17.184)

$$\{\mathbf{H}^{\{1\}}_{\text{Fermi}}\}|\Psi^{\{1\}}_k\rangle = \{E^{\{1\}}_k\}|\Psi^{\{1\}}_k\rangle \tag{17.189}$$

where $E^{\{1\}}_k$ are the eigenvalues of the matrix representation of Eq. (17.184)

$$|\Psi^{\{1\}}_k\rangle = \sum_n \{C^{\{1\}}_{nk}\}|(n)\rangle \tag{17.190}$$

Then, insert the closure relation involving these eigenkets between the exponential operator and the ket $|(n)\rangle$ to give

$$\mathbf{G}(t) \propto \sum_n\sum_k (e^{-(n+1/2)\lambda})\langle (n)| \exp\left\{ i\frac{\{\mathbf{H}^{\{1\}}_{\text{Fermi}}\}t}{\hbar} \right\} |\Psi^{\{1\}}_k\rangle\langle\Psi^{\{1\}}_k|(n)\rangle(e^{-i(n+1/2)\Omega t}) \tag{17.191}$$

Hence, using Eq. (17.189), the ACF (17.191) becomes

$$\mathbf{G}(t) \propto \sum_n\sum_k (e^{-(n+1/2)\lambda})\langle (n)|\Psi^{\{1\}}_k\rangle \exp\{i\omega^{\{1\}}_k t\}\langle\Psi^{\{1\}}_k|(n)\rangle(e^{-i(n+1/2)\Omega t})$$

or

$$\mathbf{G}(t) \propto \sum_n\sum_k |C^{\{1\}}_{nk}|^2(e^{-n\lambda}) \exp\{i\omega^{\{1\}}_k t\}(e^{-i(n+1/2)\Omega t}) \tag{17.192}$$

Now, the spectral density may then be obtained by taking the Fourier transform of the ACF (17.192) times an exponential decay $\exp(-\gamma^\circ t)$ expressing the lifetime of the excited state $|\{1\}\rangle$ of the fast mode. Hence, using Eq. (17.188), one obtains

$$\mathbf{I}(\omega) \propto \sum_n\sum_k |C^{\{1\}}_{nk}|^2 \int_{-\infty}^{+\infty} (e^{-(n+1/2)\lambda}) \exp\{i\omega^{\{1\}}_k t\}(e^{-i(n+1/2)\Omega t})e^{-i\omega t}e^{-\gamma^\circ t}\, dt \tag{17.193}$$

To get numerical values for the spectral density, it is convenient to numerically diagonalize the eigenvalue equation (17.189) for matrix representations of the Hamiltonian (17.184) by increasing progressively the matrix size and, then, to introduce the numerical eigenvalues and eigenvectors (17.190) in (17.193) until stabilization is achieved.

Figure 17.13 compares eight line shapes involving a single Fermi resonance (continuous thick lines) to the corresponding ones without it (filled line shapes). All line shapes involve the same fast mode angular frequency $\omega^\circ = 3000\,\text{cm}^{-1}$ and also the same parameters α°, Ω, ξ, and γ° (given in the figure caption) except the bending mode angular frequency ω_δ (when Fermi resonance occurs), which is varying from $2\omega_\delta = 1000\,\text{cm}^{-1}$ to $2\omega_\delta = 3800\,\text{cm}^{-1}$. It may be observed from the figure that in the range lying from $2\omega_\delta = 1400\,\text{cm}^{-1}$ to $2\omega_\delta = 3200\,\text{cm}^{-1}$, the coupling Fermi resonance $\xi\sqrt{2}\Omega = 120\,\text{cm}^{-1}$ affects the line shape showing that Fermi resonance plays

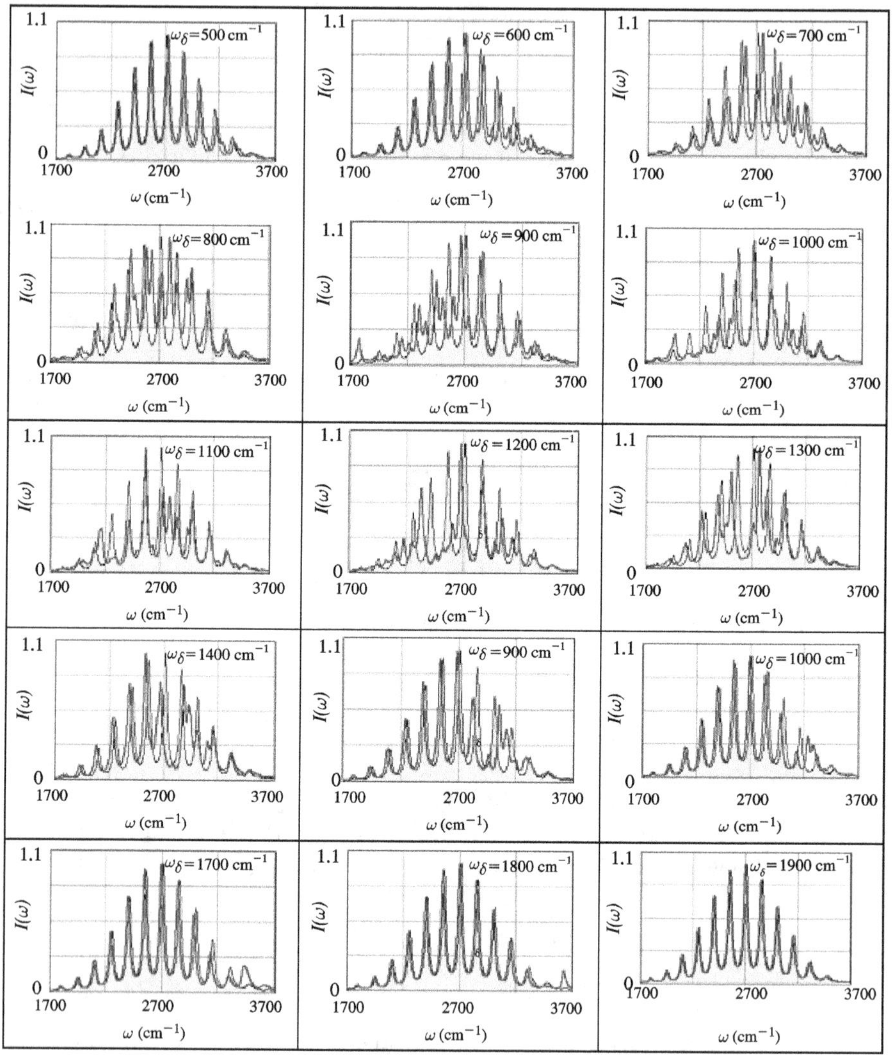

Figure 17.13 ν_{X-H} spectral densities of weak H-bonded species involving a Fermi resonance for different values of the ω_δ. $\omega^\circ = 3000\,\text{cm}^{-1}$, $\Omega = 150\,\text{cm}^{-1}$, $\alpha^\circ = 1.5$, $\xi\sqrt{2} = 0.8$, $\gamma^\circ = 0.15\,\Omega$.

a role in the physical system over a range of 1800 cm^{-1}, that is,15 times the coupling. This must be emphasized since in the absence of anharmonic coupling the Fermi resonance affects the line shape on a range that is around 0.2 times the coupling, so that it appears that the anharmonic coupling $\alpha°\Omega = 1.5 \times 150\,\text{cm}^{-1} = 225\,\text{cm}^{-1}$ increases the ability of the Fermi resonance to affect the line shapes by a factor $15/0.2 = 75$. Thus, it may be concluded that the strong anharmonic coupling drastically assists the Fermi resonance.

These results may be compared to the corresponding ones where the anharmonic coupling between the high-frequency mode and the H-bond bridge vanishes, that is, where $\alpha° = 0$, which is equivalent to considering a situation where there is only a Fermi resonance between the bending mode of angular frequency ω_δ and the fast one of angular frequency $\omega°$. These special line shapes are given in Fig. 17.14 for

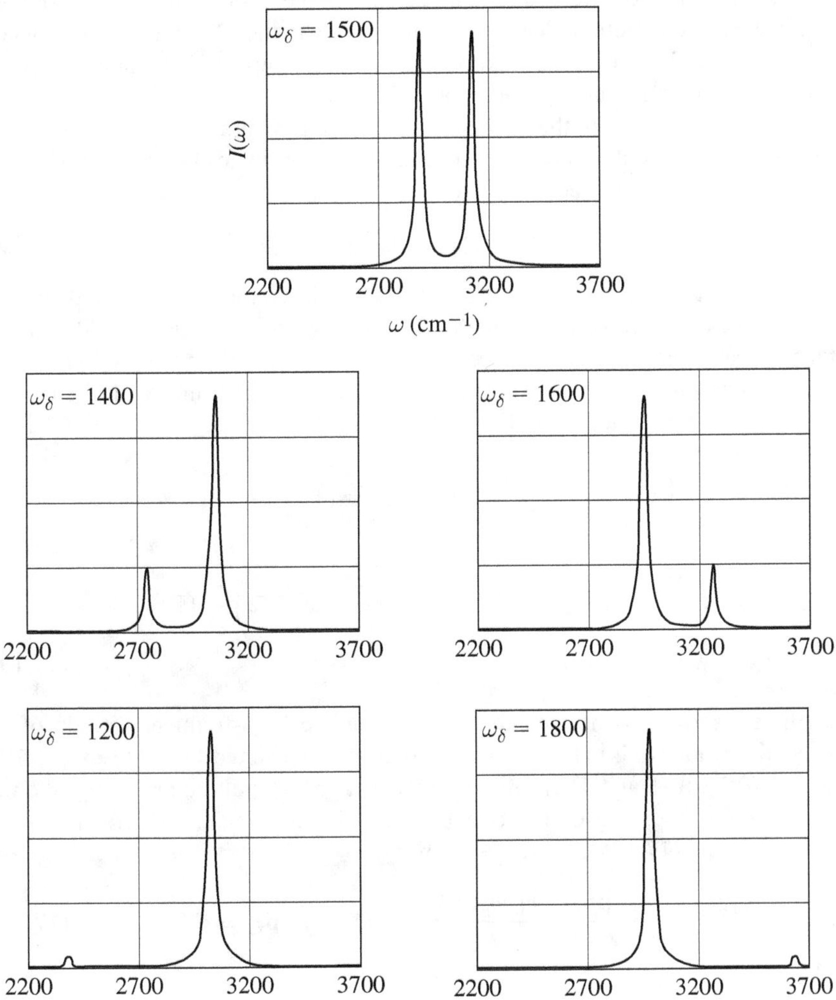

Figure 17.14 Line shapes obtained from Eq. (17.193) when the Fermi coupling is vanishing.

four line shapes of the same angular frequency $\omega^\circ = 3000\,\text{cm}^{-1}$ and the same Fermi coupling $\xi\sqrt{2}\Omega = 120\,\text{cm}^{-1}$ (as in the present figure), $2\omega_\delta$ passing from the resonant situation $2\omega_\delta = 3000\,\text{cm}^{-1}$ to nonresonant ones $2\omega_\delta = 3000\,\text{cm}^{-1} \pm 200\,\text{cm}^{-1}$ and $2\omega_\delta = 3000\,\text{cm}^{-1} \pm 600\,\text{cm}^{-1}$.

Inspection of this figure shows that when the gap between ω° and $2\omega_\delta$ is equal to $600\,\text{cm}^{-1}$, that is, around twice $\xi\sqrt{2}\Omega$, one of the doublet intensities almost vanishes.

17.8 LINE SHAPES OF H-BONDED CYCLIC DIMERS

17.8.1 Hamiltonian of dimer embedded in thermal bath

In Section 10.6 it was shown how to diagonalize the Hamiltonian of a cyclic dimer of carboxylic acid involving two H-bond bridges in the strong anharmonic coupling theory. In the present section, these results are applied to the IR spectral density where the irreversible influence of the surroundings is taken into account using the driven damped quantum harmonic oscillator formalism.

Recall that because of the symmetry of the centrosymmetric cyclic dimer, a $\widehat{\mathbf{C}}_2$ operator exists (with $\widehat{\mathbf{C}}_2^2 = 1$), which exchanges the coordinates $\mathbf{Q}_i$ of the two H-bond bridges of the cyclic dimer according to

$$\widehat{\mathbf{C}}_2\mathbf{Q}_a = \mathbf{Q}_b \qquad \widehat{\mathbf{C}}_2\mathbf{Q}_b = \mathbf{Q}_a \tag{17.194}$$

Consider the system in the strong anharmonic coupling theory and the two H-bond bridges as harmonic oscillators. Then, if these bridges are embedded within the thermal bath, each of them is described by effective Hamiltonians similar to those in (10.119) and (10.120). However, in each of them, the coupling with the thermal bath has to be incorporated, as in Eqs. (16.4) and (16.5):

$$\{\mathbb{H}^{\{0\}}\}_i = \left(\frac{\mathbf{P}_i^2}{2M} + \frac{M\Omega^2\mathbf{Q}_i^2}{2}\right) + \{\mathbb{H}^{\{\theta\}}\} + \{\mathbf{V}\}_i \qquad i = a, b \tag{17.195}$$

$$\{\mathbb{H}^{\{1\}}\}_i = \left(\frac{\mathbf{P}_i^2}{2M} + \frac{M\Omega^2\mathbf{Q}_i^2}{2} + \hbar b\mathbf{Q}_i + \hbar\omega^\circ - \alpha^{\circ 2}\hbar\Omega\right) + \{\mathbb{H}^{\{\theta\}}\} + \{\mathbf{V}\}_i \quad \text{with} \quad i = a, b \tag{17.196}$$

Here, the subscripts are related to the fact that the high-frequency mode of the corresponding moiety i is in its ground state or in its first excited state. In Eqs. (17.195) and (17.196), $\{\mathbb{H}^{\{\theta\}}\}$ and $\{\mathbf{V}\}_i$ are, respectively, the Hamiltonians of the thermal bath and of the coupling of the two bridges with it, which, in agreement with Eqs. (17.195) and (17.196), yield, respectively,

$$\{\mathbb{H}^{\{\theta\}}\} = \sum_r \left(\frac{\widetilde{\mathbf{p}}_r^2}{2m_r} + \frac{m_r\omega_r^2\widetilde{\mathbf{q}}_r^2}{2}\right) \quad \text{with} \quad [\widetilde{\mathbf{q}}_r, \widetilde{\mathbf{p}}_s] = i\hbar\delta_{rs} \tag{17.197}$$

$$\{\mathbf{V}\}_I = \sum_r \widetilde{\mathbf{q}}_r\mathbf{Q}_i\eta_r \tag{17.198}$$

In these equations, $\mathbf{P}_i$ are the conjugate momenta of the coordinates $\mathbf{Q}_i$ of the H-bond bridges of the two moieties, whereas $\mathbf{p}_i$ and $\mathbf{q}_i$ are the coordinates and the conjugate momenta of the two degenerate high-frequency modes of the two moieties and η is the coupling parameter between the cyclic dimer oscillators and those of the thermal bath. At last, Ω is the angular frequency of the H-bond bridge, η_r is an anharmonic coupling with the thermal bath, whereas ω° is the angular frequency of the two degenerate fast modes when the corresponding H-bond bridge coordinates are at equilibrium.

In the adiabatic approximation, and just as in Eq. (10.118), the Hamiltonians of each moiety of the dimer take the form of a sum of effective Hamiltonians depending on the degree of excitation of the fast mode:

$$\{\mathbb{H}_{\text{Adiab}_i}\} = \{\mathbb{H}^{\{0\}}\}_i|\{0\}_i\rangle\langle\{0\}_i| + \{\mathbb{H}^{\{1\}}\}_i|\{1\}_i\rangle\langle\{1\}_i|$$

Next, because of the degeneracy, consider an excitation of the fast mode of one moiety of the dimer. The corresponding excited state is resonant with the state where the fast mode of the other moiety is excited. Thus, a coupling may occur (*Davydov coupling*) when one of the fast mode has been excited. The Hamiltonian of the cyclic dimer involving Davydov coupling between the first excited state of the high-frequency oscillator a of one moiety, and the excited state of the oscillator b of the other moiety and vice versa, is, according to the fact that the influence of the medium is taken or not into account, given, respectively, by

$$\{\mathbb{H}_{\text{Dav}}\} = \{\mathbb{H}_{\text{Adiab}}\}_a + \{\mathbb{H}_{\text{Adiab}}\}_b + \mathbf{V}^\circ \tag{17.199}$$

Here, $\mathbf{V}^\circ$ is the Davydov coupling Hamiltonian between the first excited state of the two high-frequency oscillators:

$$\mathbf{V}^\circ = \hbar V^\circ\{|\{1\}_a\rangle\langle\{0\}_b| + |\{0\}_a\rangle\langle\{1\}_b|\}$$

$$\widehat{\mathbf{C}}_2\mathbf{V}^\circ = \mathbf{V}^\circ$$

Now, the effective Hamiltonian describing the two H-bond bridge moieties, when both the fast modes are in their ground states, is that of Eq. (10.126) plus the presence of the Hamiltonians of the bath and of its coupling with the dimer, that is,

$$\{\mathbb{H}^{\{0,0\}}\} = \{\mathbb{H}^{\{0\}}\}_a + \{\mathbb{H}^{\{0\}}\}_b + \{V\}_b + \{\mathbb{H}^{\{\theta\}}\} \tag{17.200}$$

Also, the effective Hamiltonian describing the system when the high-frequency mode a is excited, whereas the other one b remains in its ground state, is that of Eq. (10.127) in which have been incorporated the Hamiltonian of the thermal bath and that of its coupling with the dimer

$$\{\mathbb{H}^{\{1,0\}}\} = \{\mathbb{H}^{\circ\{1\}}\}_a + \{\mathbb{H}^{\circ\{0\}}\}_b + \{\mathbf{V}\}_a + \{\mathbf{V}\}_b + \{\mathbb{H}^{\{\theta\}}\} \tag{17.201}$$

while the effective Hamiltonian corresponding to the situation where the excitation is exchanged passing from the a mode to the b one, is given by

$$\{\mathbb{H}^{\{0,1\}}\} = \{\mathbb{H}^{\circ\{0\}}\}_a + \{\mathbb{H}^{\circ\{1\}}\}_b + \{\mathbf{V}\}_a + \{\mathbf{V}\}_b + \{\mathbb{H}^{\{\theta\}}\}$$

In the basis (10.124), the Hamiltonian (17.199) is

$$\{\mathbb{H}_{\text{Dav}}\} = \begin{pmatrix} \{\mathbb{H}^{\{0,0\}}\} & 0 & 0 \\ 0 & \{\mathbb{H}^{\{1,0\}}\} & \mathbf{V}^{\circ} \\ 0 & \mathbf{V}^{\circ} & \{\mathbb{H}^{\{0,1\}}\} \end{pmatrix} \tag{17.202}$$

As for the situation without thermal bath, and according to Eqs. (10.128) and (10.129), the action of the parity operator on the Hamiltonian $\mathbb{H}^{\{1,0\}}$ transforms it into $\mathbb{H}^{\{0,1\}}$ and vice versa, while this same operator does not affect the Hamiltonian $\mathbf{V}^{\circ}$.

$$\widehat{\mathbf{C}}_2\{\mathbb{H}^{\{i,j\}}\} = \{\mathbb{H}^{\{j,i\}}\}$$

$$\widehat{\mathbf{C}}_2\mathbf{V}^{\circ} = \mathbf{V}^{\circ} \tag{17.203}$$

Note that Eq. (17.203) has to be read keeping in mind Eq. (10.130), that is, the following explicit expressions hold for any entity $\mathbf{A}$ such as an operator or a ket

$$\widehat{\mathbf{C}}_2\{\mathbb{H}^{\{i,j\}}\}\mathbf{A} = \{\mathbb{H}^{\{j,i\}}\}\widehat{\mathbf{C}}_2\mathbf{A} \qquad \widehat{\mathbf{C}}_2\mathbf{V}^{\circ}\mathbf{A} = \mathbf{V}^{\circ}\widehat{\mathbf{C}}_2\mathbf{A} \tag{17.204}$$

Besides, the square of the parity operator must be unity, that is,

$$(\widehat{\mathbf{C}}_2)^2 = \widehat{\mathbf{1}} \tag{17.205}$$

17.8.2 Diagonal Hamiltonian form using symmetry

Starting from Eq. (17.202) and proceeding as in Section 10.6 dealing with Davydov coupling, when passing from Eqs. (10.125) to (10.142), one may verify, by the aid of the spinors (10.135) and (10.138),

$$\{\tilde{\boldsymbol{\beta}}^{(+)}\} = \begin{pmatrix} \{\boldsymbol{\beta}^{(+)}\} \\ +\hat{\mathbf{C}}_2\{\boldsymbol{\beta}^{(+)}\} \end{pmatrix} \quad \text{and} \quad \{\tilde{\boldsymbol{\beta}}^{(-)}\} = \begin{pmatrix} \{\boldsymbol{\beta}^{(-)}\} \\ -\hat{\mathbf{C}}_2\{\boldsymbol{\beta}^{(-)}\} \end{pmatrix} \tag{17.206}$$

that an equation of the same kind as (10.142) (which was obtained in the absence of thermal bath) holds also in the presence of a coupling with the thermal bath, so that it is possible to write in the present situation:

$$\{\mathbb{H}_{\text{Dav}}\} = \begin{pmatrix} \{\mathbb{H}^{\{0,0\}}\} & 0 & 0 \\ 0 & \{\mathbb{H}^{\{1,1\}}_{\{+\}}\} & 0 \\ 0 & 0 & \{\mathbb{H}^{\{1,1\}}_{\{-\}}\} \end{pmatrix} \tag{17.207}$$

with

$$\{\mathbb{H}^{\{1,1\}}_{\{\pm\}}\} \equiv \{\mathbb{H}^{\{1,0\}}\} \pm \mathbf{V}^{\circ}\widehat{\mathbf{C}}_2 \tag{17.208}$$

and where, owing to Eqs. (17.200) and (17.201), and due to Eqs. (17.195)–(17.198), the effective Hamiltonians appearing in Eq. (17.207) are, respectively, given by

$$\{\mathbb{H}^{\{0,0\}}\} = \left(\frac{\mathbf{P}_a^2}{2M} + \frac{M\Omega^2\mathbf{Q}_a^2}{2}\right) + \left(\frac{\mathbf{P}_b^2}{2M} + \frac{M\Omega^2\mathbf{Q}_b^2}{2}\right) + \sum_r \left(\frac{\widetilde{\mathbf{p}}_r^2}{2m_r} + \frac{m_r\omega_r^2\widetilde{\mathbf{q}}_r^2}{2}\right) \tag{17.209}$$

and

$$\begin{aligned}\{\mathbb{H}^{\{1,1\}}_{\{\pm\}}\} &= \left(\frac{\mathbf{P}_a^2}{2M} + \frac{M\Omega^2\mathbf{Q}_a^2}{2}\right) + \hbar b\mathbf{Q}_a + \hbar\omega^\circ - \alpha^{\circ 2}\hbar\Omega \\ &+ \left(\frac{\mathbf{P}_b^2}{2M} + \frac{M\Omega^2\mathbf{Q}_b^2}{2}\right) + \sum_r \left(\frac{\widetilde{\mathbf{p}}_r^2}{2m_r} + \frac{m_r\omega_r^2\widetilde{\mathbf{q}}_r^2}{2}\right) \\ &+ \sum_r \widetilde{\mathbf{q}}_r(\mathbf{Q}_a + \mathbf{Q}_b)\eta_r \pm \mathbf{V}^\circ\widehat{\mathbf{C}}_2 \end{aligned} \tag{17.210}$$

Next, in order to use the symmetry properties of the system, pass to symmetry coordinates according to

$$\mathbf{Q}_g = \left(\frac{\mathbf{Q}_a + \mathbf{Q}_B}{\sqrt{2}}\right) \quad \text{and} \quad \mathbf{Q}_u = \left(\frac{\mathbf{Q}_a - \mathbf{Q}_B}{\sqrt{2}}\right) \tag{17.211}$$

$$\mathbf{P}_g = \left(\frac{\mathbf{P}_a + \mathbf{P}_B}{\sqrt{2}}\right) \quad \text{and} \quad \mathbf{P}_u = \left(\frac{\mathbf{P}_a - \mathbf{P}_B}{\sqrt{2}}\right) \tag{17.212}$$

for which, due to Eq. (17.194),

$$\widehat{\mathbf{C}}_2\mathbf{Q}_g = \mathbf{Q}_g \quad \text{and} \quad \widehat{\mathbf{C}}_2\mathbf{Q}_u = -\mathbf{Q}_u$$

Again, in the symmetrical coordinates (17.211) and (17.212), the interaction Hamiltonians Eqs. (17.197) and (17.198) yield

$$\{\mathbf{V}\}_a + \{\mathbf{V}\}_b = \sqrt{2}\sum_r \widetilde{\mathbf{q}}_r\mathbf{Q}_g\eta_r \equiv \{\mathbb{H}_g^{\{Int\}}\}$$

while, the following relation holds for the symmetrized and nonsymmetrized squared position and momentum coordinates

$$\mathbf{P}_a^2 + \mathbf{P}_b^2 = \mathbf{P}_g^2 + \mathbf{P}_u^2 \tag{17.213}$$

$$\mathbf{Q}_a^2 + \mathbf{Q}_b^2 = \mathbf{Q}_g^2 + \mathbf{Q}_u^2 \tag{17.214}$$

Hence, within the symmetrized coordinates, the effective ground-state Hamiltonian (17.209) becomes

$$\{\mathbb{H}^{\{0,0\}}\} = \{\mathbb{H}_g^{\{0\}}\} + \{\mathbf{H}_u^{\{0\}}\} \tag{17.215}$$

with

$$\{\mathbb{H}_g^{\{0\}}\} = \{\mathbf{H}_g^{\{0\}}\} + \sum_r \left(\frac{\widetilde{\mathbf{p}}_r^2}{2m_r} + \frac{m_r\omega_r^2\widetilde{\mathbf{q}}_r^2}{2}\right) \tag{17.216}$$

$$\{\mathbf{H}_g^{\{0\}}\} = \left(\frac{\mathbf{P}_g^2}{2M} + \frac{M\Omega^2\mathbf{Q}_g^2}{2}\right) \tag{17.217}$$

and

$$\{\mathbf{H}_u^{\{0\}}\} = \left(\frac{\mathbf{P}_u^2}{2M} + \frac{M\Omega^2\mathbf{Q}_u^2}{2}\right) \tag{17.218}$$

the corresponding eigenvalue equations of the g and u effective Hamiltonians (17.224) and (17.225) being

$$\begin{aligned}
\{\mathbf{H}_g^{\{1\}}\}|\{\alpha_g^{\{1\}}\}_\nu\rangle &= \{\mathbf{E}_g^{\{1\}}\}_\nu|\{\alpha_g^{\{1\}}\}_\nu\rangle \\
\{\mathbf{H}_u^{\{+\}}\}|\{\xi^{\{+\}}\}_\mu\rangle &= \{\hbar\omega^{\{+\}}\}_\mu|\{\xi^{\{+\}}\}_\mu\rangle \\
\{\mathbf{H}_u^{\{\pm\}}\}|\{\xi^{\{-\}}\}_\mu\rangle &= \{\hbar\omega^{\{-\}}\}_\mu|\{\xi^{\{-\}}\}_\mu\rangle
\end{aligned} \tag{17.219}$$

with the last two being condensed into

$$\{\mathbf{H}_u^{\{\pm\}}\}|\{\xi^{\{\pm\}}\}_\mu\rangle = \{\hbar\omega^{\{\pm\}}\}_\mu|\{\xi^{\{\pm\}}\}_\mu\rangle \tag{17.220}$$

Now, the excited effective Hamiltonian (17.201) transforms into

$$\{\mathbb{H}^{\{1,0\}}\} = \{\mathbb{H}_g^{\{1\}}\} + \{\mathbf{H}_u^{\{1\}}\} + \hbar\omega^\circ - \alpha^{\circ 2}\hbar\Omega \tag{17.221}$$

where

$$\{\mathbf{H}_u^{\{1\}}\} = \left\{\left(\frac{\mathbf{P}_u^2}{2M} + \frac{M\Omega^2\mathbf{Q}_u^2}{2}\right) + \hbar b\frac{\mathbf{Q}_u}{\sqrt{2}}\right\} \tag{17.222}$$

$$\{\mathbb{H}_g^{\{1\}}\} = \{\mathbf{H}_g^{\{1\}}\} + \sqrt{2}\sum_r \widetilde{\mathbf{q}}_r\mathbf{Q}_g\eta_r + \sum_r\left(\frac{\widetilde{\mathbf{p}}_r^2}{2m_r} + \frac{m_r\omega_r^2\widetilde{\mathbf{q}}_r^2}{2}\right) \tag{17.223}$$

with

$$\{\mathbf{H}_g^{\{1\}}\} = \left\{\left(\frac{\mathbf{P}_g^2}{2M} + \frac{M\Omega^2\mathbf{Q}_g^2}{2}\right) + \hbar b\frac{\mathbf{Q}_g}{\sqrt{2}}\right\} \tag{17.224}$$

Next, according to Eq. (17.210) and also due to Eq. (17.222), the $u\{\pm\}$ subspace Hamiltonian components yield

$$\{\mathbf{H}_u^{\{\pm\}}\} = \left(\frac{\mathbf{P}_u^2}{2M} + \frac{M\Omega^2\mathbf{Q}_u^2}{2}\right) + \hbar b\frac{\mathbf{Q}_u}{\sqrt{2}} \pm \mathbf{V}^\circ\widehat{\mathbf{C}}_2 \tag{17.225}$$

Hence, owing to Eqs. (17.215) and (17.221), the Davydov Hamiltonian (17.207) reads

$$\begin{aligned}
\{\mathbb{H}_{\text{Dav}}\} = {} & \{\mathbb{H}_g^{\{0\}}\}\big|\{0\}_g\big\rangle\big\langle\{0\}_g\big| + \{\mathbf{H}_u^{\{0\}}\}\big|\{0\}_u\big\rangle\big\langle\{0\}_u\big| \\
& + \{\mathbb{H}_g^{\{1\}}\}\sum_\nu\big|\{\alpha_g^{\{1\}}\}_\nu\big\rangle\big\langle\{\alpha_g^{\{1\}}\}_\nu\big| + \{\mathbf{H}_u^{\{+\}}\}\sum_\mu\big|\{\xi^{\{+\}}\}_\mu\big\rangle\big\langle\{\xi^{\{+\}}\}_\mu\big| \\
& + \{\mathbf{H}_u^{\{-\}}\}\sum_\mu\big|\{\xi^{\{-\}}\}_\mu\big\rangle\big\langle\{\xi^{\{-\}}\}_\mu\big|
\end{aligned} \tag{17.226}$$

Observe that when the Davydov coupling is missing, Eqs. (17.224) and (17.225) indicate that

$$\{\mathbf{H}_u^{\{\pm\}}\} \rightarrow \{\mathbf{H}_u^{\{1\}}\}$$

so that Eq. (17.226) reduces to

$$\begin{aligned}\{\mathbb{H}_{\mathrm{Dav}}\} \rightarrow\ & \{\mathbb{H}_g^{\{0\}}\}|\{0\}_g\rangle\langle\{0\}_g| + \{\mathbf{H}_u^{\{0\}}\}|\{0\}_u\rangle\langle\{0\}_u| \\ & + \{\mathbb{H}_g^{\{1\}}\}\sum_\nu |\{\alpha_g^{\{1\}}\}_\nu\rangle\langle\{\alpha_g^{\{1\}}\}_\nu| + \{\mathbb{H}_u^{\{1\}}\}\sum_\nu |\{\alpha_u^{\{1\}}\}_\nu\rangle\langle\{\alpha_u^{\{1\}}\}_\nu|\end{aligned}$$

17.8.3 Full symmetric or antisymmetric ground and excited states

To get the IR selection rule holding for the centrosymmetric cyclic dimer, it is necessary to consider symmetrized states for the initial and final situations involved in the absorption process. Besides, it is then unnecessary to take into account the thermal bath, so that one may consider $\{\mathbf{H}_g^{\{1\}}\}$ in place of $\{\mathbb{H}_g^{\{1\}}\}$.

17.8.3.1 Some symmetry properties of states involved in the theory of cyclic dimers The eigenvectors appearing in Eqs. (17.219) and (17.220) may be expanded in the basis of the eigenkets of the harmonic Hamiltonians (17.217) and (17.218) given by the eigenvalue equations

$$\{\mathbf{H}_g^{\{0\}}\}|(n)_{\mathrm{ger}}\rangle = \hbar\Omega\left(n_{\mathrm{ger}} + \tfrac{1}{2}\right)|(n)_{\mathrm{ger}}\rangle \tag{17.227}$$

$$\{\mathbf{H}_u^{\{0\}}\}|(n)_{\mathrm{ung}}\rangle = \hbar\Omega\left(n_{\mathrm{ung}} + \tfrac{1}{2}\right)|(n)_{\mathrm{ung}}\rangle \tag{17.228}$$

leading, therefore, to

$$|\{\alpha_g^{\{1\}}\}_\nu\rangle = \sum_n \{C_{n\nu}^{\{1\}}\}|(n)_{\mathrm{ger}}\rangle \quad \text{with} \quad \{C_{n\nu}^{\{1\}}\} = \langle(n)_{\mathrm{ger}}|\{\alpha_g^{\{1\}}\}_\nu\rangle \tag{17.229}$$

$$|\{\xi^{\{\pm\}}\}_\mu\rangle = \sum_n \{B_{n\mu}^{\{\pm\}}\}|(n)_{\mathrm{ung}}\rangle \quad \text{with} \quad \{B_{n\mu}^{\{\pm\}}\} = \langle(n)_{\mathrm{ung}}|\{\xi^{\{\pm\}}\}_\mu\rangle \tag{17.230}$$

Now, observe that if the eigenkets appearing in Eq. (17.227) are all symmetric with respect to the parity operator, at the opposite, according to (10.157), those involved in Eq. (17.228) may be either symmetric or antisymmetric:

$$\hat{\mathbf{C}}_2|(n)_{\mathrm{ger}}\rangle = |(n)_{\mathrm{ger}}\rangle \qquad \text{and} \qquad \hat{\mathbf{C}}_2|(n)_{\mathrm{ung}}\rangle = (-1)^n|(n)_{\mathrm{ung}}\rangle \tag{17.231}$$

First, it is clear that since expansion coefficients cannot be affected by a symmetry operation, the action of the parity operator on the expansion of (17.229) reads

$$\hat{\mathbf{C}}_2|\{\alpha_g^{\{1\}}\}_\nu\rangle = \sum_n \{C_{n\nu}^{\{1\}}\}\hat{\mathbf{C}}_2|(n)_{\mathrm{ger}}\rangle$$

so that, due to the first equation of (17.231),

$$\hat{\mathbf{C}}_2|\{\alpha_g^{\{1\}}\}_\nu\rangle = \sum_n \{C_{n\nu}^{\{1\}}\}|(n)_{\mathrm{ger}}\rangle$$

and thus, according to the expression of the expansion of (17.229),

$$\hat{C}_2|\{\alpha_g^{\{1\}}\}_\nu\rangle = |\{\alpha_g^{\{1\}}\}_\nu\rangle \tag{17.232}$$

on the other hand, the action of the parity operator on the expansion of (17.230) yields

$$\hat{C}_2|\{\xi^{\{\pm\}}\}_\mu\rangle = \sum_n \{B_{n\mu}^{\{\pm\}}\}\hat{C}_2|(n)_{\text{ung}}\rangle$$

so that due to the last equation of (17.231)

$$\hat{C}_2|\{\xi^{\{\pm\}}\}_\mu\rangle = \sum_n \{B_{n\mu}^{\{\pm\}}\}(-1)^n|(n)_{\text{ung}}\rangle \tag{17.233}$$

For these last kets obeying (17.233) involving complex symmetry properties, it is suitable to get their g symmetric or u antisymmetric projections with the help of Eqs. (3.39) and (3.40), that is,

$$\tfrac{1}{2}(1+\hat{C}_2)\big|\{\xi^{\{+\}}\}_\mu\big\rangle = \big|\{\xi^{\{+\}}\}_\mu^g\big\rangle \tag{17.234}$$

$$\tfrac{1}{2}(1+\hat{C}_2)\big|\{\xi^{\{-\}}\}_\mu\big\rangle = \big|\{\xi^{\{-\}}\}_\mu^g\big\rangle \tag{17.235}$$

$$\tfrac{1}{2}(1-\hat{C}_2)\big|\{\xi^{\{+\}}\}_\mu\big\rangle = \big|\{\xi^{\{+\}}\}_\mu^u\big\rangle \tag{17.236}$$

$$\tfrac{1}{2}(1-\hat{C}_2)\big|\{\xi^{\{-\}}\}_\mu\big\rangle = \big|\{\xi^{\{-\}}\}_\mu^u\big\rangle \tag{17.237}$$

where the four ket varieties $\pm g$ or $\pm u$ states appearing in Eqs. (17.234)–(17.237) verify the symmetry properties:

$$\hat{C}_2\big|\{\xi^{\{+\}}\}_\mu^g\big\rangle = +1\big|\{\xi^{\{+\}}\}_\mu^g\big\rangle \quad \text{and} \quad \hat{C}_2\big|\{\xi^{\{+\}}\}_\mu^u\big\rangle = -1\big|\{\xi^{\{+\}}\}_\mu^u\big\rangle \tag{17.238}$$

$$\hat{C}_2\big|\{\xi^{\{-\}}\}_\mu^g\big\rangle = +1\big|\{\xi^{\{-\}}\}_\mu^g\big\rangle \quad \text{and} \quad \hat{C}_2\big|\{\xi^{\{-\}}\}_\mu^u\big\rangle = -1\big|\{\xi^{\{-\}}\}_\mu^u\big\rangle \tag{17.239}$$

On the other hand, observe that

$$\hat{C}_2\big|\{k\}_a\big\rangle = \big|\{k\}_b\big\rangle \quad \text{and} \quad \hat{C}_2\big|\{k\}_b\big\rangle = \big|\{k\}_a\big\rangle \quad \text{with} \quad k = 0, 1 \tag{17.240}$$

so that the action of the parity operator on the product $|\{1\}_a\rangle|\{0\}_b\rangle$ or $|\{0\}_a\rangle|\{1\}_b\rangle$ reads

$$\hat{C}_2|\{1\}_a\rangle|\{0\}_b\rangle = |\{1\}_b\rangle|\{0\}_a\rangle \quad \text{and} \quad \hat{C}_2|\{0\}_a\rangle|\{1\}_b\rangle = |\{1\}_a\rangle|\{0\}_b\rangle$$

As a consequence, one obtains, respectively, for the positive and negative linear combinations of $|\{1\}_a\rangle|\{0\}_b\rangle$ and $|\{0\}_a\rangle|\{1\}_b\rangle$:

$$\hat{C}_2\{|\{1\}_a\rangle|\{0\}_b\rangle + |\{0\}_a\rangle|\{1\}_b\rangle\}_g = +1\{|\{1\}_a\rangle|\{0\}_b\rangle + |\{0\}_a\rangle|\{1\}_b\rangle\}_g \tag{17.241}$$

$$\hat{C}_2\{|\{1\}_a\rangle|\{0\}_b\rangle - |\{0\}_a\rangle|\{1\}_b\rangle\}_u = -1\{|\{1\}_a\rangle|\{0\}_b\rangle + |\{0\}_a\rangle|\{1\}_b\rangle\}_u \tag{17.242}$$

17.8.3.2 Constructing full symmetric or antisymmetric states for the excited cyclic dimer The actions of the parity operator on the spinors (17.206) read, respectively,

$$\hat{C}_2\{\tilde{\boldsymbol{\beta}}^{(+)}\} = \begin{pmatrix} \hat{C}_2\{\boldsymbol{\beta}^{(+)}\} \\ +(\hat{C}_2)^2\{\boldsymbol{\beta}^{(+)}\} \end{pmatrix} = \begin{pmatrix} \hat{C}_2\{\boldsymbol{\beta}^{(+)}\} \\ \{\boldsymbol{\beta}^{(+)}\} \end{pmatrix} = +1\{\tilde{\boldsymbol{\beta}}^{(+)}\}$$

$$\hat{C}_2\{\tilde{\boldsymbol{\beta}}^{(-)}\} = \begin{pmatrix} \hat{C}_2\{\boldsymbol{\beta}^{(-)}\} \\ -(\hat{C}_2)^2\{\boldsymbol{\beta}^{(-)}\} \end{pmatrix} = \begin{pmatrix} \hat{C}_2\{\boldsymbol{\beta}^{(-)}\} \\ -\{\boldsymbol{\beta}^{(-)}\} \end{pmatrix} = -\begin{pmatrix} -\hat{C}_2\{\boldsymbol{\beta}^{(-)}\} \\ \{\boldsymbol{\beta}^{(-)}\} \end{pmatrix} = -1\{\tilde{\boldsymbol{\beta}}^{(-)}\}$$

Now, observe that the components $\boldsymbol{\beta}^{(\pm)}$ of these spinors are in the following connection with the states appearing in Eqs. (17.219) and (17.220):

$$\{\tilde{\boldsymbol{\beta}}^{(+)}\} \longleftrightarrow |\{\xi^{\{+\}}\}_\mu\rangle|\{\alpha_g^{\{1\}}\}_\nu\rangle \tag{17.243}$$

$$\{\tilde{\boldsymbol{\beta}}^{(-)}\} \longleftrightarrow |\{\xi^{\{-\}}\}_\mu\rangle|\{\alpha_g^{\{1\}}\}_\nu\rangle \tag{17.244}$$

Then, it is easy to get for the dimer full symmetric or antisymmetric states composed of those involved either in (17.243) or in (17.244) and from the symmetric and antisymmetric combinations of $|\{1\}_a\rangle|\{0\}_b\rangle$ and $|\{0\}_a\rangle|\{1\}_b\rangle$ rewritten in the symbolic form:

$$\left|\left\{\Phi^{\{+(1,0)\}}_{\{-(0,1)\}}\right\}^{\{-\}}_{\mu,\nu}\right\rangle_u \equiv \{|\{1\}_a\rangle|\{0\}_b\rangle - |\{0\}_a\rangle|\{1\}_b\rangle\}_u\left|\{\alpha_g^{\{1\}}\}_\nu\right\rangle\left|\{\xi^{\{-\}}\}^g_\mu\right\rangle \tag{17.245}$$

$$\left|\left\{\Phi^{\{+(1,0)\}}_{\{+(0,1)\}}\right\}^{\{-\}}_{\mu,\nu}\right\rangle_u \equiv \{|\{1\}_a\rangle|\{0\}_b\rangle + |\{0\}_a\rangle|\{1\}_b\rangle\}_g\left|\{\alpha_g^{\{1\}}\}_\nu\right\rangle\left|\{\xi^{\{-\}}\}^u_\mu\right\rangle \tag{17.246}$$

$$\left|\left\{\Phi^{\{+(1,0)\}}_{\{-(0,1)\}}\right\}^{\{+\}}_{\mu,\nu}\right\rangle_g \equiv \{|\{1\}_a\rangle|\{0\}_b\rangle - |\{0\}_a\rangle|\{1\}_b\rangle\}_u\left|\{\alpha_g^{\{1\}}\}_\nu\right\rangle\left|\{\xi^{\{+\}}\}^u_\mu\right\rangle \tag{17.247}$$

$$\left|\left\{\Phi^{\{+(1,0)\}}_{\{+(0,1)\}}\right\}^{\{+\}}_{\mu,\nu}\right\rangle_g \equiv \{|\{1\}_a\rangle|\{0\}_b\rangle + |\{0\}_a\rangle|\{1\}_b\rangle\}_g\left|\{\alpha_g^{\{1\}}\}_\nu\right\rangle\left|\{\xi^{\{+\}}\}^g_\mu\right\rangle \tag{17.248}$$

The action of the parity operator on the state (17.245), which reads

$$\hat{C}_2\left|\left\{\Phi^{\{+(1,0)\}}_{\{-(0,1)\}}\right\}^{\{-\}}_{\mu,\nu}\right\rangle_u = \hat{C}_2\{|\{1\}_a\rangle|\{0\}_b\rangle - |\{0\}_a\rangle|\{1\}_b\rangle\}_u\left|\{\alpha_g^{\{1\}}\}_\nu\right\rangle\left|\{\xi^{\{-\}}\}^g_\mu\right\rangle$$

leads, with the help of Eq. (17.242), to

$$\hat{C}_2\left|\left\{\Phi^{\{+(1,0)\}}_{\{-(0,1)\}}\right\}^{\{-\}}_{\mu,\nu}\right\rangle_u = -1\{|\{1\}_a\rangle|\{0\}_b\rangle - |\{0\}_a\rangle|\{1\}_b\rangle\}_u\hat{C}_2\left|\{\alpha_g^{\{1\}}\}_\nu\right\rangle\left|\{\xi^{\{-\}}\}^g_\mu\right\rangle$$

and, by the aid of Eq. (17.232),

$$\hat{C}_2\left|\left\{\Phi^{\{+(1,0)\}}_{\{-(0,1)\}}\right\}^{\{-\}}_{\mu,\nu}\right\rangle_u = -1\{|\{1\}_a\rangle|\{0\}_b\rangle - |\{0\}_a\rangle|\{1\}_b\rangle\}_u\left|\{\alpha_g^{\{1\}}\}^g_\nu\right\rangle\hat{C}_2\left|\{\xi^{\{-\}}\}^g_\mu\right\rangle$$

and, thus, due to the first equation of (17.239),

$$\hat{C}_2\left|\left\{\Phi^{\{+(1,0)\}}_{\{-(0,1)\}}\right\}^{\{-\}}_{\mu,\nu}\right\rangle_u = -1\left|\left\{\Phi^{\{+(1,0)\}}_{\{-(0,1)\}}\right\}^{\{-\}}_{\mu,\nu}\right\rangle_u$$

In a similar way, one would obtain

$$\hat{C}_2\left|\left\{\Phi^{\{+(1,0)\}}_{\{+(0,1)\}}\right\}^{\{-\}}_{\mu,\nu}\right\rangle_u = -1\left|\left\{\Phi^{\{+(1,0)\}}_{\{+(0,1)\}}\right\}^{\{-\}}_{\mu,\nu}\right\rangle_u$$

$$\hat{C}_2 \left| \{\Phi^{\{+(1,0)\}}_{\{-(0,1)\}}\}^{\{+\}}_{\mu,\nu} \right\rangle_u = +1 \left| \{\Phi^{\{+(1,0)\}}_{\{+(0,1)\}}\}^{\{+\}}_{\mu,\nu} \right\rangle_u$$

$$\hat{C}_2 \left| \{\Phi^{\{+(1,0)\}}_{\{+(0,1)\}}\}^{\{+\}}_{\mu,\nu} \right\rangle_u = +1 \left| \{\Phi^{\{+(1,0)\}}_{\{+(0,1)\}}\}^{\{+\}}_{\mu,\nu} \right\rangle_u$$

Again, using Eqs. (17.234)–(17.237), the states (17.245)–(17.248) read, respectively,

$$\left| \{\Phi^{\{+(1,0)\}}_{\{-(0,1)\}}\}^{\{-\}}_{\mu,\nu} \right\rangle_u = \frac{1}{2}\{|\{1\}_a\rangle|\{0\}_b\rangle - |\{0\}_a\rangle|\{1\}_b\rangle\}_u \left\{ \left|\{\alpha_g^{\{1\}}\}_\nu\right\rangle (1+\hat{C}_2) \left|\{\xi^{\{-\}}\}_\mu\right\rangle \right\}_g \tag{17.249}$$

$$\left| \{\Phi^{\{+(1,0)\}}_{\{+(0,1)\}}\}^{\{-\}}_{\mu,\nu} \right\rangle_u = \frac{1}{2}\{|\{1\}_a\rangle|\{0\}_b\rangle + |\{0\}_a\rangle|\{1\}_b\rangle\}_g \left\{ \left|\{\alpha_g^{\{1\}}\}_\nu\right\rangle (1-\hat{C}_2) \left|\{\xi^{\{-\}}\}_\mu\right\rangle \right\}_u \tag{17.250}$$

$$\left| \{\Phi^{\{+(1,0)\}}_{\{-(0,1)\}}\}^{\{+\}}_{\mu,\nu} \right\rangle_g = \frac{1}{2}\{|\{1\}_a\rangle|\{0\}_b\rangle - |\{0\}_a\rangle|\{1\}_b\rangle\}_u \left\{ \left|\{\alpha_g^{\{1\}}\}_\nu\right\rangle (1-\hat{C}_2) \left|\{\xi^{\{+\}}\}_\mu\right\rangle \right\}_u \tag{17.251}$$

$$\left| \{\Phi^{\{+(1,0)\}}_{\{+(0,1)\}}\}^{\{+\}}_{\mu,\nu} \right\rangle_g = \frac{1}{2}\{|\{1\}_a\rangle|\{0\}_b\rangle + |\{0\}_a\rangle|\{1\}_b\rangle\}_g \left\{ \left|\{\alpha_g^{\{1\}}\}_\nu\right\rangle (1+\hat{C}_2) \left|\{\xi^{\{+\}}\}_\mu\right\rangle \right\}_g \tag{17.252}$$

Now, consider the full ground state written according to

$$\left| \{\Phi^{\{0,0\}}\}^{n_{\text{ger}}}_{n_{\text{ung}}} \right\rangle = |\{0\}_a\rangle|\{0\}_b\rangle|(n)_{\text{ger}}\rangle|(n)_{\text{ung}}\rangle \tag{17.253}$$

Due to Eq. (17.240), the action of the parity operator reads successively

$$\begin{aligned}\hat{C}_2|\{0\}_a\rangle|\{0\}_b\rangle|(n)_{\text{ger}}\rangle|(n)_{\text{ung}}\rangle &= |\{0\}_b\rangle\hat{C}_2|\{0\}_b\rangle|(n)_{\text{ger}}\rangle|(n)_{\text{ung}}\rangle \\ &= |\{0\}_b\rangle|\{0\}_a\rangle\hat{C}_2|(n)_{\text{ger}}\rangle|(n)_{\text{ung}}\rangle\end{aligned}$$

which, owing to Eq. (17.231), gives successively

$$\hat{C}_2|\{0\}_a\rangle|\{0\}_b\rangle|(n)_{\text{ger}}\rangle|(n)_{\text{ung}}\rangle = |\{0\}_a\rangle|\{0\}_b\rangle|(n)_{\text{ger}}\rangle\hat{C}_2|(n)_{\text{ung}}\rangle$$

$$\hat{C}_2|\{0\}_a\rangle|\{0\}_b\rangle|(n)_{\text{ger}}\rangle|(n)_{\text{ung}}\rangle = (-1)^{n_{\text{ung}}}|\{0\}_a\rangle|\{0\}_b\rangle|(n)_{\text{ger}}\rangle|(n)_{\text{ung}}\rangle$$

so that it appears that the expressions (17.253) of the full ground states are symmetrized, being either symmetric or antisymmetric, since they obey

$$\hat{C}_2\left| \{\Phi^{\{0,0\}}\}^{n_{\text{ger}}}_{n_{\text{ung}}} \right\rangle_g = +1\left| \{\Phi^{\{0,0\}}\}^{n_{\text{ger}}}_{n_{\text{ung}}} \right\rangle_g$$

and

$$\hat{C}_2\left| \{\Phi^{\{0,0\}}\}^{n_{\text{ger}}}_{n_{\text{ung}}} \right\rangle_u = -1\left| \{\Phi^{\{0,0\}}\}^{n_{\text{ger}}}_{n_{\text{ung}}} \right\rangle_u$$

17.8.4 Matrix elements of the dipole moment operator

In Section 17.1.3 it was shown that homonuclear diatomic molecules cannot emit or absorb IR radiations because their symmetry forbids them to have dipole moment operators capable of interacting with the electromagnetic field. For centrosymmetric cyclic dimers, the influence of the symmetry on the ability for the molecular system to interact with the field is not so drastic even if it leads one to forbid some transition, leading, as we shall see, to selection rules.

Now, consider the dipole moment operator of the high-frequency modes. Roughly speaking, one may consider two symmetric and antisymmetric modes

$$\mathbf{q}_b \pm \mathbf{q}_a$$

Now, despite the $\mathbf{C}_{2v}$ symmetry, the change in the dipole moment operator induced by the two possible motions $\mathbf{q}_b \pm \mathbf{q}_a$ may be different from zero only in the $\mathbf{q}_b - \mathbf{q}_a$ motion so that the first-order expansion of this dipole moment operator with respect to the $\mathbf{q}_a$ and $\mathbf{q}_b$ must be antisymmetric with respect to the variations in $\mathbf{q}_a$ and $\mathbf{q}_b$, and thus given by

$$\hat{\mu}(\mathbf{q}_a, \mathbf{q}_b) = \mu(0) + \left(\frac{\partial \hat{\mu}}{\partial \mathbf{q}_b}\right)_{\mathbf{q}_b=0} \mathbf{q}_b - \left(\frac{\partial \hat{\mu}}{\partial \mathbf{q}_a}\right)_{\mathbf{q}_a=0} \mathbf{q}_a \tag{17.254}$$

or

$$\hat{\mu}(\mathbf{q}_a, \mathbf{q}_b) = \mu(0) + \mu^\circ(\mathbf{q}_b - \mathbf{q}_a) \tag{17.255}$$

since, for symmetry reasons

$$\left(\frac{\partial \hat{\mu}}{\partial \mathbf{q}_b}\right)_{\mathbf{q}_b=0} = \left(\frac{\partial \hat{\mu}}{\partial \mathbf{q}_a}\right)_{\mathbf{q}_a=0} = \mu^\circ$$

Equation (17.254) is the basis of the IR selection rules that will be later obtained.

To get them, consider the matrix elements of the dipole moment operator (17.255) between symmetrized excited states of the forms (17.249)–(17.252) and symmetrized ground states of the form (17.253), that is,

$$\left\langle\left\{\Phi^{\{+(1,0)\}}_{\{\pm(0,1)\}}\right\}^{\{\pm\}}_{\mu,\nu}\middle|\hat{\mu}(\mathbf{q}_a, \mathbf{q}_b)\middle|\{\Phi^{\{0,0\}}\}^{n_{\text{ger}}}_{n_{\text{ung}}}\right\rangle = \mu^\circ\left\langle\left\{\Phi^{\{+(1,0)\}}_{\{\pm(0,1)\}}\right\}^{\{\pm\}}_{\mu,\nu}\middle|(\mathbf{q}_a - \mathbf{q}_b)\middle|\{\Phi^{\{0,0\}}\}^{n_{\text{ger}}}_{n_{\text{ung}}}\right\rangle \tag{17.256}$$

which, due to Eqs. (17.249)–(17.252) and using Eq. (17.253) are

$$\left\langle\left\{\Phi^{\{+(1,0)\}}_{\{\pm(0,1)\}}\right\}^{\{\pm\}}_{\mu,\nu}\middle|\hat{\mu}(\mathbf{q}_a, \mathbf{q}_b)\middle|\{\Phi^{\{0,0\}}\}^{n_{\text{ger}}}_{n_{\text{ung}}}\right\rangle = 2\mu^\circ\sqrt{\frac{\hbar}{2m\omega^\circ}}$$

$$\langle\{\xi^{\{\pm\}}\}_\mu|\langle\{\alpha_g^{\{1\}}\}^g_\nu|\{\langle\{0\}_a|\langle\{1\}_b| - \langle\{1\}_a|\langle\{0\}_b|\}$$

$$\times\,(1 \pm \hat{\mathbf{C}}_2)(\mathbf{q}_a - \mathbf{q}_b)|\{0\}_a\rangle|\{0\}_b\rangle|(n)_{\text{ger}}\rangle|(n)_{\text{ung}}\rangle$$

To obtain these expressions, first observe that

$$\langle\{1\}_a|\langle\{0\}_b|\mathbf{q}_a|\{0\}_b\rangle|\{0\}_a\rangle = \langle\{1\}_a|\mathbf{q}_a|\{0\}_a\rangle\langle\{0\}_b|\{0\}_b\rangle$$

Then, using the normality property and after passing to Boson operators, one gets

$$\langle\{1\}_a|\langle\{0\}_b|\mathbf{q}_a|\{0\}_b\rangle|\{0\}_a\rangle = \sqrt{\frac{\hbar}{2m\omega^\circ}}(\langle\{1\}_a|\{\{\mathbf{a}\}^\dagger_a + \{\mathbf{a}\}_a\}|\{0\}_a\rangle)$$

and thus, as usual

$$\langle\{1\}_a|\langle\{0\}_b|\mathbf{q}_a|\{0\}_b\rangle|\{0\}_a\rangle = \sqrt{\frac{\hbar}{2m\omega^\circ}} \tag{17.257}$$

In like manner

$$\langle\{0\}_a|\langle\{1\}_b|\mathbf{q}_b|\{0\}_b\rangle|\{0\}_a\rangle = \sqrt{\frac{\hbar}{2m\omega^\circ}} \tag{17.258}$$

Now, observe that

$$\langle\{0\}_a|\langle\{0\}_b|\mathbf{q}_a|\{1\}_b\rangle|\{0\}_a\rangle = \langle\{0\}_a|\mathbf{q}_a|\{0\}_a\rangle\langle\{1\}_b|\{0\}_b\rangle = 0 \tag{17.259}$$

while

$$\langle\{0\}_a|\langle\{0\}_b|\mathbf{q}_b|\{0\}_b\rangle|\{1\}_a\rangle = 0 \tag{17.260}$$

Hence

$$\{\langle\{0\}_a|\langle\{1\}_b| - \langle\{1\}_a|\langle\{0\}_b|\}(\mathbf{q}_b - \mathbf{q}_a)|\{0\}_a\rangle|\{0\}_b\rangle = 2\sqrt{\frac{\hbar}{2m\omega^\circ}} \tag{17.261}$$

while

$$\{\langle\{0\}_a|\langle\{1\}_b| + \langle\{1\}_a|\langle\{0\}_b|\}(\mathbf{q}_b - \mathbf{q}_a)|\{0\}_a\rangle|\{0\}_b\rangle = 0 \tag{17.262}$$

Therefore, owing to Eqs. (17.261) and (17.262), it appears that among the matrix elements (17.256), the following ones are zero:

$$\left\langle\left\{\Phi^{\{+(1,0)\}}_{\{+(0,1)\}}\right\}^{\{+\}}_{\mu,\nu}\right|_g(\mathbf{q}_b - \mathbf{q}_a)\left|\{\Phi^{\{0,0\}}\}^{n_{\text{ger}}}_{n_{\text{ung}}}\right\rangle = 0$$

$$\left\langle\left\{\Phi^{\{+(1,0)\}}_{\{+(0,1)\}}\right\}^{\{-\}}_{\mu,\nu}\right|_u(\mathbf{q}_b - \mathbf{q}_a)\left|\{\Phi^{\{0,0\}}\}^{n_{\text{ger}}}_{n_{\text{ung}}}\right\rangle = 0$$

while the others read

$$\left\langle\left\{\Phi^{\{+(1,0)\}}_{\{-(0,1)\}}\right\}^{\{+\}}_{\mu,\nu}\right|_g(\mathbf{q}_b - \mathbf{q}_a)\left|\{\Phi^{\{0,0\}}\}^{n_{\text{ger}}}_{n_{\text{ung}}}\right\rangle$$
$$= 2\sqrt{\frac{\hbar}{2m\omega^\circ}}\langle\{\alpha_g^{\{1\}}\}_\nu|\{\langle\{\xi^{\{+\}}\}_\mu|(1 - \hat{\mathbf{C}}_2)|(n)_{\text{ung}}\rangle\}|(n)_{\text{ger}}\rangle \tag{17.263}$$

$$\left\langle\left\{\Phi^{\{+(1,0)\}}_{\{-(0,1)\}}\right\}^{\{-\}}_{\mu,\nu}\right|_g(\mathbf{q}_b - \mathbf{q}_a)\left|\{\Phi^{\{0,0\}}\}^{n_{\text{ger}}}_{n_{\text{ung}}}\right\rangle$$
$$= 2\sqrt{\frac{\hbar}{2m\omega^\circ}}\langle\{\alpha_g^{\{1\}}\}_\nu|\{\langle\{\xi^{\{-\}}\}_\mu|(1 + \hat{\mathbf{C}}_2)|(n)_{\text{ung}}\rangle\}|(n)_{\text{ger}}\rangle \tag{17.264}$$

17.8.5 Spectral density

17.8.5.1 Ingredients for spectral density Now, recall that the light intensity $\mathbf{I}(\omega)$ of angular frequency ω, absorbed or emitted by a molecular system when passing from the state $|\Psi_k\rangle$ to the state $|\Psi_l\rangle$, due to the coupling of its dipole moment operator $\widehat{\boldsymbol{\mu}}$, with the electric field $\mathbf{E}(\omega)$ at angular frequency ω, is given by Eq. (17.64), that is,

$$\mathbf{I}(\omega) = \frac{2\pi}{\hbar}(\mathbf{E}(\omega))^2\frac{1}{2\pi}\int_{-\infty}^{\infty}\sum_k\sum_l\langle\Psi_k|\boldsymbol{\rho}_B|\Psi_k\rangle\langle\Psi_k(0)|\widehat{\boldsymbol{\mu}}|\Psi_l(0)\rangle\langle\Psi_l(t)|\widehat{\boldsymbol{\mu}}|\Psi_k(t)\rangle e^{-i\omega t}dt \tag{17.265}$$

where $\boldsymbol{\rho}_B$ is the Boltzmann density operator, an expression that may be also written

$$\mathbf{I}(\omega) \propto \sum_k\sum_l\int_{-\infty}^{\infty}\mathbf{G}_{kl}(t)e^{-i\omega t}dt \tag{17.266}$$

with

$$\mathbf{G}_{kl}(t) = \rho_{kk}\{S_{kl}(0)\}\{S_{lk}(t)\} = \rho_{kk}\{S_{lk}(0)\}^*\{S_{lk}(t)\} \tag{17.267}$$

where

$$\rho_{kk} = \langle\Psi_k|\boldsymbol{\rho}_B|\Psi_k\rangle \qquad \text{and} \qquad \{S_{lk}\} = \langle\Psi_l|\widehat{\boldsymbol{\mu}}|\Psi_k\rangle$$

Now, applying these to the present situation, one has

$$|\Psi_k\rangle \rightarrow |\{\Phi^{\{0,0\}}\}_{n_{\text{ung}}}^{n_{\text{ger}}}\rangle \qquad \text{and} \qquad |\Psi_l\rangle \rightarrow \left|\left\{\Phi_{\{-(0,1)\}}^{\{+(1,0)\}}\right\}_{\mu,\nu}^{\{\pm\}}\right\rangle_x \quad \text{with} \quad x = g, u$$

$$\widehat{\boldsymbol{\mu}} \rightarrow \mu^\circ(\mathbf{q}_b - \mathbf{q}_a)$$

so that

$$\{S_{lk}\} \rightarrow \{S_{n_g n_u}^{\mu\nu\{\pm\}}\} \propto \left\langle\left\{\Phi_{\{-(0,1)\}}^{\{+(1,0)\}}\right\}_{\mu,\nu}^{\{\pm\}}\right|_g (\mathbf{q}_b - \mathbf{q}_a)\left|\{\Phi^{\{0,0\}}\}_{n_{\text{ung}}}^{n_{\text{ger}}}\right\rangle$$

where due to Eqs. (17.263) and (17.264)

$$\{S_{n_g n_u}^{\mu\nu\{\pm\}}\} \propto \langle\{\alpha_g^{\{1\}}\}_\nu|\langle\{\xi^{\{\pm\}}\}_\mu|(1 \mp \hat{\mathbf{C}}_2)|(n)_{\text{ung}}\rangle|(n)_{\text{ger}}\rangle \tag{17.268}$$

At an initial time, the complex conjugate of Eq. (17.268) reads

$$\{S_{n_g n_u}^{\mu\nu\{\pm\}}\}^* \propto \langle(n)_{\text{ger}}|\langle(n)_{\text{ung}}|(1 \mp \hat{\mathbf{C}}_2)|\{\xi^{\{\pm\}}\}_\mu\rangle|\{\alpha_g^{\{1\}}\}_\nu\rangle \tag{17.269}$$

while at time t, this same expression reads

$$\{S_{n_g n_u}^{\mu\nu\{\pm\}}(t)\} \propto$$
$$\langle\{\alpha_g^{\{1\}}\}_\nu|(e^{i\mathbf{H}_{\text{Dav}}t/\hbar})\langle\{\xi^{\{\pm\}}\}_\mu|(e^{i\mathbf{H}_{\text{Dav}}t/\hbar})(1 \mp \hat{\mathbf{C}}_2)(e^{-i\mathbf{H}_{\text{Dav}}t/\hbar})|(n)_{\text{ung}}\rangle(e^{-i\mathbf{H}_{\text{Dav}}t/\hbar})|(n)_{\text{ger}}\rangle \tag{17.270}$$

and thus, due to Eq. (17.226),

$$\{S_{n_g n_u}^{\mu\nu\{\pm\}}(t)\} \propto$$
$$\langle\{\alpha_g^{\{1\}}\}_\nu|(e^{i\mathbf{H}_g^{\{1\}}t/\hbar})\langle\{\xi^{\{\pm\}}\}_\mu|(e^{i\mathbf{H}^{\{\pm\}}t/\hbar})(1 \mp \hat{\mathbf{C}}_2)(e^{-i\mathbf{H}_u^{\{0\}}t/\hbar})|(n)_{\text{ung}}\rangle(e^{-i\mathbf{H}_g^{\{0\}}t/\hbar})|(n)_{\text{ger}}\rangle \tag{17.271}$$

While, for the diagonal elements ρ_{kk} of the Boltzmann density operator, one has

$$\rho_{kk} \rightarrow \{\rho_{n_{\text{ger}},n_{\text{ung}}}\} = \left\langle\{\Phi^{\{0,0\}}\}_{n_{\text{ung}}}^{n_{\text{ger}}}\right|\boldsymbol{\rho}_B\left|\{\Phi^{\{0,0\}}\}_{n_{\text{ung}}}^{n_{\text{ger}}}\right\rangle$$

where

$$\begin{aligned}
\{\rho_{n_{\text{ger}},n_{\text{ung}}}\} &= \frac{1}{Z}\langle(n)_{\text{ung}}|\langle(n)_{\text{ger}}|\langle\{0\}_b|\langle\{0\}_a|(e^{-\beta\mathbf{H}_{\text{Dav}}})|\{0\}_a\rangle|\{0\}_b\rangle|(n)_{\text{ger}}\rangle|(n)_{\text{ung}}\rangle \\
&= \frac{1}{Z}\langle(n)_{\text{ung}}|\langle(n)_{\text{ger}}|\langle\{0\}_b|\langle\{0\}_a|(e^{-\beta(\mathbf{H}_a^{\{0\}}+\mathbf{H}_b^{\{0\}})})|\{0\}_a\rangle|\{0\}_b\rangle|(n)_{\text{ger}}\rangle|(n)_{\text{ung}}\rangle \\
&= \frac{1}{Z}\langle(n)_{\text{ung}}|\langle(n)_{\text{ger}}|(e^{-\beta(\mathbf{H}_a^{\{0\}}+\mathbf{H}_b^{\{0\}})})\langle\{0\}_a|\{0\}_a\rangle\langle\{0\}_b|\{0\}_b\rangle|(n)_{\text{ger}}\rangle|(n)_{\text{ung}}\rangle \\
&= \frac{1}{Z}\langle(n)_{\text{ung}}|\langle(n)_{\text{ger}}|(e^{-\beta(\mathbf{H}_g^{\{0\}}+\mathbf{H}_u^{\{0\}})})|(n)_{\text{ger}}\rangle|(n)_{\text{ung}}\rangle \\
&= \frac{1}{Z_g}\frac{1}{Z_u}\langle(n)_{\text{ger}}|(e^{-\beta\mathbf{H}_g^{\{0\}}})|(n)_{\text{ger}}\rangle\langle(n)_{\text{ung}}|(e^{-\beta\mathbf{H}_u^{\{0\}}})|(n)_{\text{ung}}\rangle \\
&= (1-e^{-\lambda})\langle(n)_{\text{ger}}|(e^{-\beta(n_g+1/2)\hbar\Omega})|(n)_{\text{ger}}\rangle \\
&\quad\times(1-e^{-\lambda})\langle(n)_{\text{ung}}|(e^{-\beta(n_u+1/2)\hbar\Omega})|(n)_{\text{ung}}\rangle \\
&= (1-e^{-\lambda})(e^{-\beta(n_g+1/2)\hbar\Omega})(1-e^{-\lambda})(e^{-\beta(n_u+1/2)\hbar\Omega})
\end{aligned} \tag{17.272}$$

17.8.5.2 Autocorrelation function

17.8.5.2.1 Factorization of autocorrelation function Now, Eq. (17.267) takes the form

$$\mathbf{G}_{kl}(t) \to \{G_{n_g n_u}^{\mu\nu\{\pm\}}(t)\}$$

$$\{G_{n_g n_u}^{\mu\nu\{\pm\}}(t)\} = \{\rho_{n_{\text{ger}},n_{\text{ung}}}\}\{S_{n_g n_u}^{\mu\nu\{\pm\}}(0)\}^*\{S_{n_g n_u}^{\mu\nu\{\pm\}}(t)\} \tag{17.273}$$

Hence, comparing Eqs. (17.269), (17.271), and (17.272), Eq. (17.273) yields

$$\begin{aligned}
\{G_{n_g n_u}^{\mu\nu\{\pm\}}(t)\} &= (1-e^{-\lambda})(e^{-\beta(n_g+1/2)\hbar\Omega})(1-e^{-\lambda})(e^{-\beta(n_u+1/2)\hbar\Omega}) \\
&\quad\times\langle(n)_{\text{ger}}|\langle(n)_{\text{ung}}|(1\mp\hat{\mathbf{C}}_2)|\{\xi^{\{\pm\}}\}_\mu\rangle|\{\alpha_g^{\{1\}}\}_\nu^g\rangle\langle\{\alpha_g^{\{1\}}\}_\nu^g|(e^{i\mathbf{H}_g^{\{1\}}t/\hbar}) \\
&\quad\times\langle\{\xi^{\{\pm\}}\}_\mu|(e^{i\mathbf{H}^{\{\pm\}}t/\hbar})(1\mp\hat{\mathbf{C}}_2)(e^{-i\mathbf{H}_u^{\{0\}}t/\hbar})|(n)_{\text{ung}}\rangle(e^{-i\mathbf{H}_g^{\{0\}}t/\hbar})|(n)_{\text{ger}}\rangle
\end{aligned}$$

or

$$\{G_{n_g n_u}^{\mu\nu\{\pm\}}(t)\} = \{G_{n_g}^{\nu}(t)\}\{G_{n_u}^{\mu\{\pm\}}(t)\} \tag{17.274}$$

with

$$\begin{aligned}
\{G_{n_g}^{\nu}(t)\} &= (1-e^{-\lambda})(e^{-\beta(n_g+1/2)\hbar\Omega})\langle(n)_{\text{ger}}|\{\alpha_g^{\{1\}}\}_\nu^g\rangle \\
&\quad\times\langle\{\alpha_g^{\{1\}}\}_\nu^g|(e^{i\mathbf{H}_g^{\{1\}}t/\hbar})(e^{-i\mathbf{H}_g^{\{0\}}t/\hbar})|(n)_{\text{ger}}\rangle
\end{aligned} \tag{17.275}$$

$$\begin{aligned}
\{G_{n_u}^{\mu\{\pm\}}(t)\} &= (1-e^{-\lambda})(e^{-\beta(n_u+1/2)\hbar\Omega}) \\
&\quad\times\langle(n)_{\text{ung}}|(1\mp\hat{\mathbf{C}}_2)|\{\xi^{\{\pm\}}\}_\mu\rangle \\
&\quad\times\langle\{\xi^{\{\pm\}}\}_\mu|(e^{i\mathbf{H}^{\{\pm\}}t/\hbar})(1\mp\hat{\mathbf{C}}_2)(e^{-i\mathbf{H}_u^{\{0\}}t/\hbar})|(n)_{\text{ung}}\rangle
\end{aligned} \tag{17.276}$$

Finally, to get the spectral density for the present situation where the damping is momentarily ignored, we use Eq. (17.266), which takes on the form

$$\mathbf{I}^{\circ}(\omega) \propto \sum_{n_u}\sum_{n_g}\sum_{\mu}\sum_{\nu}\int_{-\infty}^{\infty}\{G_{n_g n_u}^{\mu\nu\{\pm\}}(t)\}e^{-i\omega t}dt$$

or, owing to Eq. (17.274)

$$\mathbf{I}^{\circ}(\omega) \propto \int_{-\infty}^{\infty}\{\sum_{n_g}\sum_{\nu}\{G_{n_g}^{\nu}(t)\}\sum_{n_u}\sum_{\mu}(\{G_{n_u}^{\mu\{+\}}(t)\} + \{G_{n_u}^{\mu\{-\}}(t)\})\}e^{-i\omega t}dt$$

a result that may be written

$$\mathbf{I}^{\circ}(\omega) \propto \int_{-\infty}^{\infty}\{\mathbf{G}_g^{\circ}(t)\}(\{\mathbf{G}^{\{+\}}(t)\} + \{\mathbf{G}^{\{-\}}(t)\})e^{-i\omega t}dt \tag{17.277}$$

with, respectively,

$$\{\mathbf{G}_g^{\circ}(t)\} = \sum_{n_g}\sum_{\nu}\{G_{n_g}^{\nu}(t)\} \tag{17.278}$$

$$\{\mathbf{G}^{\{\pm\}}(t)\} = \sum_{n_u}\sum_{\mu}\{G_{n_u}^{\mu\{\pm\}}(t)\} \tag{17.279}$$

17.8.5.2.2 The G_g°(t) autocorrelation function Now, observe that, according to Eq. (17.275), the function (17.278) in which the influence of the medium is ignored, reads

$$\{\mathbf{G}_g^{\circ}(t)\} = \sum_{n_g}\sum_{\nu}(1 - e^{-\lambda})(e^{-\beta(n_g+1/2)\hbar\Omega})$$
$$\times \langle(n)_{\text{ger}}|\{\alpha_g^{\{1\}}\}_{\nu}^{g}\rangle\langle\{\alpha_g^{\{1\}}\}_{\nu}^{g}|(e^{i\mathbf{H}_g^{\{1\}}t/\hbar})(e^{-i\mathbf{H}_g^{\{0\}}t/\hbar})|(n)_{\text{ger}}\rangle$$

Then, with the help of the closure relation on the eigenkets of $\mathbf{H}_g^{\{1\}}$, that is,

$$\sum_{\nu}|\{\alpha_g^{\{1\}}\}_{\nu}^{g}\rangle\langle\{\alpha_g^{\{1\}}\}_{\nu}^{g}| = 1$$

simplifies to

$$\{\mathbf{G}_g^{\circ}(t)\} = \sum_{n_g}(1 - e^{-\lambda})(e^{-\beta n_g\hbar\Omega})\langle(n)_{\text{ger}}|(e^{i\mathbf{H}_g^{\{1\}}t/\hbar})(e^{-i\mathbf{H}_g^{\{0\}}t/\hbar})|(n)_{\text{ger}}\rangle$$

Moreover, since the argument of the Boltzmann exponential is a scalar, it commutes with the bra so that

$$\{\mathbf{G}_g^{\circ}(t)\} = \sum_{n_g}(1 - e^{-\lambda})\langle(n)_{\text{ger}}|(e^{-\beta n_g\hbar\Omega})(e^{i\mathbf{H}_g^{\{1\}}t/\hbar})(e^{-i\mathbf{H}_g^{\{0\}}t/\hbar})|(n)_{\text{ger}}\rangle$$

or, due to the eigenvalue equation of $\mathbf{H}_g^{\{0\}}$,

$$\{\mathbf{G}_g^{\circ}(t)\} = \sum_{n_g}(1 - e^{-\lambda})\langle(n)_{\text{ger}}|(e^{-\beta\mathbf{H}_g^{\{0\}}})(e^{i\mathbf{H}_g^{\{1\}}t/\hbar})(e^{-i\mathbf{H}_g^{\{0\}}t/\hbar})|(n)_{\text{ger}}\rangle$$

which may be also written

$$\{\mathbf{G}_g^\circ(t)\} = \mathbf{tr}\{\rho_B(e^{i\mathbf{H}_g^{\{1\}}t/\hbar})(e^{-i\mathbf{H}_g^{\{0\}}t/\hbar})\} \tag{17.280}$$

with

$$\rho_B = (1 - e^{-\lambda})(e^{-\beta\mathbf{H}_g^{\{0\}}})$$

17.8.5.2.3 The $G^{\{\pm\}}$(t) autocorrelation functions Now, due to Eq. (17.276), the ACF (17.279) reads

$$\{\mathbf{G}^{\{\pm\}}(t)\} = (1 - e^{-\lambda})\sum_{n_u}\sum_{\mu}(e^{-\beta(n_u+1/2)\hbar\Omega})$$
$$\times \langle(n)_{\text{ung}}|(1 \mp \hat{\mathbf{C}}_2)|\{\xi^{\{\pm\}}\}_\mu\rangle\langle\{\xi^{\{\pm\}}\}_\mu|(e^{i\mathbf{H}_u^{\{\pm\}}t/\hbar})(1 \mp \hat{\mathbf{C}}_2)(e^{-i\mathbf{H}_u^{\{0\}}t/\hbar})|(n)_{\text{ung}}\rangle$$

Then, owing to the eigenvalue equations (17.227) and (17.220), the latter expression transforms to

$$\{\mathbf{G}^{\{\pm\}}(t)\} = (1 - e^{-\lambda})\sum_{n_u}\sum_{\mu}(e^{-\beta(n_u+1/2)\hbar\Omega})$$
$$\times \langle(n)_{\text{ung}}|(1 \mp \hat{\mathbf{C}}_2)|\{\xi^{\{\pm\}}\}_\mu\rangle\langle\{\xi^{\{\pm\}}\}_\mu|(e^{i\omega_\mu^{\{\pm\}}t})(1 \mp \hat{\mathbf{C}}_2)|(n)_{\text{ung}}\rangle(e^{-i(n_u+1/2)\Omega t}) \tag{17.281}$$

Next, owing to the last expression of (17.231), we have

$$(1 \mp \hat{\mathbf{C}}_2)|(n)_{\text{ung}}\rangle = (1 \mp (-1)^n)|(n)_{\text{ung}}\rangle = |(n)_{\text{ung}}\rangle(1 \mp (-1)^n)$$

the Hermitian conjugate of which is

$$\langle(n)_{\text{ung}}|(1 \mp \hat{\mathbf{C}}_2) = \langle(n)_{\text{ung}}|(1 \mp (-1)^n) = (1 \mp (-1)^n)\langle(n)_{\text{ung}}|$$

Hence, the ACFs (17.281) become

$$\begin{aligned}\{\mathbf{G}^{\{\pm\}}(t)\} &= (1 - e^{-\lambda})\sum_{n_u}\sum_{\mu}(e^{-\beta(n_u+1/2)\hbar\Omega}) \\ &\quad \times (1 \mp (-1)^n)\langle(n)_{\text{ung}}|\{\xi^{\{\pm\}}\}_\mu\rangle\langle\{\xi^{\{\pm\}}\}_\mu|(n)_{\text{ung}}\rangle \\ &\quad \times (1 \mp (-1)^n)(e^{i\omega_\mu^{\{\pm\}}t})(e^{-i(n_u+1/2)\Omega t})\end{aligned}$$

or, due to Eq. (17.230),

$$\begin{aligned}\{\mathbf{G}^{\{\pm\}}(t)\} &= (1 - e^{-\lambda})\sum_{n_u}\sum_{\mu}(e^{-\beta(n_u+1/2)\hbar\Omega}) \\ &\quad \times (1 \pm (-1)^{n+1})^2|B_{n\mu}^{\{\pm 1\}}|^2(e^{i\omega_\mu^{\{\pm\}}t})(e^{-i(n_u+1/2)\Omega t})\end{aligned} \tag{17.282}$$

It may be of interest to observe that the term

$$(1 \pm (-1)^{n+1})^2 \tag{17.283}$$

involved in the ACF (17.282) is the result of the IR selection rule resulting from the symmetry of the molecule. But, due to the complexity of the centrosymmetric dimers, some loss of selection rule may occur so that it may be suitable to modify it weakly by taking in place of that (17.283) the following one:

$$(1 \pm (-1)^{n+1})^2 + \eta(1 \mp (-1)^{n+1})^2 \quad \text{with} \quad \eta < 1$$

Note that when η is unity, this selection rule does not depend on n since

$$(1 \pm (-1)^{n+1})^2 + (1 \mp (-1)^{n+1})^2 = 4$$

17.8.5.2.4 The $G_g(t)$ autocorrelation function in presence of damping Now, remember that if the u part of the Hamiltonian is unaffected by the medium, that is, by the presence of the thermal bath and of its coupling with the centrosymmetric dimer, so that Eq. (17.282) holds also when this coupling is taken into account, the same is not true for the g part. Hence, in order to take into account the irreversible influence of the medium on the line shapes, it is necessary to consider for its g part an ACF taking into account the surroundings. For this purpose, it is suitable to pass from the ACF (17.280) to the following one, in a somewhat formal manner, namely:

$$\{\mathbf{G}_g(t)\} = \mathbf{tr}\{\boldsymbol{\rho}_B \mathbf{tr}_\theta\{(e^{i\mathbb{H}_g^{\{1\}}t/\hbar})(e^{-i\mathbb{H}_g^{\{0\}}t/\hbar})\}\}$$

in which the Hamiltonians (17.216) and (17.223) have replaced, respectively, those (17.217) and (17.224) occurring in the exponentials and where $\mathbf{tr}_\theta$ is the trace over the thermal bath. This expression is of the same form as that (17.136) appearing in the approach of the line shape of single H-bonded species.

Next, recall Eqs. (17.92) and (17.224) dealing, respectively, with the effective Hamiltonians $\mathbf{H}^{\{1\}}$ of the single H-bond bridge and $\mathbf{H}_g^{\{1\}}$ of the centrosymmetric cyclic H-bonded dimer, and involving, respectively, $\mathbb{H}^{\{1\}}$ and $\mathbb{H}_g^{\{1\}}$:

$$\{\mathbf{H}^{\{1\}}\} = \left(\frac{\mathbf{P}^2}{2M} + \frac{1}{2}M\Omega^2\mathbf{Q}^2\right) + b\mathbf{Q} + \hbar\omega^\circ - \alpha^{\circ 2}\hbar\Omega$$

$$\{\mathbf{H}_g^{\{1\}}\} = \left(\frac{\mathbf{P}^2}{2M} + \frac{1}{2}M\Omega^2\mathbf{Q}^2\right) + \frac{b}{\sqrt{2}}\mathbf{Q} + \hbar\omega^\circ - \alpha^{\circ 2}\hbar\Omega$$

Hence, due to the presence of the factor $1/\sqrt{2}$ in the driven part of the last effective Hamiltonian, Eq. (17.137), which is a tractable expression of the ACF (17.136), we have in the present situation

$$\mathbf{G}_g(t) \propto e^{i\omega^\circ t}e^{-i\alpha^{\circ 2}\Omega t}\mathbf{tr}\{\boldsymbol{\rho}_B \hat{\mathbf{P}}\{\exp -i\frac{b}{\sqrt{2}}\int_0^t \langle \mathbf{Q}(t')\rangle dt'\}\}$$

Hence, just as passing from Eq. (17.136)) to (17.146)

$$\mathbf{G}(t) \propto e^{i\omega^\circ t}(e^{-i2\alpha^{\circ 2}\Omega t})(e^{i\tilde{\beta}^{\circ 2}e^{-\gamma t/2}\sin\Omega t})(e^{\tilde{\beta}^{\circ 2}(\langle \mathbf{n}\rangle + 1/2)(2e^{-\gamma t/2}\cos\Omega t - e^{-\gamma t} - 1)}) \tag{17.284}$$

with

$$\tilde{\beta}^\circ = \frac{\beta^\circ}{\sqrt{2}}$$

and where α° and β° are, respectively, given by Eqs. (17.108) and (17.141), keeping in mind that $\langle \mathbf{n} \rangle$ is the thermal average of the excitation degree of the g slow mode given by Eq. (17.144).

17.8.6 Line shape

We are now able to find the IR line shape of the centrosymmetric cyclic dimer under the irreversible influence of the medium. For this purpose, it is convenient to replace Eq. (17.277) by a new one in which the damped ACF $\{\mathbf{G}_g(t)\}$ given by Eq. (17.284) takes the place of the undamped one $\{\mathbf{G}_g^\circ(t)\}$, and to multiply it by some damping resulting from the lifetime of the excited state of the fast modes of the cyclic dimer, so that

$$\mathbf{I}(\omega) \propto \mathrm{Re} \int_0^\infty \{\mathbf{G}_g(t)\}(\{\mathbf{G}^{(+)}(t)\} + \{\mathbf{G}^{(-)}(t)\})(e^{-\gamma^\circ t})e^{-i\omega t}dt \qquad (17.285)$$

which, with the help of Eqs. (17.282) and (17.284) reads

$$\mathbf{I}(\omega) = (\mu_u^\circ)^2(\mathbf{I}^+(\omega) + \mathbf{I}^-(\omega)) \qquad (17.286)$$

The components appearing in this equation are given by

$$\mathbf{I}^\pm(\omega) \propto \sum_{m_g}\sum_{n_g} \mathbf{P}_{m_g n_g} \sum_{n_u}\sum_{\mu}$$

$$e^{-\tilde{\lambda} n_u \Omega t}\{(1 \pm (-1)^{n_u+1}) + \eta^\circ(1 \mp (-1)^{n_u+1})\}^2 |B_{n_u\mu}^{\{\pm\}}|^2 (\mathbf{I}_{m_g n_g n_u \mu}^\pm(\omega))$$

with, respectively,

$$\mathbf{P}_{m_g n_g} = \frac{(1 + \langle \mathbf{n} \rangle^{m_g})\langle \mathbf{n} \rangle^{n_g} \tilde{\alpha}^{\circ 2(m_g + n_g)}}{m_g! n_g!} \qquad \tilde{\alpha}^\circ = \alpha^\circ \sqrt{2}$$

$$\mathbf{I}_{m_g n_g n_u \mu}^\pm(\omega) \propto \frac{\gamma_{m_g n_g}}{(\omega - \Omega_{m_g n_g\ n_u \mu}^\pm)^2 + (\gamma_{m_g n_g})^2}$$

In this last expression, the angular frequencies and damping parameters are given by

$$\Omega_{m_g n_g n_u \mu}^\pm = \omega^\circ - \{(m_g - n_g + n_u)\Omega - \omega_\mu^\pm\} - 2\alpha^{\circ 2}\Omega$$

$$\gamma_{m_g n_g} = (m_g + n_g)\tilde{\gamma} + \gamma^\circ \qquad \tilde{\gamma} = \gamma\sqrt{2}$$

whereas the thermal average mean number occupation is given by

$$\langle \mathbf{n} \rangle = \frac{1}{e^\lambda - 1} \quad \text{with} \quad \lambda = \frac{\hbar\Omega}{k_B T}$$

17.8.7 Limit situations

It may be of interest to examine the limiting situations of the line shape (17.285). In the first place one may observe that when the H-bond bridge damping is missing, $\mathbf{G}_g(t)$ reduces to $\mathbf{G}_g^\circ(t)$, which, due to Eq. (17.284), reads

$$\{\mathbf{G}_g^\circ(t)\} \propto e^{i\omega^\circ t}(e^{-i2\alpha^{\circ 2}\Omega t})(e^{i(\alpha^{\circ 2}/2)\sin\Omega t})(e^{(\alpha^{\circ 2}/2)(\langle\mathbf{n}\rangle+1/2)2(\cos\Omega t-1)})$$

whereas when the Davydov coupling vanishes, both $\mathbf{G}^{(+)}(t)$ and $\mathbf{G}^{(-)}(t)$ reduce to

$$\mathbf{G}^{(\pm)}(t) \to (e^{i(\alpha^{\circ 2}/2)\sin\Omega t})(e^{(\alpha^{\circ 2}/2)(\langle\mathbf{n}\rangle+1/2)2(\cos\Omega t-1)})$$

so that the line shape (17.285) reduces to

$$\mathbf{I}(\omega) \propto 4\mathrm{Re}\int_0^\infty e^{i\omega^\circ t}(e^{-i2\alpha^{\circ 2}\Omega t})(e^{i(\alpha^{\circ 2}/2)\sin\Omega t})$$
$$\times (e^{(\alpha^{\circ 2}/2)(\langle\mathbf{n}\rangle+1/2)2(\cos\Omega t-1)})(e^{i(\alpha^{\circ 2}/2)\sin\Omega t})(e^{(\alpha^{\circ 2}/2)(\langle\mathbf{n}\rangle+1/2)2(\cos\Omega t-1)})$$
$$\times (e^{-\gamma^\circ t})e^{-i\omega t}dt \qquad (17.287)$$

or, after multiplying exponentials of the same kind,

$$\mathbf{I}(\omega) \propto 4\mathrm{Re}\int_0^\infty e^{i\omega^\circ t}(e^{-i2\alpha^{\circ 2}\Omega t})(e^{i(\alpha^{\circ 2}/2)\sin\Omega t})(e^{(\alpha^{\circ 2}/2)(\langle\mathbf{n}\rangle+1/2)2(\cos\Omega t-1)})e^{-i\omega t}dt \qquad (17.288)$$

Hence, in this special situation dealing with the spectral density of two noninteracting single H-bonded species, Eq. (17.288) is exactly the same as that of (17.115) obtained for a single H-bonded species, the global intensity being, as required, twice that of a single H-bonded species.

In like manner, it would be possible to show[3] that the spectral density (17.285) reduces satisfactorily to many special situations available in the literature, specially to those obtained by:

1. Maréchal and Witkowski (see Bibliography) in the absence of damping
2. Rösch and Ratner[4] in the absence of Davydov coupling and indirect damping
3. Boulil et al.[5] in the absence of Davydov coupling and direct damping
4. Robertson and Yarwood[6] in the semiclassical limit, without Davydov coupling and direct damping

[3] O. Henri-Rousseau and P. Blaise. Infrared lineshapes of Weak Hydrogen Bonds Centrosymmetric Cyclic Dimers of Carboxylic Acids. In *Adv. Chem. Phys.*, Vol. 139. I. Prigogine and S. A. Rice (Eds.). Wiley: Hoboken, NJ, 2008, pp. 245–496. The Infrared Spectral Density of Weak Hydrogen Bonds within the Linear Response Theory. In *Adv. Chem. Phys.*, Vol. 103. Wiley: Hoboken, NJ, 1998, p. 1.

[4] N. Rösch and M. Ratner. *J. Chem. Phys.*, **61** (1974): 3344.

[5] B. Boulil, O. Henri-Rousseau, and P. Blaise. *Chem. Phys.*, **126** (1988): 263.

[6] G. Robertson and J. Yarwood. *Chem. Phys.*, **32** (1978): 267.

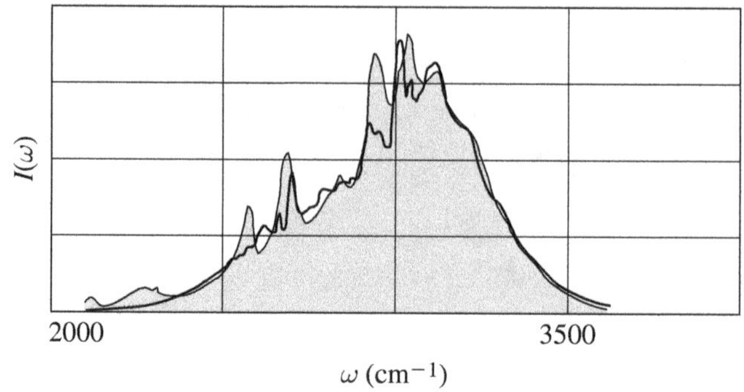

Figure 17.15 IR spectrum for the CD_3CO_2H dimer in the gas phase at room temperature. Parameters: $T = 300\,\text{K}$, $\Omega = 88\,\text{cm}^{-1}$, $\alpha^{\circ} = 1.19$, $\omega^{\circ} = 3100\,\text{cm}^{-1}$, $V^{\circ} = -1.55\Omega$, $\eta = 0.25$, $\gamma = 0.24\Omega$, $\gamma^{\circ} = 0.10\Omega$.

Figure 17.15 gives spectra for CD_3CO_2H dimer in the gas phase at room temperature. The gray spectrum represents the experimental line shapes taken from Novak et al.,[7] whereas the solid line is the theoretical line shape[8] computed by the aid of Eq. (17.286). The parameters used for the calculations are given in the caption of this figure.

BIBLIOGRAPHY

Generalities

G. Barrow. *Introduction to Molecular Spectroscopy*. McGraw-Hill: New York, 1962.

Linear response theory

R. Gordon. *Adv. Magn. Res.*, **3** (1968): 1.

Fermi resonances

A. Brodka and B. Stryczek. *Mol. Phys.*, **54** (1985): 677.
K. Fujita and M. Kimura. *Mol. Phys.*, **41** (1980): 1203.
J. L. McHale and C. H. Wang. *J. Chem. Phys.*, **73** (1980): 3601.
P. Piaggio, G. Dellepiane, R. Turbino, and L. Piseri. *Chem. Phys.*, **77** (1983): 185.
M. Schwartz and C. H. Wang. *J. Chem. Phys.*, **59** (1973): 5258.
E. Weidemann and A. Hayd. *J. Chem. Phys.*, **67** (1977): 3713.

[7] M. Haurie and A. Novák. *J. Chim. Phys.*, **62** (1965): 146.
[8] P. Blaise, M. J. Wojcik, and Olivier Henri-Rousseau. *J. Chem. Phys.*, **122** (2005): 1.

Fermi resonances involving H-bonded species

D. Chamma and O. Henri-Rousseau. *Chem. Phys.*, **229** (1998): 51.
O. Henri-Rousseau and D. Chamma. *Chem. Phys.*, **229** (1998): 37.
A. Witkowski and M.Wojcik. *Chem. Phys.*, **1** (1973): 9.

H-bonded species involving Davydov coupling

P. Blaise, M. E-A. Benmalti, and O. Henri-Rousseau. *J. Chem. Phys.*, **124** (2006): 1.
P. Blaise, M. J. Wojcik, and O. Henri-Rousseau. *J. Chem. Phys.*, **122** (2005): 1.
Y. Maréchal and A. Witkowski. *J. Chem. Phys.*, **48** (1968): 3637.

CHAPTER 18

APPENDIX

18.1 AN IMPORTANT COMMUTATOR

First, we shall explicitly write the following commutator $[\mathbf{A}, \mathbf{BC}]$ involving three operators:

$$[\mathbf{A}, \mathbf{BC}] = \mathbf{A}(\mathbf{BC}) - (\mathbf{BC})\mathbf{A}$$

which be written by adding and subtracting the same product $\mathbf{BAC}$:

$$[\mathbf{A}, \mathbf{BC}] = \mathbf{A}(\mathbf{BC}) - (\mathbf{BC})\mathbf{A} + \mathbf{BAC} - \mathbf{BAC}$$

Rearranging, we have

$$[\mathbf{A}, \mathbf{BC}] = \{\mathbf{ABC} - \mathbf{BAC}\} + \{\mathbf{BAC} - \mathbf{BCA}\}$$

On factorization one obtains

$$[\mathbf{A}, \mathbf{BC}] = (\mathbf{AB} - \mathbf{BA})\mathbf{C} + \mathbf{B}(\mathbf{AC} - \mathbf{CA})$$

which may be expressed in terms of two simple commutators:

$$\boxed{[\mathbf{A}, \mathbf{BC}] = [\mathbf{A}, \mathbf{B}]\mathbf{C} + \mathbf{B}[\mathbf{A}, \mathbf{C}]} \tag{18.1}$$

Of course, if operators $\mathbf{B}$ and $\mathbf{C}$ are the same, we have

$$\boxed{[\mathbf{A}, \mathbf{B}^2] = [\mathbf{A}, \mathbf{B}]\mathbf{B} + \mathbf{B}[\mathbf{A}, \mathbf{B}]} \tag{18.2}$$

18.2 AN IMPORTANT BASIC CANONICAL TRANSFORMATION

We start from the linear operator $\mathbf{F}(\xi)$ defined by

$$\mathbf{F}(\xi) = (e^{\xi\mathbf{A}}\mathbf{B}e^{-\xi\mathbf{A}}) \tag{18.3}$$

where ξ is a c-number and $\mathbf{A}$ and $\mathbf{B}$ two independent linear operators that do not depend on ξ and that do not commute, that is,

$$[\mathbf{A}, \mathbf{B}] \neq 0$$

Now, when ξ is vanishing, operator (18.3) simply reduces to operator $\mathbf{B}$:

$$\mathbf{F}(0) = \mathbf{B} \tag{18.4}$$

Quantum Oscillators, First Edition. Olivier Henri-Rousseau and Paul Blaise.
© 2011 John Wiley & Sons, Inc. Published 2011 by John Wiley & Sons, Inc.

By partial differentiation of Eq. (18.3) with respect to ξ, one obtains

$$\left(\frac{\partial \mathbf{F}(\xi)}{\partial \xi}\right) = \left(\frac{\partial e^{\xi \mathbf{A}}}{\partial \xi}\right) \mathbf{B} e^{-\xi \mathbf{A}} + e^{\xi \mathbf{A}} \mathbf{B} \left(\frac{\partial e^{-\xi \mathbf{A}}}{\partial \xi}\right)$$

or

$$\left(\frac{\partial \mathbf{F}(\xi)}{\partial \xi}\right) = (\mathbf{A} e^{\xi \mathbf{A}}) \mathbf{B} e^{-\xi \mathbf{A}} - e^{\xi \mathbf{A}} \mathbf{B} (\mathbf{A} e^{-\xi \mathbf{A}})$$

Since operator $\mathbf{A}$ commutes with operators which are functions of itself, we have

$$\left(\frac{\partial \mathbf{F}(\xi)}{\partial \xi}\right) = \mathbf{A}(e^{\xi \mathbf{A}} \mathbf{B} e^{-\xi \mathbf{A}}) - (e^{\xi \mathbf{A}} \mathbf{B} e^{-\xi \mathbf{A}})\mathbf{A}$$

Then, using Eq. (18.3) this result transforms to

$$\left(\frac{\partial \mathbf{F}(\xi)}{\partial \xi}\right) = \mathbf{A}\mathbf{F}(\xi) - \mathbf{F}(\xi)\mathbf{A}$$

or

$$\left(\frac{\partial \mathbf{F}(\xi)}{\partial \xi}\right) = [\mathbf{A}, \mathbf{F}(\xi)] \tag{18.5}$$

a result that, when $\xi = 0$, reduces to

$$\left(\frac{\partial \mathbf{F}(\xi)}{\partial \xi}\right)_{\xi=0} = [\mathbf{A}, \mathbf{B}] \tag{18.6}$$

Now, consider the second partial derivative with respect to ξ of (18.3)

$$\left(\frac{\partial^2 \mathbf{F}(\xi)}{\partial \xi^2}\right) = \frac{\partial}{\partial \xi}\left(\frac{\partial \mathbf{F}(\xi)}{\partial \xi}\right)$$

which, owing to Eq. (18.5), reads

$$\left(\frac{\partial^2 \mathbf{F}(\xi)}{\partial \xi^2}\right) = \frac{\partial}{\partial \xi}[\mathbf{A}, \mathbf{F}(\xi)]$$

or, since operator $\mathbf{A}$ does not depend on ξ,

$$\left(\frac{\partial^2 \mathbf{F}(\xi)}{\partial \xi^2}\right) = \left[\mathbf{A}, \left(\frac{\partial \mathbf{F}(\xi)}{\partial \xi}\right)\right]$$

so that, due to Eq. (18.5),

$$\left(\frac{\partial^2 \mathbf{F}(\xi)}{\partial \xi^2}\right) = [\mathbf{A}, [\mathbf{A}, \mathbf{F}(\xi)]]$$

and thus, when $\xi = 0$,

$$\left(\frac{\partial^2 \mathbf{F}(\xi)}{\partial \xi^2}\right)_{\xi=0} = [\mathbf{A}, [\mathbf{A}, \mathbf{B}]] \tag{18.7}$$

Next, by recurrence, one obtains from Eqs. (18.6) and (18.7)

$$\left(\frac{\partial^3 \mathbf{F}(\xi)}{\partial \xi^3}\right)_{\xi=0} = [\mathbf{A}, [\mathbf{A}, [\mathbf{A}, \mathbf{B}]]] \tag{18.8}$$

and so on. Moreover, expand operator $\mathbf{F}(\xi)$ in series around $\mathbf{F}(0)$ yielding

$$\mathbf{F}(\xi) = \mathbf{F}(0) + \left(\frac{\partial \mathbf{F}(\xi)}{\partial \xi}\right)_{\xi=0} \xi + \frac{1}{2}\left(\frac{\partial^2 \mathbf{F}(\xi)}{\partial \xi^2}\right)_{\xi=0} \xi^2 + \frac{1}{3!}\left(\frac{\partial^3 \mathbf{F}(\xi)}{\partial \xi^3}\right)_{\xi=0} \xi^3 + \cdots$$

which transforms, using Eqs. (18.4), (18.6), (18.7), and (18.8) into

$$\mathbf{F}(\xi) = \mathbf{B} + [\mathbf{A}, \mathbf{B}]\xi + \frac{1}{2}[\mathbf{A}, [\mathbf{A}, \mathbf{B}]]\xi^2 + \frac{1}{3!}[\mathbf{A}, [\mathbf{A}, [\mathbf{A}, \mathbf{B}]]]\xi^3 + \cdots$$

so that, due to Eq. (18.3), it yields

$$\boxed{e^{\xi \mathbf{A}}\mathbf{B}e^{-\xi \mathbf{A}} = \mathbf{B} + [\mathbf{A}, \mathbf{B}]\xi + \frac{1}{2}[\mathbf{A}, [\mathbf{A}, \mathbf{B}]]\xi^2 + \frac{1}{3!}[\mathbf{A}, [\mathbf{A}, [\mathbf{A}, \mathbf{B}]]]\xi^3 + \cdots} \quad (18.9)$$

18.3 CANONICAL TRANSFORMATION ON A FUNCTION OF OPERATORS

Now, consider the linear operator $\mathbf{F}(\mathbf{B})$ function of another linear operator $\mathbf{B}$, which may be expanded according to

$$\mathbf{F}(\mathbf{B}) = \sum_n C_n \mathbf{B}^n \quad (18.10)$$

and consider the following transformation of this operator:

$$e^{\xi \mathbf{A}}\mathbf{F}(\mathbf{B})e^{-\xi \mathbf{A}} = \sum_n C_n (e^{\xi \mathbf{A}}\mathbf{B}^n e^{-\xi \mathbf{A}}) \quad (18.11)$$

where ξ is a c-number and A is a linear operator. One term of the expansion appearing on the right-hand side of this last equation may be written as

$$e^{\xi \mathbf{A}}\mathbf{B}^n e^{-\xi \mathbf{A}} = e^{\xi \mathbf{A}}\mathbf{B}\mathbf{B}\cdots\mathbf{B}e^{-\xi \mathbf{A}} \quad (18.12)$$

Again, introducing between each $\mathbf{B}$ operator appearing on the right-hand side the following unity operator:

$$\mathbf{1} = e^{-\xi \mathbf{A}}e^{\xi \mathbf{A}}$$

Then, Eq. (18.12) reads

$$e^{\xi \mathbf{A}}\mathbf{B}^n e^{-\xi \mathbf{A}} = (e^{\xi \mathbf{A}}\mathbf{B}e^{-\xi \mathbf{A}})(e^{\xi \mathbf{A}}\mathbf{B}e^{-\xi \mathbf{A}})\cdots(e^{\xi \mathbf{A}}\mathbf{B}e^{-\xi \mathbf{A}})$$

Hence, the result may condensed into

$$e^{\xi \mathbf{A}}\mathbf{B}^n e^{-\xi \mathbf{A}} = (e^{\xi \mathbf{A}}\mathbf{B}e^{-\xi \mathbf{A}})^n$$

Thus, the canonical transformation (18.11) becomes

$$e^{\xi \mathbf{A}}\mathbf{F}(\mathbf{B})e^{-\xi \mathbf{A}} = \sum_n C_n (e^{\xi \mathbf{A}}\mathbf{B}e^{-\xi \mathbf{A}})^n$$

which, by comparison with Eq. (18.10), appears that this last equation yields

$$\boxed{e^{\xi \mathbf{A}}\mathbf{F}(\mathbf{B})e^{-\xi \mathbf{A}} = \mathbf{F}(e^{\xi \mathbf{A}}\mathbf{B}e^{-\xi \mathbf{A}})}$$

18.4 GLAUBER–WEYL THEOREM

Next, consider the linear operator $\mathbf{F}(\xi)$ defined by

$$\mathbf{F}(\xi) = e^{\xi\mathbf{A}}e^{\xi\mathbf{B}} \quad \text{with} \quad \mathbf{F}(0) = 1 \tag{18.13}$$

where ξ is a c-number and **A** and **B** are independent operators that do not commute:

$$[\mathbf{A}, \mathbf{B}] \neq 0$$

Now, assume that **A** and **B** are chosen so as the following double commutators are zero:

$$[\mathbf{A}, [\mathbf{A}, \mathbf{B}]] = 0 \qquad \text{and} \qquad [\mathbf{B}, [\mathbf{A}, \mathbf{B}]] = 0 \tag{18.14}$$

Then the partial derivative of $\mathbf{F}(\xi)$ with respect to ξ

$$\left(\frac{\partial \mathbf{F}(\xi)}{\partial \xi}\right) = \left(\frac{\partial e^{\xi\mathbf{A}}}{\partial \xi}\right)e^{\xi\mathbf{B}} + e^{\xi\mathbf{A}}\left(\frac{\partial e^{\xi\mathbf{B}}}{\partial \xi}\right)$$

is

$$\left(\frac{\partial \mathbf{F}(\xi)}{\partial \xi}\right) = \mathbf{A}e^{\xi\mathbf{A}}e^{\xi\mathbf{B}} + e^{\xi\mathbf{A}}\mathbf{B}e^{\xi\mathbf{B}} \tag{18.15}$$

Again, introduce within the last right-hand term of Eq. (18.15), the following unity operator:

$$1 = e^{-\xi\mathbf{A}}e^{\xi\mathbf{A}}$$

to get

$$\left(\frac{\partial \mathbf{F}(\xi)}{\partial \xi}\right) = \mathbf{A}e^{\xi\mathbf{A}}e^{\xi\mathbf{B}} + e^{\xi\mathbf{A}}\mathbf{B}(e^{-\xi\mathbf{A}}e^{\xi\mathbf{A}})e^{\xi\mathbf{B}}$$

or

$$\left(\frac{\partial \mathbf{F}(\xi)}{\partial \xi}\right) = \mathbf{A}(e^{\xi\mathbf{A}}e^{\xi\mathbf{B}}) + (e^{\xi\mathbf{A}}\mathbf{B}e^{-\xi\mathbf{A}})(e^{\xi\mathbf{A}}e^{\xi\mathbf{B}})$$

so that, comparing Eq. (18.13),

$$\left(\frac{\partial \mathbf{F}(\xi)}{\partial \xi}\right) = \{\mathbf{A} + (e^{\xi\mathbf{A}}\mathbf{B}e^{-\xi\mathbf{A}})\}\mathbf{F}(\xi) \tag{18.16}$$

Furthermore, observe that owing to Eq. (18.9), the right-hand central term of this last equation takes the form

$$e^{\xi\mathbf{A}}\mathbf{B}e^{-\xi\mathbf{A}} = \mathbf{B} + [\mathbf{A}, \mathbf{B}]\xi + \frac{1}{2}[\mathbf{A}, [\mathbf{A}, \mathbf{B}]]\xi^2 + \frac{1}{3!}[\mathbf{A}, [\mathbf{A}, [\mathbf{A}, \mathbf{B}]]]\xi^3 + \cdots$$

which, comparing the assumption (18.14), reduces to

$$e^{\xi\mathbf{A}}\mathbf{B}e^{-\xi\mathbf{A}} = \mathbf{B} + [\mathbf{A}, \mathbf{B}]\xi$$

Thus, the derivative (18.16) becomes

$$\left(\frac{\partial \mathbf{F}(\xi)}{\partial \xi}\right) = (\mathbf{A} + \mathbf{B} + [\mathbf{A}, \mathbf{B}])\xi\mathbf{F}(\xi) \tag{18.17}$$

the integration of which between $\xi = 0$ and $\xi = \xi°$ reads

$$\int_0^{\xi°} \frac{d\mathbf{F}(\xi)}{\mathbf{F}(\xi)} = (\mathbf{A}+\mathbf{B})\int_0^{\xi°} d\xi + [\mathbf{A},\mathbf{B}]\int_0^{\xi°} \xi\, d\xi$$

so that

$$\ln\left(\frac{\mathbf{F}(\xi°)}{\mathbf{F}(0)}\right) = (\mathbf{A}+\mathbf{B})\xi° + \frac{1}{2}[\mathbf{A},\mathbf{B}]\xi^{\circ 2}$$

or, after passing to the exponentials,

$$\mathbf{F}(\xi°) = \mathbf{F}(0)e^{(\mathbf{A}+\mathbf{B})\xi° + [\mathbf{A},\mathbf{B}]\xi^{\circ 2}/2}$$

Moreover, because of Eq. (18.14), the exponential operators may be written as the product of two exponentials involving operators that do not commute (which may be verified expanding the exponentials), according to

$$\mathbf{F}(\xi°) = \mathbf{F}(0)e^{(\mathbf{A}+\mathbf{B})\xi°}e^{[\mathbf{A},\mathbf{B}]\xi^{\circ 2}/2} \tag{18.18}$$

Now, owing to Eq. (18.13), and since Eq. (18.18) holds for any $\xi°$, we have

$$\boxed{e^{\xi\mathbf{A}}e^{\xi\mathbf{B}} = e^{(\mathbf{A}+\mathbf{B})\xi}e^{[\mathbf{A},\mathbf{B}]\xi^2/2} \quad \text{with} \quad [\mathbf{A},[\mathbf{A},\mathbf{B}]] = 0 \quad \text{and} \quad [\mathbf{B},[\mathbf{A},\mathbf{B}]] = 0} \tag{18.19}$$

This equation is named the *Glauber* or also the *Glauber–Weyl relation.* Again, multiplying the two members of the Glauber relation by the same exponential involving the scalar $[\mathbf{A},\mathbf{B}]$ yields

$$e^{\xi\mathbf{A}}e^{\xi\mathbf{B}}e^{-[\mathbf{A},\mathbf{B}]\xi^2/2} = e^{(\mathbf{A}+\mathbf{B})\xi}e^{+(1/2)[\mathbf{A},\mathbf{B}]\xi^2}e^{-[\mathbf{A},\mathbf{B}]\xi^2/2}$$

a result that, due to

$$e^{[\mathbf{A},\mathbf{B}]\xi^2/2}e^{-[\mathbf{A},\mathbf{B}]\xi^2/2} = 1$$

transforms to a new equivalent form of the Glauber–Weyl relation (18.19):

$$e^{(\mathbf{A}+\mathbf{B})\xi} = e^{\xi\mathbf{A}}e^{\xi\mathbf{B}}e^{-[\mathbf{A},\mathbf{B}]\xi^2/2}$$

18.5 COMMUTATORS OF FUNCTIONS OF THE P AND Q OPERATORS

18.5.1 Commutators of observables as functions of $\mathbf{P}_k$ and $\mathbf{Q}_k$

As expected from the above conclusion, the noncommutativity properties of observables, which are very general, play a fundamental role in the knowledge of the measurement of a physical variable. In order to find how the observables work in a quantum mechanical sense, it is necessary to determine the commutators $[\mathbf{Q},\mathbf{F}(\mathbf{P})]$ and $[\mathbf{P},\mathbf{F}(\mathbf{Q})]$.

First, seek the commutator of $\mathbf{Q}$ and $\mathbf{P}^2$, keeping in mind Eq. (18.2), that is,

$$[\mathbf{A}, \mathbf{B}^2] = [\mathbf{A}, \mathbf{B}]\mathbf{B} + \mathbf{B}[\mathbf{A}, \mathbf{B}]$$

which reads

$$[\mathbf{Q}, \mathbf{P}^2] = [\mathbf{Q}, \mathbf{P}]\mathbf{P} + \mathbf{P}[\mathbf{Q}, \mathbf{P}] \tag{18.20}$$

Next, due to the basic quantum mechanical commutator

$$[\mathbf{Q}, \mathbf{P}] = i\hbar$$

the left-hand side of Eq. (18.20) becomes

$$[\mathbf{Q}, \mathbf{P}^2] = 2i\hbar\mathbf{P} \tag{18.21}$$

Moreover, consider the following commutator $[\mathbf{Q}, \mathbf{P}^3]$ for which, applying Eq. (18.1),

$$[\mathbf{A}, \mathbf{BC}] = [\mathbf{A}, \mathbf{B}]\mathbf{C} + \mathbf{B}[\mathbf{A}, \mathbf{C}]$$

with

$$\mathbf{A} = \mathbf{Q} \qquad \mathbf{B} = \mathbf{P}^2 \qquad \mathbf{C} = \mathbf{P}$$

leading to

$$[\mathbf{Q}, \mathbf{P}^2\mathbf{P}] = [\mathbf{Q}, \mathbf{P}^2]\mathbf{P} + \mathbf{P}^2[\mathbf{Q}, \mathbf{P}] \tag{18.22}$$

or, comparing Eqs. (18.20) and (18.21),

$$[\mathbf{Q}, \mathbf{P}^3] = (2i\hbar\mathbf{P})\mathbf{P} + \mathbf{P}^2 i\hbar = 3i\hbar\mathbf{P}^2$$

Then, from Eqs. (18.21) and (18.22), one obtains by recurrence

$$\boxed{[\mathbf{Q}, \mathbf{P}^n] = n(i\hbar)\mathbf{P}^{n-1}} \tag{18.23}$$

Furthermore, consider the commutator of $\mathbf{Q}$ with some function $\mathbf{F}(\mathbf{P})$, which may be expanded as

$$\mathbf{F}(\mathbf{P}) = \sum_n f_n \mathbf{P}^n \tag{18.24}$$

where the f_n are the expansion coefficients. Then, due to (18.24), this commutator reads

$$[\mathbf{Q}, \mathbf{F}(\mathbf{P})] = (i\hbar)\sum_n n f_n \mathbf{P}^{n-1} \tag{18.25}$$

Now, the partial derivative of Eq. (18.24) with respect to $\mathbf{P}$ yields

$$\left(\frac{\partial \mathbf{F}(\mathbf{P})}{\partial \mathbf{P}}\right) = \sum_n n f_n \mathbf{P}^{n-1} \tag{18.26}$$

so that, by identification of Eqs. (18.25) and (18.26),

$$\boxed{[\mathbf{Q}, \mathbf{F}(\mathbf{P})] = (i\hbar)\left(\frac{\partial \mathbf{F}(\mathbf{P})}{\partial \mathbf{P}}\right)} \tag{18.27}$$

In like manner for the commutators of **P** with different powers of **Q**, or with some function **F**(**Q**), we have

$$\boxed{[\mathbf{P}, \mathbf{Q}^n] = -n(i\hbar)\mathbf{Q}^{n-1}} \tag{18.28}$$

$$\boxed{[\mathbf{P}, \mathbf{F}(\mathbf{Q})] = -(i\hbar)\left(\frac{\partial \mathbf{F}(\mathbf{Q})}{\partial \mathbf{Q}}\right)} \tag{18.29}$$

18.6 DISTRIBUTION FUNCTIONS AND FOURIER TRANSFORMS

18.6.1 Fourier analysis and Bessel–Parseval relation

Consider a periodic function of the form

$$f(x+L) = f(x)$$

which may be expanded as

$$f(x) = \sum_{n=-\infty}^{\infty} C_n e^{ik_n x} \tag{18.30}$$

with

$$k_n = \frac{2\pi}{L} n$$

Now, multiplying the function $f(x)$ given by Eq. (18.30) by $e^{-ik_m x}$ and integrating between x_0 and $x_0 + L$ reads

$$\int_{x_0}^{x_0+L} f(x) e^{-ik_m x} dx = \sum_{n=-\infty}^{\infty} C_n \int_{x_0}^{x_0+L} e^{i(k_n - k_m)x} dx$$

Since

$$\int_{x_0}^{x_0+L} e^{i(k_n - k_m)x} dx = 0 \quad \text{if} \quad n \neq m \tag{18.31}$$

$$\int_{x_0}^{x_0+L} e^{i(k_n - k_m)x} dx = L \quad \text{if} \quad n = m \tag{18.32}$$

we have

$$\int_{x_0}^{x_0+L} f(x) e^{-ik_m x} dx = C_m L$$

so that the coefficients of the expansion (18.30) are simply

$$C_m = \frac{1}{L} \int_{x_0}^{x_0+L} f(x) e^{-ik_m x}\, dx$$

Next, integrating the modulus of the function (18.30) between x_0 and $x_0 + L$, that is,

$$\int_{x_0}^{x_0+L} |f(x)|^2 dx = \sum_{n=-\infty}^{\infty} \sum_{m=-\infty}^{\infty} C_n^* C_m \int_{x_0}^{x_0+L} e^{i(k_n - k_m)x}\, dx$$

yields, comparing Eqs. (18.31) and (18.32),

$$\int_{x_0}^{x_0+L} |f(x)|^2\, dx = \sum_{n=-\infty}^{\infty} |C_n|^2 L$$

which, may be written equivalently in a more symmetric form called the *Bessel–Parseval relation*:

$$\sum_{n=-\infty}^{\infty} |C_n|^2 = \frac{1}{L} \int_{-L/2}^{L/2} |f(x)|^2\, dx \tag{18.33}$$

Next, if one passes from the function $f(x)$ to the function $f(k)$,

$$f(k) = \sum_{n=-\infty}^{\infty} C_n e^{ik_n L}$$

which is periodic in the interval $-L/2$ to $+L/2$:

$$f\left(k + \frac{2\pi}{L}\right) = f(k)$$

where one obtains in place of Eq. (18.33)

$$\sum_{n=-\infty}^{\infty} |C_n|^2 = \frac{L}{2\pi} \int_{-\pi/L}^{\pi/L} |f(k)|^2\, dk \tag{18.34}$$

18.6.2 Fourier transforms

18.6.2.1 Definition The Fourier transforms $f(k)$ of any continuous function $f(x)$, and its inverse, are by definition given by

$$f(k) = \frac{1}{\sqrt{2\pi}} \int_{-\infty}^{+\infty} f(x) e^{-ikx}\, dx \tag{18.35}$$

$$f(x) = \frac{1}{\sqrt{2\pi}} \int_{-\infty}^{+\infty} f(k) e^{ikx}\, dk \tag{18.36}$$

where x is a length. Their 3D generalization reads

$$\mathbf{F}(\mathbf{k}) = \left(\frac{1}{2\pi}\right)^{3/2} \int \mathbf{F}(\mathbf{r}) e^{-i\mathbf{k}\cdot\mathbf{r}} d^3\mathbf{r} \quad \text{and} \quad \mathbf{F}(\mathbf{r}) = \left(\frac{1}{2\pi}\right)^{3/2} \int \mathbf{F}(\mathbf{k}) e^{i\mathbf{k}\cdot\mathbf{r}} d^3\mathbf{k} \tag{18.37}$$

where the integral over $d^3\mathbf{r}$ stands for

$$\int \mathbf{F}(\mathbf{r}) e^{-i\mathbf{k}\cdot\mathbf{r}} d^3\mathbf{r} = \int_{-\infty}^{+\infty}\int_{-\infty}^{+\infty}\int_{-\infty}^{+\infty} \mathbf{F}(x, y, z) e^{-i(k_x x + k_y y + k_z z)} dx\, dy\, dz$$

with a similar expression for that over $d^3\mathbf{k}$.

Next, if the length x is replaced by the time t and the value k by an angular frequency ω, Eqs. (18.35) and (18.36) read

$$I(\omega) = \frac{1}{\sqrt{2\pi}} \int_{-\infty}^{+\infty} G(t) e^{-i\omega t}\, dt \tag{18.38}$$

$$G(t) = \frac{1}{\sqrt{2\pi}} \int_{-\infty}^{+\infty} I(\omega) e^{+i\omega t}\, d\omega$$

where $I(\omega)$ plays the role of $f(k)$ and $G(t)$ that of $f(x)$. In quantum mechanics the variable k is replaced by the momentum p, and the Fourier transforms (18.35) and (18.36) take, respectively, the forms

$$\Psi(p) = \frac{1}{\sqrt{2\pi\hbar}} \int_{-\infty}^{+\infty} \Psi(x) e^{-ipx/\hbar}\, dx \tag{18.39}$$

$$\Psi(x) = \frac{1}{\sqrt{2\pi\hbar}} \int_{-\infty}^{+\infty} \Psi(p) e^{ipx/\hbar}\, dp \tag{18.40}$$

the only difference being the supplementary presence of $1/\sqrt{\hbar}$ in front of the integrals in Eqs. (18.39) and (18.40). That is required in order to render the product $\Psi(p)\Psi(x)$ dimensionless just as the dimensionless product $f(k)f(x)$, because the dimension of the product $dx\, dp$ is $\hbar$.

18.6.2.2 First property We start from Eq. (18.39), and replace p by $(p - p^\circ)$. Then, Eq. (18.39) reads

$$\Psi(p - p^\circ) = \frac{1}{\sqrt{2\pi\hbar}} \int_{-\infty}^{+\infty} \Psi(x) e^{-i(p-p^\circ)x/\hbar}\, dx$$

or, after factorizing the exponential,

$$\Psi(p - p^\circ) = \frac{1}{\sqrt{2\pi\hbar}} \int_{-\infty}^{+\infty} \{\Psi(x) e^{ip^\circ x/\hbar}\} e^{-ipx/\hbar}\, dx$$

In a like manner as Eq. (18.38)

$$I(\omega - \omega^\circ) = \frac{1}{\sqrt{2\pi}} \int_{-\infty}^{+\infty} \{G(t) e^{i\omega^\circ t}\} e^{-i\omega t}\, dt$$

18.6.2.3 Second property Now, the complex conjugate of the Fourier transform (18.39) is

$$\Psi(p)^* = \frac{1}{\sqrt{2\pi\hbar}} \int_{-\infty}^{+\infty} \Psi(x)^* e^{ipx/\hbar}\, dx \tag{18.41}$$

Now, if $\Psi(x)$ is real,

$$\Psi(x)^* = \Psi(x)$$

so that Eq. (18.41) simplifies to

$$\Psi(p)^* = \frac{1}{\sqrt{2\pi\hbar}} \int_{-\infty}^{+\infty} \Psi(x) e^{ipx/\hbar}\, dx \tag{18.42}$$

hence, comparing Eqs. (18.42) and Eq. (18.39), we have

$$\boxed{\Psi(p)^* = \Psi(-p) \quad \text{if} \quad \Psi(x)^* = \Psi(x)} \tag{18.43}$$

Now, if $\Psi(x)$ is purely imaginary,

$$\Psi(x)^* = -\Psi(x)$$

Equation (18.41) reduces to

$$\Psi(p)^* = -\frac{1}{\sqrt{2\pi\hbar}} \int_{-\infty}^{+\infty} \Psi(x) e^{ipx/\hbar}\, dx$$

so that, by comparison of Eq. (18.42) with Eq. (18.39), we have

$$\boxed{\Psi(p)^* = -\Psi(-p) \quad \text{if} \quad \Psi(x)^* = -\Psi(x)}$$

18.6.2.4 Parseval–Plancherel transformation Starting from

$$\int_{-\infty}^{+\infty} |\Psi(x)|^2\, dx = \int_{-\infty}^{+\infty} \Psi(x)^* \Psi(x)\, dx$$

and using Eq. (18.40) for $\Psi(x)$ appearing on the right-hand side of this equation, we have

$$\int_{-\infty}^{+\infty} |\Psi(x)|^2\,dx = \int_{-\infty}^{+\infty} \Psi(x)^* \left\{ \frac{1}{\sqrt{2\pi\hbar}} \int_{-\infty}^{+\infty} \Psi(p) e^{ipx/\hbar}\,dp \right\} dx$$

or

$$\int_{-\infty}^{+\infty} |\Psi(x)|^2\,dx = \int_{-\infty}^{+\infty} \Psi(p)\,dp \left\{ \frac{1}{\sqrt{2\pi\hbar}} \int_{-\infty}^{+\infty} \Psi(x)^* e^{ipx/\hbar}\,dx \right\}$$

and thus, using the complex conjugate of Eq. (18.39),

$$\int_{-\infty}^{+\infty} |\Psi(x)|^2\,dx = \int_{-\infty}^{+\infty} \Psi(p)\Psi(p)^*\,dp$$

so that

$$\boxed{\int_{-\infty}^{+\infty} |\Psi(x)|^2\,dx = \int_{-\infty}^{+\infty} |\Psi(p)|^2\,dp} \tag{18.44}$$

that is the *Parseval–Plancherel equation* relating the squared modulus of any complex function $\Psi(x)$ to that of its Fourier transform $\Psi(p)$.

18.6.2.5 Fourier transforms of divergences and rotationals

18.6.2.5.1 Fourier Transforms of $\nabla \cdot \mathbf{f}(\mathbf{r})$ Start from the Fourier transform (18.36) of $f(k)$ leading to $f(x)$:

$$f(x) = \frac{1}{\sqrt{2\pi}} \int_{-\infty}^{+\infty} \{f(k)\} e^{ikx}\,dk$$

By derivation with respect to x, this expression reads

$$\left(\frac{\partial f(x)}{\partial x} \right) = \frac{1}{\sqrt{2\pi}} \frac{\partial}{\partial x} \int_{-\infty}^{+\infty} \{f(k)\} e^{ikx}\,dk$$

or, since the x partial derivative commutes with the integral and due to the fact that the Fourier transform $f(k)$ does not depend on x

$$\left(\frac{\partial f(x)}{\partial x} \right) = \frac{1}{\sqrt{2\pi}} \int_{-\infty}^{+\infty} \{f(k)\} \frac{\partial}{\partial x} (e^{ikx})\,dk$$

that leads to

$$\left(\frac{\partial f(x)}{\partial x} \right) = \frac{1}{\sqrt{2\pi}} \int_{-\infty}^{+\infty} \{ikf(k)\} e^{ikx}\,dk \tag{18.45}$$

this result shows that if $f(k)$ is the Fourier transform of $f(x)$, then $ikf(k)$ is the Fourier transform of $\partial f(x)/\partial x$, allowing one to write for the inverse transformation

$$\{ikf(k)\} = \frac{1}{\sqrt{2\pi}} \int_{-\infty}^{+\infty} \left\{ \frac{\partial f(x)}{\partial x} \right\} e^{-ikx}\, dx \tag{18.46}$$

Then, passing to the 3D generalization, one obtains the following generalization of the 1D Eqs. (18.45) and (18.46):

$$\boxed{i\mathbf{k}\cdot\mathbf{f}(\mathbf{k}) = \left(\frac{1}{2\pi}\right)^{3/2} \int \nabla\cdot\mathbf{f}(\mathbf{r}) e^{-i\mathbf{k}\cdot\mathbf{r}}\, d^3\mathbf{r}} \tag{18.47}$$

the inverse Fourier transform of which is

$$\boxed{\left(\frac{1}{2\pi}\right)^{3/2} \int \mathbf{k}\cdot\mathbf{f}(\mathbf{k}) e^{i\mathbf{k}\cdot\mathbf{r}}\, d^3\mathbf{k} = i\nabla\cdot\mathbf{f}(\mathbf{r})} \tag{18.48}$$

with, of course,

$$\nabla\cdot\mathbf{f}(\mathbf{r}) = \frac{\partial f_x(x,y,z)}{\partial x} + \frac{\partial f_y(x,y,z)}{\partial y} + \frac{\partial f_z(x,y,z)}{\partial z}$$

$$\mathbf{k}\cdot\mathbf{f}(\mathbf{k}) = k_x f_x(k) + k_y f_y(k) + k_z f_z(k)$$

Now, observe that taking in place of k the scalar momentum p, and keeping in mind Eqs. (18.39) and (18.40) dealing with the quantum mechanics Fourier transforms, Eqs. (18.45) and (18.46) read, respectively,

$$\boxed{\frac{1}{\sqrt{2\pi\hbar}} \int_{-\infty}^{+\infty} \frac{p}{\hbar} f(p) e^{ipx/\hbar}\, dx = i\hbar\left(\frac{\partial f(x)}{\partial x}\right)} \tag{18.49}$$

$$\boxed{\frac{1}{\hbar} pf(p) = \frac{1}{\sqrt{2\pi\hbar}} i \int_{-\infty}^{+\infty} \left(\frac{\partial f(x)}{\partial x}\right) e^{-ipx/\hbar}\, dx} \tag{18.50}$$

18.6.2.5.2 Fourier transform of $\nabla \times \mathbf{f}(\mathbf{r})$ Furthermore, look at the Fourier transform of the rotational of $\mathbf{f}(\mathbf{r})$, that is, $\nabla \times \mathbf{f}(\mathbf{r})$, the x component of which reads

$$\{\nabla \times \mathbf{f}(\mathbf{r})\}_x = \left(\frac{\partial f_z(x,y,z)}{\partial y}\right) - \left(\frac{\partial f_y(x,y,z)}{\partial z}\right) \tag{18.51}$$

Now, consider the following Fourier transform:

$$f_z(x,y,z) = \left(\frac{1}{2\pi}\right)^{3/2} \int_{-\infty}^{+\infty}\int_{-\infty}^{+\infty}\int_{-\infty}^{+\infty} \{f_z(k_x,k_y,k_z)\} e^{i(k_x x + k_y y + k_z z)}\, dk_x\, dk_y\, dk_z$$

Its y partial derivative reads

$$\left(\frac{\partial f_z(x,y,z)}{\partial y}\right) = \left(\frac{1}{2\pi}\right)^{3/2} \int_{-\infty}^{+\infty}\int_{-\infty}^{+\infty}\int_{-\infty}^{+\infty} \{f_z(k_x,k_y,k_z)\}\frac{\partial}{\partial y}e^{i(k_x x+k_y y+k_z z)}\,dk_x\,dk_y\,dk_z$$

or

$$\left(\frac{\partial f_z(x,y,z)}{\partial y}\right) = \left(\frac{1}{2\pi}\right)^{3/2} \int_{-\infty}^{+\infty}\int_{-\infty}^{+\infty}\int_{-\infty}^{+\infty} i\{k_y f_z(k_x,k_y,k_z)\}e^{i(k_x x+k_y y+k_z z)}\,dk_x\,dk_y\,dk_z$$

showing that the Fourier transform of the y partial derivative of $f(x,y,z)$ is $k_y f(k_x,k_y,k_z)$ so that, using a similar result for the z partial derivative, one gets

$$\left(\frac{\partial f_z(x,y,z)}{\partial y}\right) - \left(\frac{\partial f_y(x,y,z)}{\partial z}\right) = \left(\frac{1}{2\pi}\right)^{3/2}$$

$$\times \int_{-\infty}^{+\infty}\int_{-\infty}^{+\infty}\int_{-\infty}^{+\infty} i\Big\{(k_y f_z(k_x,k_y,k_z) - k_x f_y(k_x,k_y,k_z))\Big\}e^{i(k_x x+k_y y+k_z z)}\,dk_x\,dk_y\,dk_z \quad (18.52)$$

Then, due to the fact that

$$k_y f_z(k_x,k_y,k_z) - k_x f_y(k_x,k_y,k_z) = \{\mathbf{k}\times\mathbf{f}(\mathbf{k})\}_x$$

and owing to Eq. (18.51), it appears that Eq. (18.52) reads

$$\{\nabla\times\mathbf{f}(\mathbf{r})\}_x = \left(\frac{1}{2\pi}\right)^{3/2} \int_{-\infty}^{+\infty}\int_{-\infty}^{+\infty}\int_{-\infty}^{+\infty} i\{\mathbf{k}\times\mathbf{f}(\mathbf{k})\}_x e^{i(k_x x+k_y y+k_z z)}\,dk_x\,dk_y\,dk_z$$

or

$$\{\nabla\times\mathbf{f}(\mathbf{r})\}_x = \left(\frac{1}{2\pi}\right)^{3/2} \int i\{\mathbf{k}\times\mathbf{f}(\mathbf{k})\}_x e^{i(\mathbf{k}.\mathbf{r})}\,d^3\mathbf{k}$$

which, with the help of similar equations for the y and z components of the cross products, yields

$$\{\nabla\times\mathbf{f}(\mathbf{r})\} = \left(\frac{1}{2\pi}\right)^{3/2} \int i\mathbf{k}\times\mathbf{f}(\mathbf{k})e^{i(\mathbf{k}.\mathbf{r})}\,d^3\mathbf{k}$$

showing that $\{\nabla\times\mathbf{f}(\mathbf{r})\}$ is the Fourier transform of $i\mathbf{k}\times\mathbf{f}(\mathbf{k})$ leading, therefore, for the inverse Fourier transform

$$\boxed{\left(\frac{1}{2\pi}\right)^{3/2} \int \{\nabla\times\mathbf{f}(\mathbf{r})\}e^{-i\mathbf{k}\cdot\mathbf{r}}\,d^3\mathbf{r} = i\mathbf{k}\times\mathbf{f}(\mathbf{k})} \quad (18.53)$$

18.6.3 Dirac distribution

18.6.3.1 Definition The Dirac distribution denoted $\delta(x - x^\circ)$ may be introduced via

$$f(x^\circ) = \int_{-\infty}^{+\infty} \delta(x - x^\circ) f(x)\,dx \quad (18.54)$$

where $f(x)$ is any function of x and $f(x^\circ)$ its expression when $x = x^\circ$. Observe that $\delta(x - x^\circ)$ acts in Eq. (18.54) as if

$$\delta(x - x^\circ) = 1 \quad \text{if} \quad x = x^\circ \qquad \text{and} \qquad \delta(x - x^\circ) = 0 \quad \text{if} \quad x \neq x^\circ$$

We remark, however, that the dimension of the Dirac distribution $\delta(x - x^\circ)$ is that of the inverse of x because of the presence of the differential term dx involved in the integral appearing on the right-hand side of Eq. (18.54), defining the Dirac distribution. We observe also that the Dirac distribution may be viewed as the generalization to continuous functions of the Kronecker symbol $\delta_{i,j}$ used for discrete variables, corresponding to

$$\delta_{kl} = 1 \quad \text{if} \quad k = l \qquad \text{and} \qquad \delta_{kl} = 0 \quad \text{if} \quad k \neq l$$

In the special case where $x^\circ = 0$, Eq. (18.54) simplifies to

$$f(0) = \int_{-\infty}^{+\infty} \delta(x) f(x)\, dx$$

18.6.3.2 Limiting expressions for the Dirac distributions Many limiting expressions $\delta^\varepsilon(x)$ of the Dirac distribution $\delta(x)$ exist, chosen in such a way as

$$\delta(x) = \lim_{\varepsilon \to 0^+} \{\delta^\varepsilon(x)\} \tag{18.55}$$

where ε is a small positive parameter having the same dimension as x, which has to be vanishing. One of the limiting expressions $\delta^\varepsilon(x)$ is the *rectangular function* defined by

$$\delta^\varepsilon(x) = \frac{1}{\varepsilon} \quad \text{if} \quad -\frac{\varepsilon}{2} < x < \frac{\varepsilon}{2} \qquad \text{and} \qquad \delta^\varepsilon(x) = 0 \quad \text{otherwise}$$

Its graph has a width ε, a height $1/\varepsilon$, and thus an area equal to unity.

Among other limiting expressions $\delta^\varepsilon(x)$ of the Dirac distribution, one may quote the following:

$$\delta^\varepsilon(x) = \frac{1}{2}\left(\frac{e^{-|x/\varepsilon|}}{\varepsilon}\right)$$

$$\delta^\varepsilon(x) = \left(\frac{e^{-x^2/\varepsilon^2}}{\varepsilon}\right)$$

$$\delta^\varepsilon(x) = \frac{1}{\pi}\left(\frac{\varepsilon}{x^2 + \varepsilon^2}\right) \tag{18.56}$$

$$\delta^\varepsilon(x) = \frac{1}{\pi}\left(\frac{\sin(x/\varepsilon)}{x}\right)$$

$$\delta^\varepsilon(x) = \frac{1}{\pi}\varepsilon\left(\frac{\sin^2(x/\varepsilon)}{x^2}\right) \tag{18.57}$$

One may verify that all these limiting distributions have, as required, the same x^{-1} dimension.

18.6.3.3 An integral representation for the Dirac distribution Consider the following Fourier transform of the Dirac distribution $\delta(x - x^\circ)$, which according to Eq. (18.39) is

$$F(p) = \frac{1}{\sqrt{2\pi\hbar}} \int_{-\infty}^{+\infty} \delta(x - x^\circ) e^{-ipx/\hbar}\, dx \qquad (18.58)$$

Now, taking

$$f(x) = e^{-ipx/\hbar}$$

in Eq. (18.54), the integral appearing on the right-hand side of Eq. (18.58), simplifies to

$$F(p) = \frac{1}{\sqrt{2\pi\hbar}} e^{-ipx^\circ/\hbar} \qquad (18.59)$$

Next, consider the inverse Fourier transform of Eq. (18.58), which according to Eq. (18.40) is

$$\delta(x - x^\circ) = \frac{1}{\sqrt{2\pi\hbar}} \int_{-\infty}^{+\infty} F(p) e^{ipx/\hbar}\, dp$$

Then, using Eq. (18.59) for $F(p)$, we have

$$\delta(x - x^\circ) = \frac{1}{\sqrt{2\pi\hbar}} \frac{1}{\sqrt{2\pi\hbar}} \int_{-\infty}^{+\infty} e^{-ipx^\circ/\hbar} e^{ipx/\hbar}\, dp$$

Hence, we have the important result

$$\boxed{\delta(x - x^\circ) = \frac{1}{2\pi\hbar} \int_{-\infty}^{+\infty} e^{ip(x - x^\circ)/\hbar}\, dp} \qquad (18.60)$$

Observe that, by changing the momentum p into the modulus of the wave vector k, Eq. (18.60) may be also written

$$\delta(x - x^\circ) = \frac{1}{2\pi} \int_{-\infty}^{+\infty} e^{ik(x - x^\circ)}\, dk \qquad (18.61)$$

and that, passing from the x variables to the angular frequencies ω, and from k to the time t,

$$\delta(\omega - \omega^\circ) = \frac{1}{2\pi} \int_{-\infty}^{+\infty} e^{it(\omega - \omega^\circ)}\, dt \qquad (18.62)$$

18.6.4 An important integral

One has sometimes to find expressions of the form

$$F(\omega^\circ) = \int_{-\infty}^{+\infty} f(\omega)\left(\int_0^{+\infty} e^{-i(\omega-\omega^\circ)\tau}\, d\tau\right) d\omega \tag{18.63}$$

To solve this integral, it is convenient to transform it into

$$F(\omega^\circ) = \int_{-\infty}^{+\infty} f(\omega)\left(\lim_{\varepsilon\to 0^+}\left\{\int_0^{\infty} e^{-i(\omega-\omega^\circ)\tau} e^{-\varepsilon\tau}\, d\tau\right\}\right) d\omega$$

or

$$F(\omega^\circ) = \int_{-\infty}^{+\infty} f(\omega)\left(\lim_{\varepsilon\to 0^+}\left\{\int_0^{\infty} e^{-i(\omega-\omega^\circ+i\varepsilon)\tau}\, d\tau\right\}\right) d\omega$$

with ε real, so that after integration, one gets

$$F(\omega^\circ) = -i\int_{-\infty}^{+\infty} f(\omega)\left\{\lim_{\varepsilon\to 0^+}\left\{\frac{1}{(\omega-\omega^\circ)+i\varepsilon}\right\}\right\} d\omega$$

with

$$\left(\frac{1}{(\omega-\omega^\circ)+i\varepsilon}\right) = \left(\frac{(\omega-\omega^\circ)-i\varepsilon}{((\omega-\omega^\circ)+i\varepsilon)((\omega-\omega^\circ)-i\varepsilon)}\right) = \left(\frac{(\omega-\omega^\circ)-i\varepsilon}{(\omega-\omega^\circ)^2+\varepsilon^2}\right)$$

so that $F(\omega^\circ)$ may be decomposed into two integrals according to

$$\boxed{\begin{aligned} F(\omega^\circ) = &-i\int_{-\infty}^{+\infty} f(\omega)\left\{\lim_{\varepsilon\to 0^+}\left(\frac{\omega-\omega^\circ}{(\omega-\omega^\circ)^2+\varepsilon^2}\right)\right\} d\omega \\ &-i\int_{-\infty}^{+\infty} f(\omega)\left\{\lim_{\varepsilon\to 0^+}\left(\frac{-i\varepsilon}{(\omega-\omega^\circ)^2+\varepsilon^2}\right)\right\} d\omega \end{aligned}} \tag{18.64}$$

Now, keeping in mind Eqs. (18.55) and (18.56) allows one to write the Dirac distribution $\delta(x)$ according to

$$\delta(\omega-\omega^\circ) = \lim_{\varepsilon\to 0^+}\left\{\frac{1}{\pi}\left(\frac{\varepsilon}{(\omega-\omega^\circ)^2+\varepsilon^2}\right)\right\}$$

Hence, it may be concluded that the second term of the right-hand side of Eq. (18.64) reads

$$\boxed{\lim_{\varepsilon\to 0^+}\left(\frac{i\varepsilon}{(\omega-\omega^\circ)^2+\varepsilon^2}\right) = i\pi\delta(\omega-\omega^\circ)} \tag{18.65}$$

On the other hand, taking $x = (\omega - \omega^\circ)$ in the first integral of Eq. (18.64), it is suitable to decompose this integral into three parts according to

$$\lim_{\varepsilon\to 0^+} \int_{-\infty}^{+\infty} f(\omega)\left(\frac{(\omega-\omega^\circ)}{(\omega-\omega^\circ)^2+\varepsilon^2}\right) d\omega = \lim_{\varepsilon\to 0^+}\lim_{\eta\to 0^+}$$

$$\left\{\int_{-\infty}^{-\eta} f(x)\left(\frac{x}{x^2+\varepsilon^2}\right) dx + \int_{-\eta}^{+\eta} f(x)\left(\frac{x}{x^2+\varepsilon^2}\right) dx + \int_{+\eta}^{+\infty} f(x)\left(\frac{x}{x^2+\varepsilon^2}\right) dx\right\} \tag{18.66}$$

Next, look at the limit of the second term of the right-hand side of this last equation involving the integration over x between $-\eta$ and $+\eta$. Since $\eta \to 0^+$, that means that the integrand x is very near $x = 0$ so that $f(0)$ may be taken in place of $f(x)$ and then displaced in front of the integral sign, leading thus to

$$\lim_{\varepsilon\to 0^+}\lim_{\eta\to 0^+}\left\{\int_{-\eta}^{+\eta} f(x)\left(\frac{x}{x^2+\varepsilon^2}\right) dx\right\} = f(0)\lim_{\varepsilon\to 0^+}\lim_{\eta\to 0^+}\left\{\int_{-\eta}^{+\eta}\left(\frac{x}{x^2+\varepsilon^2}\right) dx\right\} \tag{18.67}$$

Now, observe that the right-hand integral of (18.67) between $+\eta$ and $-\eta$ is zero because involving an odd integrand

$$\lim_{\varepsilon\to 0^+}\lim_{\eta\to 0^+}\left\{\int_{-\eta}^{+\eta} f(x)\left(\frac{x}{x^2+\varepsilon^2}\right) dx\right\} = 0$$

As a consequence, Eq. (18.66) simplifies to

$$\lim_{\varepsilon\to 0^+}\left\{\int_{-\infty}^{+\infty} f(x)\left(\frac{x}{x^2+\varepsilon^2}\right) dx\right\}$$

$$= \lim_{\varepsilon\to 0^+}\lim_{\eta\to 0^+}\left\{\int_{-\infty}^{-\eta} f(x)\left(\frac{x}{x^2+\varepsilon^2}\right) dx + \int_{+\eta}^{+\infty} f(x)\left(\frac{x}{x^2+\varepsilon^2}\right) dx\right\} \tag{18.68}$$

Besides, observe that in the limit where $\varepsilon \to 0^+$

$$\lim_{\varepsilon\to 0^+}\left\{\int_{-\infty}^{-\eta} f(x)\left(\frac{x}{x^2+\varepsilon^2}\right) dx + \int_{+\eta}^{+\infty} f(x)\left(\frac{x}{x^2+\varepsilon^2}\right) dx\right\}$$

$$= \left\{\int_{-\infty}^{-\eta} f(x)\left(\frac{1}{x}\right) dx + \int_{+\eta}^{+\infty} f(x)\left(\frac{1}{x}\right) dx\right\} \tag{18.69}$$

Hence, coming back from x to $\omega - \omega^{\circ}$, Eq. (18.68) reduces to the following expression:

$$\boxed{\lim_{\varepsilon \to 0^+} \left\{ \int_{-\infty}^{+\infty} f(\omega) \frac{(\omega - \omega^{\circ})}{(\omega - \omega^{\circ})^2 + \varepsilon^2} d\omega \right\} = \left\{ \int_{-\infty}^{+\infty} f(\omega) \mathcal{P}\left(\frac{1}{(\omega - \omega^{\circ})} \right) d\omega \right\}} \quad (18.70)$$

which is named the *Cauchy principal part* $\mathcal{P}(1/x)$:

$$\left\{ \int_{-\infty}^{+\infty} f(x) \mathcal{P}\left(\frac{1}{x} \right) dx \right\} = \lim_{\eta \to 0^+} \left\{ \int_{-\infty}^{-\eta} f(x) \left(\frac{1}{x} \right) dx + \int_{+\eta}^{+\infty} f(x) \left(\frac{1}{x} \right) dx \right\}$$

As a consequence of Eqs. (18.65) and (18.70), Eq. (18.64) yields

$$\boxed{F(\omega^{\circ}) = -i \left\{ \int_{-\infty}^{+\infty} f(\omega) \mathcal{P}\left(\frac{1}{(\omega - \omega^{\circ})} \right) d\omega \right\} d\omega - \int_{-\infty}^{+\infty} f(\omega) \pi \delta(\omega - \omega^{\circ})\, d\omega} \quad (18.71)$$

18.7 LAGRANGE MULTIPLIERS METHOD

Consider a function $f(x, y)$ of the variables x and y. These variables are not independent but relate via an equation of the form

$$\Phi(x, y) = cst \quad (18.72)$$

Since the function $\Phi(x, y)$ relating x and y is a constant, its differential is zero, that is,

$$d\Phi(x, y) = 0$$

Now, to get the extrema of the function $f(x, y)$ one has to impose the condition

$$df(x, y) = 0$$

Hence, the total differentials of the functions $f(x, y)$ and $\Phi(x, y)$ are zero so that one may write, respectively,

$$df(x, y) = \left(\frac{\partial f(x, y)}{\partial x} \right) dx + \left(\frac{\partial f(x, y)}{\partial y} \right) dy = 0 \quad (18.73)$$

$$d\Phi(x, y) = \left(\frac{\partial \Phi(x, y)}{\partial x} \right) dx + \left(\frac{\partial \Phi(x, y)}{\partial y} \right) dy = 0 \quad (18.74)$$

To find the extrema of the function $f(x, y)$ subjected to the constraint (18.72), it is convenient to combine the two equalities (18.73) and (18.74) by adding to the first, the last times a λ parameter, which may be according to

$$\left\{ \left(\frac{\partial f(x, y)}{\partial x} \right) dx + \left(\frac{\partial f(x, y)}{\partial y} \right) dy \right\} + \lambda \left\{ \left(\frac{\partial \Phi(x, y)}{\partial x} \right) dx + \left(\frac{\partial \Phi(x, y)}{\partial y} \right) dy \right\} = 0 \quad (18.75)$$

Now, the λ value may be chosen in such a way as

$$\left(\frac{\partial f(x,y)}{\partial y}\right) + \lambda\left(\frac{\partial \Phi(x,y)}{\partial y}\right) = 0 \tag{18.76}$$

which implies that

$$\lambda = -\frac{\left(\frac{\partial f(x,y)}{\partial y}\right)}{\left(\frac{\partial \Phi(x,y)}{\partial y}\right)}$$

However, this last equation is unnecessary. In fact, this is exactly the point of the Lagrange multipliers, and one may avoid this expression by simply concluding from the imposed Eq. (18.76) that Eq. (18.75) implies

$$\left(\frac{\partial f(x,y)}{\partial x}\right) + \lambda\left(\frac{\partial \Phi(x,y)}{\partial x}\right) = 0 \tag{18.77}$$

Then, one may observe that Eqs. (18.76) and (18.77) are the two equations that would be written from a function $F(x, y)$ of the form

$$F(x,y) = f(x,y) + \lambda\Phi(x,y) \tag{18.78}$$

if one imposes the condition

$$dF(x,y) = 0 \tag{18.79}$$

since its zero total differential is

$$dF(x,y) = \left\{\left(\frac{\partial f(x,y)}{\partial x}\right) + \lambda\left(\frac{\partial \Phi(x,y)}{\partial x}\right)\right\}dx + \left\{\left(\frac{\partial f(x,y)}{\partial y}\right) + \lambda\left(\frac{\partial \Phi(x,y)}{\partial y}\right)\right\}dy = 0$$

an expression that is equivalent to Eq. (18.75).

Thus, to find the extrema of some function $f(x, y)$ subjected to a constraint of the form (18.72), it is convenient to define a function $F(x, y)$ of the form (18.78) and then set equal to zero its total differential according to Eq. (18.79), so that Eqs. (18.76) and (18.77) are satisfied. Hence, one has the three equations (18.72), (18.76), and (18.77) for three unknown, that is the values of x° and y° allowing the function $f(x, y)$ to be an extremum, and also the value of λ (the *Lagrange multiplier parameter*) introduced through Eq. (18.78), when defining the function $F(x, y)$.

18.8 TRIPLE VECTOR PRODUCT

This section is devoted to the calculation of the triple vectorial product $\vec{\mathbf{A}} \times (\vec{\mathbf{B}} \times \vec{\mathbf{C}})$. Figure 18.1 represents the different vectors involved in the calculation of this triple vectorial product.

A vector equation is true independently of the coordinate system. Thus, we can simplify our work by choosing our coordinate system intelligently in such a way so

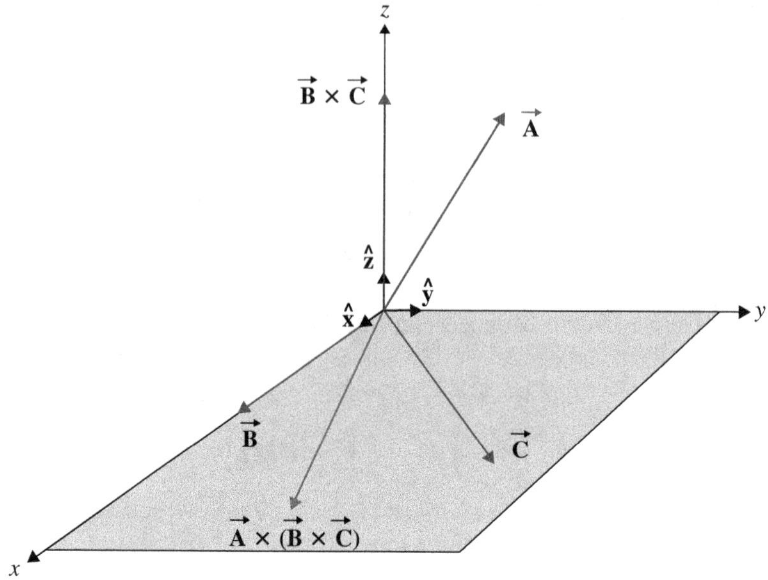

Figure 18.1 Triple vectorial product $\vec{\mathbf{A}} \times (\vec{\mathbf{B}} \times \vec{\mathbf{C}})$.

that the demonstration is the simplest. Thus, choose the referential for the vectors $\vec{\mathbf{A}}$, $\vec{\mathbf{B}}$, and $\vec{\mathbf{C}}$ according to

$$\vec{\mathbf{A}} = A_x\hat{\mathbf{x}} + A_y\hat{\mathbf{y}} + A_z\hat{\mathbf{z}} \tag{18.80}$$

$$\vec{\mathbf{B}} = B_x\hat{\mathbf{x}} \tag{18.81}$$

$$\vec{\mathbf{C}} = C_x\hat{\mathbf{x}} + C_y\hat{\mathbf{y}} \tag{18.82}$$

where $\hat{\mathbf{x}}$, $\hat{\mathbf{y}}$, and $\hat{\mathbf{z}}$ are the unit vectors along the x, y, and z directions obeying

$$\hat{\mathbf{x}} \cdot \hat{\mathbf{x}} = \hat{\mathbf{y}} \cdot \hat{\mathbf{y}} = \hat{\mathbf{z}} \cdot \hat{\mathbf{z}} = 1 \tag{18.83}$$

$$\hat{\mathbf{x}} \times \hat{\mathbf{x}} = \hat{\mathbf{y}} \times \hat{\mathbf{y}} = \hat{\mathbf{z}} \times \hat{\mathbf{z}} = 0 \tag{18.84}$$

$$\hat{\mathbf{x}} \times \hat{\mathbf{y}} = \hat{\mathbf{z}} \qquad \hat{\mathbf{y}} \times \hat{\mathbf{z}} = \hat{\mathbf{x}} \qquad \hat{\mathbf{z}} \times \hat{\mathbf{x}} = \hat{\mathbf{y}} \tag{18.85}$$

$$\hat{\mathbf{y}} \times \hat{\mathbf{x}} = -\hat{\mathbf{z}} \qquad \hat{\mathbf{z}} \times \hat{\mathbf{y}} = -\hat{\mathbf{x}} \qquad \hat{\mathbf{x}} \times \hat{\mathbf{z}} = -\hat{\mathbf{y}} \tag{18.86}$$

Using Eqs. (18.81) and (18.82) dealing with $\vec{\mathbf{B}}$ and $\vec{\mathbf{C}}$ leads to

$$\vec{\mathbf{B}} \times \vec{\mathbf{C}} = B_x\hat{\mathbf{x}} \times (C_x\hat{\mathbf{x}} + C_y\hat{\mathbf{y}}) = B_xC_x(\hat{\mathbf{x}} \times \hat{\mathbf{x}}) + B_xC_y(\hat{\mathbf{x}} \times \hat{\mathbf{y}})$$

or, using the first properties of (18.84) and (18.85)

$$\vec{\mathbf{B}} \times \vec{\mathbf{C}} = B_xC_y\hat{\mathbf{z}}$$

Next, using the expression of $\vec{\mathbf{A}}$ appearing in Eq. (18.80), the triple product reads

$$\vec{\mathbf{A}} \times (\vec{\mathbf{B}} \times \vec{\mathbf{C}}) = A_xB_xC_y(\hat{\mathbf{x}} \times \hat{\mathbf{z}}) + A_yB_xC_y(\hat{\mathbf{y}} \times \hat{\mathbf{z}}) + A_zB_xC_y(\hat{\mathbf{z}} \times \hat{\mathbf{z}})$$

Then, owing to the equations of interest appearing in (18.84)–(18.86), it simplifies to

$$\vec{\mathbf{A}} \times (\vec{\mathbf{B}} \times \vec{\mathbf{C}}) = -A_x B_x C_y \widehat{\mathbf{y}} + A_y B_x C_y \widehat{\mathbf{x}}$$

Next, adding and substracting the same term $A_x B_x C_x \widehat{\mathbf{x}}$ on the right-hand side yields

$$\vec{\mathbf{A}} \times (\vec{\mathbf{B}} \times \vec{\mathbf{C}}) = -A_x B_x C_x \widehat{\mathbf{x}} - A_x B_x C_y \widehat{\mathbf{y}} + A_y B_x C_y \widehat{\mathbf{x}} + A_x B_x C_x \widehat{\mathbf{x}}$$

or

$$\vec{\mathbf{A}} \times (\vec{\mathbf{B}} \times \vec{\mathbf{C}}) = -A_x B_x (C_x \widehat{\mathbf{x}} + C_y \widehat{\mathbf{y}}) + (A_y C_y + A_x C_x) B_x \widehat{\mathbf{x}} \tag{18.87}$$

Now, observe that, according to Eqs. (18.80)–(18.82), it appears that

$$A_x B_x = \vec{\mathbf{A}} \cdot \vec{\mathbf{B}} \qquad \text{and} \qquad (A_x C_x + A_y C_y) = \vec{\mathbf{A}} \cdot \vec{\mathbf{C}}$$

Hence, due to the expressions of $\vec{\mathbf{B}}$ and $\vec{\mathbf{C}}$ appearing in Eqs. (18.81) and (18.82), the triple product (18.87) takes the form

$$\boxed{\vec{\mathbf{A}} \times (\vec{\mathbf{B}} \times \vec{\mathbf{C}}) = (\vec{\mathbf{A}} \cdot \vec{\mathbf{C}})\vec{\mathbf{B}} - (\vec{\mathbf{A}} \cdot \vec{\mathbf{B}})\vec{\mathbf{C}}} \tag{18.88}$$

18.9 POINT GROUPS

18.9.1 Symmetry and groups

18.9.1.1 Symmetry elements and symmetry operations For molecules involving symmetry properties, it is necessary to distinguish between symmetry elements and symmetry operations. The symmetry elements, their symbols, and their corresponding symmetry operations are given in Eq. (18.89):

Symbol	Symmetry Element	Symbol	Symmetry Operation
E	Identity	$\mathbf{R}_E$	No change
σ	Symmetry plane	$\mathbf{R}_\sigma$	Reflection through σ
i	Center of symmetry	$\mathbf{R}_i$	Inversion through i
C_p^k	Axis of symmetry	$\mathbf{R}_{C_p}$	Rotation about C_p by $2\pi/p$
C_p	Axis of symmetry	$\mathbf{R}_{C_p^k}$	Rotation about C_p by $2\pi k/p$
S_p	Rotation–reflection axis	$\mathbf{R}_{S_p}$	C_p rotation followed by σ reflection

(18.89)

Note that to a given symmetry axis, several kinds of rotation may correspond. Although there is only a finite number of fashions to meet the different symmetry elements, each of these possible combinations forms a *point group* because all the symmetry operations that are associated with the symmetry elements involved in these combinations act as elements of a mathematical group and leave fixed in space a point of the molecule belonging to this point group (in contrast with space groups dealing with crystals where a translation of a unit cell leads to a new equivalent disposition in the crystal).

In the following we shall limit ourselves only to the very simple examples of point groups, which are C_{2v} and C_{3v}, the symmetry elements of which appear in (18.90):

Point Group	Element	Operation	Examples
C_{2v}	$E, C_2, \sigma_v, \sigma'_v$	$R_E, R_{C_2}, R_{\sigma_v}, R_{\sigma'_v}$	H_2O
C_{2h}	E, C_2, σ_h, i	$R_E, R_{C_2}, R_{\sigma_h}, R_i$	C_2H_4
C_{3v}	$E, C_3, C_3^2, \sigma_v, \sigma'_v, \sigma''_v$	$R_E, R_{C_3}, R_{C_3^2}, R_{\sigma_v}, R_{\sigma'_v}, R_{\sigma''_v}$	NH_3

(18.90)

18.9.1.2 Symmetry operations as a group As it has been said, the symmetry operations of a given point group obey the properties of a mathematical group, which are now of interest to recall. A set of elements $P, Q, R, S \ldots$ form a group if all the following conditions are satisfied:

1. A multiplication rule exists according to which the multiplication of two elements of a group leads to another element of this group, that is, for instance
$$PQ = R \qquad PR = S$$
2. The multiplication rule obeys an associative law, that is,
$$P(QR) = (PQ)R$$
3. The set of elements of a group contains an identity element E such as for all other elements R of the group
$$ER = RE = R$$
4. To every element R of the group, there exists an inverse R^{-1}, which belongs also to this group, according to
$$RR^{-1} = R^{-1}R = E$$

Besides, a *class* of a group is a subset of its elements $P, Q \cdots S$ which remain members of this class when subjected to the following operation involving all members R of the group:

$$R^{-1}PR \qquad R^{-1}QR \cdots R^{-1}PSR \tag{18.91}$$

Now, as it may be verified, the multiplications of the different symmetry operations of the $\mathbf{C}_{2v}$ and $\mathbf{C}_{2h}$ point groups verify the group multiplication rule as summarized in (18.92):

$\mathbf{C}_{2v}$	E	C_2	σ_v	σ'_v
E	E	C_2	σ_v	σ'_v
C_2	C_2	E	σ'_v	σ_v
σ_v	σ_v	σ'_v	E	C_2
σ'_v	σ'_v	σ_v	C_2	E

$\mathbf{C}_{2h}$	E	C_2	i	σ_h
E	E	C_2	i	σ_h
C_2	C_2	E	σ_h	i
i	i	σ_h	E	C_2
σ_h	σ_h	i	C_2	E

(18.92)

Underline the presence of the identity operation E and also the fact that the inverse requirement is satisfied.

Moreover, for the $\mathbf{C}_{3v}$ point group, the symmetry operations C_3 and C_3^2 belong to the same class, whereas the symmetry operations σ_v, σ_v', and σ_v'' belong to another class, which may be verified from (18.93) for C_3 and C_3^2 by looking at all the transformations of the kind (18.91) involving all the symmetry elements of the point group:

$$\begin{array}{lllllll}
(E)^{-1} \times C_3 \times (E) & = & C_3 & \text{and} & (E)^{-1} \times C_3^2 \times (E) & = & C_3^2 \\
(C_3)^{-1} \times C_3 \times (C_3) & = & C_3 & \text{and} & (C_3)^{-1} \times C_3^2 \times (C_3) & = & C_3 \\
(C_3^2)^{-1} \times C_3 \times (C_3^2) & = & C_3^2 & \text{and} & (C_3^2)^{-1} \times C_3^2 \times (C_3^2) & = & C_3^2 \\
(\sigma_v)^{-1} \times C_3 \times (\sigma_v) & = & C_3^2 & \text{and} & (\sigma_v)^{-1} \times C_3^2 \times (\sigma_v) & = & C_3 \\
(\sigma_v')^{-1} \times C_3 \times (\sigma_v') & = & C_3^2 & \text{and} & (\sigma_v')^{-1} \times C_3^2 \times (\sigma_v') & = & C_3 \\
(\sigma_v'')^{-1} \times C_3 \times (\sigma_v'') & = & C_3^2 & \text{and} & (\sigma_v'')^{-1} \times C_3^2 \times (\sigma_v'') & = & C_3
\end{array} \tag{18.93}$$

In the same way, it would be found easily that the three symmetry operations σ_v, σ_v', and σ_v'' belong to the same class and that E forms alone by itself a third class.

18.9.2 Group representations resulting from several symmetry transformation

To visualize how the components of the rotation and translation vectors transform under the action of symmetry operations, look at the simple examples of the $\mathbf{C}_{2v}$ and $\mathbf{C}_{3v}$ point groups.

18.9.2.1 C_{2v} point group By inspection of Fig. 18.2 it appears that the $\mathbf{C}_{2v}$ symmetry operations on the components T_x, T_y, and T_z of the translation vector $\mathbf{T}$ and on the components R_x, R_y, and R_z of the rotation vector $\mathbf{R}$, are given in (18.94):

$$\begin{array}{llll}
ET_x = T_x & C_2T_x = -T_x & \sigma_vT_x = T_x & \sigma_v'T_x = -T_x \\
ET_y = T_y & C_2T_x = -T_y & \sigma_vT_y = -T_y & \sigma_v'T_y = T_y \\
ET_z = T_z & C_2T_z = T_z & \sigma_vT_z = T_z & \sigma_v'T_z = T_z \\
ER_x = R_x & C_2R_x = -R_x & \sigma_vR_x = R_x & \sigma_v'R_x = R_x \\
ER_y = R_y & C_2R_x = -R_y & \sigma_vR_y = R_y & \sigma_v'R_y = -R_y \\
ER_z = R_z & C_2R_z = R_z & \sigma_vR_z = -R_z & \sigma_v'R_z = -R_z
\end{array} \tag{18.94}$$

From these results one may observe that there is not any symmetry operation that modifies the translation vector T_z, since all the symmetry operations affect T_z by the unity factor. Besides, R_z conserves its sign under the E and C_2 operations, whereas it changes under the σ_v and σ_v' ones. Furthermore, both R_y and T_x and both R_x and

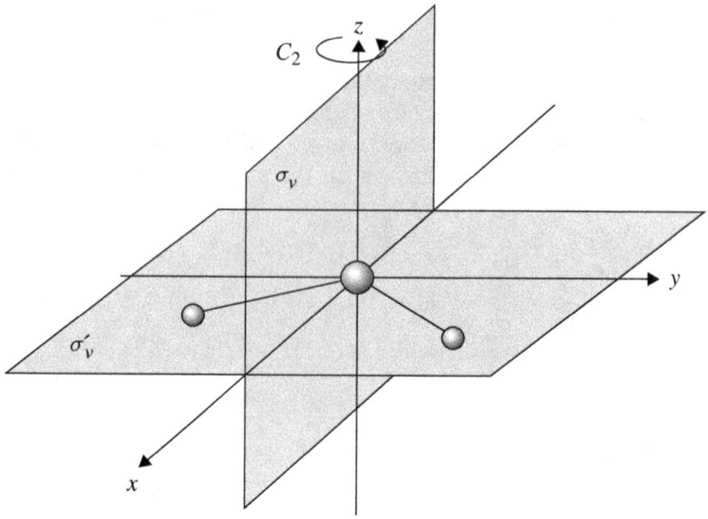

Figure 18.2 Symmetry elements for a C_{2v} molecule.

T_y conserve or change their sign in the same way under all the symmetry operations, which is summarized in (18.95):

Rot and Trans	E	C_2	σ_v	σ_v'
T_z	+1	+1	+1	+1
R_z	+1	+1	−1	−1
R_y, T_x	+1	−1	+1	−1
R_x, T_y	+1	−1	−1	+1

(18.95)

18.9.2.2 C_{3v} point group Now, passing to the $\mathbf{C}_{3v}$ point group, look at Fig. 18.3 in which are depicted the transformations of the components of the translation and rotation vectors under the symmetry elements of this group.

The two clockwise operations C_3 and C_3^2 along the Z coordinate appearing in this figure Involve the rotation angles $\theta = -120°$ and $\theta = -240°$. Besides, a rotation around the Z axis by the angle θ induces the following linear transformation on the Cartesian displacement and rotation coordinates:

$$\mathbf{C}(\theta)\begin{pmatrix} X \\ Y \end{pmatrix} = \begin{pmatrix} \cos\theta & -\sin\theta \\ \sin\theta & \cos\theta \end{pmatrix}\begin{pmatrix} X \\ Y \end{pmatrix} \tag{18.96}$$

where X and Y stand either for T_x and T_y or for R_x and R_y. That implies that under rotations of $\theta = -120°$ corresponding to the symmetry operation $\mathbf{C}_3$, and of $\theta = -240°$ corresponding to the symmetry operation $\mathbf{C}_3^2$, the X and Y components are mixed so that the treatment of the $\mathbf{C}_{3v}$ point group becomes more complex than that of the $\mathbf{C}_{2v}$. Now, keeping in mind that the Z components of the rotation and translation vectors cannot be modified by such rotations, since they are around this Z axis, the action of

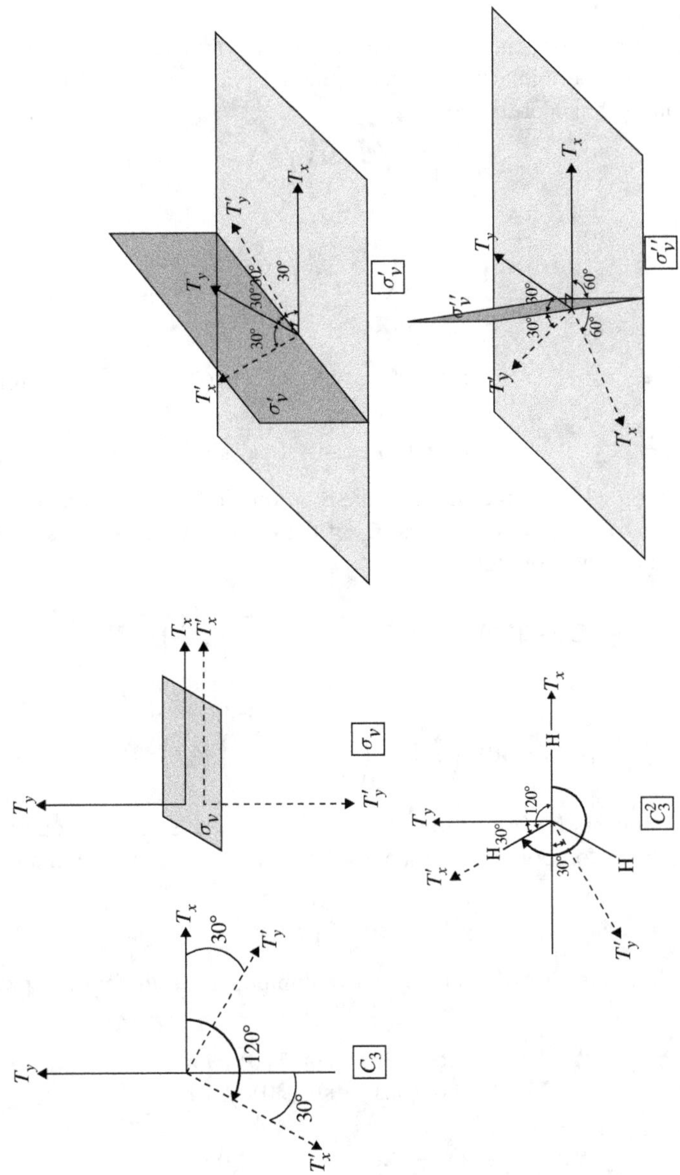

Figure 18.3 The C_{3v} symmetry operations. (See color insert.)

the different symmetry operations are simply given by

$$\begin{array}{llllll} \mathbf{E}T_z = T_z & \mathbf{C}_3 T_z = T_z & \mathbf{C}_3^2 T_z = T_z & \sigma_v T_z = T_z & \sigma_v' T_z = T_z & \sigma_v' T_z = T_z \\ \mathbf{E}R_z = R_z & \mathbf{C}_3 R_z = R_z & \mathbf{C}_3^2 R_z = R_z & \sigma_v R_z = -R_z & \sigma_v' R_z = -R_z & \sigma_v' R_z = -R_z \end{array} \tag{18.97}$$

Besides, if the identity operation **E** cannot modify anything and thus the X and Y components, the σ_v operation, which acts in the XZ plane, does not changes X but transforms Y into $-Y$, according to

$$\mathbf{E}\begin{pmatrix} X \\ Y \end{pmatrix} = \begin{pmatrix} 1 & 0 \\ 0 & 1 \end{pmatrix}\begin{pmatrix} X \\ Y \end{pmatrix} \tag{18.98}$$

$$\sigma_v\begin{pmatrix} X \\ Y \end{pmatrix} = \begin{pmatrix} 1 & 0 \\ 0 & -1 \end{pmatrix}\begin{pmatrix} X \\ Y \end{pmatrix} \tag{18.99}$$

Moreover, observe that

$$\cos(-120^\circ) = -\tfrac{1}{2} \qquad -\sin(-120^\circ) = \tfrac{\sqrt{3}}{2} \qquad \sin(-120^\circ) = -\tfrac{\sqrt{3}}{2} \text{ for } \mathbf{C}_3$$

$$\cos(-240^\circ) = -\tfrac{1}{2} \qquad -\sin(-240^\circ) = -\tfrac{\sqrt{3}}{2} \qquad \sin(-240^\circ) = \tfrac{\sqrt{3}}{2} \text{ for } \mathbf{C}_3^2$$

Hence, owing to Eq. (18.96), the action of the rotation on the translation and rotation vector components T_x and T_y and R_x and R_y of the rotation vectors act as a whole in the same way on the two components:

$$\{\mathbf{C}_3(-120)\}\begin{pmatrix} X \\ Y \end{pmatrix} = \begin{pmatrix} -\frac{1}{2} & \frac{\sqrt{3}}{2} \\ -\frac{\sqrt{3}}{2} & -\frac{1}{2} \end{pmatrix}\begin{pmatrix} X \\ Y \end{pmatrix} \tag{18.100}$$

$$\{\mathbf{C}_3^2(-240)\}\begin{pmatrix} X \\ Y \end{pmatrix} = \begin{pmatrix} -\frac{1}{2} & -\frac{\sqrt{3}}{2} \\ \frac{\sqrt{3}}{2} & -\frac{1}{2} \end{pmatrix}\begin{pmatrix} X \\ Y \end{pmatrix} \tag{18.101}$$

Furthermore, it appears by inspection of Fig. 18.3 that the symmetry operation σ_v does not affect the X component but changes the Y one into its inverse, namely,

$$\sigma_v\begin{pmatrix} X \\ Y \end{pmatrix} = \begin{pmatrix} 1 & 0 \\ 0 & -1 \end{pmatrix}\begin{pmatrix} X \\ Y \end{pmatrix} \tag{18.102}$$

whereas the action of σ_v' and σ_v'' on the X and Y components transform, respectively, according to

$$\sigma_v'\begin{pmatrix} X \\ Y \end{pmatrix} = \begin{pmatrix} \cos(120) & \sin(120) \\ \cos(30) & \sin(30) \end{pmatrix}\begin{pmatrix} X \\ Y \end{pmatrix}$$

$$\sigma_v''\begin{pmatrix} X \\ Y \end{pmatrix} = \begin{pmatrix} \cos(-120) & \sin(-120) \\ \cos(150) & \sin(150) \end{pmatrix}\begin{pmatrix} X \\ Y \end{pmatrix}$$

Hence, using, respectively, for $\boldsymbol{\sigma}_v'$ and $\boldsymbol{\sigma}_v''$ the trigonometric relations

$$\begin{array}{llll} \cos 120^\circ = -\frac{1}{2} & \sin 120^\circ = -\frac{1}{2} & \cos 30^\circ = \frac{\sqrt{3}}{2} & \sin 30^\circ = \frac{1}{2} \\ \cos(-120^\circ) = -\frac{1}{2} & \sin(-120^\circ) = -\frac{\sqrt{3}}{2} & \cos(150^\circ) = -\frac{\sqrt{3}}{2} & \sin(150^\circ) = \frac{1}{2} \end{array}$$

they yield

$$\sigma_v' \begin{pmatrix} X \\ Y \end{pmatrix} = \begin{pmatrix} -\frac{1}{2} & \frac{\sqrt{3}}{2} \\ \frac{\sqrt{3}}{2} & \frac{1}{2} \end{pmatrix} \begin{pmatrix} X \\ Y \end{pmatrix} \tag{18.103}$$

$$\sigma_v'' \begin{pmatrix} X \\ Y \end{pmatrix} = \begin{pmatrix} -\frac{1}{2} & -\frac{\sqrt{3}}{2} \\ -\frac{\sqrt{3}}{2} & \frac{1}{2} \end{pmatrix} \begin{pmatrix} X \\ Y \end{pmatrix} \tag{18.104}$$

All the transformations in Eqs. (18.97)–(18.104) are summarized in (18.105):

E	C_3	C_3^2	σ_v	σ_v'	σ_v''	Rot, Trans
1	1	1	+1	+1	+1	T_z
1	1	1	−1	−1	−1	R_z
$\begin{matrix} 1 & 0 \\ 0 & 1 \end{matrix}$	$\begin{matrix} -\frac{1}{2} & +\frac{\sqrt{3}}{2} \\ -\frac{\sqrt{3}}{2} & -\frac{1}{2} \end{matrix}$	$\begin{matrix} -\frac{1}{2} & -\frac{\sqrt{3}}{2} \\ +\frac{\sqrt{3}}{2} & -\frac{1}{2} \end{matrix}$	$\begin{matrix} +1 & 0 \\ 0 & -1 \end{matrix}$	$\begin{matrix} -\frac{1}{2} & +\frac{\sqrt{3}}{2} \\ +\frac{\sqrt{3}}{2} & +\frac{1}{2} \end{matrix}$	$\begin{matrix} -\frac{1}{2} & -\frac{\sqrt{3}}{2} \\ -\frac{\sqrt{3}}{2} & +\frac{1}{2} \end{matrix}$	$R_x, R_y,$ T_x, T_y

(18.105)

Now, use the fact that the traces of the 2×2 transformation matrices involved in (18.105) are given, respectively, by

$$\mathbf{tr} \begin{pmatrix} 1 & 0 \\ 0 & 1 \end{pmatrix} = 2 \qquad \mathbf{tr} \begin{pmatrix} -\frac{1}{2} & +\frac{\sqrt{3}}{2} \\ -\frac{\sqrt{3}}{2} & -\frac{1}{2} \end{pmatrix} = \mathbf{tr} \begin{pmatrix} -\frac{1}{2} & -\frac{\sqrt{3}}{2} \\ \frac{\sqrt{3}}{2} & -\frac{1}{2} \end{pmatrix} = -1$$

$$\mathbf{tr} \begin{pmatrix} 1 & 0 \\ 0 & -1 \end{pmatrix} = \mathbf{tr} \begin{pmatrix} -\frac{1}{2} & \frac{\sqrt{3}}{2} \\ \frac{\sqrt{3}}{2} & \frac{1}{2} \end{pmatrix} = \mathbf{tr} \begin{pmatrix} -\frac{1}{2} & -\frac{\sqrt{3}}{2} \\ -\frac{\sqrt{3}}{2} & \frac{1}{2} \end{pmatrix} = 0$$

Then, replacing in (18.105) the 2×2 matrices by their corresponding traces, one obtains the result shown in (18.106):

Rot and Trans	E	C_3	C_3^2	σ_v	σ_v'	σ_v''
T_z	1	1	1	1	1	1
R_z	1	1	1	−1	−1	−1
R_x, R_y, T_x, T_y	2	−1	−1	0	0	0

(18.106)

Next, one may observe by inspection of (18.106) that the symmetry operations C_3 and C_3^2 belonging to the same class of the point group affect in the same way all the translation and rotation vectors, and so for the three symmetry operations σ_v, σ_v', and σ_v'' belonging to another class of this group. That allows one to compact (18.106) according to

Rot and Trans	E	$2C_3$	$3\sigma_v$
T_z	1	1	1
R_z	1	1	−1
R_x, R_y, T_x, T_y	2	−1	0

(18.107)

where the effects of C_3 and C_3^2 and those of σ_v, σ_v', and σ_v'' are condensed in the same columns.

18.9.3 Some representation of the C_{2v} and C_{3v} point group

The different fashions according to which the components of the rotation and translation vectors transform under the different symmetry operation of a given group, denoted Γ_i, may be considered as different representations of the group. For leading for the C_{2v} point group, (18.95) reads as follows:

C_{2v}	E	C_2	σ_v	σ_v'
Γ_1	1	1	1	1
Γ_2	1	1	−1	−1
Γ_3	1	−1	1	−1
Γ_4	1	−1	−1	1

C_{2v}	E	C_2	σ_v	σ_v'
E	E	C_2	σ_v	σ_v'
C_2	C_2	E	σ_v'	σ_v
σ_v	σ_v	σ_v'	E	C_2
σ_v'	σ_v'	σ_v	C_2	E

(18.108)

where the multiplication table of this point group appearing in (18.92) has also been added. Now, one may observe for each Γ_i row that multiplying the number characterizing one of the symmetry operations by that characterizing another symmetry transformation yields the number characterizing the symmetry operations resulting from the product of the first symmetry operation by the last one. That is given in (18.109) for some multiplication involving E, C_2, σ_v, and σ_v':

	$E\sigma_v = \sigma_v$	$C_2\sigma_v = \sigma_v'$	$C_2\sigma_v' = \sigma_v$	$\sigma_v'\sigma_v = C_2$
Γ_1	$(1)(1)=1$	$(1)(1)=1$	$(1)(1)=1$	$(1)(1)=1$
Γ_2	$(1)(-1)=-1$	$(1)(-1)=-1$	$(1)(-1)=-1$	$(-1)(-1)=1$
Γ_3	$(1)(1)=1$	$(-1)(1)=-1$	$(-1)(-1)=1$	$(-1)(1)=-1$
Γ_4	$(1)(-1)=-1$	$(-1)(-1)=1$	$(-1)(1)=-1$	$(1)(-1)=-1$

(18.109)

18.9.3.1 The C_{2v} and C_{3v} point group tables The notations A, B, and E are generally used for the Γ_i representations according to A for nondegenerate representations symmetric with respect to the principal symmetry axis, B for the nondegenerate representation, which is antisymmetric with respect to the principal symmetry axis, and E for degenerate representations. Besides, for the A and B situations, supplementary subscripts 1 and 2 and g and u are used in the following situations in order to complete the symmetry properties:

A_1	for the full symmetric representations
A_2	for situations that are antisymmetric with respect to symmetry planes
B_1 and B_2	for situations B, which are either symmetric or antisymmetric with respect to symmetry planes

A_g and A_u for situations A, which are, respectively, either symmetric or antisymmetric with respect to the inversion center

B_g and B_u for situations A, which are, respectively, either symmetric or antisymmetric with respect to the inversion center

Owing to these conventions, the tabular data on the left in (18.108) dealing with the $\mathbf{C}_{2v}$ point group takes the following form:

$\mathbf{C}_{2v}$	E	C_2	σ_v	σ_v'	Rot and Trans
Γ_{A_1}	1	1	1	1	T_z
Γ_{A_2}	1	1	−1	−1	R_z
Γ_{B_1}	1	−1	1	−1	R_y, T_x
Γ_{B_2}	1	−1	−1	1	R_x, T_y

(18.110)

Now, observe that the four numbers appearing in the four rows of (18.110) act as components of orthogonal vectors since

$$
\begin{aligned}
(\Gamma_{A_1}, \Gamma_{A_2}) &= (1 \times 1) + (1 \times 1) + (1 \times -1) + (1 \times -1) = 0 \\
(\Gamma_{A_1}, \Gamma_{B_1}) &= (1 \times 1) + (1 \times -1) + (1 \times 1) + (1 \times -1) = 0 \\
(\Gamma_{A_1}, \Gamma_{B_2}) &= (1 \times 1) + (1 \times -1) + (1 \times -1) + (1 \times 1) = 0 \\
(\Gamma_{A_2}, \Gamma_{B_1}) &= (1 \times 1) + (1 \times -1) + (-1 \times 1) + (-1 \times -1) = 0 \\
(\Gamma_{A_2}, \Gamma_{B_2}) &= (1 \times 1) + (1 \times -1) + (-1 \times -1) + (-1 \times 1) = 0 \\
(\Gamma_{B_1}, \Gamma_{B_2}) &= (1 \times 1) + (-1 \times -1) + (1 \times -1) + (-1 \times 1) = 0
\end{aligned}
\tag{18.111}
$$

On the other hand, owing to the above convention, (18.106) dealing with the $\mathbf{C}_{3v}$ point group reads:

$\mathbf{C}_{3v}$	E	C_3	C_3^2	σ_v	σ_v'	σ_v''	Rot and Trans
Γ_{A_1}	1	1	1	1	1	1	T_z
Γ_{A_2}	1	1	1	−1	−1	−1	R_z
Γ_E	2	−1	−1	0	0	0	R_x, R_y, T_x, T_y

(18.112)

Here also, as for the $\mathbf{C}_{2v}$ point-group table, one may observe that the six rows of (18.111) act also as orthogonal vectors since the following relations are satisfied:

$$
\begin{aligned}
(\Gamma_{A_1}, \Gamma_{A_2}) &= (1 \times 1) + (1 \times 1) + (1 \times 1) + (1 \times -1) + (1 \times -1) + (1 \times -1) = 0 \\
(\Gamma_{A_2}, \Gamma_E) &= (1 \times 2) + (1 \times -1) + (1 \times -1) + (1 \times -0) + (-1 \times 0) + (-1 \times 0) = 0 \\
(\Gamma_{A_1}, \Gamma_E) &= (1 \times 2) + (1 \times -1) + (1 \times -1) + (1 \times -0) + (1 \times 0) + (1 \times 0) = 0
\end{aligned}
\tag{18.113}
$$

Besides, in the same way, (18.107), which is the contraction of (18.112) by considering only the different classes, reads:

$$
\begin{array}{c|cccl}
\mathbf{C}_{3v} & E & 2C_3 & 3\sigma_v & \text{Rot and Trans} \\
\hline
\Gamma_{A_1} & 1 & 1 & 1 & T_z \\
\Gamma_{A_2} & 1 & 1 & -1 & R_z \\
\Gamma_E & 2 & -1 & 0 & R_x, R_y, T_x, T_y
\end{array}
\tag{18.114}
$$

which reveals also that each of its three rows acts as orthogonal vectors according to:

$$
\begin{aligned}
(\Gamma_{A_1}, \Gamma_{A_2}) &= (1 \times 1) + 2(1 \times 1) + 3(1 \times -1) = 0 \\
(\Gamma_{A_2}, \Gamma_E) &= (1 \times 2) + 2(1 \times -1) + 3(-1 \times 0) = 0 \\
(\Gamma_{A_1}, \Gamma_E) &= (1 \times 2) + 2(1 \times -1) + 3(1 \times 0) = 0
\end{aligned}
\tag{18.115}
$$

18.9.4 Obtaining reducible group representation

18.9.4.1 C_{2v} matrix representation obtained from the H_2O Cartesian coordinates One may obtain other representations of a given point group to which belongs a molecule by studying the results of the action of its different symmetry operations on the molecular atomic coordinates. As an illustration, apply that to a triatomic molecule such as H_2O.

In this case, the action of the symmetry operations **R** over the atomic coordinates $\mathbf{q}_{H_1}$, $\mathbf{q}_{H_2}$, and $\mathbf{q}_O$ of the two hydrogen atoms and of the atomic oxygen reads

$$
\begin{pmatrix} [\tilde{q}_{H_1}] \\ [\tilde{q}_{H_2}] \\ [\tilde{q}_O] \end{pmatrix} = \boxed{\mathbf{D(R)}} \begin{pmatrix} [q_{H_1}] \\ [q_{H_2}] \\ [q_O] \end{pmatrix}
$$

where

$$
[q_K] = \begin{pmatrix} (q_x)_K \\ (q_y)_K \\ (q_z)_K \end{pmatrix} \quad \text{and} \quad [\tilde{q}_{At}] = \begin{pmatrix} (\tilde{q}_x)_K \\ (\tilde{q}_y)_K \\ (\tilde{q}_z)_K \end{pmatrix} \quad \text{with} \quad K = H_1, H_2, O
$$

whereas **D(R)** appearing in the linear transformations is the transformation matrix corresponding to the symmetry operation R. Of course, that corresponding to the identity operation **E**, which does not modify anything, reads

$$
\boxed{\mathbf{D(E)}} = \begin{pmatrix} \boxed{\mathbf{1}} & \boxed{\mathbf{0}} & \boxed{\mathbf{0}} \\ \boxed{\mathbf{0}} & \boxed{\mathbf{1}} & \boxed{\mathbf{0}} \\ \boxed{\mathbf{0}} & \boxed{\mathbf{0}} & \boxed{\mathbf{1}} \end{pmatrix}
\tag{18.116}
$$

where the 3×3 square matrices are given by

$$
\boxed{\mathbf{1}} = \begin{pmatrix} 1 & 0 & 0 \\ 0 & 1 & 0 \\ 0 & 0 & 1 \end{pmatrix} \quad \text{and} \quad \boxed{\mathbf{0}} = \begin{pmatrix} 0 & 0 & 0 \\ 0 & 0 & 0 \\ 0 & 0 & 0 \end{pmatrix}
\tag{18.117}
$$

In a similar way, the transformation matrices corresponding to other symmetry operations may be found to be

$$\boxed{\mathbf{D(C_2)}} = \begin{pmatrix} \boxed{\mathbf{0}} & \boxed{\mathbf{A}} & \boxed{\mathbf{0}} \\ \boxed{\mathbf{A}} & \boxed{\mathbf{0}} & \boxed{\mathbf{0}} \\ \boxed{\mathbf{0}} & \boxed{\mathbf{0}} & \boxed{\mathbf{A}} \end{pmatrix} \tag{18.118}$$

$$\boxed{\mathbf{D}(\sigma_v)} = \begin{pmatrix} \boxed{\mathbf{0}} & \boxed{\mathbf{B}} & \boxed{\mathbf{0}} \\ \boxed{\mathbf{B}} & \boxed{\mathbf{0}} & \boxed{\mathbf{0}} \\ \boxed{\mathbf{0}} & \boxed{\mathbf{0}} & \boxed{\mathbf{B}} \end{pmatrix} \tag{18.119}$$

$$\boxed{\mathbf{D}(\sigma_v')} = \begin{pmatrix} \boxed{\mathbf{C}} & \boxed{\mathbf{0}} & \boxed{\mathbf{0}} \\ \boxed{\mathbf{0}} & \boxed{\mathbf{C}} & \boxed{\mathbf{0}} \\ \boxed{\mathbf{0}} & \boxed{\mathbf{0}} & \boxed{\mathbf{C}} \end{pmatrix} \tag{18.120}$$

with, respectively,

$$\boxed{\mathbf{A}} = \begin{pmatrix} -1 & 0 & 0 \\ 0 & -1 & 0 \\ 0 & 0 & 1 \end{pmatrix} \quad \boxed{\mathbf{B}} = \begin{pmatrix} 1 & 0 & 0 \\ 0 & -1 & 0 \\ 0 & 0 & 1 \end{pmatrix} \quad \boxed{\mathbf{C}} = \begin{pmatrix} -1 & 0 & 0 \\ 0 & 1 & 0 \\ 0 & 0 & 1 \end{pmatrix} \tag{18.121}$$

Now, observe that the traces of the five 3×3 matrices appearing in (18.117) and (18.121) are, respectively,

$$\mathbf{tr}\boxed{\mathbf{0}} = 0 \qquad \mathbf{tr}\boxed{1} = 3 \qquad \mathbf{tr}\boxed{\mathbf{A}} = -1 \qquad \mathbf{tr}\boxed{\mathbf{B}} = \mathbf{tr}\boxed{\mathbf{C}} = 1$$

so that the traces of the matrix representations (18.116), (18.118), (18.119), and (18.120) of the different symmetry operations of the group defined by

$$\{\chi_{(R)}^{\Gamma^\circ}\} = \mathbf{tr}\boxed{\mathbf{D(R)}}$$

yield, respectively,

$$\begin{aligned} \{\chi_{(E)}^{\Gamma^\circ}\} &= \mathbf{tr}\boxed{\mathbf{D}(E)} = 3\ \mathbf{tr}\boxed{1} = 9 \\ \{\chi_{(C_2)}^{\Gamma^\circ}\} &= \mathbf{tr}\boxed{\mathbf{D}(C_2)} = \mathbf{tr}\boxed{\mathbf{A}} = -1 \\ \{\chi_{(\sigma_v)}^{\Gamma^\circ}\} &= \mathbf{tr}\boxed{\mathbf{D}(\sigma_v)} = \mathbf{tr}\boxed{\mathbf{B}} = 1 \\ \{\chi_{(\sigma_v')}^{\Gamma^\circ}\} &= \mathbf{tr}\boxed{\mathbf{D}(\sigma_{v'})} = 3\ \mathbf{tr}\boxed{\mathbf{C}} = 3 \end{aligned}$$

From these results dealing with the characters of the different symmetry transformations, one obtains the following description of the corresponding Γ° representation of H_2O:

$\mathbf{C_{2v}}$	E	C_2	σ_v	σ_v'
Γ°	9	−1	1	3

(18.122)

18.9.4.2 General method to obtain a group representation There is a general method to get the characters of the representative matrices of the symmetry operators appearing in the representation of any point group by aid of the study of the transformation of the atomic coordinates of a molecule without searching to know these matrices. That is because in such studies diagonal elements that only interest the characters, that is, the trace of the matrices, occur only if the symmetry operations leave the position of atoms unchanged. That gives a possibility to get the character contributions of the symmetry elements of a point group by looking only at the atoms that the symmetry operations leave unchanged. The tabular data in (18.123) gives these contributions for different symmetry operations:

Operations	E	σ	i	C_2	C_3^1	C_3^2	C_4^1	C_4^3	C_6^1	C_6^5
χ contributions	3	1	−3	−1	0	0	1	1	2	2

(18.123)

That may be verified for H_2O, since the results (18.122) proceed directly from (18.123); and (18.123) opens the possibilities to get the reducible representation Γ° of a molecule belonging to a given point group.

18.9.5 From reducible to irreducible representations

Keeping in mind that there are many matrix representations $\boxed{\mathbf{D}(R_r)}$ of one R_r symmetry operation of a point group such as those involved in the Γ° representations obtained by aid of the action of the symmetry operators on the atomic coordinates of H_2O, it appears of interest to know that there are matrices $\boxed{\mathbf{C}}$ allowing to transform all the different matrices $\boxed{\mathbf{D}(R_r)}$ corresponding to the different R_r symmetry operations into new ones $\boxed{\tilde{\mathbf{D}}(R_r)}$ having the same block forms for all the symmetry transformations R_r of the point group, according to

$$\boxed{\tilde{\mathbf{D}}(R_r)} = \boxed{\mathbf{C}}^{-1} \boxed{\mathbf{D}(R_r)} \boxed{\mathbf{C}} \tag{18.124}$$

with, for instance

$$\boxed{\tilde{\mathbf{D}}(R_r)} = \begin{pmatrix} \boxed{\mathbf{D}^{\Gamma_k}(R_r)} & \boxed{\mathbf{0}} & \boxed{\mathbf{0}} & \cdots \\ \boxed{\mathbf{0}} & \boxed{\mathbf{D}^{\Gamma_k}(R_r)} & \boxed{\mathbf{0}} & \\ \boxed{\mathbf{0}} & \boxed{\mathbf{0}} & \boxed{\mathbf{D}^{\Gamma_m}(R_r)} & \\ \cdots & \cdots & \cdots & \boxed{\mathbf{D}^{\Gamma_l}(R_r)} \end{pmatrix} \tag{18.125}$$

In these last expressions, the diagonal blocks, which may be the same, are squared matrices of dimensions $k \times k$, $m \times m$, $l \times l, \ldots$, whereas the off-diagonal blocks are zero rectangular matrices. The consequence of (18.125) is that this matrix representation may be decomposed according to

$$\boxed{\tilde{\mathbf{D}}(R_r)} = a_{\Gamma_k} \boxed{\mathbf{D}^{\Gamma_k}(R_r)} + a_{\Gamma_l} \boxed{\mathbf{D}^{\Gamma_l}(R_r)} + a_{\Gamma_m} \boxed{\mathbf{D}^{\Gamma_m}(R_r)} \tag{18.126}$$

where the coefficients a_{Γ_k}, a_{Γ_l}, and a_{Γ_m} are the number of times the corresponding matrix appears in the block form (18.125), that is,

$$a_{\Gamma_k} = 2 \qquad a_{\Gamma_m} = 1 \qquad a_{\Gamma_l} = 1$$

As a consequence of the linear transformation (18.124) it becomes possible to break a *reducible* representation Γ of the form

	R_1	R_2	R_3	...
Γ°	$\boxed{\mathbf{D}(R_1)}$	$\boxed{\mathbf{D}(R_2)}$	$\boxed{\mathbf{D}(R_3)}$	...

in the same block form according to

	R_1	R_2	R_3	...
Γ_k	$\boxed{\mathbf{D}^{\Gamma_k}(R_1)}$	$\boxed{\mathbf{D}^{\Gamma_k}(R_2)}$	$\boxed{\mathbf{D}^{\Gamma_k}(R_3)}$	...
Γ_l	$\boxed{\mathbf{D}^{\Gamma_l}(R_1)}$	$\boxed{\mathbf{D}^{\Gamma_l}(R_2)}$	$\boxed{\mathbf{D}^{\Gamma_l}(R_3)}$	...
Γ_m	$\boxed{\mathbf{D}^{\Gamma_m}(R_1)}$	$\boxed{\mathbf{D}^{\Gamma_m}(R_2)}$	$\boxed{\mathbf{D}^{\Gamma_m}(R_3)}$	...
...	...	...	...	...

One says that the reducible representation Γ° has been reduced into a set of reduced representations Γ_k. Now, if the block matrices corresponding to the $\Gamma_k, \Gamma_l, \Gamma_m, \ldots$ representations cannot be reduced in turn through new similarity transformations of the form (18.124), these reduced representations are said to be *irreducible*. Next, if the block matrices labeled by k appear a_k times, and if the block matrices labeled by l appear a_l times, in the full transformation matrices (18.125), the reducible representation Γ° may be written

$$\Gamma^\circ = a_{\Gamma_k}\Gamma^k + a_{\Gamma_l}\Gamma^l + \cdots \tag{18.127}$$

It must be emphasized that the transformations (18.124) conserve the matrix traces, that is,

$$\mathbf{tr}\boxed{\mathbf{D}(R_r)} = \mathbf{tr}\boxed{\tilde{\mathbf{D}}(R_r)}$$

Therefore, owing to the block form (18.126) it reads

$$\mathbf{tr}\boxed{\tilde{\mathbf{D}}(R_r)} = \sum_k a_{\Gamma_k}\mathbf{tr}\boxed{\mathbf{D}^{\Gamma_k}(R_r)}$$

so that

$$\{\chi^{\Gamma^\circ}(R_r)\} = \sum_k a_{\Gamma_k}\{\chi^{\Gamma_k}(R_r)\} \tag{18.128}$$

with

$$\{\chi^{\Gamma^\circ}(R_r)\} = \mathbf{tr}\boxed{\tilde{\mathbf{D}}(R_r)} \qquad \text{and} \qquad \{\chi^{\Gamma_k}(R_r)\} = \mathbf{tr}\boxed{\mathbf{D}^{\Gamma_k}(R_r)}$$

18.9.6 Great orthogonality theorem

The *great orthogonality theorem* states that if Γ_1 and Γ_2 are two nonequivalent irreducible unitary representations characterized, respectively, by matrices $\mathbf{D}^{\Gamma_1}(R_r)$ and $\mathbf{D}^{\Gamma_2}(R_r)$ of dimension n_μ and n_ν, their matrix elements obey

$$\sum_r \{D_{ik}^{\Gamma_\mu}(R_r)\}\{D_{jm}^{\Gamma_\nu}(R_r^{-1})\} = \left(\frac{g}{n_\mu}\right)\delta_{\mu\nu}\,\delta_{ij}\,\delta_{km}$$

Here, g is the order of the group, the sum runs over all the symmetry operations R_r of the group, and the μ and ν identify the irreducible representations Γ_μ and Γ_ν. The reduced transformation matrix being unitary so that

$$\boxed{\{\mathbf{D}^{\Gamma_\mu}(R_r^{-1})\}} = \boxed{\{\mathbf{D}^{\Gamma_\mu}(R_r)\}}^{-1} = \boxed{\{\mathbf{D}^{\Gamma_\mu}(R_r)\}}^{\dagger}$$

the great orthogonality theorem reads

$$\sum_r \{D_{ik}^{\mu}(R_r)\}\{D_{jm}^{\nu}(R_r)^*\} = \left(\frac{g}{n_\mu}\right)\delta_{\mu\nu}\delta_{ij}\delta_{km} \tag{18.129}$$

Applied to the $\mathbf{C}_{2v}$ group, where $g=4$, and $n_{\Gamma_1} = n_{\Gamma_2} = 1$, the great orthogonality theorem (18.129) reads for $\Gamma_\mu = \Gamma_{A_1}$ and $\Gamma_\nu = \Gamma_{A_2}$

$$\begin{aligned}\{D_{ik}^{\Gamma_{A1}}(E)\}\{D_{ik}^{\Gamma_{A2}}(E)\} &+ \{D_{ik}^{\Gamma_{A1}}(C_2)\}\{D_{ik}^{\Gamma_{A2}}(C_2)\}\\ &+ \{D_{ik}^{\Gamma_{A1}}(\sigma_v)\}\{D_{ik}^{\Gamma_{A2}}(\sigma_v)\} + \{D_{ik}^{\Gamma_{A1}}(\sigma_{v'})\}\{D_{ik}^{\Gamma_{A2}}(\sigma_v')\} = 0\end{aligned}$$

and, for $\Gamma_\mu = \Gamma_\nu = \Gamma_{A_2}$

$$\begin{aligned}\{D_{ik}^{\Gamma_{A2}}(E)\}\{D_{ik}^{\Gamma_{A2}}(E)\} &+ \{D_{ik}^{\Gamma_{A2}}(C_2)\}\{D_{ik}^{\Gamma_{A2}}(C_2)\}\\ &+ \{D_{ik}^{\Gamma_{A2}}(\sigma_v)\}\{D_{ik}^{\Gamma_{A2}}(\sigma_v)\} + \{D_{ik}^{\Gamma_{A2}}(\sigma_{v'})\}\{D_{ik}^{\Gamma_{A2}}(\sigma_v')\} = 4\end{aligned}$$

As it appears, the g matrix elements indexed by i and k values, that is, for the $\mathbf{C}_{2v}$ group, the four elements

$$\{D_{ik}^{\Gamma_{A2}}(E)\}, \{D_{ik}^{\Gamma_{A2}}(C_2)\}, \{D_{ik}^{\Gamma_{A2}}(\sigma_v)\}, \{D_{ik}^{\Gamma_{A2}}(\sigma_v')\}$$

act as the components of a g-dimensional vector. Thereby, since there are $n_{\Gamma_1}^2$ such sets of matrix elements corresponding to $i = 1, 2 \ldots n_{\Gamma_1}$, and $k = 1, 2 \ldots n_{\Gamma_1}$, that implies that there are as many vectors as there are irreducible representations that are mutually orthogonal.

Next, applied to the diagonal elements of the matrices $\mathbf{D}^{\Gamma_\mu}(R)$ and $\mathbf{D}^{\Gamma_\nu}(R)$, the great orthogonality theorem (18.129) yields

$$\sum_r \{D_{ii}^{\Gamma_\mu}(R_r)\}\{D_{jj}^{\Gamma_\nu}(R_r)^*\} = \left(\frac{g}{n_\mu}\right)\delta_{\mu\nu}\delta_{ij}\delta_{ij} = \left(\frac{g}{n_\mu}\right)\delta_{\mu\nu}\delta_{ij}$$

Then, performing the trace over the two involved matrices, according to

$$\sum_r \sum_{i=1}^{n_\mu} \{D_{ii}^{\Gamma_\mu}(R_r)\} \sum_{j=1}^{n\nu} \{D_{jj}^{\Gamma_\nu}(R_r)^*\} = \left(\frac{g}{n_\mu}\right)\delta_{\mu\nu}\sum_{i=1}^{n_\mu}\sum_{j=1}^{n_\nu}\delta_{ij}$$

that is,

$$\sum_r \mathbf{tr}\boxed{\mathbf{D}^{\Gamma_\mu}(R_r)}\,\mathbf{tr}\boxed{\mathbf{D}^{\Gamma_\nu}(R_r)}^\dagger = \left(\frac{g}{n_\mu}\right)\delta_{\mu\nu}\sum_{i=1}^{n_\mu}\sum_{j=1}^{n_\nu}\delta_{ij}$$

reads

$$\sum_r \{\chi^{\Gamma_\mu}(R_r)\}\{\chi^{\Gamma_\nu}(R_r)\} = \left(\frac{g}{n_\mu}\right)\delta_{\mu\nu}\sum_{i=1}^{n_\mu}\sum_{j=1}^{n_\nu}\delta_{ij} \tag{18.130}$$

with

$$\{\chi^{\Gamma_1}(R_r)\} = \mathbf{tr}\boxed{\mathbf{D}^{\Gamma_1}(R_r)} \tag{18.131}$$

Then, owing to the properties of the Kronecker symbol, leading to

$$\sum_{i=1}^{n_{\Gamma_1}}\sum_{j=1}^{n_{\Gamma_2}}\delta_{ij} = n_\mu$$

where n_μ is the dimension of the irreducible representation Γ_μ, it appears that Eq. (18.130) simplifies to

$$\sum_r \{\chi^{\Gamma_\mu}(R_r)\}\{\chi^{\Gamma_\nu}(R_r)\}^* = g\delta_{\mu\nu} \tag{18.132}$$

showing that a reduced representation of a point group has g components

$$\{\chi^{\Gamma_\mu}(R_1)\},\{\chi^{\Gamma_\mu}(R_2)\},\ldots,\{\chi^{\Gamma_\mu}(R_r)\}$$

which act as the components of a g-dimensional vector, and that such vectors corresponding to different nonequivalent irreducible representations are orthogonal. As it may be verified, the set of Eqs. involved in (18.111) and (18.115) are illustrations of the very general expression (18.132).

18.9.7 Irreducible representation analysis

Now, in order to find the values of the coefficients involved in Eq. (18.127), that is, the number of times the irreducible representation appears in a reducible representation Γ°, start from Eq. (18.128), that is,

$$\{\chi^{\Gamma^\circ}(R_r)\} = \sum_\mu a_{\Gamma_\mu}\{\chi^{\Gamma_\mu}(R_r)\}$$

Then, multiply both members of this equation by $\chi_{\Gamma_\nu(R)}$, and sum over all the symmetry operations of the point group to give

$$\sum_r \{\chi^{\Gamma_\nu}(R_r)\}\{\chi^{\Gamma^\circ}(R_r)\} = \sum_r\sum_\mu a_{\Gamma_\mu}\{\chi^{\Gamma_\mu}(R_r)\}\{\chi^{\Gamma_\nu}(R_r)\} \tag{18.133}$$

then, owing to Eq. (18.132), that is,

$$\sum_r \{\chi^{\Gamma_\mu}(R_r)\}\{\chi^{\Gamma_\nu}(R_r)\} = g\,\delta_{\mu\nu}$$

Eq. (18.133) yields

$$\sum_r \{\chi^{\Gamma_\nu}(R_r)\}\{\chi^{\Gamma^\circ}(R_r)\} = g \sum_\mu a_{\Gamma_\mu} \delta_{\mu\nu}$$

or, after simplification using the Kronecker symbol

$$\sum_r \{\chi^{\Gamma_\nu}(R_r)\}\{\chi^{\Gamma^\circ}(R_r)\} = g a_{\Gamma_\nu}$$

so that the weight of the irreducible representation Γ_ν inside the reduced representation Γ° reads

$$\boxed{a_{\Gamma_\nu} = \frac{1}{g} \sum_r \{\chi^{\Gamma_\nu}(R_r)\}\{\chi^{\Gamma^\circ}(R_r)\}} \tag{18.134}$$

18.10 SCIENTIFIC AUTHORS APPEARING IN THE BOOK

Balmer

Johann Jakob Balmer (1825–1898), Swiss mathematician and mathematical physicist, whose major contribution (made in 1885) was an empirical formula for the visible spectral lines of the hydrogen atom.

Bessel

Friedrich Bessel (1784–1846), German mathematician and astronomer, systematizer of the Bessel functions, was a contemporary of Carl Gauss, also a mathematician and astronomer.

Bloch

Felix Bloch (1905–1983), Swiss born and American physicist, who with Edward Purcell were awarded the 1952 Nobel Prize for "their development of new ways and methods for nuclear magnetic precision measurements." His doctoral thesis established the quantum theory of solids, using Bloch waves to describe the electrons. He studied with Pauli in Zürich, Bohr in Copenhagen, and Fermi in Rome. In 1933, immediately after Hitler came to power, he left Germany, emigrating and in 1939, became a naturalized citizen of the United States. After the war he concentrated on investigations into nuclear induction and nuclear magnetic resonance. In 1946 he proposed the Bloch equations, which determine the time evolution of nuclear magnetization.

Bogolyubov

Nikolaï Bogolyubov (1899–1980), Russian and Ukrainian mathematician and theoretical physicist known for a significant contribution to quantum field theory, classical and quantum statistical mechanics, and to the theory of dynamical systems.

Bohr

Niels Bohr (1885–1962), Danish physicist, was at the origin of the old quantum mechanics for which he received the Nobel Prize in Physics in 1922, by theoretically reproducing the experimental line spectra of atomic hydrogen. Becoming a leader for the new physicists working in quantum mechanics, he contributed to what is known as the *Copenhagen interpretation of quantum mechanics* and may be considered as one of the most influential physicists of the twentieth century.

Boltzmann

Ludwig Boltzmann (1844–1906), Austrian physicist, was one of the most important advocates for atomic theory when that scientific model was still highly controversial with the majority of physicists. He is famous for his founding contributions in the fields of statistical mechanics and statistical thermodynamics. His physical ideas are at the origin of the basic Planck hypothesis.

Born

Max Born (1882–1970), German-born physicist and mathematician, was instrumental in the development of quantum mechanics with his young assistants Heisenberg, Pauli, and Franck, all future Nobel prizes winners in the 1920s. He also made contributions to solid-state physics and optics. Born won the 1954 Nobel Prize in Physics.

Bose

Satyendra Bose (1894–1974), Indian physicist from Bangladesh, specializing in mathematical physics is known for his work on quantum mechanics in the early 1920s, providing the foundation for Bose–Einstein statistics and the theory of the Bose–Einstein condensate for which he is honored as the namesake of the boson.

Brillouin

Léon Brillouin (1899–1969), French physicist, made contributions to quantum mechanics, radio wave propagation in the atmosphere, solid-state physics, and information theory relating statistical entropy to information.

Cauchy

Augustin-Louis Cauchy (1789–1857), French mathematician, early pioneer of analysis, started the project of formulating and proving the theorems of infinitesimal calculus in a rigorous manner, gave also several important theorems in complex analysis, and initiated the study of permutation groups in abstract algebra. Profound mathematician, Cauchy exercised a great influence over his contemporaries and successors.

Compton

Arthur Compton (1892–1962), American physicist, Nobel Price of Physics (1927) with Charles Wilson, for his discovery of the Compton effect dealing with the fact that electromagnetic radiation can, in certain circumstances, show a corpuscular behavior (photon), later head of the laboratory where the first controlled nuclear chain reaction was performed, played an important role in the study of the atomic bomb.

Condon

Edward Condon (1902–1974), American nuclear physicist, during World War II participated in the development of radar and the atom bomb.

Coulomb

Charles de Coulomb (1736–1806), French physicist, is best known for developing Coulomb's law, dealing with the attraction and repulsion of electrostatic forces and has given his name to SI unit of charge, the coulomb.

Coulson

Charles Coulson (1910–1974), English theoretical physicist worked in mathematics, physics, theoretical chemistry, chemistry, and molecular biology.

Broglie (de)

Louis Duke de Broglie (1892–1987), French mathematician and physicist. He was awarded the Nobel Prize in Physics in 1929 for his discovery of the wave nature of electrons. Influenced by the work of Einstein, he asserted that, in the same way that waves can behave as particles, particles can behave like waves of a wavelength (the de Broglie's wavelength), a hypothesis that was confirmed by Davisson's experiments on the diffraction of electrons. Schrö dinger used the ideas of De Broglie to formulate wave mechanics. Unsatisfied by the probabilistic interpretation of quantum physics, de Broglie attempted after 1950 to restore a strictly deterministic formulation of this theory, but his ideas remained isolated in the community of physicists, except for Einstein and Schrödinger.

Davisson

Clinton Davisson (1881–1958), American physicist, won the 1937 Nobel Prize in Physics for his discovery of electron diffraction in 1927. Davisson shared the Nobel Prize with George Paget Thomson, who independently discovered electron diffraction at about the same time as Davisson.

Davydov

Alexander Davydov (1912–1994), Ukrainian physicist, contributed to theory of absorption, scattering, and dispersion of the light in molecular crystals. He predicted the phenomenon known as *Davydov splitting* in 1948. He also worked on the theory of collective excited states in spherical and nonspherical nuclei.

Dirac

Paul Dirac (1902–1984), English physicist and mathematician, co-recipient with Schrödinger of the Nobel Prize in Physics 1933 for the discovery of new and useful forms of atomic theory, one of the "fathers" of quantum mechanics in its axiomatic form. He predicted later the existence of antimatter through his relativistic formulation of quantum mechanics by the aid of four components wavefunctions. He was the first to introduce coherent states.

Dyson

Freeman Dyson (1923–), British-born American physicist and mathematician, contributed particularly to the foundations of quantum electrodynamics in 1948 in which time ordering is required.

Ehrenfest

Paul Ehrenfest (1880–1933), Austrian physicist, contributed to the physical debate at the origin of the physics at the beginning of the twentieth century and wrote with his wife Tatiana a famous article in 1911 on the foundations of statistical mechanics.

Einstein

Albert Einstein (1879–1955), German physicist, was awarded the Nobel Price in Physics (1921) for his interpretation of the photoelectric effect. He is at the origin of the special theory of relativity (1905) and of its generalization (1915) according to which the gravitational forces are simple consequence of the space curvature by masses. Einstein also contributed to statistical mechanics and has also introduced in theoretical spectroscopy the concept of induced and spontaneous absorption coefficients. Einstein rejected until his death the official statistical interpretation of quantum mechanics by the Copenhagen School.

Fermi

Enrico Fermi (1901–1954), Italian physicist, was awarded the Nobel Prize in Physics (1938) for his demonstration of the existence of new radioactive elements produced by neutron bombardment and his discovery of nuclear reactions induced by slow neutrons.

Fokker

Adriaan Fokker (1887–1972), Dutch physicist, derived the Fokker–Planck equation along with Planck and made several contributions to special and general relativity.

Fourier

Joseph Fourier (1768–1830), French mathematician and physicist, best known for initiating the investigation of Fourier series and their application to problems of heat transfer. The Fourier transform and Fourier's law are also named in his honor.

Franck

James Franck (1882–1964), German physicist, received the Nobel Prize in Physics, mostly for his work in 1912–1914, which included the Franck–Hertz experiment, an important confirmation of the Bohr model of the atom.

Galileo

Galileo Galilei (1564–1642), Italian physicist, astronomer, and philosopher, played a major role in the scientific revolution and may be considered as the "father of modern physics." His more pioneering work was the experimental study of the motion of uniformly accelerating objects.

Gauss

Johann Carl Friedrich Gauss (1777–1855), German mathematician and astronomer, contributed significantly to many fields, including number theory, statistics, analysis, differential geometry, geodesy, geophysics, electrostatics, astronomy, and optics. Sometimes referred to as the Princeps mathematicorum, Gauss had a remarkable influence in many fields of mathematics and science and is ranked as one of history's most influential mathematicians.

Gegenbauer

Leopold Bernhard Gegenbauer (1849–1903), Austrian mathematician, had many mathematical interests such as number theory, function theory, and the theory of integration, but he was chiefly an algebraist. He is remembered for the Gegenbauer polynomials, a class of orthogonal polynomials.

Germer

Lester Germer (1896–1971), American physicist, proved with Davisson in 1927 the wave-particle duality of matter and thus supported the theoretical work of De Broglie.

Glauber

Roy Glauber (1925–), American theoretical physicist, awarded one half of the 2005 Nobel Prize in Physics "for his contribution to the quantum theory of optical coherence," with the other half shared by John L. Hall and Theodor W. Hänsch. In this work, published in 1963, he created a model for photodetection and explained the fundamental characteristics of different types of light, such as laser light involving coherent state. His theories are widely used in the field of quantum optics.

Hamilton

Sir William Hamilton (1805–1865), Irish physicist, astronomer, and mathematician, made important contributions to classical mechanics, optics, and algebra. His studies of mechanical and optical systems led him to discover new mathematical concepts and techniques, his greatest contribution being the reformulation of Newtonian mechanics, now called Hamiltonian mechanics. This work has proven central to the modern study of classical field theories and to the development of quantum mechanics. In mathematics, he is also known as the inventor of quaternions.

Hausdorff

Felix Hausdorff (1868–1942), German mathematician, is considered to be one of the founders of modern topology and contributed significantly to functional analysis.

Heisenberg

Werner Heisenberg (1901–1976), German physicist, was awarded the 1932 Nobel Prize in Physics for the creation of quantum mechanics for which he made foundational contributions and is particularly known for asserting the uncertainty principle of quantum theory. He was a fervent advocate of the Copenhagen statistical interpretation of quantum mechanics. In addition, he made important contributions to nuclear physics, quantum field theory, and particle physics. along with Max Born and Pascual Jordan, set forth the matrix formulation of quantum mechanics in 1925.

Helmoltz

Hermann von Helmholtz (1821–1894), German scientist, made significant contributions to several widely varied areas of modern science: in physiology and psychology for his works on vision, in physics for his theories on the conservation of energy, on chemical thermodynamics, and on the mechanical foundations of thermodynamics.

Hermite

Charles Hermite (1822–1901), French mathematician, did research on number theory, quadratic forms, invariant theory, orthogonal polynomials, elliptic functions, and algebra. Hermite polynomials, Hermite interpolation, Hermite normal form, Hermitian operators, and cubic Hermite splines are named in his honor. One of his students was Henri Poincaré.

Jeans

Sir James Jeans (1877–1946), English physicist, astronomer, and mathematician, wrote books that made him fairly well known as an expositor of the revolutionary scientific discoveries of his day, especially in relativity and physical cosmology.

Jordan

Pascual Jordan (1902–1980), German theoretical and mathematical physicist, made significant contributions to quantum mechanics and quantum field theory, through the mathematical form of matrix mechanics, together with Heisenberg, Born, and Von Neumann, and developed canonical anticommutation relations for fermions. While the Jordan algebra he invented is no longer employed in quantum mechanics, it has found other mathematical applications.

Kirchhoff

Gustav Kirchhoff (1824–1887), German physicist, contributed to the fundamental understanding of electrical circuits, spectroscopy, and the emission of black-body radiation by heated objects. He coined the term "black-body" radiation in 1862, and two sets of independent concepts in both circuit theory and thermal emission are named "Kirchhoff's laws" after him.

Kronecker

Leopold Kronecker (1823–1891), German mathematician and logician, argued in the spirit of the Pythagorians that arithmetic and analysis must be founded on "whole numbers."

Lagrange

Joseph Lagrange (1736–1813), north Italian–born French mathematician, astronomer, and physicist, made significant contributions to all fields of analysis and to number theory. Lagrange's treatise on analytical mechanics, written in Berlin and first published in 1788, offered the most comprehensive treatment of classical mechanics since Newton and formed the basis for the development of mathematical physics in the nineteenth century.

Laguerre

Edmond Laguerre (1834–1886), French mathematician, worked in the areas of geometry and complex analysis and investigated orthogonal Laguerre polynomials.

Lamb

Willis Lamb (1913–2008), Americain physicist, won Nobel Prize in Physics (1955) for his discoveries concerning the fine structure of the hydrogen spectrum.

Langevin

Paul Langevin (1872–1946), French physicist, noted for his work on paramagnetism and diamagnetism, also did much to spread the theory of relativity in France and created what is now called the twin paradox.

Liouville

Joseph Liouville (1809–1882), French mathematician, worked in a number of different fields in mathematics, including number theory, complex analysis, differential geometry, and topology but also mathematical physics and even astronomy. He is remembered particularly for the Liouville theorem, a nowadays rather basic result in complex analysis. In number theory, he was the first to prove the existence of transcendental numbers.

Lorentz

Hendrik Lorentz (1853–1928), Dutch physicist, shared the 1902 Nobel Prize in Physics with Pieter Zeeman for the discovery and theoretical explanation of the Zeeman effect. He also derived the transformation equations subsequently used by Einstein in his special theory of relativity.

Markov

Andrei Markov (1856–1922), Russian mathematician, is best known for his work on theory of stochastic processes known as Markov chains.

Maxwell

James Clerk Maxwell (1831–1879), Scottish physicist. His most important achievement was classical electromagnetic theory, synthesizing all previously unrelated observations, experiments, and equations of electricity, magnetism, and even optics into a consistent theory. His set of equations—Maxwell's equations—demonstrated that electricity, magnetism, and even light are all manifestations of the same phenomenon: the electromagnetic field. From that moment on, all other classic laws or equations of these disciplines became simplified cases of Maxwell's equations. Maxwell's work in electromagnetism has been called the "second great unification in physics," after the first one carried out by Isaac Newton.

Morse

Philip Morse (1903–1985), American physicist, worked on physical vibrations. He became later administrator and pioneer of operations research in World War II.

Neumann

John von Neumann (1903–1957), Austro-Hungarian-born American mathematician, made major contributions to a vast range of fields, including set theory, functional analysis, quantum mechanics, ergodic theory, continuous geometry, economics and game theory, computer science, numerical analysis, hydrodynamics (of explosions), and statistics, as well as many other mathematical fields. He was a pioneer of the application of operator theory to quantum mechanics, in the development of functional analysis. With Teller he has studied the key steps involved in the nuclear physics of thermonuclear reactions and hydrogen bombs.

Newton

Sir Isaac Newton (1643–1727), English physicist, mathematician, astronomer, natural philosopher, alchemist, and theologian. His 1687 publication, the *Principia*, is considered to be among the most influential books in the history of science, laying the groundwork for most of classical mechanics. In this work, Newton described universal gravitation and the three laws of motion, which dominated the scientific view of the physical universe for the next three centuries. Newton showed that the motions of objects on Earth and of celestial bodies are governed by the same set of natural laws by demonstrating the consistency between Kepler's laws of planetary motion and his theory of gravitation, thus removing the last doubts about heliocentrism and advancing the scientific revolution. In mathematics, Newton shares the credit with Leibniz for the development of the differential and integral calculus. He also demonstrated the generalized binomial theorem, developed Newton's method for approximating the roots of a function, and contributed to the study of power series.

Parseval

Marc-Antoine Parseval (1755–1836), French mathematician, most famous for what is now known as Parseval's theorem, which presaged the unitarity of the Fourier transform.

Planck

Max Planck (1858–1947), German physicist, Nobel Price in Physics (1918), is considered to be the founder of the quantum theory, and thus one of the most important physicists of the twentieth century.

Poisson

Siméon-Denis Poisson (1781–1840), French mathematician and physicist, worked in pure mathematics, applied mathematics, mathematical physics, and rational mechanics, and celestial mechanics.

Pauli

Wolgang Pauli (1900–1958), Austrian physicist, one of the pioneers of quantum physics, when working with Born and Heisenberg, received later the Nobel Prize in Physics in 1945 for his "decisive contribution through his discovery of a new law of Nature, the exclusion principle or Pauli principle." In 1930 he postulated the existence of the neutrino.

Rabi

Isidor Rabi (1898–1988), Galician-born American physicist, was awarded the Nobel Prize in 1944 for his discovery of nuclear magnetic resonance. He worked with Bohr, Heisenberg, and Pauli.

Raylegh

John Rayleigh (1842–1919), English physicist, discovered with Ramsay the element argon, for which he earned the Nobel Prize in Physics in 1904. He also discovered the phenomenon now called Rayleigh scattering, explaining why the sky is blue, and predicted the existence of the surface waves now known as Rayleigh waves. He anticipated the catastrophic character for statistical mechanics of the black-body radiation law (which he called the ultraviolet catastrophe).

Sackur

Otto Sackur (1880–1914), German physical chemist, is known for the development of the Sackur–Tetrode equation, which he developed independently of Tetrode.

Schrödinger

Erwin Schrödinger (1887–1961), Austrian physicist, received the Nobel Price in Physics in 1933 for his contributions to quantum mechanics, especially the Schrödinger equation. In 1926, Schrödinger gave in a first paper a "derivation" of the wave equation for time-independent systems and showed that it gave the correct energy eigenvalues for the hydrogenlike atom. In a second paper submitted just four weeks later, he solved the quantum harmonic oscillator, the rigid rotor, and the diatomic molecule and gave a new derivation of the Schrödinger equation. In a third paper of the same year, he showed the equivalence of his approach to that of Heisenberg and gave the treatment of the Stark effect. In a fourth paper he showed how to treat problems in which the system changes with time, as in scattering problems.

Stefan

Joseph Stefan (1835–1893), Slovenian-born Austrian, physicist and mathematician, is best known for originating a physical power law in 1879 stating that the total radiation from a black body is proportional to the fourth power of its thermodynamic temperature *T*. Stefan deduced the law from experimental measurements made by the Irish physicist Tyndall. In 1884 the law was derived theoretically in the framework of thermodynamics by Stefan's student Ludwig Boltzmann and hence is known as the Stefan–Boltzmann law.

Taylor

Brook Taylor (1685–1731), English mathematician. Added a new branch to the higher mathematics, now designated the "calculus of finite differences." Among other ingenious applications, he used it to determine the form of movement of a vibrating string, by him first successfully reduced to mechanical principles. The same work contained the celebrated formula known as Taylor's theorem, the importance of which remained unrecognized until 1772, when Lagrange, realizing its powers, termed it "the main foundation of differential calculus."

Tetrode

Hugo Tetrode (1895–1931), Dutch theoretical physicist, contributed to statistical physics, early quantum theory, and quantum mechanics. He corresponded with Einstein, Lorentz, and Ehrenfest on quantum mechanics and wrote several influential papers on quantum mechanics that introduced key ideas that were fruitful in the formulation of the Wheeler–Feynman time symmetric theory. Tetrode developed independently from Sackur the Sackur–Tetrode equation, a quantum mechanical expression of the entropy of an ideal gas.

Weyl

Hermann Weyl (1885–1955), German mathematician, who was one of the most influential mathematicians of the twentieth century, developed research of major significance for theoretical physics as well as pure disciplines, including number theory dealing with space, time, matter, philosophy, logic, symmetry, and the history of mathematics. He was one of the first to conceive of combining general relativity with the laws of electromagnetism.

Wigner

Eugene Wigner (1902–1995), American physicist of Hungarian origin, received the Nobel Prize in Physics in 1963 "for his contributions to the theory of the atomic nucleus and the elementary particles, particularly through the discovery and application of fundamental symmetry principles." Wigner is important for having laid the foundation for the theory of symmetries in quantum mechanics as well as for his research into the structure of the atomic nucleus and for his several mathematical theorems.

Young

Thomas Young (1773–1829), English genius admired by Herschel and Einstein, who is famous for having partly deciphered Egyptian hieroglyphs (specifically the Rosetta Stone) before the definitive work of Champollion, made notable scientific contributions to the fields of light, solid mechanics, energy, physiology, vision, musical harmony language, and Egyptology.

INDEX

Quantum Oscillators, First Edition. Olivier Henri-Rousseau and Paul Blaise.
© 2011 John Wiley & Sons, Inc. Published 2011 by John Wiley & Sons, Inc.

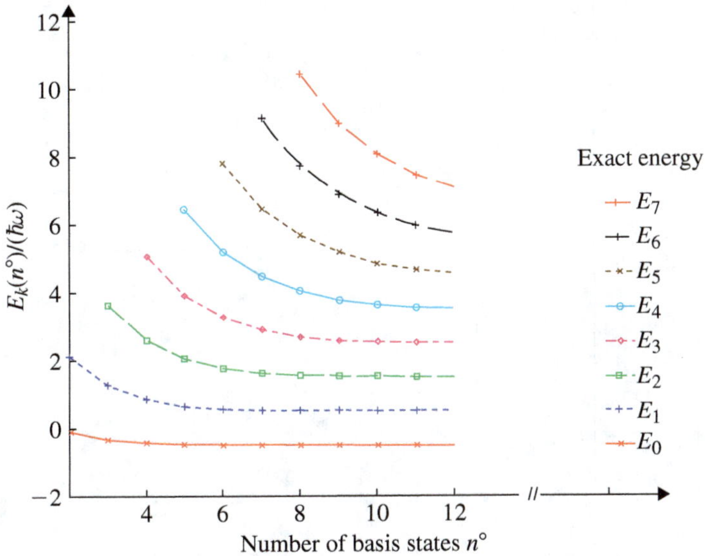

Figure 6.3 Stabilization of the energy of the eight lowest eigenvalues $E_k(n^\circ)/\hbar\omega^\circ$ with respect to n°.

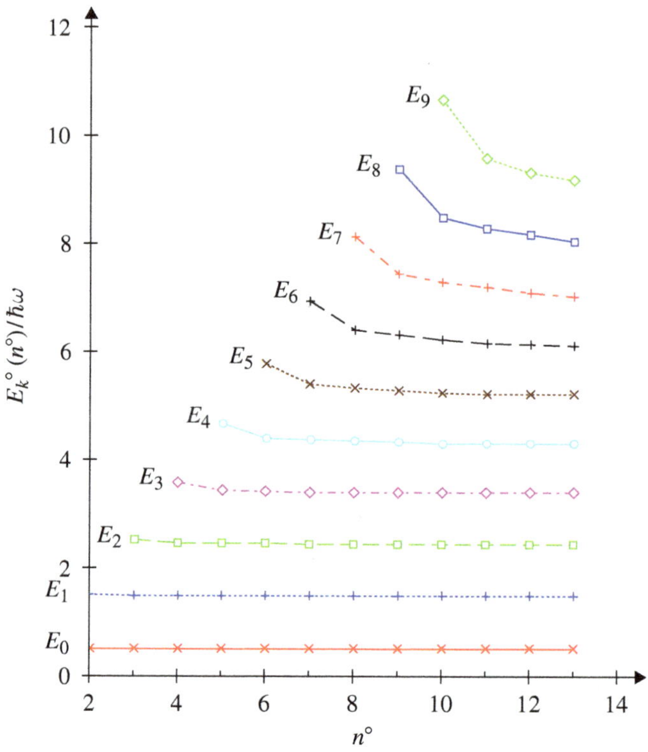

Figure 9.2 Progressive stabilization of the eigenvalues appearing in Eq. (9.50) with the dimension n° of the truncated matrix representation ($\eta = -0.017$).

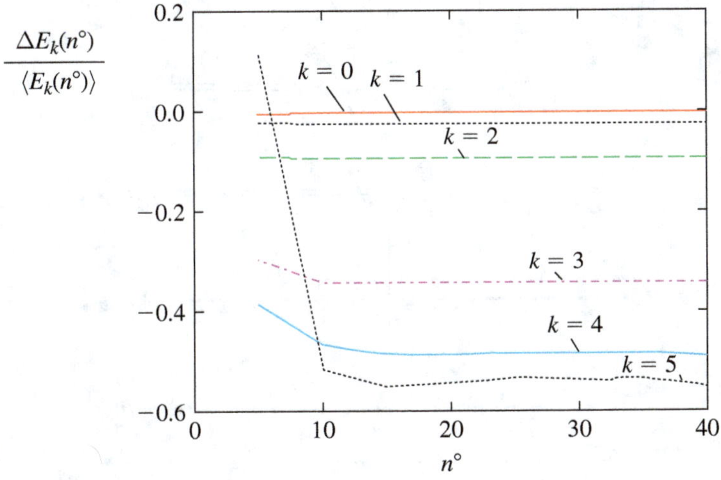

Figure 9.3 Relative dispersion of the difference between the energy levels and the virial theorem.

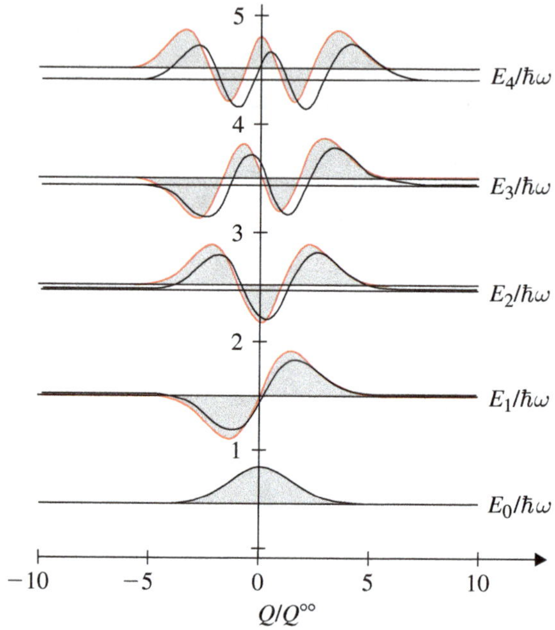

Figure 9.4 Five lowest wavefunctions $\Psi_k(\xi)$ of the Morse Hamiltonian compared to the five symmetric or antisymmetric lowest wavefunctions $\Phi_n(\xi)$ of the harmonic Hamiltonian. The length unit is $Q^{oo} = \sqrt{h/2mw}$.

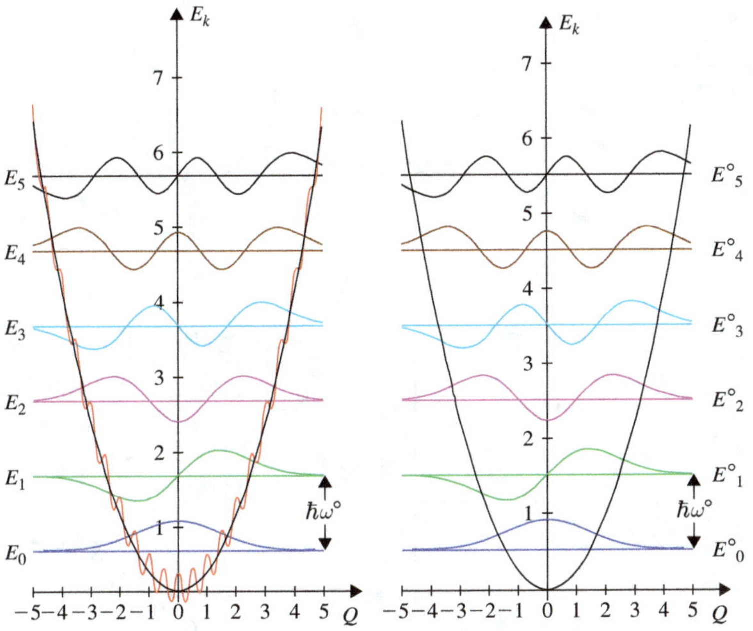

Figure 9.7 Comparison between the energy levels calculated by Eq. (9.100) and the wavefunctions obtained by Eq. (9.101) and the energy levels and the wavefunctions of the harmonic oscillator.

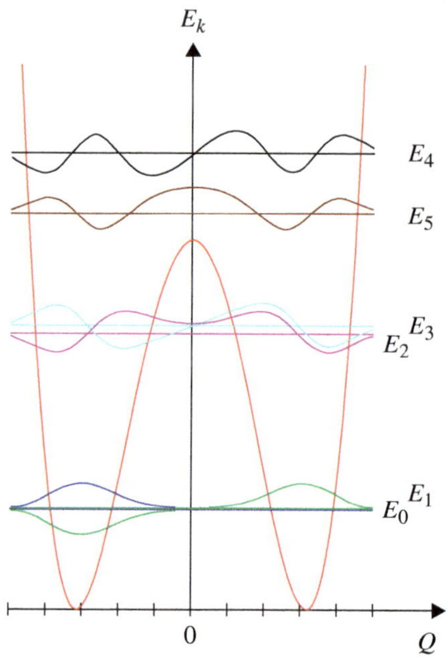

Figure 9.11 Representation of the six lowest wavefunctions and the corresponding energy levels for symmetrical double-well potential.

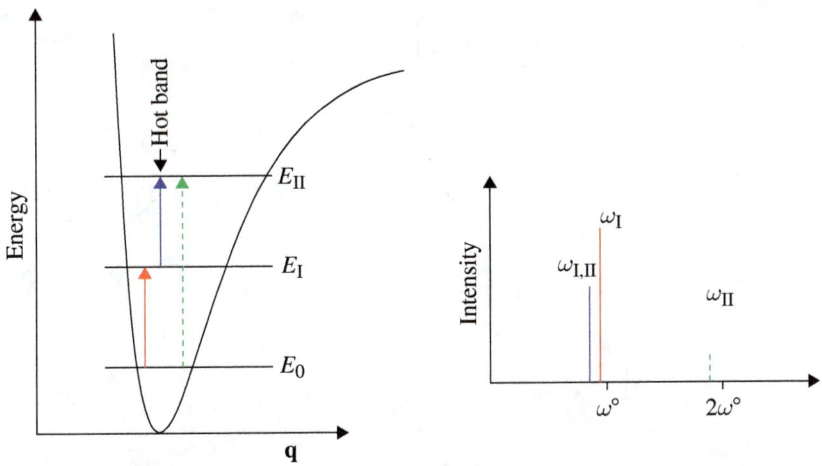

Figure 17.3 Appearance of a hot band in the IR spectrum of a Morse oscillator.

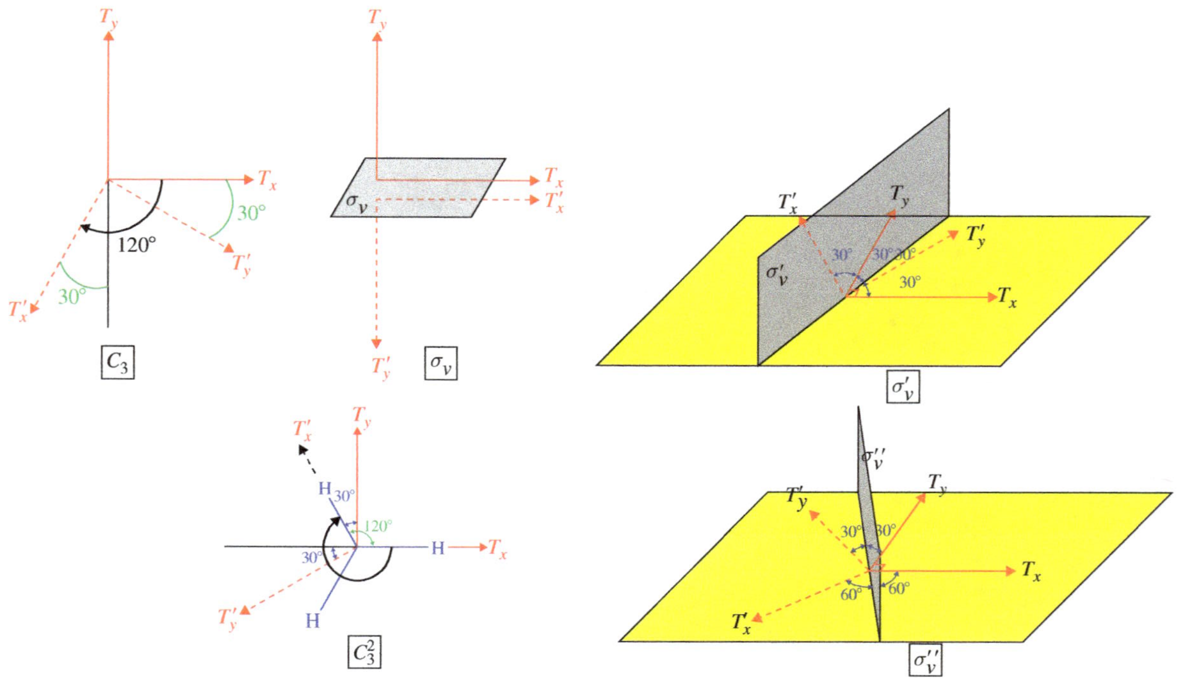

Figure 18.3 The C_{3v} symmetry operations.

www.ingramcontent.com/pod-product-compliance
Lightning Source LLC
LaVergne TN
LVHW011537200926
848484LV00001B/275